W9-AXP-245

Constructing and Manufacturing

WOOD PRODUCTS

Constructing and Manufacturing WOOD PRODUCTS

DR. WAYNE H. ZOOK
Chairman, Department of Industrial Technology
University of North Dakota
Grand Forks, North Dakota

Formerly:
Associate Professor
Department of Industrial Technology
Illinois State University
Normal, Illinois

McKNIGHT PUBLISHING COMPANY
Bloomington, Illinois

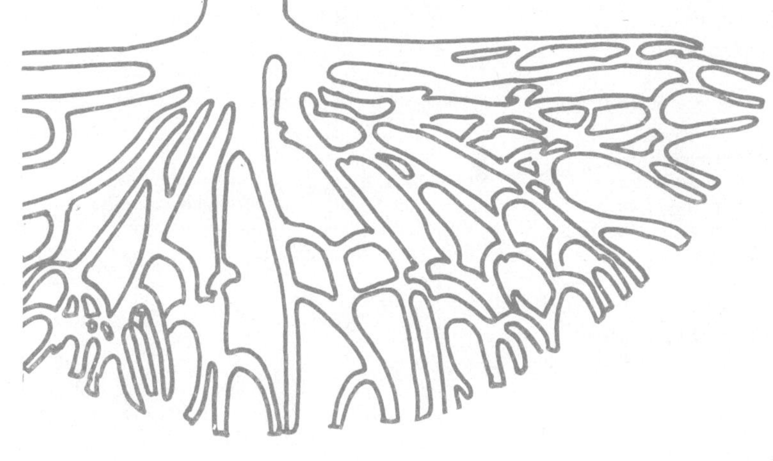

FIRST EDITION

Lithographed in U.S.A.

McKNIGHT

COPYRIGHT 1973

PUBLISHING COMPANY/BLOOMINGTON, ILLINOIS 61701

Ronald E. Dale, Vice President-Editorial, would like to acknowledge the skills and talents of the following people and organizations in the preparation of this publication

Donna M. Faull
Production Editor

Bettye King
Copy Editor

Elizabeth Purcell
Art Editor

Mary Catherine Fairfield
Proofreader

R. Scott Jones
Staff Artist

Howard Davis
Line Artist — Photographer

Nadine Fred
Researcher — Typist

William McKnight, III
Manufacturing

Jeff Boswell
Dallas, Texas
Cover Artist

Gorman's Typesetting
Bradford, Illinois
Compositor

R. R. Donnelley & Sons Co.
Chicago, Illinois
Preproduction — Printing

Northwestern Engraving Co.
Menasha, Wisconsin
Four Color Preproduction

Library of Congress
Card Catalog Number: 72-82079

SBN: 87345-048-5

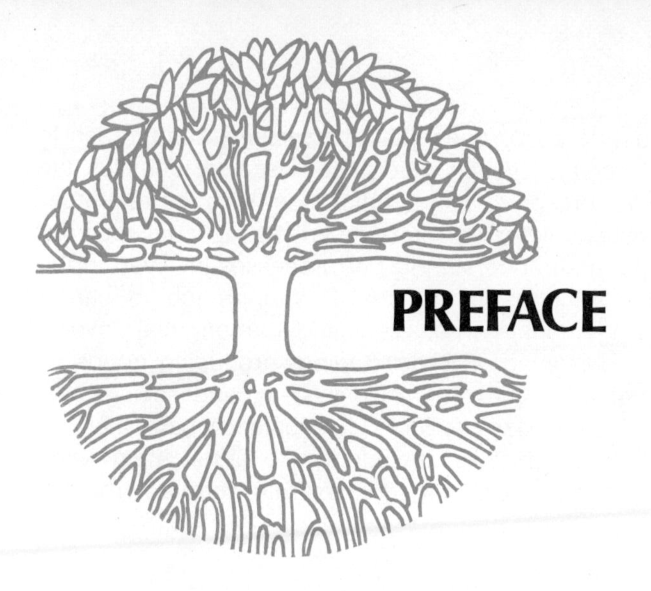

PREFACE

The economy of our nation depends on a variety of enterprises. Our youth are involved in personal development toward assuming adult responsibilities. Adult responsibilities could be summarized into two main generalizations. Each adult is, **first,** a consumer of the outputs of enterprises in terms of goods and services and, **secondly,** he may be a human resource as an input to an enterprise for the production, creation, management, and distribution of the goods or services. Using this as a general premise, Dr. Wayne Zook conceptualized this textbook to help students gain a better understanding of the construction and manufacturing enterprises that use wood as a basic material. The author's initial overview of career opportunities and functional relationships to our society should be of assistance to students when making tentative career selections that meet their life style goals. Before a student can begin identifying tentative career choices to fulfill his own personal needs, he should understand:

- how an enterprise utilizes materials, establishes standard sizes and shapes of materials;
- what the advantages and disadvantages of materials are in terms of physical properties;
- the variety of research and development techniques needed to gain a knowledge that can be used as a basis for decision-making;
- what the financial factors are that influence decision making;
- what steps are involved in preparing to produce goods or services; and
- the variety of production processes that are available for constructing or manufacturing consumer products.

The author unfolds the major concepts of a systematic approach to producing goods and services by an enterprise. He also identifies the importance of people as a resource to the enterprise. With the reinforcement of appropriate psychomotor activities in an instructional environment, students will acquire an understanding of processes involved in the construction and manufacturing enterprises. This approach to career education should assist students when selecting careers that involve managing, researching, designing and developing; producing goods or providing services; controlling the enterprise system; and selling, servicing, and distributing consumer products. It should prepare them to begin to qualify for entry-level skills. It should also assist them when making consumer decisions for goods or services from wood enterprises.

Ronald E. Dale
Vice President - Editorial
McKnight Publishing Company
Bloomington, Illinois

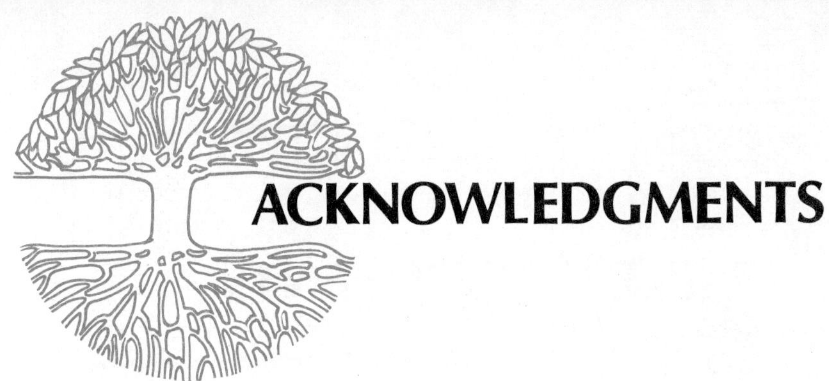

ACKNOWLEDGMENTS

The author wishes to especially thank **Mr. Howard Davis** for his assistance in the preparation of drawings, photographs, and illustrations used in this book. The assistance provided by **Dr. Gary Weede, Mr. James S. Stinson** and **Mrs. Nadine Fred** was also greatly appreciated.

The author also sincerely appreciates the cooperation and courtesy of the following companies who supplied photographs and illustrations used in the book:

Adjustable Clamp Company
American Forest Industries, Inc.
American Forest Institute
American Forest Products
American Steel and Wire Company
Amerock Corporation
Arvids Iraids Multi Clamps
Baird Manufacturing Company
Biltbest Windows
Binks Spray Systems
Black Brothers Company, Inc.
Boice Crane
Brandt Furniture
Buck Brothers, Inc.
California Redwood Association
Clark Equipment Company
H. H. Connally
The Conn. Valley Mfg. Company
Culley Engineering & Mfg. Company
Delta Div., Rockwell Mfg. Company
DeVilbiss Company
Dewalt Div., American Machine & Foundry Company
Fairfield Engineering and Manufacturing Company
J. A. Fay and Egan Company
Forest Products Laboratory
Georgia Pacific
Greenlee Bros. & Company
Greenlee Tool Company
Gulf Oil Company
Hardwood Plywood Manufacturers Association
Harloc Products Corporation
Hedrick-Blessing
Independent Lock Company
Independent Nail Corporation
Irwin Auger Bit Company
Jasper Cabinet Company
Keller Furniture
Kemp Furniture Industries
Kroehler Furniture Mfg. Company
Lumtape Corporation
Millers Falls Company
Milwaukee Electric Tool Company
National Forest Products
National Homes Corporation
Nicholson File Company
Norton Coated Abrasive Division
H. K. Porter Company, Inc., Henry Disston Div.
Powermatic Machine Company
Power Tool Div., Rockwell Mfg. Company
Richmond Cedar Works
Rockwell Mfg. Company
Rohm and Haas Company
B. M. Root Company
RUVO Automatic Corporation
Sand-O-Flex
Selig Manufacturing Company
Senco Products, Inc.
Simonds Saw & Steel Company
Southern Forest Products Association
Stanley Tools
Union Twist Drill Company
Universal Chief
U. S. Forest Service
Vega Industries, Inc.
Walker-Turner Div., Rockwell Mfg. Company
Western Wood Products Association
Westinghouse
Weyerhaeuser Company
Winchester-Western Div.,
Olin-Mathieson Chemical Corp.
Winnebago Industries
Workrite Products Company

TABLE OF CONTENTS

Chapter 10 Principle of Cutting Holes 210

Chapter 11 Shaping and Forming Irregular Shapes 233

Chapter 12 Forming Cylindrical Shapes 258

Chapter 13 Assembly and Fastening of Parts 276

Chapter 14 Product Finishing 322

Chapter 15 Elements of Mass Production 355

Chapter 16 Cabinet and Furniture Construction 363

Chapter 17 Foundry Patternmaking 382

Chapter 18 Carpentry and Building Construction . . . 388

CHAPTER 1
THE WOOD INDUSTRY

INTRODUCTION

Our daily lives are influenced by the wood industry. The wood industry provides many of the goods or products which we use each day. The industry provides employment for hundreds of thousands of people. This employment is essential to the well-being of our country.

Wood as a material influences our *economy.* The economy is the financial system of production, distribution, and consumption of *goods and services.* Wood products are an important part of our economy. It is important that you understand as much as possible about the wood industry. The trees of our forests provide us with the raw material for thousands of products. Paper, charcoal, various oils and solvents, plastics, construction materials, and adhesives are only a few of the products derived from wood, Fig. 1-1. Our expanding technology is constantly finding new uses for wood.

Wood is a *natural* resource. The forests are unlike most other natural resources. They can be replenished as the trees are cut. Each year new trees are planted and cultivated to replace those which have been used to serve our needs for wood products,

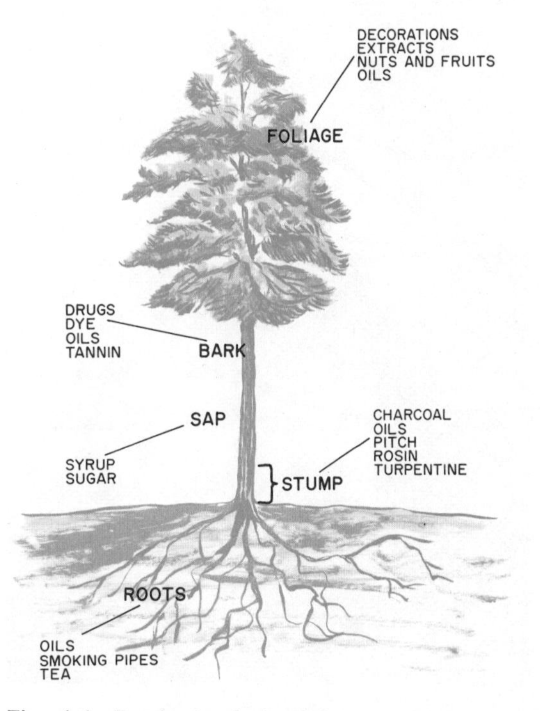

Fig. 1-1. Products derived from various parts of a tree.

Fig. 1-2. The replenished forests will always be an important part of our nation's economy.

The wood industry offers many *career opportunities* to you. As you study the wood industry, you will become aware of many careers which may interest you. This course may help you to make a career choice. It is important for you to know as much about as many jobs as you can before making a career choice. Choosing your occupation will be one of the most important decisions you will make in your life. You should select a career which fits your (1) personality, (2) mental ability, and (3) physical ability. As you study about the wood industry, you may discover a career opportunity for which you are fitted in these three ways. This course will familiarize you with what people do in their occupations or jobs. You will experience what some of the people do in their jobs as you perform the class activities in this course.

Wood as a material has many advantages for structural uses. It is lightweight, but very strong. Wood is easily shaped and formed with tools and machines. It can be fastened together quite easily with nails, screws, and glue. Expensive and complicated processes are not necessary for fastening wood. Wood can easily be combined with other materials for structural purposes.

Fig. 1-2. Trees are planted to replenish native forests. (Kimberly-Clark Corp.)

Wood has always been a favorite material for home furnishings because of its beauty. Wood is a good insulator because it is porous. One chief reason why most of our furniture is constructed from wood is that wood feels warm to the touch when compared to other materials such as metal or plastic.

The natural beauty of wood has always appealed to people for decorative purposes. It has been impossible to chemically recreate all of the desirable properties of wood. Wood has a distinct smell, touch, and sense of beauty. There are many species or kinds of wood and each kind has different properties. Therefore, a particular type of wood is generally available for a specific purpose.

Wood, like all other materials, also has some disadvantages. Wood can decay if it is not treated with paint when exposed to the weather. It is also relatively soft and can be dented and gouged. Wood expands in size when it becomes wet. However, these problems or disadvantages can be overcome if you are knowledgeable about wood and woodworking. The advantages are greater than the disadvantages, therefore, wood remains a favorite structural material.

When subcomponents such as 2x4's and components such as windows are needed in large quantities, they are mass-produced in a *manufacturing plant.* Manufacturing provides many of the components for construction. *Construction is the process of assembling all of the manufactured components on a specific building site.* This means that the construction industry is dependent on the selection of the construction site. The transportation industry moves the raw materials and the manufactured components to the construction site, and it also transports the manufactured products to the consumer.

The service industry supports the construction, transportation, and manufacturing industries by providing special services such as consulting, maintenance, research and development, record keeping, sales, and accounting. All of these industries are dependent on each other. A business relationship is formed when an organization can benefit by cooperating and serving other organizations to accomplish their separate goals of making a profit.

WOOD INFLUENCES OUR ECONOMY

Wood influences our economy because it is part of the (1) *construction,* (2) *manufacturing,* (3) *service,* and (4) *transportation* industries. The *Gross National Product* (GNP) is a means of expressing the contribution of a particular industry to the total economy of a country. The GNP is the total value of all of the products and services produced in a country. The GNP of the wood industry is given as the number of dollars spent in the United States for products and services provided by the industry, Fig. 1-3.

The GNP of the United States has increased each year. This increase is due to the growing population and because people have purchased more products and services than they did the previous year. A comparison of the GNP for the wood and other industries would reveal the importance of the wood industry to the nation.

CONSTRUCTION

Large quantities of wood and wood products are consumed by the construction industry. The construction industry can be defined as the process of building a structure on a site. Figure 1-4 shows a house being constructed on the property where it will remain. This would be regarded as a part of the construction industry. In addition to houses, other structures such as bridges, large office buildings, schools, churches, and similar structures would be considered construction.

Wood is a vital material upon which the construction industry is dependent. Can you imagine building a house without using any wood or wood products? With today's construction methods, it would be very difficult. Think of all of the people who make their living in the construction industry. Since wood is used so much in the entire construction industry, it is important for these people to understand wood.

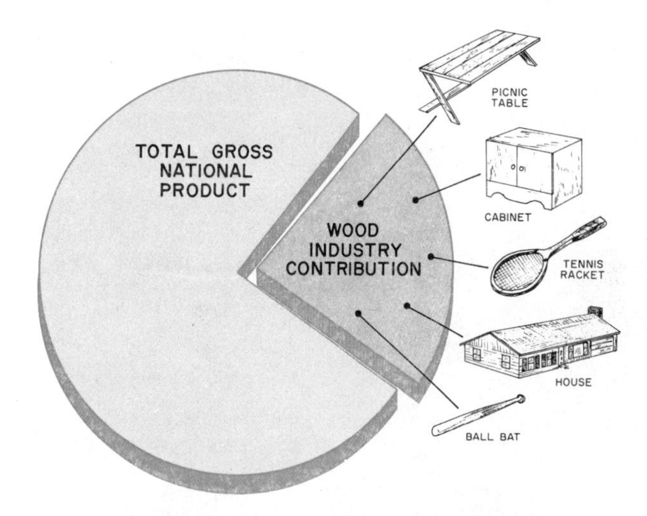

Fig. 1-3. The wood industry contributes to the gross national product.

Fig. 1-4. Building a house on a site is one example of construction.

MANUFACTURING

The wood industry is a part of manufacturing. *Manufacturing* is the process of combining and altering the shapes of raw materials into a finished product. That is, a product which can be sold to people who have a desire to own and use it. The value of raw materials is increased through the manufacturing process, Fig. 1-5. For example, lumber is sold to a furniture manufacturing company which produces furniture you may someday buy. The lumber is the raw material in this case, and the furniture is the finished product. The completed furniture is the *output* of the manufacturing industry, Fig. 1-6.

Wood as a material is processed many times. The log is cut in the forest and sawed into lumber at a mill. This is manufacturing. The lumber may then be manufactured into a kitchen cabinet. The cabinet then becomes a *component part* (one of the many subparts used to manufacture a product) for the interior construction of your home. The altering and combining of materials into a product is regarded as *manufacturing.* The same piece of wood can be used at different stages of manufacturing as a raw material or subassembly, Fig. 1-7.

The manufacturing process requires the services of many people, Fig. 1-8. The first

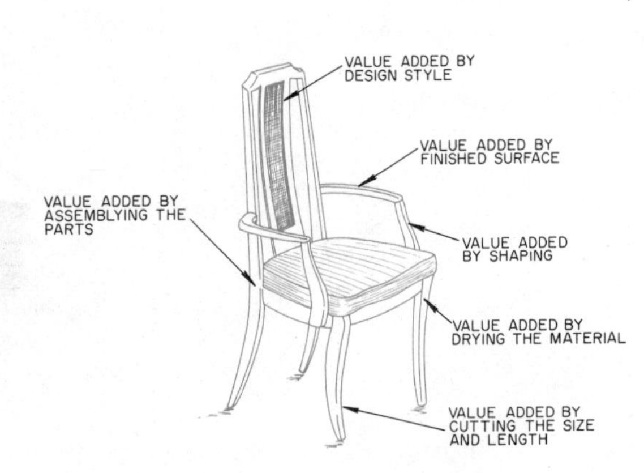

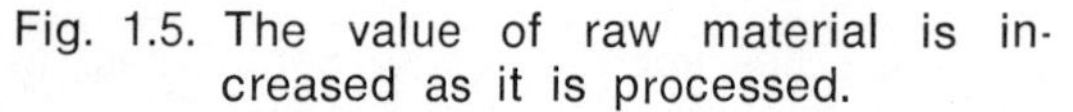
Fig. 1.5. The value of raw material is increased as it is processed.

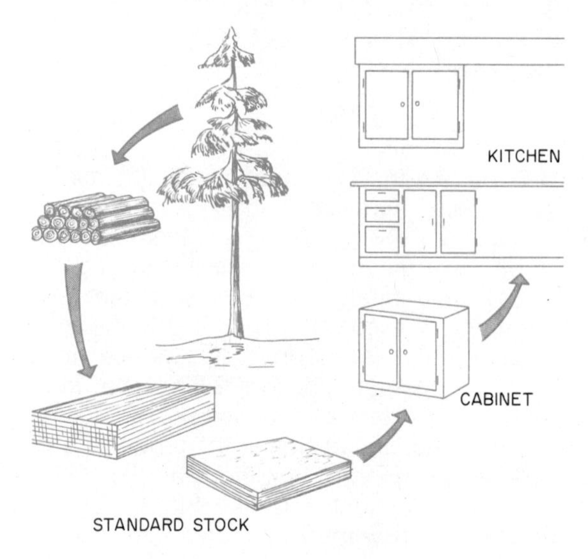

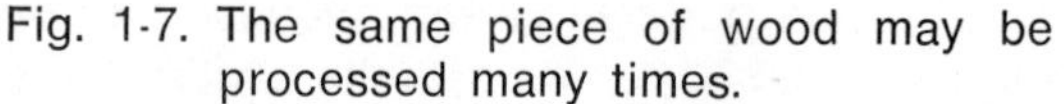
Fig. 1-7. The same piece of wood may be processed many times.

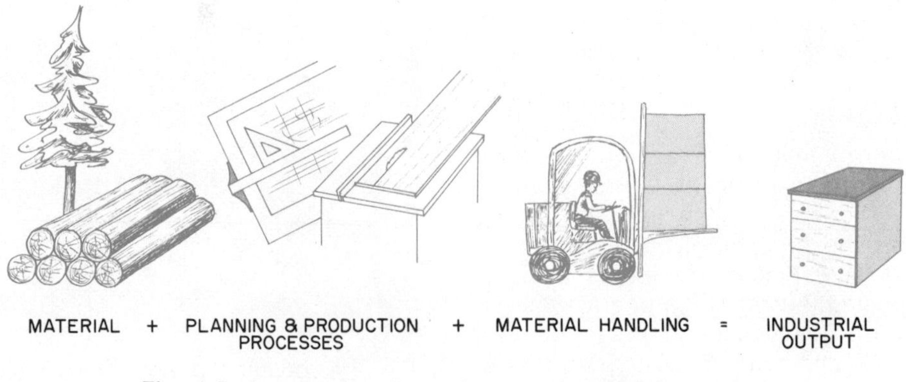

Fig. 1-6. A completed product is industrial output.

concern of manufacturing is to *identify a consumer need;* that is, to identify a product the consumer will buy. The need is then given to the designer who *designs a product to meet the need at the lowest cost.* The management personnel then plans how the raw materials can be transformed into a finished product. This phase is called *production planning.* Management then *tools up for production.* Jobs are planned for different workers to produce the product. *Production control systems* are developed to insure the quality and quantities of products produced.

SERVICE

You have now begun to explore the contributions of wood to our economic system. The next dimension is all of the *services* which are rendered by people throughout the nation to support the manufacturing and construction industries. The people in service industries sell only their services. They do not produce a product as do the manufacturing and construction industries.

A repairman is employed for a short time to repair a broken machine in a manufacturing or construction industry. His only job is to restore the machine so it operates properly, Fig. 1-9.

A team of experts may be hired by a manufacturing company to evaluate the means of production. After a thorough evaluation of the existing production system, the team will make recommendations for improving the efficiency of the company. These people are called *industrial engineers.* If the suggestions are good and put into practice, the company's output will be increased. The industrial engineers have sold their services to make a profit. The manufacturing company will also make a profit by using this service. The industrial engineers were hired because they could increase the company's profit. If the time to produce a product can be reduced with the same

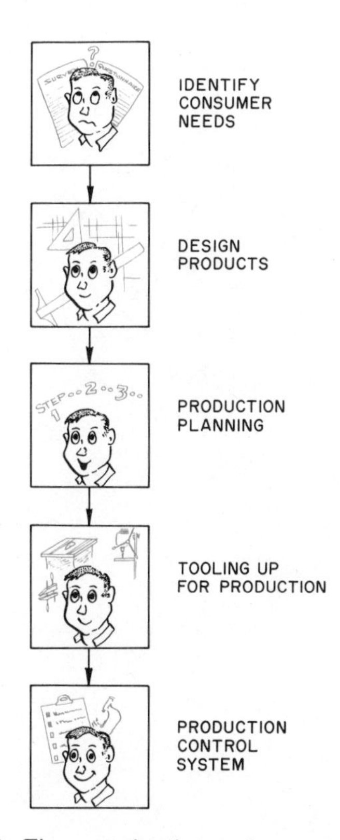

Fig. 1-8. The production of a product requires the services of many people.

Fig. 1-9. A machinery repairman restoring a machine to operating condition.

equipment, a greater profit will result for the company. Profit is the money the company has left after paying its bills or costs.

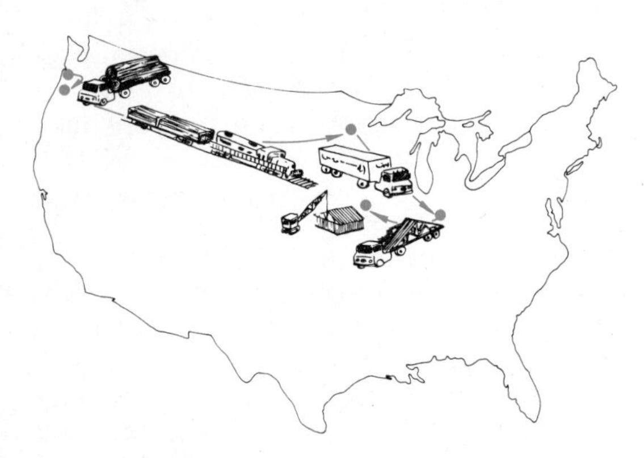

Fig. 1-10. Wood may be transported many times before it reaches the consumer.

When the profit becomes too low, a company can no longer afford to exist. The services which individuals and groups of individuals sell must benefit the company. If the services are of no value to the company, they will not be purchased.

The service industries and wood industries are dependent upon one another. Services which are sold may include machine repair, safety inspections, computer services, radio and television advertising, and many others.

TRANSPORTATION

After the logs are cut in the forest, they are moved (transported) many times before the finished product reaches the consumer. The transportation process is a large industry in itself. It includes the use of railways,

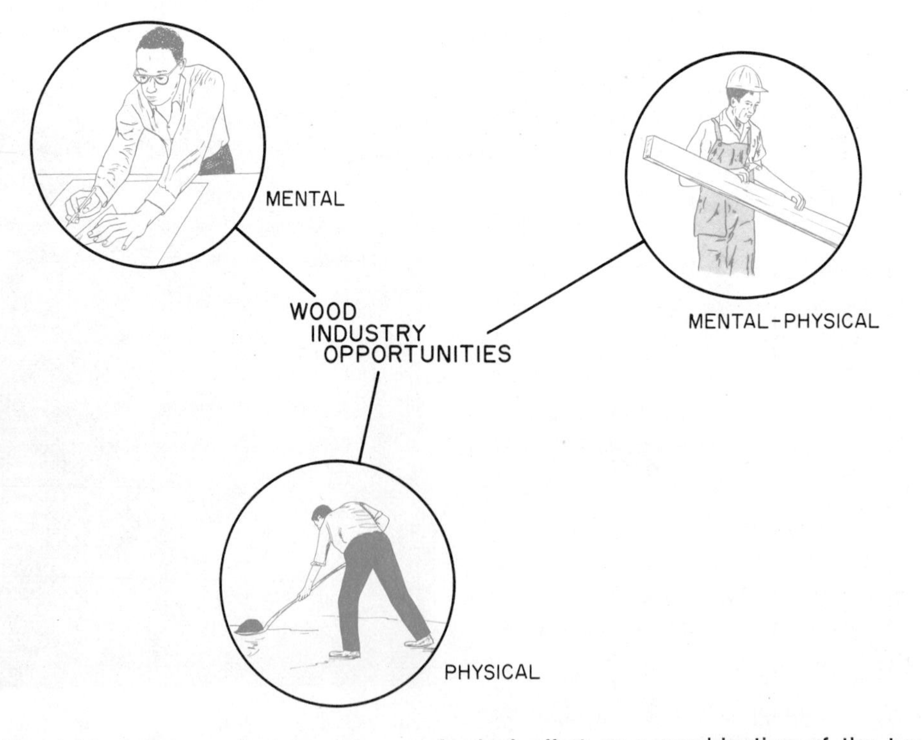

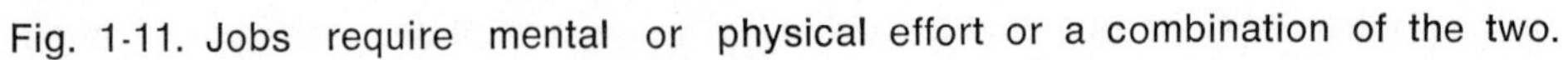

Fig. 1-11. Jobs require mental or physical effort or a combination of the two.

trucks, ships, airlines, and many special carriers. The transportation industry is also dependent upon the wood industry.

Let us follow what may likely be the transportation involved in moving the log from the forest to you, the consumer. The log is cut, for example, in Oregon and moved from the forest to the sawmill on a truck. At the sawmill, the log is sawn into *standard* lumber sizes. The lumber is then loaded on a railroad car and shipped to a window factory in Minnesota. The lumber is processed in the manufacturing plant into window frames. This manufacturer then sells the window units to a housing manufacturing company in Indiana. The windows are shipped from Minnesota to Indiana by railroad. The windows are installed in a wall section for a house. The completed wall section is then loaded on a truck and shipped to a contractor in Iowa who erects the house on a building site, Fig. 1-10.

The lumber is moved many times from one manufacturing plant to the next. *Each time the value of the boards cut from the logs has increased* because of transportation, labor, and processing costs. The industries involved could not exist without the services of the transportation industry.

CAREER OPPORTUNITIES

The woodworking industry provides many *occupational opportunities.* An occupation is the type of work a person does to earn a living. An occupation and a job are the same. The occupations which exist in the wood industry vary. The type of work you would perform could require great physical efforts or mental efforts, Fig. 1-11. There are also jobs which require both physical and mental efforts.

Before you can make a wise occupational choice, you must first consider your abilities and interests. The job must fit your *personality, mental abilities, and physical abilities.* Certain jobs require different personal traits and abilities. You are the only one who can determine if you will like a particular job. However, before you can make this personal assessment, you must know what the job requires. The purpose of this section of the book is to familiarize you with some of the job opportunities in the wood industry.

You will need to determine how much training you will need to enter the jobs in the industry. Some of the jobs will require a college education. Other jobs in the industry require that you become highly skilled with tools and machines. You can become a skilled worker or craftsman working on the job and going to a special trade school. A craftsman must enjoy working with his hands, tools, and ideas.

There are many sales jobs in the wood industry. This type of job may or may not require a college education. However, sales work requires a person with the ability to communicate well with other people. A salesman is like a teacher; he must explain and demonstrate his products. A salesman's personality must include the enjoyment of meeting and talking with people, Fig. 1-12.

Fig. 1-12. A salesman meets and talks to people. (Georgia Pacific)

As you study about the jobs in the wood industry, ask yourself if you would like that type of work. Before deciding on a job, make sure that you have the necessary qualifications. Compare your mental and physical abilities with what the job demands. Some jobs require that you travel and be away from home a lot. You should also be concerned about the job future. It is important to select a job in which you can "grow," Fig. 1-13. The pay or salary for various types of work will also differ.

The jobs in the wood industry can be divided into groups or clusters. Let's start by dividing the jobs according to the types or classification of work. The job classifications are as follows: unskilled, semiskilled, skilled, semiprofessional, and professional, Fig. 1-14.

Fig. 1-13. It is important to select a job which offers advancement.

UNSKILLED OCCUPATIONS

An *unskilled* occupation is one which requires little or no special training to perform the job. These jobs involve the handling and moving of materials. The unskilled jobs often involve heavy physical work, Fig. 1-15.

Fig. 1-15. An unskilled worker performing a job task.

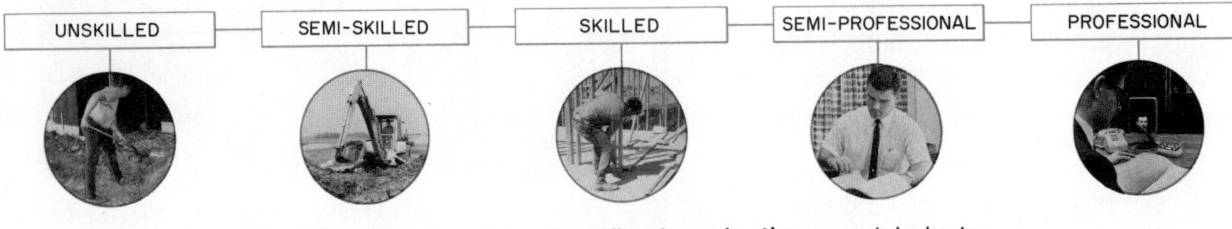

Fig. 1-14. Common job classifications in the wood industry

Unskilled workers are often referred to as *laborers*. An unskilled worker usually does not have special training.

SEMISKILLED OCCUPATIONS

The *semiskilled* occupational classification is the largest in the labor force. This job classification requires some special training to perform the work tasks. The training is often provided on the job. However, the school shop can often provide the needed training. Many of the semiskilled workers perform as machine operators, Fig. 1-16.

SKILLED OCCUPATIONS

A *skilled* worker is also called a skilled *craftsman* or skilled *tradesman*. A skilled craftsman must be able to perform all of the tasks which are involved in his trade. A craftsman can learn his trade in many ways. The apprenticeship method is one of the most common methods. An *apprentice* learns his trade from a master craftsman. He works on the job with the craftsman and learns as he works. The apprentice is paid less wages, but it increases as he learns the trade. In addition to working on the job, he must attend classroom instruction in the evenings, Fig. 1-17. The apprenticeship period consists of four years of on-the-job training.

The apprentice becomes a *journeyman* after successfully completing four years of training. The journeyman must continue to learn on the job as new methods are introduced into the industry. He becomes highly skilled as he continues to work in his trade.

A high school education is usually required to become an apprentice. You will also be required to pass some special tests to be admitted to an apprenticeship program.

Some skilled craftsmen learn the trades informally. They may learn the trade as helpers or unskilled workers on the job.

There are also many good trade schools in which you can learn a trade. What you learn in this course will help you if you desire to become a skilled craftsman.

There are many skilled crafts in the wood industry. The most common ones include *carpentry, cabinetmaking,* and *patternmaking*. Each of these trades require special training and skills. Further information about

Fig. 1-16. A machine operator is a semiskilled worker. (Norton Coated Abrasive Division)

Fig. 1-17. Apprentices receiving classroom instruction.

skilled trades may be found in the *OCCUPATIONAL OUTLOOK HANDBOOK*.[1]

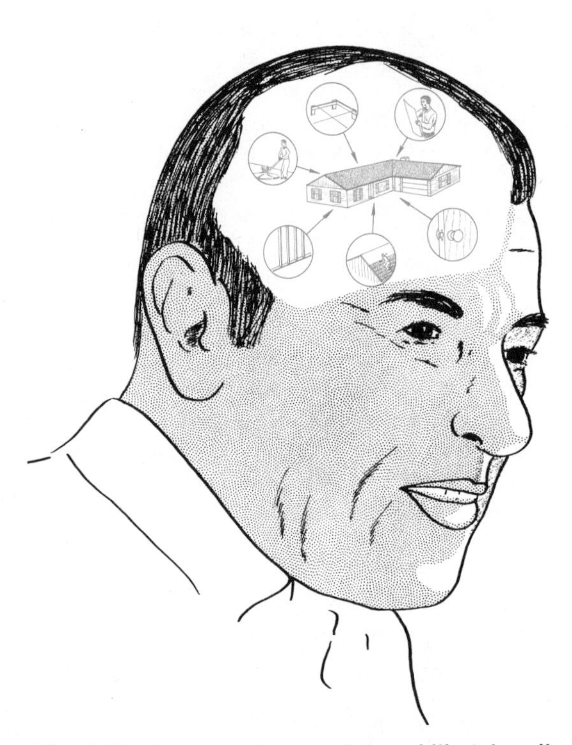

Fig. 1-18. A carpenter must be skilled in all job tasks of the trade.

Fig. 1-19. Forming carpenters erecting wood frames in which concrete will be placed.

[1]U.S. Department of Labor, Bureau of Labor Statistics, *Occupational Outlook Handbook* (Washington, D.C.: U.S. Government Printing Office 1968-1969 Edition) 759 pages. A revised edition is printed every two years.

Carpentry. Carpentry is the largest single skilled trade in the United States. It is reasonable to believe that due to increased demands for housing and all types of construction that this will remain true. A carpenter is a skilled craftsman. He must be able to perform all of the building tasks common to the construction of wood frame buildings, Fig. 1-18. Some carpenters *specialize* in the trade; that is, they only do certain types of carpentry tasks. For example, you may desire to be a *framing carpenter.* This person builds the frames for the structure and does not work on the structure to completion. Framing carpentry is heavy work and is usually all outside work. There are also *finishing carpenters.* These carpenters start on the job after the frame is constructed. They install the doors and trim. It is inside

Fig. 1-20. A cabinetmaker must be very skillful and work to close tolerances. (Norton Coated Abrasive Division)

work most of the time. Some carpenters spend most of their time building forms in which concrete is poured. They are called *forming carpenters.* They usually work on large construction jobs such as commercial buildings or bridges, Fig. 1-19. This is heavy work and is normally outside. All carpenters must be able to read drawings.

Cabinetmaking. A cabinetmaker is a specialized type of carpenter. He is highly skilled with woodworking machines and tools. His job is building cabinets to be installed in houses and commercial buildings. He does his work inside, Fig. 1-20. The cabinetmaker must be able to join and fit wood very accurately.

Patternmaking. Patternmaking is a very specialized craft in the wood industry. The patternmaker constructs patterns from wood for making metal castings. Figure 1-21 shows a wood pattern and a metal casting which was made from that pattern.

The patternmaker is highly skilled in the use of hand and power tools. In addition to working to close tolerances, he must be able to read complicated blueprints. The selection and proper use of wood is an important consideration to the patternmaker. Figure 1-22 shows a patternmaker working on a pattern.

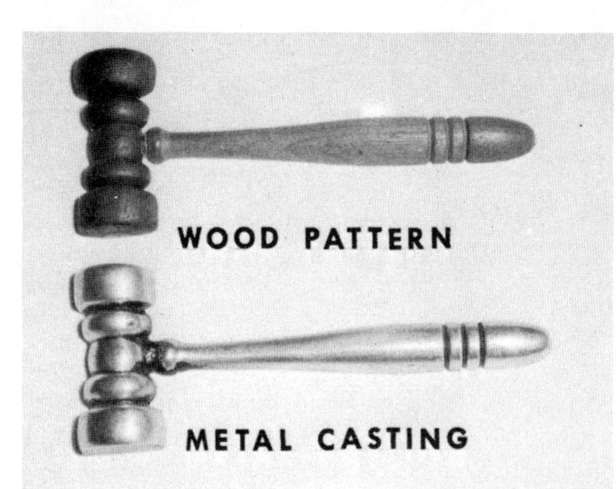

Fig. 1-21. A metal casting poured from a wood pattern.

SEMIPROFESSIONAL OCCUPATIONS

There are many semiprofessional occupations which require a knowledge of the wood industry. The broad knowledge of the wood industry you would need as a semiprofessional worker can be learned in school and on the job. Let us look at some of the semi-professional occupation job descriptions and the type of work involved. A semiprofessional occupation will probably not involve much physical labor.

Retail lumber salesman. A retail lumber salesman would be an example of a semiprofessional occupation. A sales person in a retail lumber company would be involved in selling building materials. He must constantly work with customers, therefore, he must like to meet people. Most of the products he will be selling are building materials. He must know about different kinds of wood and what they are used for in construction. He also needs a knowledge of hardware and related construction materials.

The retail lumber salesman does a lot of his work from drawings the customer provides. He figures the kind, amounts, and

Fig. 1-22. A patternmaker constructs the pattern from which a metal casting will be made.

Fig. 1-23. A retail lumber salesman helps a customer.

Fig. 1-24. Industry is dependent upon the draftsman to prepare the final drawing which will be used to produce the product. (Georgia Pacific)

Fig. 1-25. A forester protects and conserves the forests to provide recreation for the public. (U.S. Forest Service)

costs of the material needed from the drawings, Fig. 1-23.

Draftsman. A draftsman would prepare the final drawing from an architect's specifications. He has to know how the structure is put together or constructed. He must be a skillful draftsman to prepare the final drawings, Fig. 1-24. A draftsman generally receives his training in a trade school or vocational school.

PROFESSIONAL OCCUPATIONS

A college education is necessary for most of the professional occupations. The college training may be in many different fields such as: forestry, chemistry, marketing, architecture, business, design, and many others. The related training in many cases gives the person the skills needed to function in the wood industry. Let's look at some of the specific professional occupations.

Forester. A forester is trained to work in forest management, conservation, and research. He is concerned with providing forests to produce more lumber and also recreational opportunities. The forester is also trained in management and conservation of the forest. He decides what trees should be cut for lumber each year, as well as supervising the planting of new trees. He also protects the forest from fire through planned programs, Fig. 1-25. He works out-of-doors a lot of the time.

Fig. 1-26. Chemists are employed as researchers to find new and improved uses for wood. (Georgia Pacific)

Fig. 1-27. A forester supervising the planting of new trees. (Weyerhaeuser Company)

A forester may also specialize in research. This involves the discoveries of new uses for trees and better methods of growing trees. Many professional foresters are employed in the Forest Products Laboratory. This laboratory is for research and is located in Madison, Wisconsin. You can request information from the laboratory about problems which you cannot solve related to wood. Many chemists are employed in research activities at the Forest Products Laboratories, Fig. 1-26.

Fig. 1-28. A modern forest fire detection center includes a microwave relay tower, a short-wave radio antenna, and an overhead helitack plane, as well as the lookout tower. (American Forest Institute)

Many professional foresters devote their efforts to replenishing the forests after they are cut. This involves the selection, planting, and caring for the young trees, Fig. 1-27.

The forests have many enemies. Each year thousands of acres of trees are destroyed. Among the enemies are insects and forest fires. To prevent the wasteful loss of valuable trees, foresters also watch for fires from fire towers or from helicopters, Fig. 1-28. Many people make their living protecting the forests from fires. Early detection of fire is important. The prevention and control of forest fires is important to the wood industry.

Engineers. Engineers are constantly discovering new methods of using wood in construction. Engineers are scientists. These people work in laboratories with laboratory equipment. Often it is necessary to test the strength of different types of wood, Fig. 1-29. This type of research is needed to determine how large beams must be to support various structures. This information is used to build safe, durable buildings. Engineers must have a college education.

Fig. 1-29. An engineer testing the strength of a beam by adding metal weights on loading yokes. (Forest Products Laboratory)

Fig. 1-30. An architect checking the progress of a structure.

Architects. An architect designs buildings constructed of wood. He also inspects the work as it is being performed by the construction personnel. It is his responsibility to make sure the building is properly constructed, Fig. 1-30. The architect determines the quality and type of materials which will be used in the structure. He must be knowledgeable of the materials that can be used for the various parts. The architect is concerned with the appearance and the durability of the structure, so he must also have a background in design. Figure 1-31 shows a structure designed by an architect which is both beautiful and structurally strong. An architect must have a college education and be licensed to practice.

Furniture designers. A furniture designer is somewhat like an architect. Figure 1-32 shows a piece of furniture a furniture designer has planned to serve varied needs of the consumer. The designer must also understand the methods used in furniture construction. It is important for the designer to study about wood as a material and the processes used to create a finished product. A designer usually has special training.

Real estate broker. A real estate broker sells houses. He should know how houses are constructed and designed to sell houses to future owners. He must know

Fig. 1-31. A building designed by an architect for public use. (Western Wood Products Association)

how to appraise or judge the price of a house. The price of a house is based on size, location, and type of construction. Figure 1-33 shows a real estate broker selling a house to the future homeowner. If you like to meet people and have a knowledge of house construction, you may want to consider becoming a real estate broker. Both men and women are employed in the real estate business.

A real estate broker may not always have a college education. However, he needs special training and must be licensed to sell real estate.

Bankers. A large amount of a bank's money is loaned to people to buy houses in which to live. A banker must be knowledgeable of the wood industry, since much of the bank's money is invested in real estate. He must be able to recognize quality construction. The banker must also appraise the value of a house. His appraisal must be accurate to determine how much money to lend a customer on a particular house, Fig. 1-34. The banker's job requires a broad knowledge of the wood industry such as you may learn in this course. He must also be able to read blueprints for new houses.

Fig. 1-33. A real estate broker showing a house to a prospective buyer.

Fig. 1-32. A piece of furniture designed by a furniture designer to fit into contemporary furnishings. (Brandt Furniture)

Fig. 1-34. A banker who lends money to clients to purchase homes must be knowledgeable about house construction.

Fig. 1-35. A contractor supervising a job must know all phases of construction.

Fig. 1-36. A carpenter is directed by the contractor to assemble the construction as designed.

Banks serve the community by lending money. The bank makes a profit by charging *interest*. Interest is a charge to the borrower for using the money. The rate of interest is expressed as a percentage. For example, you may borrow $100 at an 8% annual interest rate. This means it will cost you $8.00 to use the $100 for one year.

A banker meets the public each day in his work. As a banker he must also be able to evaluate people to determine if they are a good risk when lending money. If you want to be a banker, a college education will help. This course, combined with other experiences, will give you a broad knowledge of construction and the wood industry.

Management personnel. There are many professional occupations in management. There are many types of management positions in the wood industry. Contractors are managers — their job is to make the decisions on a construction job. They manage the workers and decide when a particular job will be performed. A contractor also bids (decides how much he will charge for his company's services) on jobs. He must always have his workers and materials at the right place at the right time, Fig. 1-35. A contractor owns the company and the equipment to perform the services. He may often buy the services of other professionals such as architects, engineers, and accountants.

Many unskilled, semiskilled, and skilled workers are employed by contractors, Fig. 1-36.

Executives. Large companies hire executives to run the total business. An executive may be the president or any of the other officers in the company. The executives make the important decisions which affect the company. The decisions of the company executives are put into practice by the employees at lower levels. The executives must be well-informed so that they will make decisions which will benefit the company, Fig. 1-37.

Executives usually have a college education. They are trained in business administration. The executives of a wood manufacturing industry must know about wood to make important decisions. A poor decision would lose the company money.

Personnel management. Large companies have people in jobs called personnel management. People employed in personnel management hire all of the employees. They must determine what jobs need to be filled. If you were in personnel, you would determine what type of employee would best perform the job. You would then interview people until you found the person with the needed qualifications. Tests are often given to prospective employees to determine their ability to do a specific job. The personnel management officers interview the prospective employees, Fig. 1-38.

The people in personnel management must hire people who will benefit the company. They are always looking for people who will become more valuable to the company as they work.

Teachers. Industrial Arts teachers are professional people in the wood industry. If you enjoy helping people, you may like to teach, Fig. 1-39. A teacher has a college education. You can learn about the qualifications and opportunities in teaching from your teacher.

After you have made a careful study of the wood industry, you will probably find there are many jobs you would like. As you try to choose a job, continue to assess your own abilities and interests. Your lifetime occupation is an important decision. You can learn more about the job from your teacher, parents, and guidance counselor. Before making your final decision, talk to people who work at the job. Ask them what they do and how you can qualify yourself for this type of job.

Fig. 1-37. Major decisions are made by company executives.

Fig. 1-38. An interview of a prospective employee is made by the personnel manager.

Fig. 1-39. Many people are employed to teach students about the wood industry.

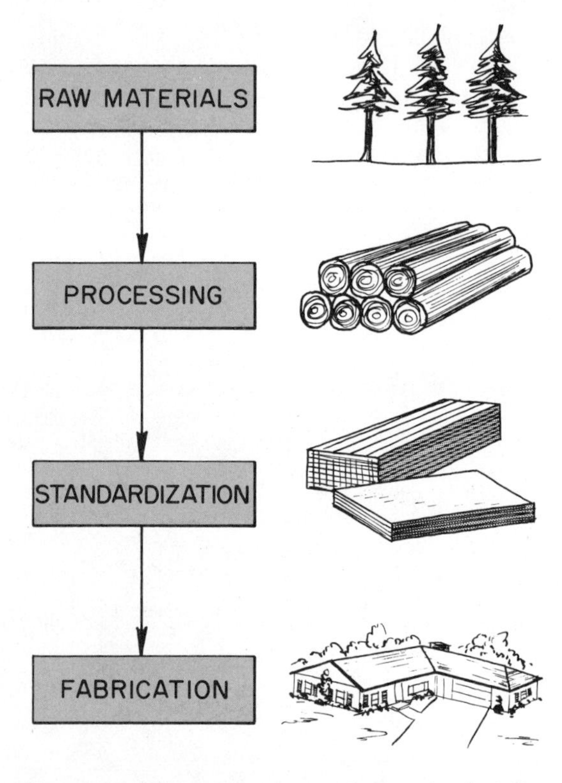

Fig. 1-40. The sequence of events in the wood industry.

THE WOODWORKING INDUSTRY

The woodworking industry includes those industries which manufacture or construct a product from wood. The products produced are as varied as the jobs involved in producing the products. The list of products is endless and consists of several thousands. The industry consists of a sequence of events. The events include (1) *raw material processing*, (2) *standardizing,* and (3) *fabricating*, Fig. 1-40.

RAW MATERIALS

The *raw material* in the wood industry is a mature tree standing in the forest. The first step toward product production is the selecting and marking of the trees for harvest, Fig. 1-41, by a forester. The marked

Fig. 1-41. Each year a forester marks the trees which are ready to be harvested. (California Redwood Association)

Fig. 1-42. Special loading and hauling equipment is used to move the trees from the forest to the sawmill. (Southern Forest Products Association)

trees are then cut and prepared to go to the sawmill. The logs are loaded on special trucks and hauled to the mill, Fig. 1-42. At the mill, the logs are sawed into lumber or plywood, Fig. 1-43.

The logs are sawed into *standard sizes.* That is, they are cut to a uniform thickness, width, and length. This is the first step in the standardization of materials.

The sawed lumber is placed in huge ovens called *kilns*, Fig. 1-44. The kilns are heated and force the moisture from the freshly-sawed boards. A large portion of the sawed lumber is planed smooth on four sides and prepared for shipment to the next industry. Most of the lumber is shipped on railroad cars. The lumber is often bundled and wrapped in large packages, Fig. 1-45. These packages make it possible to load and unload the lumber with machines. The wrapping material is waterproof and protects the lumber from the weather.

Fig. 1-44. A drying kiln removes the moisture from the freshly-sawn lumber.

STANDARD MATERIALS

The logs are cut into standard size materials at the sawmill. The standard sizes simplify mass marketing of the structural materials. The manufacturer knows what is

Fig. 1-43. Logs being sawn into lumber. (Simonds Saw and Steel Co.)

Fig. 1-45. The lumber is wrapped and packaged at the sawmill.

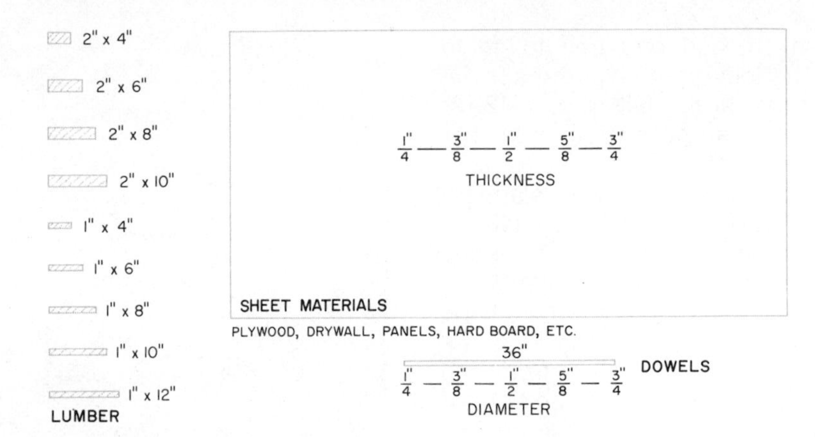

Fig. 1-46. Standard lumber and sheet sizes available in the wood industry.

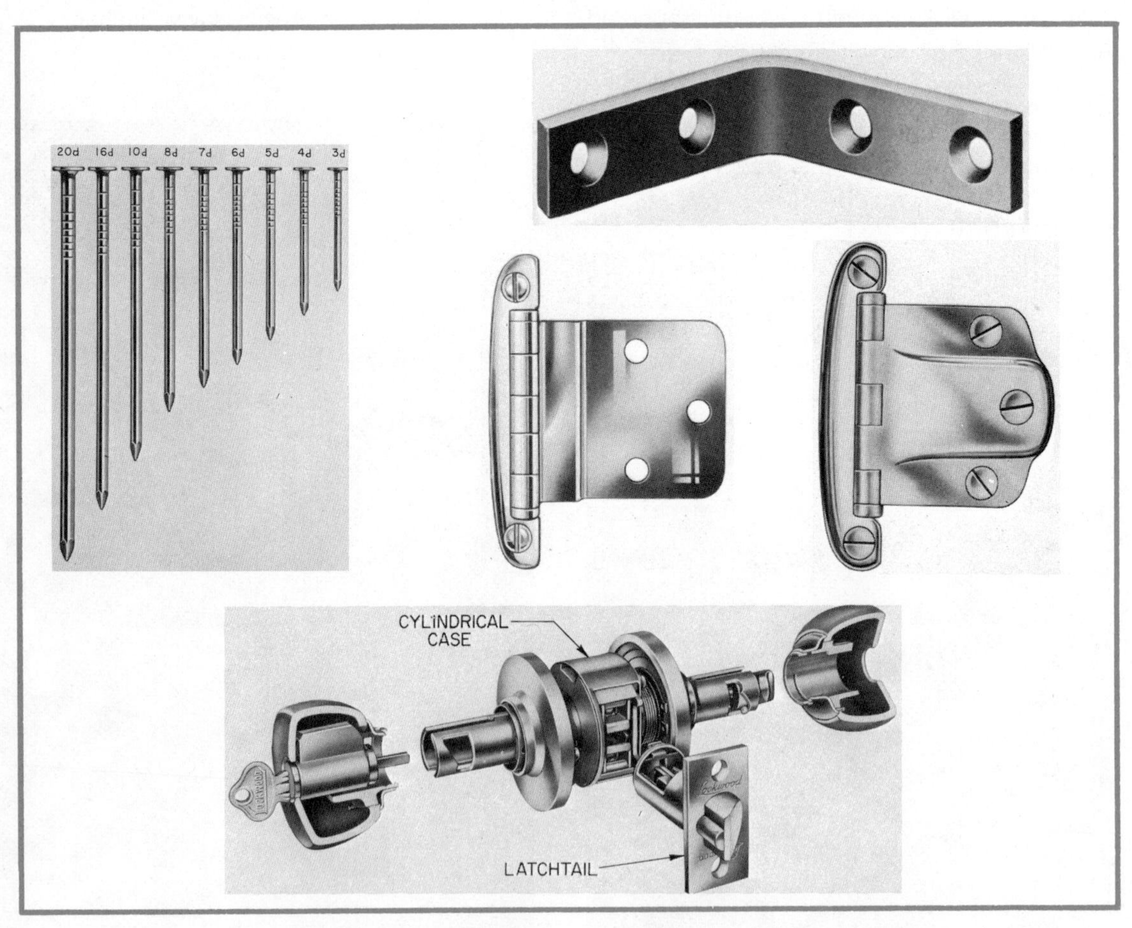

Fig. 1-47. An assortment of standard materials available to the wood industry. (Independent Lock Co.; American Steel & Wire Co.; and Stanley Tools)

available and can design his production system to utilize the materials. Figure 1-46 shows some of the standard lumber and wood sheet material sizes available.

There are other manufacturing companies which produce standard component parts. For example windows are manufactured in standard sizes. The main reasons for this standardization are interchangability and replacement factors. An architect can specify a window type and size and be sure that the window will be available when it is ordered.

Standard materials can save the manufacturing company money. The manufacturing company has a choice of (1) buying standard materials which are available from other suppliers, or (2) to produce their own. If the cost is greater to produce their own standard material, they will choose to buy the pre-cut material.

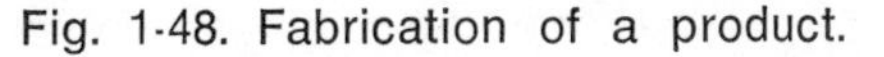

In our society of mass production manufacturing and product specialization, many companies produce only standard parts which are used by other companies. Some companies produce hardware such as hinges. However, this same company does not produce the cabinet doors on which they are eventually installed. Other companies producing doors do not manufacture hinges, and so on, Fig. 1-47.

When manufacturing companies need component parts which are not a line of their specialization, they generally will purchase them. The manufacturing companies can find sources of needed standard material from sales literature, catalogs, and representatives.

FABRICATION

The fabrication process is that of forming, shaping, and combining materials into consumable finished products, Fig. 1-48. *It should be understood that the finished product for one industry may be a standard material or component part for another industry,* Fig. 1-49.

Manufacturing industries are closely controlled and monitored organizations. They are a managed production system.

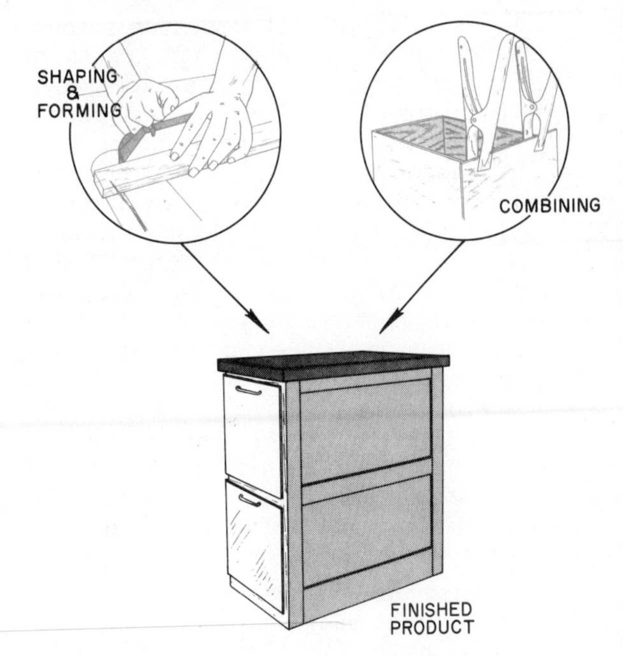

Fig. 1-48. Fabrication of a product.

Fig. 1-49. A kitchen cabinet is a component part for the construction industry, but is the finished product of a cabinetmaker.

The system for product production consists of *management practices, production practices* and *personnel practices.* The combined efforts of these divisions result in the efficient fabrication of a product, Fig. 1-50.

Management. Management practices are those concerned with the overall operation of the plant. The management division determines what will be produced, how it will be produced, and the quantity to be produced. This level is responsible for all major decisions influencing the operation of the company.

Production. This division of the industry is responsible for changing the form and shape of the raw material to add value to the materials. The products produced in the production practice division are those which are marketed.

Personnel. This division of the industry is concerned with coordinating the employees and the jobs. The purpose is to fit people to the jobs for the greatest worker output and worker satisfaction. Workers who enjoy their work are more productive.

PROBLEM 1

It is important that you tentatively select several occupations and then explore them before you actually accept a job. Working in school using the materials, processes, and tools is one way to test the realistic basis for your choice. However, you should also become acquainted with people involved in various jobs and talk with them about the work in which you are interested. It is likely there is a person in your neighborhood, school, church, or a friend of your family who works in the type of job in which you are interested.

Sometime during this course you should interview at least two people and ask them questions about the job. You will need to contact them in advance and make an

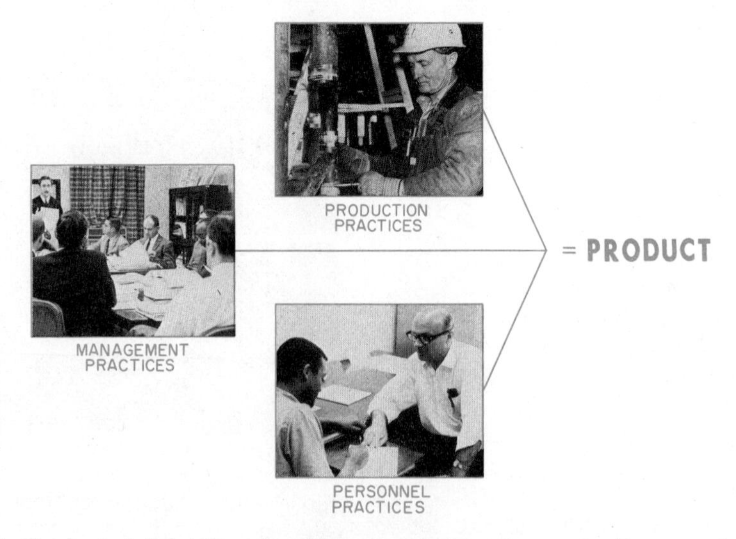

Fig. 1-50. Product fabrication involves management, production, and personnel practices.

appointment for the interview. Study the following interview guide and add any questions you would like to ask prior to the interview, then use this form to help you conduct an interesting and productive interview.

QUESTIONS FOR CAREER INTERVIEWS

1). How did you gain a knowledge of the wood industry (from school, on the job training, or through your work experience)?
2). What does your job involve? Do you need a special skill such as a carpenter? Are job opportunities available for qualified people? Do you meet or work with other people?
3). What type of education and training would be helpful to enter an occupation in the wood industry?
4). What are the working conditions for your job? Is the work seasonal? Can you work year around? Do you travel? Do you work inside most of the time? Is your work safe?
5). How does your income compare with other jobs in the community?
6). If you were my age, what type of a job would you prepare for in the wood industry?
7). Where do you apply for a job?

Take notes during the interview and then write up a report about the job. Do you like the type of work that the person you interviewed does? Why?

PROBLEM 2

Study the photographs of the wood products shown here. Identify all of the occupations that would have been involved with the creation of these products from raw materials to delivery to the consumer.

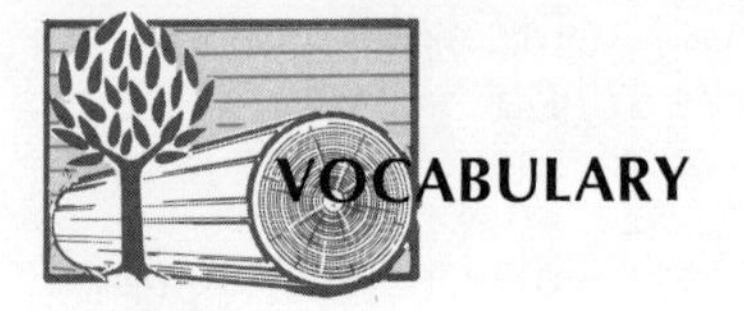

Economy
Manufacturing
Construction
Transportation
Service industry
Gross National Product (GNP)
Output
Component
Subassembly
Consumer

Occupations
- Unskilled
- Semiskilled
- Skilled
- Semiprofessional
- Professional

Fabrication
Management
Production
Personnel

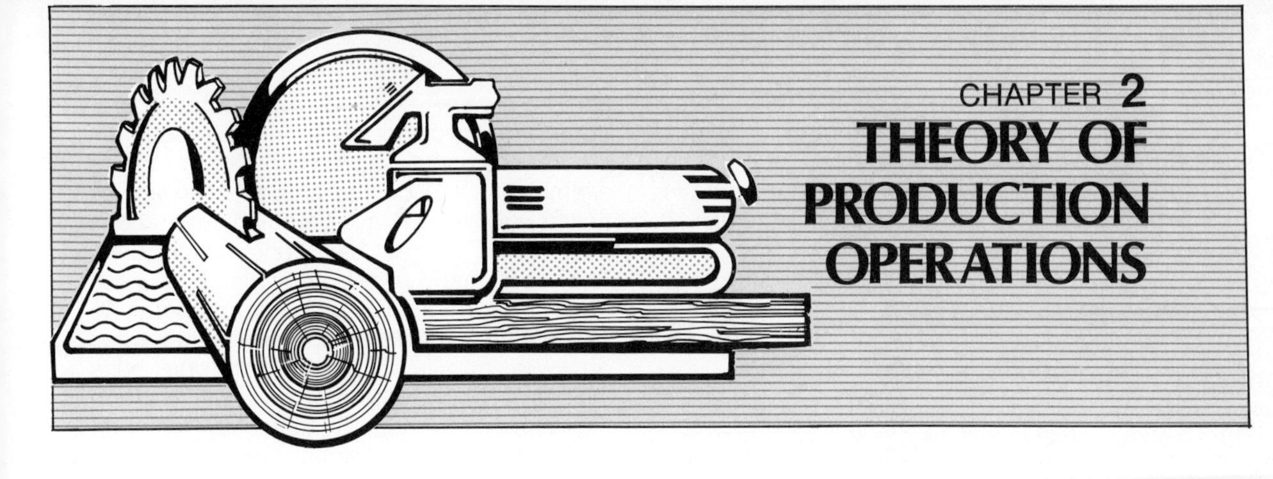

CHAPTER 2 THEORY OF PRODUCTION OPERATIONS

INTRODUCTION

Industry is based on *profit.* A company must make a profit to remain in business. Each production operation or process *adds value to a product.* Manufacturing industries attempt to *reduce the number of operations required* to produce their product. This results in a direct manufacturing savings and the product is less expensive to the consumer. If a product becomes overpriced, the potential consumer will either go without or find a substitute. For example, if plywood wall paneling becomes too expensive, the consumer may buy wallpaper to decorate a wall in his home. The wallpaper is a substitute for his first choice. As you can see, if the price becomes too high for the consumer, the manufacturer will lose his market. If he cannot sell his product, his company can no longer exist.

You will learn that there are many ways products can be produced. Industry must use the most efficient processes to produce a product which will make it competitive on the consumer market. You, as a consumer, might select one of the chairs shown in Fig. 2-1. If the quality of all of the products is the same, you will no doubt select the least expensive. An industry's production costs determines how much you will have to pay for a product. Industry knows it must keep its price competitive.

Fig. 2-1. Competitive products from which a consumer may select. (Kroehler)

The purpose of this chapter is to introduce you to the common production operations used in the wood industry. All of these processes are used for producing various wood products. The processes include:

1. Separating — the process of reducing the size of stock by cutting or shearing
2. Planing—the process of smoothing and reducing the size of stock by the removal of small pieces.
3. Cutting holes — the process in which a hole is cut by a revolving tool held at a right angle to the tool holder
4. Shaping of irregular parts — forming of a part which does not have all right angle corners

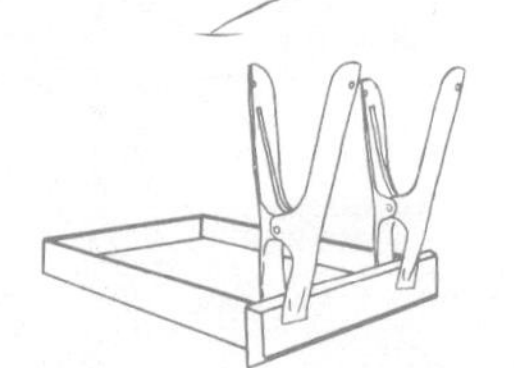

5. Assembly of parts — the process of positioning and fastening component parts into a finished product

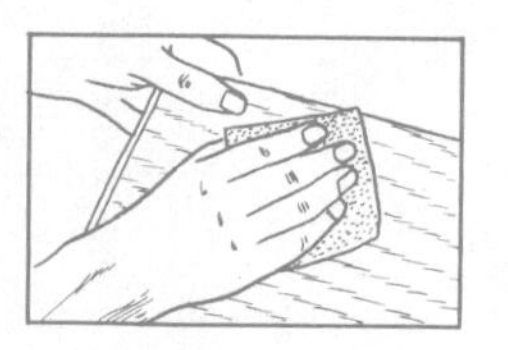

6. Conditioning of materials — the process of changing the properties of a material for better utilization.

7. Finishing — the process of applying decorative and protective coatings to a product.

As you become more involved in the study of the wood industry, you will have an opportunity to perform many of these operations. You first need to know what operations are involved in product production before designing and constructing one. For example, if you design a product in a particular way, it can save you many operations and will result in less construction time. There are many methods by which you can perform a particular production operation. The method you use depends on the types of tools and machines which you have available. Many different tools will perform the same operation. It is usually faster to use a hand tool when performing special production operations that are not repetitive such as trimming a door or cutting a special length of wood. However, repetitive operations such as cutting rafters or wall studs should be done on a machine so the cutting time is reduced by a mass production machine setup.

For example, there are many processes you could use to produce the product shown in Fig. 2-2. The purpose of this chapter is to show you the different tools and machines which will perform the *same operations*.

After you learn what tools and machines will produce the desired results, you then must make the selection decision. For example, if you only have one small piece of wood to separate, it may be faster to use

Fig. 2-2. Many different processes can be used to produce this product.

a handsaw. However, if you have many of them to cut, a machine which would do the same job would be a better choice. You will have to make decisions like this throughout this course. These decisions will be easy if you have information on which to base your decisions. This chapter will give you some of this information.

You will begin to think as people in decision-making positions in industry think. You will start asking yourself questions like this: *What is the best way to save time and money and still have a quality product*?

After you learn what the common production processes involve, other chapters in this book will show you how to perform each of the operations.

Hand tools have an economic purpose. Some operations are not performed very often, and you cannot afford to invest in expensive power equipment for occasional use. For example, a carpenter could not afford to invest in a shaping machine if he only had occasion to need it once a year— he would use a hand tool. The hand tool would require more time to perform the job; however, it would be less expensive for him to take more time rather than to invest in expensive equipment. If he used the machine every day, it would save him time, and therefore, he could make more money. These are the types of decisions industry must make. The decision will be made according to what will result in the greatest profit for the company.

PRINCIPLE OF SEPARATING

Separating is one of the most basic operations in the production of goods. *Separating* is the process by which one piece is removed from another to change the shape or size. This process is commonly referred to as a *sawing operation* because this method is most often used. However, there are two basic types of separating methods used in the wood industry. They are (1) *shearing* and (2) *sawing*. Shearing is a means of separating stock *without a loss of material.* For example, a log is rotary cut into thin sheets of material called veneer for manufacturing plywood. The thin sheets are removed from the log without any loss of material, Fig. 2-3.

Sawing *results in a loss of material* as the pieces are separated. The portion of the board through which the saw passes falls out as fine sawdust. This results in wasted material. Each separating cut with a saw results in at least 1/8″ of wasted stock, Fig. 2-4. Many of the expensive woods cut into

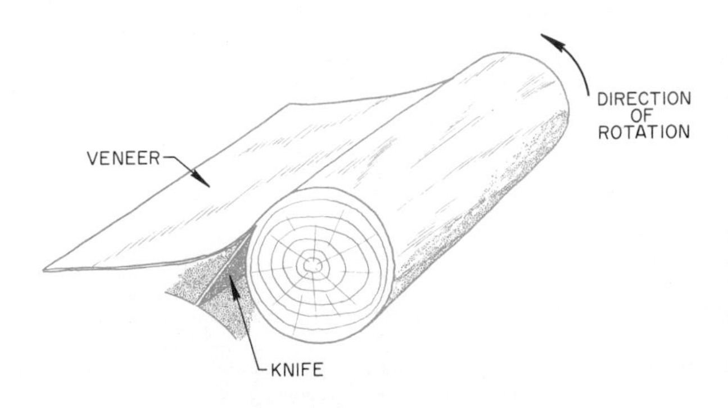

Fig. 2-3. Veneer is separated from the log by a shearing process.

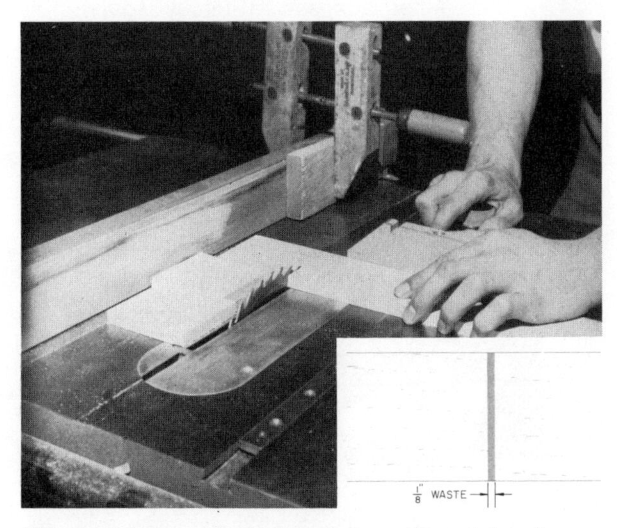

Fig. 2-4. Each separating cut wastes at least 1/8″ stock.

Fig. 2 5. Separating a tree from its roots by sawing. (National Forest Products)

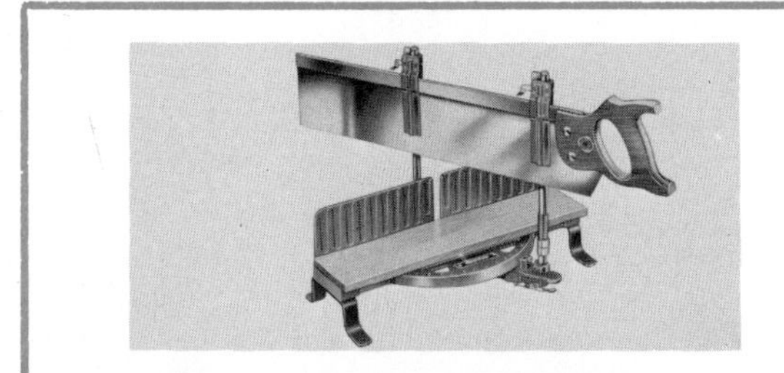

A. Miter box saw. (Stanley Tools)

B. Portable electric circular saw.

C. Dovetail saw. (H. K. Porter Co., Inc., Henry Disston Div.)

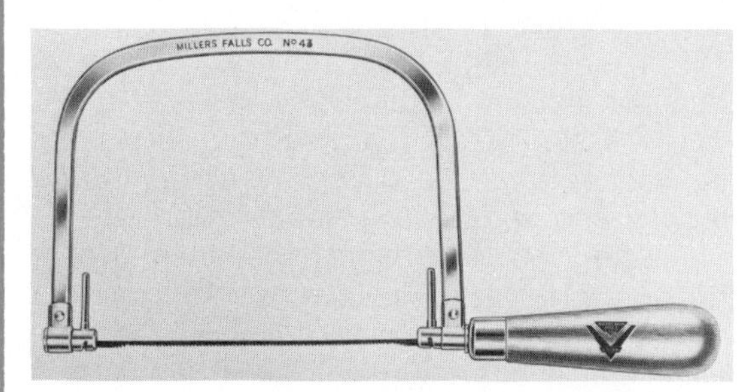

D. Coping saw. (Millers Falls Co.)

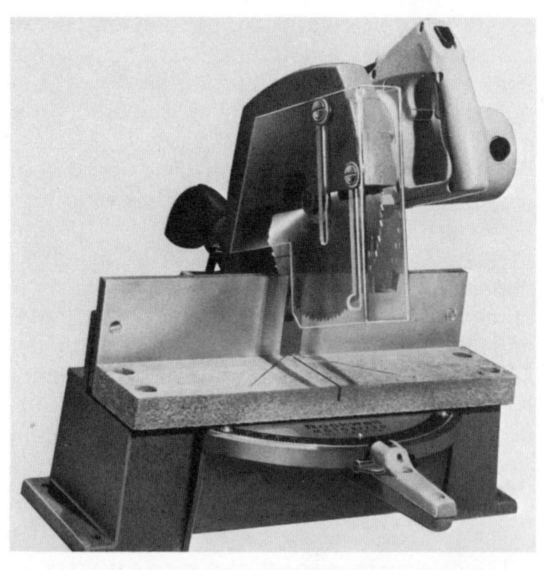

E. Motorized miter box saw. (Rockwell Mfg. Co.)

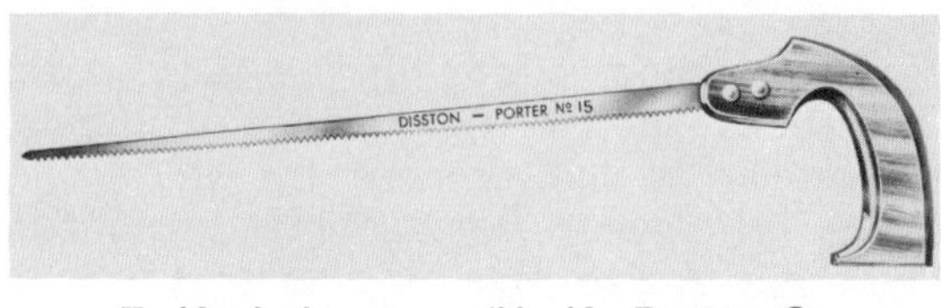

F. Keyhole saw. (H. K. Porter Co., Inc., Henry Disston Div.)

Fig. 2-6. A selection of sawing tools and machines.

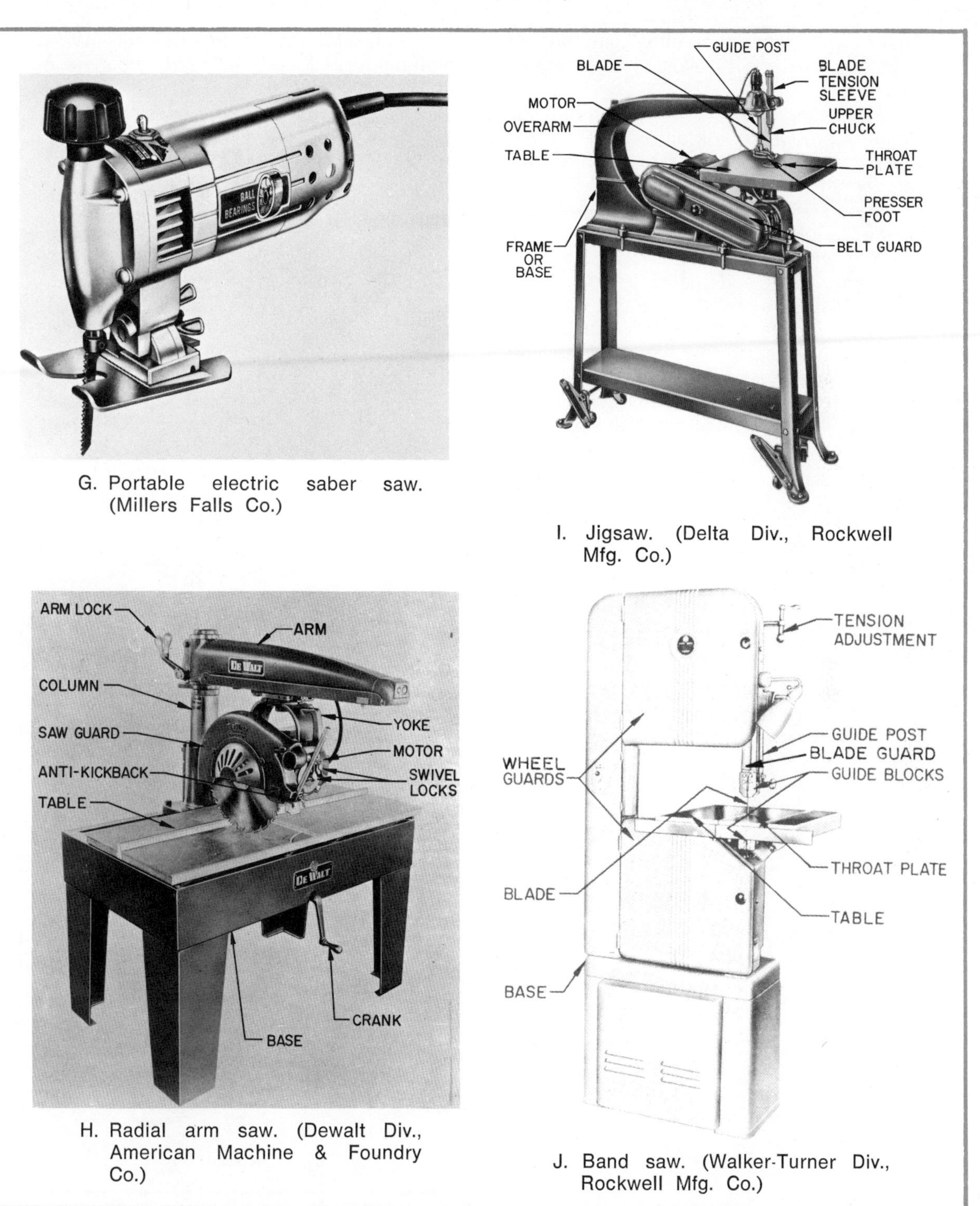

G. Portable electric saber saw. (Millers Falls Co.)

I. Jigsaw. (Delta Div., Rockwell Mfg. Co.)

H. Radial arm saw. (Dewalt Div., American Machine & Foundry Co.)

J. Band saw. (Walker-Turner Div., Rockwell Mfg. Co.)

veneer are separated into thin veneer sheets by shearing to avoid this excessive material waste.

Separating may be used at nearly all stages of production. As an example, a tree is separated from its roots to become raw material for other products, Fig. 2-5.

Separating, as a production process, can be performed with many types of tools and machines. Figure 2-6 shows a few of the sawing tools and machines which are used to separate wood.

The principle of separation remains the same regardless of the types of sawing tools and machines used, Fig. 2-7. As the teeth enter the wood, small chips are removed. The section of wood the width of the saw blade which is removed is called the *kerf,* Fig. 2-8. The kerf is wider than the back of the saw blade because the teeth are *set.* The set of the teeth prevents the saw from binding due to friction as the blade enters the board.

The shape of the saw blade determines the purpose of a particular saw, Fig. 2-9. A wide blade is used for straight cuts. This blade is rigid and will follow the straight line. A narrow blade is designed for cutting curves because it will turn easily. The blade must be very narrow if it must follow a sharp curve. Some saws have pointed blades which are designed for inserting through small holes for making internal cuts. When separating stock, *first evaluate the type of cut you will be making, and then select a saw designed for this type of work.*

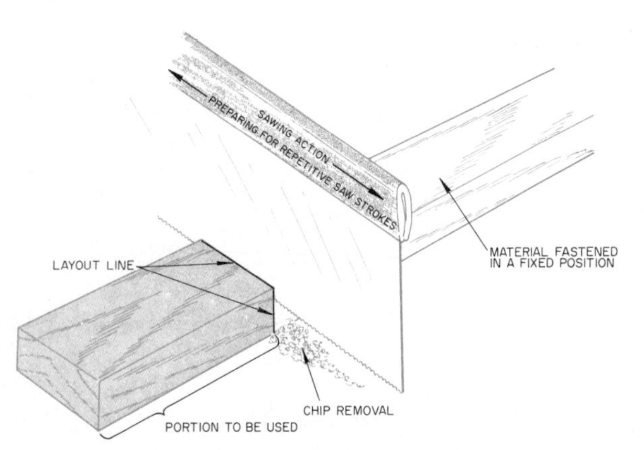

Fig. 2-7. Basic principle of separating wood with a saw.

HAND TOOLS FOR SEPARATING WOOD

The hand tools for separating wood each have a particular purpose. Proper

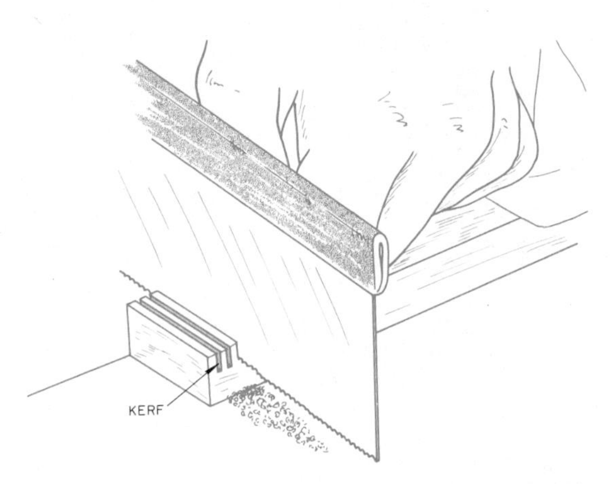

Fig. 2-8. The kerf is the section of wood removed by the saw.

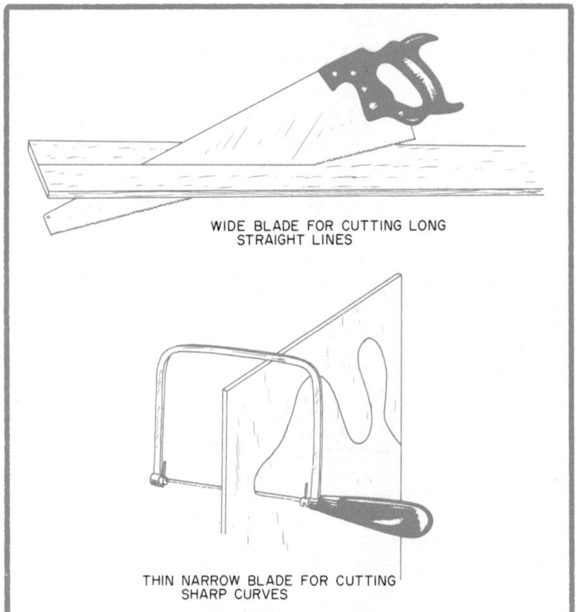

Fig. 2-9. The shape of the blade determines its purpose.

tool selection results in better and faster work.

Crosscut saws. Crosscut saws are used for cutting across the grain of a board to reduce the length. The more teeth the saw has, the smoother the cut. A saw with fewer teeth cuts faster. Figure 2-10 shows a crosscut saw and how it is used.

Ripsaws. Ripsaws are used for separating a board to the desired width. Figure 2-11 shows a ripsaw being used to separate a board.

Coping saws. Coping saws are used for cutting curves and irregular shapes. The blade is replacable and is held in the frame, Fig. 2-12. The blade can be placed through a hole to make internal cuts on a board.

Compass saws. Compass saws have a tapered blade which comes to a point on the end. This saw is used to cut an inside hole on a board. The blade will also follow a curve because it is narrow. Figure 2-13

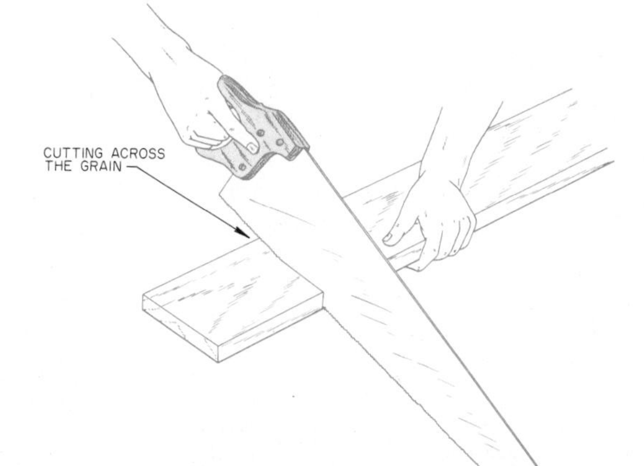

Fig. 2-10. A crosscut saw.

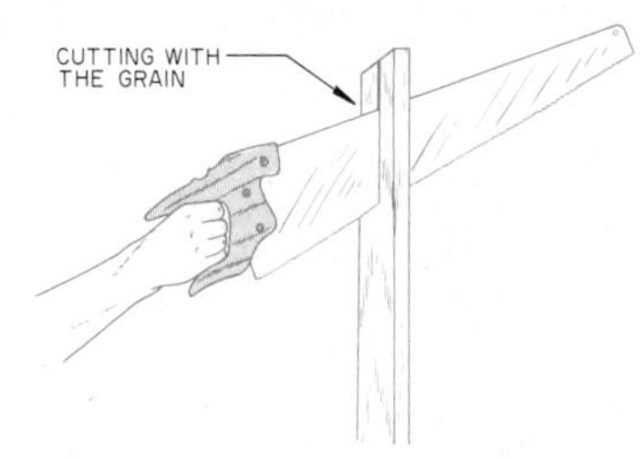

Fig. 2-11. A ripsaw.

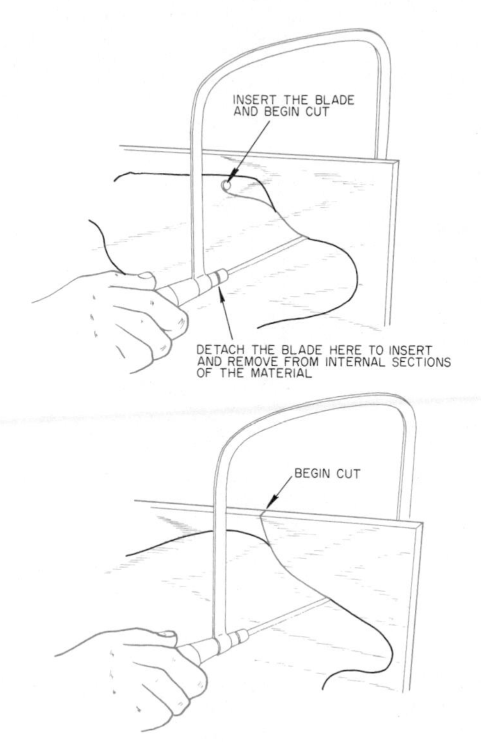

Fig. 2-12. A coping saw.

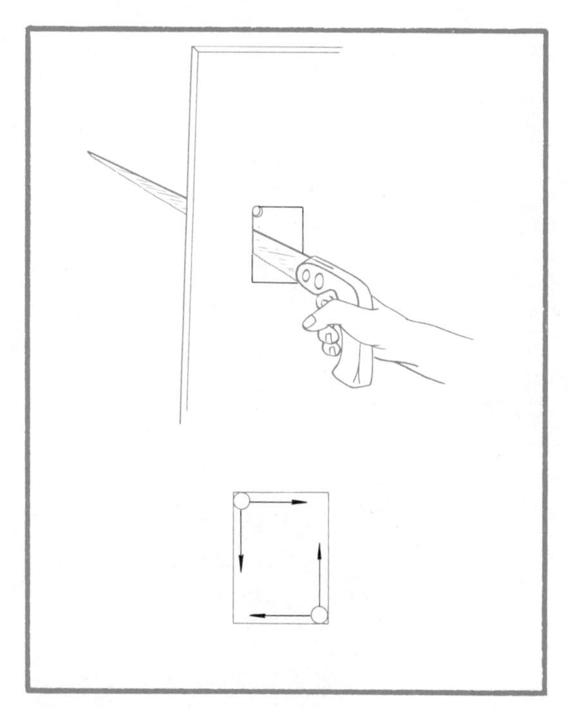

Fig. 2-13. A compass saw.

shows the compass saw being used to make an internal cut and an external cut. The hole must be bored to start the end of the blade for internal cutting.

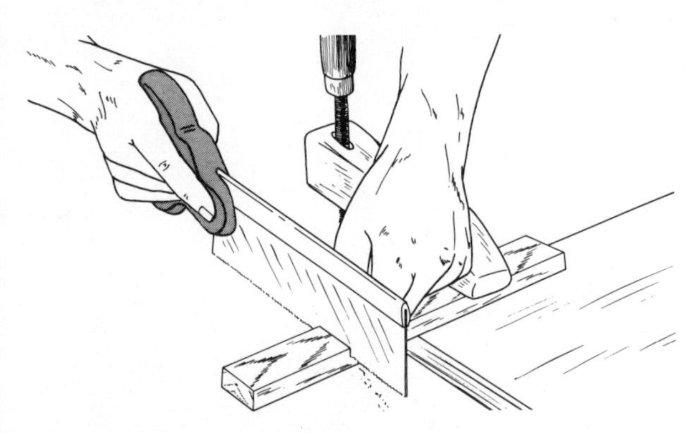

Fig. 2-14. A backsaw.

Fig. 2-15. A dovetail saw.

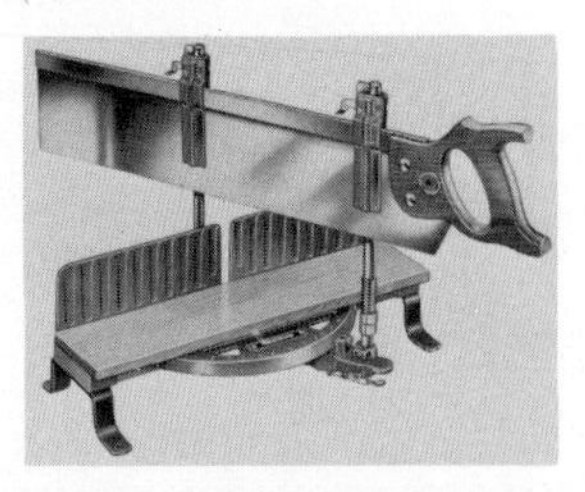

Fig. 2-16. A miter box saw. (Stanley Tools)

Backsaws. Backsaws, Fig. 2-14, have teeth like a crosscut saw. This saw is often used for cutting joints. The blade is very thin and removes a narrow kerf. The thin blade is reinforced with a steel back on the top of the blade. This saw will produce a very smooth cut.

Dovetail saws. Dovetail saws, Fig. 2-15, are similar in appearance to backsaws. The handle and blade are the major differences. The dovetail saw is also used for cutting joints. The teeth are very fine and produce a smooth cut.

Miter box saws. Miter box saws are used in combination with a metal frame, Fig. 2-16. The frame helps to guide the saw as the cut is made. The saw can be turned to any desired angle for cutting miter joints and other angular cuts. The saw used with a miter box is a large backsaw designed to fit the frame.

PORTABLE POWER SAWS

There are several types of portable power saws used for separating wood. The portable power saws are powered by an electric motor. Portable power saws produce high quality work and speed the operation. The two most common portable power saws often used are the portable jigsaw and circular saws.

Portable jigsaw. A portable jigsaw, Fig. 2-17, is also called a saber saw. This is a power machine designed to do work

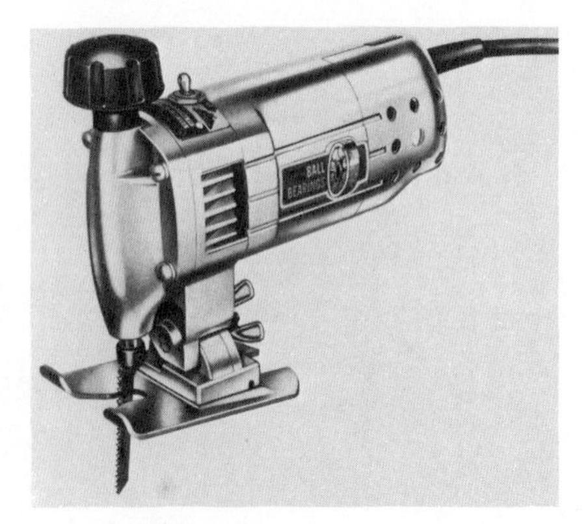

Fig. 2-17. A portable power jigsaw. (Millers Falls Co.)

similar to that of a coping saw. This saw is limited to light work and is not designed for crosscutting and ripping large stock.

Portable circular saws. Portable circular saws, Fig. 2-18, are generally used on the construction site by carpenters. They are excellent machines for cutting large stock to width and length. They are not designed for performing close tolerance finish work.

STATIONARY SAWING MACHINES

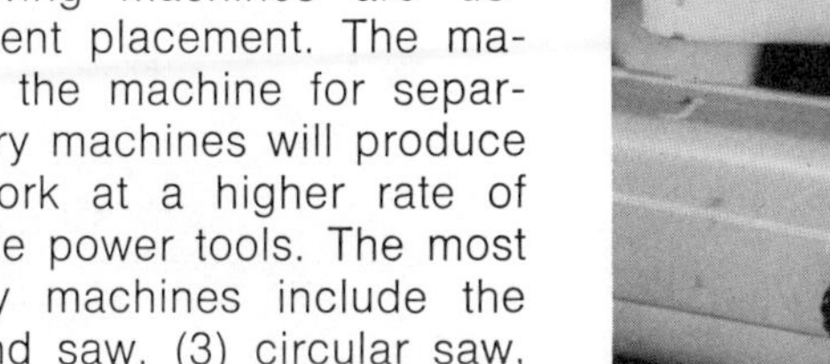

Stationary sawing machines are designed for permanent placement. The material is moved to the machine for separating. The stationary machines will produce close tolerance work at a higher rate of speed than portable power tools. The most common stationary machines include the (1) jigsaw, (2) band saw, (3) circular saw, (4) radial arm saw, and (5) power miter saws.

Jigsaws. Jigsaws are used for cutting curves and irregular shapes. The blade moves up and down as it operates. The blade is clamped in place at both ends and can be easily replaced. Figure 2-19 shows a jigsaw being used for cutting a curve on a piece of stock. The blade can be loosened on one end and placed through a hole in the board for making internal cuts, Fig. 2-20.

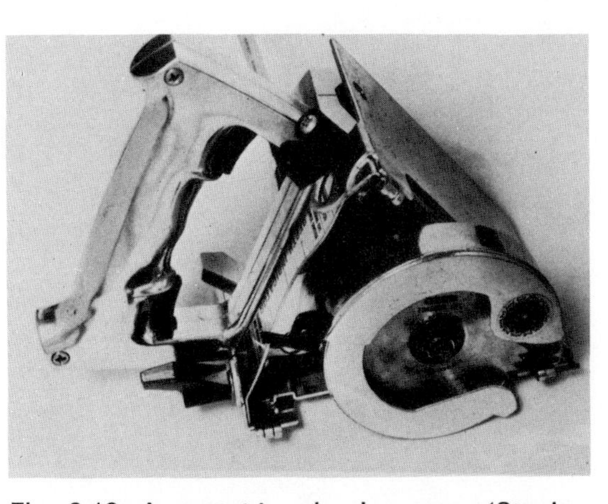

Fig. 2-18. A portable circular saw. (Stanley Tools)

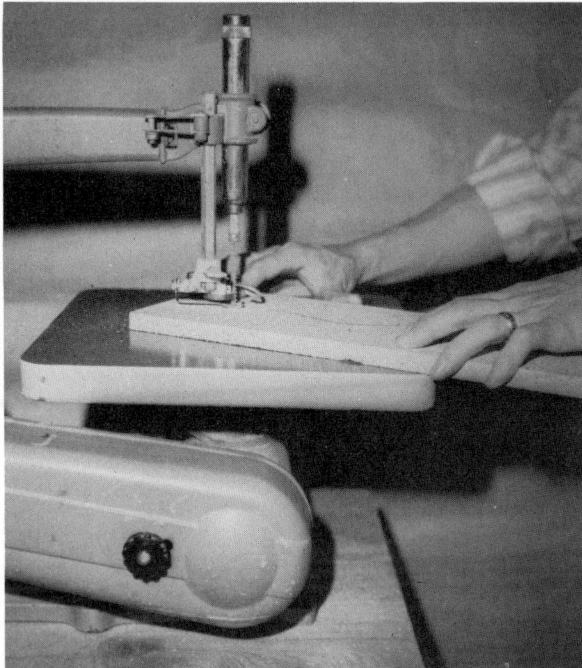

Fig. 2-19. Cutting stock on a jigsaw.

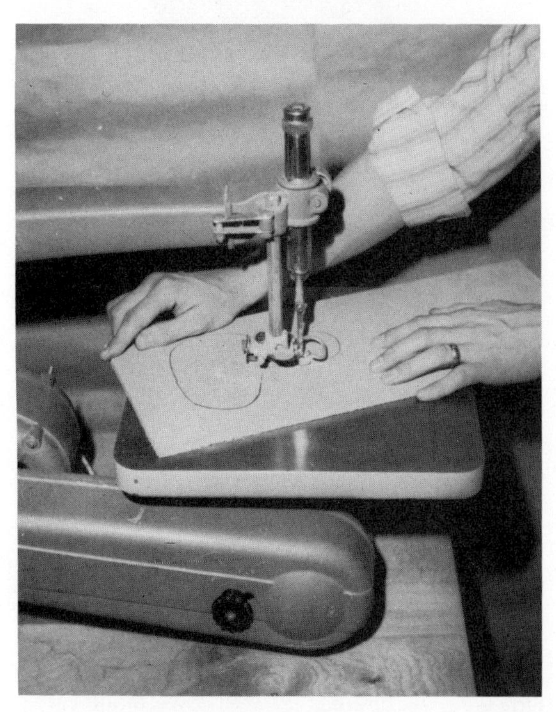

Fig. 2-20. Making an internal cut on a jigsaw.

Band saws. Band saws, Fig. 2-21, are used for cutting large curves and irregular shapes. The blade in the band saw is one continuous band which runs on the wheels. The band saw can also be used for cross-cutting and ripping stock.

Circular saws. Circular saws, Fig. 2-22, are the most-used production machines in the wood processing industry. The blade is mounted on a shaft called an *arbor*. The machine is equipped with a crosscutting support called a *miter gauge*. This slides in a slot machined into the table top. The miter gauge supports the stock as it is passed through the revolving blade, Fig. 2-23.

A straight edge called a fence is clamped in position on the table top to guide the stock as it passes through the blade, Fig. 2-24. The circular saw is used for making angular cuts, tapers, joints and many other uses, Fig. 2-25.

Fig. 2-21. A band saw

Fig. 2-23. Cutting stock to length on a circular saw.

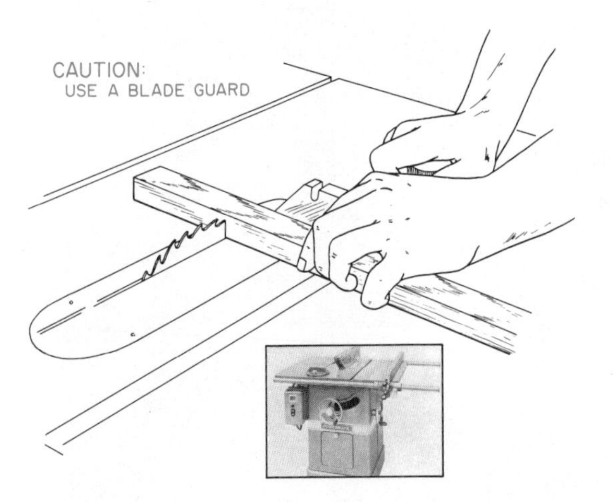

Fig. 2-22. A circular saw. (Powermatic Machine Co.)

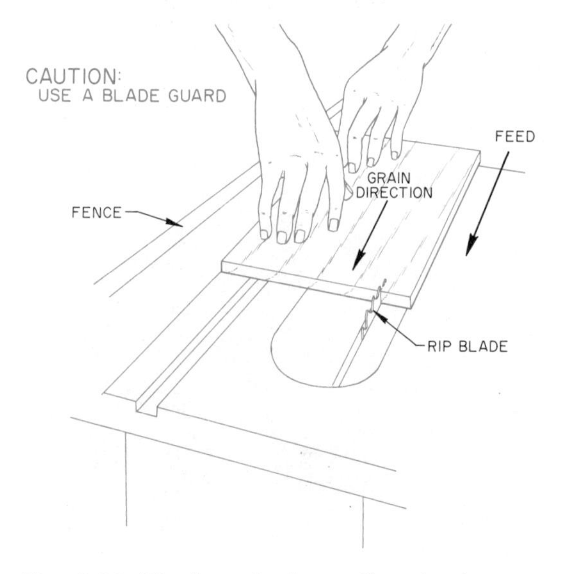

Fig. 2-24. Ripping stock on the circular saw using a fence to guide the stock.

Radial arm saws. Radial arm saws such as shown in Fig. 2-26 are also general-purpose saws. This saw will perform many of the operations which can be done on a circular saw. The stock is placed on the table and the blade is pulled through the stock for most work. However, the head of the machine is rotated 90° for ripping operations. The stock is then pushed through the blade for this and similar operations, Fig. 2-27.

Power miter saws. Power miter saws such as shown in Fig. 2-28 are designed for installing trim in houses and other structures. An accurate 45° angle can be cut for constructing mitered joints. This saw can also be used for light cutoff operations.

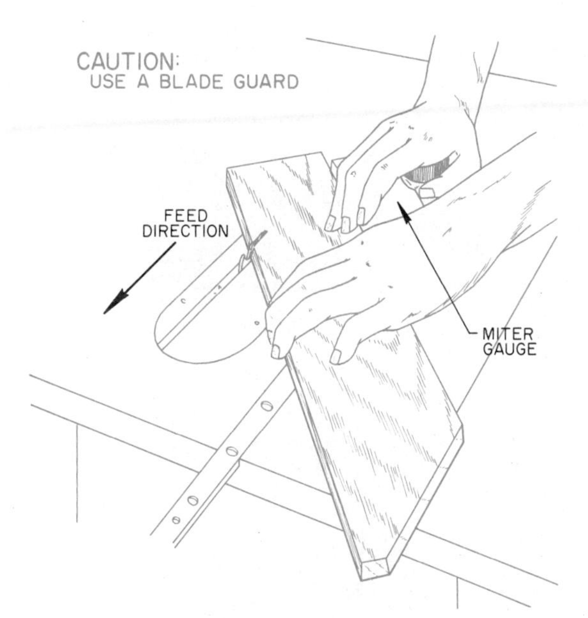

Fig. 2-25. Making an angular cut on the circular saw.

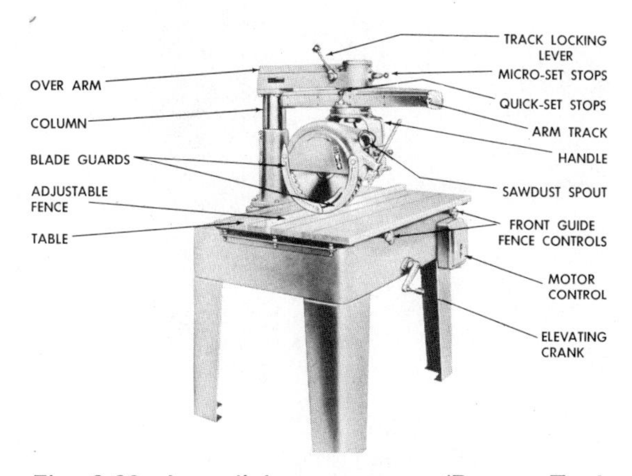

Fig. 2-26. A radial arm saw. (Power Tool Div., Rockwell Mfg. Co.)

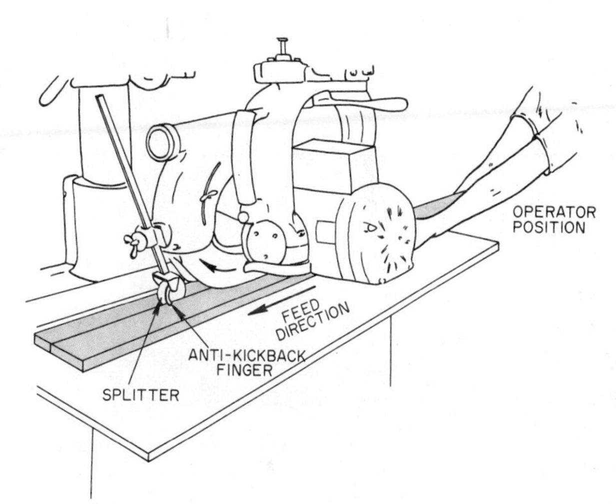

Fig. 2-27. Ripping stock on the radial arm saw.

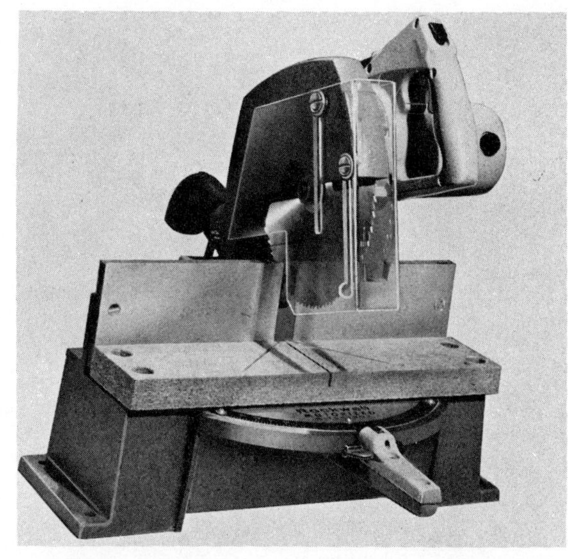

Fig. 2-28. A power miter saw. (Rockwell Mfg. Co.)

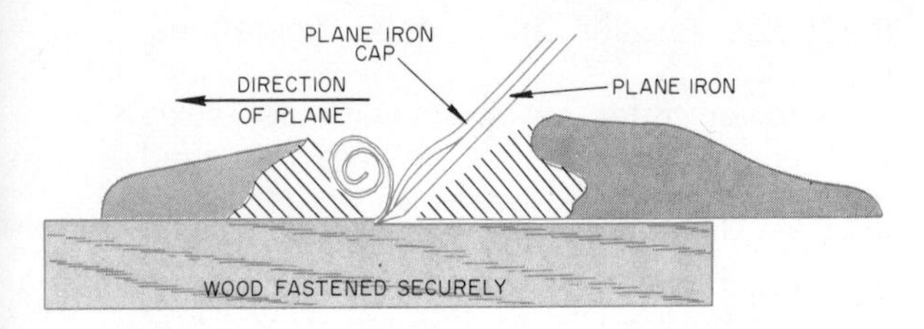

Fig. 2-29. Cutting action of a hand plane.

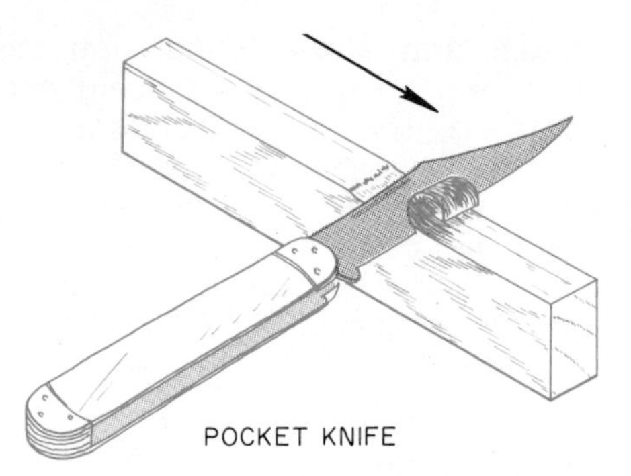

Fig. 2-30. A pocket knife cuts with a wedging action like a plane.

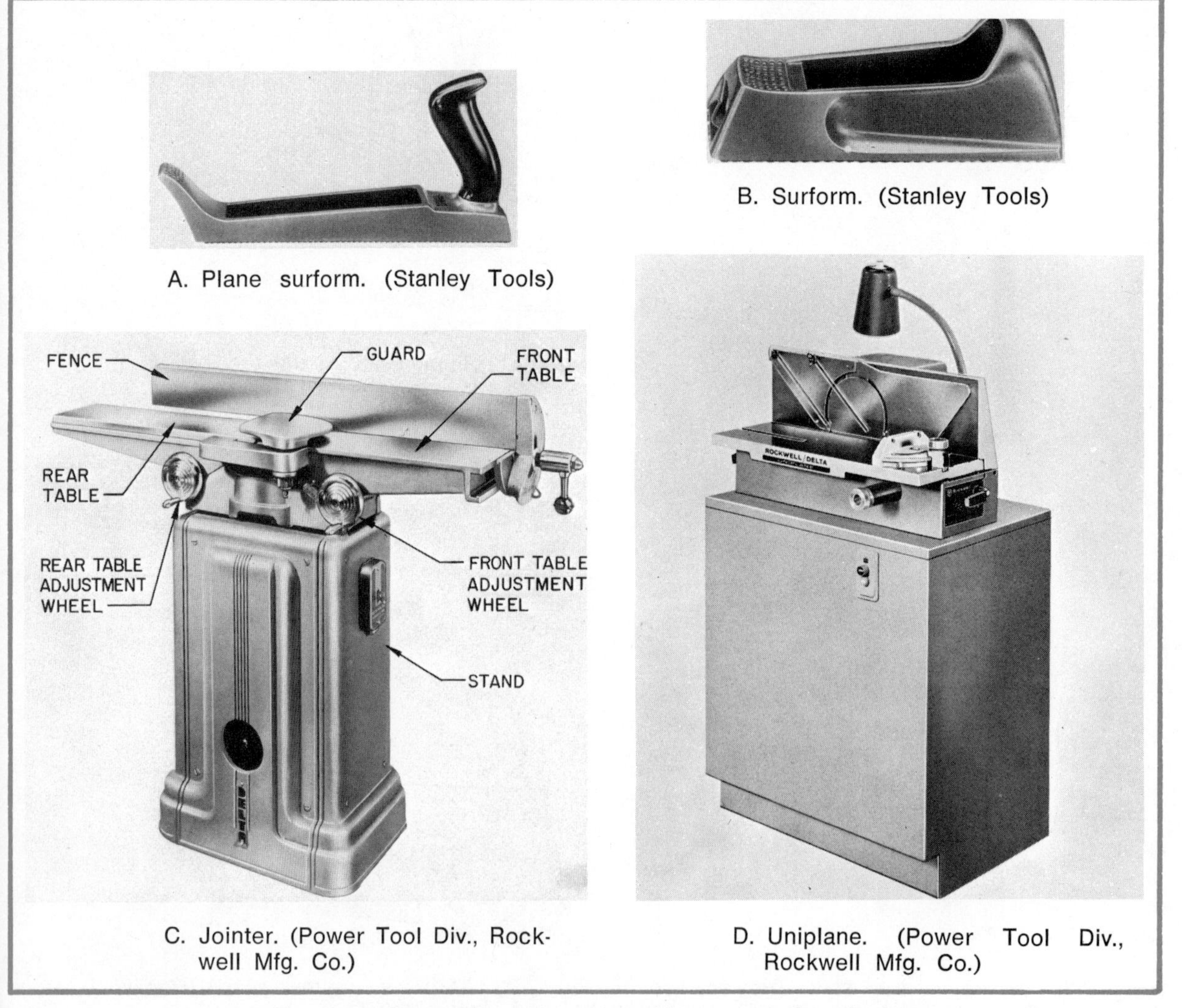

A. Plane surform. (Stanley Tools)

B. Surform. (Stanley Tools)

C. Jointer. (Power Tool Div., Rockwell Mfg. Co.)

D. Uniplane. (Power Tool Div., Rockwell Mfg. Co.)

Fig. 2-31. Common tools and machines used for planing.

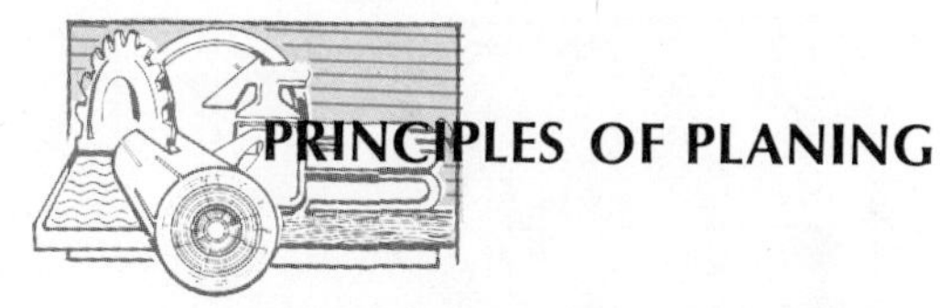

PRINCIPLES OF PLANING

Planing serves two main purposes: (1) to reduce the stock in size and (2) to smooth the surface. Stock is planed with edge-cutting tools. These are tools which have cutting blades shaped like inclined planes, Fig. 2-29. The cutting edge shears the wood fibers. A pocket knife produces a shearing cut, Fig. 2-30. The cutting principle remains the same for all edge-cutting tools and machines.

Edge-cutting tools include hand planes, chisels, power driven planes, and surfacing machines. Detailed instruction for planing and smoothing stock is given in Chapter 10. Figure 2-31 shows a few of the many hand tools commonly used for planing surfaces.

HAND PLANING TOOLS

There are many special-purpose planing tools. It is important that you select the correct tool for the particular operation. The various tools will be classified as hand planes, chisels, and power machines.

Jack planes. Jack planes are the most common planes used in woodworking. The jack plane is regarded as a general-purpose plane, Fig. 2-32. If you learn to use this plane, you will be able to assemble and operate nearly every type of plane. The jack plane is about 14″ long. This plane is used for rough planing, smoothing surfaces, edges, and ends and reducing stock to size.

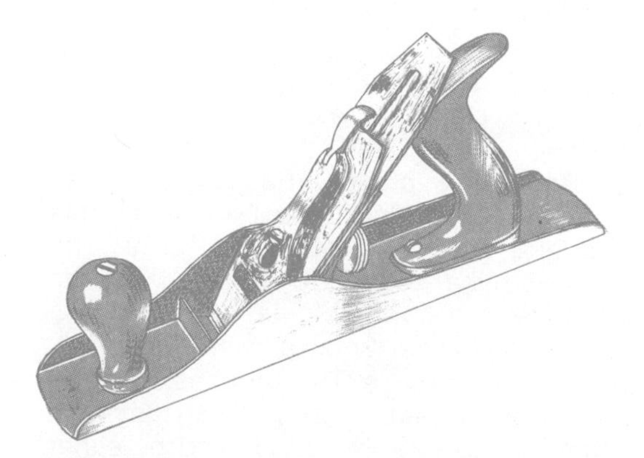

Fig. 2-32. A jack plane. (Stanley Tools)

Smooth planes. Smooth planes are like the jack planes except in size. Smooth planes are shorter in length. The smooth plane is about 10″ long, Fig. 2-33. It is used for many of the same purposes as the jack plane.

Jointer planes. Jointer planes are longer than the jack plane. It measures about 22″ in length, Fig. 2-34. This plane is used for smoothing and trueing long edges.

The jack plane, smooth plane, and jointer plane all have *double-plane iron assemblies*, Fig. 2-35. The cutting edge

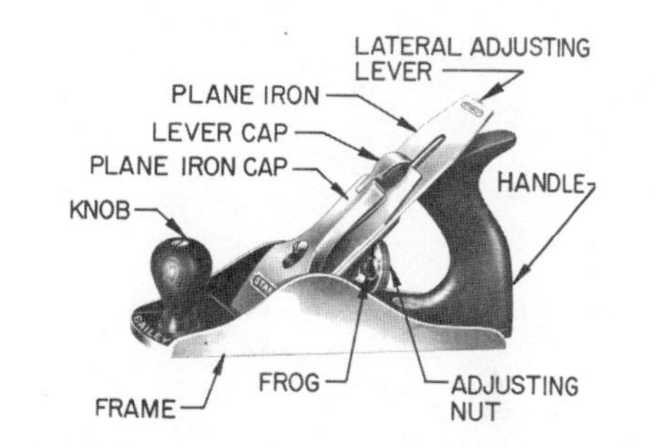

Fig. 2-33. A smoothing plane. (Stanley Tools)

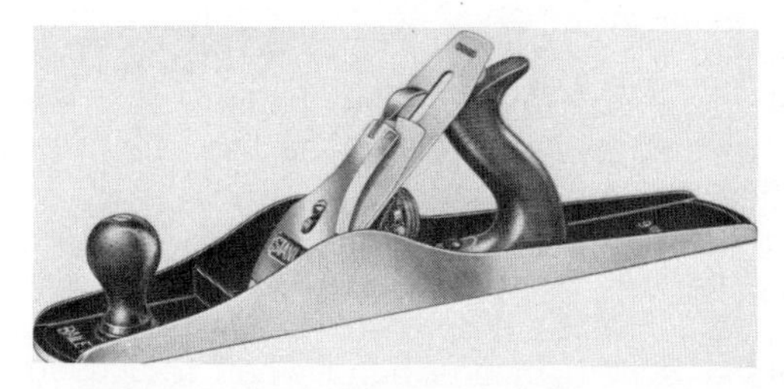

Fig. 2-34. A jointer plane. (Stanley Tools)

Fig. 2-35. A double plane iron assembly.

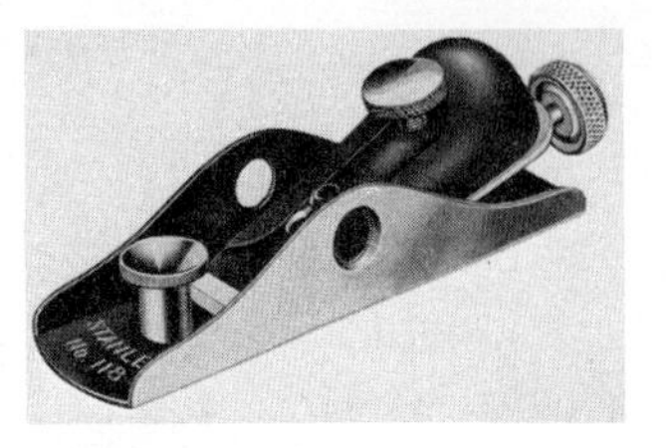

Fig. 2-36. A block plane. (Stanley Tools)

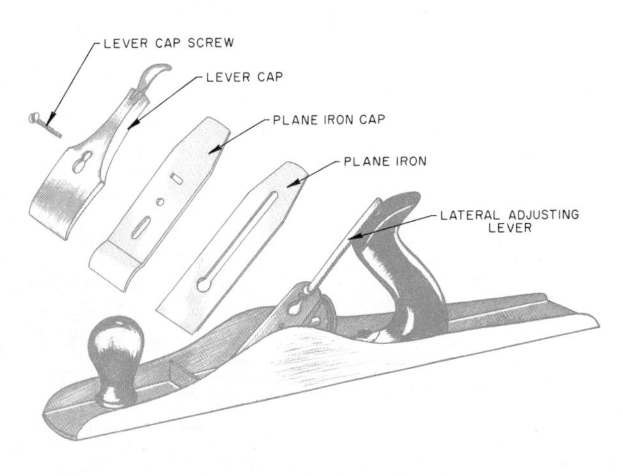

Fig. 2-37. Single plane iron assembly.

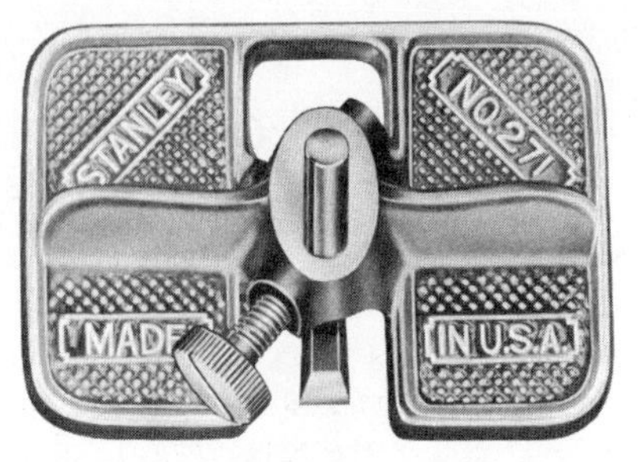

Fig. 2-38. A router plane. (Stanley Tools)

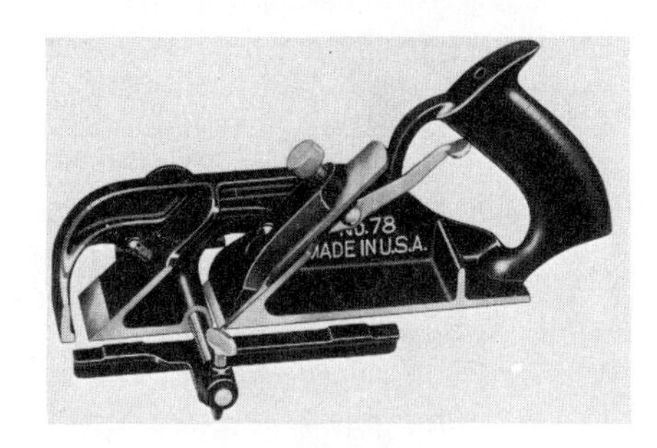

Fig. 2-39. A rabbeting plane.

Fig. 2-40. A spokeshave.

consists of a blade called a *plane iron* and a *plane iron cap*. The two parts are held together by means of a cap screw. The plane iron cap causes the shavings to break and curl out of the plane as they are sheared from the wood, Fig. 2-37.

Block planes. Block planes are the smallest of the planes. They range in size from 4″ to 8″ long, Fig. 2-36. The block plane is held in one hand when it is used. It is commonly used to plane end grain. Carpenters often use this tool for fitting siding and molding around doors and windows.

The block plane has a single-plane iron. The plane iron is set at a low angle in the plane.

Router planes. Router planes are special-purpose tools used for constructing dado joints, Fig. 2-38. The cutter or bit can be lowered as the cut is made. The sides of the dado are first cut with a backsaw and the excess stock is removed with the router plane.

Rabbeting planes. Rabbeting planes are used for cutting rabbet joints on the ends and edges of stock, Fig. 2-39. The fence and the cutter are adjusted for the desired size of rabbet.

Spokeshaves. Spokeshaves, Fig. 2-40, are used like a plane. This tool has a single-plane iron assembly. It is used to form irregular shapes. This tool is used by either pulling or pushing.

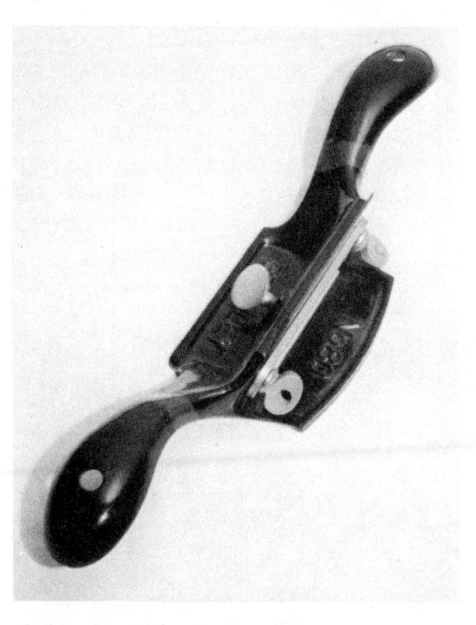

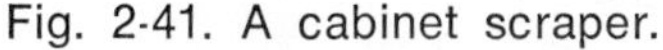

Fig. 2-41. A cabinet scraper.

Cabinet scrapers. Cabinet scrapers are used much like a spokeshave, but the blade is different. It has a "hooked" edge and is a scraping tool, Fig. 2-41. It is used to remove small amounts of stock. The cabinet scraper is generally used to smooth the surface just before sanding.

HAND CHISELS

Hand chisels cut stock very much like planes. The main difference is that they are guided entirely by the hands. There are many types of chisels, but the two most common are: (1) the tang chisel and (2) the socket chisel. Chisels are most often used for joining and fitting pieces together after they have been planed to size.

Tang chisels. Tang chisels are constructed so that part of the chisel called the tang enters the handle, Fig. 2-42. It is used to remove excess stock. It is used mainly as a hand paring chisel and is not pounded with a mallet.

Socket chisels. Socket chisels are constructed with the handle entering part of the chisel, Fig. 2-43. This tool is usually more durable and can withstand light pounding.

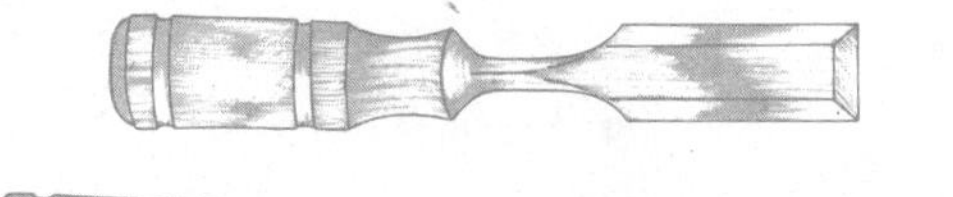

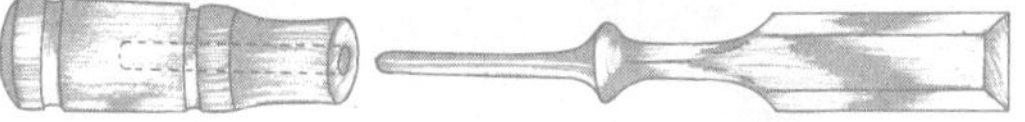

Fig. 2-42. A tang chisel.

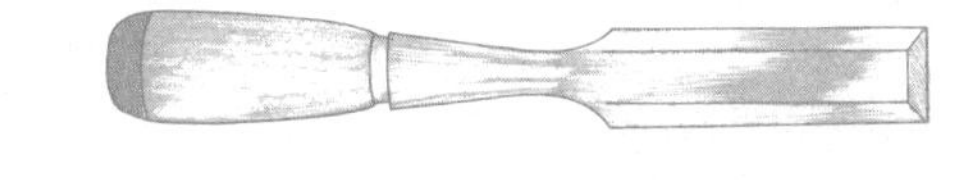

Fig. 2-43. A socket chisel.

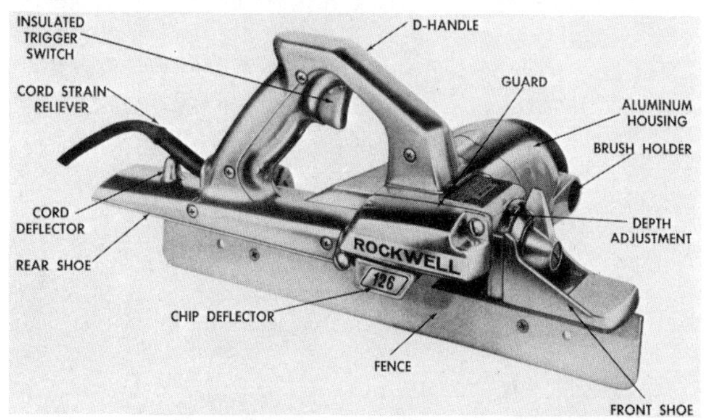

Fig. 2-44. A standard portable power plane. (Power Tool Div., Rockwell Mfg. Co.)

PORTABLE POWER-DRIVEN PLANES

Portable power-driven planes are generally used for smoothing surfaces and edges. However, they can also be used for planing tapers, bevels, chamfers, and end grain. The portable power planes are used for such jobs as planing doors and truing edges for joining. The two basic types of portable power planes are the (1) standard electric plane and the (2) power block plane. Chapter 10 provides detailed instruction for using edge cutting tools.

Standard electric planes. Standard electric planes are general-purpose tools similar to the hand jack plane, Fig. 2-44.

This plane is used for reducing stock to size and preparing stock to be joined together.

Power block planes. Power block planes can be compared to the hand block plane. It is smaller and lighter than the standard electric plane, Fig. 2-45. It is held in one hand and is used on short surfaces and edges.

STATIONARY POWER PLANES

Stationary planing machines are designed for permanent placement. The work is taken to the machine for processing. Stationary machines usually have greater capacities for work and will produce high-quality work. The three most used stationary planing machines are the (1) *jointer*, (2) *surface planer*, and (3) *uniplane*.

Uniplanes. Uniplanes are used for surfacing and squaring stock. They are especially useful for planing small pieces, Fig. 2-46.

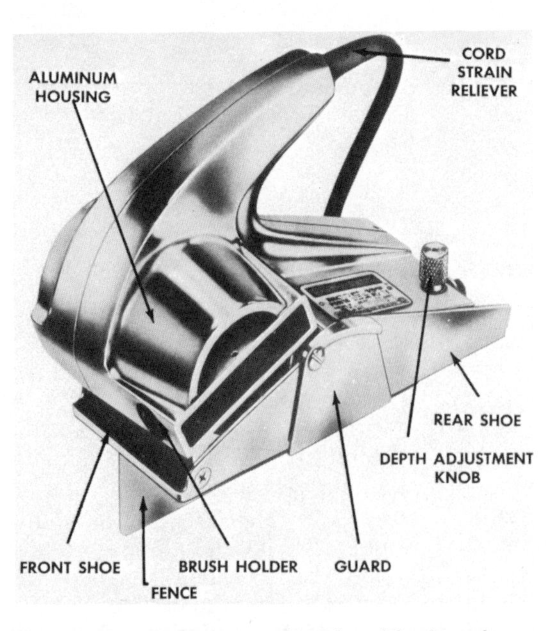

Fig. 2-45. A power electric block plane. (Power Tool Div., Rockwell Mfg. Co.)

Jointers. Jointers are generally used for truing and smoothing the edges and surfaces of boards, Fig. 2-47. In addition to these operations, the jointer will also perform rabbeting and tapering operations.

Surface planers. Surface planers (sometimes called thickness planers or surfacers)

Fig. 2-46. A uniplane. (Power Tool Div., Rockwell Mfg. Co.)

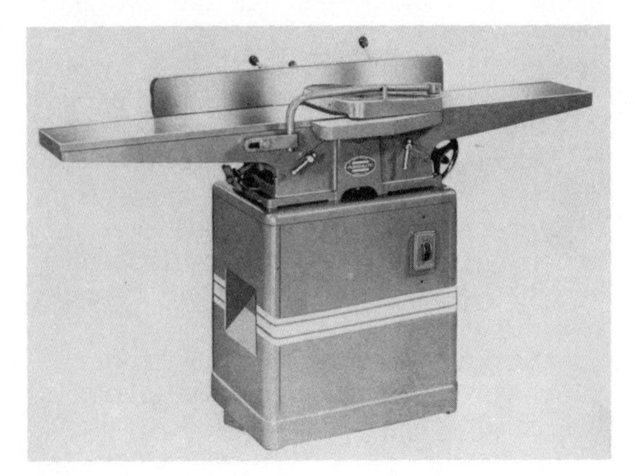

Fig. 2-47. A jointer.

are single-purpose machines, Fig. 2-48. They will produce a level surface as they reduce the stock to thickness.

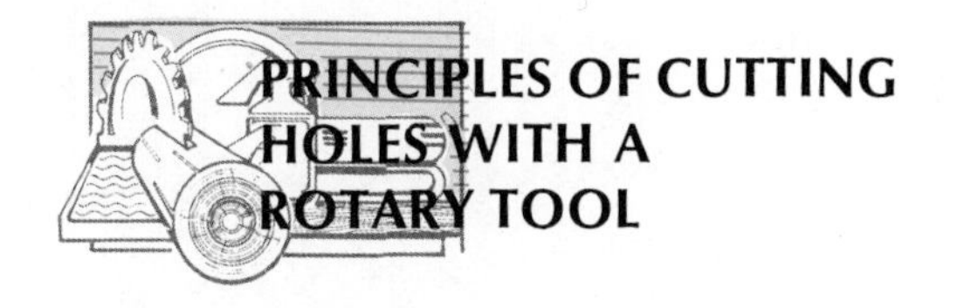

PRINCIPLES OF CUTTING HOLES WITH A ROTARY TOOL

Cutting holes is a general term used to refer to the process of drilling and boring. The cutting of holes as a production process refers to the *linear control of a rotary tool,* Fig. 2-49.

The making of holes is an essential production operation. The hole-cutting processes are used for the assembly of wood parts into finished products, Fig. 2-50. The holes are used for the installation of mechanical fasteners and the joining of wood to wood.

The hole-cutting processes are generally divided into two categories. They are (1) *drilling* and (2) *boring*. Holes larger than ¼″ in diameter are *bored,* and those less than ¼″ are *drilled.* Various hand tools and

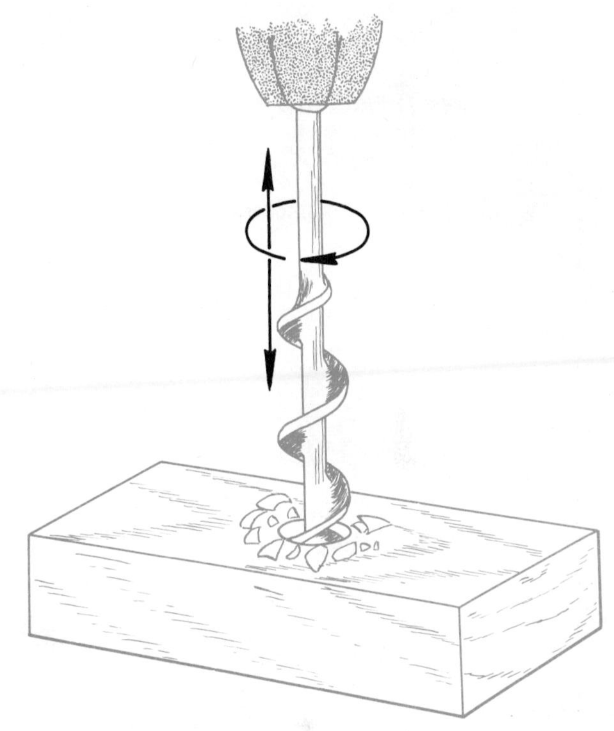

Fig. 2-49. Cutting holes with linear control of a rotary tool.

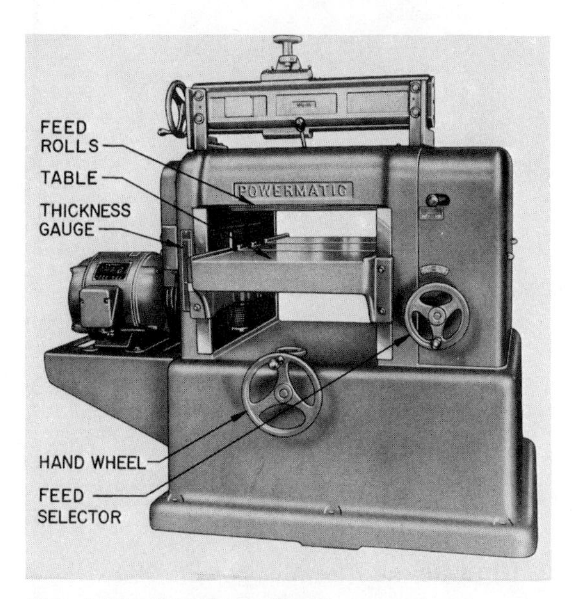

Fig. 2-48. A surface planer.

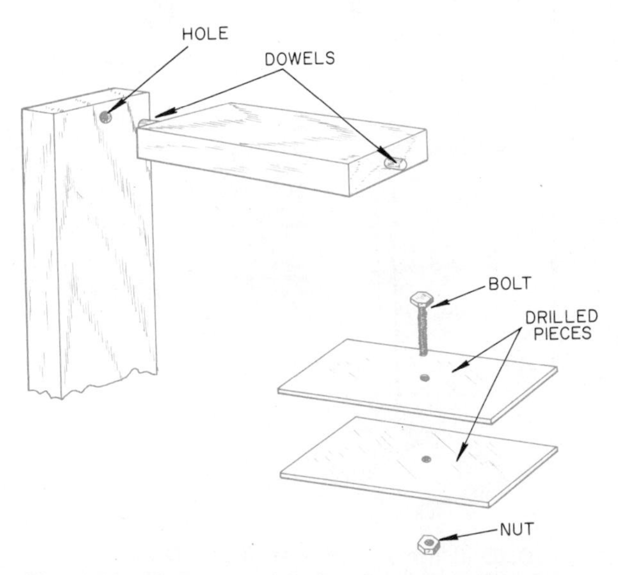

Fig. 2-50. The use of holes for assembly of parts.

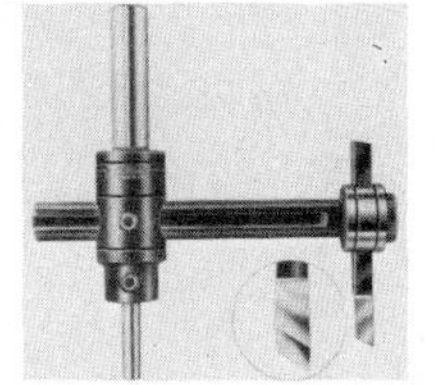

A. Fly cutter. (Stanley Tools)

B. Hole saw. (Millers Falls Co.)

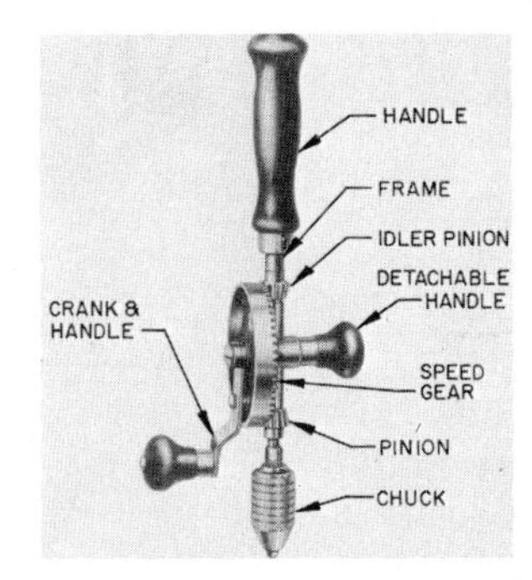

E. Hand drill. (Stanley Tools)

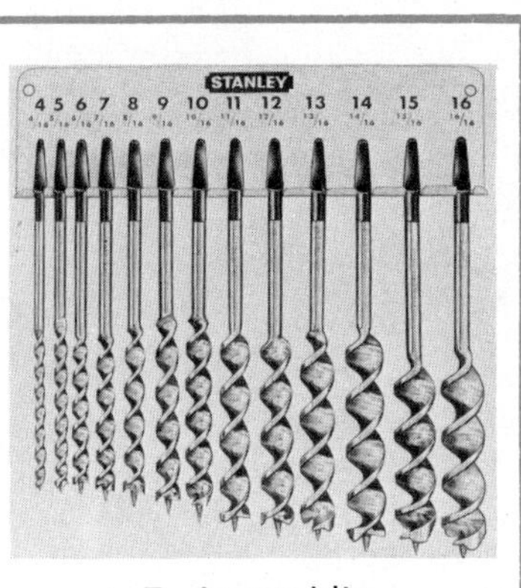

F. Auger bits. (Stanley Tools)

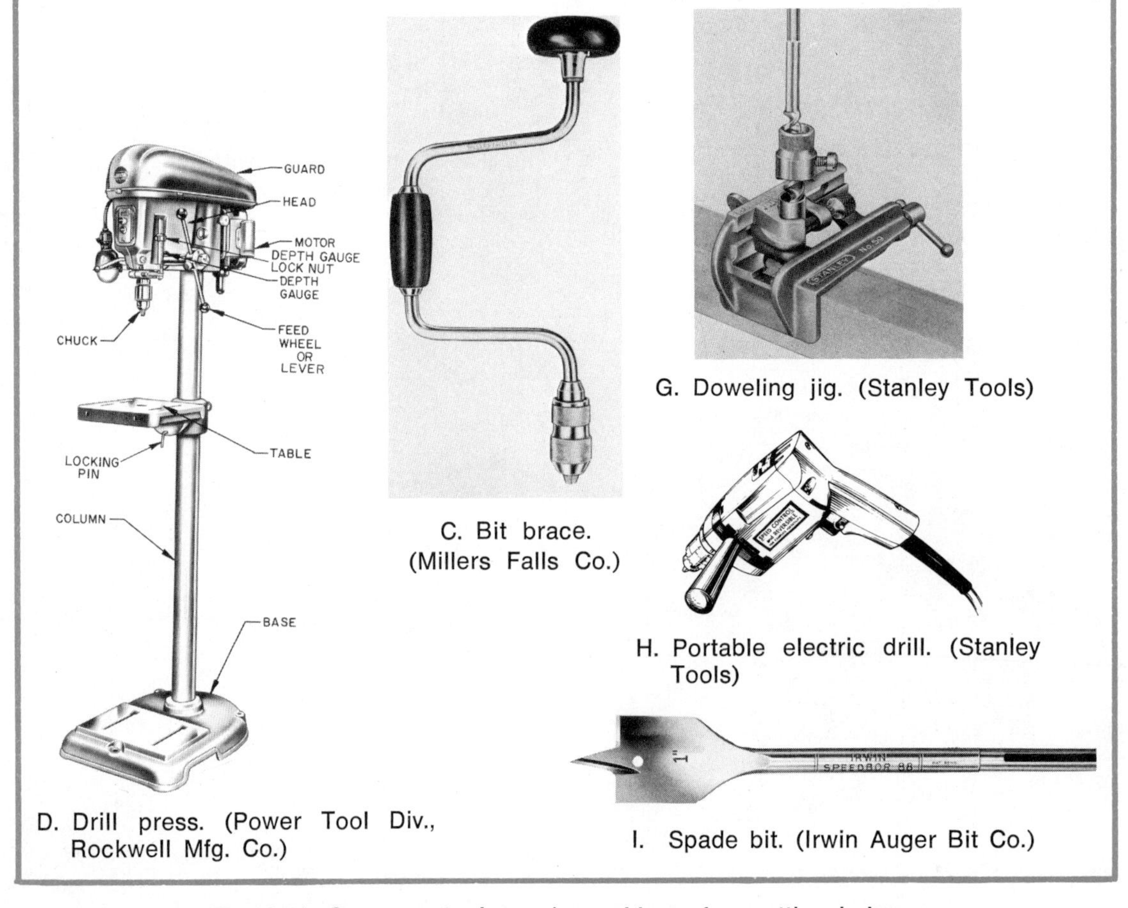

C. Bit brace. (Millers Falls Co.)

D. Drill press. (Power Tool Div., Rockwell Mfg. Co.)

G. Doweling jig. (Stanley Tools)

H. Portable electric drill. (Stanley Tools)

I. Spade bit. (Irwin Auger Bit Co.)

Fig. 2-51. Common tools and machines for cutting holes.

power machines can be used for cutting holes, Fig. 2-51.

HAND TOOLS FOR CUTTING HOLES

A large assortment of hand tools can be used for cutting holes in wood, Fig. 2-52.

Hand drills. Hand drills are used for drilling holes as large as 1/4″ in diameter. Twist drills ranging in size from 1/16″ to 1/4″ by thirty-seconds (1/32) or sixty-fourths (1/64) are installed in the drill chuck. Figure 2-53 shows the hand drill ready for use.

Automatic push drills. Automatic push drills such as shown in Fig. 2-54 are designed for drilling small holes rapidly. It is a handy tool for drilling small holes for nails and screws. The bits can be changed when different size holes are needed.

Ratchet braces. Ratchet braces are designed for boring holes larger than 1/4″ in diameter. The square shank of a bit is placed in the chuck for boring. Figure 2-55 shows a brace. The brace will hold auger bits, Forstner bits, and expansion bits, Fig. 2-56.

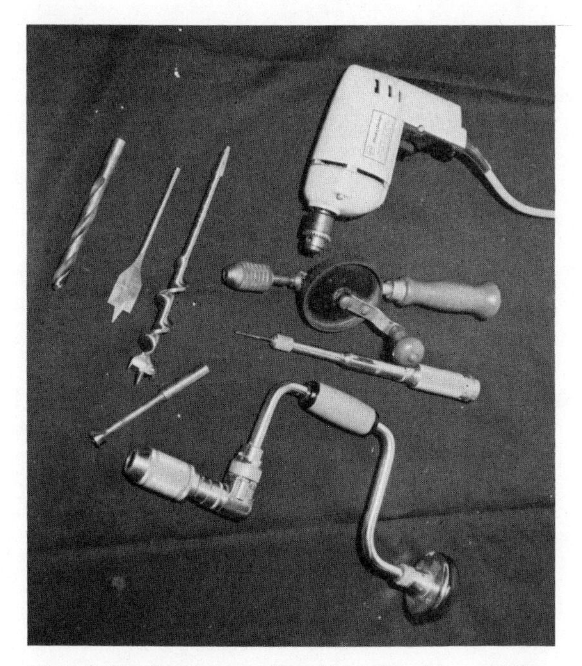

Fig. 2-52. Common hand tools for cutting holes.

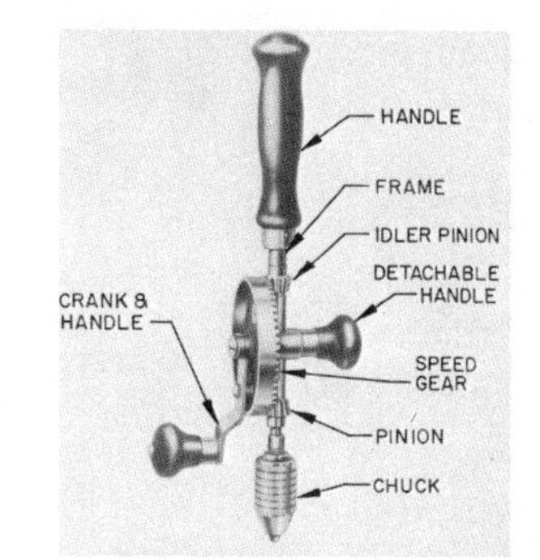

Fig. 2-53. A hand drill. (Stanley Tools)

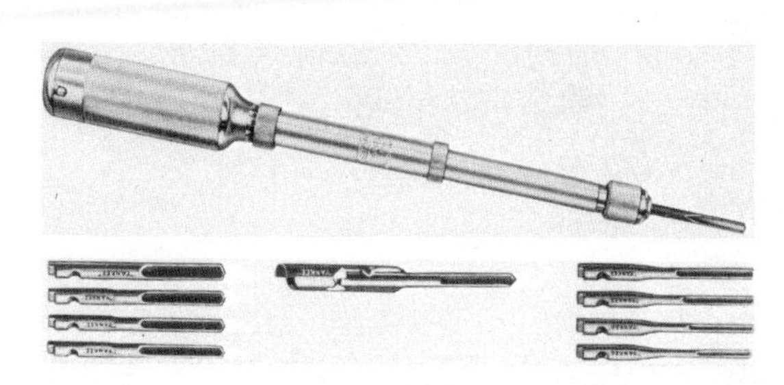

Fig. 2-54. An automatic push drill and bits.

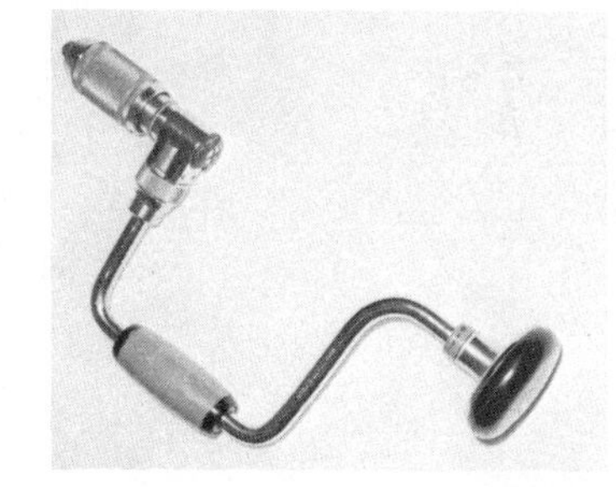

Fig. 2-55. A brace.

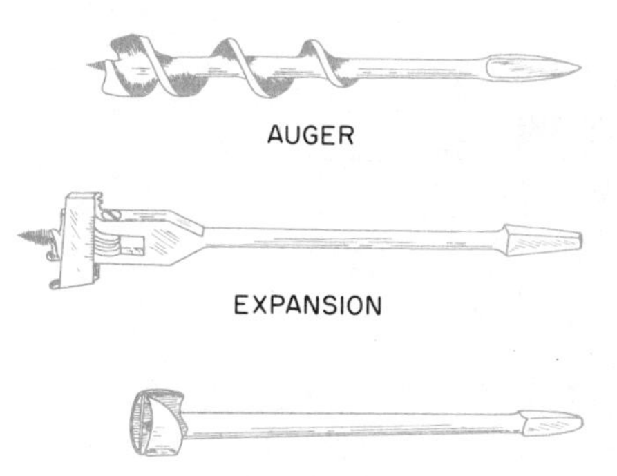

Fig. 2-56. Bits commonly used in a brace.

PORTABLE POWER TOOLS FOR CUTTING HOLES

Portable electric drills are among the most versatile tools used in the wood industry, Fig. 2-57. Most portable electric drills have either a ¼″ or ⅜″ chuck. The diameter of the bit shank cannot exceed the maximum chuck opening.

STATIONARY DRILLING AND BORING MACHINES

Stationary hole-cutting machines are generally called *drill presses.* A drill press will perform accurate work at a high rate of speed. Drill presses can be set up as a production machine, Fig. 2-58. The drill press can be equipped with special holding devices to locate and secure the work to the table.

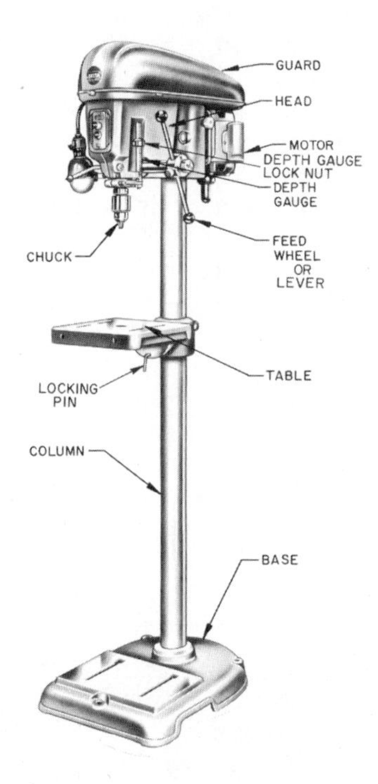

Fig. 2-58. A drill press. (Power Tool Div., Rockwell Mfg. Co.)

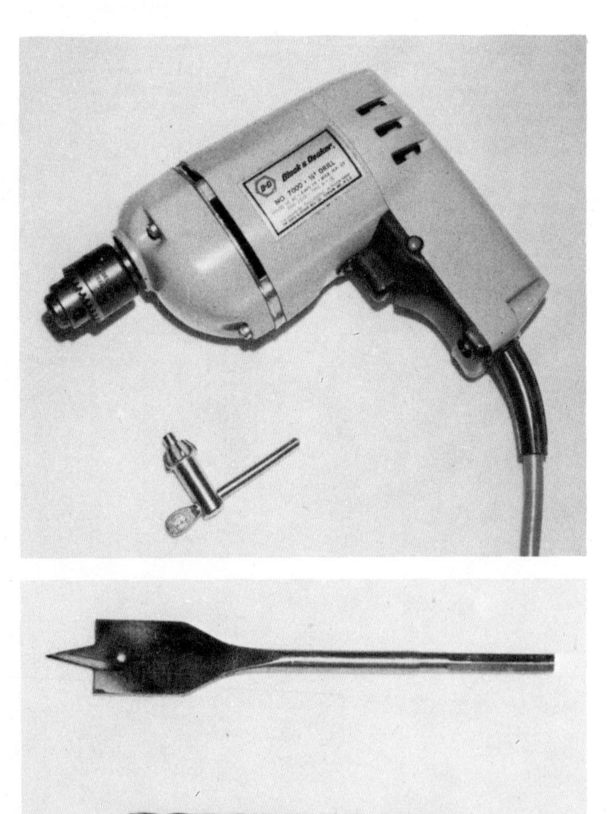

Fig. 2-57. Portable electric drill and bits commonly used to cut holes.

Fig. 2-59. Products consisting of irregular shapes.

PRINCIPLE OF SHAPING AND FORMING IRREGULAR SURFACES

Manufactured products often consist of irregularly shaped parts, Fig. 2-59. An irregularly shaped part is one which does not have all right-angled corners, Fig. 2-60. A number of hand tools and machine processes can be used to form the irregular parts. Detailed instruction for shaping irregular parts is presented in Chapters 12 and 13 of this book.

HAND TOOLS FOR FORMING IRREGULAR PARTS

Gouges, files, and rasps can be used in forming the irregular shapes.

Gouges. Gouges are edge-cutting tools similar to chisels except the blades are curved or rounded, Fig. 2-61. A gouge the shape of the desired contour is selected. It can be forced into the work or pounded lightly with a mallet. Large amounts of wood can be removed quickly with a sharp gouge. They can be used to form both inside and outside curves.

Files and rasps. Files and rasps work well for forming a contour on the outside of a part. They are available in many shapes, Fig. 2-62.

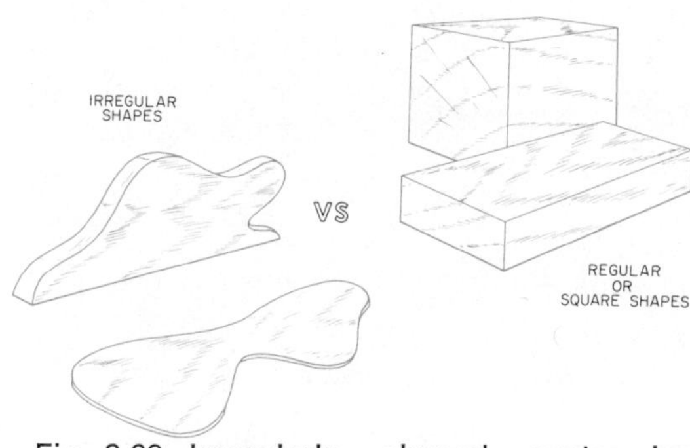

Fig. 2-60. Irregularly shaped parts have corners which are not right angles.

PORTABLE POWER TOOLS FOR FORMING IRREGULAR SHAPES

A router such as shown in Fig. 2-63 is used to form an irregularly shaped edge on a piece of stock. The router is equipped with a chuck to hold the router *bit*. The tool

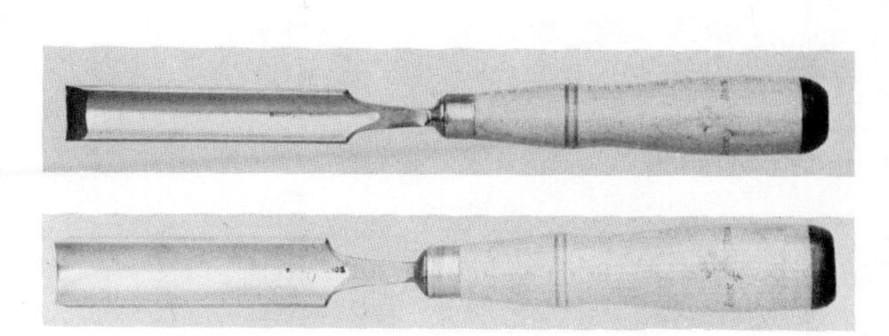

Fig. 2-61. Gouges. (Buck Brothers, Inc.)

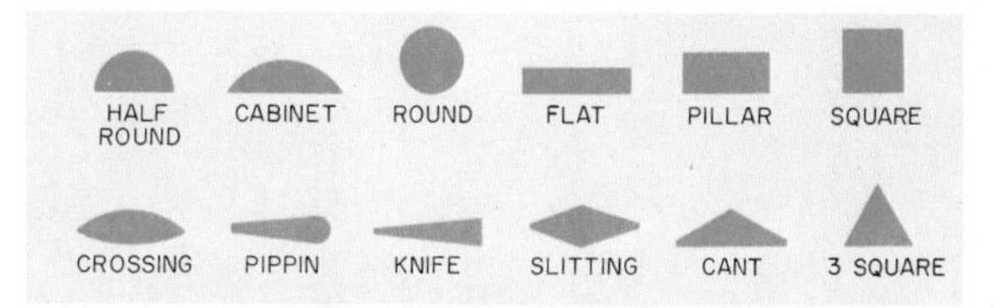

Fig. 2-62. Common shapes of files and rasps. (Nicholson File Co.)

Fig. 2-63. A router. (Rockwell Mfg. Co.)

bits are available in many shapes, Fig. 2-64. The router spindle turns at a high rate of speed. The router bit is advanced into the stock as the cut is made, Fig. 2-65.

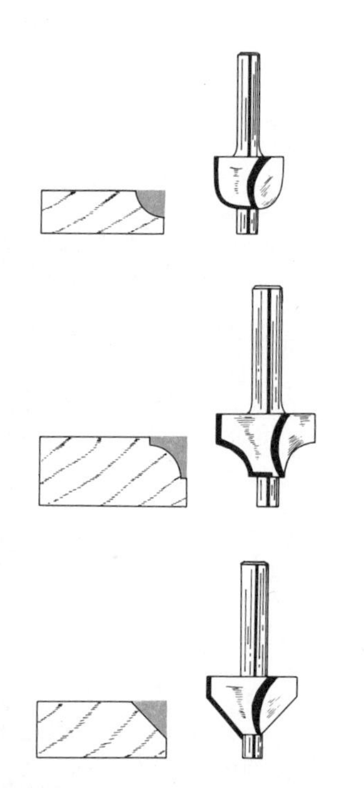

Fig. 2-64. Common shapes of router bits.

Fig. 2-65. Advancing the router into the stock as the cut is performed. (Stanley Tools)

STATIONARY POWER TOOLS FOR FORMING IRREGULAR SHAPES

Irregularly shaped work can be formed on stationary machines such as a *spindle shaper* and *lathe.*

Spindle shapers. Spindle shapers are very similar to a router except the work is advanced through the machine, Fig. 2-66. The spindle shaper is a high production machine. The cutter selection is mounted

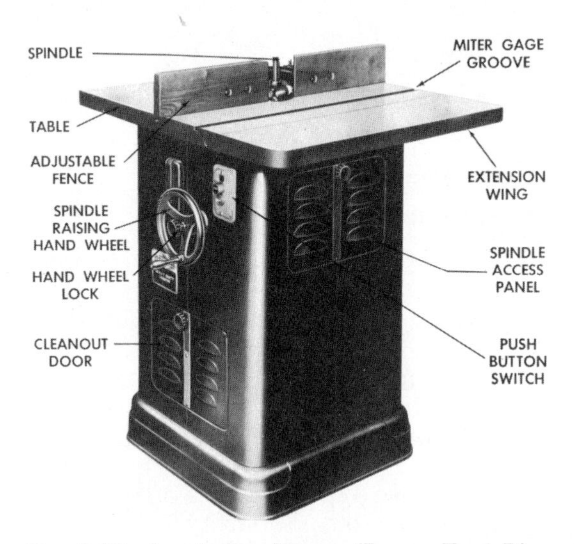

Fig. 2-66. A spindle shaper. (Power Tool Div., Rockwell Mfg. Co.)

Fig. 2-67. A table with irregularly shaped legs which have been turned on a lathe. (Kemp Furniture Industries)

on the shaper spindle. The cutter revolves at a very high speed.

Lathes. Lathes are used for forming irregular shapes such as table legs, Fig. 2-67. The lathe is a machine on which the stock is usually turned between two centers, Fig. 2-68. The stock revolves and the tool is held in your hand and forced into the workpiece. Different types of cutting tools are used to form the shapes, Fig. 2-69.

BENDING WOOD TO FORM IRREGULAR SHAPES

Sometimes irregular shapes are formed by bending wood, Fig. 2-70. There are three techniques commonly used for bending and forming wood. They are (1) lamination, (2) steaming or soaking, and (3) kerfing.

LAMINATION

Lamination is a forming process whereby thin strips of wood are put into plies (layers) and glued together in the desired

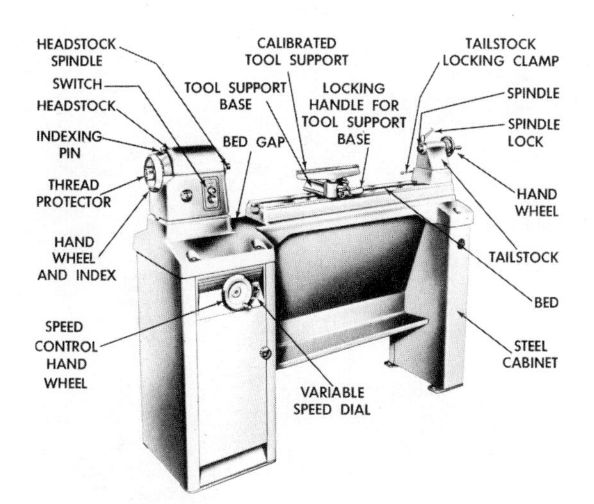

Fig. 2-68. A wood lathe. (Power Tool Div., Rockwell Mfg. Co.)

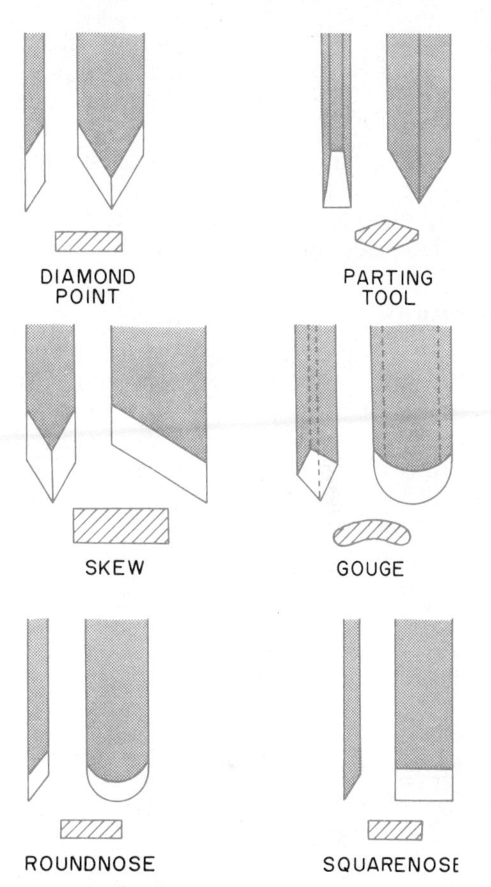

Fig. 2-69. Common lathe tools for turning irregular shapes.

Fig. 2-70. Wood parts which have been formed by bending.

shape, Fig. 2-71. The plies are clamped together in a *forming mold* and allowed to dry into the desired shape. The plies usually run parallel to the bend.

The lamination form is constructed in two sections. The form must match properly so that even pressure is applied to the entire part being shaped.

STEAMING

Steaming or soaking is a technique by which solid wood can be bent and formed to shape. Mechanical pressure and additional moisture is added to the wood. The moisture allows the wood to become more flexible. After the wood has become pliable due to this additional moisture, it is bent into shape. The wet wood is left clamped in the form until it is dry, Fig. 2-72.

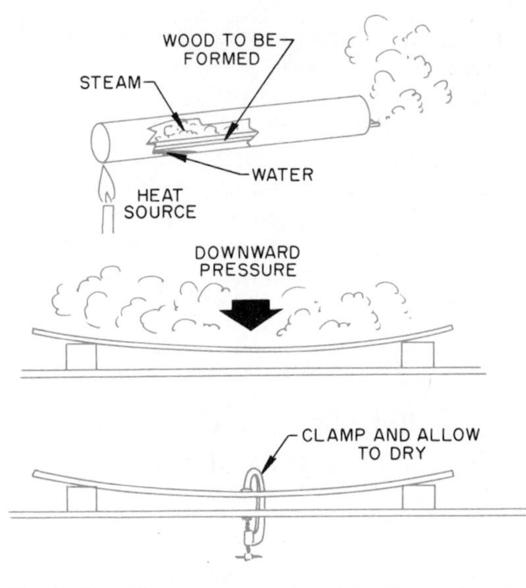

Fig. 2-72. The process for bending steamed wood.

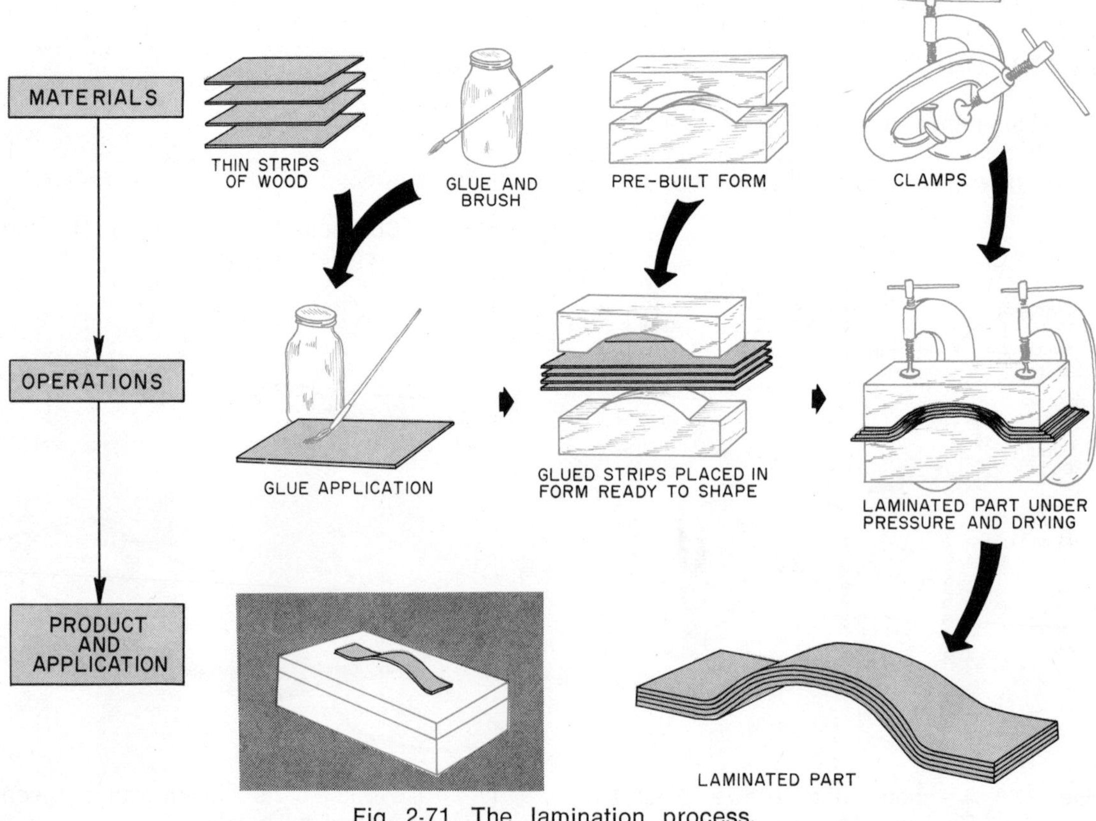

Fig. 2-71. The lamination process.

KERFING

Kerfing is also a technique used for bending wood, Fig. 2-73. As you recall, a kerf is the section of wood the width of the saw blade which is removed as sawdust. Kerf bending is performed by sawing several kerfs about three-fourths of the way through the thickness of a board on the side which will not show when assembled.

The saw kerfs close when the board is bent. The strength of the kerfed board can be increased by filling the kerfs with a mixture of glue and fine sawdust. The glue also helps hold the board in the bent position. Kerf bending is often used for constructing round tables, Fig. 2-74.

PRINCIPLE OF ASSEMBLING PARTS

In the wood industry, the assembly of multiple parts is carefully planned. Product assembly requires planning so that all of the components (parts) will be ready for assembly into the finished product. A manufacturing company attempts to avoid excess inventory of component parts. As you recall, a manufacturer adds value to a product each time he processes it or handles it. Therefore, *the manufacturer wants to put his product on the market as soon as possible after investing his money in processing.*

Most woodworking products consist of several components. The difficulty of the process is dependent upon the kind of article and the number of parts. These are the factors which determine the extent to which the process must be planned. The product shown in Fig. 2-75 would require a lot of planning. The drawer is only one of many components.

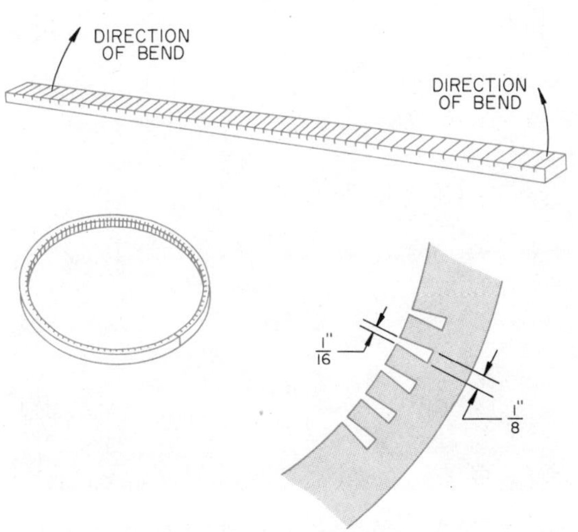

Fig. 2-73. The kerfing process for bending wood.

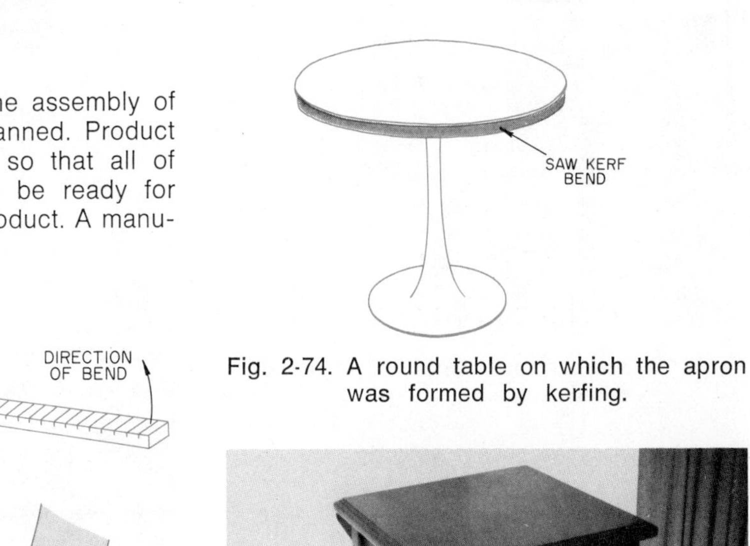

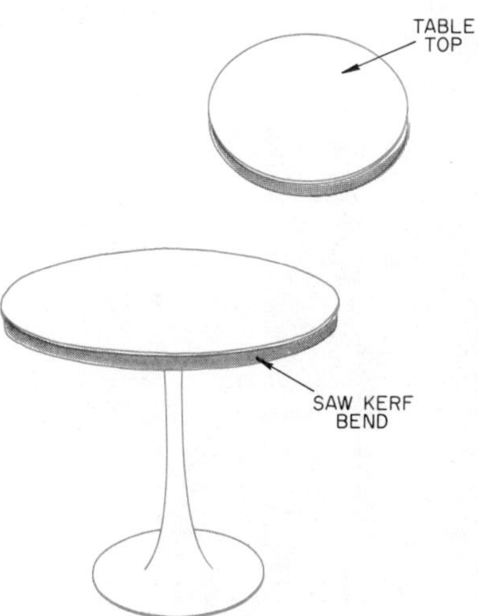

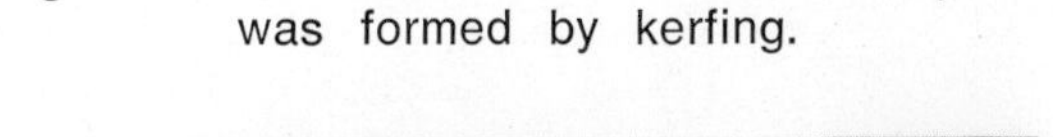

Fig. 2-74. A round table on which the apron was formed by kerfing.

Fig. 2-75. A product consisting of many parts.

An assembly line is used in most large wood industries, Fig. 2-76. The components are added to the product as it progresses down the line.

Industry during the assembly process is primarily concerned with fastening techniques. As you know, wood is a very easy material to fasten. There are many methods of fastening wood. The methods can be categorized in the two broad classifications of (1) *mechanical* fastening and (2) *chemical* fastening. A combination of mechanical and chemical fastening is often used.

Fig. 2-76. An assembly line for a prefabricated home.

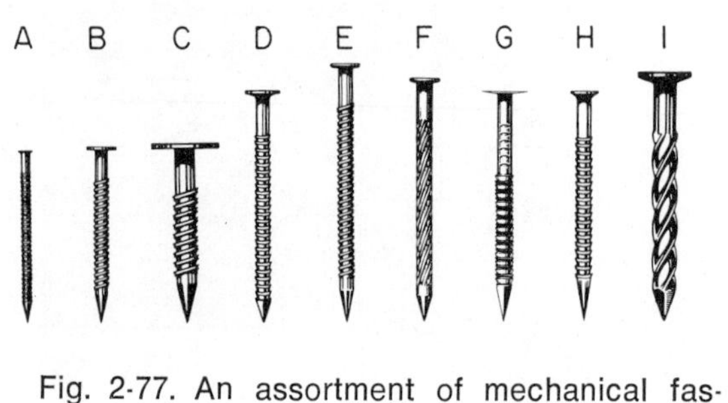

Fig. 2-77. An assortment of mechanical fasteners.

Fig. 2-78. A house being assembled and fastened with nails. (Western Wood Products Association)

Fig. 2-79. Assembling a product with an automatic stapling gun. (Senco Products, Inc.)

MECHANICAL FASTENING

Mechanical fastening refers to metal fasteners such as nails, screws, bolts, staples, and other special-purpose fasteners, Fig. 2-77.

Nails are the most common type of mechanical fasteners. The *nail* is a satisfactory fastening device for many purposes, such as house construction and packing crates. Nails have sufficient holding strength and do not distract from the appearance of most finished products, Fig. 2-78.

In woodworking industries where production lines are used for assembly, nails are rapidly being replaced with *staples.* Staples installed automatically with machines have reduced assembly time and assembly cost, Fig. 2-79.

The greatest disadvantage for all mechanical fasteners is appearance on some products. For this reason, the exterior of quality furniture is never assembled with mechanical fasteners. The lack of strength is also a reason for using other methods than mechanical fasteners. Often a combination of mechanical and chemical fastening is used for greater strength. It also requires time to prepare the automatic equipment for operation.

CHEMICAL FASTENING

Chemical fastening refers to the use of various types of *glues* and *adhesives* to bond surfaces together. Chemical fastening is often used because it has greater strength and better appearance since it does not show on the surface.

There are several adhesives and glues available. Each type of glue has certain characteristics for specific purposes. For example, some dry fast, are less expensive, ready to use, waterproof, will bond unlike materials, plus other qualities.

Adhesives are in two general categories. They are (1) waterproof and (2) non-waterproof. Waterproof glue can be used outside where it will be exposed to water. A boat would be built with waterproof glue. Non-waterproof glue will not withstand exposure to water. Its uses are limited to products which will remain indoors. Furniture is usually assembled with non-waterproof glue. Figure 2-80 shows a variety of glues and adhesives which are available. The selection and use of glues and adhesives is presented in detail in Chapter 14. The greatest disadvantage of glue is the time required for applying and drying. The production rate can be increased for some types of glue when heat is applied.

SURFACE PREPARATION FOR FASTENING

Regardless of the fastening technique used, the surfaces to be joined together must be properly prepared. This preparation is called *joinery.* The selection of the proper type of joint is important. The main factors industry considers in selecting joints are (1) *cost,* (2) *strength,* and (3) *appearance.* The least expensive joint which has ample strength and satisfactory appearance for the product is selected. The cost of a joint increases as the time to produce it increases. *Excessive construction time reduces industry profit.*

Fig. 2-80. A variety of glues and adhesives used for fastening wood products.

Certain joints are more satisfactory for some purposes than others, Fig. 2-81. Detailed instruction on joint construction is given in Chapter 14.

PRINCIPLE OF CONDITIONING MATERIALS

The quality of the finished product is largely dependent upon the quality of the raw material from which it is constructed. The raw material in the wood industry is lumber. The quality of lumber is dependent upon the *conditioning process.*

METHODS OF SAWING LUMBER

Conditioning processes involve various methods of cutting and drying wood. Every attempt is made in the conditioning process *to utilize the best characteristics of the material.* There are two basic ways wood is cut at the sawmill. They are (1) plain or flat sawing and (2) quarter sawing.

Plain-sawn lumber is cut parallel to the annual growth rings of the log, Fig. 2-82. A *quarter-sawn* board is cut at an angle (between 30° and 70°) to the annual growth rings. Figure 2-83 shows the sawing method for quarter sawing lumber. Quarter-sawn wood is more stable; that is, it does not *warp* as much as plain-sawn lumber.

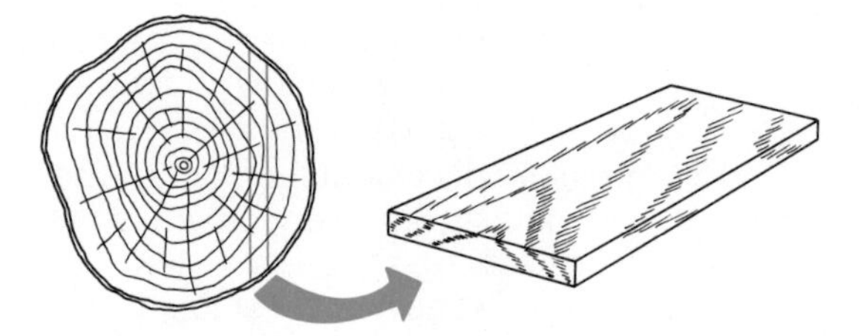

Fig. 2-82. Cutting plain sawn lumber from a log.

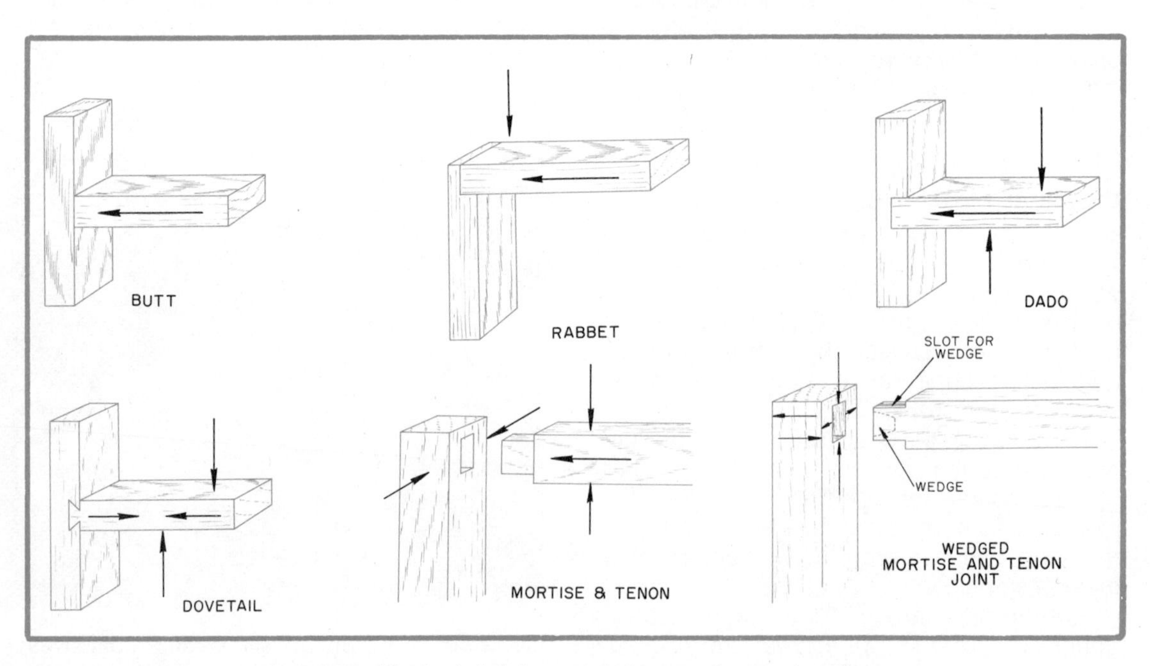

Fig. 2-81. Typical joints used in wood construction.

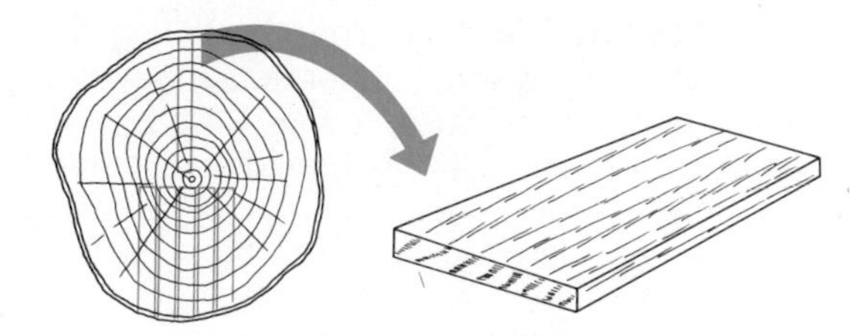

Fig. 2-83. Cutting quarter sawn lumber from a log.

METHODS OF DRYING LUMBER

Lumber must be dried before it has any value for most commercial uses. That is, the water or *moisture* must be removed. The moisture content in wood is expressed as a percentage. Normally wood which is used for cabinet manufacture is dried to 6-10% moisture content. When the moisture present in the air increases, wood will absorb moisture. This causes it to expand slightly in size. As the moisture present in the air decreases, wood gives off moisture and begins to shrink.

Lumber can be dried by two different methods: (1) air drying, and (2) kiln drying.

Air-dried lumber. Air-dried lumber is exposed to the open air *until the moisture evaporates*. Fresh sawn lumber is stacked with spaces called *stickers* between each layer of boards for drying. A cover is placed over the stack to protect it from rain and the direct sun, Fig. 2-84. Notice the stickers between each layer of boards. The stickers provide for air circulation between each of the boards.

The time required to air dry lumber depends upon the climate, and may exceed a year. Local weather conditions during a particular season determine the drying time. Of course, the wood will dry very slowly during rainy weather. The minimum moisture content in most air-dried lumber is about 12% to 18%.

Uses for air-dried lumber. The uses for air-dried lumber are limited. It is seldom used for furniture and cabinetmaking due to the high moisture content. If air-dried lumber was used for furniture, it would shrink excessively because the atmosphere in cold climates is dry in your house during the winter months. The excessive shrinkage would cause the wood to crack.

Air-dried lumber is primarily used for the construction of packing crates, pallets, and buildings.

Advantages of air-dried lumber. The chief advantage of air-dried lumber is it is usually less expensive. There is no processing cost other than storage and stacking involved when lumber is allowed to dry naturally. Air-dried lumber which is used for construction of exterior buildings should be dried in the same locality that it is used for the proper moisture content. This will eliminate the likelihood of the wood expanding or contracting after it is placed in a structure, Fig. 2-85.

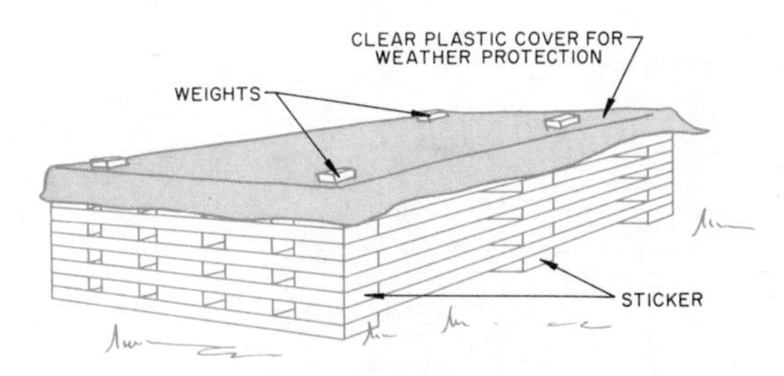

Fig. 2-84. Lumber stacked for air drying.

Fig. 2-85. Air-dried lumber is dried to the same moisture content as the locality in which it will be used.

Fig. 2-86. A drying kiln.

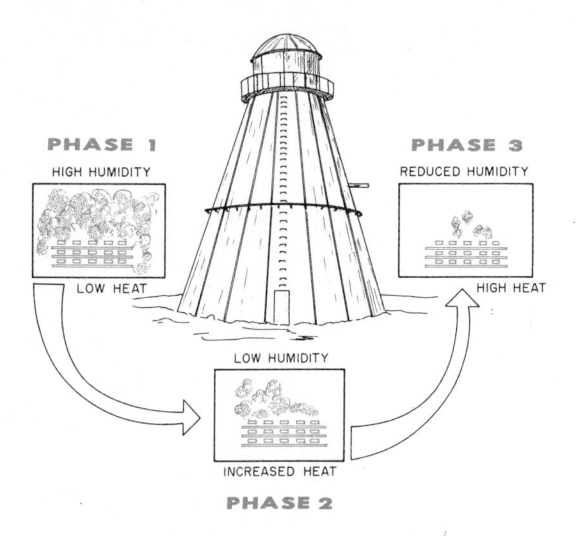

Fig. 2-87. The kiln-drying process.

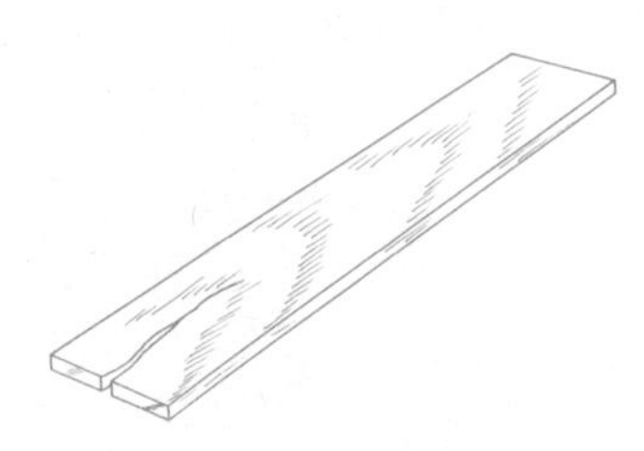

Fig. 2-88. A check on the end of a board created by drying stress.

Kiln-dried lumber. The kiln-drying process is used when lumber must be dried to a *specific moisture content in a short time.* In kiln drying, the humidity and temperature is carefully *controlled,* Fig. 2-86. The lumber is stacked like for air drying and then pulled into the kiln.

The kiln drying process consists of three phases (Fig. 2-87):

1. In the first phase, the lumber is subjected to low heat and high humidity.
2. During the second phase, the humidity is lowered and the temperature is increased.
3. The third phase of drying consists of low humidity and high heat.

During the entire drying process, the air inside the kiln is circulated with electric fans. The circulating air helps the wood to dry evenly. The kiln-drying process reduces drying time for 1″ lumber to 10% moisture content in three or four days. The moisture content is reduced to about 5% for most lumber. After the lumber has been removed from the kiln, the moisture content may increase slightly due to high humidity in the air.

Advantages of kiln drying. Kiln drying produces high-quality lumber in a short time. The defects which are common to air-dried lumber are reduced or eliminated by kiln drying. The most common defect is *checking* due to stress created by uneven drying, Fig. 2-88. The stresses occur when the outside of the board shrinks faster than the inside portion.

FINISHING

Finishing is generally the final process in the production of a product. A finish is applied to the final product for two main purposes. The finish (1) decorates and (2) protects and preserves the product.

Industry is also concerned with giving the product "consumer appeal." A decorative finish can give the product this consumer appeal, Fig. 2-89. In addition to decorating the product, the finish is also protecting the wood.

Wood is a cellular material and tends to absorb moisture from the air, Fig. 2-90. When excessive amounts of moisture are absorbed by the wood, decay may occur. A finish forms a barrier on the surface of the wood and prevents the absorption of moisture, thus helping to preserve and lengthen the life of wood. Paint is applied to the outside of a house to protect the wood from excess moisture. The paint coating also improves the appearance of the house.

Fig. 2-89. Finishes decorate and protect wood products.

KINDS OF FINISHES

Finishes can be classified as either (1) opaque or (2) transparent. *Opaque* finishes are those which contain pigments. Paint such as you would apply to your house is an opaque finish. This type of finish covers the material completely so you cannot see the wood. The opaque finish is generally used over less expensive woods which do not have much natural beauty. The opaque finish both protects and decorates. It is not a means of enhancing the natural beauty of the wood.

Transparent (clear) finishes are used to finish most furniture. The natural beauty of the wood shows through the transparent finish. Wood is a beautiful material and this is one reason it is selected for furniture construction. Transparent finishes protect the surface of the wood from moisture and dirt, and at the same time, enhance the natural beauty of the wood.

Transparent coatings and treatments. Transparent finishes can be classified as either (1) coatings, or (2) penetrating treatments.

Coating finishes are those which increase in thickness on the surface with each additional application. They are sometimes called "buildup" finishes because each coating results in a thicker surface buildup.

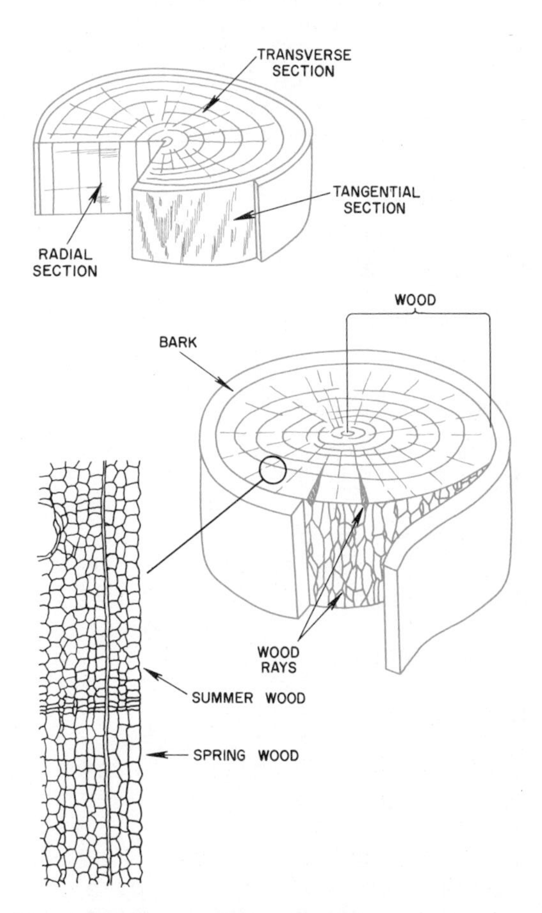

Fig. 2-90. The cellular structure of wood.

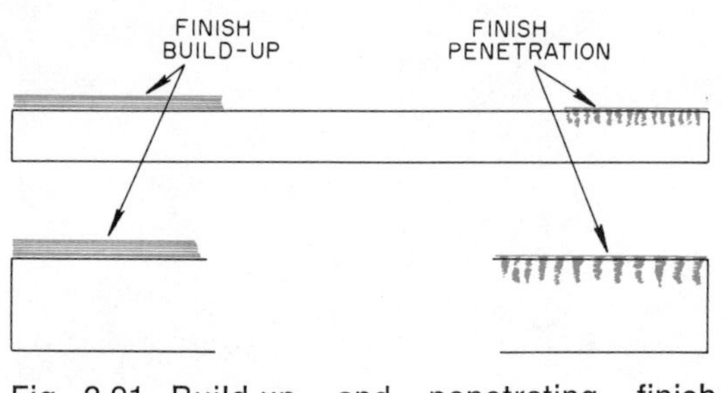

Fig. 2-91. Build-up and penetrating finish treatments.

Treatment finishes penetrate into the pores of the wood. There is very little surface coating developed with treatment finishes. Figure 2-91 shows a comparison of the two types of treatments. Both finishes serve to preserve the wood and enhance the grain figuration.

Color treatments. Wood is often colored or stained differently than its natural color. This process is referred to as *staining.* Inexpensive woods are stained to imitate more expensive woods. The woodworking industry relies upon this technique to give products consumer appeal. The grain figuration of many of the fine cabinet woods can also be enhanced by means of color staining.

FINISH APPLICATION

The primary concern of industry when applying finish is a quality job at a reasonable cost. The type of product and the type of finish determines how the finish will be applied. The most common methods used include: (1) air spray; (2) rollers; and (3) brushes.

Air spraying. Air spraying of finishing materials is the most popular industrial technique, Fig. 2-92. This is a fast way to apply the finish and results in a smooth, even coating. Both opaque and transparent finishes are applied with the air spray method.

Roller applicators. Roller applicators are often used for applying paint. This method is faster than brushing and does not require the preparation time necessary with air spray application.

Fig. 2-92. A product being finished by air-spraying. (DeVilbiss Co.)

Fig. 2-93. Brushing finish on a small product.

Brushes. Brushes of various types are used for applying finishes to small surfaces. The brushing method is satisfactory for home use, but is too time consuming for industrial use, Fig. 2-93.

SELECTING A FINISH

Finishes are selected according to *product requirements, material cost,* and *application costs.* Manufacturing industries will sometimes conduct tests to determine the effectiveness of a finish. Time required to apply a finish, and the time required for a finish to dry are prime factors for material selection. Since drying time is a cost factor, most industries have selected quick-drying finishes. A quick-drying finish is less likely to be contaminated by dust which can ruin a finish. It also reduces costs because more products can be finished in a given period of time.

The number of applications of finish required is also important. Each application is an additional operation and increases the product's cost.

The purpose of this chapter was to provide you with information about processes common to the wood industry. These processes included: (1) separating, (2) planing, (3) cutting holes, (4) shaping of irregular parts, (5) assembly of parts, (6) conditioning of materials, and (7) finishing. You also realize now that there are many ways to perform the same processes. Every product presents many construction challenges and you must make decisions as to the best methods of solving the problems.

Many of these decisions will be automatic because of limited tools and machines to perform the processes. In these cases, you must use what is available. You will often have to use hand tool processes, because to own expensive machines for occasional use would be poor economics.

Let us select a typical product such as shown in Fig. 2-94 which you may sometime want to construct. Now let us consider the various production processes which may be used for the construction of this product. First, consider all of the separating processes and the tools and machines which may be used. Separating can be done at the same time for all of the shelves and ends since they are all the same width. These parts can be ripped to width using a hand ripsaw, stationary circular saw, portable circular saw, or a radial arm saw. The various parts can be cut to length using a hand crosscut saw, portable circular saw, stationary circular saw, or a radial arm saw.

The front member which extends from the bottom shelf to the floor can be cut by the same methods described for the ends and shelves.

The decorative trim on the top can be cut to the irregular shape with a coping saw, portable jigsaw, stationary jigsaw, or a band saw.

The stock can be reduced to thickness and smoothed with a jack plane or smoothing plane. The operation can be performed faster using a jointer and a surface planer if machines are available.

The shelves can be attached to the ends by means of a dado joint, Fig. 2-95.

Fig. 2-94. A typical product involving many processes.

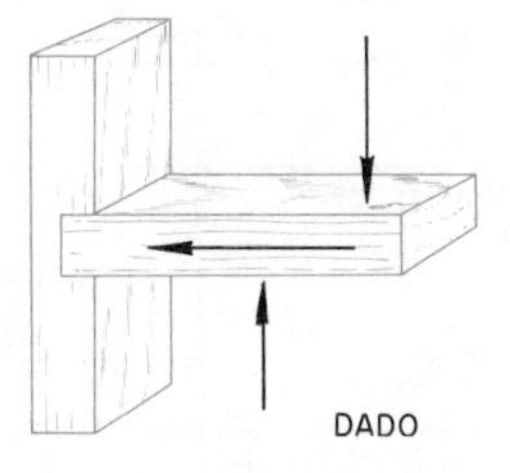

Fig. 2-95. Shelves can be assembled with dado joints.

The joint can be cut using a backsaw and chisel, a router plane, router with a straight bit, circular saw or radial arm saw.

The parts can be assembled using an adhesive (glue). The hardboard for the back can be cut to size using a saw designed for cutting straight lines. The back can be attached with small nails and glue.

PROBLEM

Study the three products shown here. Identify the tools used for each process listed in the example chart below. Sequence the processes as they would take place on each product. (DO NOT WRITE IN YOUR TEXTBOOK)

Separating
Assembly
Conditioning
Finishing
Profit
Kerf
Set
Bit
Lamination

Product A. (Selig Manufacturing Co.)

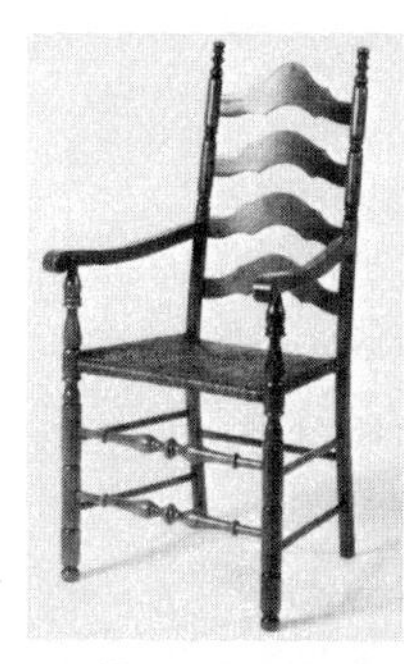

Product B. (Kemp Furniture Industries)

Product C. (Kemp Furniture Industries)

Methods Of	PRODUCT A		PRODUCT B		PRODUCT C	
	Tools	Sequence	Tools	Sequence	Tools	Sequence
Separating						
Planing						
Cutting Holes with Rotary Tools						
Shaping & Forming Irregular Shapes						
Bending Wood to Form Irregular Shapes						
Assembling Parts						
Conditioning Materials						
Finishing						

EXAMPLE CHART

CHAPTER 3 STANDARD MATERIALS

INTRODUCTION

A manufacturer of a product can reduce the number of production operations by using *standard materials*. A standard material is a part, component, or assembly of a (1) *standard size*, (2) *shape*, or (3) *quality* that is available as a stock item. Some manufacturing companies specialize in the production of standard material, Fig. 3-1. These products are purchased by other manufacturing companies and combined into a consumable product.

When a company considers a product for production, it considers how the product can be produced and what materials can be used. The company's *industrial engineers* will generally make a study of the standard materials which are available. The company will determine if they will *make or buy the parts* from another company. Let us consider a residential house manufacturing company. It will need windows to install in the structure. The company will study the problem and then decide whether to buy the manufactured (ready-made) window units or manufacture their own. If the manufacturing costs are as great to produce their own windows, they will probably buy them from another company.

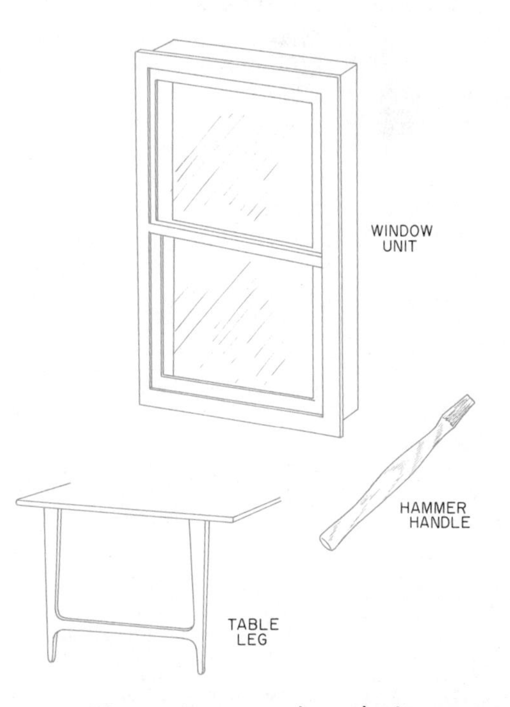

Fig. 3-1. These three wood products are examples of products produced by manufacturing companies to serve as components for other companies.

Remember, industry and business is based on the profit motive. An industry will not manufacture a particular product if it is not profitable. It costs industry money for production equipment and labor each time it performs a production operation on a product. Since fewer production operations usually mean a lower cost of production, industry is concerned with methods of reducing the number of operations required for a given product. The material from which the product is produced influences the number of production operations, Fig. 3-2. The large sheet of plywood in the tennis table saved numerous production operations in the construction of the game table top.

Some manufacturing industries have *specialized* in the production of standard materials. These companies have perfected their production methods by specializing. Specialized companies can produce large volumes of various standard materials and supply them to other manufacturers at a cost that is lower because of the high volume of production. A company that specializes in the production of nails can supply many companies that use them rather than have each company manufacture only the amount of nails they use.

It is important that you understand that each type of manufacturer performs a service to other manufacturers. For example, a carpenter could not afford to make the plywood he uses. However, the plywood manufacturing company is dependent upon carpenters as customers to buy the plywood and they specialize in making the plywood. The carpenter can buy the plywood cheaper than he can make it.

Standard materials mean different things to different industries. For example, standard materials to a sawmill would be logs cut from trees, Fig. 3-3. Standard material to a furniture industry would be boards directly from the sawmill. A carpenter would consider a prehung door unit such as shown in Fig. 3-4 as standard material. As you see, the standard material produced in one industry is often the raw material for the next industry, from which it will begin to develop a different product, Fig. 3-5.

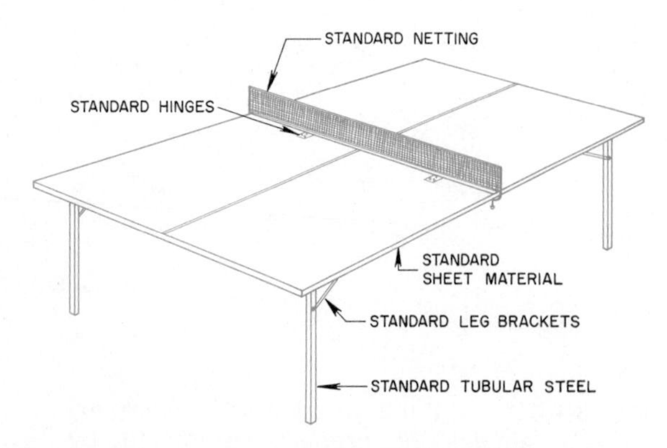

Fig. 3-2. A product may be manufactured from several different materials.

Fig. 3-3. Sawn logs are standard materials for sawmill operations.

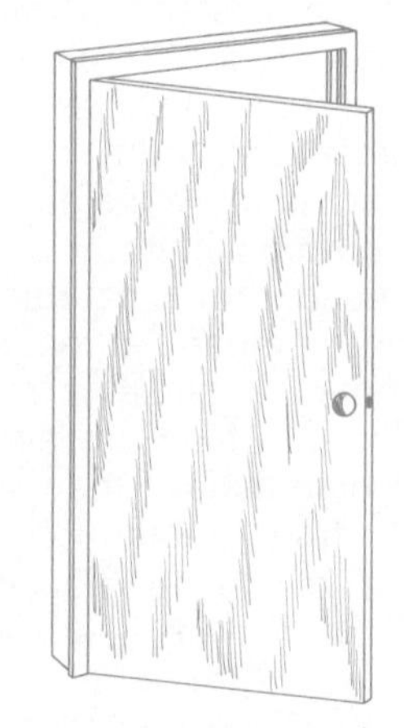

Fig. 3-4. A pre-hung door unit is standard material for a carpenter.

Fig. 3-5. Veneer is raw material for many furniture industries. (American Forest Products)

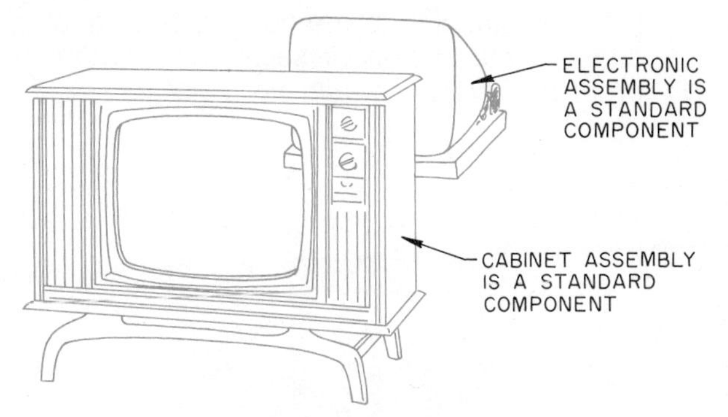

Fig. 3-6. The electronic assembly of a television is a standard component around which a television cabinet must be designed.

Standard materials influence product design because the final cost of the product is dependent on the cost of materials. A product can be produced at a lower cost if standard materials can be used for manufacturing. There are many types of standard materials available to the wood industry ranging from lumber and plywood to various types of special-purpose hardware. Lumber and plywood are available in standard sizes and qualities. A uniform grading system is used to insure equal quality for the consumer from different manufacturers of standard materials. The grades are standardized for plywood, lumber, and similar materials.

The manufacturing process for man-made sheet materials, i.e. plywood and hardboard, makes it possible to meet the manufacturer's needs. The size and shape of the material can be more varied than lumber.

STANDARD MATERIALS AND PRODUCT DESIGN

Standard materials influence product design. The cost to manufacture an essential part of a product may require that the entire product is designed around the part. For example, a furniture manufacturing company which produces television cabinets must design the cabinet to fit the television component, Fig. 3-6. The minimum size of the cabinet must be such that the television will fit inside. The cabinet must be designed so the television can be removed for servicing. Vents must be placed in the proper locations for the television speakers. The vent opening must also be the correct size. In addition, the cabinet must be adequately ventilated to prevent overheating when the television is operating. This is only one example of how standard component parts influence product design.

It is important to become familiar with the standard materials which can be used *before you design a product from wood.* The size of the standard material will often influence the design of the finished product. Sometimes you will design a product because the plywood panel you will buy is a certain width or length. An architect must first know what size windows are available *before specifying a given size.* If he does not use a standard size, it will be *more costly* to have special windows made to a particular size. Standard parts generally mean more volume to the manufacturer. With large-volume production, a manufacturer can save money by buying large quantities of material. Large-volume buying reduces costs and increases profits. *Without profit the industry would not exist.*

Fig. 3-7. Different standard materials are selected for various products. (Keller Furniture)

A lower cost lumber may be used to frame a house and expensive wood may be used for fine furniture construction.

STANDARD MATERIALS AVAILABLE TO THE WOOD INDUSTRY

The wood industries are dependent on many other industries for standard materials. Major materials supplied include lumber, sheet materials such as plywood, hardboard, and particle board, chemical and mechanical fasteners, hardware, and various wood component parts. All types of standard materials can be specified by size, quantity, and quality when ordering. The specifications insure you of receiving the exact material you need.

STANDARD LUMBER

Dimension lumber is a standard material which is available in a variety of sizes and grades. Lumber is divided into two classifications: (1) hardwoods, and (2) softwoods. Hardwoods are produced by leaf-bearing trees. Softwoods are produced by needle-bearing trees. The grading systems and standard sizes of lumber classified as hardwood and softwood differ. The grading systems differ because most softwoods are used for structural purposes. The requirements for structural lumber are different than those for cabinet construction. Most of the quality hardwood is used for cabinet and furniture construction. The grading system is different because wood of a different quality is needed for each type of work, Fig. 3-7.

Actual and nominal lumber sizes. Lumber sizes for both hardwoods and softwoods are usually stated with the (1) nominal or (2) actual (dressed) size. The nominal size is the actual size before the board is planed smooth or *dressed.* The nominal size of a 2X4 piece of lumber is exactly 2″ thick and 4″ wide while the actual (dressed) size is 1½″ thick and 3½″ wide, Fig. 3-8. Most lumber is surfaced or planed at the sawmill on two sides or four sides. This is called *dressed* lumber. The planing process reduces the thickness or width of lumber to what is called the actual size.

When lumber is planed or dressed smooth on all four sides, it is referred to as "surfaced on four sides." This is abbreviated as S4S. If the boards are surfaced on only two sides, it is abbreviated as S2S. S2S lumber is surfaced on the face sides.

Softwood is available in standard thicknesses, widths, and lengths, Table 1. The most common nominal thickness available in retail lumber companies are 1″, 2″, and 3″. The standard lengths of softwoods available start at 4′ and range up to 18′ long. The lumber is stocked in length intervals of 2′. For example, you could purchase lumber in any of the following standard lengths: 4′, 6′, 8′, 10′, 12′, 14′, 16′, and 18′. This means if you needed a board 5½′ long, you would need to purchase one 6′ long.

Table 1
Standard Softwood Lumber Sizes

Thickness		Width		Length*
¾″	X	3½″	X	6′ - 20′
¾″	X	5½″	X	6′ - 20′
¾″	X	7¼″	X	6′ - 20′
¾″	X	9¼″	X	6′ - 20′
¾″	X	11¼″	X	6′ - 20′
1½″	X	3½″	X	6′ - 20′
1½″	X	5½″	X	6′ - 20′
1½″	X	5½″	X	6′ - 20′
1½″	X	7¼″	X	6′ - 20′
1½″	X	9¼″	X	6′ - 20′
1½″	X	11¼″	X	6′ - 20′

*Standard lengths are: 6′, 8′, 10′, 12′, 14′, 16′, 18′, 20′. Lengths from 20′ - 26′ are available at slightly higher cost.

The size of lumber is always stated in the following order: (1) *thickness*, (2) *width*, and (3) *length*. You would list the sizes of lumber on an order as shown in Fig. 3-9.

GRADES OF SOFTWOOD LUMBER

The lumber grading standards are regulated by the U. S. Department of Commerce. This insures the consumer that the lumber he buys meets the standards for a particular grade. The standards have been changed several times. The 1970 grading rules differ greatly from those of previous years.

As a consumer of lumber, you are interested in purchasing the least expensive lumber which will serve your purpose. Most local lumber retail companies anticipate the material demands of a particular community and inventory these materials. There are so

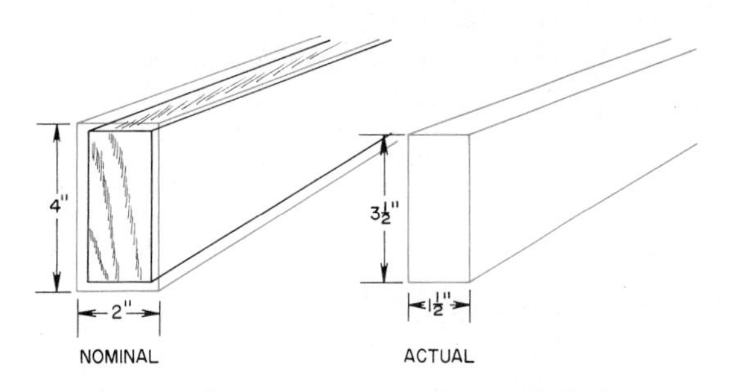

Fig. 3-8. Comparative sizes of nominal and actual sized softwood.

ESTIMATE

SIMONSON LUMBER AND HARDWARE — Nº 3890
FARGO - GRAND FORKS - GRAFTON — NORTH DAKOTA

Date November 1, 1971
Owner Wayne H. Zook — Phone 775-2458
Address 501 34th Avenue South
Contractor Davis — Phone 775-8542
Address Grand Forks
For Recreation Room

	No. Pcs.	DESCRIPTION	Feet	Price	Amount	Price
1	5	1" x 4" x 8' #2 Common Pine				
2	2	1" x 6" x 10' C-Select Pine				
3	10	1" x 12" x 8' #1 Common Pine				
4	3	1" x 12" x 6' #2 Common Pine				
5						
6						
7						
8						
9						
10						
11						
12						
13						
14						
15						
16						
17						
18						
19						
20						
21						
22						
23						

Fig. 3-9. An order giving lumber listed by thickness, width, and length.

many grades of lumber that a retail company cannot afford to inventory each of the items. A lumber company must stock some of the various grades and also an assortment of sizes. If a company stocked every dimensional size and grade, the investment in inventory would probably reduce the company's profit.

Lumber can be classified by its *appearance* and by its *strength.* The best appearing grades are generally the highest grades and most expensive. Lumber which is used to construct the frame of a house must be strong, but the appearance is secondary. The lumber is covered by other finishing material and is not visible in the finished house. The important factor when buying framing lumber is that it be strong and retain its shape.

Stress grading of lumber. Much of the lumber intended for framing purposes is mechanically graded according to its strength rather than appearance. This is called "stress grading." Each piece of lumber is graded according to its minimum strength, Fig. 3-10. A roll at the outfeed end of the machine stamps the stress rating on the end of the board. The rating is determined by the amount the board bends or deflects under a given load, Fig. 3-11. The instrument in the center records the amount of deflection or bending. The correct grade stamp is placed on the board automatically as the board passes through. The stress grading method is rather new and improvements are constantly being made. The amount of load or force placed on the board is in accordance with the thickness, width, and length of board. For example, a 2X4 is not loaded as much as a 2X12 board.

Stress grading insures the consumer of a given strength factor if the board is properly installed on the job.

The strength of lumber is determined by the species (kind) and basic characteristics. The nature of the grain and various defects influence the strength of wood. The quality of lumber is determined by the amount of clear cutting which a board contains. This means the amount of the board which is free of knots and other defects.

Common defects in softwood lumber. The most common defects in lumber are knots, Fig. 3-12. The knots are caused by

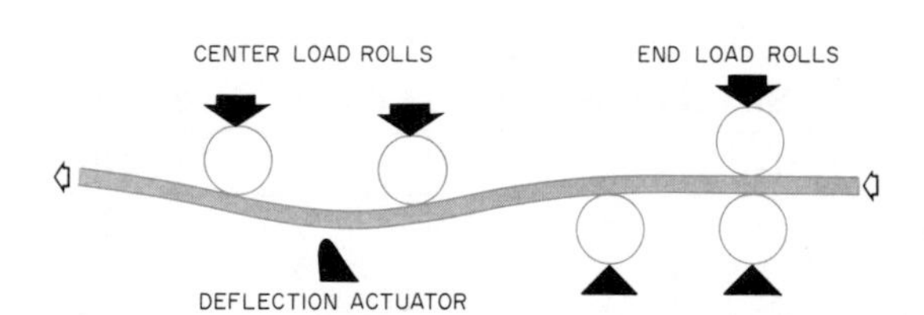

Fig. 3-11. The principle of stress grading.

Fig. 3-10. A stress grading machine. (Forest Products Laboratory)

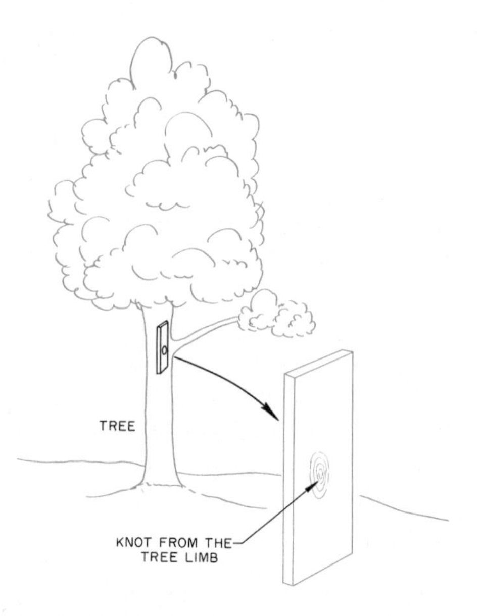

Fig. 3-12. A knot defect.

the growth of a limb on the live tree. Many of the low limbs on a tree drop off as it grows and the point of separation is covered by new layers of growth. Usually the lumber with less defects is located in the trunk of the tree where there are fewer limbs. The shape of a knot as it appears on a board is determined by the direction the log was held when sawed into boards, Fig. 3-13.

In addition to knots affecting the grading of lumber, other factors are considered. They include checks, splits, shakes, pitch pockets, wane, and warpage.

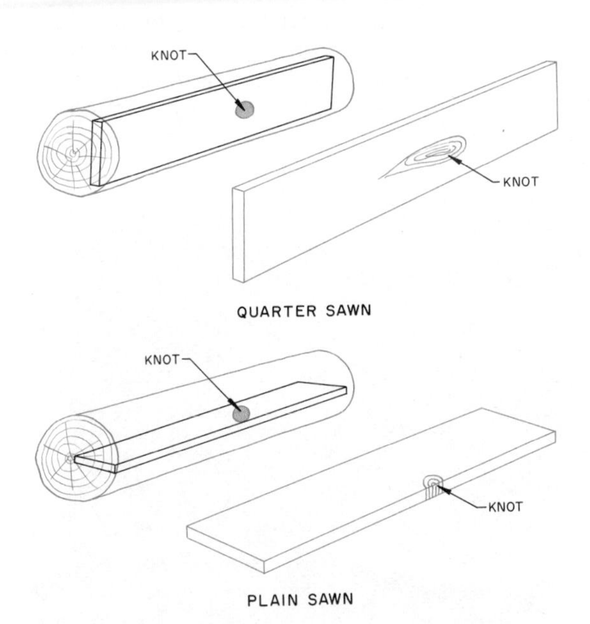

Fig. 3-13. The position of the log during sawing determines the shape of a knot defect.

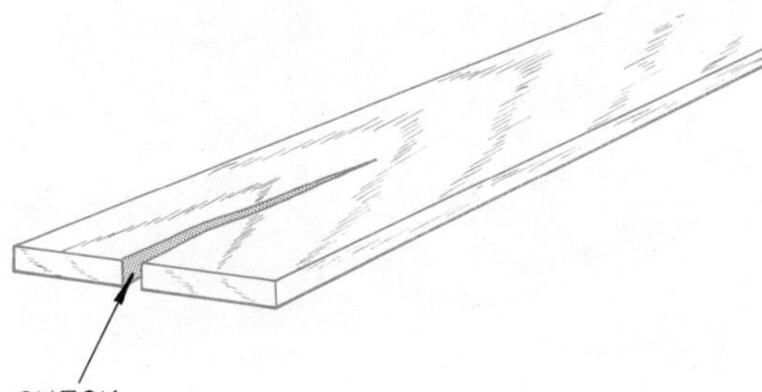

Fig. 3-14. A check is a common defect in sawn boards.

A *check* is a split in a board parallel to the grain, Fig. 3-14. The check occurs due to drying stress. Sometimes defects similar to checks are caused when warped lumber is surfaced. The pressure of the planer rollers flaten the warped boards, causing it to split.

A *shake* is a defect which also runs parallel to the grain, Fig. 3-15. The shake is a defect in which the annual rings in the board appear to be separated.

A *pitch pocket* is an accumulation of resins in a hollow in the wood. The affect of a pitch pocket on the grade of lumber depends on its size.

A board which still has some bark on the edge is regarded as defective. This type of defect is called *wane.* It is defective because the full width of the board cannot be used.

Warp is a variation in a board from a flat plane or surface. There are numerous types of warps. Common types of warps are cups, crooks, bows, and twists, Fig. 3-16. Twists are also called winds.

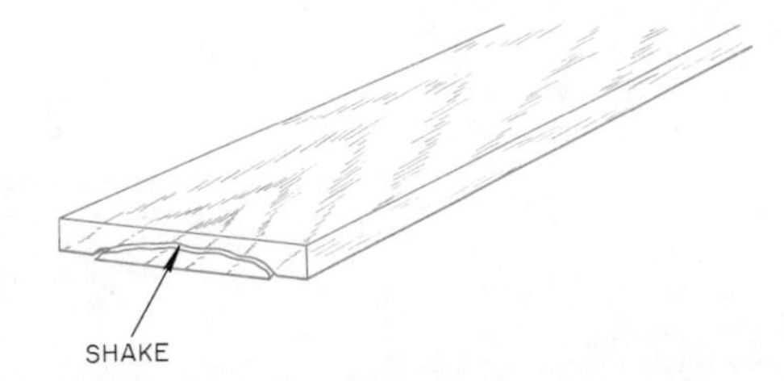

Fig. 3-15. A shake defect.

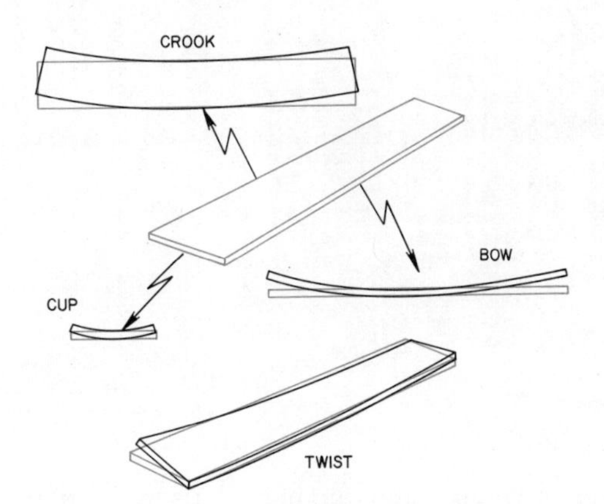

Fig. 3-16. Common types of warpage.

The product will determine the quality of lumber needed. The grade will indicate the appearance or the number of defects in the lumber. Boards can be purchased which are free of visual defects. This lumber would be the most expensive. It is more economical to select lumber according to its use. For example, if strength is not a major factor, it is more economical to use the lower grades in products where it will not show. The framework of a cabinet could be lower grade material if it does not show.

When ordering lumber for a product, you can depend upon the services of your local lumber retail dealer. He will supply the grade of lumber and dimensional size necessary for the product which you are constructing.

Grades of softwood lumber. The finest quality of 1″ nominal-sized softwood lumber is called *selects.* This grade of lumber would be specified if lumber is required with few, if any, defects. There are three grades of selects which include: (1) B & BTR, (2) C, and (3) D. The best grade of select is B & BTR.

Select lumber would be used for interior walls, woodwork, furniture, cabinets, and similar purposes.

A wide range of lumber called *commons* is available. This grade has more defects (such as knots) than the select grades. There are five common board grades ranging from *one* to *five.* The number *three* is the most widely used and is most often stocked by retail lumber companies. The knots are generally tight and will not fall out. The number three common boards are used for shelving, paneling, siding, fences, and many industrial uses.

Larger dimensions ranging from 2X4's to 2×12's are graded somewhat differently, and are referred to as light framing lumber. This lumber is generally used for framing in

Fig. 3-17. Framing lumber is used in house construction.

Fig. 3-18. A package of precut studs ready for the carpenter to assemble on the job.

house construction and is classified as *construction, standard,* and *utility* grades. The construction grade is the highest quality.

Light framing lumber has defects such as knots. The main concern for this type of lumber is strength and not appearance. Figure 3-17 shows framing lumber being used for house construction. It provides a strong frame, but it will eventually be covered.

Studs, which are the vertical support members in light frame construction, are graded as one grade. The grading of studs is limited to factors which affect the strength and fastening qualities. Studs can be purchased precut to the exact length, Fig. 3-18. This saves the carpenter on the job one time-consuming operation. This is another example of an available standard material.

You should now realize there are many grades of softwood lumber available. It is important to reduce cost so the grade of lumber is matched to the product produced. Your local lumber dealer will help you, and you should use this service.

Extensive information on the grading of lumber is available from the Forest Products Laboratory in Madison, Wisconsin.

STANDARD HARDWOOD LUMBER

The standards used for the grading of hardwoods have been adopted by the American Hardwood Lumber Association. The grading system is based on the proportion of a board which can be cut into smaller pieces which are clear on one side. The smaller pieces cut from larger boards are usually referred to as *cuttings.*

Standard sizes of hardwood lumber. The standard sizes of hardwoods are usually different from the softwoods. The hardwoods are generally more expensive, and to cut all of the wood to given sizes would result in excessive waste. For this reason, some hardwoods are supplied in uneven lengths such as 7′ and 9′, etc. When possible, they are cut to even lengths such as 6′, 8′, and 10′, etc.

The widths in which most hardwoods are available will be specified as a *minimum width.* An example would be to specify the order as 6″ wide and wider. Figure 3-19 shows a sample order for hardwood lumber.

Hardwood lumber may also be ordered as *random lengths and widths.* This means that you will receive lumber of varying sizes. However, you also specify in random ordering the minimum widths and lengths you will accept, Fig. 3-20.

The standard lengths of hardwoods range from 4′ to 16′ in intervals of 1′. The

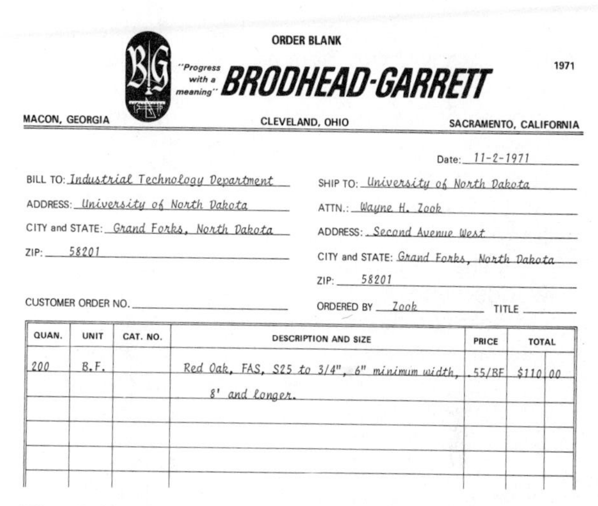
ORDER BLANK

"Progress with a meaning" BRODHEAD-GARRETT 1971

MACON, GEORGIA CLEVELAND, OHIO SACRAMENTO, CALIFORNIA

Date: 11-2-1971

BILL TO: Industrial Technology Department
ADDRESS: University of North Dakota
CITY and STATE: Grand Forks, North Dakota
ZIP: 58201

SHIP TO: University of North Dakota
ATTN.: Wayne H. Zook
ADDRESS: Second Avenue West
CITY and STATE: Grand Forks, North Dakota
ZIP: 58201

CUSTOMER ORDER NO.
ORDERED BY Zook TITLE

QUAN.	UNIT	CAT. NO.	DESCRIPTION AND SIZE	PRICE	TOTAL
200	B.F.		Red Oak, FAS, S25 to 3/4", 6" minimum width, 8' and longer.	.55/BF	$110.00

Fig. 3-19. A sample order for hardwood lumber.

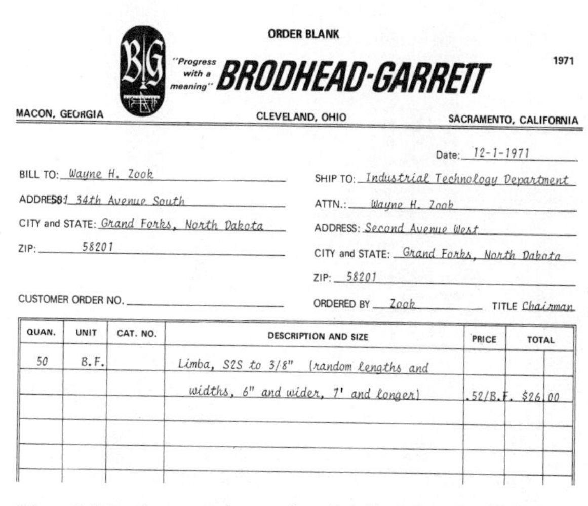
ORDER BLANK

"Progress with a meaning" BRODHEAD-GARRETT 1971

MACON, GEORGIA CLEVELAND, OHIO SACRAMENTO, CALIFORNIA

Date: 12-1-1971

BILL TO: Wayne H. Zook
ADDRESS: 581 34th Avenue South
CITY and STATE: Grand Forks, North Dakota
ZIP: 58201

SHIP TO: Industrial Technology Department
ATTN.: Wayne H. Zook
ADDRESS: Second Avenue West
CITY and STATE: Grand Forks, North Dakota
ZIP: 58201

CUSTOMER ORDER NO.
ORDERED BY Zook TITLE Chairman

QUAN.	UNIT	CAT. NO.	DESCRIPTION AND SIZE	PRICE	TOTAL
50	B.F.		Limba, S2S to 3/8" (random lengths and widths, 6" and wider, 7' and longer)	.52/B.F.	$26.00

Fig. 3-20. A sample order for random widths and lengths of hardwood lumber.

standard thicknesses of rough sawn hardwood lumber start at 3/8″ and range to 3/4″ by 1/8″ intervals. Hardwoods from 3/4″ thick to 1 1/2″ thick are sold in thickness intervals of 1/2″.

A hardwood log is cut into widths and thicknesses which will produce the most lumber with the least amount of waste. Excessive waste at any stage of production forces an increase in product cost to the consumer.

Grades of hardwood lumber. The best grade of hardwood lumber is *Firsts,* and the next grade is *Seconds.* These two grades are usually combined and are called *Firsts and Seconds,* written FAS. Hardwoods graded as Firsts must yield a minimum of 91 2/3% clear material (free of defects). Seconds must yield 83 1/3% cuttings. First and Second grades must be boards not less than 6″ wide and 8′ long.

The next grade classification of hardwoods is called *Selects.* This grade classification permits small pieces of clear lumber. Selects may be a minimum of 4″ wide and 6′ long. The back side of selects may also have more imperfections than permitted in Firsts and Seconds.

No. 1 Common and *No. 2 Common* are grades below the Select grade. Both Common grades require a minimum size board of 3″ wide and 4′ long. No. 1 Common must yield 66 2/3% clear face cuttings. No. 2 Common must yield 50% clear face cuttings. A smaller minimum size clear cutting is permitted in the common grade than in the FAS and Select grades.

MAN MADE SHEET MATERIALS

Man-made sheet materials are the result of continuous research in the woodworking industry. Wood scientists and wood technologists have constantly sought better ways to use wood. Man-made sheet material has added a new dimension to the woodworking industry. Man-made sheet material has remedied many of the limitations of wood as a structural material.

Manufactured materials make it possible for the builder to specify sizes unknown before to the industry. Now the builder can specify exact quality and size in building materials. Before manufacturing, the size of man's materials was determined by the size of the tree.

There are numerous types of man-made sheet materials available. The most common are plywood, hardboard, and particle board.

THE MANUFACTURE OF PLYWOOD AND VENEER

Plywood is fabricated sheet material consisting of layers of wood panels bonded together. Plywood consists of three or more plies (layers) of veneer with each ply at a right angle to the adjacent ply. Plywood is always constructed in odd numbers of plies, Fig. 3-21. The use of odd numbers of plies at right angles to one another stabilizes the sheet. The balanced construction of plywood tends to keep the sheet flat when the moisture content changes.

Note the direction of the grain in each layer in Fig. 3-21. The plies on each side of the center which go in the same direction are of equal thickness to maintain a stress balance.

The outer plies always have the grain running in the same direction. The inner plies are called the *core.* Most plywood is manufactured with a veneer core; some plywood is constructed with a lumber core as shown in Fig. 3-22.

The logs which are cut into veneer for plywood production are carefully selected. The best logs are selected for veneer and are used for the face ply on high grade plywood panels. The use of thin veneer makes it possible to better utilize a limited amount of fine cabinet wood.

The logs selected for veneer are often called "peeler" logs. The rotary method is used most often for cutting veneer from the log. For rotary cutting, the log is mounted

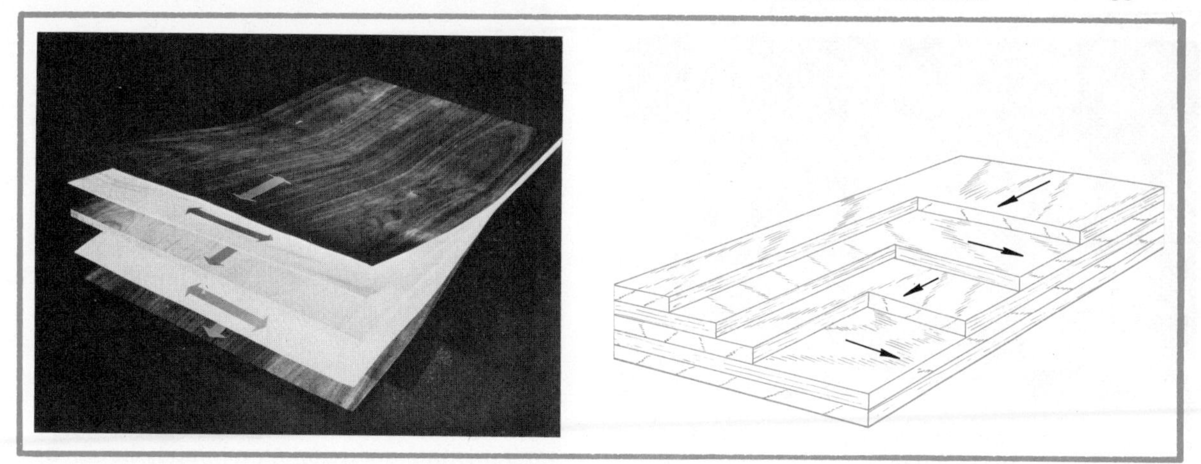

Fig. 3-21. A cross section of a typical piece of 5-ply plywood. (American Forest Institute)

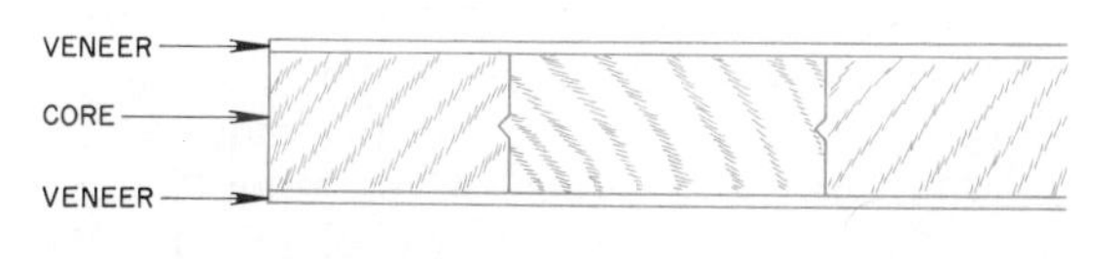

Fig. 3-22. A sectional view of lumber core plywood.

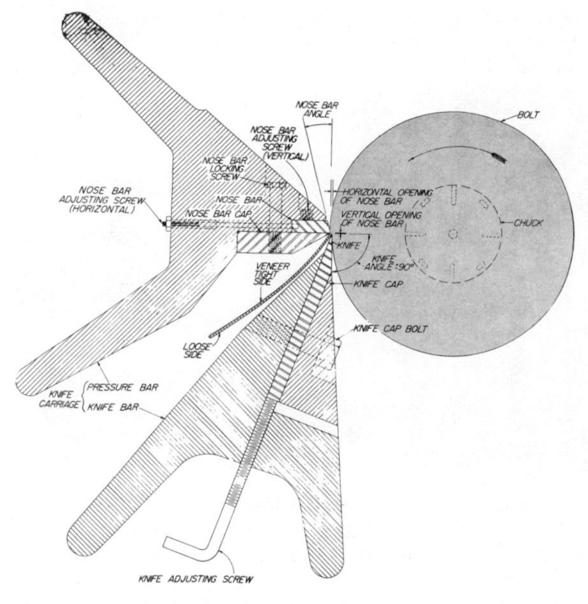

Fig. 3-24. Cross-sectional detail of a rotary lathe carriage.

Fig. 3-23. Rotary cutting of veneer.

in a large lathe and rotated against a razor-sharp knife, Fig. 3-23.

Rotary cutting produces a continuous length of veneer the same width as the length of the log. As the log becomes smaller, the knife advances toward the log. This insures a uniform thickness for the veneer. Figure 3-24 shows the cross-sectional detail of the rotary lathe carriage.

Fig. 3-25. One of the production processes for manufacturing plywood.

Fig. 3-26. A veneer drier for drying to desired moisture content. (Hardwood Plywood Manufacturers Association)

Fig. 3-27. Visual grading of veneer.

The production of a sheet of plywood involves many processes, Fig. 3-25. After passing through the rotary veneer lathe, the veneer is cut into the required lengths and widths.

The veneer is dried to the necessary moisture content in a large drier as shown in Fig. 3-26. The sheets of veneer are fed through at a constant speed on the conveyor system. The moisture content after drying is usually within the range of 4% to 6%. The dried sheets of veneer are graded in much the same manner as dimensional lumber, Fig. 3-27. The graded sheets are stacked and used for either face, back, or core plies. When the sheets of veneer are dried and graded, defects such as knots are cut out with a die. The die-cut holes are filled with solid wood patches.

The graded sheets are fed through a glue applicator and a film of adhesive is applied on both sides of the sheets. The adhesive-coated sheets are stacked with alternate sheets of dry veneer to form the complete plywood panel, Fig. 3-28. The plies of veneer are stacked with the grain in opposite directions. The panel consisting of an odd number of plies is placed in a huge press equipped with heated platens for the glue to cure. The cured plywood

Fig. 3-28. Adhesives are applied to veneer sheets. (Hardwood Plywood Manufacturers Association)

panel is passed through saws which trim and square the completed panel. Finish sanding is the final operation in the manufacturing of the plywood panel. The complete panel is inspected, graded, and prepared for shipment.

The majority of the veneer for plywood is cut by the rotary process. However, some of the finer cabinet woods used in the manufacture of plywood are plain-sliced or quarter-sliced. Plain and quarter-sliced veneers often result in a more interesting grain figuration, Fig. 3-29.

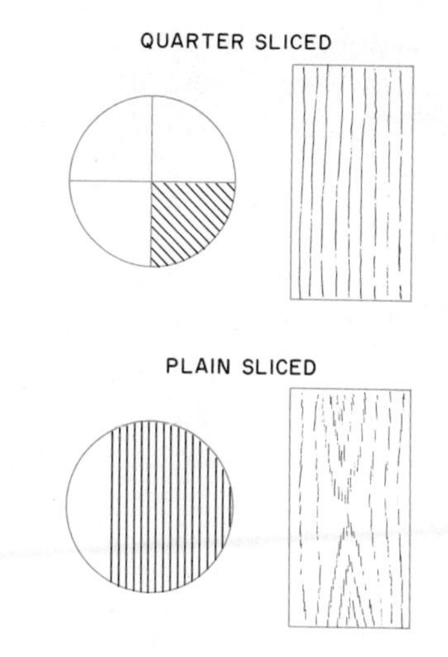

Fig. 3-29. Common methods used for slicing veneer.

Hardwood and softwood plywood. Plywood is available in two broad classifications which are (1) hardwood and (2) softwood. Douglas fir is one of the most common species used for softwood plywood. However, other species such as western hemlock, redwood, white fir, redwood, pine, and others are also used. The hardwood plywoods are manufactured from a wide variety of species.

Interior and exterior plywood. In both hardwoods and softwoods a variety of grades and types are manufactured. Grade refers to the quality of the veneer used in the sheet construction. Type refers to the moisture resistance of the glue used to bond the panels together. Type is classified as either exterior or interior plywood.

Interior plywood will withstand occasional wetting followed by thorough drying without losing strength. Interior plywood is usually bonded with glue having a soybean base. The exterior plywood will retain its shape and strength when exposed to water and other exterior weather elements. The only difference between interior and exterior plywood is the resistance of the glue to moisture.

Advantages of plywood. The structure of plywood gives it properties superior to those of solid dimension lumber. The cross banding of the plies gives it dimensional stability and some strength characteristics which are greater than that of solid wood.

One of the greatest advantages of plywood is its availability in large sheet form. The large sheets *reduce construction time.* Production costs less because less assembly time is required when large sheets are used. Plywood can also be bent and formed to conform to irregular shapes. Like solid lumber, plywood is available in many grades and species. To insure the buyer of the quality needed for a particular product, a standard grading system has been developed. The grading system for plywood manufactured from softwood and hardwood differs.

Grades of plywood. The grade of plywood is determined by the quality of the veneer used on the face side and back side of the sheet. Softwood plywood is graded somewhat differently than hardwood. The quality of the face and back veneers is designated by the letters A, B, C, and D for softwood plywood. The best grade of softwood plywood is A and the poorest grade is D.

Softwood plywood graded A-A would have A grade veneer on both the face and back sides. A-D plywood is made of a face veneer of A quality and a back of D quality. The interior plies of each panel are usually of one of the lower grades. Table 2 gives the common grades of plywood and some of their uses.

The grading system for plywood was established by the Department of Commerce and the plywood industry. The American Plywood Association helped to establish the rules for grading softwood plywood. Plywood graded under these rules is stamped "DFPA" which stands for "Division for Product Approval."

Hardwood plywood is graded 1, 2, 3, and 4. A hardwood plywood panel graded 1-2 would mean that the face veneer is good and is carefully matched. The back panel on the 1-2 grade would be good but the veneer panels would not be carefully matched. Plywood graded as 3 would have some defects and would be patched. However, other than appearance, the panel would be structurally sound.

In addition to the four common grades of hardwood plywood, special grades can be ordered. It is possible to order plywood sequencially matched from one sheet to the next. The use of this type of matched plywood is limited to elaborate architectural structures due to the cost.

Plywood is often designated by G1S (good one side) or G2S (good two sides). This usually means that the face side is of the high grade and the back is of the lower grade.

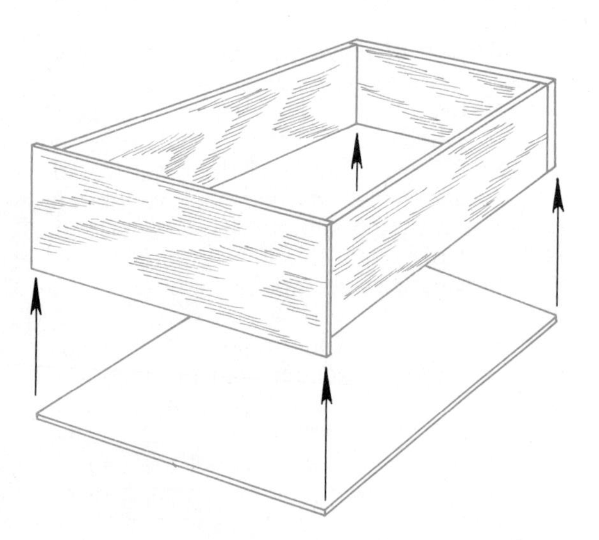

Fig. 3-30. Hardboard is accepted and used for drawer bottoms and furniture backs.

Standard sizes of plywood and how it is sold. Plywood is available in a variety of sizes. The most common size sheet of plywood is 4′ wide and 8′ long. The thickness of plywood varies from 1/8″ to thickness of over 1″. The thicknesses commonly available from retail lumberyards are 1/4″, 3/8″, 1/2″, 5/8″, and 3/4″ in the standard 4′ × 8′ panel. Other sheet sizes can also be ordered depending upon the species and the grade.

Fig. 3-31. The interior walls of this room are being finished with hardboard wall paneling. (Hardwood Plywood Manufacturers Association and Georgia-Pacific Corp.)

Plywood is sold by the square foot. It is most economical to purchase plywood in full sheet sizes. The grade of plywood will be determined by the purpose for which it will be used. Lower grades can be used for products which are painted. The paint tends to cover many of the small defects.

When only one side of the plywood is exposed, the back side can be of a less expensive grade. Purchasing higher quality material than the product demands is a waste of money.

MANUFACTURE OF HARDBOARD SHEET MATERIAL

Hardboard is a manufactured panel consisting of wood fibers. The wood fibers are bonded together with a natural substance called *lignin* which exists in all woods. The wood fibers are bonded together through the utilization of the natural lignin and extreme pressure. Hardboard is commonly manufactured from sawmill residue and low-quality logs which are unsuitable for lumber usage. It is very economical, and can be used for construction of products which will be painted, Fig. 3-30.

Large quantities of wall paneling are made from hardboard, Fig. 3-31. A special finishing technique is used to make it appear like wood grain. Figure 3-32 shows hardboard as a material for siding on the outside of a house.

Fig. 3-32. A house with hardboard siding. (National Forest Products)

Hardboard is also available in a vast array of decorative shapes and patterns. Pegboard is a hardboard material which has been widely accepted for decorative and utility purposes, Fig. 3-33.

Hardboard is manufactured from the sawmill residue. This residue was a costly material for sawmills to discard before the manufacturing process for hardboard was discovered. Through scientific development,

Fig. 3-33. Pegboard as interior wall covering. (U. S. Plywood, Div. of U. S. Plywood-Champion Papers, Inc.)

Table 2

Common Grades of Plywood and Their Uses.

	Symbol indicating grade	Common Uses	Common Thicknesses					
			1/4	3/8	1/2	5/8	3/4	1
INTERIOR TYPES	A-A INT - DFPA*	Interior uses where both sides will be exposed. Used for cabinets, built-ins, furniture and similar uses.	X	X	X	X	X	X
	A-B INT - DFPA	Used where the appearance of the second side is not as important as the first. Both sides are smooth and solid.	X	X	X	X	X	X
	A-D INT - DFPA	Used where the appearance of one side is important. Commonly used for shelving, partitions, paneling and built-ins.	X	X	X	X	X	X
	B-B INT - DFPA	Both sides may be plugged. Provides smooth paintable surface.	X	X	X	X	X	X
	B-D INT - DFPA	One side is plugged and paintable. Back side is not patched and is usually not exposed.	X	X	X	X	X	X
	Decorative Panels	Grooved, brushed, rough sawn faces and used for decorative effect on interior walls, cabinets and built-ins.	X	X	X			
EXTERIOR TYPES	A-A EXT - DFPA	Used where both sides will be exposed. Common uses include fences, displays, boats, etc.	X	X	X	X	X	X
	A-B EXT - DFPA	Used for many of the same purposes as A-A.	X	X	X	X	X	X
	A-C EXT - DFPA	For exterior uses where only one side shows. Typical uses include siding, soffits, farm buildings, fences, etc.	X	X	X	X	X	X
	HDO EXT - DFPA	High density plywood with resin-fiber overlay. Used for concrete forms, acid tanks, and counter tops. It is abrasive resistant.		X	X	X	X	X
	Marine EXT - DFPA	It is made only from Douglas fir or western larch. Commonly used for boat hulls and other purposes where it will be subjected to water.	X	X	X	X	X	X
	C-C EXT - DFPA	Unsanded outer surfaces bonded with water resistant glue. Used for subflooring and roof decking.	X	X	X	X	X	

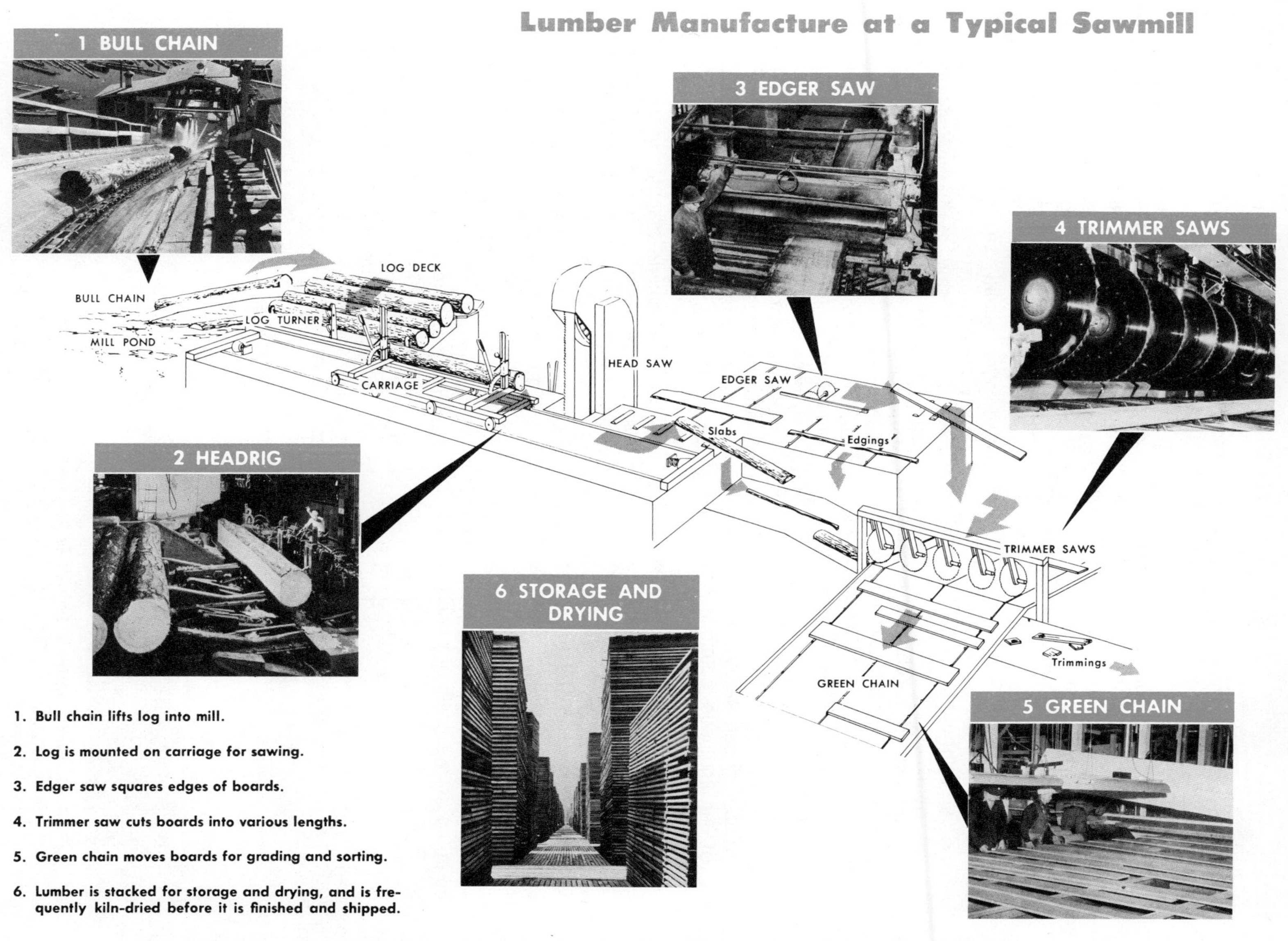

Fig. 3-34. The manufacturing process for hardboard. (American Forest Products Industries, Inc.)

a waste material has become a profit-producing material for the wood industry, Fig. 3-34.

Kinds of hardboard. Hardboard is classified as either *standard or tempered*. Standard hardboard is somewhat softer than tempered hardboard and is less resistant to moisture and abrasion. Tempered hardboard is manufactured by impregnating the wood fibers with drying oils and adhesive additives. The result is a manufactured sheet with greater wearing qualities and high moisture resistance.

Advantages of hardboard. Hardboard is a lightweight and rather inexpensive material. It is available in large sheets and can easily be cut to size. It does not expand and contract like wood, and the size remains constant as the moisture in the atmosphere changes.

Hardboard can easily be finished with any of the methods commonly used for wood.

Standard size of hardboard and how it is sold. Hardboard is available in many sizes and thicknesses. The thicknesses most often stocked by retail lumber yards are 1/8″ and 1/4″ in sheet sizes of 4′ x 8′. Tempered and standard hardboard are available in 1/10″, 1/8″, 3/16″, 1/4″, and 5/16″ thicknesses in a range of sizes starting with 1′ × 4′. The price for hardboard is computed on the basis of price per square foot.

Fig. 3-35. A piece of particle board as it is being produced. (Southwest Forest Industries)

MANUFACTURE OF PARTICLE BOARD SHEET MATERIAL

Particle board is a man-made structural material consisting of wood chips and flakes in combination with adhesive. Particle board is the result of research in the woodworking industry to make better use of forest products. Particle board is constructed from some of the sawmill residue and poor quality saw logs which would otherwise be wasted, Fig. 3-35.

The first step in the manufacture of particle board is to chip the wood into uniform flake size. The particles are combined with an adhesive and placed in a hot press under high pressure. Particle board can be produced for interior or exterior use depending upon the adhesive. The manufacturing process of particle board is very similar to hardboard production.

Advantages of particle board. Particle board is very stable and is affected very little by humidity changes. It is used extensively in furniture manufacturing. Particle board can also be formed into desired shapes other than flat sheet form, Fig. 3-36. It is also used in house construction as *underlayment.* This is the second layer of the floor directly under the finished floor covering material such as carpet and tile.

Particle board has a very hard surface, but can be cut and formed easily with woodworking tools and machines.

Standard sizes of particle board and how it is sold. Particle board can be purchased in standard panel sizes. The most common panel size is 4′ × 8′. Standard thicknesses in sizes ranging from ⅛″ to 1″

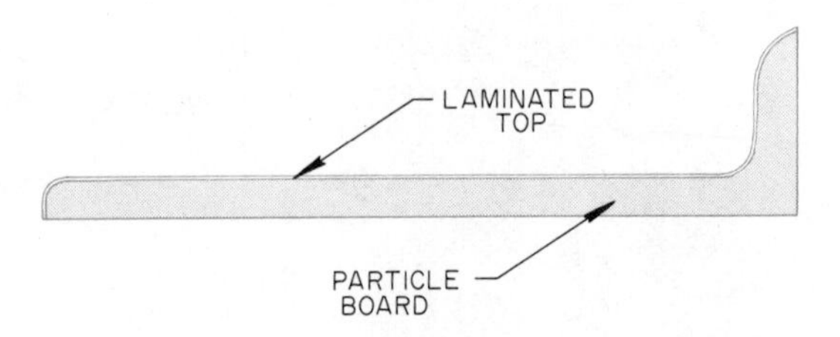

Fig. 3-36. A counter top made from formed particle board and laminated.

are available. The cost for particle board is computed by the price per square foot.

STANDARD HARDWARE

The selection of hardware available is extremely varied, so only a few of the basic types will be covered in this chapter. Items commonly referred to as hardware include hinges, pulls, latches, locks, and door tracks, to mention only a few. It is important to be familiar with the basic types of hardware and their uses as they relate to product design and production. Hardware contributes to the appearance of a product as well as its usefulness. Hardware is usually not attached to the product until the finish has been applied. However, the type of hardware needed should be determined during the product-planning stage of production. Hardware for furniture is available in numerous styles to harmonize with the period represented, Fig. 3-37.

Fig. 3-37. An Early American desk with hardware designed for that furniture period. (Kemp Furniture Industries)

DOOR MECHANISMS

Numerous types of hardware assemblies can be used for swinging and sliding doors. Doors are generally equipped with a knob or handle and a device to hold the door shut. Various types of door hardware are shown in the following figures.

Butt hinges. This is one of the most common types of hinges, Fig. 3-38. Butt hinges are available with either loose pins or solid pins, Fig. 3-39. The butt hinge requires the cutting of a gain in the rail of the cabinet or door jamb, Fig. 3-40.

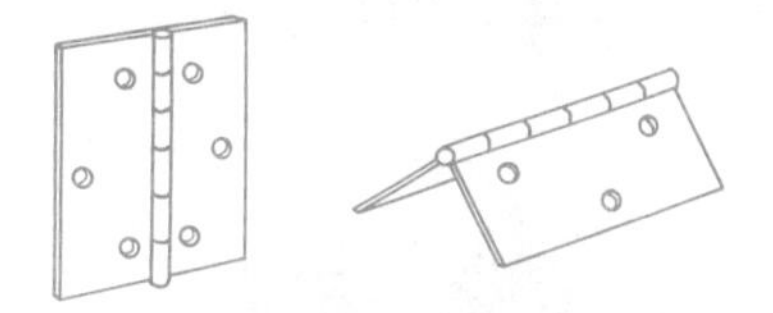

Fig. 3-38. A butt hinge.

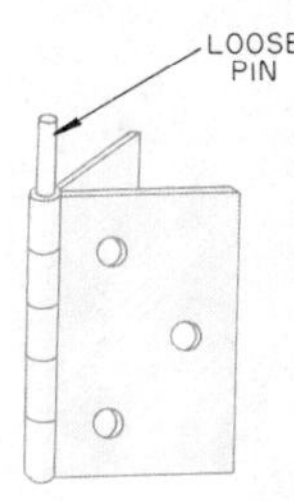

Fig. 3-39. A butt hinge with a removable pin.

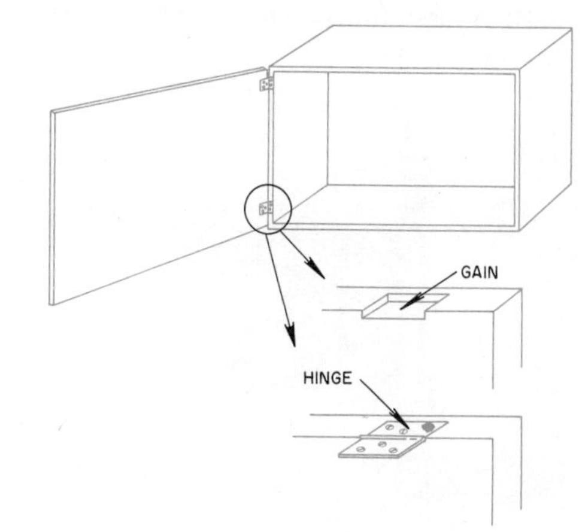

Fig. 3-40. A gain is cut to receive the butt hinge leaf.

Surface hinges. This hinge is often used because it is easy to install. The surface hinge does not require the cutting of a gain. It can also add a decorative effect, Fig. 3-41.

Semiconcealed hinges. This hinge is used to install overlapping doors, Fig. 3-42. A gain is not cut to install this hinge. It works well with plywood doors because the screws are installed in the flat grain. Only one-half of the hinge shows after it is installed.

Door pulls. A door pull is both functional and decorative. Figure 3-43 shows a group of door pulls. They can serve as a handle and also a latch.

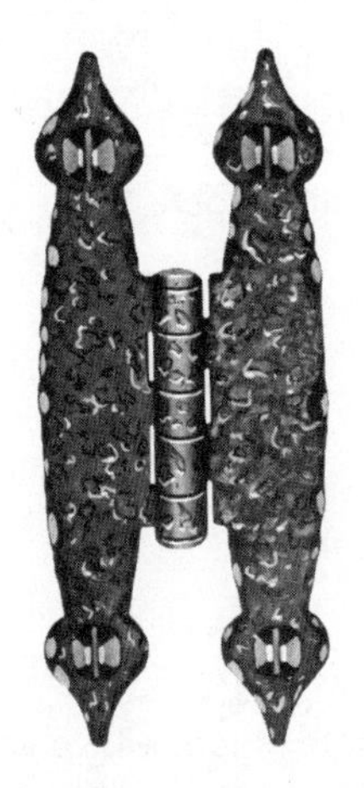

Fig. 3-41. A decorative surface hinge. (Amerock Corp.)

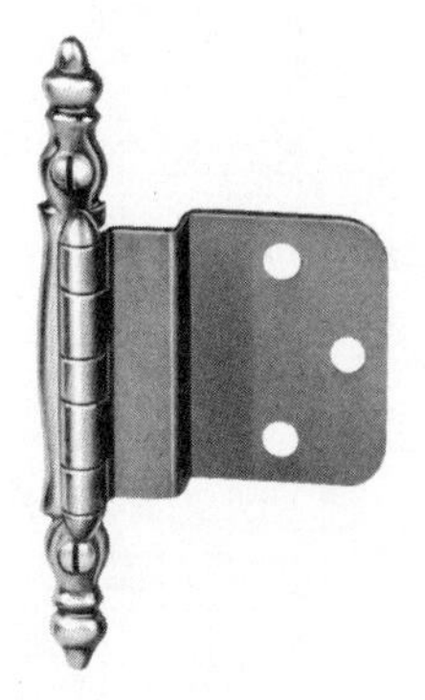

Fig. 3-42. A semi-concealed hinge. (Amerock Corp.)

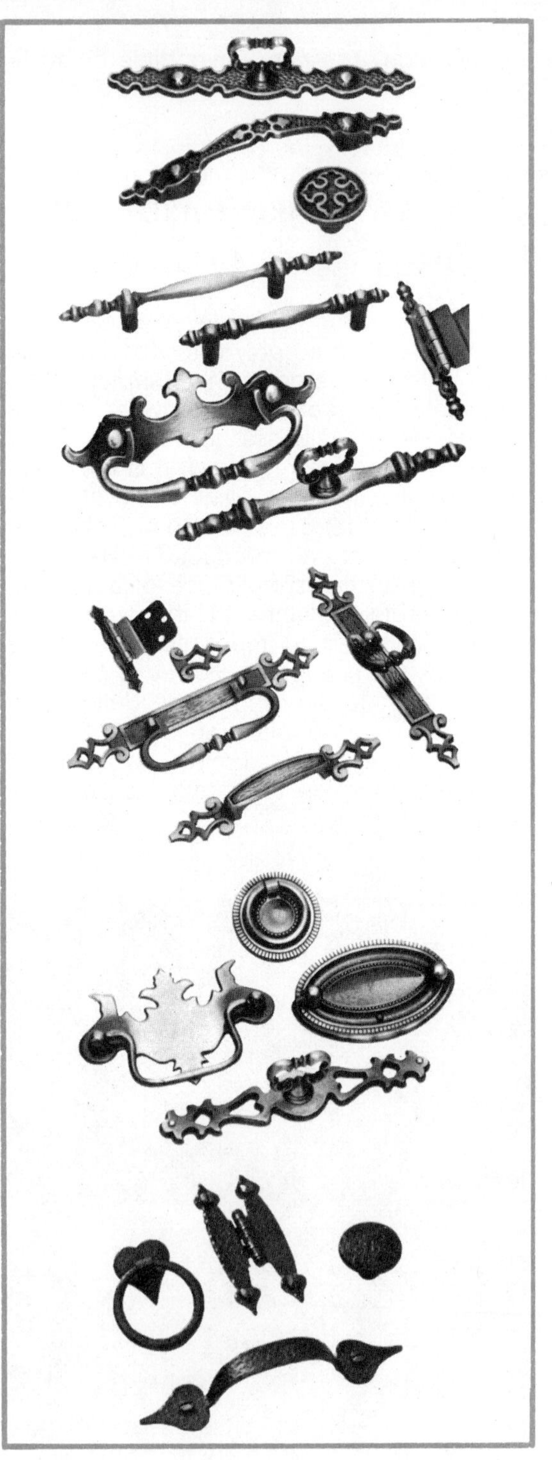

Fig. 3-43. Decorative door pulls. (Amerock Corp.)

Sliding door track. Sliding doors are installed when there is not room for the doors to swing. Sliding doors can also be a design feature. Figure 3-44 shows hardware which can be installed on various types of closet and cabinet doors.

DRAWER MECHANISMS

Cabinet drawers are equipped with guides so they will open easily and remain in alignment. There are numerous commercial drawer guides which can be purchased, Fig. 3-45. Guides are often constructed as a part of the cabinet. Details for constructing guides are covered in a later chapter.

Fig. 3-44. Hardware for cabinet doors.

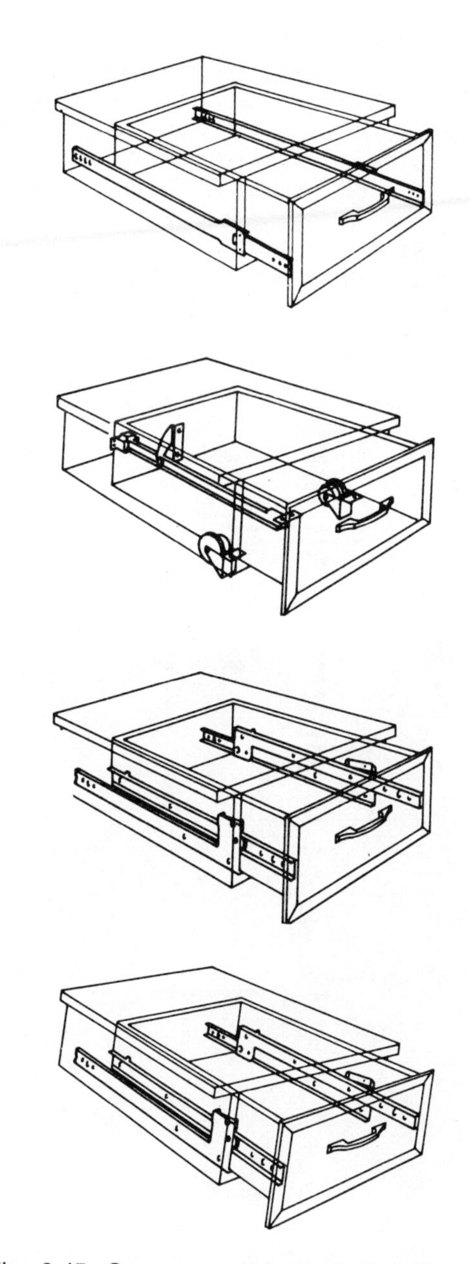

Fig. 3-45. Common mechanical drawer guide systems.

THE INFLUENCE OF STANDARDIZATION ON MANUFACTURING AND PRODUCT AVAILABILITY

The standardization of materials has influenced product design and production operations. Industry first makes a thorough study of what can be used for material to produce a new product. Decisions are made whether it is less costly to manufacture their own parts or to buy them and only do assembly in their own plant.

The product cost is generally reduced as the number produced increases. This usually makes it less expensive to buy standard parts from other companies specializing in the manufacture of a particular part. For example, consider all of the parts used in the manufacture of the piece of furniture shown in Fig. 3-46. This product requires hinges, fasteners, adhesives, plywood, and lumber. You immediately realize that the manufacturer purchases many of the standard parts. The company can purchase the necessary hardware at less cost than it could possibly manufacture its own.

You will soon have to make these same decisions when you produce a product. Undoubtedly you will purchase many standard materials because it will be faster and less expensive.

A complete description of the material on the order form will insure you of receiving the desired material. The supplier is as anxious to supply the material as you are to receive it, Fig. 3-47. As you can see, the supplier can furnish the same lumber rough or dressed. However, the additional processing costs you money. You must decide if this additional processing is worth the extra money. Buying this additional service saves you labor and equipment investment.

Fig. 3-46. A manufactured piece of furniture requiring many standard materials. (Brandt)

ORDER BLANK

"Progress with a meaning" BRODHEAD-GARRETT 1971

MACON, GEORGIA — CLEVELAND, OHIO — SACRAMENTO, CALIFORNIA

Date: 11-2-1971

BILL TO: Industrial Technology Department
ADDRESS: University of North Dakota
CITY and STATE: Grand Forks, North Dakota
ZIP: 58201

SHIP TO: University of North Dakota
ATTN.: Wayne H. Zook
ADDRESS: Second Avenue West
CITY and STATE: Grand Forks, North Dakota
ZIP: 58201

CUSTOMER ORDER NO.
ORDERED BY Zook TITLE

QUAN.	UNIT	CAT. NO.	DESCRIPTION AND SIZE	PRICE	TOTAL
1 pr.	pr.	811	Drop leaf table hinge	.75	.75
2	1	11405	Ball Knob (maple)	.30	.60
3	1		5/16" x 2 1/2" Hanger bolts	.03	.09
10	1	1914	Steel table leg brace, size #3	.26	2.60
1	gal		lacquer thinner	2.85	2.85
1	pint		4 lb. cut white shellac	.70	.70
2	sheet		1/4" x 4' x 8' standard hardboard	2.50	5.00

Fig. 3-47. Furnish the supplier with a complete description to order materials.

ACTIVITIES

PROBLEM

Identify all of the standard materials used in the construction of the house shown in Fig. 3-48.

Fig. 3-48. A completed house constructed from standard materials.

VOCABULARY

Standard material
Plywood
Stress grading
Hardwood
Softwood
Particle board
Nominal size
Actual size
S4S
FAS
G1S

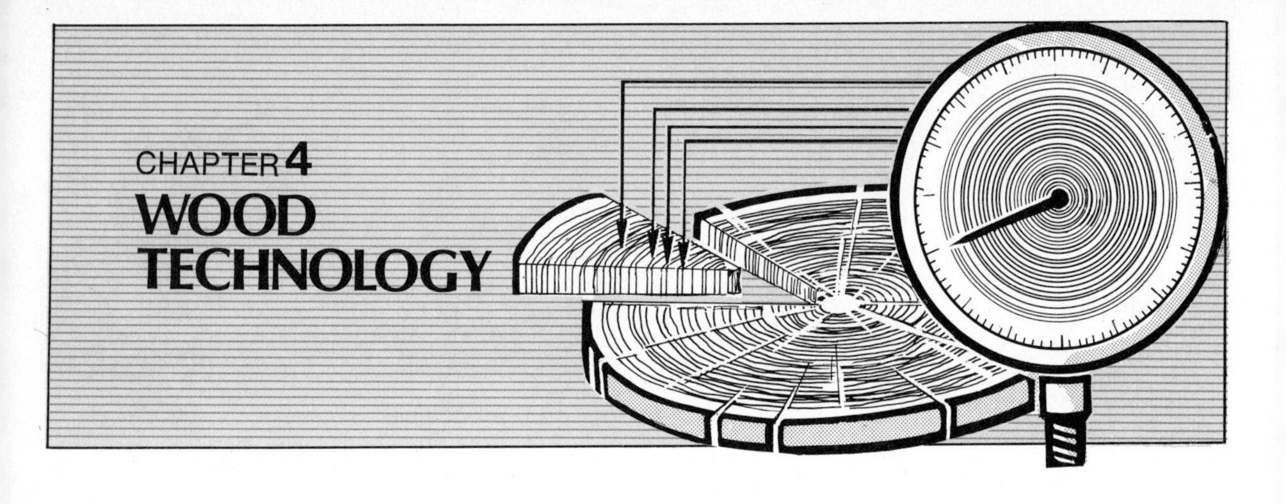

INTRODUCTION

Wood technology is the scientific study of the characteristics of wood. The knowledge gained through science is used in the manufacturing of wood products. You will need to understand some of the characteristics of wood to successfully design and manufacture wood products effectively.

The *physical properties* of wood are studied in wood technology. There are several different types of woods. When you know the advantages and disadvantages of different species of wood, you can *select* those which best serve your needs. It is also important to understand *how moisture affects wood.* Moisture is the greatest factor which must be considered in woodworking.

You should also learn *how to identify different kinds of wood.* Once you become familiar with the common species of woods, you will be able to identify different kinds by *visual inspection.* Weight, color, and grain structure are identifying factors for most common wood species.

Furniture designers, architects, and other construction personnel associated with manufactured product development have a knowledge of wood as a structural material. The direction of the grain, for example, affects the position in which it is placed in the structure.

The quality of lumber cannot be changed very easily while it is growing. Trees can be planted and protected as seedlings but after that, it is a *natural process.* The characteristics of wood cannot be changed much after it is cut into lumber. Some woods are much harder or stronger than others. You must have a knowledge of wood to determine which is the best for a particular purpose.

While it is important to select functional species of wood, you must also consider the characteristic of wood expansion and contraction. Wood tends to *expand and contract across the grain* with only slight change in the direction of the grain. The amount of moisture content affects this characteristic and it must be considered during the design stage, Fig. 4-1. Wood tends to split much easier with (in the direction of) the grain than across the grain, Fig. 4-2. This factor must also be considered when designing with wood.

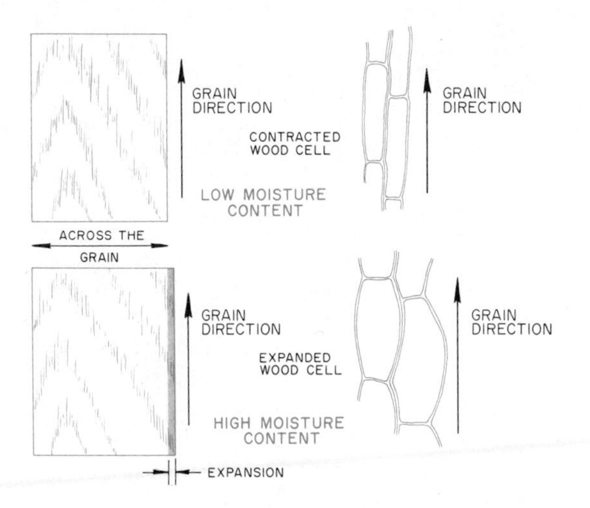

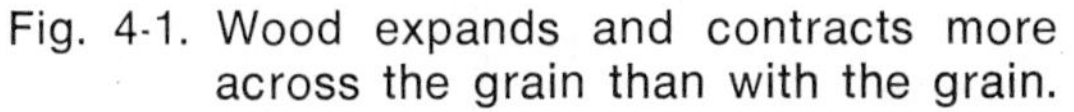

Fig. 4-1. Wood expands and contracts more across the grain than with the grain.

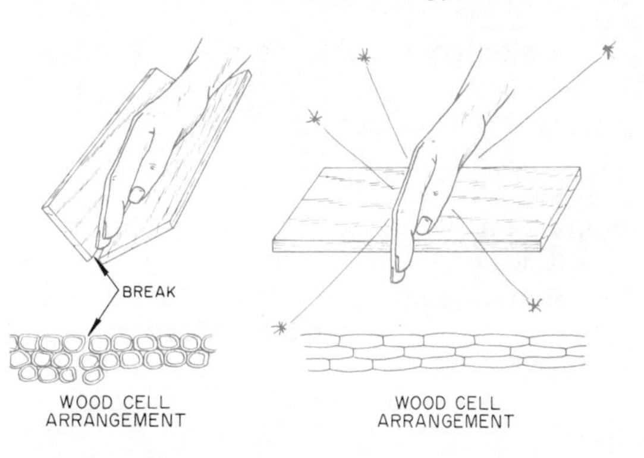

Fig. 4-2. Wood splits more easily across the grain than with the grain.

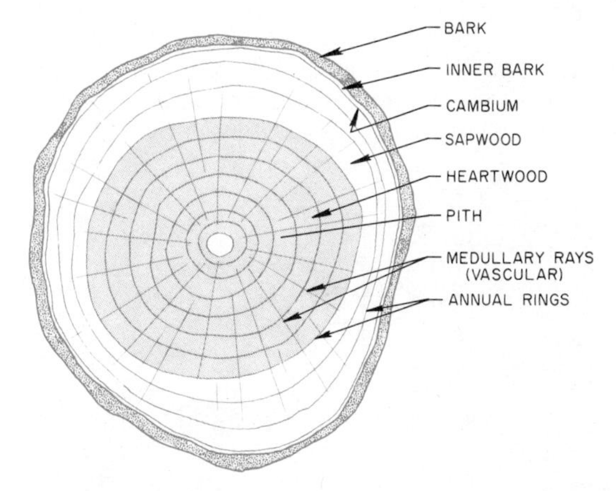

Fig. 4-3. Cross section of a tree.

PARTS OF A TREE

To truly understand the characteristics of wood, you must understand *the growth process of a tree.* The tree trunk is made up of *layers*. Each layer serves a specific function in the growth of the tree. Refer to the cross-section of a tree in Fig. 4-3 as you study the major parts of a tree and their function:

Bark. The outer layer of the tree trunk is a protective layer called *bark*. It is composed of dead, dry tissue and serves to protect the inner, living tissues. Extensive use has been made of bark for landscaping purposes.

Inner bark. Directly beneath the outer bark is a layer known as the inner bark. It is a soft, moist layer that carries food from the leaves to the growing parts of the tree.

Cambium. The cambium layer forms wood and bark cells. This greenish layer is microscopically thin. The cambium layer is responsible for all new growth. Growth occurs by means of cell division.

Sapwood. This is the light-colored wood beneath the bark. It carries sap from the roots to the leaves. It also stores excess food. The widest layers of sapwood are found in the fastest growing trees. The sapwood varies in thickness among different trees and the number of growth rings it contains.

Heartwood. This layer is made up of dead or inactive cells. They are formed from sapwood. The heartwood is found as the sapwood nearest the center of the tree becomes inactive. It is often dark in color because the cavities contain deposits of various materials. The heartwood serves the growing tree by providing rigid support.

Pith. This is the soft tissue about which the first wood growth takes place in newly formed twigs. In a tree trunk, the pith is the small central core which is darker in color. Most branches originate at the pith.

Medullary rays. These rays connect the various layers from the pith to the bark. They carry food across the section of the tree. These rays can only be seen in a few species of trees. In quarter sawn lumber from oak and sycamore trees, the rays show up as flakes.

Annual rings. In many species which grow in temperate climates, there is a difference between the wood formed early and the wood formed late in the growing season. This change in growth rate forms an *annual growth ring.* In tropical climates where there is very little change in the seasons, the marked differences in annual rings may not appear — the trees tend to grow at the same rate throughout the year. When differences in growth rate do appear, they produce annual growth rings. The age of the tree can be determined by counting the concentric annual growth rings.

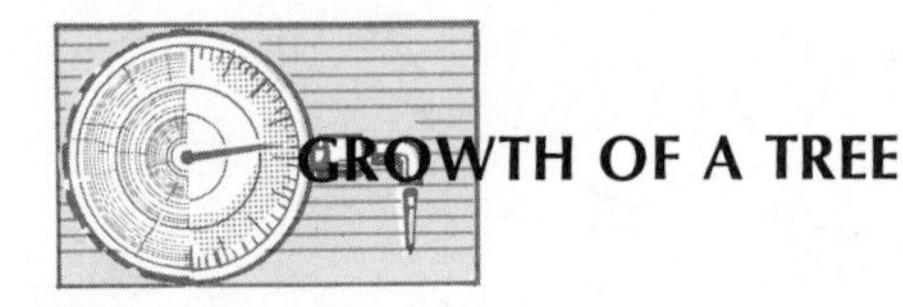

GROWTH OF A TREE

Trees, like all other plant life, grow due to the *photosynthesis* process. Photosynthesis in its simplest form is the process whereby water is absorbed by the roots of the tree and is transported through the leaves, where it combines with carbon dioxide from the air, and oxygen is given off. The sunlight changes these elements to food in the form of carbohydrates which nourish the tree and stimulate growth.

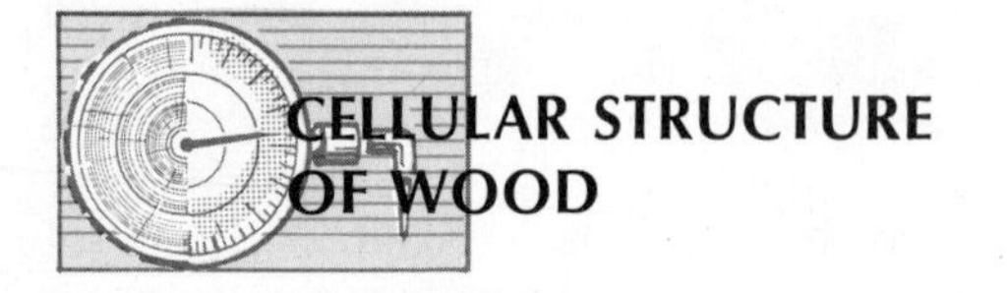

CELLULAR STRUCTURE OF WOOD

Wood, like every living organism, is made up of tiny units called *cells.* These cells are of various sizes and shapes and are tightly bound together by *lignin,* which is the natural cohesive substance found in wood. Most cells are a pipe-like, elongated structure with pointed ends. The cells overlap each other in a staggered arrangement. These cells are called *fibers* or *tracheids.* Most cells run parallel to the length of wood, but some, such as the medullary rays, run at right angles to the wood. Fibers in hardwood are about 1/25″ long. Hardwoods also contain some large cells called *vessels.* Vessels are the main arteries for the movement of sap.

Fibers in softwoods are called *tracheids.* The fibers in softwoods are usually 1.8″ to 1.3″ long. Cells called *parenchyma* are found in both soft and hardwoods. These cells are used to store food.

CONDITIONING OF WOOD

Wood must be *conditioned* before it is of any value for structural uses. Conditioning means to cut the logs to the desired lumber size and to remove the excess moisture. To understand how moisture is removed, you must know how it is contained in the cellular structure of the wood material.

MOISTURE CONTENT IN WOOD

It has been stated previously that wood is a cellular material. When a tree is living prior to cutting, the moisture content accounts for over one-half of the total weight. Before wood can be used commercially, a large percent of the moisture must be removed. The two most common methods of drying are *air drying* and *kiln drying.* These drying methods were discussed in Chapter 2. The moisture content is expressed as the percentage of oven-dry weight. Figure 4-4 shows two common methods of determining the moisture content of wood. For most purposes, a meter such as shown in Fig. 4-5 is used to measure the moisture content.

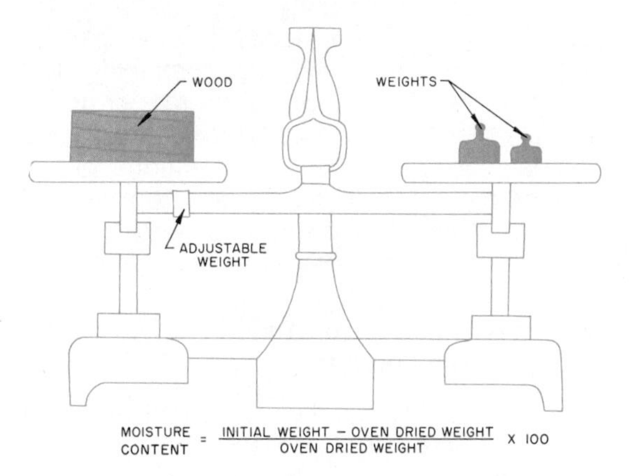

A. Oven drying technique.

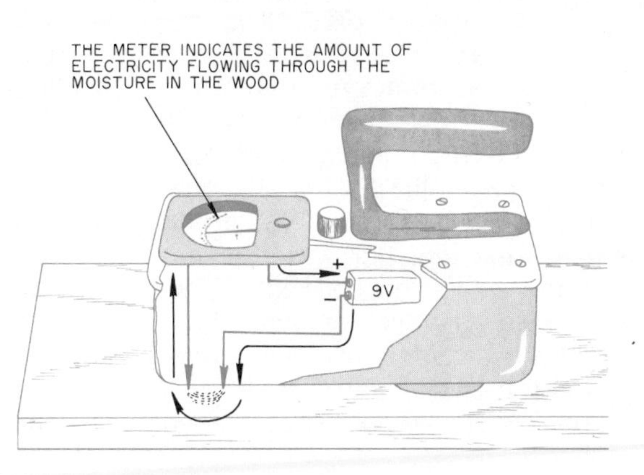

B. Moisture meter method. (Moisture Register Company)

Fig. 4-4. Methods of determining moisture content.

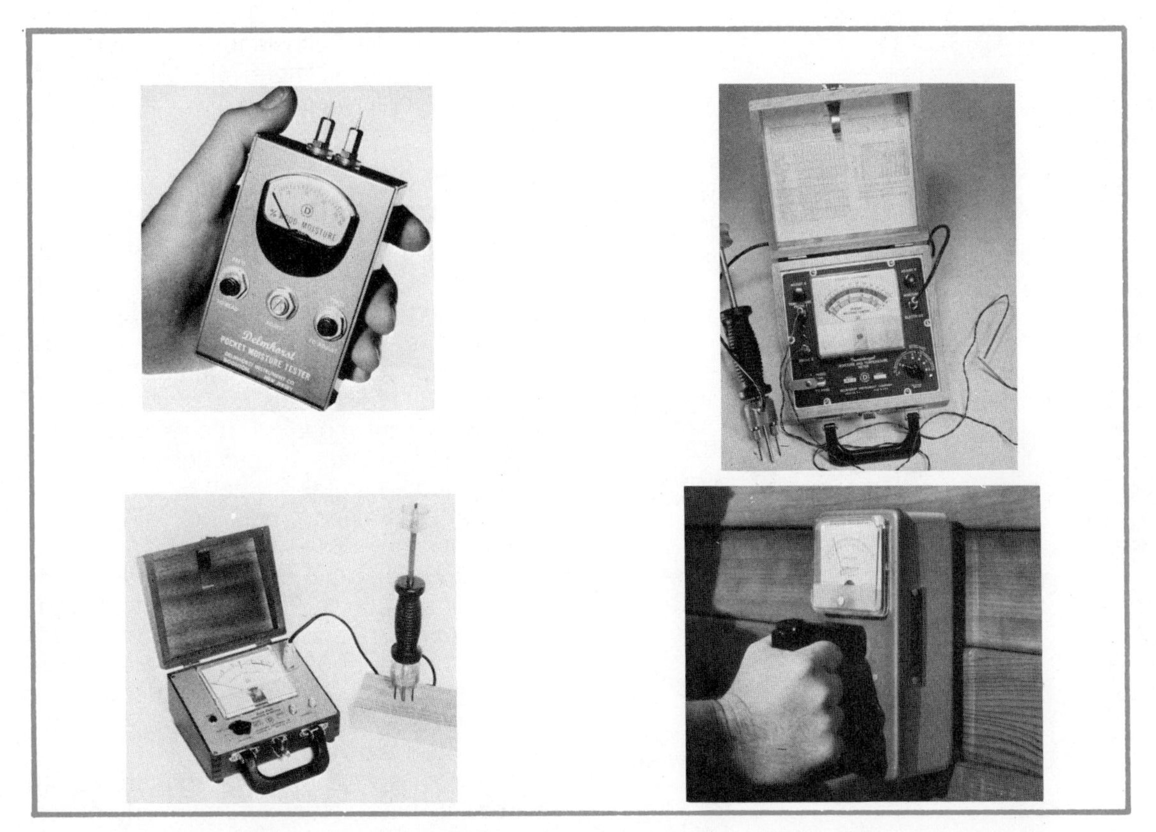

Fig. 4-5. A variety of resistance-type moisture meters which show the moisture content instantly. (Dilmhorst Instrument Co.)

Moisture is present in both the cell cavities and the walls of the cells. When moisture is removed from wood cells, the cells themselves empty first. The walls of the cells still contain moisture. This condition is called the *fiber saturation point.* Wood does not start to shrink until this point has been reached, Fig. 4-6. The moisture contained within the cells is called *free water.*

Moisture is removed from wood to reduce chances of decay and make the wood a more stable material. If wood is kept constantly submerged in water or constantly dry, it will not decay. If wood can be kept dry enough, it will last indefinitely. The two main factors that affect the rate of decay are (1) moisture and (2) temperature. Wood exposed to warm, humid conditions deteriorates more rapidly than wood kept in cool or dry areas. When the moisture content exceeds 20%, decay will begin.

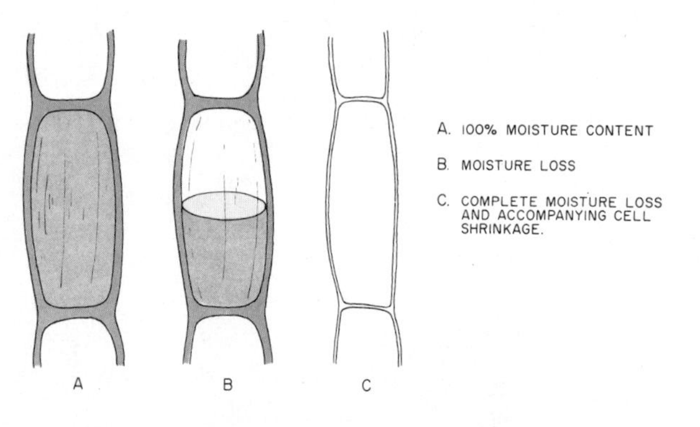

Fig. 4-6. Wood begins to shrink when the moisture content drops below the fiber saturation point.

THE EFFECTS OF MOISTURE ON WOOD STRENGTH

It can be generally stated that as wood dries, its strength increases. The increase in strength does not begin until the fiber-saturation point is reached. However, not all strength properties of wood increase with the reduction of moisture content. Some properties such as toughness or shock resistance actually decrease as the wood dries. Toughness is dependent on both strength and pliability. Therefore, although dried wood can withstand a greater load, it will break or fail more rapidly when bent than will green wood.

METHODS OF SAWING LOGS INTO LUMBER

There are two ways in which lumber can be cut from logs. They are: (1) plain sawing and (2) quarter sawing, Fig. 4-7. If the wood is cut tangent to the annual rings, it is called *"plain sawed"* lumber in hardwoods and *"flat-grained"* or *"slash-grained"* lumber in softwoods. If the wood is cut radially to the rings or parallel to the rays, it is called "quarter-sawed" in hardwoods and "edge-grained" or "vertical-grained" lumber in softwoods.

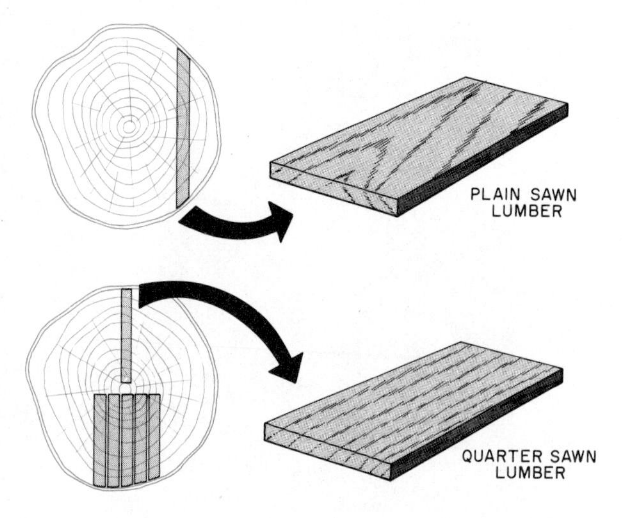

Fig. 4-7. Comparison of plain sawn and quarter sawn lumber.

For commercial use, lumber with rings at angles of 45° to 90° with the surface is called *quarter-sawed,* and lumber with rings at angles of 0° to 45° with the surface is called *plain-sawed.* Quarter-sawed or edge-grained lumber is not cut strictly parallel with the rays.

Either plain-sawed or quarter-sawed lumber is satisfactory for many purposes. However, each does have certain advantages.

Plain-sawed lumber. Plain-sawed lumber is less expensive because less time and waste are involved in sawing the log. The grain patterns are more conspicuous in plain-sawed lumber. Plain-sawed wood does not collapse as easily in drying. It also shrinks and expands less in thickness as the moisture content changes.

Quarter-sawed lumber. Quarter-sawed or edge-grained lumber results in less shrinkage and swelling in width. It also twists and cups less than plain-sawed. Other advantages include less surface checks and splits and more even wearability. The raised grain of quarter-sawed lumber caused by the separation in the annual rings is not as pronounced as plain-sawed. Figuration from rays, interlock grain, and wavy grain are brought out more conspicuously. Quarter-sawed wood exposes the edge of the annual ring which results in better wear.

Rotary cutting. Rotary cutting is used in cutting veneer. It is a process in which continuous sheets of flat-grained veneer are cut by revolving the log against a knife, Fig. 4-8. Most veneer is cut by this method because it is less costly.

PHYSICAL PROPERTIES OF WOOD

Every kind of wood is different from other species. These differences can be

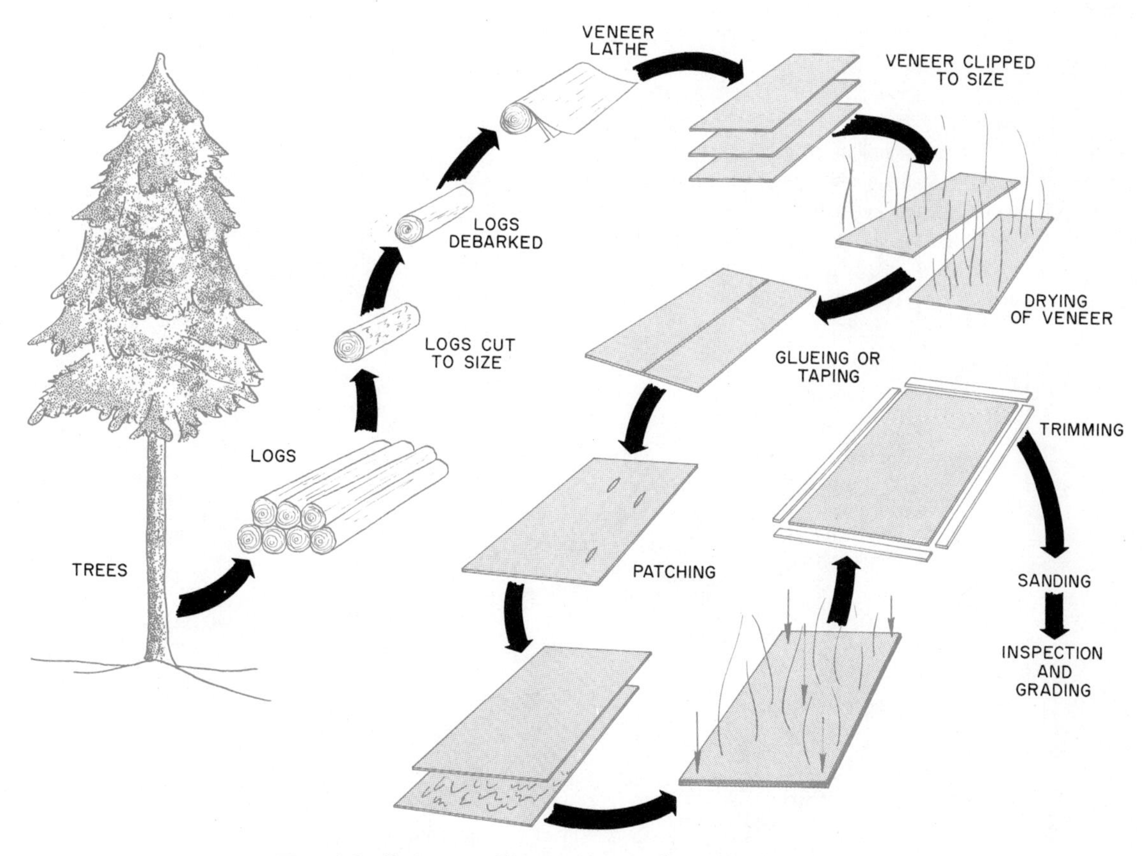

Fig. 4-8. Rotary cutting process for veneer.

referred to as the *physical properties.* The properties can be measured as a means of comparing different species of wood. Common properties which are generally considered in wood technology are:

1. Specific gravity
2. Density
3. Weight
4. Grain texture
5. Hardness
6. Tensile strength
7. Compressive resistance
8. Cleavage resistance
9. Stiffness
10. Shock resistance
11. Shear resistance

Specific gravity. Specific gravity is the relative density of a substance compared with *water.* Wood will float in water because the specific gravity of wood is less than that of water. Steel will not float because its specific gravity is greater than that of water, Fig. 4-9. The value for specific gravity is computed by dividing the weight of a substance by an equal volume of pure water. Pure water has a specific gravity of one (1.0). This means that if the mass of one pint of pure water weighed one pound, and

Fig. 4-9. Specific gravity of wood.

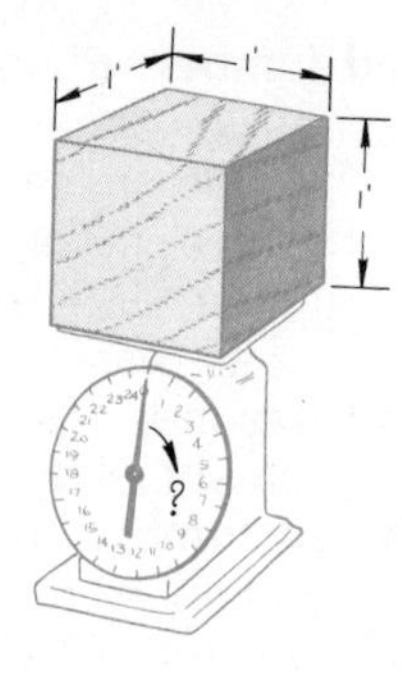

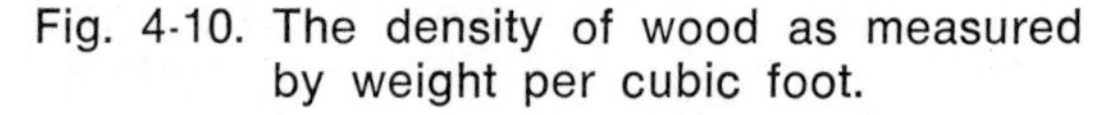

Fig. 4-10. The density of wood as measured by weight per cubic foot.

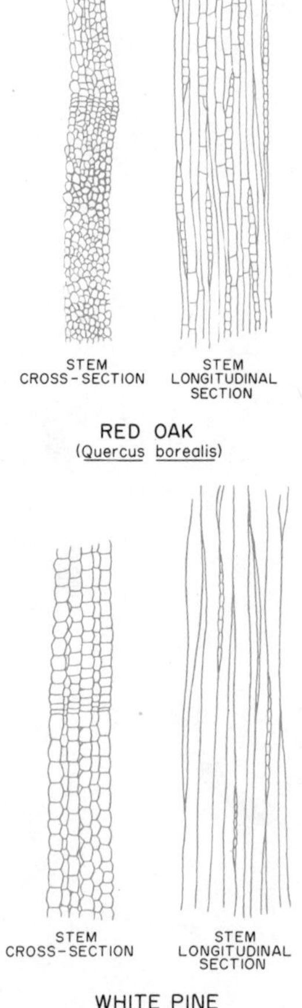

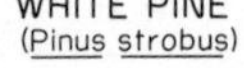

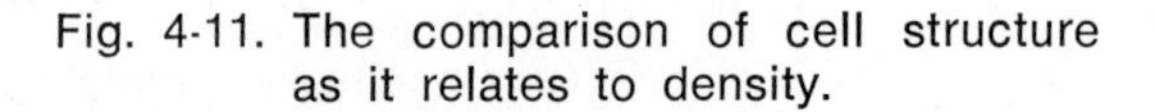

Fig. 4-11. The comparison of cell structure as it relates to density.

if the same volume (size) of mass is lighter than one pound, it will float. It is for this reason any material with a specific gravity of greater than 1.0 will sink in water.

Density and weight. Density is the weight of a given volume of wood. The unit volume of wood is *one cubic foot.* The density of wood is referred to as the *weight per cubic foot,* Fig. 4-10. There is a high relationship between the density of wood and its strength. Very dense wood has more woody material in its structure. This means that the cell walls are thicker and therefore stronger, Fig. 4-11.

The density or weight of wood is measured when the moisture content is at a given level. The weight of wood is often compared *when* the moisture content of all the samples is at 12%.

Grain texture. All woods have a grain pattern, Fig. 4-12. Some woods may have a more attractive and pronounced pattern than others. The various patterns are created by the relationship of the annual rings to the wood fibers.

Some woods are very *porous.* These woods have large cells. When a porous wood is cut, it has a distinct cell pattern. Woods which have large cell structures are referred to as *open grained* woods; those with less distinct cell patterns are referred to as *close grain* woods. Oak, mahogany, and ash are examples of open grained woods and pine, cherry, and maple are examples of close grain woods. Figure 4-13 shows the grain texture of two common commercial woods.

Hardness. Hardness of wood can be determined by resistance to indentation, Fig. 4-14. The hardness is measured by the amount of force required to imbed half of a steel ball with a diameter of .444″ into the wood surface. This test reveals the resistance that the cellular structure of the wood has to crushing when subjected to a load. There is a direct relationship between the hardness and density of wood — the more dense woods are also harder.

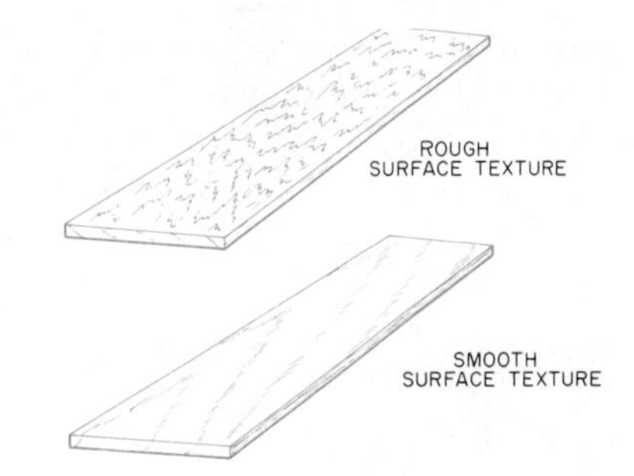

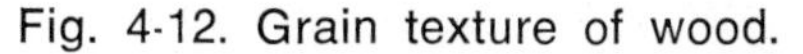

Fig. 4-12. Grain texture of wood.

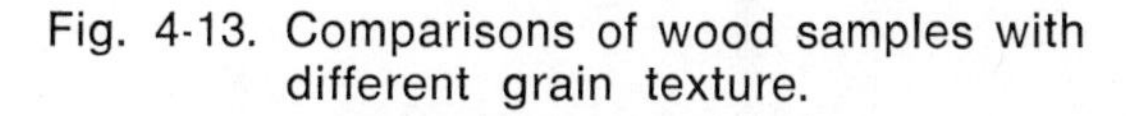

Fig. 4-13. Comparisons of wood samples with different grain texture.

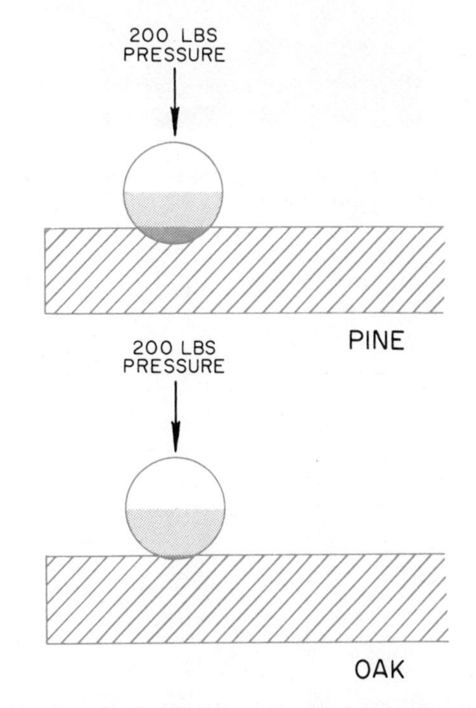

Fig. 4-14. Comparison of the hardness of wood.

Figure 4-15 shows a piece of wood being tested for hardness. The sample specimen should be 2″ square and 6″ long. The steel ball is advanced slowly into the stock at a rate of approximately ¼″ per minute.

The hardness factor is important for flooring and furniture which may be dented from daily use. Harder woods are not scratched or dented as easily as softer woods. Figure 4-31 shows the hardness values of common structural woods.

Tensile strength. Tensile strength is the *resistance of wood to forces* which attempt to pull it apart parallel to the grain, Fig. 4-16. The force is applied in opposite directions. When testing the tensile strength of wood specimens, the center portion of cross-section is reduced. Figure 4-17 shows the specimen in a laboratory testing machine. The resistance is recorded as the P.S.I. (pounds per square inch) of force required to fracture or break the specimen.

Compressive resistance. This term refers to the ability of wood to *resist crushing* when forces are applied from opposite

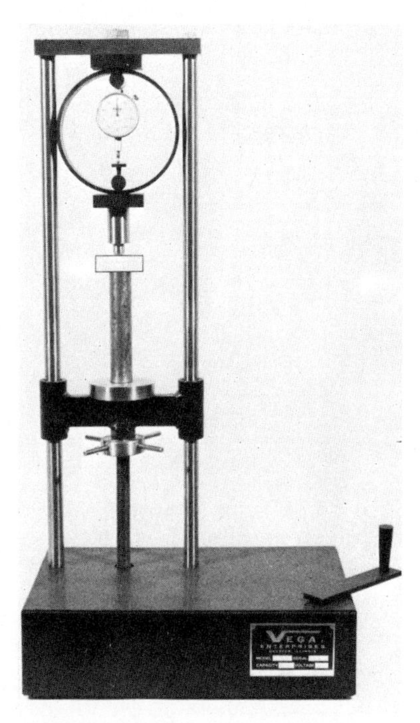
Fig. 4-15. A wood hardness tester. (Vega Enterprises)

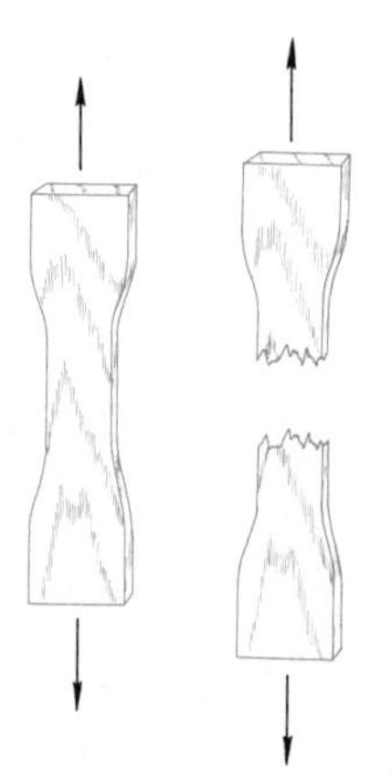
Fig. 4-16. Tensile strength of wood parallel to the grain.

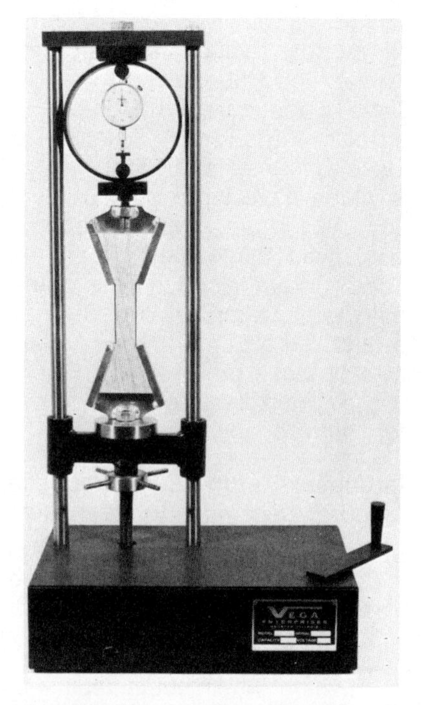
Fig. 4-17. Testing tensile strength of wood with a laboratory test machine. (Vega Enterprises)

ends towards the center, Fig. 4-18. The compressive resistance of wood is an important factor in construction. For example, the studs which support the weight of a house are under a constant compressive load, Fig. 4-19.

The compressive strength of a small specimen is being tested in Fig. 4-20. The specimen is placed in the machine and the load is slowly increased until the specimen fails. This test is generally performed with the force applied parallel to the direction of the grain.

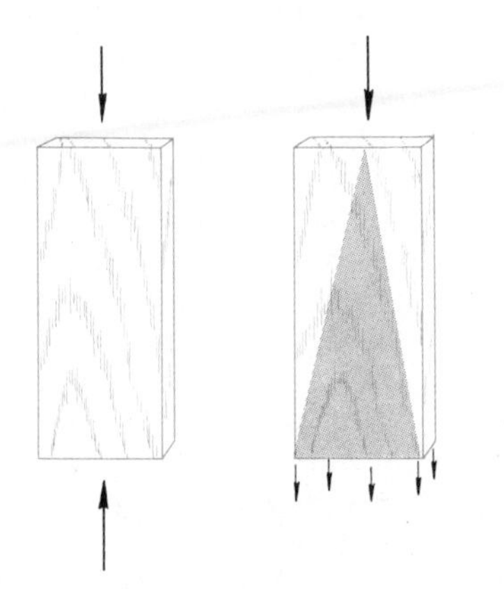

Fig. 4-18. Compression resistance of wood parallel to the grain.

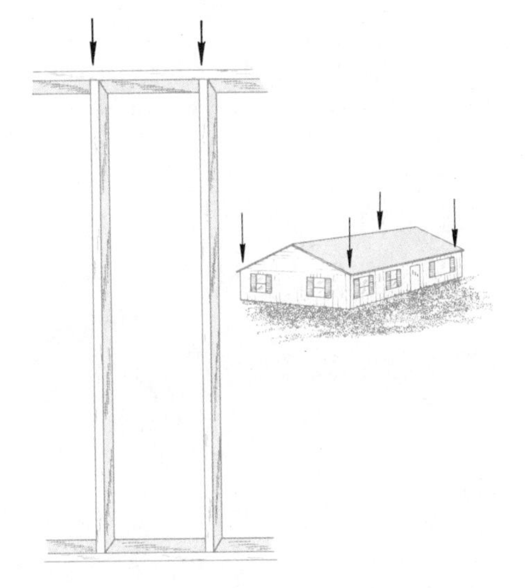

Fig. 4-19. Studs in a house must resist compression.

Cleavage resistance. Cleavage is the *resistance of wood to splitting parallel to the grain* when forces are applied in opposite directions, Fig. 4-21. The rafter

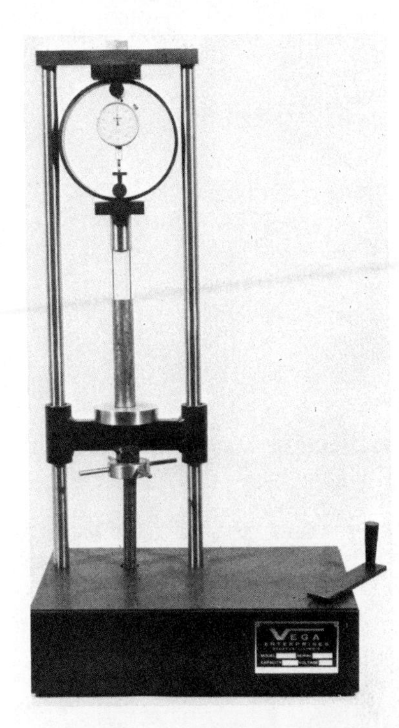

Fig. 4-20. Testing the compressive resistance of a specimen of wood in a laboratory test machine. (Vega Enterprises)

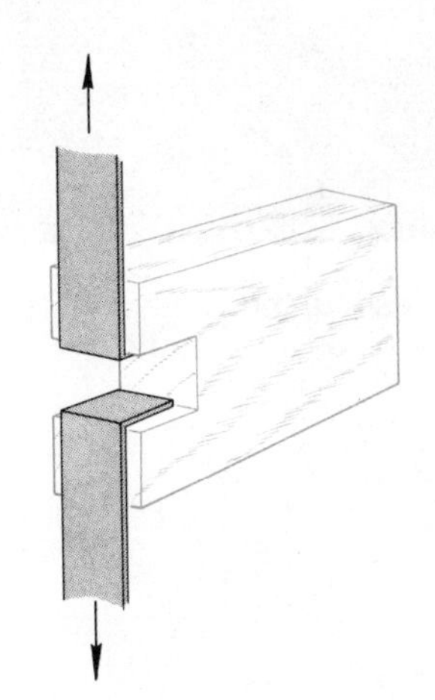

Fig. 4-21. Cleavage is the resistance of wood to splitting parallel to the grain.

shown in Fig. 4-22 must have resistance to splitting. This would be a form of cleavage resistance. The specimen shown in the machine in Fig. 4-23 is being tested for cleavage strength. A notch is cut in the end of the specimen and jaw attachments are placed in the notch. Slow, even pressure is applied to the specimen.

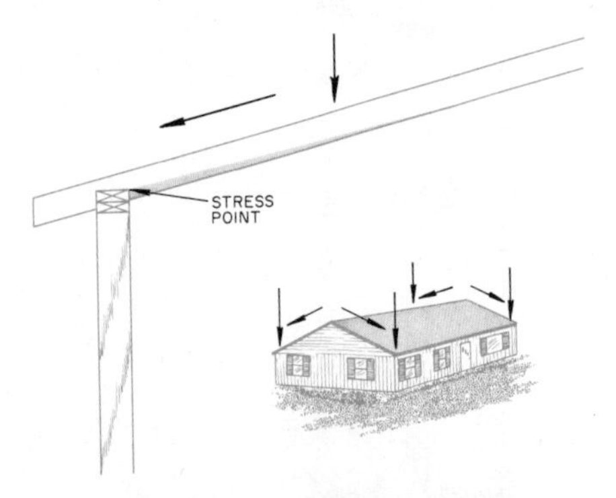

Fig. 4-22. Rafters must resist splitting.

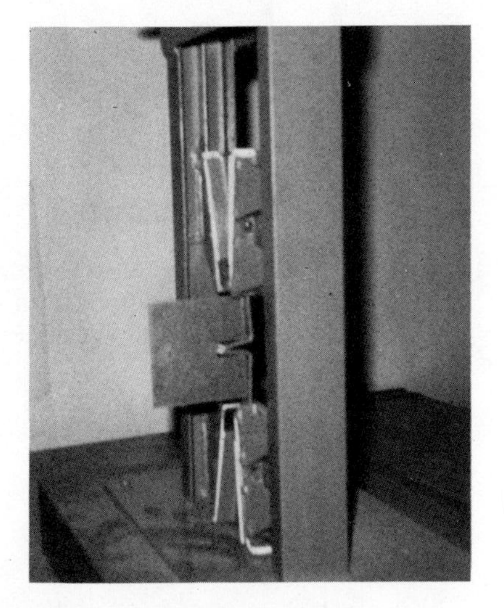

Fig. 4-23. Testing the cleavage resistance of a piece of wood.

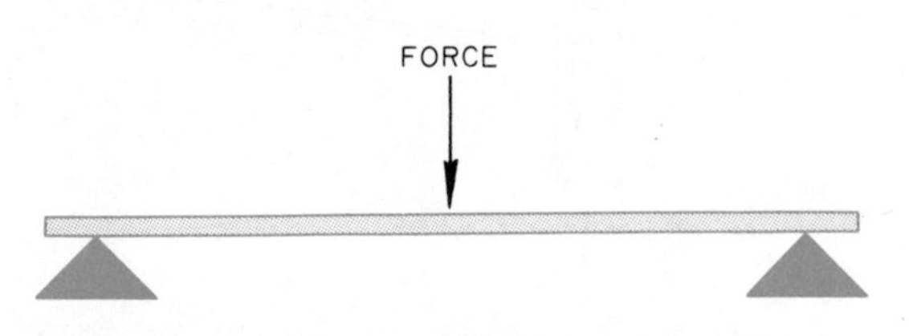

Fig. 4-24. Stiffness is the resistance of a piece of wood to bending.

Stiffness. Bending resistance and stiffness can be considered the same. The amount of bending which occurs when a beam is loaded is called *deflection*, Fig. 4-24. The amount of deflection is measured in the center of the beam. The load which is applied is given as p.s.i.

Stiffness is an important characteristic in building construction. The floor joists shown in Fig. 4-25 must be stiff or they will bend when they are loaded.

Stiffness involves the *tensile and compressive strength* of a structural member. The upper portion of the member is subjected to a compressive load and the lower portion to a tensile force. A neutral axis is present in the approximate center. The type of force is just the opposite on either side of the neutral axis.

Figure 4-26 shows a test specimen being tested for the amount of deflection under varying loads. The force is increased slowly until the desired load is placed or the specimen fails.

Shock resistance. This is sometimes called impact bending. It is a load suddenly applied, Fig. 4-27. When a bat hits a baseball, it is a form of impact bending. An object falling to the floor is an impact load. To test a piece of wood, a weight is dropped from a pre-determined height, Fig. 4-28.

Fig. 4-25. Floor joists must be stiff.

The weight is gradually dropped from an increasing height until the force causes the specimen to fracture.

Shear strength. This is the resistance of wood to forces exerted which cause one part of the wood to slide along the other part, Fig. 4-29. The test specimen is placed in a machine such as shown in Fig. 4-30.

Fig. 4-26. Testing the deflection of a wood specimen.

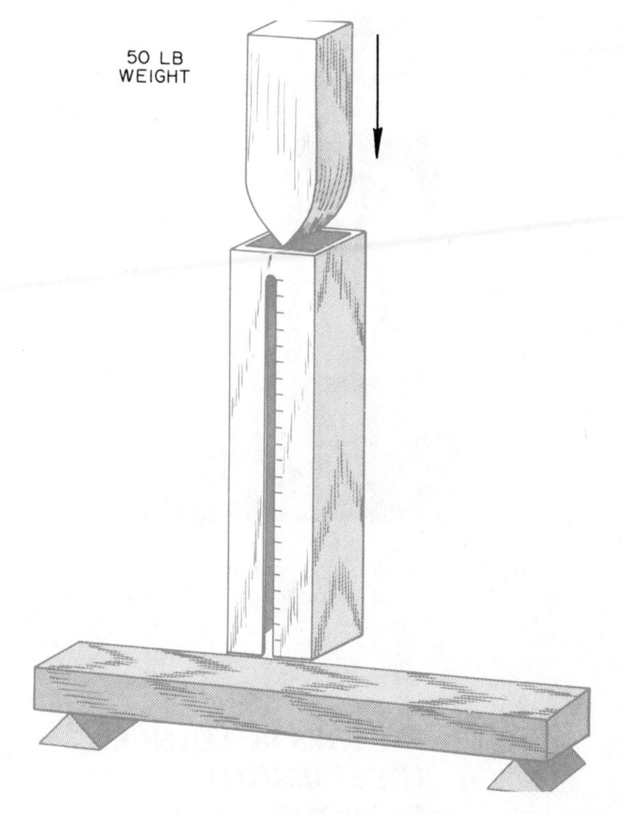

Fig. 4-28. Testing the impact resistance of a wood specimen.

Fig. 4-27. Shock resistance is the ability to remain stable when a sudden load is applied.

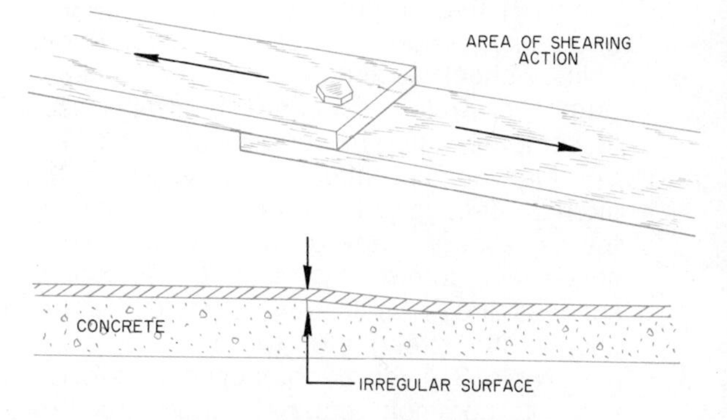

Fig. 4-29. Shear resistance of wood.

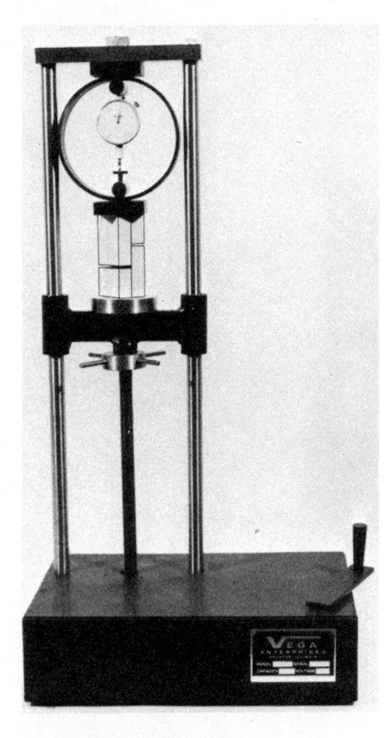

Fig. 4-30. Testing the shear resistance of a wood specimen. (Vega Enterprises)

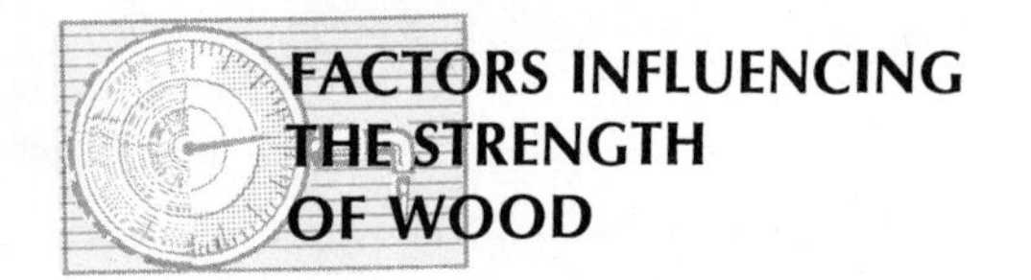

FACTORS INFLUENCING THE STRENGTH OF WOOD

The two most important factors influencing the strength of wood are *defects* and *moisture content.* Defects include knots, splits, checks, and others. For testing purposes, a clear specimen is always used to compare different species of wood. The moisture content affects nearly all of the strength factors. All samples must contain the same percentage of moisture to make a successful comparative test. Other factors such as fire, insects, and climate also influence the strength of wood.

Table 3 gives a composite of properties of common cabinet and structural woods. For the purpose of equalizing the structural comparison, the moisture content of all of the specimens were the same.

TYPES OF STRUCTURAL LOADING

The load applied on wood may be either *dead* or *live.* The amount of load is generally expressed in pounds of weight. A *dead load* on a part or member of a structure includes its own weight and it is the stationary portion of the structure such as floors, ceilings, and walls. A *live load* is people, furniture, equipment, etc. There are three types of live loads which include *static, repeated,* and *impact.*

Static loads. A static load is one which is applied slowly and remains constant, Fig. 4-31. This type of loading is used for testing beams for deflection or bending. A load is applied slowly and evenly. A large piano resting on the floor of your living room is a static load.

Repeated loads. A repeated load is often called a fatigue load. This is when a structure is loaded time after time and placed under a stress. The repeated loading of a bathtub with water is an example. The load applied to the parallel bars in a gymnasium or loads applied to stairs would be examples of repeated loading, Fig. 4-32.

Impact loading. An impact load is a sudden force applied to a structure. A book dropped on the top of your desk is an impact load, or the ball hitting a paddle as shown in Fig. 4-33. If the book remained on the desk top, it would then become a static load. Table 3 shows the comparative resistance of common structural woods to impact loading.

Fig. 4-31. An example of static load.

Fig. 4-32. A repeated load.

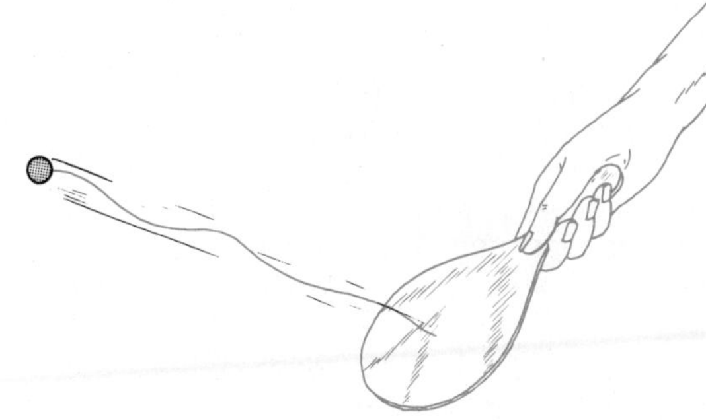

Fig. 4-33. An impact load.

Table 3

The Comparative Strength Values of Common Woods
(All Specimens at 12% M.C.)

Name	Weight/ Cu. Ft. in Lbs.	Grain Open-Closed	Hard or Soft Wood	Hardness (pounds)	Impact Resistance (Inches/50# weight dropped)	Compressive Resistance Parallel to Grain	Shear Resistance Parallel to Grain	Common Commercial Uses
Cedar, Western Red	23	C	S	450	17	5020	860	siding, shingles, fencing
Pine, Sugar	25	C	S	430	18	4770	1050	patterns, furniture
Pine, White	25	C	S	380	23	5620	900	construction crates, patterns
Redwood	28	C	S	630	19	6150	940	fencing, siding, framing studs, exterior trim
Fir, Douglas	33	C	S	710	31	7430	1160	plywood, framing lumber, flooring
Gum, Sweet	34	C	H	850	32	6320	1600	furniture, veneer, interior wood trim
Cherry, Black	35	C	H	950	29	7100	1700	furniture, patterns, veneer
Walnut, Black	38	0	H	1010	34	7580	1370	furniture, veneer woodware, plywood
Ash, White	42	0	H	1320	43	7410	1950	furniture, veneer handles, sporting goods, woodware
Elm, Rock	44	0	H	1320	56	7050	1920	bent wood, furniture, framing
Maple, Sugar	44	C	H	1450	39	7830	2330	furniture, veneer patterns, flooring, handles
Oak, Red	44	0	H	1210	42	6760	1910	furniture, woodware, trim flooring
Birch, Yellow	46	C	H	1260	55	8170	1880	furniture, flooring, veneer woodware
Oak, White	48	0	H	1300	37	7440	1320	furniture, trim barrels
Hickory	50	0	H	1750	67	9210	2430	bent wood, sporting goods, furniture

WOOD IDENTIFICATION

There are certain characteristics which make each species of wood different than others. The color, weight, and grain structure are used for visual identification. Many woods may appear about the same color, but either the grain structure or the weight will usually be different. As you become more familiar with common structural woods, you will be able to readily identify different species. Some of the most common woods are shown in the color pages following page 372. You can compare the grain structure and the color of actual samples in the photos to help you determine the different kinds.

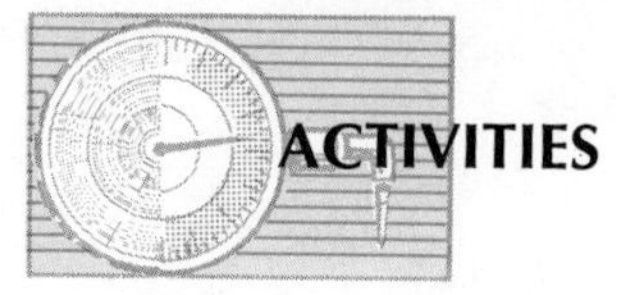

ACTIVITIES

PROBLEM

Using a testing machine or shop-made apparatus, compare the strength of different kinds of wood. If testing equipment is not available, study the testing principles and then develop equipment or tools to perform simple material tests. Compare high- and low - density woods and determine the strength differences.

VOCABULARY

Wood technology
Bark
Cambium
Medullary ray
Sapwood
Heartwood
Lignin
Moisture content
Specific gravity
Density
Hardness
Tensile strength
Compressive strength

CHAPTER 5

PRODUCT SELECTION, ENGINEERING, & PROTOTYPE CONSTRUCTION

INTRODUCTION

Before you begin to produce your first product, you will need to follow the same planning and production procedure as a manufacturer. If there is no demand for the product, industry cannot afford to produce it. Manufacturers must *sell* what they produce and make a profit, or the organization will cease to exist. A survey is made to determine the salability of a product. This is done by questioning potential customers, Fig. 5-1. You will first have to determine if there is a need for the product. This is called *consumer demand.* This will be followed by *research* and *development* (R and D). R and D is the process of determining what is the best method of producing the product. The result of accurate R and D is a product which appeals to the consumer. This involves making a consumer demand study. If a satisfactory product has already been developed at a low cost, then the consumer demand may be satisfied. However, most existing products can be improved. This is where your R and D may begin. An

Fig. 5-1. A survey is made to determine if a product has consumer appeal.

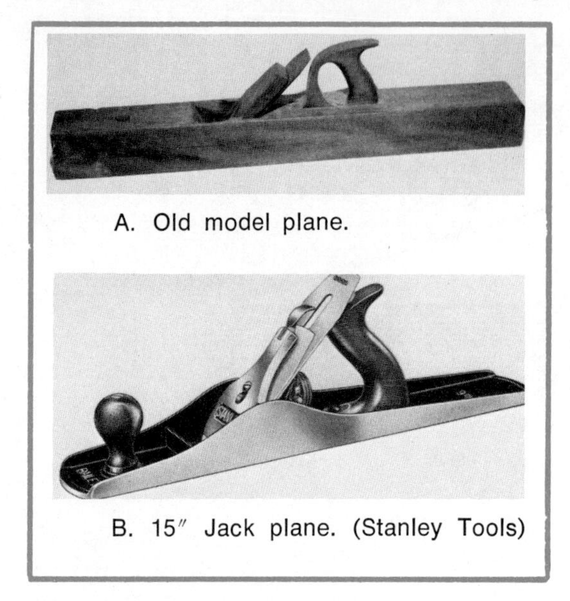

A. Old model plane.

B. 15″ Jack plane. (Stanley Tools)

Fig. 5-2. These planes show how products can be improved through R. and D.

example of product improvement is shown in Fig. 5-2. Each of the planes will cut and smooth wood. However, R and D did not stop with the first one, because someone like yourself thought they could improve the method of cutting wood.

As you research and develop new products in your study of the wood industry, you will construct a *prototype.* A prototype is *one* completed product which is constructed as a means of determining the success of your idea. Industry does this same thing. A prototype is developed and perfected *before* it is produced by means of mass-production techniques.

There are several stages through which research and development progresses before the prototype is constructed. You must design the product. During this stage, you will draw your ideas freehand, select your materials, develop a list and cost (bill) of

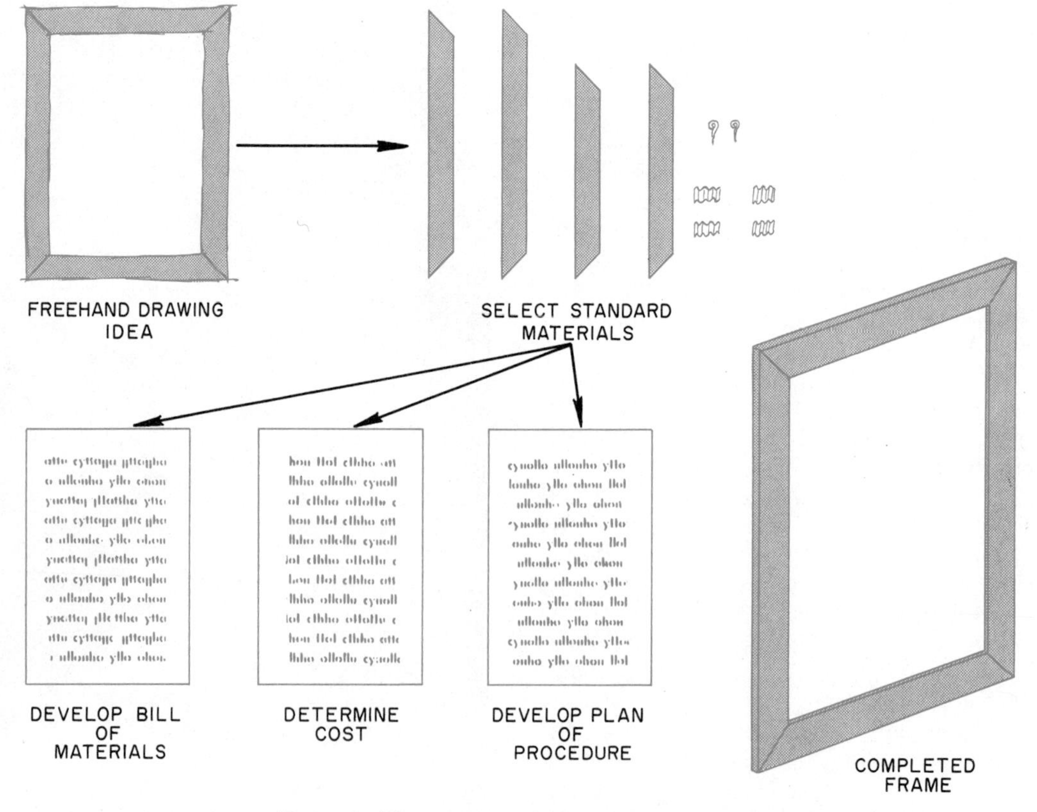

Fig. 5-3. The stages of R and D progression.

materials, determine labor costs, and develop a plan of procedure for constructing the prototype, Fig. 5-3.

The design team in industry must obtain the final approval of *management* before the product can be produced in quantity. As a student, you will need the approval of your instructor before starting to construct your prototype.

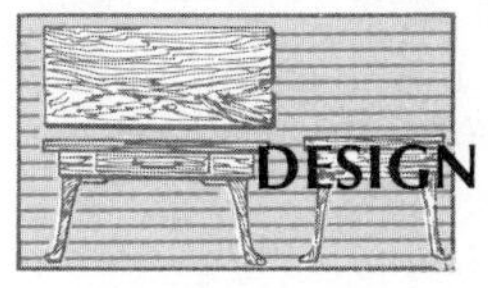

DESIGN CONSIDERATIONS

The three most essential design factors are *appearance, function,* and *ease of construction.* If any of these elements are lacking, the product will probably never succeed. Design is often a compromise. The reason for this is that construction and function will limit appearance. Products are often designed with outstanding appearance characteristics, but are found impossible or impractical to produce. The difficulties encountered in construction may cause the product to be so expensive that no one would buy it.

APPEARANCE

A product must have a pleasing appearance to sell in a competitive market. There are certain accepted design factors which affect the appearance of a product. Among the most basic are *lines, shapes, proportion, texture, color,* and *balance.*

Lines. The shape of the line can influence the mood of a product. Lines can be curved, flowing, or straight, Fig. 5-4. The lines are combined into basic shapes to form objects.

Shapes. Lines can be combined in their shapes to form squares, triangles, rounds, or circles, Fig. 5-5. All of the basic shapes may be combined into a single product.

Proportion. The designer is concerned with proper and pleasing proportions. Proportion refers to the relationship of the size of one part to another as it affects the whole. The ancient Greeks devised the "Golden Mean Rectangle" centuries ago as a relationship of proportion. In its simplest application, it is interpreted as having a size relationship of 5-to-8 proportions, Fig. 5-6.

Texture. Texture is the contrast in the surface of the material. Wood has a natural texture caused by the grain. The texture of the surface of wood can be emphasized by the finishing method.

Color. Color is an important part of design. Wood in its natural state has a wide variation in color among different species. The natural color of wood can be changed by bleaching and staining. Wood can also be painted to achieve a desired color.

Fig. 5-4. The shape of lines influence the mood of a product.

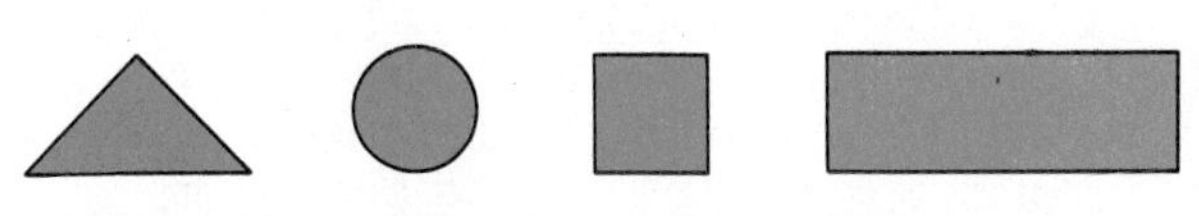

Fig. 5-5. Lines form shapes in design.

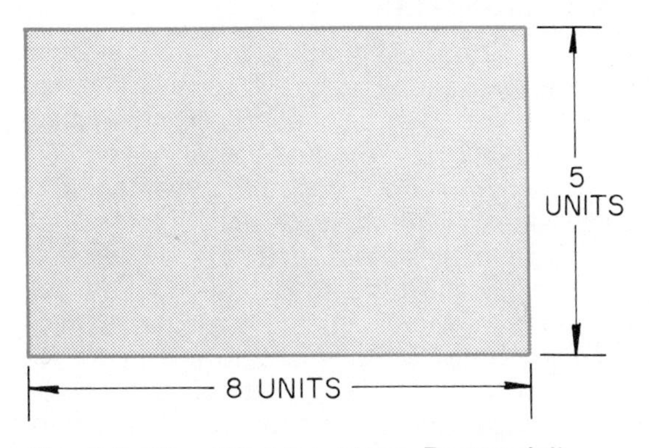

Fig. 5-6. The "Golden Mean Rectangle".

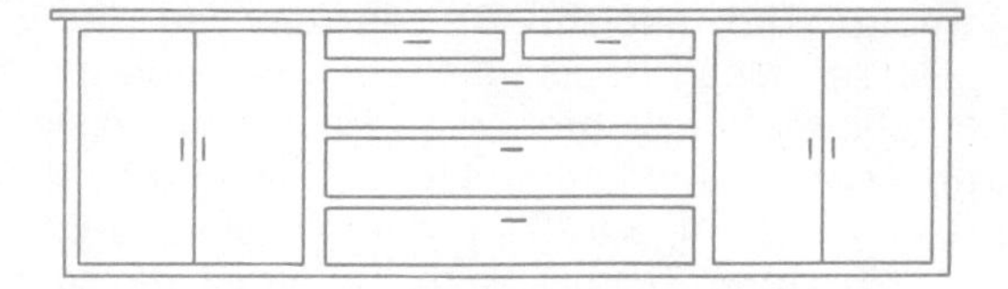

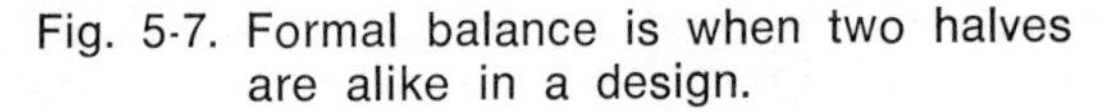

Fig. 5-7. Formal balance is when two halves are alike in a design.

Fig. 5-8. Informal balance achieved through placement of accessories.

Balance. Balance is a quality of optical relationship. Balance can be achieved by making both halves of an object exactly alike. This is called *formal balance,* Fig. 5-7. *Informal balance* can also lend interest to a design. Figure 5-8 shows the principle of informal balance through the placement of accessories on a piece of furniture.

MATERIAL SELECTION

Standard materials play an important role in product design. It is usually less expensive to design a product utilizing existing parts than to produce them yourself. Standard materials and the advantages of using them were presented in Chapter 3. As you recall, the decision of (1) *to make or* (2) *to buy* must be made. If you can buy cheaper than making, the wise decision is to buy. Determining sources and costs for standard materials is a part of the research and development process. The time spent in material selection will result in reduced

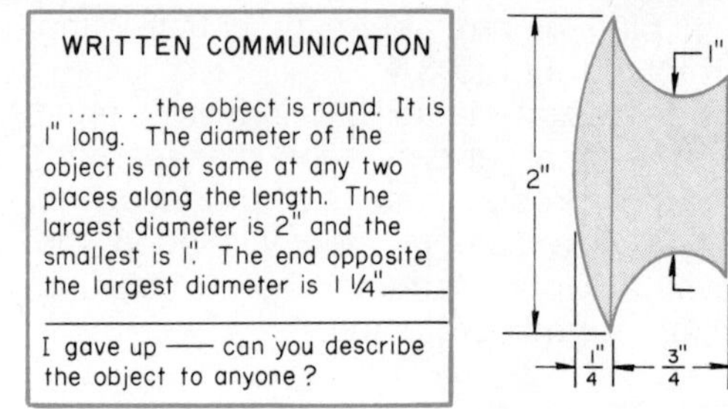

Fig. 5-9. Comparison of the use of written and graphic communication.

cost. The number of production operations can be greatly reduced by careful material selection. Remember, each production operation *adds value* to the product, and it is a cost to the manufacturer. Therefore, the *value added* to a material (wood, metal, paper, etc.) represents a financial investment for the manufacturer and he must consider this investment when pricing the finished product.

INDUSTRIAL COMMUNICATION

Drawings serve as the bases by which design ideas are communicated to other people. Drawing is a universal language. A simple drawing is meaningful to anyone, just as a photograph has meaning. The drawing serves as a means by which the designer records his ideas for the prototype. The drawing is a record which is later used by the production workers to reproduce the product in quantity.

Drawings are used in industry instead of written instructions because there is less chance of misinterpreting a drawing, Fig. 5-9. The manufacturing industries rely on drawings as a means of communication. These drawings are referred to as *working drawings*, Fig. 5-10. A working drawing provides all of the necessary information for the production of a product. The drawings show the *shape, size, type of material,* and the *special details.* The drawing conveys the information by means of lines, symbols, and dimensions.

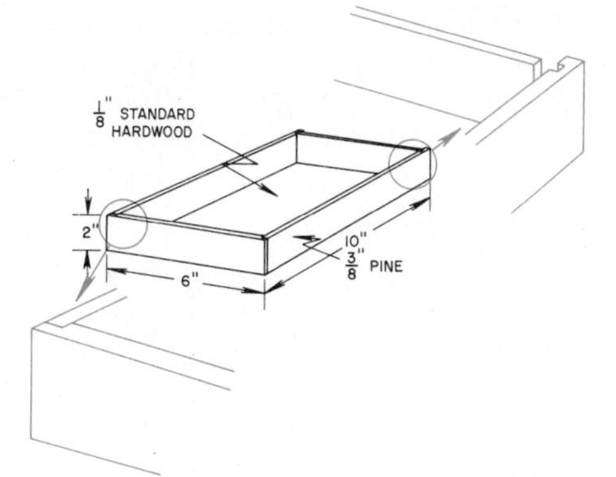

Fig. 5-10. A working drawing shows size, shape, type of material, and special detail.

The drawings are produced by draftsmen. The craftsmen working in the production shop must be able to read the drawings. This is sometimes called blueprint reading.

DEVELOPMENT OF PRODUCT DRAWINGS

The development of an idea into a finished product involves many stages. First, there must be a need for the product. The need progresses to a tentative solution to the problem. The solution is developed into a "*rough sketch.*" Alternate solutions are drawn free-hand. The best solution is developed into a working drawing. A prototype is made from the working drawings. (A prototype is a full-size model.) The prototype is evaluated to determine if it meets the needs or solves the problems. The drawings are then modified in accordance with the design changes, Fig. 5-11.

FREEHAND DRAWINGS

The first ideas the designer has for a product are developed as freehand drawings or sketches. The purpose of this drawing is to convey an idea. A sketch requires only a minimum of time to prepare.

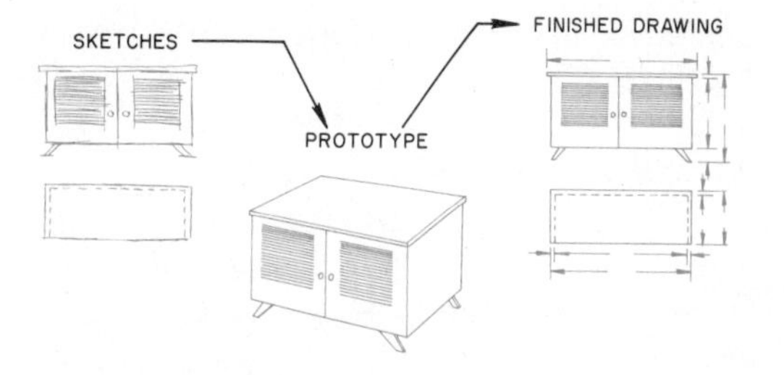

Fig. 5-11. Steps of development from an idea through placing a product in production through use of drawings.

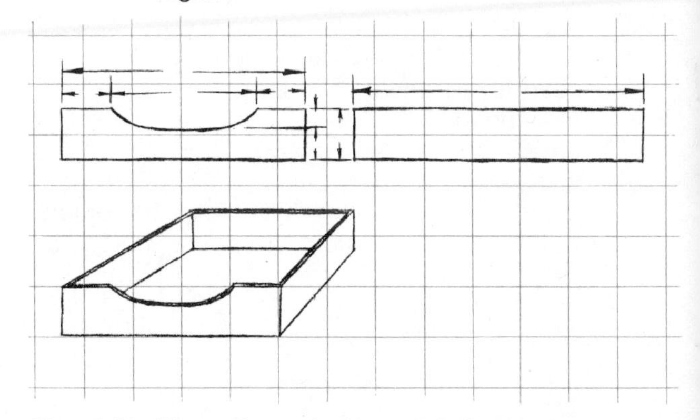

Fig. 5-12. The first design solutions are developed on grid paper by means of freehand drawing.

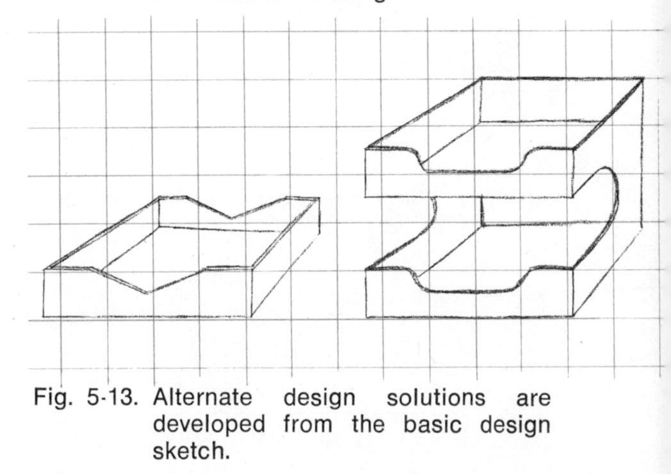

Fig. 5-13. Alternate design solutions are developed from the basic design sketch.

Grid paper makes it possible to prepare quality sketches with proper proportions, Fig. 5-12. Each square equals 1/4″. Alternate problem solutions can be developed from the basic design sketch, Fig. 5-13. The best design solution is developed into a working drawing.

WORKING DRAWINGS

Working drawings are also called *orthographic projection* drawings. Three views of the object are usually shown in most working drawings. The top, front, and side views are the ones most often shown if three views are represented. The views are drawn as though you were looking straight into the object from each of the directions, Fig. 5-14.

Several types and weights of lines are used to represent various parts of a working drawing, Fig. 5-15. Heavy lines represent the outline of the object. They are *object lines.* Dashed lines, called *hidden lines,* are used to represent changes in the object which are not visible from the particular view.

Dimensions placed on the drawing give the exact size of the object. Special notes may also be written on the drawing when additional information is needed, Fig. 5-16.

PICTORIAL REPRESENTATION

A drawing which appears like a picture may be used to show the product design. The object may be sketched as either an *isometric* or *oblique* drawing. The pictorial drawings are often combined with working drawings to represent an object, Fig. 5-17. An isometric drawing is constructed on an axis. Oblique drawings represent the object with one true view for the front. The advantage of pictorial representation is that it gives a quick image of the product to the reader.

DRAWING TO SCALE

An object is usually too large to draw full-size on a piece of paper. Large objects

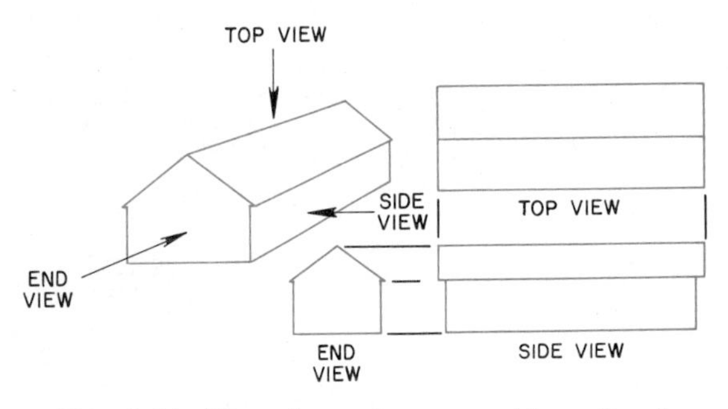

Fig. 5-14. The views for a working drawing.

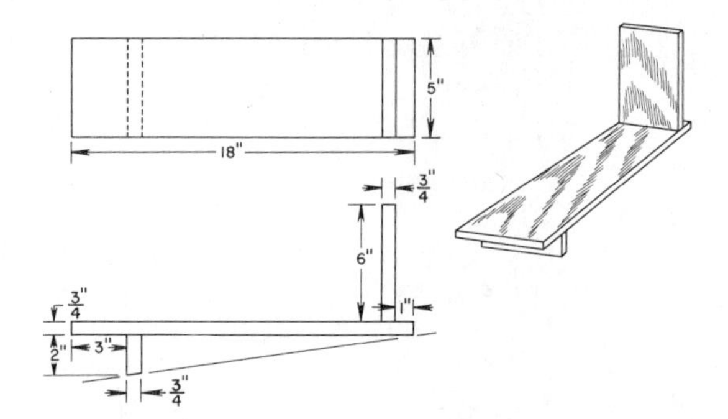

Fig. 5-16. A complete working drawing.

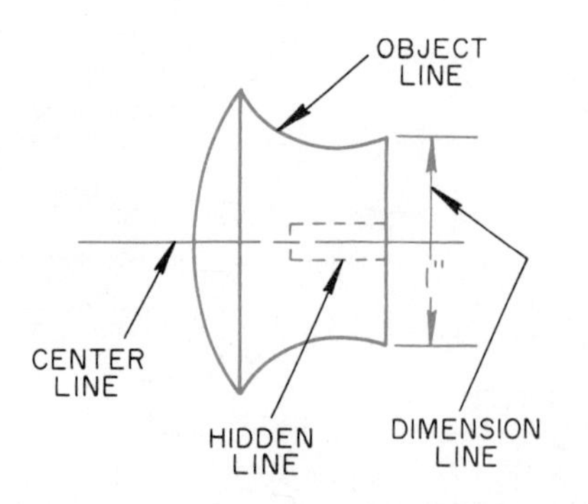

Fig. 5-15. Types and weights of lines represent various parts of a drawing.

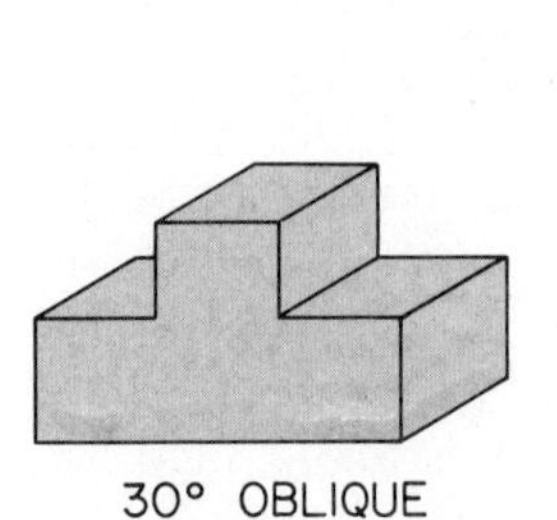

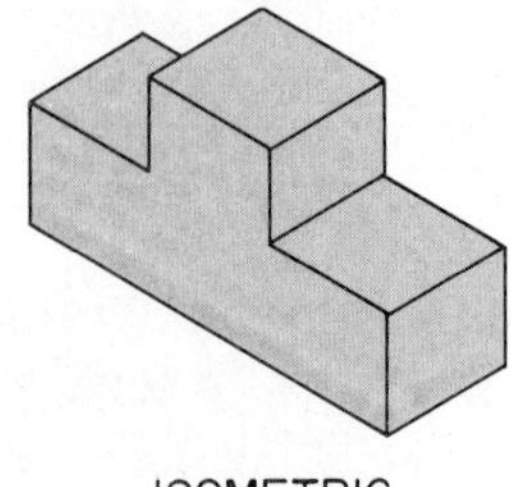

Fig. 5-17. An isometric and an oblique pictorial drawing.

are sometimes drawn one-fourth or one-half scale. An object 10″ long would be drawn 5″ long if it were drawn one-half scale. The 10″ part would be drawn 2½″ long if the one-fourth scale were used. Sometimes detailed parts are drawn larger than the rest of the object. In the wood industry, joint assemblies may be drawn as enlarged details, Fig. 5-18.

BILL OF MATERIALS

After the drawings are complete, a *bill of materials* is made before construction is started. A bill of materials is a complete list of all the lumber, hardware, finishing supplies, and the cost of each item needed to build the product. It includes the complete description of the materials. The thickness, width, and length of each piece of lumber is shown on the bill of materials. Figure 5-19 shows a typical form which may be used to record the materials. The bill of materials shown in Fig. 5-19 is for the coffee table shown in Fig. 5-20.

A bill of materials will insure you that the materials will be available once you start the prototype only if you check with the supplies department. In addition, you will also know the exact cost of the product when completed. If the cost is excessive, changes can be made in the materials selected for construction. You should consider construction material alternatives. If less expensive materials will not affect the completed product, they should be selected.

Manufacturers usually try to get the highest possible quality for the lowest possible cost. For this reason, the prototype is studied carefully for cost reduction. Most of the decisions a manufacturer makes are based on economics (cost of material, labor, handling, etc.).

BILL OF MATERIALS

NO. OF PIECES	SIZE			NAME OF PART	STANDARD MATERIAL
	T	W	L		
1	3/4″	24″	36″	TOP	BIRCH PLYWOOD - G1S
1	3/4″	24″	36″	BOTTOM	FIR PLYWOOD, A-D
2	2 1/2″	2 1/2″	14″	HANDLES	BIRCH
3	2″	2″	7″	LEGS	BIRCH
10	3/4″	2″	7″	SPACE BLOCKS	WHITE PINE
20	1/2″	1/2″	2″	GLUE BLOCKS	WHITE PINE
1	2″	2″	8″	SPOUT	BIRCH
2			20″	RIBS	#9 GALVANIZED WIRE
300			1/2″	DECORATIVE STUDS	UPHOLSTERY TACKS
1		18″	10′	SIDES	PLASTIC UPHOLSTERY MATERIAL

Fig. 5-19. Bill of Materials form.

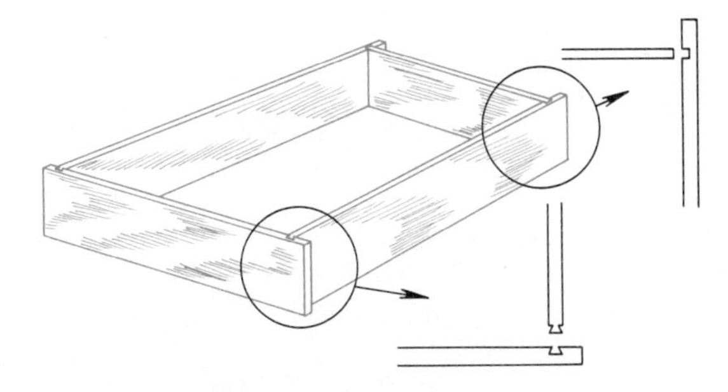

Fig. 5-18. Detailed enlargement of joints for a working drawing.

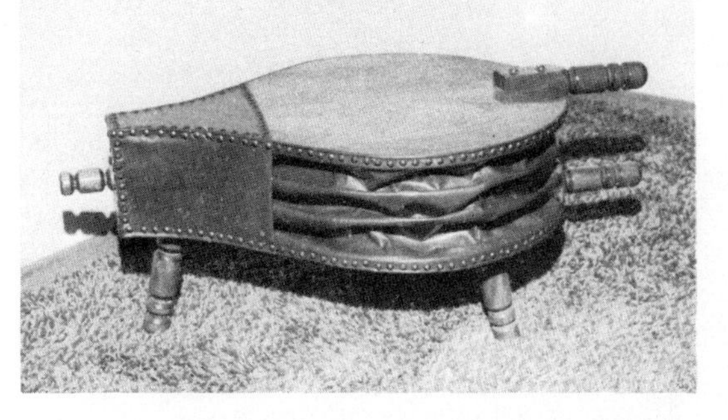

Fig. 5-20. A coffee table produced from the bill of materials in Fig. 5-19.

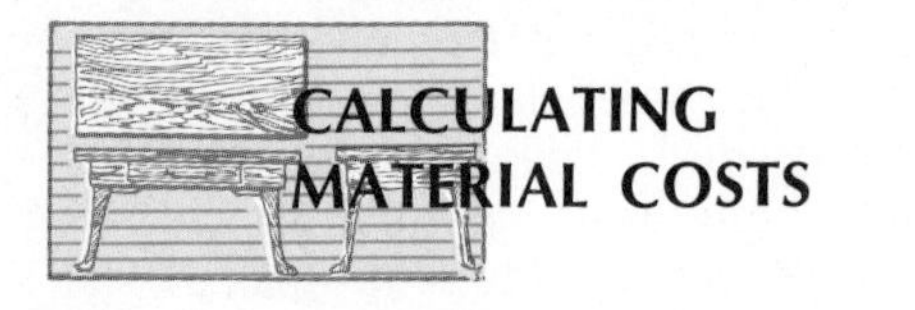

CALCULATING MATERIAL COSTS

Various types of structural materials are sold and measured by different units. Just as eggs are sold by the dozen, gloves by the pair, etc., different structural materials are also sold by special units. Materials are sold by the board foot, square foot, linear foot, pound, pair or as single units of one.

Board foot. Lumber is sold in a unit measure called a board foot. A board foot is a piece of lumber 1″ thick, 12″ wide, and 12″ long and equals 144 cu. in., Fig. 5-21. Lumber which is less than 1″ thick is figured as 1″ thick. Lumber which is over 1″ in thickness is figured according to its actual size. The following formulas are used to calculate the board footage of a piece of stock when the dimensions are all in inches.

$$\text{Bd. Ft.} = \frac{T'' \times W'' \times L''}{144} \text{ (All in inches)}$$

For example, a board ¾″ thick, 6″ wide and 24″ long would equal one board foot. The ¾″ is figured as 1″. The sizes of the stock are placed in the formula.

$$\frac{1'' \times 6'' \times 24''}{144} = \frac{144''}{144} = 1 \text{ Bd. Ft.}$$

The following formula is used if the length of the stock is given in feet. For example, a board ¾″ thick, 3″ wide, and 6′ long would equal 1½ board feet.

$$\text{Bd. Ft.} = \frac{T \text{ (In.)} \times W \text{ (In.)} \times L \text{(ft.)}}{12} \text{ (length in ft.)}$$

$$\frac{1'' \times 3'' \times 6}{12} = \frac{18}{12} = 1\tfrac{1}{2} \text{ Bd. Ft.}$$

The cost of lumber is given as the cost per board foot.

Square footage. Man-made sheet material such as plywood, particle board, and hardboard are sold by the square foot, Fig. 5-22. A square foot is a unit which measures 12″ × 12″. The thickness will depend upon the material and the specified thickness. The cost is calculated by the square foot. For example, a sheet of plywood ¼″ thick, 4′ wide, and 8′ long would contain 32 square feet.

Linear footage. Materials such as moldings, cove, quarter-round, and similar materials are sold by the linear foot, Fig. 5-23. The price varies according to the size and quality. Enough material is purchased to cover the desired area plus enough for cutting allowances.

Pound. Nails are generally sold by the pound unit. Powdered fillers and similar items may also be sold in a container by weight. The weight may not always be one pound units.

Pair. Most hinges are sold by the pair. It is seldom that hinges will be installed

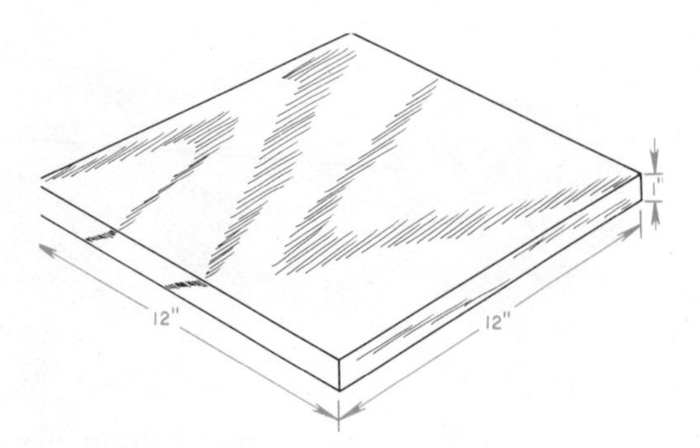

Fig. 5-21. A board foot of lumber.

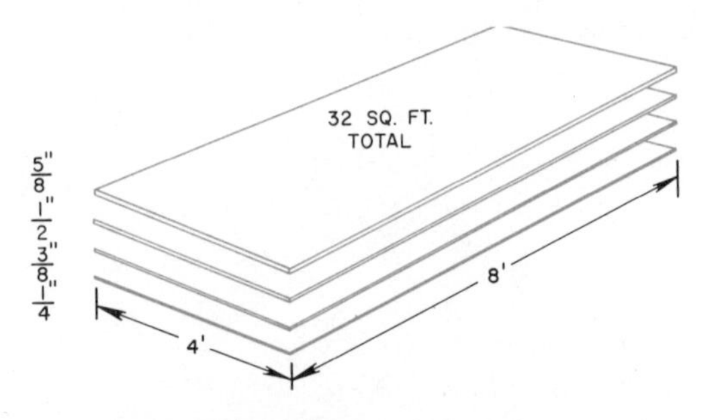

Fig. 5-22. Sheet material is sold by the square foot.
The thickness has no bearing on the number of square feet in sheet material.

unless two or more are used to support a swinging door. It is customary to have matching hinges. Items of this type are priced according to size, style, quality, etc.

Units of one. Hardware items such as pulls, catches, lid supports, etc. are sold in units of one. Many other types of hardware and fasteners can also be purchased in units of one from local hardware stores.

ORDER SPECIFICATIONS

Lumber orders are written including the size, grade, and species. The exact material description is called a specification. The surface, whether planed-smooth or rough-sawed, is included in the specifications. The size is written with the thickness, width and length given in that order. For example, a piece 2″ thick, 4″ wide, and 8′ long is written 2″×4″×8′.

The specifications must be exact to insure delivery of the materials desired from the supplier. The order specifications are taken from the bill of materials. You will need to convert your bill of materials needs to standard sizes to order materials. For example, you may need several small pieces of plywood; however, you will order a standard sheet which will provide the number of cuttings you need.

PLAN OF PROCEDURE

A plan sheet including each step or operation for the completion of the product should be developed. This part of the plan sheet is called the *plan of procedure.* A plan of procedure will provide a solution to the problems which will develop as the product is produced. A similar plan is developed in industry for the production of a product. It includes each operation and the machine or tools needed to perform the operation. Careful consideration is given to the sequence of the operations.

The plan of procedure for producing a product is similar to a game plan for a ballgame. The plan must be followed after it is developed, or it will be of no value. A planning sheet helps to plan the total procedure, such as the plan of procedure in Fig. 5-24

Fig. 5-23. The base and decorative molding installed around the outer walls of this room are sold by the linear foot.

STUDENT PLAN SHEET

Name__________ Product__________

Date Started __________ Date Completed __________

Bill of Materials

No. of Pieces	Materials Finished Size			Part	Type of Material	Number of Bd./Sq. Ft.	Cost/Unit	Total Cost
	T	W	L					
2	3/8″	2″	12″	Sides	White Pine	1/3 Bd. Ft.	.30 Bd. Ft.	.10
2	3/8″	2″	9 1/4″	Ends	White Pine	1/5 Bd. Ft.	.30 Bd. Ft.	.06
1	1/8″	9″	11 3/4″	Bottom	Hardwood	3/4 Sq. Ft.	.15 Sq. Ft.	.11
								.27

Plan of Procedure

Tools: Crosscut Saw, Ripsaw, Backsaw, Chisel, Plane, Try Square, Bench Rule, Coping Saw, File

1. Select stock and rough cut to size for sides and ends
2. Plane sides and ends smooth and square
3. Cut the contour in the front end piece
4. Smooth the contour cut
5. Cut rabbet joints for corners
6. Cut groove in side and end pieces for bottom
7. Cut hardboard bottom to size
8. Make trial assembly
9. Sand parts smooth
10. Glue and assemble
11. Finish sand and repair any defect
12. Apply finish

Fig. 5-24. A complete plan of procedure sheet.

developed for the desk organizer shown in Fig. 5-25. The plan of procedure can best be developed by carefully studying the drawings for the product.

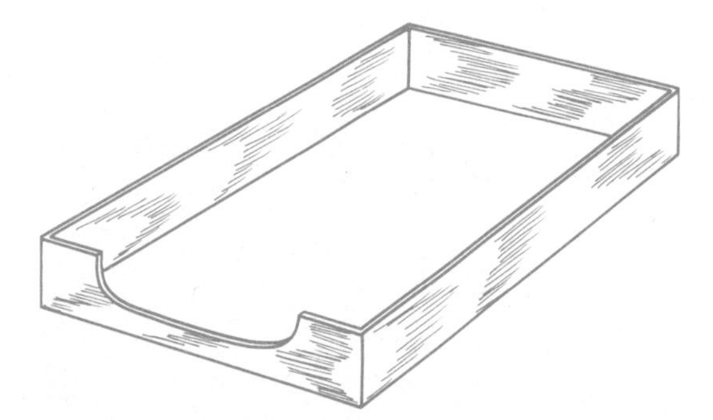

Fig. 5-25. A desk organizer to be produced from the plan of procedure in Fig. 5-24.

Fig. 5-26. A production schedule was developed before production began on this product. (Norton Coated Abrasive Division)

PREPARING TO PRODUCE PRODUCTS IN QUANTITY

The research, development, and prototype construction processes are common industrial functions before large quantities of a product are produced. After the prototype is constructed and evaluated, the necessary design changes are made before production is started.

The production schedule is then prepared. Consideration is given to how each component will be produced, Fig. 5-26. The primary concern is to schedule production in such a manner so the product can be produced at the lowest possible cost.

As you design and produce your prototype, think about how you could produce large quantities in the least amount of time.

ACTIVITIES

PROBLEM

Select a product for which you would like to develop a *prototype.* As you know, this will require research and development (R and D) before you can start designing. The major design characteristics which you must keep in mind are: (1) appearance, (2) function, (3) ease of construction, and (4) cost.

After completing the R and D for the product, prepare freehand drawings for the prototype. The freehand drawings represent the design alternatives from which you will select the best one. A detailed working drawing should be prepared from the best design alternative.

Before you start to produce your prototype, the following preparations should be completed.

1. Select a product to design
2. Prepare design alternatives
3. Prepare working drawing for the best design
4. Select standard materials
5. Prepare a bill of materials
6. Write order specifications
7. Develop a plan of procedure (review production operations in Chapter 2)

Research and development
Prototype
Oblique
Board footage
Isometric
Bill of materials
Specifications

CHAPTER 6 LAYOUT TOOLS AND PROCEDURES

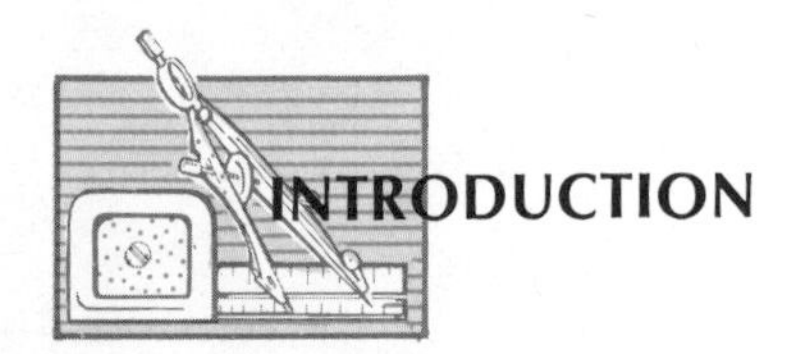

INTRODUCTION

Loss of time and misuse of materials is the enemy of any profit-making organization. To safeguard against increased cost due to excessive waste, proper measurement and layout methods must be used. *Layout* is the process of marking the stock for cutting. To insure accurate measurements, the proper tools must be used. Measuring tools, like all other tools, are designed for certain purposes. Layout tools with straight edges are designed for measuring and marking straight lines. Longer measuring tools should be selected to measure long distances.

Tools called squares are used to mark stock which will have a right angle to adjacent edges, Fig. 6-1. There are many sizes and types of squares. Special tools are used for laying out circles and irregular shapes.

In the wood industry, the unit of one inch is the basic measurement. The dimensions used in measuring include inches or the following fractions of inches: halves, quarters, eighths, sixteenths, and thirty-seconds, Fig. 6-2.

Accurate measurements are necessary in all work. Equally important is the selection of the proper measuring tool. A sharp

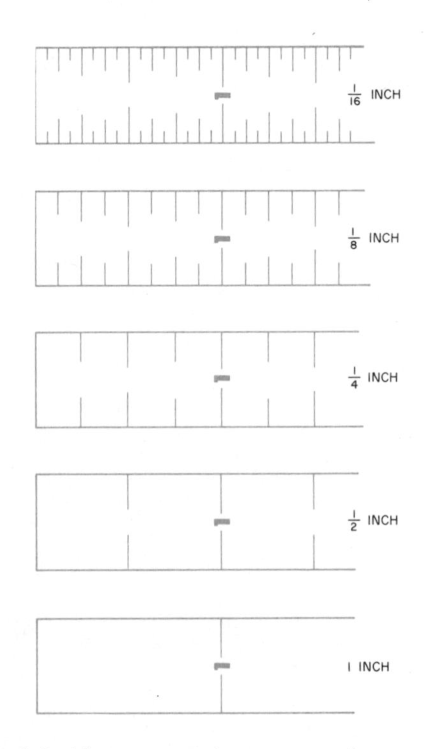

Fig. 6-2. Measurements commonly used in the wood industry.

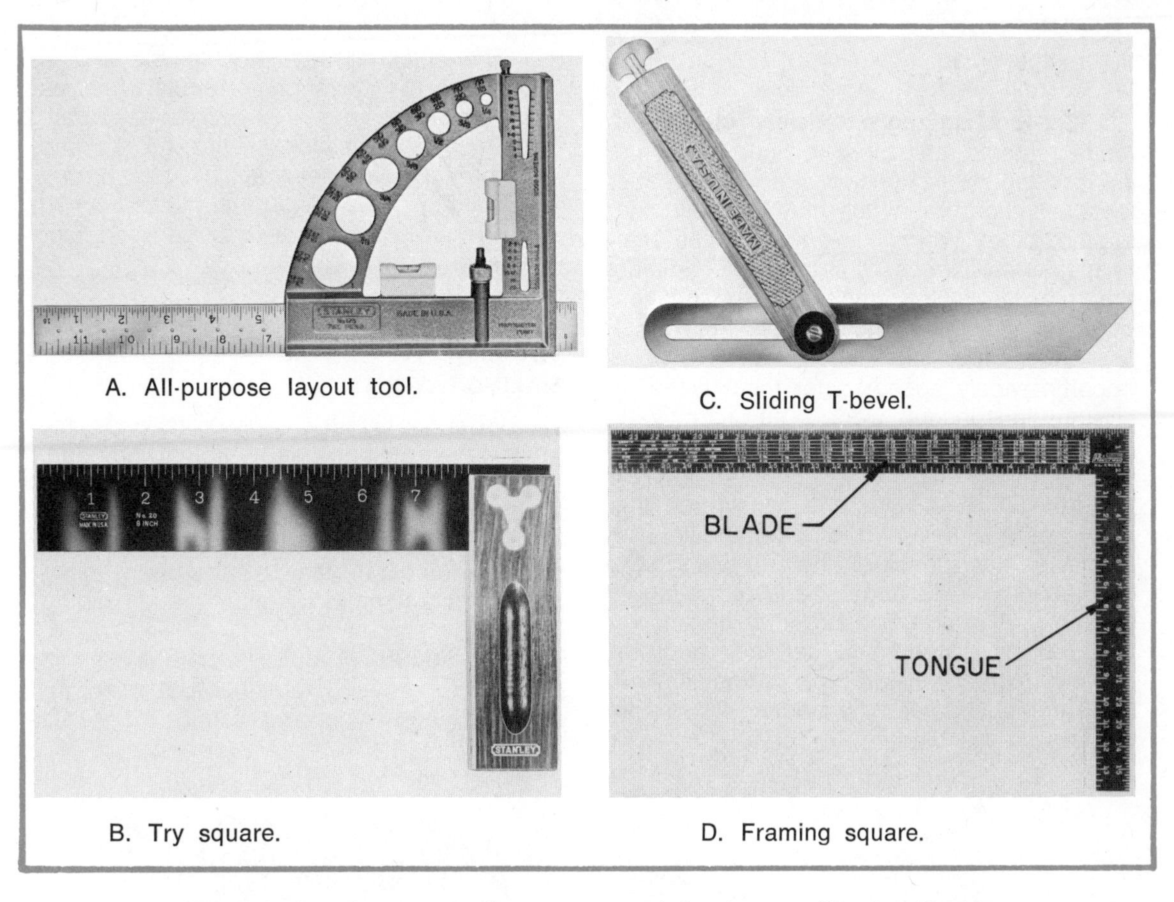

A. All-purpose layout tool.

B. Try square.

C. Sliding T-bevel.

D. Framing square.

Fig. 6-1. An assortment of squares used for layout. (Stanley Tools)

pencil should be used for accurately marking stock to length.

STRAIGHT-LINE MEASURING TOOLS

Straight-line measuring tools are used for measuring distances. The most common straight-line measuring tools include the bench rule, six-foot zig-zag extension rule, and steel tape. All three are used in measuring, but each has advantages for particular types of work.

Bench rule. A bench rule is shown in Fig. 6-3. A bench rule is a wooden rule with a brass protection rule cap on either end. It may be 12″, 24″, or 36″ long. The bench rule is best used for measuring small pieces of stock and marking short distances. It may also be used to adjust dividers. It provides an accurate measurement. One side of the rule is graduated in sixteenths of an inch, and the other side is graduated

Fig. 6-3. A bench rule. (Stanley Tools)

in eighths of an inch. Figure 6-4 shows a layout being made using a bench rule.

Six-foot zig-zag extension rule. Exact measurements are more difficult with the six-foot zig-zag extension rule due to the fact that slight variations may occur in the folding joints. When using the zig-zag rule to measure distances greater than 2′, place the rule flat on the stock. Use the rule on edge to measure distances less than 2′.

Steel tape. Steel tapes, Fig. 6-5, have become widely accepted for most types of straight-line layout work. They are very accurate and are easy to carry on the job. The tape is available in lengths of 6′, 8′, 10′, 12′, 16′, 50′, and 100′. It is considered the best tool for measuring long distances. The steel tape has a hook on the end which can be placed over the edge or end of the stock so one person can measure long distances, Fig. 6-6. The steel tape can be used for making accurate inside measurements. The measurement is read by adding 2″ to the reading on the blade.

The steel tape is very flexible and is often used for measuring irregular lines as well as straight lines. It is not very accurate when used to measure sharp curves and only provides an *estimate* of curved distances, Fig. 6-7. Accurate measurement and layout of curved lines is discussed later in this chapter.

SQUARING TOOLS FOR MAKING LAYOUTS

When a layout line is placed on the face of the stock at a right angle to an edge, a square is used, Fig. 6-8. There are many kinds of squares which, like other tools, are designed for particular uses. Common layout tools include the try square, framing square, combination square, and sliding T-bevel.

Try square. The try square shown in Fig. 6-9 is a very versatile tool used for squaring, measuring, and testing. It consists

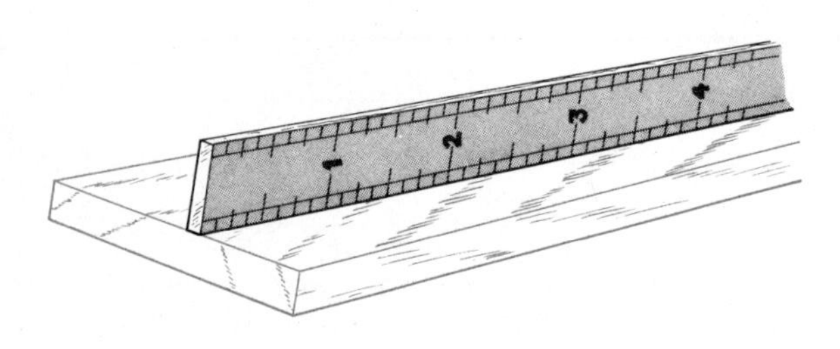

Fig. 6-4. The bench rule is held on edge to make an accurate layout.

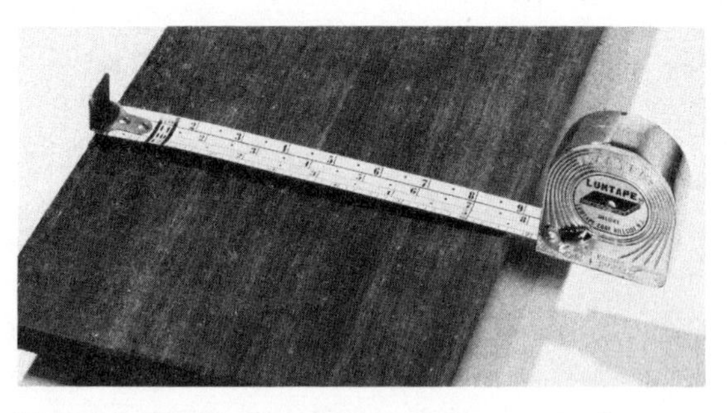

Fig. 6-6. Making a layout with a steel tape. (Lumtape Corp.)

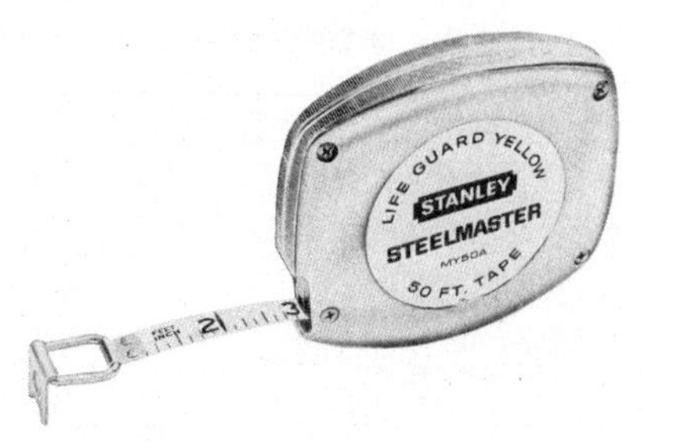

Fig. 6-5. A steel layout tape. (Stanley Tools)

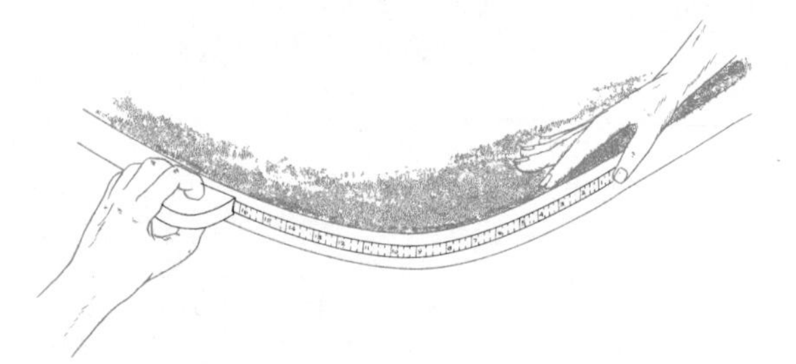

Fig. 6-7. Measuring a curved surface with a steel tape.

of a metal blade and a wood or metal handle. The try square is used to square lines across the face or edge of stock. It is also used for checking ends, edges, and surfaces for squareness, Fig. 6-10.

Framing square. The framing square shown in Fig. 6-11 is a large steel square consisting of a *blade* and *tongue.* It is sometimes called a carpenter's square. Tables are stamped on the square and are used in such operations as stair and rafter layout, Fig. 6-12.

The framing square is also used to check for squareness and for squaring lines across wide stock. A large square should always be used for squaring long lines across wide stock, Fig. 6-13. This square is often used to check large cabinets for

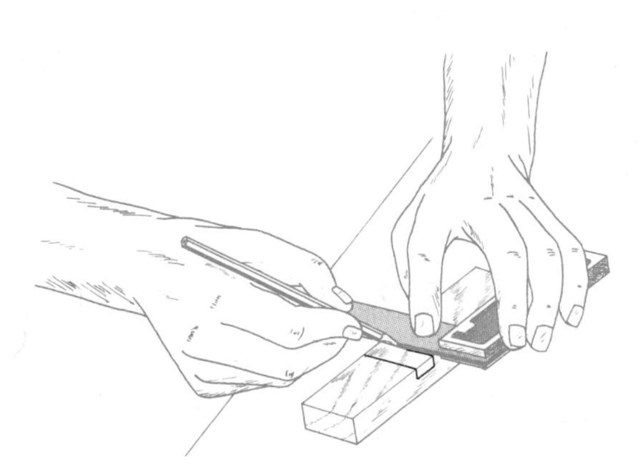

Fig. 6-8. Using a square to place a line on the stock face at a right angle to the edge.

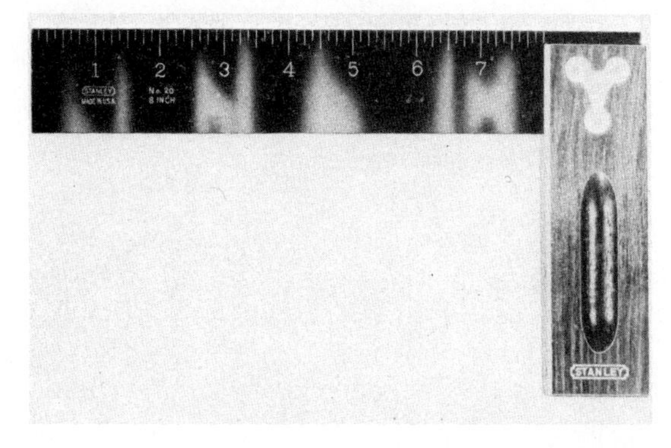

Fig. 6-9. A try square. (Stanley Tools)

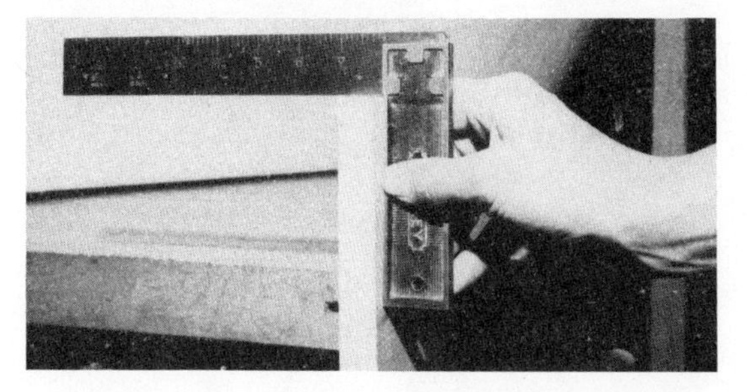

Fig. 6-10. Checking a board for squareness with a try square.

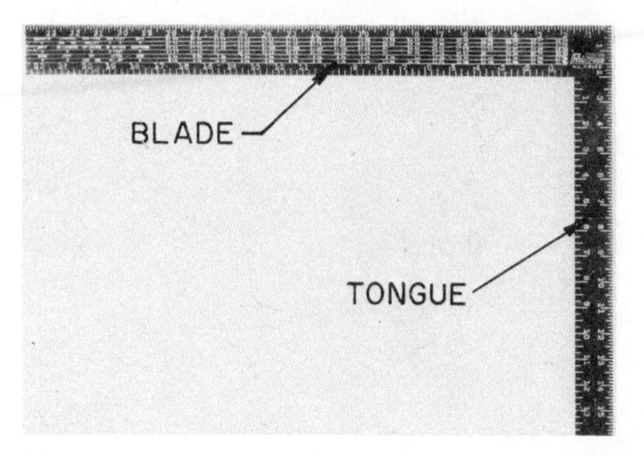

Fig. 6-11. A framing square. (Stanley Tools)

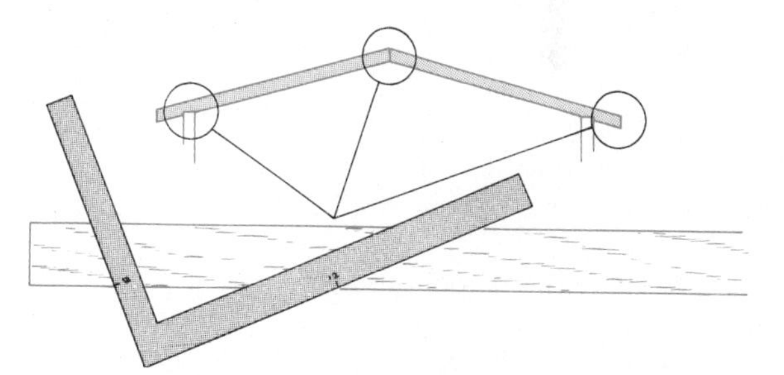

Fig. 6-12. Using the framing square to lay out angles of a rafter.

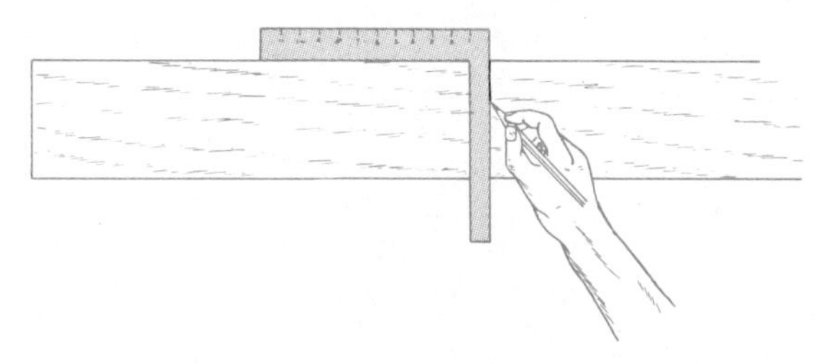

Fig. 6-13. Using a framing square to place a right angle line on the wide surface of the stock.

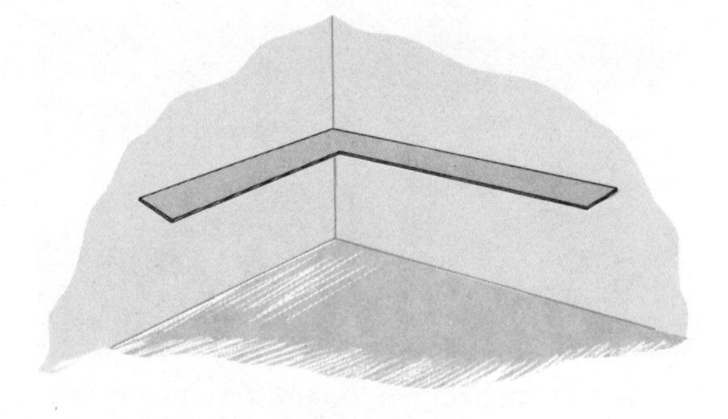

Fig. 6-14. Checking the interior of a cabinet for squareness using a framing square.

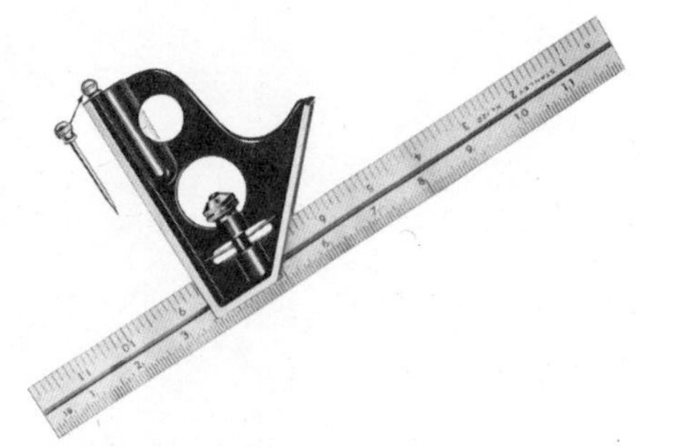

Fig. 6-15. A combination square. (Stanley Tools)

squareness during the assembly process, Fig. 6-14.

Combination square. A combination square is shown in Fig. 6-15. It is called a combination square because it can be used as a try square, depth gauge, level, miter square, and scriber. This square consists of a blade and a handle. The blade slides along the handle or head. The level and scriber are in the head. The level is used to determine if a surface is level or if it is plumb. A *level* surface is horizontal and a *plumb* surface is vertical, Fig. 6-16. The combination square can also be used for checking either inside or outside corners for squareness, Fig. 6-17.

The combination square is also designed for measuring and marking 45° angle miters, Fig. 6-18. The head can be moved along the blade to any desired distance and used as a marking gauge, Fig. 6-19.

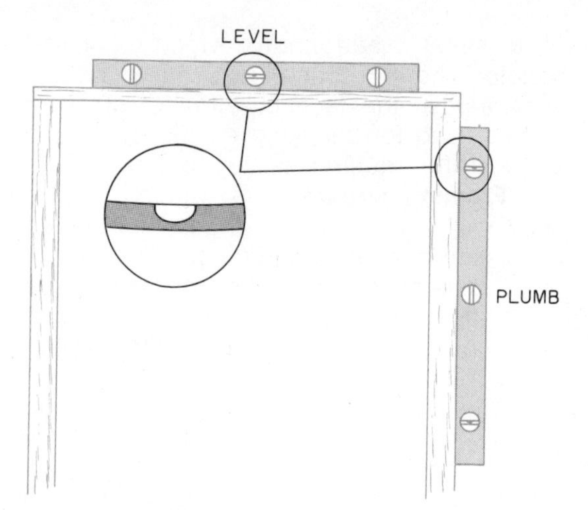

Fig. 6-16. A horizontal surface is referred to as level and a vertical surface as plumb.

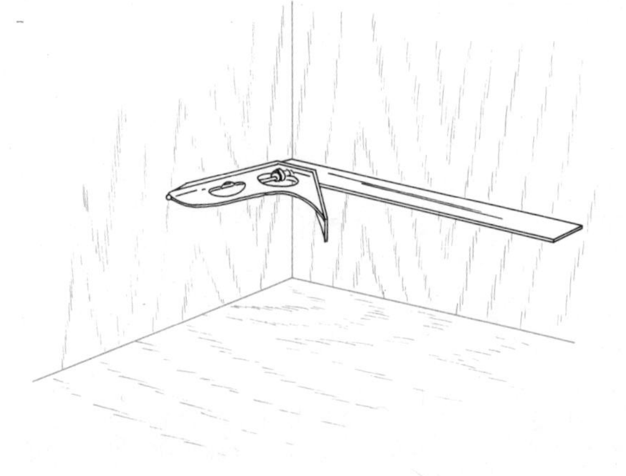

Fig. 6-17. Using a combination square to check an inside corner for squareness.

Sliding T-bevel. The sliding T-bevel, such as shown in Fig. 6-20, is used in the same manner as most other squares. The blade can be adjusted to any desired angle in the handle. A protractor is commonly used to set the exact angle.

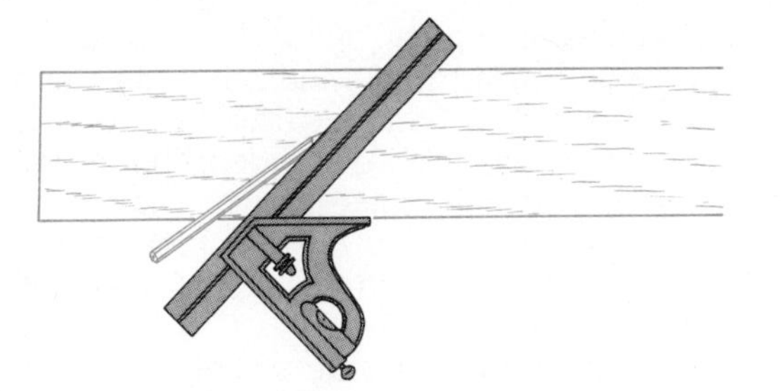

Fig. 6-18. Marking a 45° angle with a combination square.

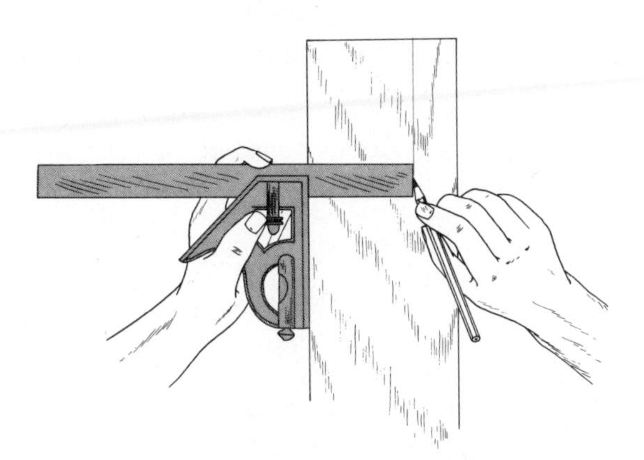

Fig. 6-19. Gauging a line with a combination square.

The sliding T-bevel is used to measure or transfer any angle between 0° and 180°, Fig. 6-21. It can also be used to check or test a miter cut. The sliding T-bevel is commonly used for transferring an angle from one piece of stock to another.

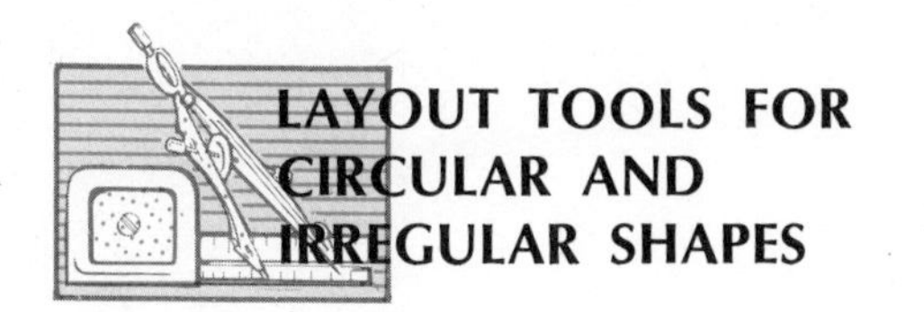

LAYOUT TOOLS FOR CIRCULAR AND IRREGULAR SHAPES

Additional special tools are needed for laying out irregular shapes. The irregular shapes can be developed directly on the stock or on paper to be transferred to the stock. Geometric shapes can usually be drawn directly on the stock. Irregular designs are generally drawn on paper to provide a full-size pattern, and then transferred to the stock. Figure 6-22 shows an irregularly shaped layout compared to a

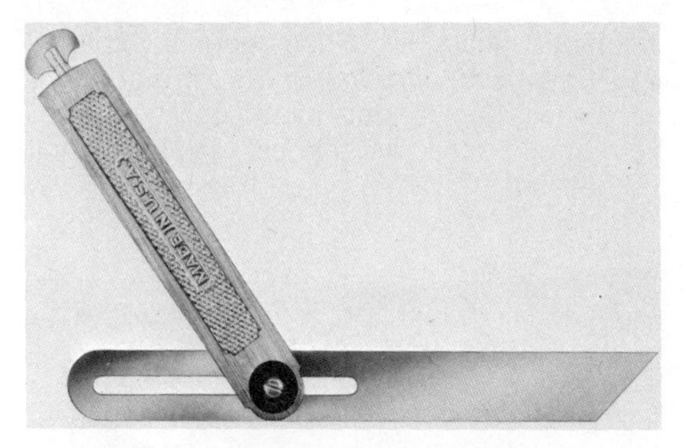

Fig. 6-20. A sliding T-bevel. (Stanley Tools)

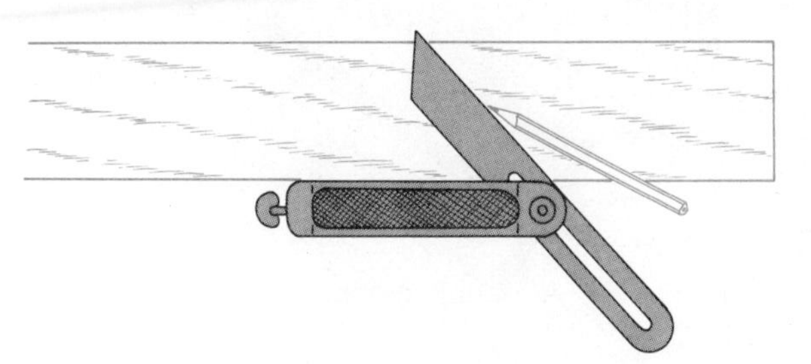

Fig. 6-21. Using a sliding T-bevel to mark an angle on the stock.

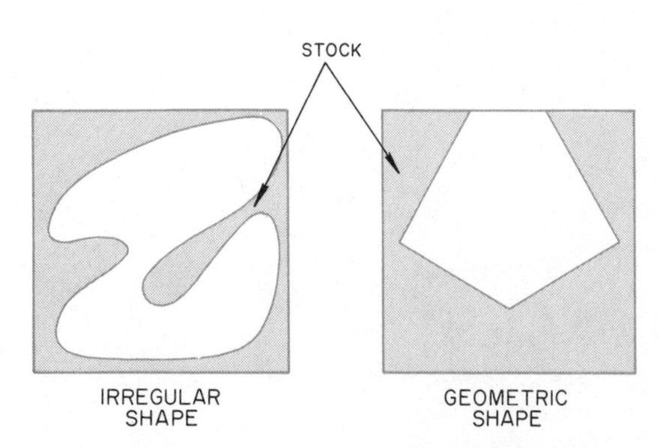

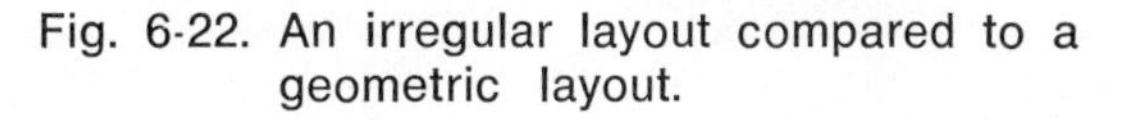

Fig. 6-22. An irregular layout compared to a geometric layout.

geometric shape. Common geometric shapes are hexagons and octagons.

The special tools commonly used for laying out irregular shapes include pencil compasses, dividers, and trammel points.

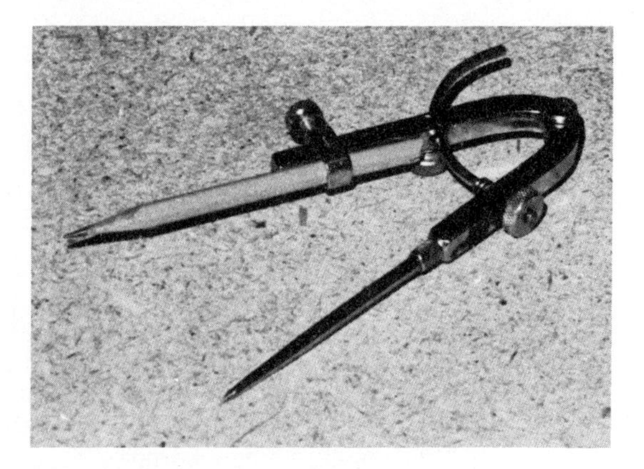

Fig. 6-23. A pencil compass.

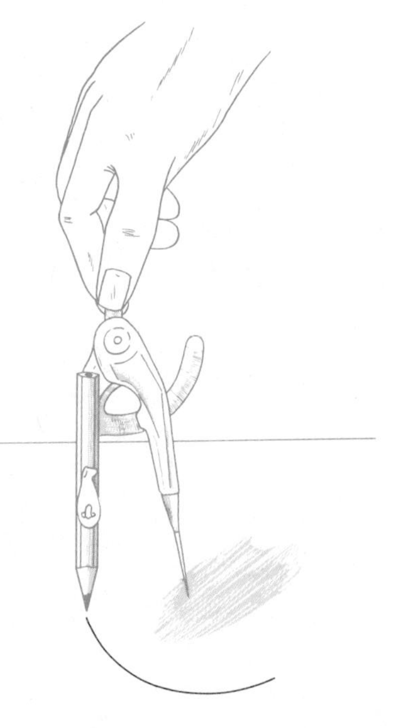

Fig. 6-24. Drawing a circle with a pencil compass.
The sharp point is held stationary in the exact center of the desired circle.

Pencil compass. A pencil compass, Fig. 6-23, is used for laying out circles and various rounded corners. The size of the compass needed is dependent upon the radius of the circle being drawn. The center of the circle is located on the stock from a dimension given on the working drawings for the product, Fig. 6-24.

Dividers. The dividers are used for dividing space equally, transferring measurements, and scribing arcs. The dividers are set using a bench rule as shown in Fig. 6-25. One leg of the divider is set on the inch mark of the rule, and the other leg is opened to the desired measurement on the rule. Then the thumbscrew is locked to hold the legs the correct distance apart. The dividers are different from the compass in that both legs have sharp metal points.

The dividers are set to equal one-half of the diameter of the circle to be drawn. The diameter of the circle should always be given on the working drawing. To scribe or scratch the circle, one leg is placed over the center of the circle and the dividers are tipped at a slight angle. The circle is scribed from left to right.

Trammel points. Trammel points, Fig. 6-26, are used for laying out large circles as shown in Fig. 6-27. The trammel points consist of two metal points that are fastened

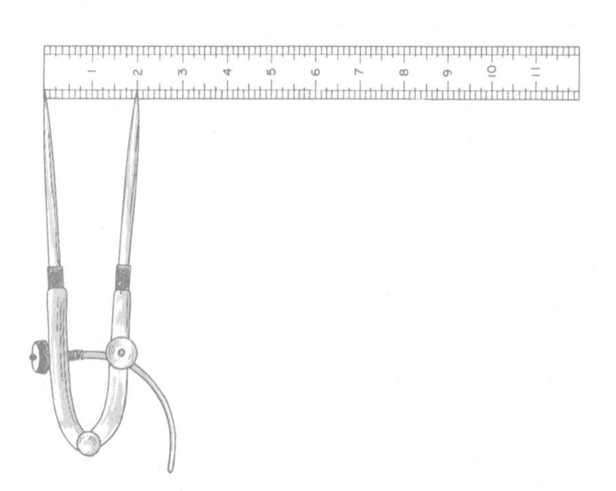

Fig. 6-25. Setting a divider to the desired size using a bench rule.

to a long bar of wood or metal. In addition to laying out large circles and arcs, the trammel points can also be used for laying out distances between two points.

LAYOUT FOR ROUNDED CORNERS

Many products have rounded corners. Figure 6-28 shows the procedure for locating the center of the radius of a rounded corner. The radius of the arc is determined from the drawing. The radius can be marked for cutting with either the dividers or pencil compass.

LAYOUT FOR A HEXAGON

A hexagon is a six-sided figure with all sides and all angles equal. To layout a hexagon, find the length of one side. Set a compass or dividers to the measurement equal to one side. Draw a circle with this radius. Begin at any point on the circle, and without changing the setting, draw a series of arcs, moving the point to the place where the preceding arc has intersected the circle. The last arc will intersect the circle at the first point made by the compass. Join the points with a straight-edge, Fig. 6-29.

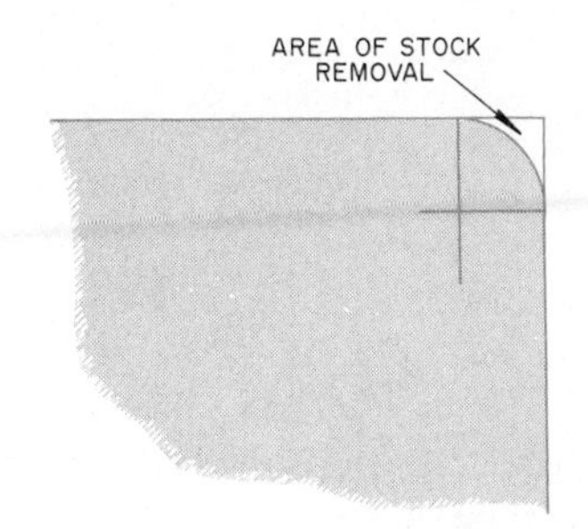

Fig. 6-28. Layout for a rounded corner.

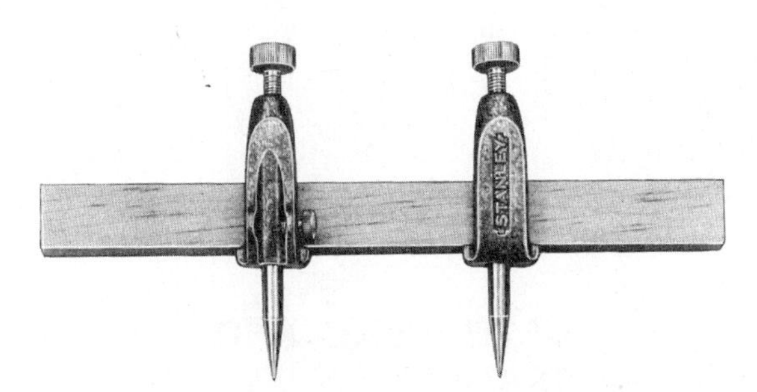

Fig. 6-26. A set of trammel points. (Stanley Tools)

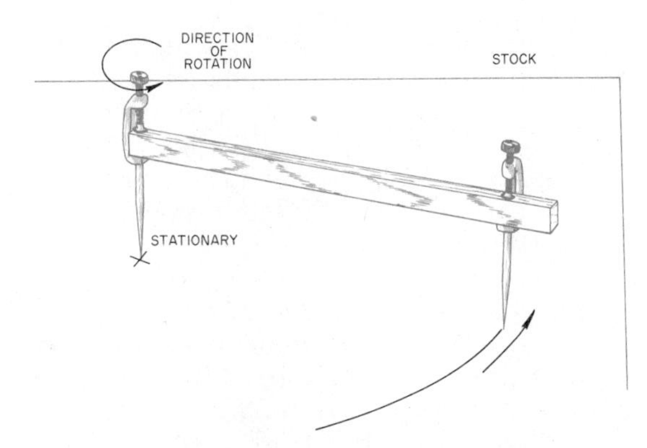

Fig. 6-27. Laying out a large circle using trammel points.

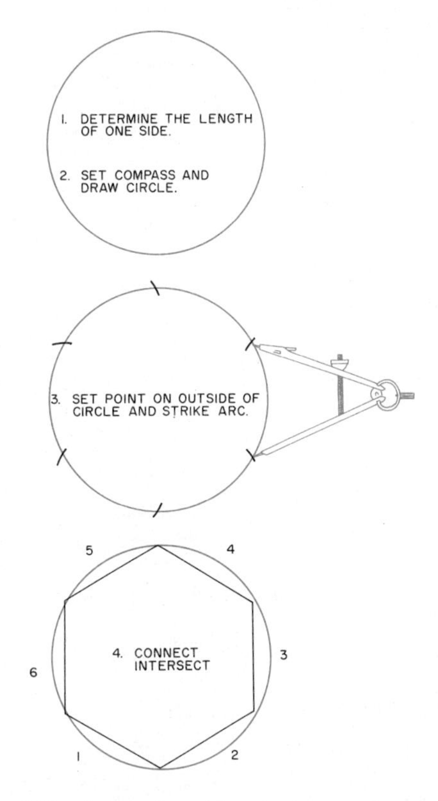

Fig. 6-29. Procedure for laying out a hexagon.

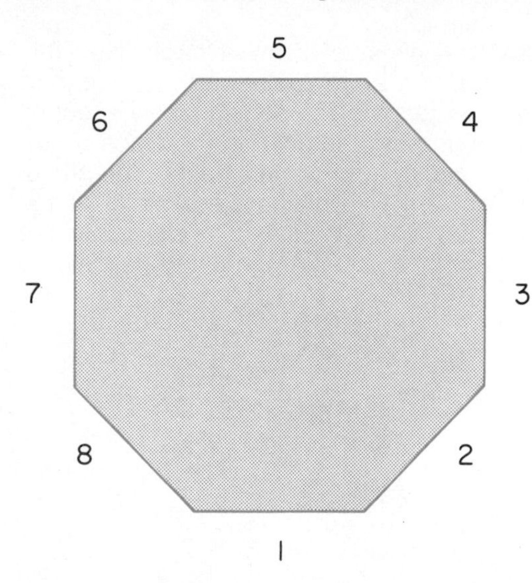

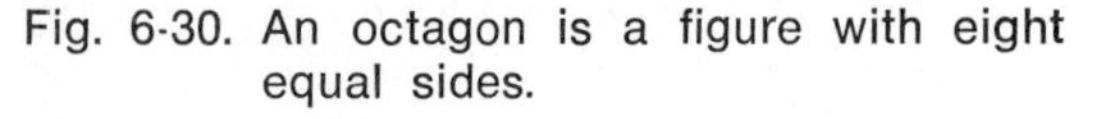

Fig. 6-30. An octagon is a figure with eight equal sides.

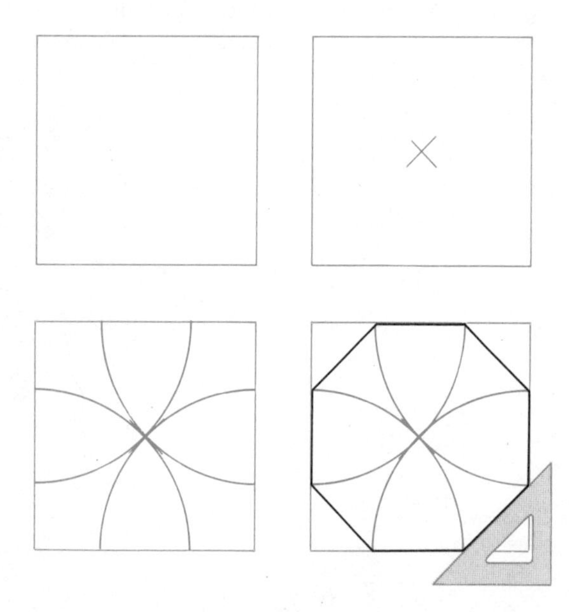

Fig. 6-31. Procedure for laying out an octagon.

LAYOUT FOR AN OCTAGON

An octagon is an eight-sided figure with all sides of equal length and angles of each line the same, Fig. 6-30. The layout for an octagon is made by first determining the distance across the octagon, Fig. 6-31. A square is constructed with sides the same length as the distance across the octagon. The compass is set equal to one-half the diagonal distance across the square, Fig. 6-31, and arcs are drawn using the same procedure as for a hexagon. The points at which the arcs meet or intersect are connected with a straightedge.

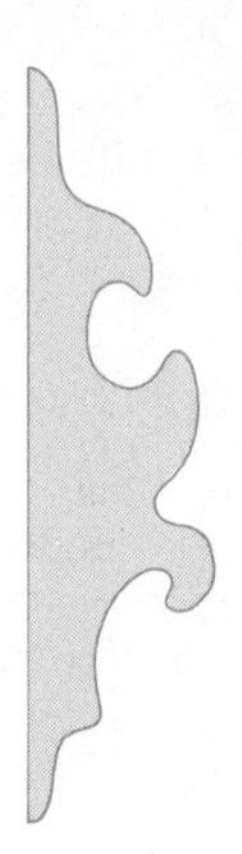

Fig. 6-32. An irregular shape to be transferred to a pattern.

DEVELOPING AND TRANSFERRING IRREGULAR SHAPES

It is sometimes necessary to transfer a pattern for irregularly shaped parts to the stock. Figure 6-32 shows a pattern which must be transferred. Sometimes the patterns will need to be enlarged, as they are usually drawn one-half or one-fourth the actual size on the working drawing.

The patterns are enlarged by the *square* method. If the outline of the shape is not drawn on squared paper on the working drawing, they are added over the top of the drawing, Fig. 6-33.

When enlarging an irregularly shaped drawing to the actual size, one inch squares are drawn on a large piece of wrapping

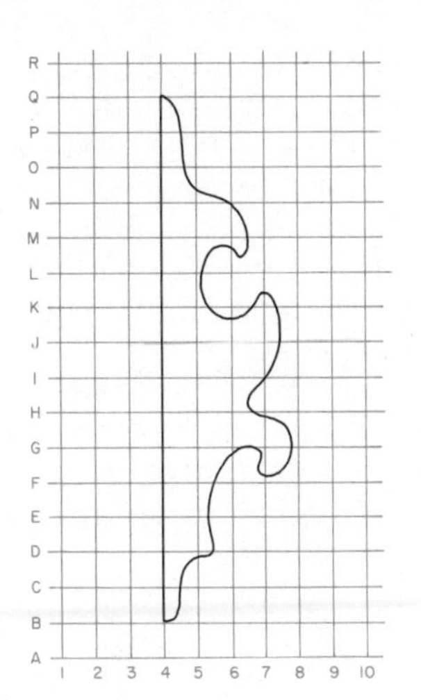

Fig. 6-33. Squares drawn for the irregular shape.

paper. If the squares are already drawn on the original plan, each horizontal line is labeled A, B, C, D, etc. from the bottom to the top. Each vertical line is labeled with a number from left to right 1, 2, 3, 4, etc. The lines on the paper for the full-sized drawing are labeled in exactly the same manner.

The design is transferred from the original pattern to the full scale enlargement by transferring points. A point is located on the original and the corresponding point is marked on the enlargement sheet. This is continued until all of the points have been located. The points are then connected on the enlargement, Fig. 6-34.

The enlarged pattern can be transferred to the stock by one of two methods. One method is to cut out the pattern, tape it in place, and trace around it. The other method is to place a piece of carbon paper between the pattern and the stock and then trace the design.

TRANSFERRING IRREGULAR SHAPES WITH TEMPLATES

When the same irregular shape will be transferred to several different pieces of stock, time can be saved by using a *tem-*

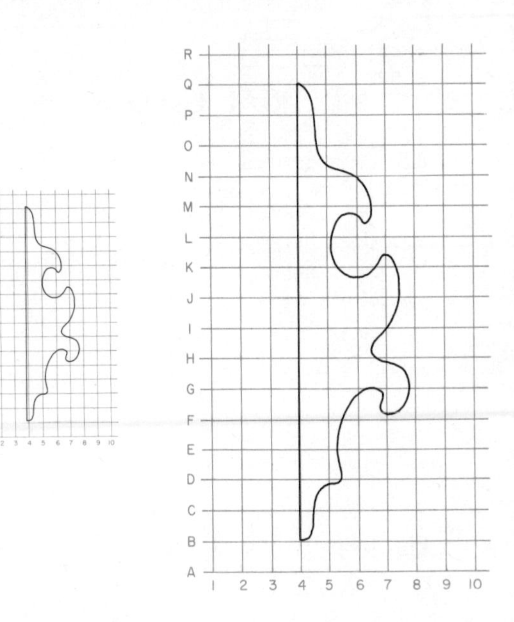

Fig. 6-34. Transferring the enlarged or reduced shape by the squares method.

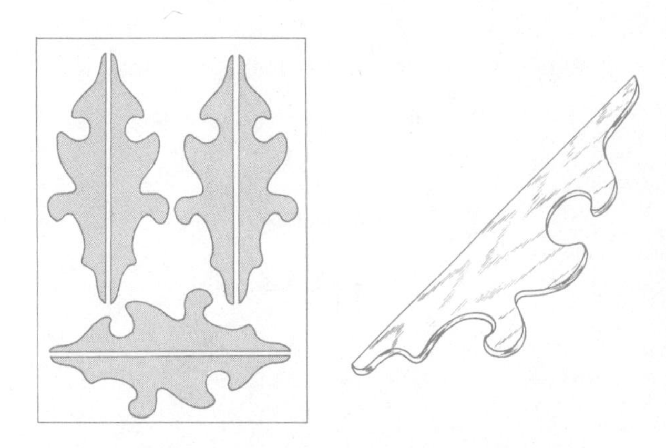

Fig. 6-35. A template developed on rigid stock for an irregular shape that may be duplicated many times.

plate. A template is a piece of rigid material on which the shape of the irregular pattern has been cut. Plywood or tempered hardboard is often used for making templates, Fig. 6-35. A template saves time and insures greater accuracy when transferring designs to the stock.

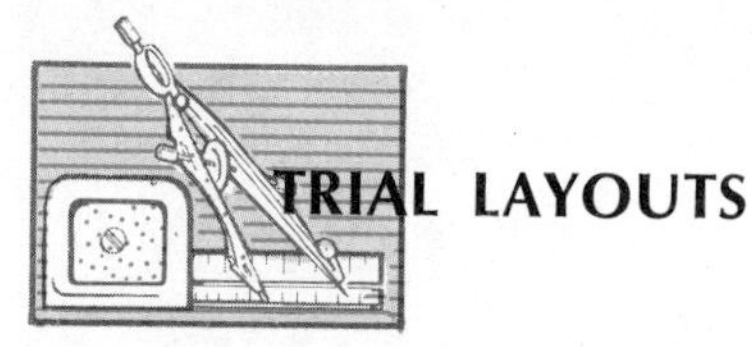

TRIAL LAYOUTS

As you recall, the main concern in the wood industry is to reduce costs. Whenever material is wasted, the costs will go up.

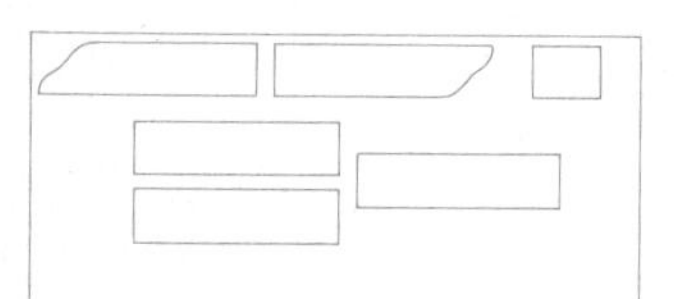

Fig. 6-36. A poor layout such as this results in material waste.

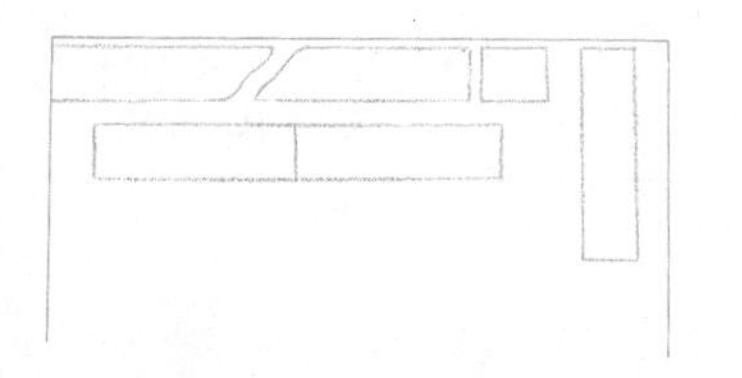

Fig. 6-37. A trial layout made on stock with a piece of chalk.

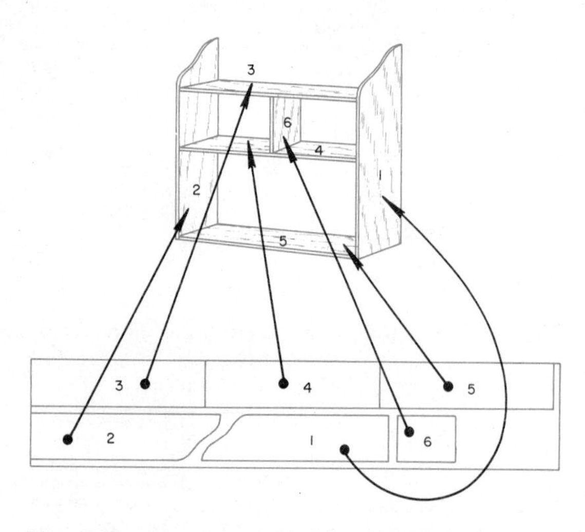

Fig. 6-38. Careful attention must be given to the direction of the grain during layout.

Long vertical parts run with the grain.

Careful layout before cutting will insure minimum waste. You should always make a trial layout before cutting stock so that you will get the best possible utilization of the standard materials, Fig. 6-36.

A trial layout can be made on the stock with a piece of chalk. Chalk is used since it can easily be removed if a change must be made. Figure 6-37 shows a trial layout for a product which will be constructed from the large plywood sheet. Notice how the irregularly shaped sides are reversed on the sheet to save material.

Careful attention must be given to the direction of the grain for various parts. For example, the sides of the bookcase shown in Fig. 6-38 should have the grain running in a vertical direction. However, the board on the bottom has the grain running in a horizontal direction. If incorrectly cut, this would cause waste and increased production costs.

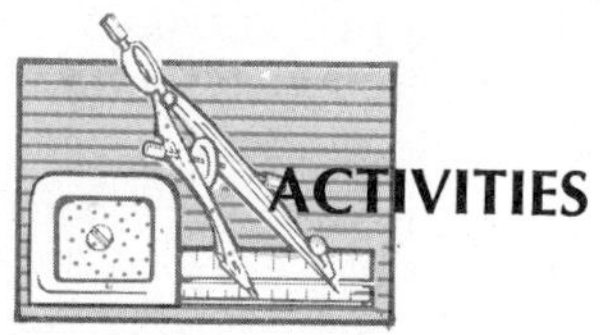

ACTIVITIES

PROBLEM

Using a large sheet of wrapping paper, make a trial layout for the prototype which you have designed. Cut the paper to the size representing the standard stock to be used for constructing your prototype. A piece of chalk for marking the layout will allow you to make any necessary changes as you develop it.

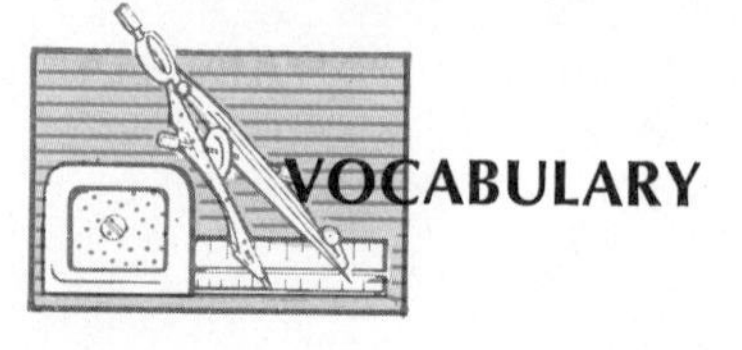

VOCABULARY

Layout
Bench rule
Zig-zag rule
Steel tape
Trammel points
Compass
Sliding T-bevel
Combination square
Template

CHAPTER 7 SAFETY AND ACCIDENT PREVENTION

INTRODUCTION

Safety and accident prevention have become great concerns of industry today. *Accidents are costly.* They cost both employer and employee. The expense to the employee is in two forms — it causes him *human suffering* and *financial loss* due to

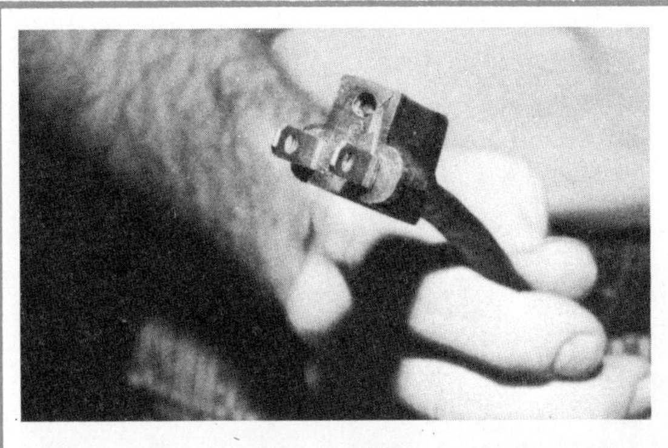

B. Unsafe condition — broken plug.

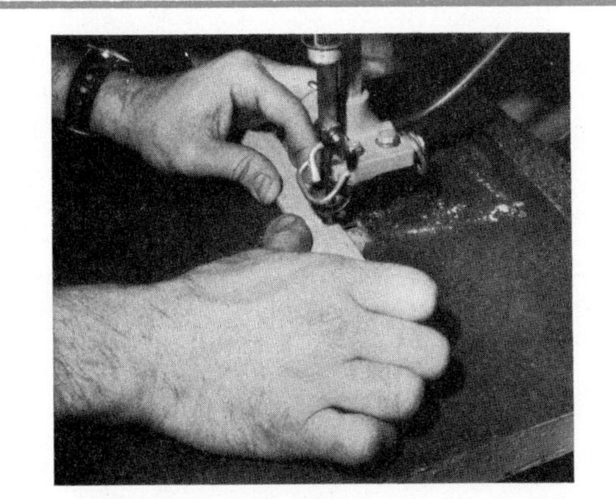

A. Unsafe act — placing the finger in front of blade.

C. Unsafe condition due to carelessness—workers leaving bricks and boards in positions where they may fall on workers.

Fig. 7-1. An accident occurs due to an unsafe act or an unsafe condition.

his inability to work. It costs industry through the loss of skilled employee's services and the medical expenses. Most industries have insurance which pays the medical expenses. However, as the accident rate increases, the cost of the industry's insurance premium also increases. Increased insurance rates mean increased production costs for the industry.

Most industrial accidents can be prevented. An accident is the result of an *unsafe condition* or an *unsafe act.* An accident may occur when either or both of the causes are present, Fig. 7-1. The unsafe condition is usually a physical problem. The problem may be an unprotected machine or a worker improperly dressed for the job, Fig. 7-2. The unsafe act is usually a form of carelessness on the part of the worker.

Until about 50 years ago, very little attention was given to industrial accident prevention. In 1912, laws were passed called *workman's compensation laws.* These laws required industries to pay the medical expenses of the employees injured on the job. This caused industries to become conscious of accident prevention and make special efforts to prevent accidents.

Insurance companies, industries, and special agencies such as the National Safety Council have worked together to organize *accident-prevention programs.* An accident-prevention program is a plan to reduce accidents by reducing hazards.

The continued success of a safety program depends to a large extent on the foreman. The foreman must start the plan of action in cooperation with the safety engineer. The foreman must enforce the plan of action and make sure it is followed by his workers at all times. He must plan safety, as well as production, into all jobs.

Fig. 7-2. The cause of an accident may be an unguarded machine.

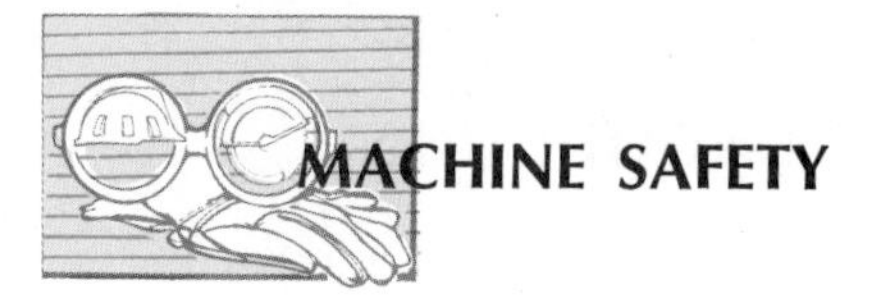

MACHINE SAFETY

Machines can be one of the safest elements in an industry. When a machine is kept in working order and is properly operated, it will generally not be the cause of an accident. *It is the operator who generally causes the accident.* Most industrial machines are equipped with special guards and shields to protect the operator. It is when the guards are removed and the machine is incorrectly operated that accidents occur. Figure 7-3 shows some of the common guards and shields designed for industrial woodworking machines to protect the worker.

SAFEGUARDING THE WORKER

It is not always possible to remove all safety hazards. However, it is possible to protect the worker from these hazards by

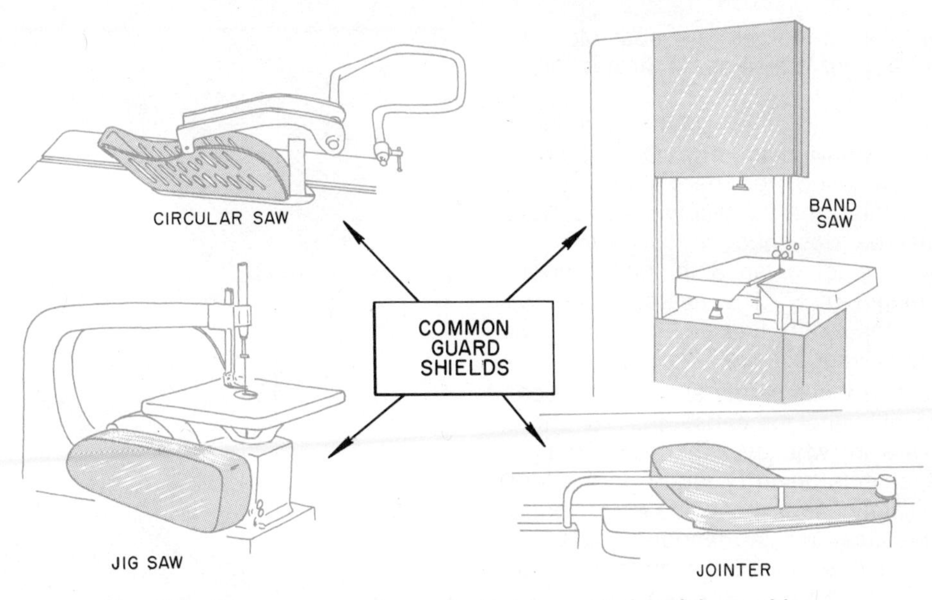

Fig. 7-3. Common guards and shields on industrial machines.

means of special clothing such as eye glasses and goggles, shoes, boots, hard hats, face shields, gloves, and respirators. Safety specialists evaluate working conditions and attempt to determine the potential safety hazards. Special efforts are made to reduce the possibility of accidents. For example, if the worker is threatened by falling objects, he is protected by special hats and shoes.

The jobs in a woodworking industry can be inspected to determine the hazards. The hazards with which the worker comes in contact are recorded. Then all of the hazards are removed, covered, or the worker is equipped with special equipment to safeguard him against the hazard.

SPECIAL SAFETY EQUIPMENT

All personal safety equipment is designed to protect a portion of the worker's body. Common safety equipment includes the following:

Hats. Hard hats are worn on jobs where there is a likelihood of objects falling and striking a worker, Fig. 7-4. Ideally, the possibility of falling objects should be eliminated.

Fig. 7-4. Hard hats are worn on some jobs to protect the workers from falling objects.

However, this is not always possible, and the hard hat furnishes some protection to the worker.

Hats, like all other safety equipment, must meet certain standards. One standard specification used for hard hats is the amount of force the hat will withstand. Most hats must be designed so they will withstand the impact of an 8 lb. ball dropped from a height of 5′. The hat cannot break or dent to the extent that any portion would touch the worker's head, Fig. 7-5.

Shoes. Special safety shoes have been designed with steel toe caps, Fig. 7-6. These shoes are worn whenever there is a danger of heavy objects being dropped on the feet.

The safety shoes must also meet certain safety specifications. A standard commonly used is that the toes of the shoe must be

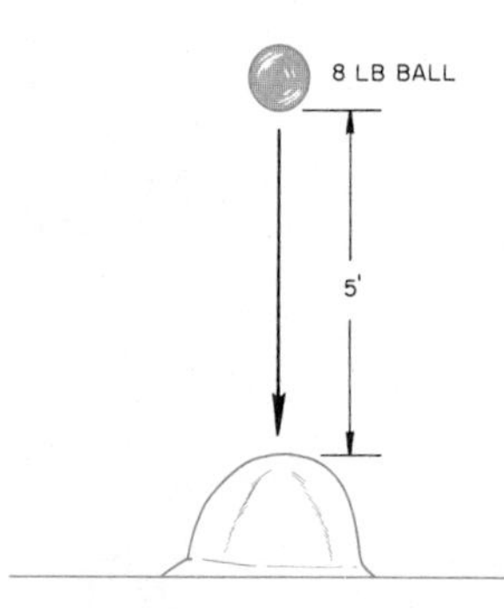

Fig. 7-5. Hard hats must be designed to withstand specific tests for quality.

Fig. 7-6. Shoes designed to protect the wearer from falling objects.

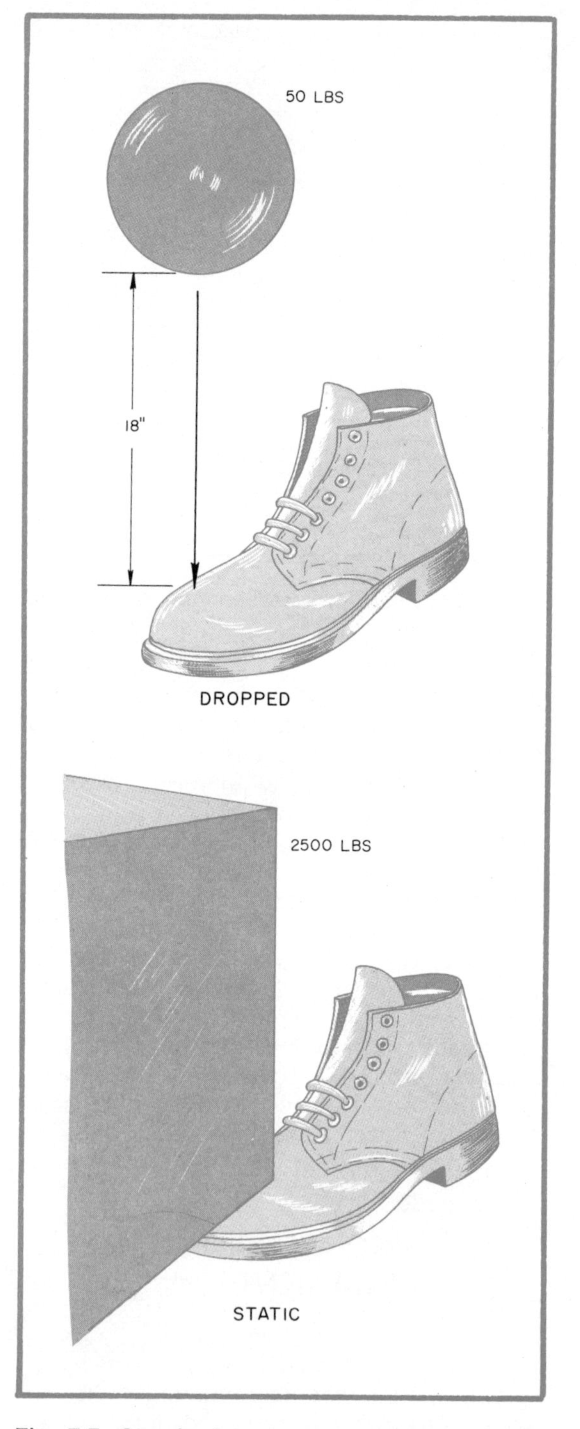

Fig. 7-7. Standard tests are applied to safety shoes to determine the level of protection.

designed to withstand the impact of a 50 lb. load dropped from a height of 18″. The toe of the shoe cannot deform under the impact test to the extent that it touches the worker's toe under the shoe, Fig. 7-7.

Protective glasses and goggles. Special eye protections such as shown in Fig. 7-8 are worn whenever there is a danger of flying particles, dust, or chemical burns. The glass lens must meet safety specifications to insure that the glass will not shatter and cut the eye when hit by a flying object. One standard specification for glasses is a test in which a 1/8″ steel ball is dropped from a height of 50″ and strikes the lens. To meet the specification, the glass lens must withstand the blow without shattering, Fig 7-9.

Protective gloves. Certain jobs in the wood industry require the workers to come in contact with strong chemicals. A safeguard against serious chemical burns can be the wearing of plastic or rubber gloves, Fig. 7-10. The type of glove to provide the

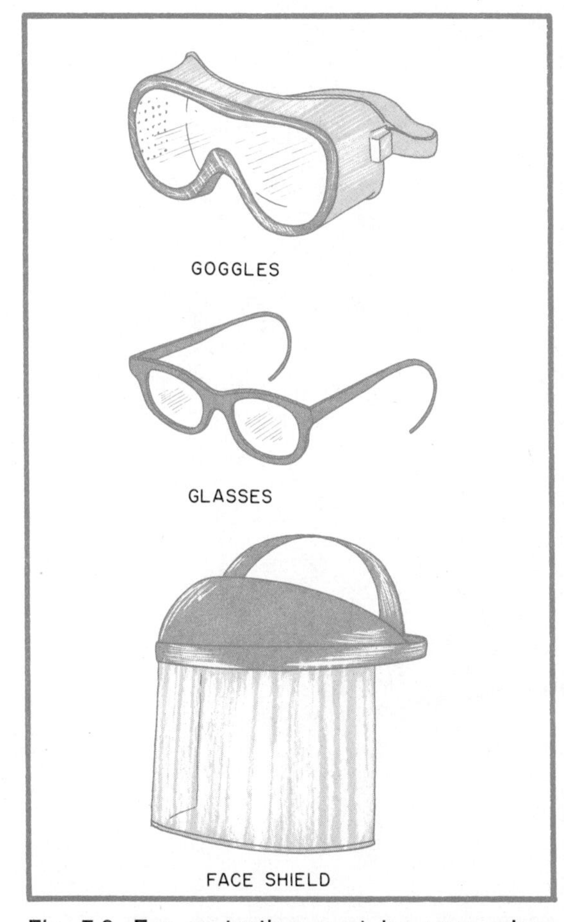

Fig. 7-8. Eye protection must be worn when there is danger of injury from flying chips or other harmful particles.

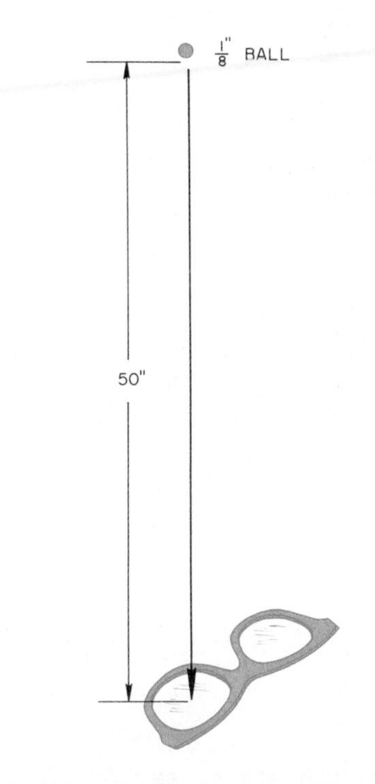

Fig. 7-9. Eye glass lens must meet the quality through standard tests.

Fig. 7-10. Gloves designed to protect workers from strong chemicals.

necessary protection is determined by the type of chemical solution being used. The manufacturer of the chemical will suggest the type of protection necessary for the safety of the worker.

Safety devices for the hands. There is a constant danger to the worker whenever operating certain sawing and planing machines. The level of danger can be reduced by using specially designed *push sticks,* Fig. 7-11. The push stick serves as an extension of the hand and eliminates the need for the hand coming close to the blades.

There are many hazards in any industrial plant. Most accidents can be avoided if a worker is alert. Each individual must practice looking for unsafe conditions and situations. By being alert at all times, a worker can reduce accidents. There are special safety considerations that all workers must be aware of to prevent accidents.

Respirator. The respirator is designed to prevent the worker from breathing harmful substances, Fig. 7-12. A respirator is

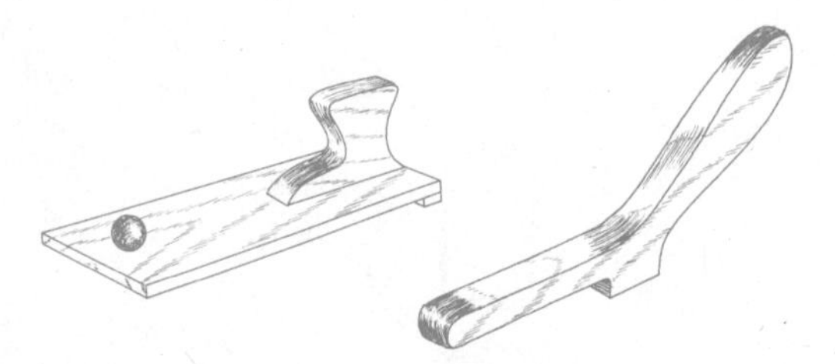

Fig. 7-11. Push sticks used to protect the hands when feeding stock into machines.

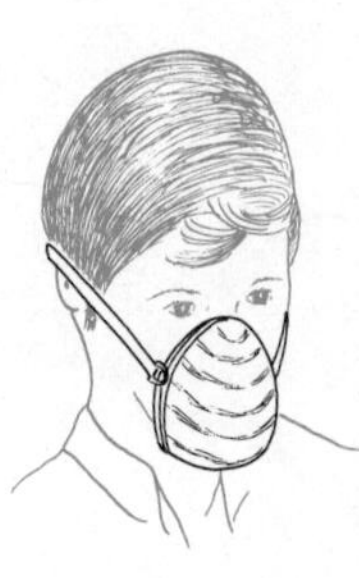

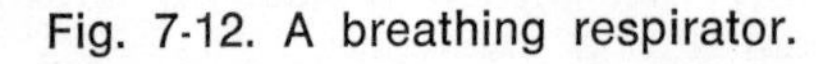

Fig. 7-12. A breathing respirator.

Fig. 7-13. Method of lifting heavy objects to avoid serious injury.

used whenever there is a chemical hazard, such as in some types of spray finishing operations, or if there is a high dust content in the air.

SPECIAL SAFETY CONSIDERATIONS

LIFTING AND CARRYING

Many accidents occur when lifting heavy objects. Many of the serious accidents which occur from lifting can be avoided by using proper lifting and carrying methods. Figure 7-13 shows the proper sequencial method for lifting a heavy object. The procedure for lifting and carrying is as follows:

1. Stand close to the heavy object you are to lift. Place your feet firmly on the floor, slightly apart.
2. *Bend the knees* and grasp the object.
3. Raise the object by straightening the legs.
4. Move with the object only when the body is in a vertical position.
5. When carrying the heavy object, turn corners by shifting the direction of your feet. Avoid twisting your body.
6. Reverse the lifting procedure to lower the object. *BEND THE KNEES.*
7. Always have a helper for moving extremely heavy objects.

SLIPS AND FALLS

Each year, slips and falls head the list of on-the-job injuries, causing many minor injuries and permanent disabilities. The major reason for this type of injury is the collection of debris scattered about the work area or liquid spilled on the floor. All of these hazards can be eliminated by the cooperation of everyone on the job. Scrap material should be piled in a designated area and not allowed to accumulate in the work area. Spilled water and other liquid materials should be cleared up immediately.

A work area which is clean and free of safety hazards is also a more efficient place to work. The time required to maintain a clean and orderly work area will result in greater work output and less personal injuries.

FIRE

Fire is one of the most costly industrial hazards. It results in serious injuries and property loss. Fire can best be controlled by eliminating the factors which support fire. The three essential factors which must be present to cause and support fire are: (1) Fuel, (2) Oxygen, and (3) Ignition temperature. The removal of any one of the essential factors for the support of the fire will immediately extinguish it.

Industry's primary concern in the control of fire loss is to make certain that the three essential elements never exist at the same time. Safety inspectors and engineers are employed by many industries to safeguard against fire loss. The danger of fire loss can be minimized through preventative measures. Common industry (or school shop) fire prevention measures include the following:

1. Avoid the accumulation of rubbish and combustible materials.
2. Install fire extinguishers and other fire-control equipment on the job.
3. Ground all electrical equipment.
4. Report all fire hazards to your supervisor or your teacher.
5. Keep all inflammable liquids in sealed containers. Store them away from any heat source.
6. Always have proper ventilation wherever combustible materials are being applied.
7. Provide closed containers for combustible materials such as oily rags, etc.
8. Conduct routine safety inspections.
9. Eliminate or repair all defective electrical equipment.
10. Avoid the use of heaters which will create a fire hazard.

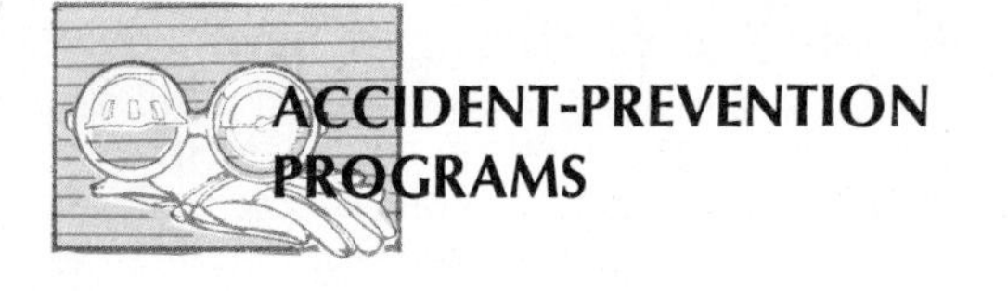

ACCIDENT-PREVENTION PROGRAMS

Many industries employ full-time professional *safety engineers.* A safety engineer is a safety specialist and works in cooperation with management and production personnel. To be effective, the accident-prevention program must be a planned course of action. All factors which make up a job for an individual must be considered. This includes tools, machines, methods, materials, and people. Any one of these elements could cause an accident. A safety engineer considers all factors which may cause accidents to the workers. He then finds a solution to remove the factors which may result in a personal accident. The proper solution to the problem may require the cooperation of the management and production personnel. Sometimes it is necessary to develop a new method of performing production operations to eliminate an accident hazard.

Constant inspections combined with safety program enforcement can help lead to a safer working environment. It is usually easier to plan safety control programs for machines than people, because the actions of people are more difficult to predict. It is more often the machine operator, and not the machine, which causes the accident. Safety must become a habit and the responsibility of everyone.

Whenever accidents do occur, regardless of how minor the injury, the cause should be determined. It will no doubt prove to be either an unsafe condition or unsafe act. In either case, a solution to the problem must be found. An accident report should always be filed even for minor accidents.

Special attention will be directed to safe machine and tool usage throughout this book. Safe work habits are good habits — to become a habit, they must be practiced.

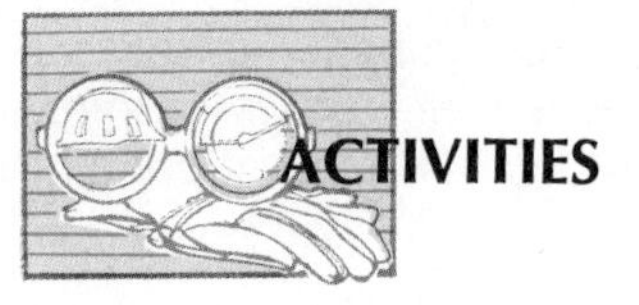

ACTIVITIES

PROBLEM 1

Your instructor may ask you to form groups of three to five students to become an inspection team. You will be assigned one of the following inspections. One student from each team will be asked to report his team's findings.

A. Inspect the floors, walls, ceilings, material storage areas, and passageways for unsafe conditions.

B. Inspect all hand or portable power tools and equipment for unsafe conditions.

C. Inspect all stationary power equipment for unsafe conditions.

D. Inspect the availability and conditions of all equipment or clothing for the protection of eyes, ears, hands, legs, feet, and other parts of the body.

E. Inspect all combustible materials, exit areas, and plan of procedure for emergencies for availability and safety practices.

PROBLEM 2

After the safety inspections have been completed, discuss various methods of reducing or eliminating all existing hazards. Determine what protective devices such as face shields, machine guards, etc. should be added in the laboratory. Also, determine the approximate cost of each suggested safety improvement.

VOCABULARY

Hazard
Safety practice
Safety inspection
Unsafe condition
Accident
Unsafe act
Accident-prevention program
Personal safety equipment
Safety engineer

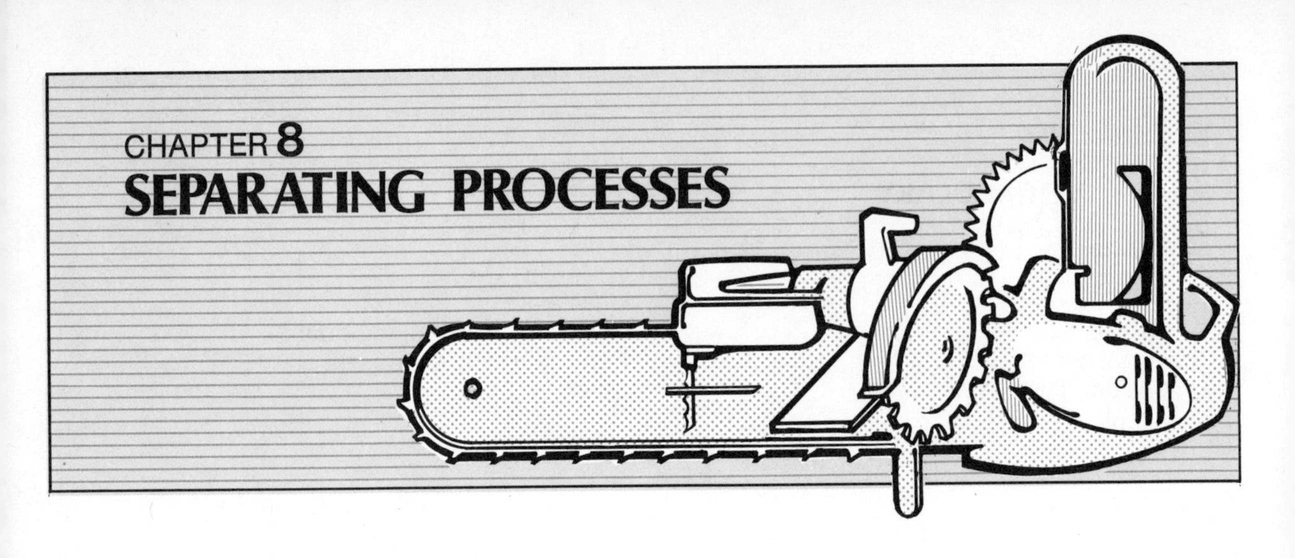

CHAPTER 8
SEPARATING PROCESSES

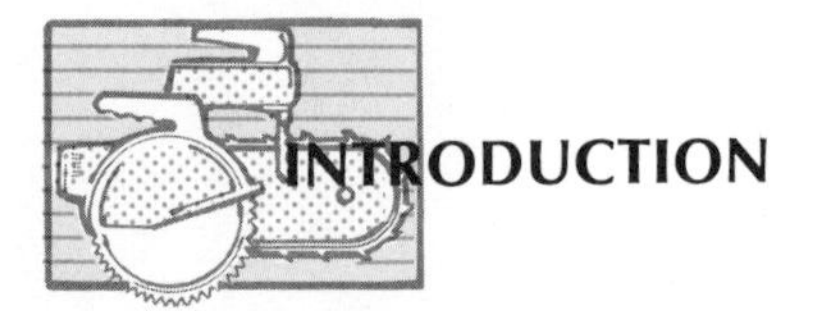

INTRODUCTION

Separating is a major production process. Separating is the process of cutting wood to a predetermined length, width, or thickness. Sawing of wood is a standard separating process. The separation of wood stock can be done by means of numerous tools and machines. In Fig. 8-1, the board is being separated by means of a handsaw.

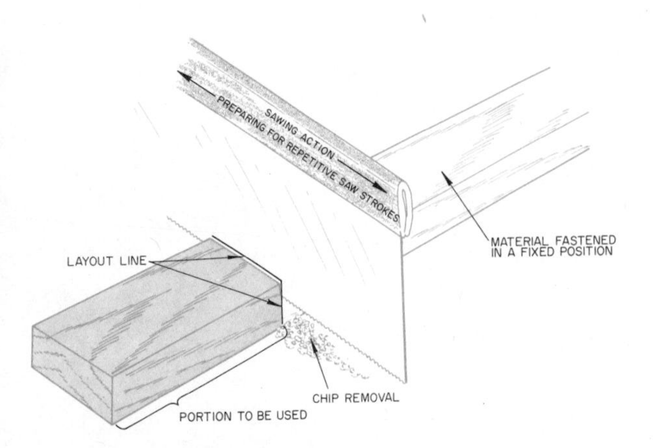

Fig. 8-1. Sawing a board is a separating process.

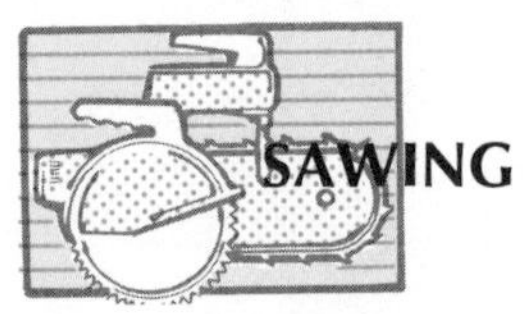

SAWING

This section will deal with the sawing method of separation. Regardless of the complexity of the tool or machine, the sawing principle remains basically the same. The major difference in the techniques used is the *time required to perform the task.* The teeth of a saw are classified as either *rip, crosscut,* or a *combination of the two.* Whichever saw is used, the process remains that of severing the wood fibers. The crosscut saw is used to cut across the width of the board. The ripsaw is used to cut the length of the board or parallel to the grain. Notice the direction of the grain and the position of the saw in Fig. 8-2.

The hand crosscut and ripsaw are used to cut or separate stock into the rough sizes. Other types of handsaws or power saws are often used for finish cuts and preparing stock for joining by chemical or mechanical fastening.

The crosscut and ripsaw are the two most common types of handsaws. Both of these saws cut on the *push stroke.* An understanding of the cutting principles of the rip and crosscut saw will apply to all handsaws. Figure 8-3 shows the parts of a handsaw.

The blade of a handsaw is tapered from the teeth to the back, Fig. 8-4. The purpose of this taper is to give clearance to the saw as it enters the stock. The back of the saw becomes thicker as it nears the handle. This is to prevent the saw from bending as cutting pressure is applied.

The length of the blade is given as that distance from the point to the butt. Crosscut saw blades range in length from 20″ to 26″. The usual length of a ripsaw blade is 26″.

TEETH AND POINTS PER INCH

The size of the teeth in a saw is given by the term *points to the inch.* The number of points per inch is one more than the number of teeth per inch, Fig. 8-5. The more points per inch on a saw, the smoother the cut. However, a saw which has more points per inch cuts slower. Crosscut saws normally have from 8 to 12 points per inch. An 8-point saw is a general-purpose saw. Ripsaws have 5½ or 6 points per inch.

The alternate teeth of a handsaw are bent in opposite directions. This bending of

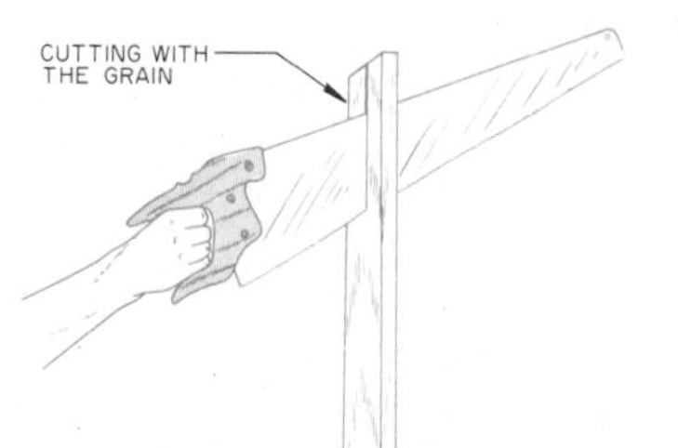

A. Ripping.

B. Crosscutting.

Fig. 8-2. The direction of the grain for separating.

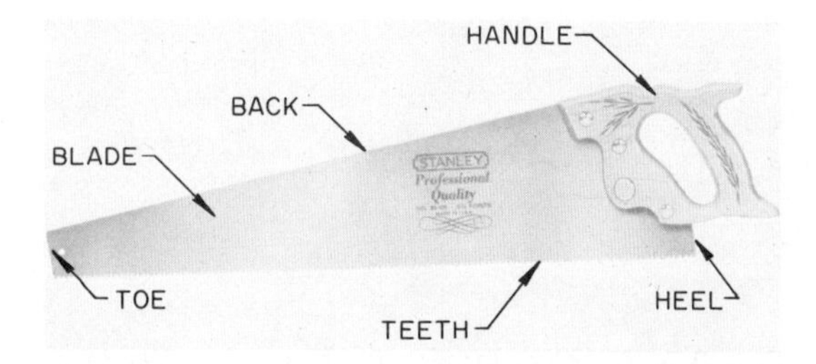

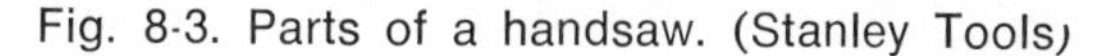
Fig. 8-3. Parts of a handsaw. (Stanley Tools)

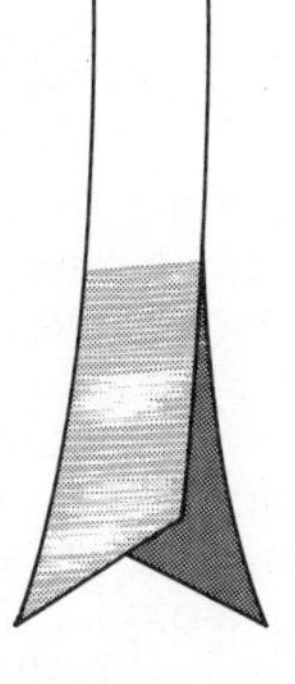

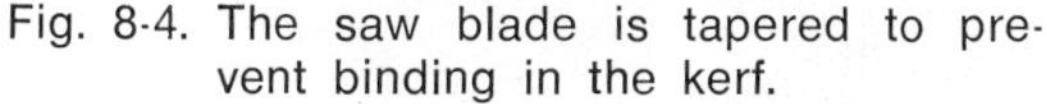
Fig. 8-4. The saw blade is tapered to prevent binding in the kerf.

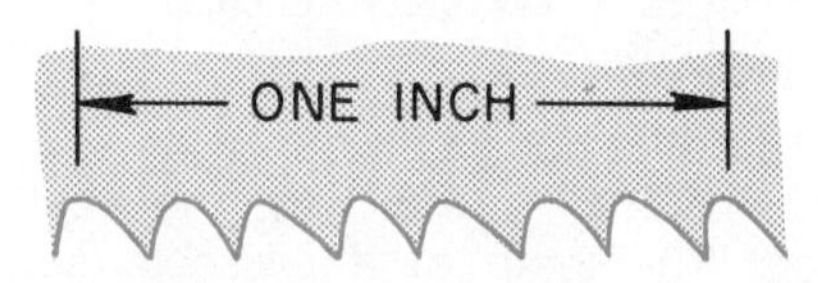

Fig. 8-5. The points per inch on a handsaw.

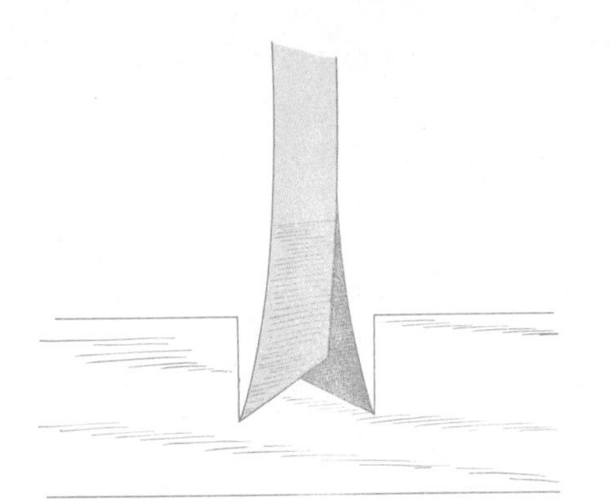

Fig. 8-6. A handsaw must be set to prevent it from binding.

Fig. 8-7. A circular saw blade is hollow ground to provide clearance and prevent binding.

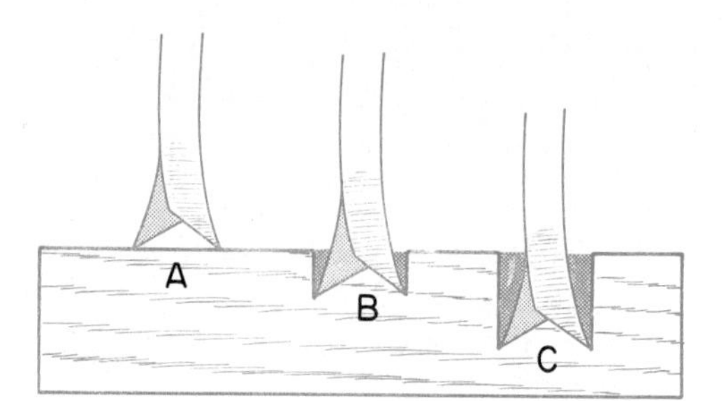

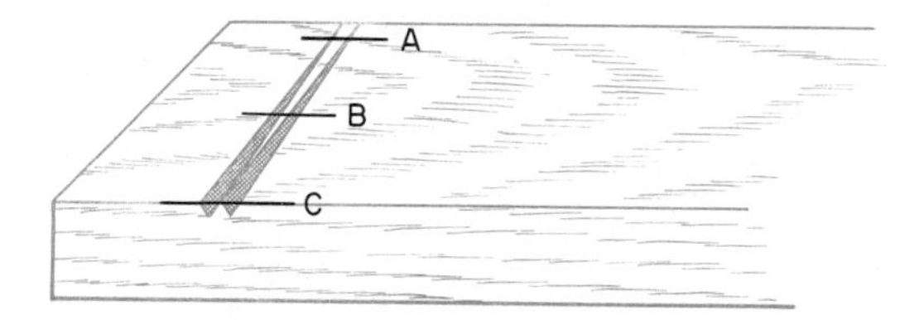

Fig. 8-8. The cutting action of a crosscut saw.

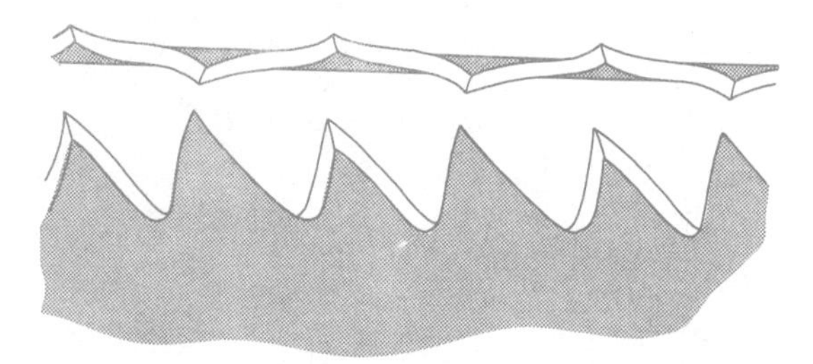

Fig. 8-9. The shape of crosscut saw teeth.

the teeth is referred to as *set*, Fig. 8-6. The set in the saw allows more stock removal in the separating process. The stock which is removed is referred to as the *kerf*.

Machine saws also require a blade that cuts a kerf wider than the blade. A circular saw blade which does not cut a wide enough kerf will overheat and scorch the wood. A circular saw blade may have teeth which are set like a handsaw or be hollow-ground, Fig. 8-7. A hollow-ground blade is thickest on the points of the teeth and thinner in the center.

DIFFERENCES BETWEEN CROSSCUT SAWS AND RIPSAWS

The difference between a crosscut saw and a ripsaw is the way the teeth are shaped. The teeth of a crosscut saw can be compared to a series of little knives. The outside edges of the teeth on a properly set saw shear the fibers on the sides of the kerf The center portion of the teeth removes the sheared fibers in the form of sawdust, Fig. 8-8.

Notice how the teeth remove the stock to form the kerf. Also ample clearance is provided due to the set.

The knife-like teeth on the crosscut saw are beveled on each side. The front of the teeth are shaped at a 15° angle to a line perpendicular to the row of teeth. The back of the teeth are shaped at a 45° angle to the row of teeth. The sides of the teeth are also filed on a bevel of about 25°, Fig. 8-9.

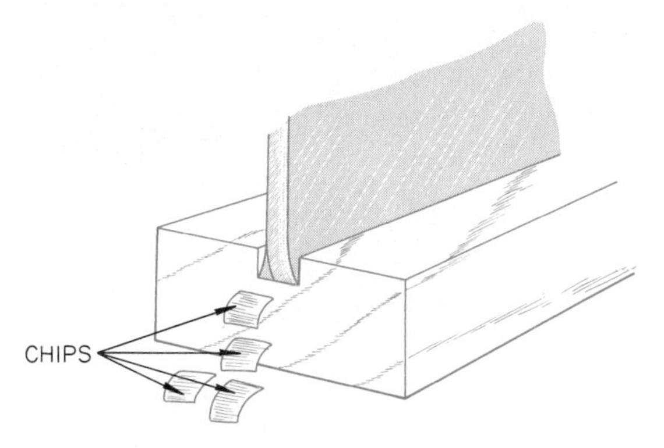

Fig. 8-10. Stock being removed from the kerf with a ripsaw.

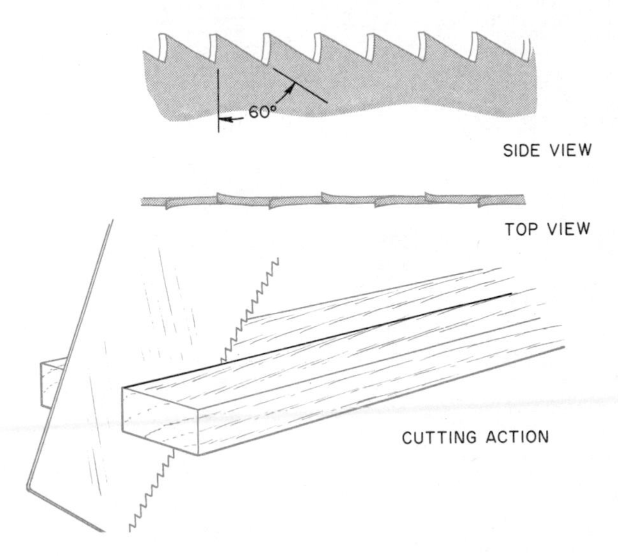

Fig. 8-11. The shape and cutting action of ripsaw teeth.

The ripsaw has teeth which are filed to cut like a series of chisels. The front of the teeth are straight across, and the wood fibers are chiseled from the stock as the board is separated, Fig. 8-10. The teeth of the ripsaw are set similar to those of a crosscut saw.

The angle of the ripsaw teeth is shown in Fig. 8-11. The front cut edge of the teeth are filed approximately straight across.

The principle of separating stock with a saw is similar regardless of the type of machine or tool used. Better work can be done in less time if the proper tool or machine is used for the job. The following pages outline the correct tool and technique to use for separating stock.

CUTTING WITH HANDSAWS

Remember, the type of saw to use will depend on the job. The crosscut saw is used to cut stock to length; the ripsaw is used to cut stock to width. However, before cutting in either direction, it is necessary to lay out the stock. After selecting stock suitable for the job, determine the best method for cutting. From your product plan, determine the most economical cutting layout. Look for any defects in the stock which will affect the layout. You will need to make allowances for cutting and smoothing of the stock.

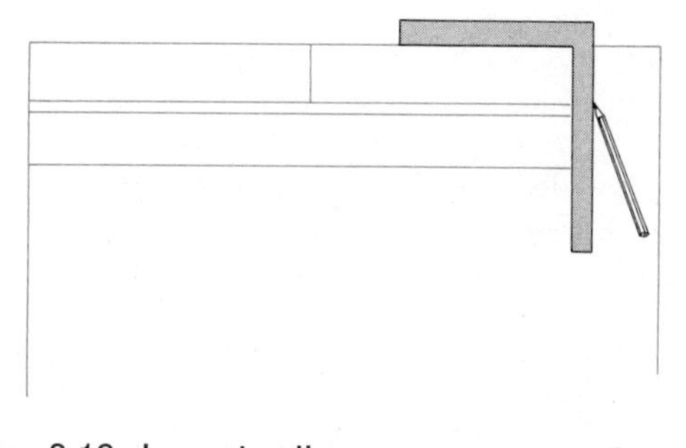

Fig. 8-12. Layout allowances are made for cutting stock.

LAYOUT FOR ROUGH CUTTING

The layout for rough cutting should include allowances for smoothing and planing as follows: (1) 1/8″ additional thickness; (2) 3/8″ additional on each edge; and (3) 1″ additional length. Figure 8-12 shows a typical cutting layout on the stock, which can also be done with a bench rule and chalk. The chalk can be removed easily if the cutting layout needs to be changed. It is helpful to label each of the parts as they are marked on the stock. Always make sure that the grain is running in the proper direction.

After a satisfactory rough layout is made, the final cutting layout is drawn with

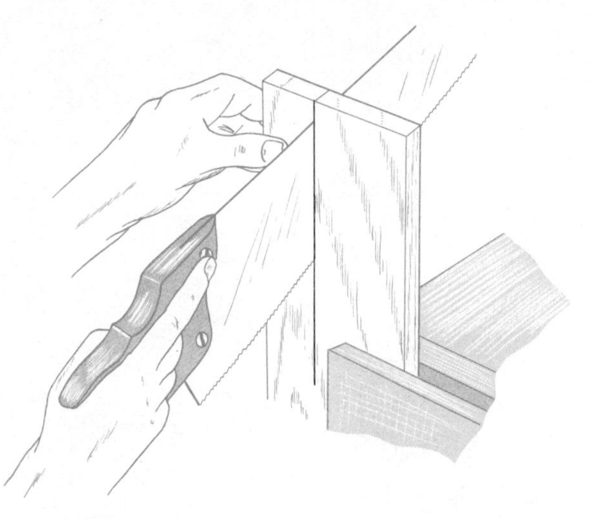

Fig. 8-13. Stock secured in a bench vise.

a framing square and a sharp pencil. Remember, when you cut a board a specific length, some of the stock is removed as sawdust. For this reason, make sure that the saw is cutting on the side of the line which is waste. If care is not taken, the result may be a board that is too short to use. An 8- or 10-point saw is used for cutting the stock to length. A ripsaw with 5½ points per inch is good for separating the stock to the proper width.

It is sometimes easier to support small pieces of stock in a vise to rip or crosscut. The board should be positioned up or down (vertically) for ripping. The line to be cut should extend to the left of the vise, Fig. 8-13.

CUTTING WITH A BACKSAW

The backsaw is used when a fine cut is desired. The teeth on the backsaw are filed as on a crosscut saw. The backsaw is a very fine-toothed crosscut saw. It usually has 14 points per inch. The blade is very thin, and the back is reinforced with a band of steel. It is called a backsaw because of the steel band, Fig. 8-14.

The backsaw makes a very smooth cut. It is used when an accurate cut is necessary.

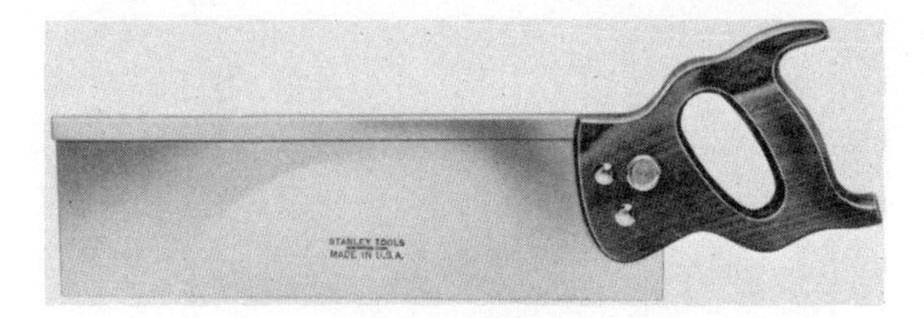

Fig. 8-14. A backsaw. (Stanley Tools)

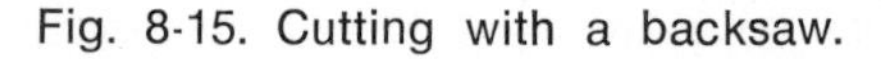

Fig. 8-15. Cutting with a backsaw.

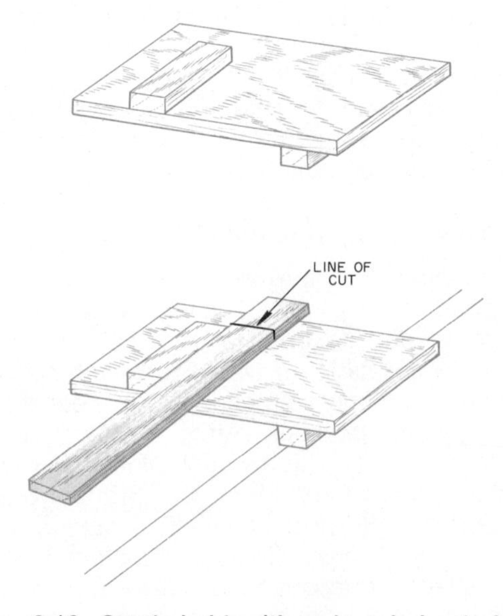

Fig. 8-16. Stock held with a bench hook for cutting with a backsaw.

The backsaw is commonly used for cutting joints and squaring stock, and is used in a horizontal position. The backsaw cut is started in the same manner as other handsaw cuts. The saw is then gradually lowered until it is in a horizontal position, Fig. 8-15.

Secure stock in position for cutting. The stock to be cut can be held in a horizontal position by several methods. The bench hook is often used to secure the work. In addition to holding the work, the bench hook also serves as a guide to aid in following the line, Fig. 8-16. The stop is equipped with a cleat on the top and bottom so it may be reversed for either a right- or left-handed person. The cleat is about 2″ shorter than the width of the board to which it is fastened. The cleat both holds the work and provides a guide against which the saw can be held.

For special work, or if a bench hook is not available, other methods can be used to hold the stock. It should be kept in mind that the extra time required to make a smooth, square cut will eliminate time otherwise needed for planing and squaring a rough cut. A clamp and a straightedge work well as holding devices to use with the backsaw, Fig. 8-17. Place scrap stock under the piece being cut to avoid damaging the bench top. The straightedge must be aligned with the cutting line so the cut or kerf will come out of the waste stock. When cutting with a guide, place pressure on the side of the saw with your free hand so it will cut squarely up and down as well as follow the line, Fig. 8-18.

It is also likely that the stock may be held in a bench vise to cut with a backsaw.

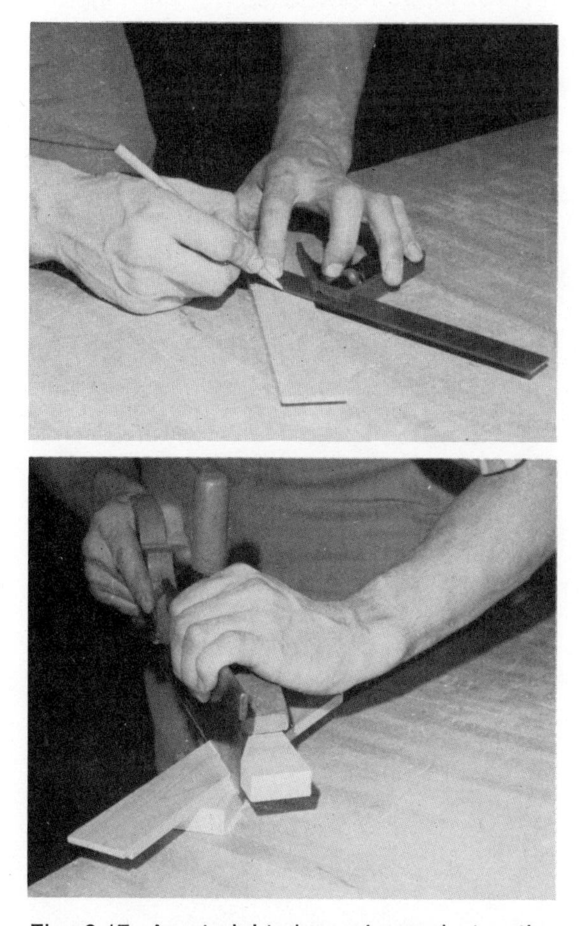

Fig. 8-17. A straightedge clamped to the stock serves as a guide for cutting with the backsaw.

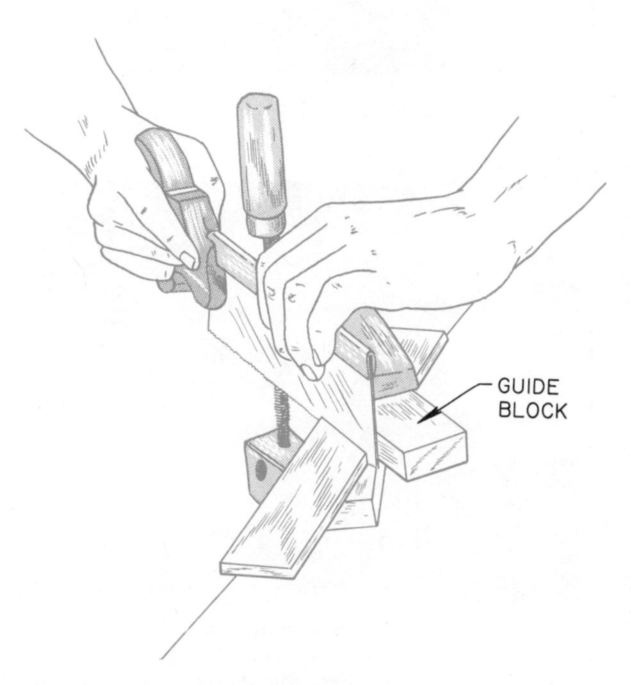

Fig. 8-18. Pressure is exerted against the side of the backsaw to hold it firmly against the guide block.

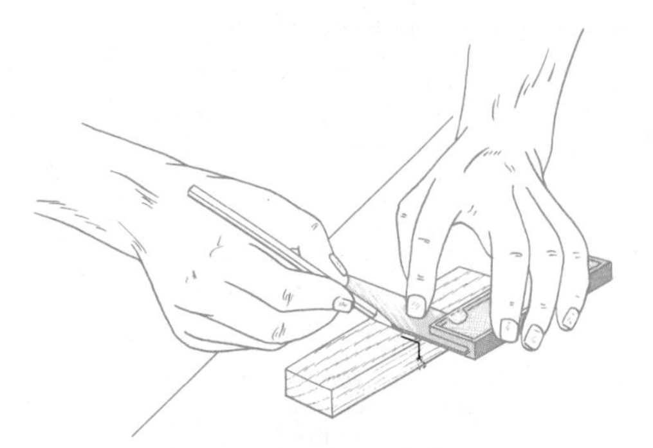

Fig. 8-19. The layout line should be made on the top and side of the stock to insure a straight cut with the backsaw.

A. The jig holds the stock in cutting position.

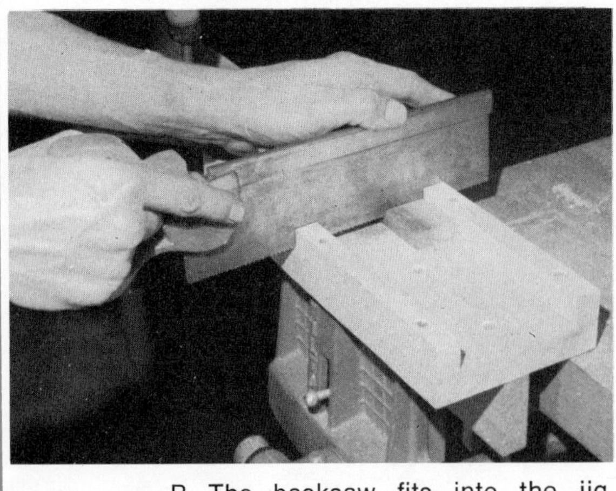

B. The backsaw fits into the jig and the cut can be made through the stock.

Fig. 8-20. A special cutting jig.

Regardless of the holding method, it is best to mark the stock on both the top and the edges. This is done with a sharp pencil and a try square. If the cut does not go completely through the stock, also mark the desired depth. Figure 8-19 shows a piece of stock being laid out for cutting. Notice that the layout is marked for a cut only one-half way through the stock.

The saw should follow the vertical lines on the sides at the same time it is following the horizontal line on top. This will assure you of a square, accurate cut.

When several pieces are to be cut with the backsaw, a simple jig will speed the operation and insure greater accuracy, Fig. 8-20. A jig is a device used for holding the work in place while cutting or forming and is sometimes used when assembling parts. A jig is usually built to serve a particular purpose, and can be constructed to reduce the amount of layout required when duplicate parts are made.

CUTTING WITH A DOVETAIL SAW

Dovetail saws are often used when a very smooth surface is needed on the end of small stock. The dovetail saw is similar to a backsaw in appearance, but is smaller, the teeth are finer, and the blade is thinner, Fig. 8-21. Due to the thin blade, the back is also reinforced with a band of steel. It is also used in the same manner as the backsaw. Most of the same cutting jigs and set-ups can be used with either a backsaw or dovetail saw.

CUTTING WITH A MITER BOX SAW

Many of the cuts made with a backsaw can also be made with a miter saw and miter box, Fig. 8-22. A miter saw is a large

Fig. 8-21. A dovetail saw. (H. K. Porter, Inc., Henry Disston Div.)

backsaw. Miter saws range in size from 24″ to 28″. The miter box may be either a simple wooden box or a commercial metal miter box, Fig. 8-23. The first type of miter box is satisfactory for limited use where a square cut or 45° angle is desired. The metal miter box has more uses because the cut can be adjusted to various angles. Common settings on a metal miter box are: 12°, 22½°, 30°, 36°, and 45° angles. The miter joint is often used in the construction of picture frames.

CUTTING IRREGULAR SHAPES

It is often necessary to cut irregular shapes in woodworking. The coping saw and compass saws are used for this purpose, Figs. 8-24 and 8-25. A saw with a thin

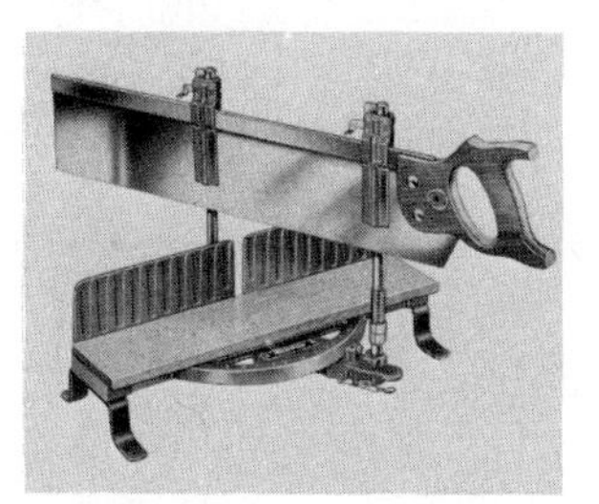

Fig. 8-22. A miter box and saw. (Stanley Tools)

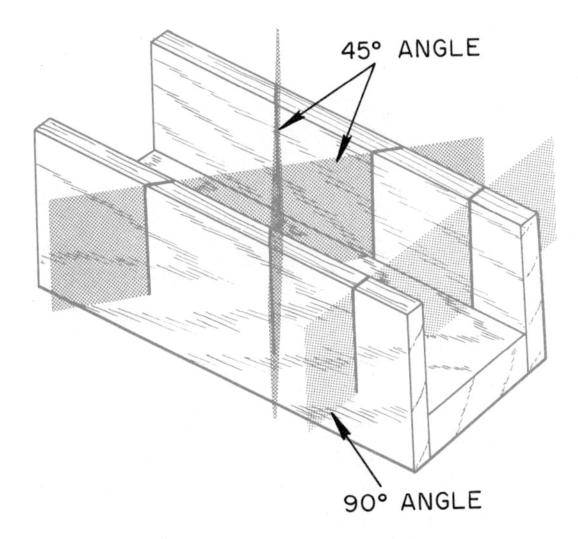

Fig. 8-23. A shop-constructed miter box.

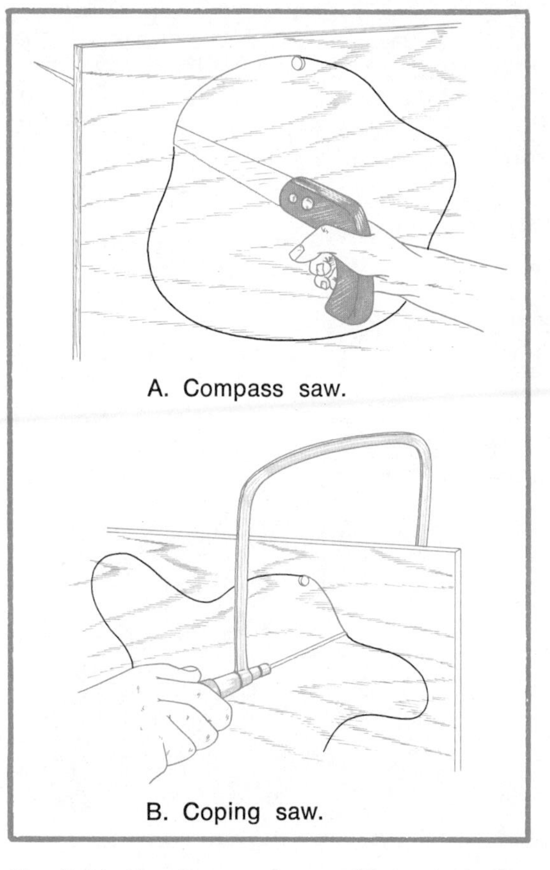

Fig. 8-24. Handsaws for cutting irregular shapes.

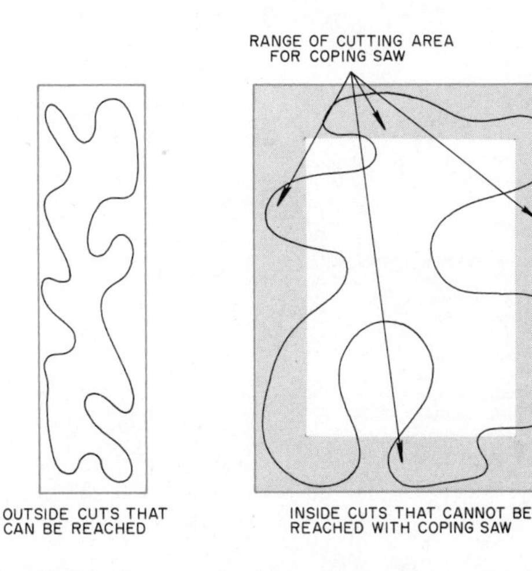

Fig. 8-25. Some inside irregular cuts are made with a compass saw because a coping saw has a limited range.

blade is used to follow the curved edges and sharp corners. A thin blade will turn more easily in the work than a wide blade. The design of a saw blade determines the purpose for which it was intended.

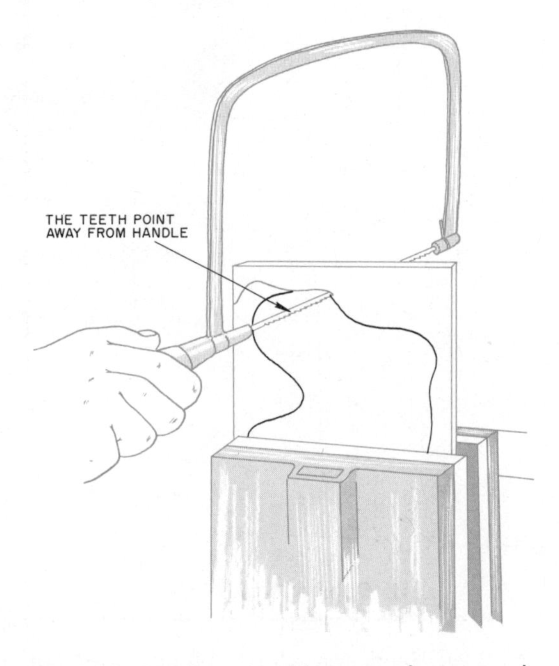

Fig. 8-26. Stock cut with a coping saw is held in a vise with the teeth of the saw pointing away from the handle.

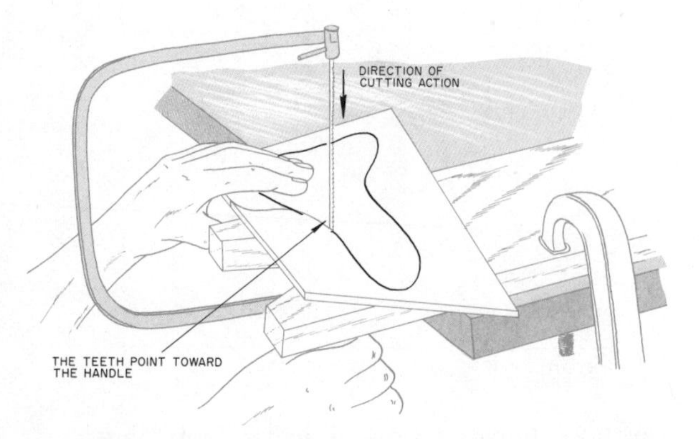

Fig. 8-27. Stock cut with a coping saw and a saw bracket.

CUTTING WITH A COPING SAW

The coping saw consists of a frame and a changeable blade. The blade is held in position by the spring tension created in the steel frame. The blade can be adjusted to cut at various angles to the frame. The handle contains an adjustment for changing the angle and adding tension to the blade. There are numerous types of blades which can be used in the coping saw. Most blades have fine teeth — a blade with 16 points per inch is good for general use. Thick stock can be cut with a blade with fewer teeth per inch. The teeth are shaped like those of a ripsaw.

The stock to be cut with a coping saw can be supported in a vise, Fig. 8-26. When the stock is held in a vise, the blade should be inserted in the frame with the teeth pointed *away* from the handle. The cut is made with the saw as you follow the line. Use light, even pressure on the cutting stroke.

A saw bracket can also be used for supporting work when cutting with a coping saw, Fig. 8-27. The blade is inserted in the frame with the teeth pointed *towards* the handle. The downward stroke is the cutting stroke. The pressure is released as the saw is moved to the up position. The saw handle is gripped with the right and the stock is supported by the left hand. The saw bracket is useful when cutting thin stock. As you cut around a sharp corner, use extra strokes and turn the blade slowly.

Inside cuts can be made by first boring a hole and inserting the blade through the hole. The blade is then replaced in the frame. Figure 8-28 shows the inside of a design being cut by means of a coping saw. After the cut is finished, remove the blade from the frame.

CUTTING WITH A COMPASS SAW

The compass saw is used to cut large circles and irregular shapes. It has a thin, tapered blade from 10″ to 14″ long. A hole is bored in the waste stock near the cutting line to insert the blade through the stock to start the cut. The blade is designed so the wide portion may be used for cutting large curves, and the narrow part of the blade is suitable for following short curves. The length of the stroke will vary depending upon the sharpness of the curve. Light pressure must be used when cutting to avoid bending the blade. The handle is designed so the blade may be reversed to allow for easier cutting in close quarters. The handle is twisted slightly to follow the layout on the work. The blade normally has about 8 points per inch for general use.

You now know there are many methods of separating wood. Some methods are better suited for a particular job than other methods. Whichever method is used, the process remains much the same. The quality of the job is dependent upon the selection and proper usage of the correct tool.

CUTTING WITH PORTABLE POWER SAWS

Technology is always changing. As a result of technology, new and improved methods of processing materials have evolved. Machines are gradually replacing hand tools for many tasks. This unit will discuss portable power saws commonly used to separate wood. A portable power tool is usually powered by an electric motor. The discussion in this unit will be limited to the portable circular saw and saber saw. Both of these tools have found wide acceptance in the processing of wood. These portable tools are capable of performing many of the separation processes which would otherwise be done with hand tools. As you become more knowledgeable of the wood industry, it will be evident that both hand tools and power tools are essential in wood processing.

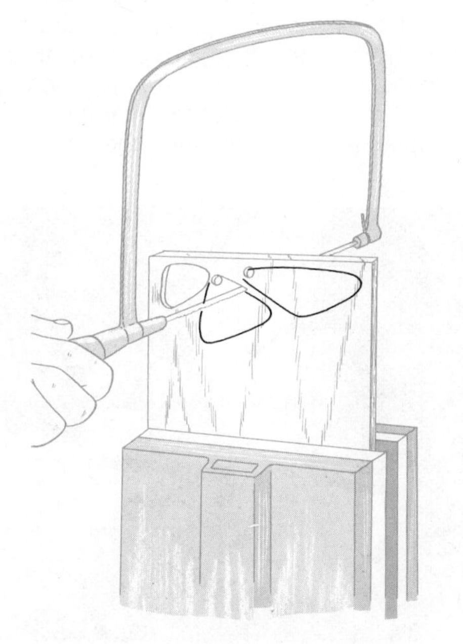

Fig. 8-28. Cutting inside curves with a coping saw.

PORTABLE CIRCULAR SAW

The portable circular saw has become widely accepted by carpenters in the construction industry. The tool has proven to be a great labor- and time-saver. It is basically used for rough framing in construction. The portable circular saw has limited use for making accurate finish cuts. You should become familiar with how the saw operates and its parts before you operate the tool, Fig. 8-29.

The size of the portable circular saw is determined by the maximum sized blade which can be used with the tool. Portable

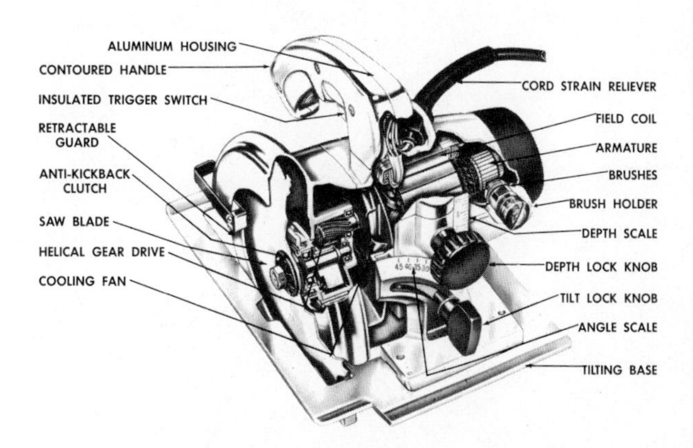

Fig. 8-29. Parts of the portable circular saw. (Power Tool Div., Rockwell Mfg. Co.)

circular saws range in size from 4½″ to 12″. The most common size for general use is 7″. With the blade at a right angle to the base, a 7″ saw will cut to a depth of 2 5/16″. When the blade is set at a 45° angle, the depth of cut is 1 13/16″. The motor ratings range from 1/16 horsepower to 1½ horsepower, depending on the blade size.

The portable circular saw cuts just the opposite of a hand saw: it cuts from the edge closest to the operator to the side furthest away; and the hand saw cuts from the furthest side toward the operator. In addition, the portable circular saw cuts from the bottom of the stock toward the top, and the hand saw cuts from the top to the bottom. The difference in cutting action and direction requires that the stock be placed differently for the two cutting methods. The stock is placed with the finished side up for the hand processes as the saw cuts from the top. With the portable circular saw, the stock is cut face side down, Fig. 8-30. This procedure avoids chipping on the finished side.

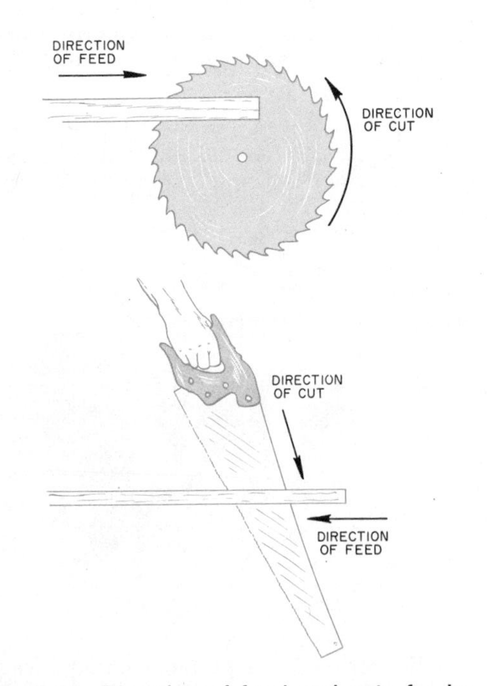

Fig. 8-30. Direction of feed and cut of a hand saw and a circular saw.

The portable circular saw has many uses — among the most common are crosscutting and ripping. In addition, the saw may be used for special cutting such as beveling and internal cutting. A combination saw blade is commonly used for all types of cutting. It combines the characteristics of both ripping and crosscutting blades. The combination blade reduces the need for changing blades for different types of work.

Operation of the portable circular saw. The saw is usually guided freehand along a layout line. The depth of the cut should be adjusted so that it extends about ⅛″ below the lower surface of the board. Crosscutting is one of the most common uses for the portable circular saw, Fig. 8-31. It can either be performed freehand or guides can be clamped to the saw or stock for this operation. After placing a layout line on the stock, position the work securely on a bench or sawhorse. Grip the saw firmly and place it on the layout line. Operate the trigger switch with the forefinger of the same hand. If the stock is securely clamped, the other hand can be used to hold the knob on the top-front of the saw. If stock is long, it may be held on the sawhorse with the knee and free hand. As the saw progresses through the work, the free hand may be needed to support the cutoff. Before starting, check all adjustments to be sure they are tight.

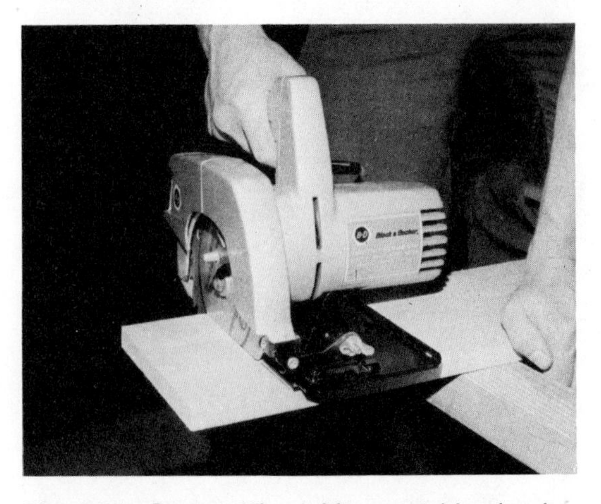

Fig. 8-31. Crosscutting with a portable circular saw.

For straight cutting, the adjustable saw base must be set at zero. Make sure the power cord will not become tangled as you are cutting. *CAUTION: Make all adjustments with the power disconnected.* The cut can be made with either a combination or cross-cut blade.

The accuracy with which the portable circular saw can be used for crosscutting is increased through the use of guides. A portable table such as shown in Fig. 8-32 also increases the accuracy of the saw. This device is especially helpful for beginners. With experience, the saw can usually be guided along a line for an accurate cut. The protractor guide can be set at the desired angle for making angle cuts across the grain.

The ripping operation can be performed with either a rip blade or a combination blade. The rip blade may cut faster and give a smoother cut. The stock being cut must be held securely in place for ripping. Sawhorses work well for most ripping operations, Fig. 8-33. The stock must be supported so that it does not bend downward from the weight of the saw. Caution must be taken not to cut into the sawhorses as you cut. The stock may be ripped by placing a line on the stock as in crosscutting, or a guide may be used, Fig. 8-34. The saw base is usually equipped with an adjustable ripping fence. The guide can be used if the edge of the stock is straight and parallel to the layout line. If a rip fence cannot be used, a straightedge can be clamped to the surface of the stock. The base of the saw will follow along the straightedge. The cutting procedure is basically the same for ripping and crosscutting. The stock may

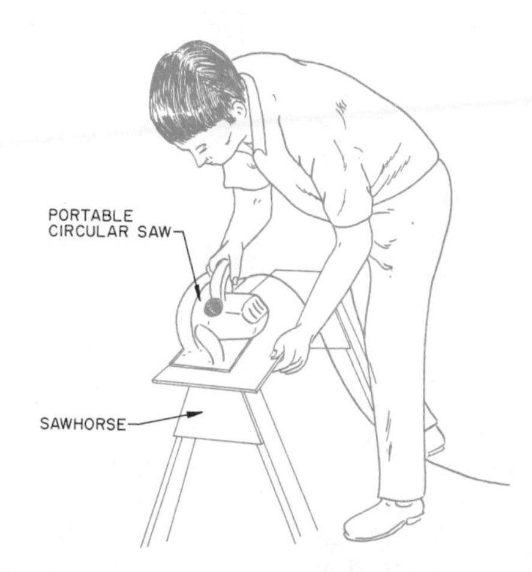

Fig. 8-33. A saw horse used to support stock for cutting operations.

Fig. 8-32. Crosscutting with the portable circular saw mounted in a special table.

Fig. 8-34. Ripping stock to a layout line.

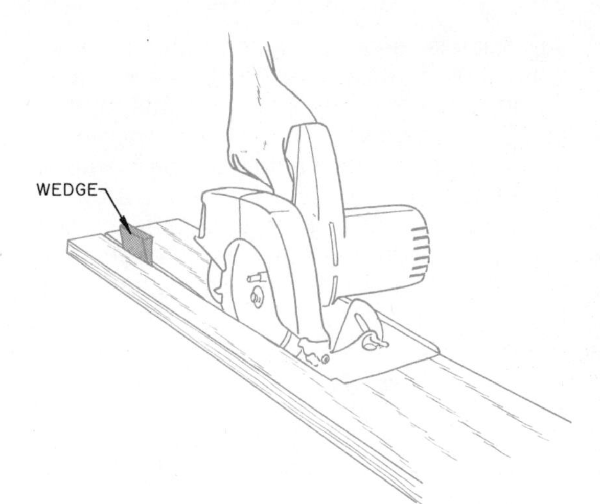

Fig. 8-35. Use of a wedge in the kerf of stock being ripped.

Fig. 8-36. A portable circular saw being used to cut rafters. (Milwaukee Electric Tool Co.)

tend to pinch close on the saw kerf during the ripping process. When this happens, a small wedge can be placed in the kerf, Fig. 8-35.

The portable circular saw is often used to cut material which is already assembled and fastened. For example, roofing and sheathing is sometimes cut to length after it is nailed in place. Rafters are sometimes cut with the portable circular saw, Fig. 8-36.

Safety considerations for the portable circular saw. Regardless of the type of cut being made with the portable circular saw, caution must be taken for its safe operation. In addition to other safety rules, observe the following when operating the portable circular saw:

1. After completing the cut, hold the saw until the blade has stopped rotating.
2. When changing blades and adjusting the saw, make sure the power cord is disconnected.
3. The stock must be supported well.
4. The depth of the blade should be 1/8″ greater than the thickness of the stock.
5. Guide the saw with both hands whenever possible.
6. The blade should be sharp and have the proper amount of set.
7. Wear goggles to protect the eyes from flying chips.
8. Allow the saw to gain top speed before starting the cut.
9. Do not bind the saw in the stock, as a "kickback" may result.
10. Use the correct blade for the particular type of work.

If these safety rules are followed, sawing accidents can be eliminated. Remember — accidents are *generally* the result of incorrect tool usage.

CUTTING WITH A SABER SAW

Another type of portable electric saw which has found wide acceptance is the *saber saw.* The saber saw is often referred to as a bayonet saw, Fig. 8-37. The basic parts of the saw are shown in the picture. The saber saw is normally used for light work for separating irregular shapes. The saber saw cuts with an up-and-down stroke. The length of the stroke for different models may vary but most fall into a range of about 1/2″. The cutting action occurs on the upward stroke, therefore the finish side of the

Fig. 8-37. A saber saw. (Power Tool Div., Rockwell Mfg. Co.)

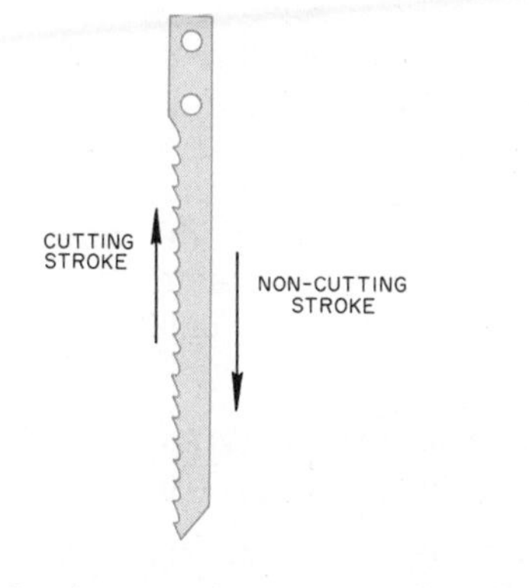

Fig. 8-38. Cutting action and direction of teeth on the blade of a saber saw.

stock being cut should be placed down. This will prevent chipping and splitting on the finish side. The blade is placed in the saw with the teeth pointing towards the base, Fig. 8-38.

The blade of the saber saw makes about 4500 strokes per minute. There are various types of blades which may be used for cutting wood. The number of teeth per inch range from 6 to 12. A blade with 10 teeth per inch is a satisfactory all-purpose blade for general use. The thickness of the stock is one factor used to determine the number of teeth per inch on the blade. There should be at least four teeth in contact with the wood during the separating operation. The operator's manual should be referred to before mounting blades in the saw. The method for mounting the blades

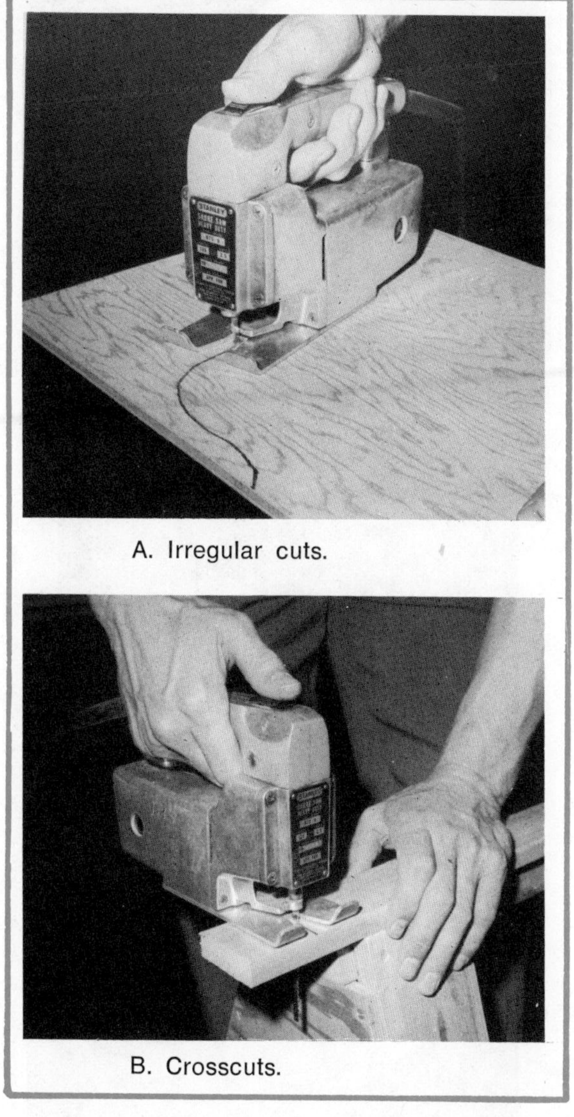

A. Irregular cuts.

B. Crosscuts.

Fig. 8-39. Cuts made with a saber saw.

will vary slightly from model to model. The operator's manual will also give routine maintenance procedures.

The saber saw can be used freehand to cut along a layout line. In addition, guides can be used to insure greater accuracy. When cutting circles, a guide can be used. For cutting irregular shapes such as curves, the saw is usually guided freehand along a layout line, Fig. 8-39.

Fig. 8-40. Cutting a bevel with a saber saw with a tilting base. (Power Tool Div., Rockwell Mfg. Co.

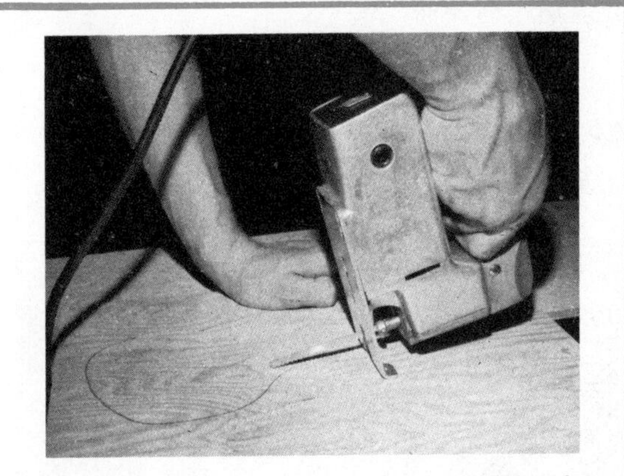

A. Starting the cut by tilting the saw.

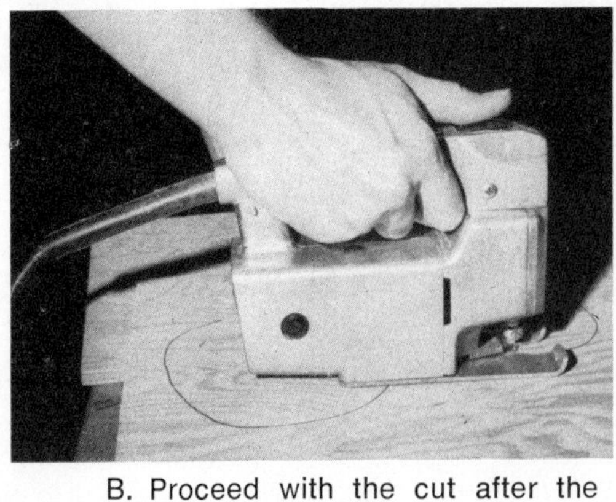

B. Proceed with the cut after the entire base rests on the stock.

Fig. 8-41. A plunge cut with a saber saw.

The base on most saber saws tilts to the desired angle for making bevel cuts, Fig. 8-40. A fence may be used to cut a straight line parallel to an edge. The fence attaches to the base of most saws. It is also possible to clamp a straightedge to the stock to aid in cutting a straight line.

The saber saw can also be used for internal cutting on flat stock. It is sometimes referred to as *plunge cutting.* It is possible to start the cut without first drilling a hole to make a plunge cut, Fig. 8-41. The saw should be tilted on the front of the base with the blade clearing the stock. Start, and gradually lower; the base as the hole is cut. After the entire base is in contact with the stock, proceed with the cut. Make sure that the saw cut is started in the waste stock. On some material it may be advisable to drill a hole to start the intended cut.

The saber saw is often used by carpenters, plumbers, and electricians to cut holes for the installation of utilities, Fig. 8-42.

SEPARATING STOCK WITH STATIONARY POWER SAW

Stationary power tools are those to which the stock is moved rather than moving the machine to the stock. Stationary machines are designed and used for produc-

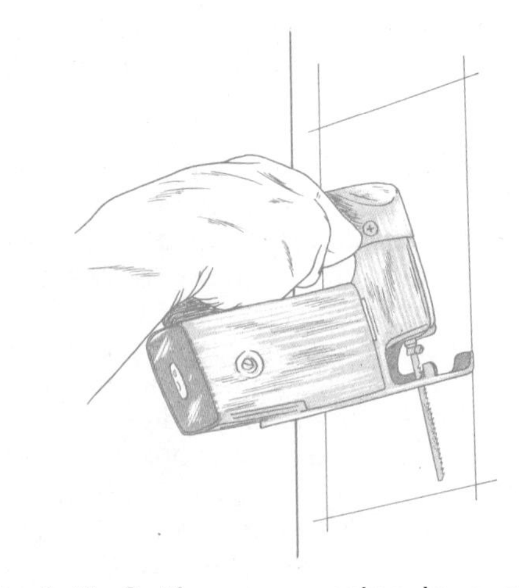

Fig. 8-42. Cutting an opening in a subfloor for a register.

tion work. Usually stationary tools will provide more accuracy for cutting.

JIGSAW

The jigsaw is one of the simplest and safest power tools for separating stock. The jigsaw blade moves up and down, which is called reciprocation action. The jigsaw is mostly used for cutting irregular shapes. It is often called a scroll saw, Fig. 8-43. The size of the jigsaw is given as the distance from the blade to the overarm. This distance is called the throat opening. The 24″ model jigsaw is a very popular size. A 24″ jigsaw should cut to the center of a 48″ circle. The cutting speed can be varied by means of the four-step cone pulley. The speeds usually vary between 600 to 1800 strokes per minute. The motion is converted from rotary to reciprocating motion by means of a pitman, Fig. 8-44. The lower chuck is driven up and down by the pitman. The blade is held tight by means of a spring in the upper chuck attached to the tension sleeve and bar, Fig. 8-45. The tension sleeve consists

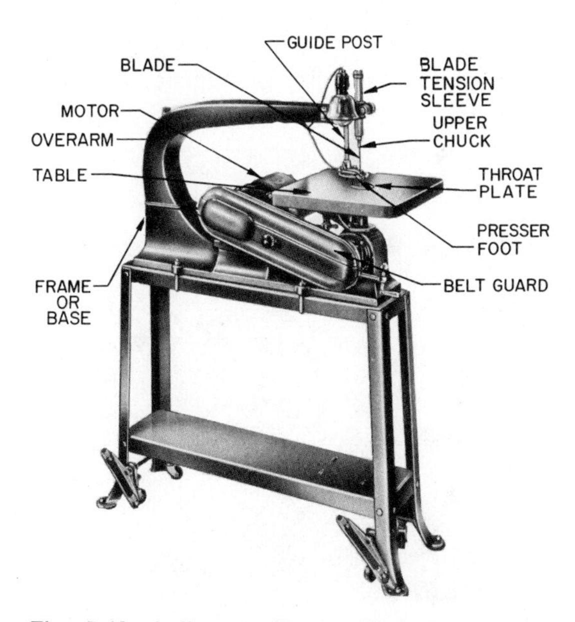

Fig. 8-43. A jigsaw. (Power Tool Div., Rockwell Mfg. Co.)

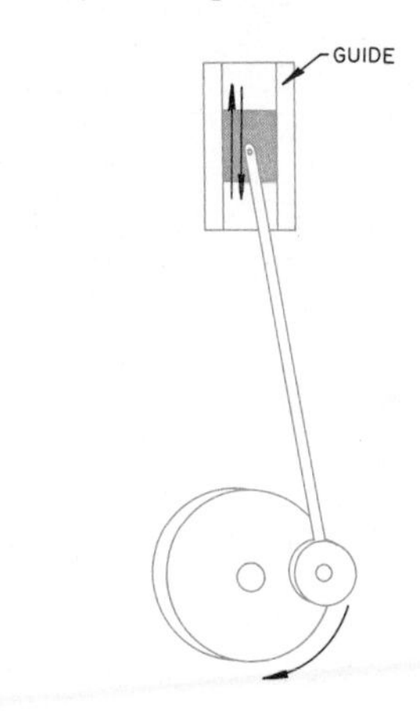

Fig. 8-44. Pitman drive mechanism of a jigsaw.

A. Upper chuck.

B. Lower chuck.

Fig. 8-45. Blades held in position in the jigsaw by means of the upper and lower chucks.

of a box which moves up and down with each stroke of the lower chuck. This sleeve can be moved up and down to provide proper tension on different length blades.

The tables on most jigsaws can be tilted to provide for bevel cutting. The blade passes through a hole in the table from the lower chuck to the upper chuck. A slotted aluminum disc called a throat plate or table insert is placed in the hole after the blade is installed. The guide assembly includes a spring hold-down. This hold-down prevents the stock from lifting off the table with each upward stroke during the cut.

Blades are normally held in position in the jigsaw by means of an upper and lower chuck, Fig. 8-43. The chucks are designed to allow the blade to be clamped either at a right angle to the overarm frame or in line with the frame. It is also possible to remove the overarm on many models of jigsaws to increase the capacity of the machine. With the overarm removed, a lower blade guide is installed to support the blade, Fig. 8-46. A sabersaw blade is then used in the jigsaw.

Installing blades in the jigsaw. Jigsaw blades are available in 3″, 5″, and 6″ lengths. The most common length is 5″. The number of teeth per inch will range from 7 to 20. On thin stock, a blade with more teeth per inch is required. Thicker stock should be cut with a blade having less teeth per inch. A general rule is that there should be three or four teeth in contact with the stock while it is being cut.

The throat plate is first removed when installing a blade in the jigsaw. Turn the saw by hand to position the lower chuck in its highest position. Place the blade in the center of the lower chuck with the teeth *pointing downward.* Tighten the thumbscrew with the fingers — *do not use pliers to tighten the chuck.* The blade should extend into the chuck about 1/2″. Make sure that the blade is straight up and down, Fig. 8-47.

Lower the upper chuck and place it over the blade. Place tension on the blade by raising the tension sleeve, Fig. 8-48. The chuck should be about 1/2″ to 3/4″ below the tension sleeve when the lower chuck is in its highest position. The saw should be turned by hand to check the blade tension.

Since the blade is narrow and flexes very easily, it must be supported by the guide assembly. The guide assembly prevents the blade from bending backwards or to the side when cutting pressure is applied. The guide support consists of a roller support and a slotted disc or adjustable guide

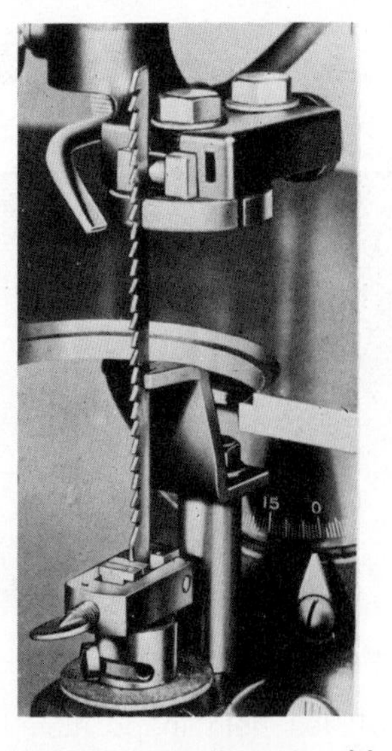

Fig. 8-46. A jigsaw equipped with a saber cutting attachment.

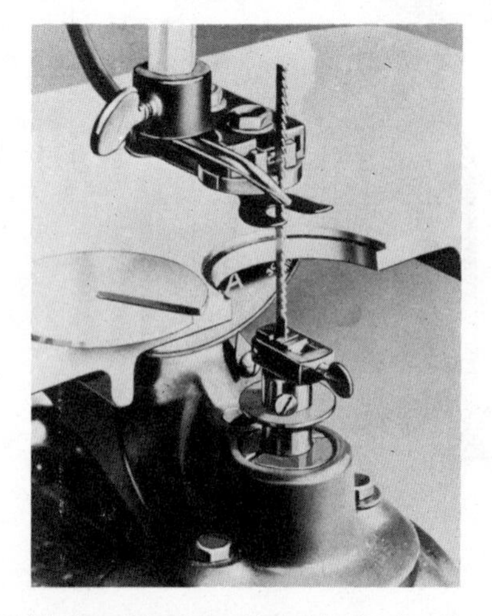

Fig. 8-47. Positioning the blade in the lower chuck of the jigsaw.

blocks, Fig. 8-49. The roller support should make very little contact when the machine is turned by hand. The side guides should be set so that the blade runs free. Lower the guide post and apply light pressure to the stock with hold-down spring. If the blade is placed too deeply into the guide supports, the set will be knocked out of the blade. The adjustment and maintenance procedures may differ slightly with different models, and the operator's manual should be consulted for specific adjustments.

Cutting with the jigsaw. The layout lines should be made on the stock with a sharp pencil. A template will be helpful for the layout if several pieces are to be cut in the same shape and size. The irregular cuts are normally the last operations after the stock has been squared and smoothed. A blade of the proper size should be selected for the work. After the machine is properly adjusted, push the work into the blade with light, even forward pressure. Start the cut in the waste stock and advance towards the layout line. If the machine is properly adjusted, you will not need to force the work into the blade. *Never* cut the entire line away, but stay outside of the line. The line will be needed as a reference line for smoothing after cutting. If the curves are not too sharp, you can follow all the way around in one continuous cut, Fig. 8-50. If

Fig. 8-48. Placing tension on the blade of the jigsaw.

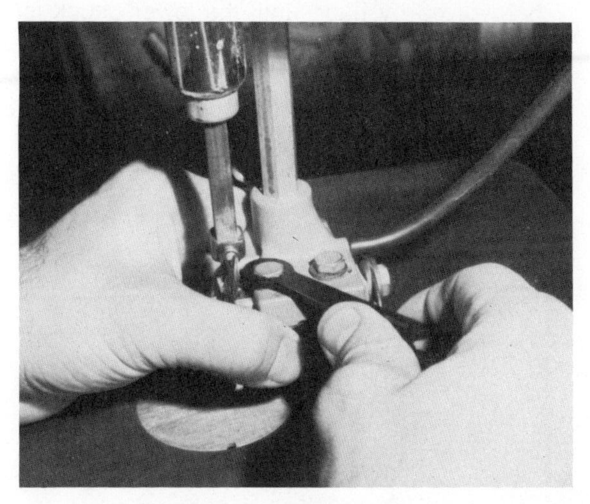

Fig. 8-49. Guide assembly adjustment on the jigsaw.

Fig. 8-50. Cutting a long, even curve on the jigsaw.

the layout has sharp curves, it will be necessary to make relief cuts. This will prevent blade breakage and the need to back out of long cuts, Fig. 8-51. It is helpful to drill small holes for cutting sharp curves. The holes will provide room to turn and change direction.

For internal cutting, a hole must be drilled or bored in the waste stock through which the blade will extend. The blade is loosened from the top chuck. The machine is revolved by hand until the lower chuck is in the lowest position. The blade is then placed through the hole in the stock and replaced in the upper chuck, Fig. 8-52. The adjustments for internal cutting and external cutting are the same. If numerous internal cuts are to be made, a saber attachment can be used at the same time. The saber sawing technique uses only the lower chuck under the table and a special guide attachment, Fig. 8-53. A shorter, heavier blade is used for saber cutting to prevent its bending when the cutting pressure is applied.

Many kinds of work can be performed on the jigsaw that cannot be done on other machines. The table of the jigsaw can be set at an angle to cut a bevel on either straight or irregular work. The table is often tilted to do inlay work (also referred to as parqueting). Two contrasting colors of wood are fixed together for inlaying. A brad in each corner in the waste stock is often used, Fig. 8-54. The design is placed on the top layer. A very small hole is drilled to pass the blade through the stock. A blade such as a jeweler's blade is used to reduce the kerf size. With the table set at a slight angle,

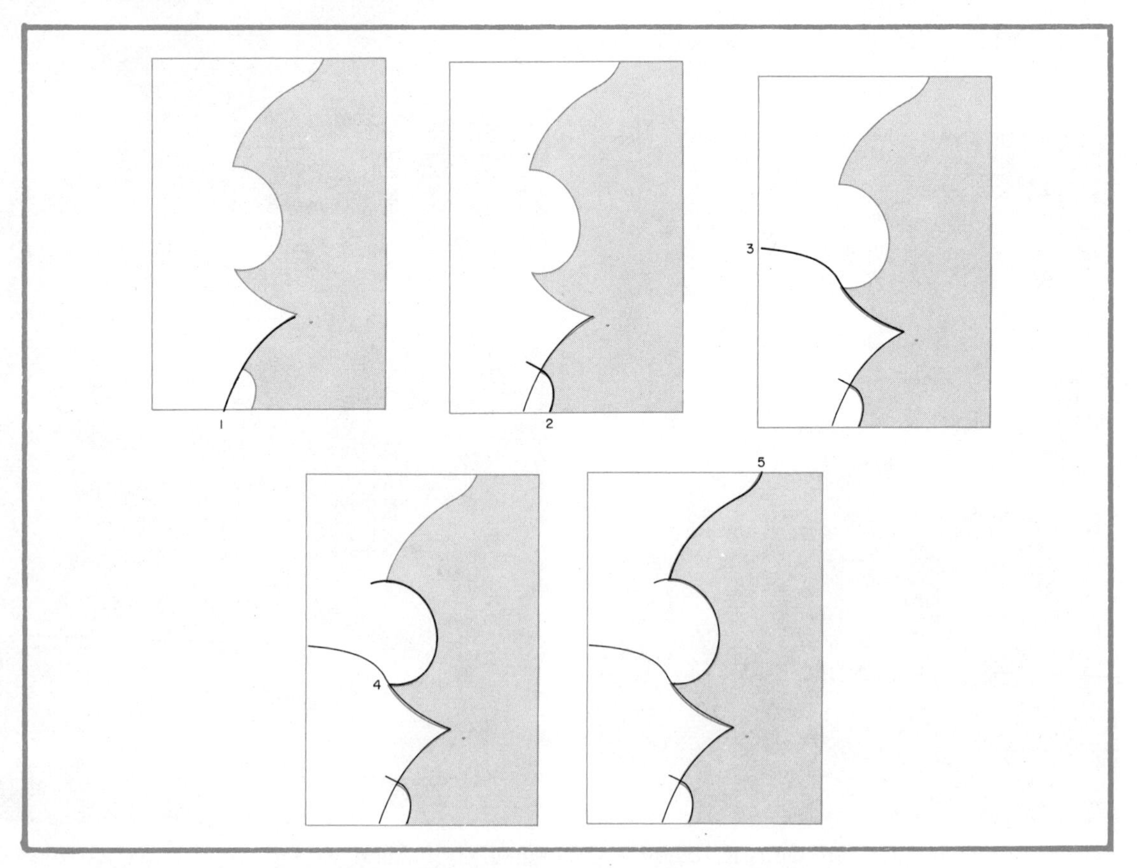

Fig. 8-51. Relief cuts simplify the cutting of sharp curves.

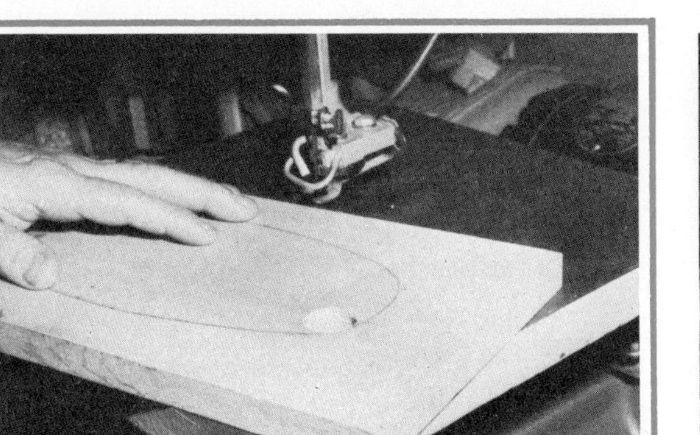

A. A hole is drilled in the waste stock near the layout line.

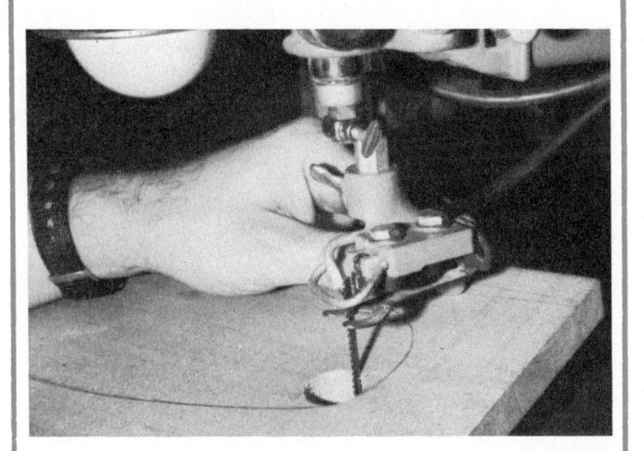

B. The blade is inserted through the hole and retightened in the upper chuck.

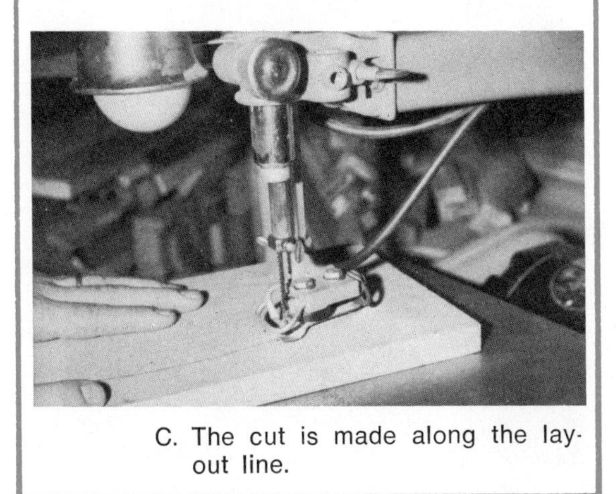

C. The cut is made along the layout line.

Fig. 8-52. Internal cutting with a jigsaw.

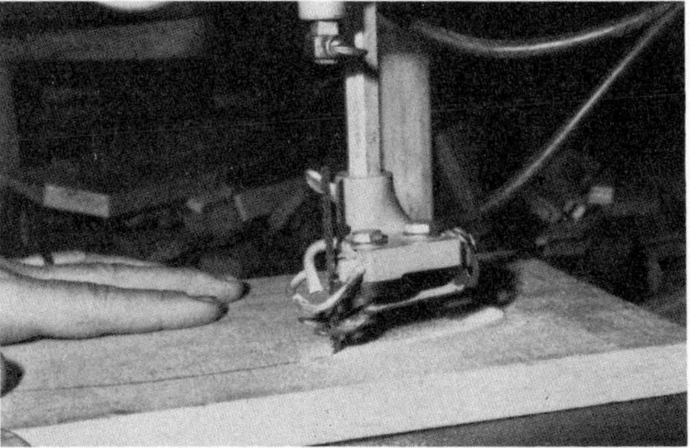

A. Blade is held in the lower chuck only.

B. A special attachment guides the blade during the cut.

Fig. 8-53. Saber sawing on the jigsaw.

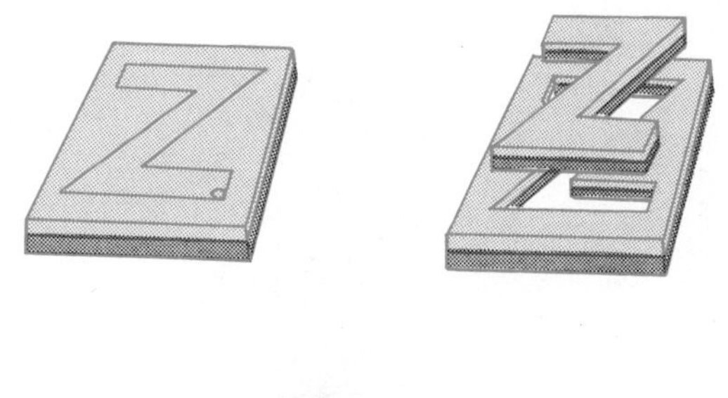

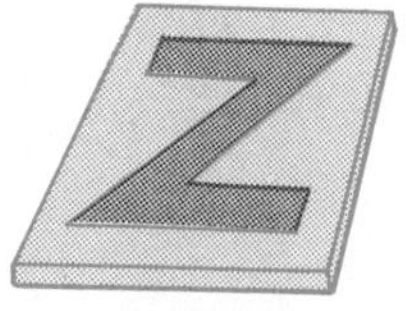

Fig. 8-54. Placement of stock for a simple inlay.

proceed with the cutting operation. All of the cutting must be done on the same side of the blade. When the cut is complete, the pieces will interchange. The angle at which the stock was cut will cause the kerf to close when the pieces are interchanged.

BAND SAW

The band saw, Fig. 8-55, consists of two vertical wheels spaced apart on which a flexible steel band travels. The band of steel has ripsaw-shaped teeth filed on the edge. The wheels are covered with rubber tires on which the continuous band travels. The wheels are equipped with adjustments and guides that cause the blade to track or remain in alignment. The band saw is designed for cutting irregular shapes and curves. The band saw is used for heavier work than is commonly performed on the jigsaw. The band saw is often used for some types of straight cutting.

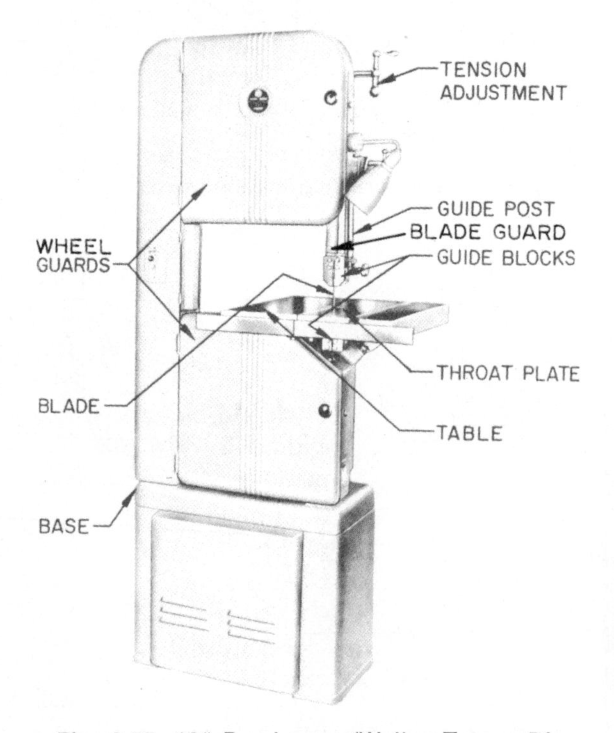

Fig. 8-55. 16″ Band saw. (Walker-Turner Div., Rockwell Mfg. Co.)

The power source is connected to the lower wheel. The upper wheel is adjustable to vary the blade tension and to track the blade. The sliding bar, Fig. 8-55, moves up and down and is equipped with a ball bearing blade support and guide block. A roller support and set of guide blocks are also attached beneath the table. The guide blocks and roller supports give the needed support to the blade during the cutting process, so they must be correctly aligned. This adjustment procedure is included later in this chapter under blade installation. The table can be adjusted to any desired angle for bevel cutting, and equipped with a ripping fence and a miter gauge, Fig. 8-56.

The size of the band saw is determined by the diameter of the wheels, ranging from 10″ to 42″. The most common general-purpose machines range from 14″ to 24″. The size of the table is proportionate to the wheel size — a 14″ saw would have a table approximately 14″ square. The saw table can be tilted to a 45° angle in one direction. Bevel and diagonal cuts can be made with the table tilted. The cutting speed for band saws is expressed as surface feet per minute (SFM). If the wheel diameter and revolutions per minute (RPM) are given, the SFM can be determined with the following formula:

$$SFM = \frac{D \times 3.14 \times RPM}{12}$$

Fig. 8-56. A fence and miter gauge in place on a band saw.

Some band saws are equipped so the speed may be controlled. The SFM should be in the range of 1500 to 3000 for cutting wood.

Blade selection. The proper selection and installation of the blade is essential for proper operation. Blades are available in widths including ⅛″, 3/16″, ¼″, ⅜″, ½″, ⅝″, and ¾″ for general-purpose use. Blades can be purchased in the correct length with the ends welded together for a particular machine. They can also be purchased as a continuous roll in lengths of 100′ or more. The rolls are then cut to the desired blade length and the ends are welded together to form a blade. A special welding machine as shown in Fig. 8-57 is used to connect the ends. The width of the blade is determined by the nature of the work: a narrow blade is used to cut sharp curves; and a wider blade is best for straight cutting. If a blade is too wide for cutting small curves and circles, the blade may be damaged during the operation. A band saw blade is likely to break if it is twisted or bent. Figure 8-58 shows the correct width blade to use for cutting a given radius. The thickness of the blade should be in relationship to the diameter of the wheels, that is, the saws with smaller wheels will require a thinner blade. The constant flexing of the blade over a wheel will tend to fatigue the metal if a thick blade is used on a small wheel.

The two most common types of band saw blades for cutting wood are the standard blade and the skip tooth, Fig. 8-59. The skip tooth is a hardened steel blade, with the appearance of every other tooth missing. This is why it is called a skip tooth blade. General-purpose band saw blades for cutting wood will have 4 or 5 points per inch.

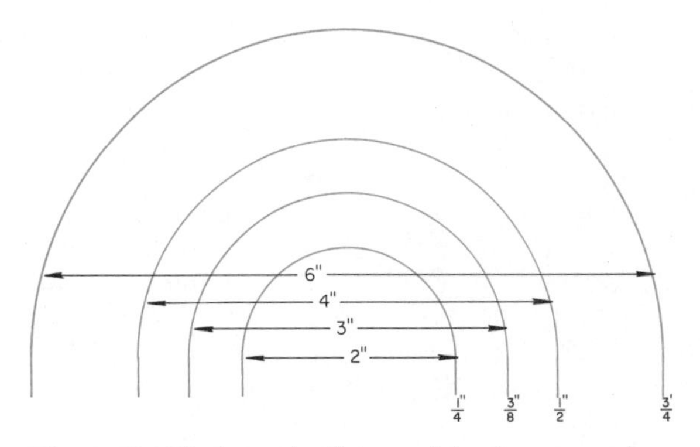

Fig. 8-58. Blade selection guide for curved and irregular cutting.

Fig. 8-57. Band saw welding machine.

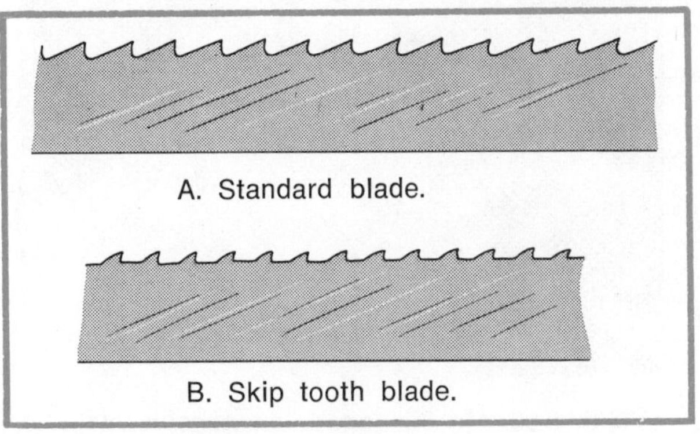

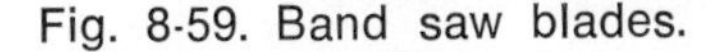
Fig. 8-59. Band saw blades.

Sometimes it is necessary to cut a short curve with a wide blade. The danger of damaging the blade can be reduced by making relief cuts, or by making a number of tangent cuts, Fig. 8-60.

Installation of blades. Before replacing or changing a band saw blade, make certain that the power is disconnected. Open or remove the upper and lower wheel guards, release the blade tension by turning the tension control knob, and loosen the upper and lower guide blocks and move the roller bearing supports back from the blade. Then remove the alignment pin from the front of the table and throat insert plate. Carefully remove the blade from the wheels. Place the replacement blade on the wheels, making certain that the teeth point downward toward the front of the table. Tighten the tension control adjustment to place slight tension on the blade. Most machines have an indicator for tension for specific blade widths. Install the throat plate and table alignment pin.

Revolve the machine by turning the upper wheel by hand. Adjust the tracking mechanism until the blade runs on the center of the wheels, then move the guide blocks to a position slightly behind the gullets of the blade. The blade should have about 1/64″ clearance between the blade and guide blocks on each side. Move the roller support wheels to within 1/64″ of the back of the blade. A piece of heavy kraft wrapping paper can be used as a spacing material, Fig. 8-61. Revolve the machine by hand to check all clearances and alignments. Replace the safety shields. *Check with your teacher before operating the machine.*

The blades are stored in coils when not in use, usually in a safe, hanging position. The procedure shown in Fig. 8-62 can be used for coiling the blades.

Common band saw cuts. The band saw can be used for many types of common cuts. There is no danger of a kickback. Rough stock can be cut to width and length. When working with warped stock, place the stock so it will not rock on the table. It is best to place the concave side of the stock on the table to prevent blade damage. The cuts may be made freehand for rough cutting stock. Much of the warp can be removed from stock by ripping it into narrower pieces on the band saw. The blade guide should not be in excess of ¼″ above the stock. The proper placement of the guide gives additional support to the blade and reduces the danger for the operator. It is best to stand to the right of the blade, support the stock in the right hand, and feed it with the left hand. Always make sure the machine has gained top speed before starting the cut. The rate of feed is determined by the kind and size of stock, blade, and the operating speed of the saw.

Fig. 8-60. Making short curve cuts with a wide blade.

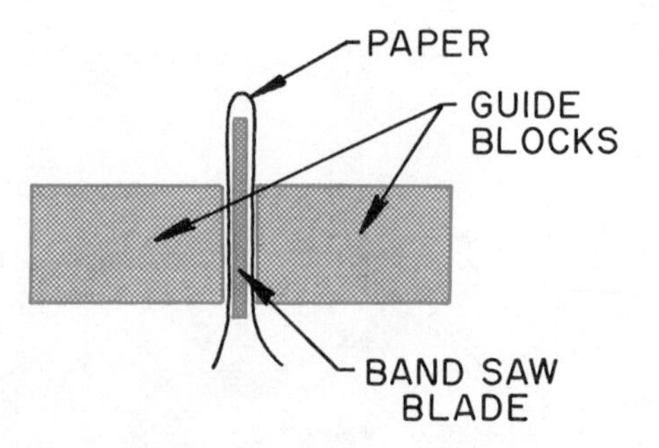

Fig. 8-61. Paper can be used to set the clearance for proper blade alignment.

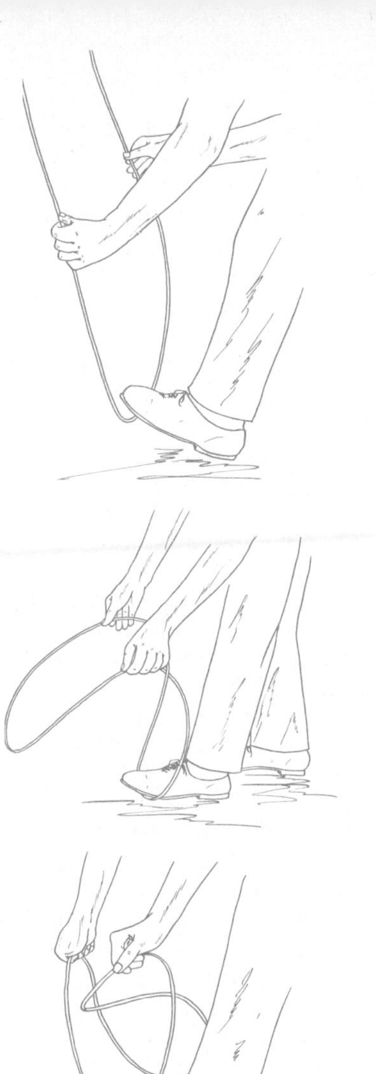

Fig. 8-62. Coiling band saw blades for storage.

Straight cuts on the band saw. The band saw miter gauge should be used for crosscutting, and the fence used for ripping. Before cutting with either the fence or miter gauge, the stock must be surfaced flat on at least one edge and face surface. To insure accurate cutting, the blade must be sharp with proper set and proper alignment.

The band saw will not produce a perfectly smooth cut. It is necessary to leave an allowance for smoothing the surface. Always cut on the waste side of the stock, leaving the layout line on irregular stock as a reference for smoothing operations.

The method for cutting an irregular shape should be planned before the cut is started. Planning will avoid the possibility of being trapped into backing the blade out of the cut, Fig. 8-63. Always make the short cuts, then the long cuts. This reduces the need to back the blade out over long distances.

A narrow cut can be done in several ways. It can be simplified if a hole is bored near the bottom of the cut, part A in Fig. 8-64. Most of the stock can be removed by

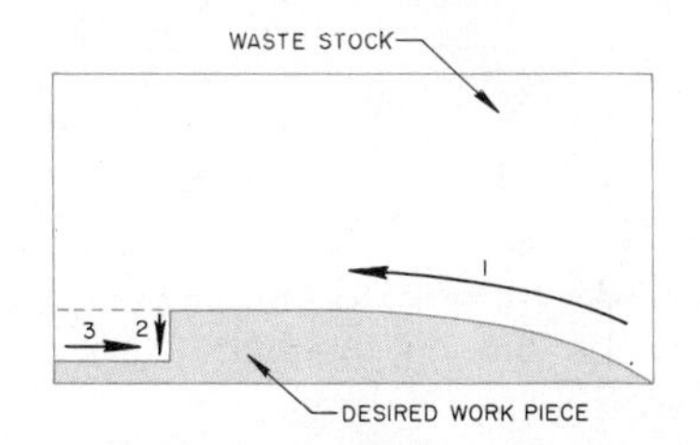

Fig. 8-63. Plan the cutting sequence prior to making cuts on the band saw.

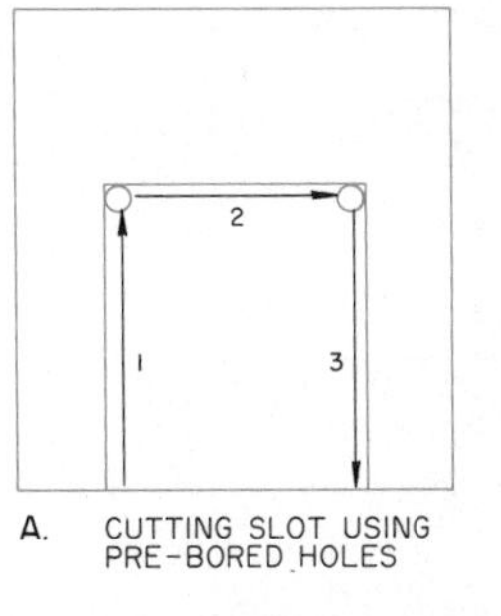

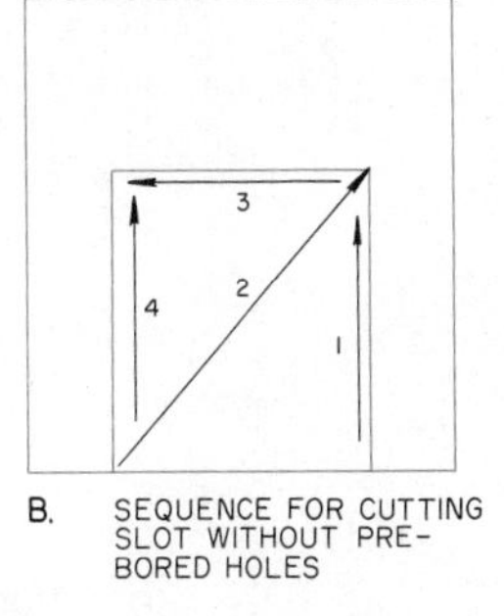

Fig. 8-64. Cutting a square slot on the band saw.

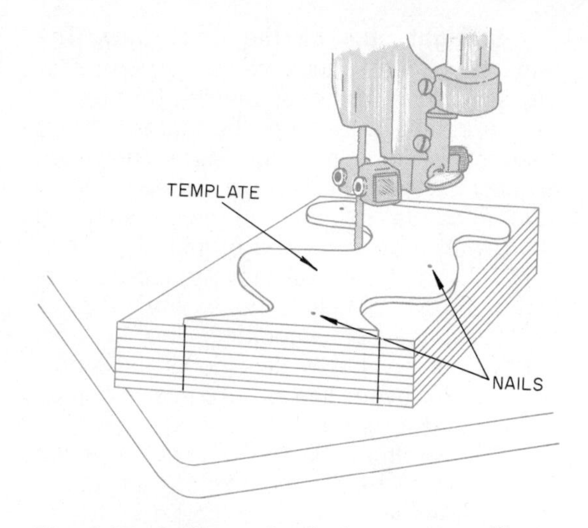

Fig. 8-65. Cutting duplicate parts with a template.

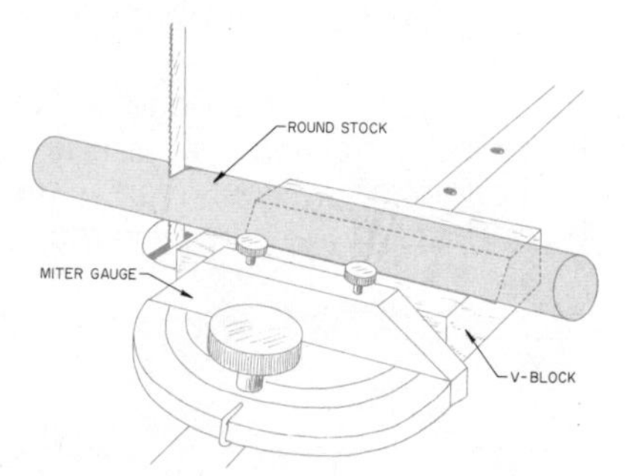

Fig. 8-66. A V-block to hold round stock to be cut on the band saw.

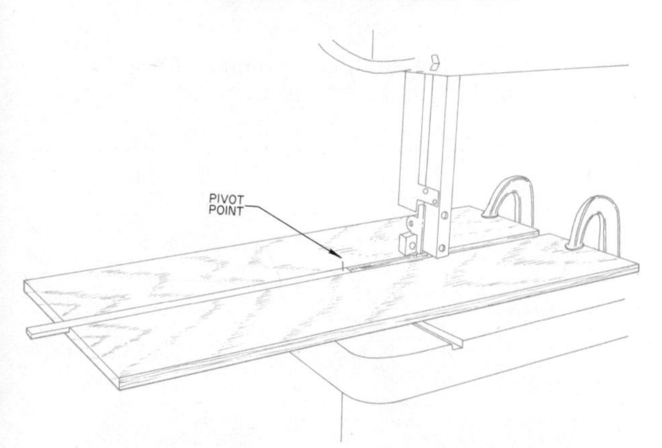

Fig. 8-67. An adjustable jig for cutting duplicate pieces of round stock on a band saw.

making two cuts and the remainder can be removed with several nibbling-type cuts. When the hole is not drilled, the cut can be made as shown in part B of Fig. 8-64. The bottom portion of the slot must also be removed with nibbling cuts.

Cutting duplicate parts on the band saw. The time required to cut duplicate parts can be reduced by cutting several at one time. This will reduce the number of separate layouts and insure uniformity between parts. The design will determine which method to use.

When the pads (stacks of stock to be cut) are held together with nails, they are placed in the waste stock away from the cutting line, Fig. 8-65. If the product is designed to have holes in it, the parts can be predrilled and padded together for cutting with dowels. The box provides not only a means of holding the pad together, but also serves as a template to eliminate the need for repeated layout.

Duplicate pieces can also be produced by first cutting the shape on a thick piece of stock. The stock is then ripped to the desired thickness.

Cutting round stock on the band saw. Round stock is often cut on the band saw. To avoid the danger of the round stock turning, the stock is held in a V-block, Fig. 8-66. If numerous pieces of the same size are to be cut, a jig can be constructed, such as shown in Fig. 8-67. A square piece of stock with a hole to accommodate the desired size of rod is used. The jig is held against the cutoff gauge to insure a square cut. If a jig is not available, the round stock can be clamped in a hand screw.

Cutting circles on the band saw. Circles can be cut on the band saw freehand, but the freehand work is usually shaped and smoothed by sanding. When a number of circular pieces are needed, the use of a *circle jig* will simplify the job, and produce the desired number of pieces more accurately, Fig. 8-68. The circle jig consists of a pivot on which the work is placed and advanced into the blade. The jig is constructed with plywood and serves as an auxiliary table clamped to the table of the machine. The jig can be made adjustable by adding a slide on which the pivot joint is attached. To cut a perfect circle, the pivot point must be at a right angle to the blade and aligned with the cutting edge of the teeth. The adjustable beam slides in a groove cut in the plywood table. Permanent cleats will insure the correct position of the jig on the table. A cleat can also be cut to fit in the miter gauge slot to insure alignment. A pin or screw tip is placed as a pivot at the end of the sliding beam. A flat head screw placed close to the beam will lock it in the desired position. The pivot pin is moved and locked at a point equal to the radius of the desired circle. The stock is placed on the pivot and rotated slowly into the blade.

Cutting compound shapes. Stock which is cut to an irregular shape from two or more sides is called compound cutting. Certain types of furniture legs are shaped by compound cutting on the band saw. Carvings and wood sculptures are rough-cut to shape. The design is drawn on two sides, Fig. 8-69. After the design is transferred to the stock, one side is cut. The waste stock which was removed is replaced either by spot-gluing or nailing it back in place. Nails must be placed so they will not interfere with the cut.

Resawing on the band saw. Resawing is the process of reducing stock held on edge into two or more pieces. The band saw is used because the blade is thin and reduces waste. A guide or pivot block such as shown in Fig. 8-70 is needed to insure an even and uniform cut. The guide block is clamped to the saw table the desired distance from the blade. A pivot block allows the operator to turn the stock if the blade leaves the line. The guide support should clear the work by about ¼″.

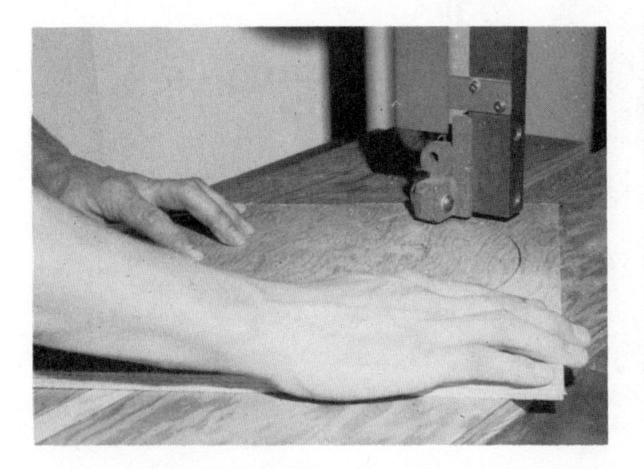

Fig. 8-68. Cutting circles with a jig on the band saw.

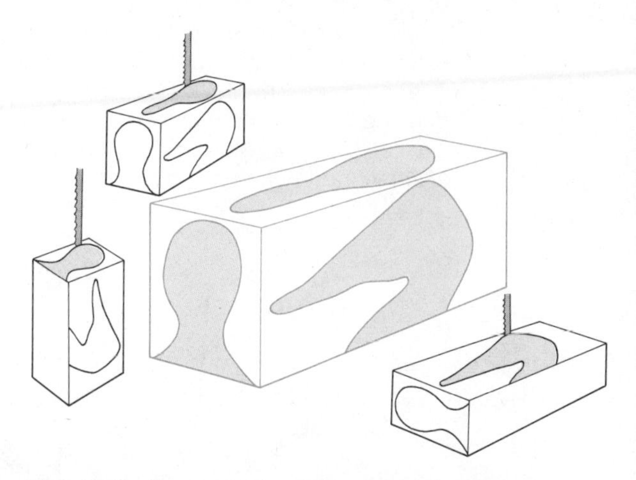

Fig. 8-69. Compound sawing on a band saw.

Fig. 8-70. Resawing stock using a pivot guide block.

Safety considerations when operating the band saw.

1. Disconnect the power to make adjustments and change blades.
2. Keep guards and wheel shields in place when operating the machine.
3. Revolve the machine by hand after adjusting or changing blades.
4. Use the correct blade for the job.
5. Be certain the blade is sharp and has proper set.
6. The electrical installation should be grounded.
7. A push stick should be used when passing stock close to the blade.
8. Do not twist the blade when cutting a small radius.
9. Allow the saw to gain top speed before starting the cut.
10. Plan the cutting procedure *before* starting the cut.
11. The upper guide support should be ½″ above the stock for cutting.
12. Use a V-block or other holding device to cut round stock.
13. After shutting the saw off, remain at the machine until the blade has stopped.

CIRCULAR SAWS

The circular saw, also called a table saw, is one of the most essential and versatile separating machines for wood. Due

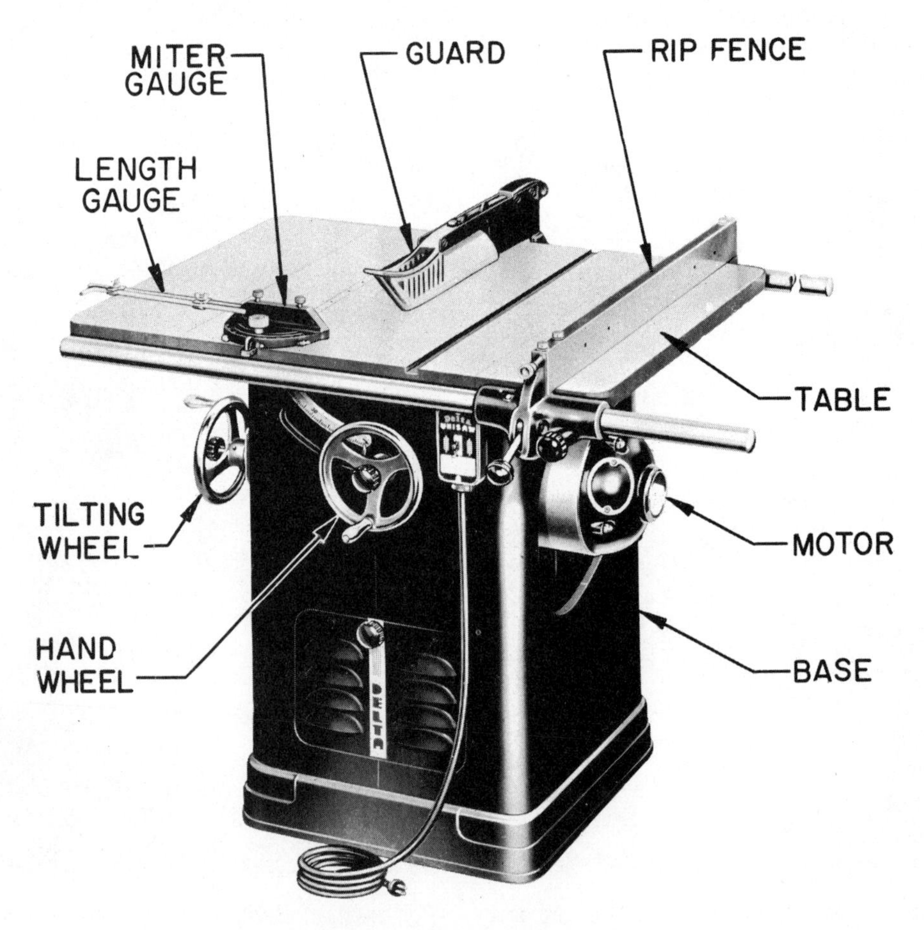

Fig. 8-71. Tilting arbor table saw. (Delta Div., Rockwell Mfg. Co.)

to its versatility in performing a wide range of operations, the circular saw is sometimes called a variety saw. The two most basic operations are crosscutting and ripping. Figure 8-71 shows a 10″, tilting arbor circular saw and the operating parts of the machine. The basic parts consist of an arbor, motor, and blade attached to a yoke-type frame. The yoke can be raised, lowered, and turned at an angle for bevel cutting. A hand wheel located on the front of most machines raises and lowers the blade for the depth cut. A ripping fence clamps to the graduated guide bar located on the front of the table. The rip fence guides the stock for ripping cuts, and the miter gauge used for crosscutting slides in a parallel slot on either side of the blade. The table is equipped with a throat plate through which the blade projects. The throat plate is removed when changing blades and it is replaced by a special insert when the dado head is used. The arbor is a shaft usually having left-hand threads and two flange collars. The blade is mounted between the collars. The arbor nut which holds the blade and collars in place is turned clockwise to loosen. The flange collars give additional support to the blade. The saw blade is tilted to an angle by tilting the arbor in the yoke mounting.

The circular saw should also be equipped with a guard and splitter, Fig. 8-72. These are designed to protect the operator from the hazard of injury due to contact with the blade or from a kickback. A *kickback* occurs when stock is thrown out of the saw by the blade. The guard is a basket which covers the exposed blade, and the splitter is a piece of steel attached directly behind the blade. It projects above the table and prevents the kerf from closing and binding the blade. The splitter is usually equipped with anti-kickback fingers which grip the wood to prevent it from being thrown back toward the operator, Fig. 8-72.

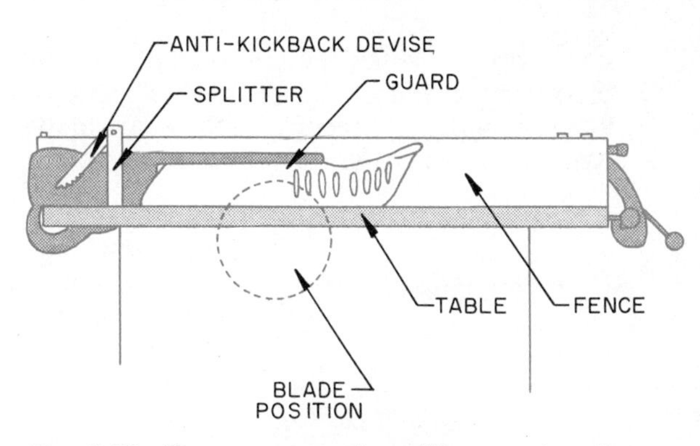

Fig. 8-72. The saw guard, splitter, and anti-kickback device.

Sizes and types of circular saws. The size of a circular saw is determined by the largest blade the saw arbor will carry. Saws are available in a wide range of sizes from 8″ to 20″. The 8″ - 14″ saw is satisfactory for most shop and production work.

Most modern circular saws have tilting arbors. That is, the table remains horizontal and the blade is adjusted to an angle. A less expensive model may have a tilting table. On this model, the blade remains vertical and the table is tilted to the desired angle. Bevel cutting can be performed on either style. The tilting arbor is more convenient for the operator and provides greater accuracy.

Some saws, called universal saws, are equipped with two arbors. A blade is mounted on each arbor, but only one blade can be raised into cutting position at a time. With a universal saw, the operator can switch from ripping to crosscutting without changing blades.

Selection of circular saw blades. The circular saw is used mainly for ripping and/or crosscutting. If a production operation calls for ripping of stock, rip blades are installed on the saw. The same is true of a production operation of crosscutting stock — crosscut blades are installed. However, when both types of cutting are performed on the same machine during a production process, a *combination* blade is often used. A combination blade combines both crosscut teeth and rip teeth on one blade, eliminating the need for constant changing of blades. Figure 8-73 shows the difference in the shape of the teeth on the three common types of blades. A rip blade has teeth which are filed straight across similar to small chisels, whereas the teeth of a crosscut blade are filed to a point.

The rip and crosscut blades are usually spring-set to provide clearance as the blade advances into the wood. To set the blade, alternate teeth are bent in opposite directions. Combination blades are hollow-ground and tapered from the outer edge to the center. Either method of providing clearance results in a saw kerf wider than the blade. For some types of work requiring a smooth cut, a *planer* blade may be used. The planer blade resembles the combination blade in appearance. It produces a very smooth cut when ripping, crosscutting, and mitering. The planer blade can greatly reduce or eliminate sanding, thus it should be used only when a high-quality finish cut is desired.

Special purpose blades are also available, Fig. 8-74. The teeth on the plywood blade are designed to reduce splitting and chipping and produce a smooth cut. Carbide tip blades are used in the wood industry for production cutting. The carbide

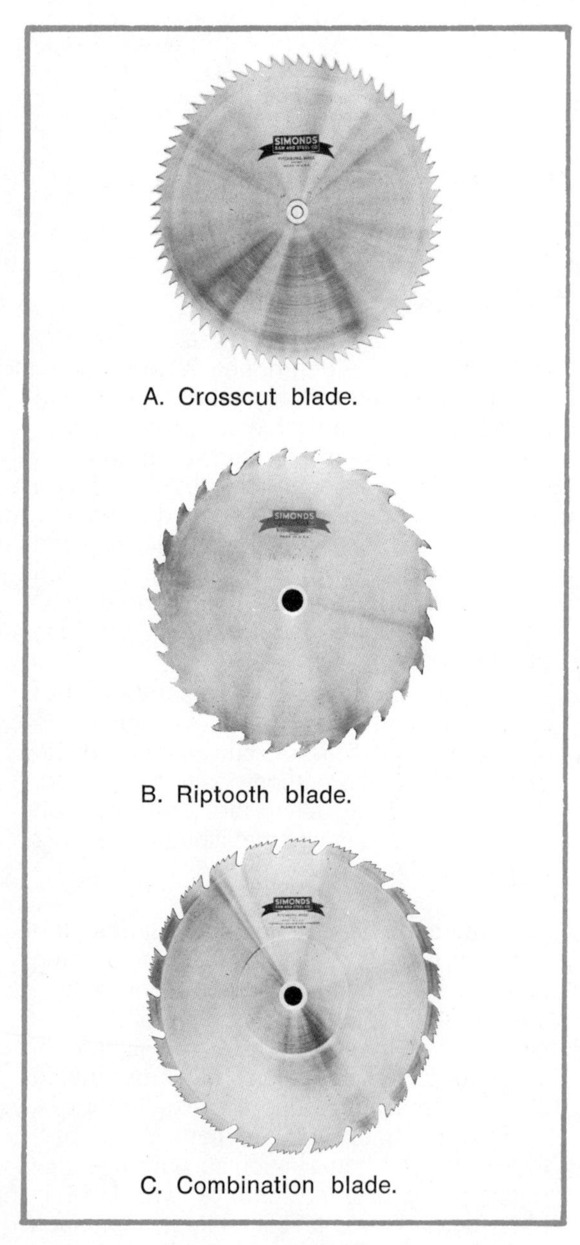

A. Crosscut blade.

B. Riptooth blade.

C. Combination blade.

Fig. 8-73. Circular saw blades. (Simonds Saw & Steel Co.)

A. Carbide tipped blade. B. Plywood blade.

Fig. 8-74. Special purpose blades for the circular saw.

tip blade will remain sharp much longer than ordinary blades. They are not affected by materials such as hardboard and plastic laminates which dull other blades.

The thickness (gauge) of a blade is determined by the diameter — larger blades are thicker to prevent vibration. If a blade is excessively thick, it will produce an excessively wide kerf and result in wasted stock. The number of teeth in blades also varies. A blade with a greater number of teeth will give a smoother cut, but must be fed slower.

When purchasing a saw blade, consideration should be given to the kind, arbor size, gauge, diameter, and the number of teeth.

Conditioning of saw blades. The quality of work produced on the circular saw will be affected by the condition of the blade. A dull blade is dangerous to use, as it can cause a kickback, or because it is dull, it may overheat and warp. Blades should be kept free of pitch and gum, which is more difficult with dull blades. The buildup can be removed with prepared pitch remover, kerosine or fine steel wool.

Most saw teeth can be filed to restore the original cutting ability of the blade, but are normally sent to an experienced craftsman for sharpening, Fig. 8-75.

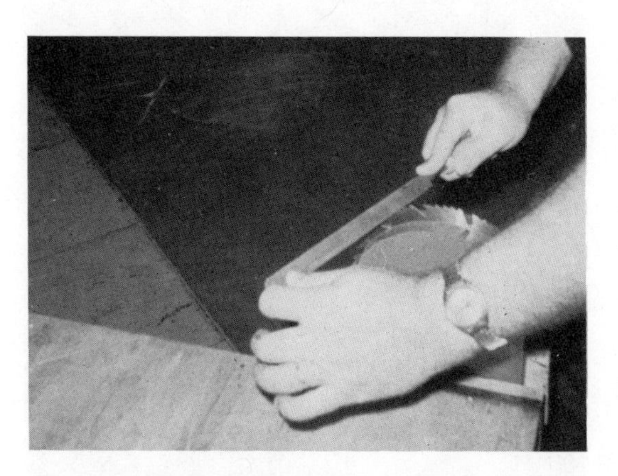

Fig. 8-75. Skilled craftsmen recondition saw blades which have become dulled from usage.

Changing circular saw blades.

1. Shut off the power leading to the saw.
2. Remove the throat plate from the table.
3. Raise the blade to its highest position.
4. Hold a block of wood against the teeth of the blade. The block may be wedged between the blade and the table, Fig. 8-76.
5. Loosen the arbor nut by turning it in the same direction the blade revolves when operating. Most saws are equipped with a left-hand thread on the arbor. Use the correct size wrench to remove the arbor nut, flange, and the old blade. Do not lay the blades on the saw table, as the blade may be damaged.

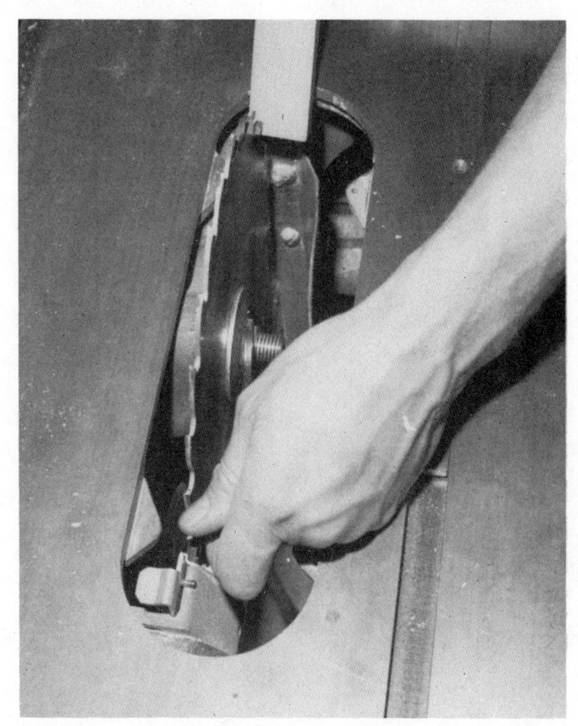

Fig. 8-76. Removing a blade from a circular saw.

6. Place the new blade on the arbor, with the trademark on the blade in the upward position. Replace the flange and arbor nut. The points of the teeth point toward the operator.
7. Tighten the arbor nut with the wrench. Hold a block of wood against the back of the blade to lock the blade while tightening.
8. Replace the throat plate.

Crosscutting on the circular saw. Crosscutting is the process of separating a board across the grain using a crosscut or a combination blade. Before stock can be cut on the circular saw, one edge must be straight and one surface must be flat. The miter gauge is used to cut a board to length on the circular saw. If only a few cuts are to be made, a layout line is drawn on the board with a square. After drawing the line, mark the side of the line which will be cut. An "X" placed on the waste side of the cutting line will indicate the side of the line on which to cut. The miter gauge is usually placed on the left side of the blade. The blade should extend ¼" above the stock. The miter gauge should be checked for alignment with the blade. A framing square placed against the blade and the miter gauge as shown in Fig. 8-77 can be used to set the blade and miter gauge at a 90° angle to each other. The square should be in contact with two teeth on opposite sides of the blade. The teeth used for the alignment must point in the same direction. Place the stock on the table and align the line to be cut with the blade. The fence should be removed, or at least moved completely out of the way. Start the saw motor and allow it to come to top speed. Hold the stock firmly against the miter gauge with the left hand. The guard and the splitter should be in place. Push the miter gauge with the right hand and advance the stock into the blade, Fig. 8-78. If the stock is too short to support with the miter gauge, it should be cut by hand.

Crosscutting duplicate pieces. If numerous pieces of the same length are needed, time can be saved and accuracy increased by means of a special setup or stop. A stop block can be clamped to the miter gauge as shown in Fig. 8-79. A straight piece of stock is attached to the

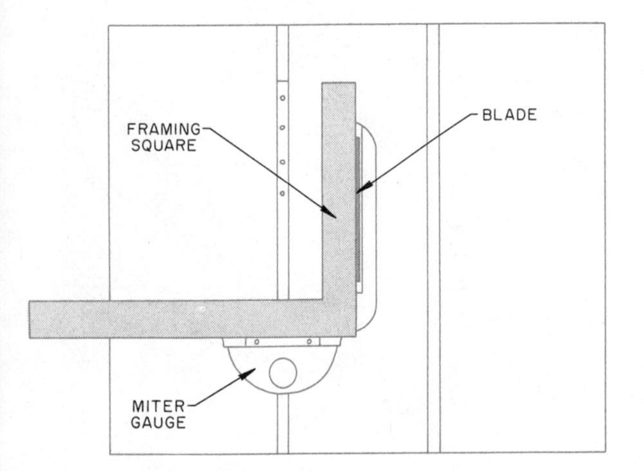

Fig. 8-77. A framing square placed against the blade and the miter gauge to check squareness.

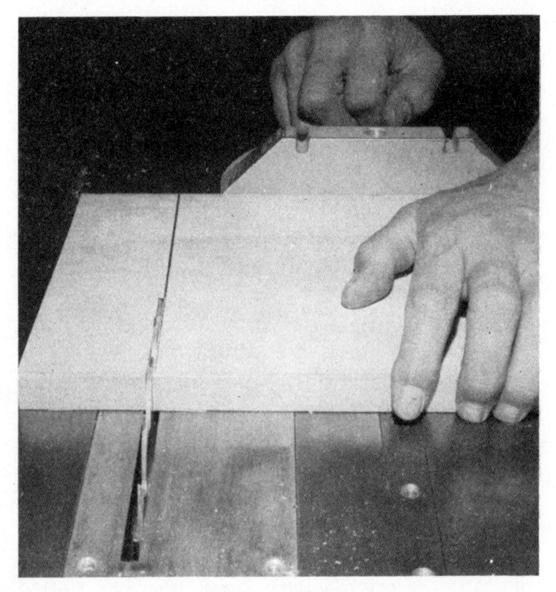

Fig. 8-78. Crosscutting on the circular saw.

miter gauge by means of screws. A block of wood is clamped to the straight piece of stock at the desired distance from the blade. The guard and splitter should be used, but remove the fence from the table for this operation. The stock to be cut is first squared on one end. This end is then placed against the stop block and the cut is made. After each cut is made, advance the remaining stock against the block until all the pieces are cut.

Duplicate cutting can be performed with a *stop rod* attached to the miter gauge, Fig. 8-80. This technique can be used if the pieces are 6″ or longer. The stop rod can be adjusted to the desired length of up to about 12″. It is extended from the miter gauge on the side opposite the blade. The fence should be removed from the saw, but the guard and splitter should be used for this operation. Make sure that the metal stop rod clears the blade before making the cut.

A *clearance block* or stop block is sometimes used for making duplicate cuts, Fig. 8-81. It is clamped to the fence behind the blade near the front of the table. The thickness of the clearance block must be great enough so the diagonal measurement of the piece to be cut is less than the distance between the fence and blade. The

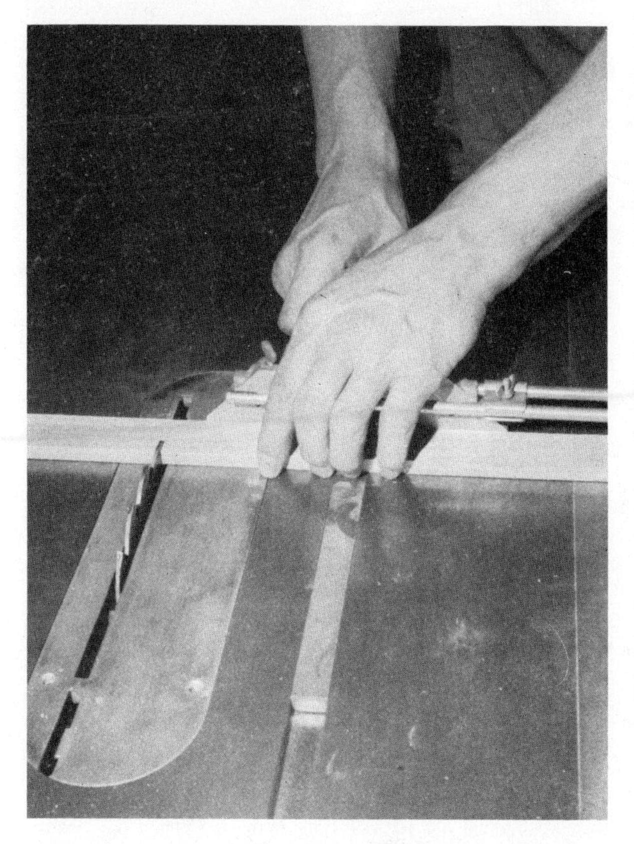

Fig. 8-80. Crosscutting duplicate pieces using a stop rod.

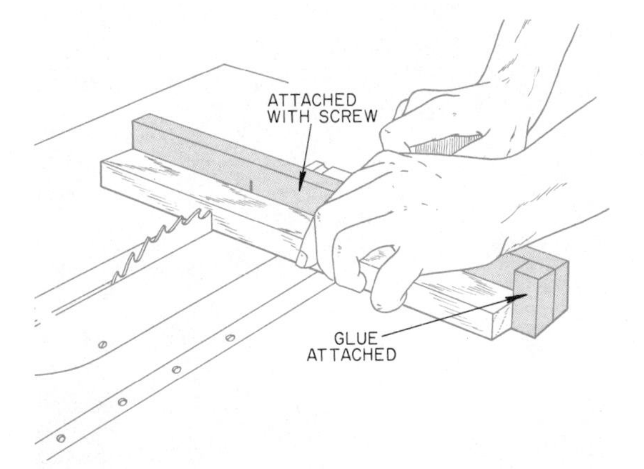

Fig. 8-79. Cutting duplicate pieces on the circular saw using a stop block attached to the miter gauge.

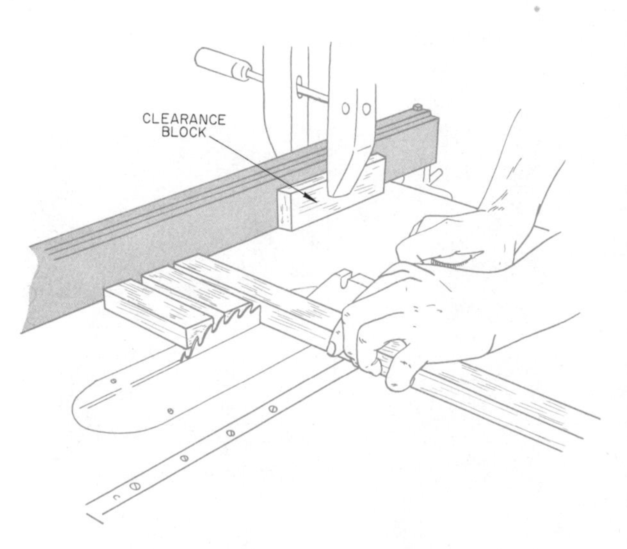

Fig. 8-81. Cutting duplicate pieces to length using a clearance block.

length of the desired piece is measured from the clearance block to a tooth set toward the fence. The stock must first be squared on one end, and this end is then placed against the miter gauge as the stock is cut. Keep the guard and splitter in place. The process is repeated for the number of pieces needed.

The cutting of duplicate parts may sometimes be performed between the fence and the blade. *However, the miter gauge must also be placed between the fence and the blade,* Fig. 8-82, and the guard and splitter should be used.

Ripping stock on the circular saw. Ripping is the process of reducing the width of a board by sawing. One edge of the stock must be straight and one surface flat to rip on the circular saw. A rip or combination blade is installed on the saw, with the blade extending about 1/4″ above the stock being ripped. The splitter and guard must be used if possible.

The fence is always used for ripping operations, usually placed to the right of the blade. The fence is set the desired distance from the blade using the scale on the saw table or by measuring with a rule, Fig. 8-83. When measuring with a rule, measure from the fence to a tooth set toward the fence. *Always stand to one side of the circular saw blade to perform ripping operations.*

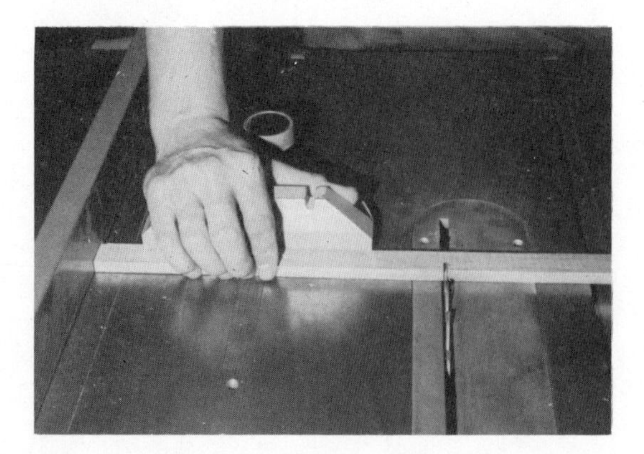

Fig. 8-82. Cutting duplicate pieces with the miter gauge between the fence and the blade.

After starting the saw, advance the board into the blade with a firm, even pressure to make the cut, Fig. 8-84. The straight edge of the stock is placed against the fence and the flat surface is against the table. The left hand holds the board against the fence as it is pushed with the right hand. If the distance between the blade and the fence is less than 4″, a push stick should be used.

Fig. 8-83. Positioning the fence on a circular saw with a rule for ripping operations.

Fig. 8-84. Ripping stock on the circular saw.

The part of the board which is passed between the fence and blade must be controlled until it is completely clear of the blade. A *kickback* will likely result if the board is between the fence and blade and is not pushed clear of the blade. A long board should be supported as it passes through the saw, so a helper should assist the operator. Whenever possible, place the wide portion of the board between the blade and the fence.

Ripping narrow strips on the circular saw. Naturally, a narrow strip is more difficult and dangerous to rip than a wide board. A push stick must always be used for ripping narrow stock, and sometimes two push sticks will be used for the operation, Fig. 8-85. The guard cannot be used for ripping narrow stock, thus additional care must be taken. Make certain the splitter is in place, and start the cut using the push stick in the left hand to hold the stock against the fence. As the end of the stock nears the front of the table, grasp a push stick in the right hand. Guide the board between the fence and the blade with the push stick to complete the cut. *Remember, a push stick should always be used when ripping narrow stock.*

Resawing on the circular saw. Resawing is the process of ripping a board held on edge into two or more thinner pieces, Fig. 8-86. The height of the blade is set to slightly over one-half the width of the stock being resawed. A *feather board,* as shown in Fig. 8-87 holds the stock against the fence. After the first cut has been made, the stock is turned end for end and the second cut is made. The same

Fig. 8-85. Ripping narrow stock using a push stick.

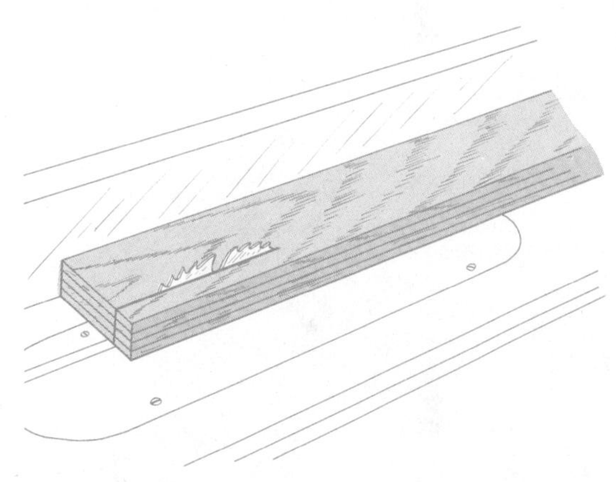

Fig. 8-86. The second cut made in the resawing of a piece of stock into several narrow strips.

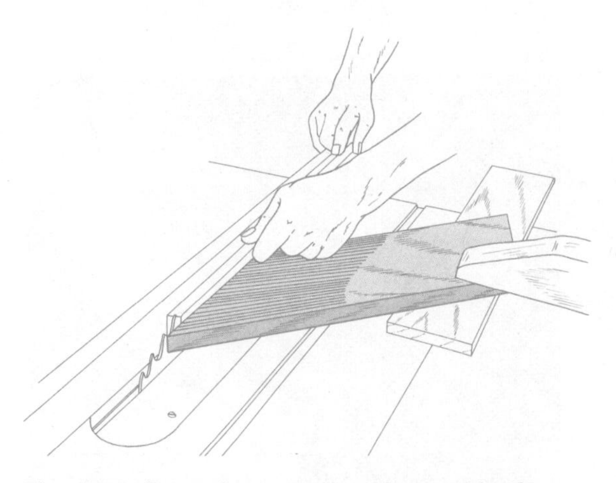

Fig. 8-87. Resawing stock using a feather board support.

surface of the board is held against the fence for both cuts.

If the stock is narrow, the operation can be performed as one cut. Stock wider than twice the maximum cutting depth of the blade must be finished with a hand saw or on the band saw. The guard cannot be used for this operation, so additional caution should be exercised. Make certain the splitter is in place and set slightly below the top of the blade.

Miter cutting is performed similar to crosscutting. The miter gauge is set to the desired angle. The crosscut or combination blade is used for mitering, Fig. 8-88.

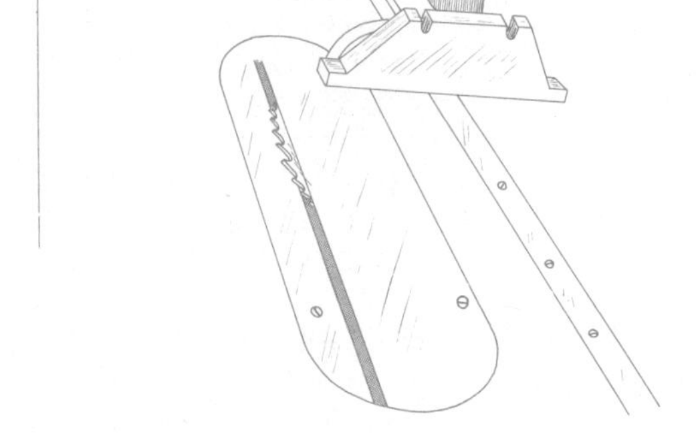

Fig. 8-89. Blade position and set up for cutting bevels and chamfers on the circular saw.

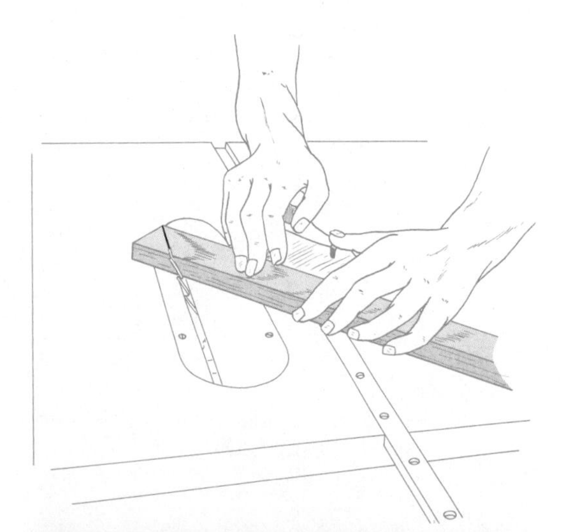

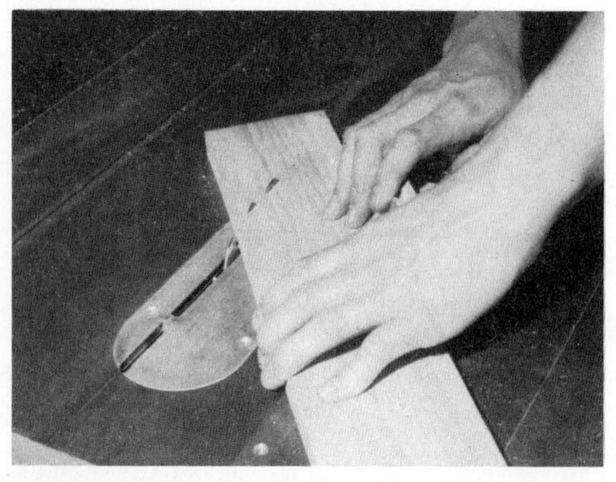

Fig. 8-88. Miter cutting on the circular saw.

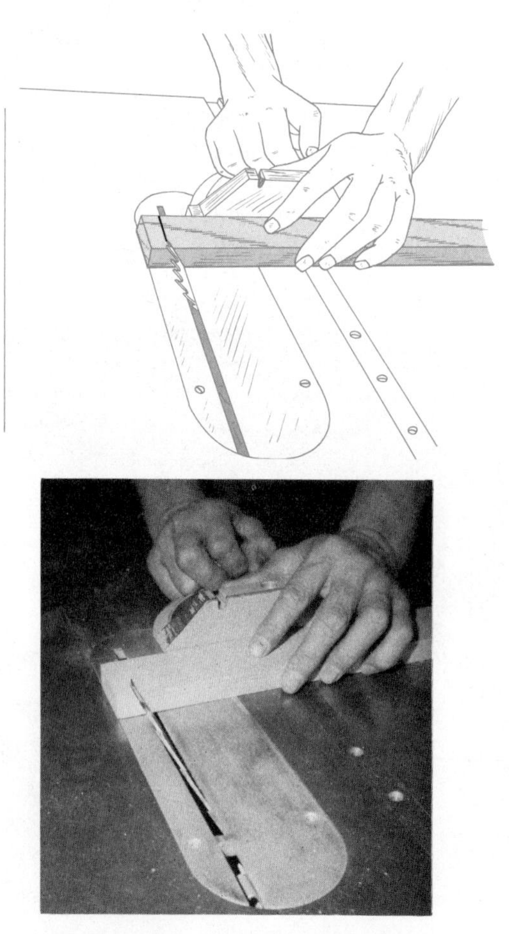

Fig. 8-90. Beveling stock across the grain on the circular saw.

Bevel and chamfer cutting on the circular saw. The blade can be tilted to an angle to crosscut or rip stock on a bevel. The fence is used to cut a bevel on stock in the direction of the grain. The blade is set to the desired angle and adjusted to the correct height. The fence is set so the blade tilts away from it, Fig. 8-89.

A bevel or chamfer can be cut across the grain by tilting the blade and advancing the stock into the blade with the miter gauge, Fig. 8-90. The procedure is similar to straight crosscutting. Hold the stock firmly to the miter gauge so it will not creep toward the blade. Make sure the blade clears the guard. Compound angles are cut by tilting the blade and setting the miter gauge at the desired angle, Fig. 8-91.

Cutting a taper on the circular saw. A special jig such as shown in Fig. 8-92 is used for cutting tapers on the circular saw. The jig consists of a straight board and two evenly-spaced notches. The width of the notches will depend on the desired taper and the length of the stock being cut. The stock is placed on the notch nearest the fence for the first cut. The stock is then turned over and placed on the next notch for the second cut. The jig and the stock are pushed along the fence to make the cut. A push stick should be used to hold the stock in the jig. The splitter can be left in place on the saw.

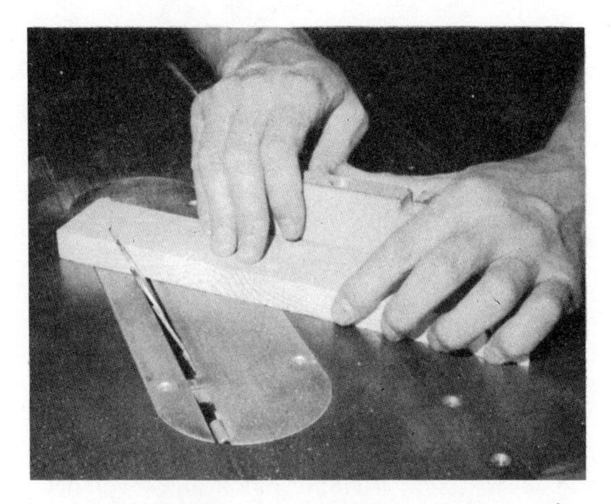

Fig. 8-91. Cutting a compound angle on the circular saw.

Cutting grooves, rabbets, and dado cuts on the circular saw. Grooves, rabbets, and dado cuts may be cut with either a standard saw blade or a dado head, Fig. 8-93.

If only a limited number of joints are to be cut, it is often easier to use a single blade. When a single blade is used to cut a rabbet on the edge of a piece of stock, the depth and width is laid out on the end of the stock. The fence is positioned the correct distance from the blade and the blade is raised to the required height. The first cut is made on the face side of the board, then the blade and fence are positioned for the second cut. The two cuts intersect to form a square corner. Whenever possible, the fence should be placed so

Fig. 8-92. A jig for cutting tapers on the circular saw.

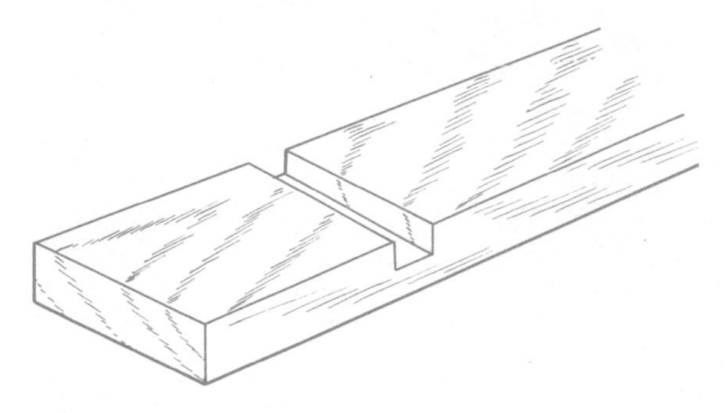

Fig. 8-93. A rabbet cut.

the waste piece falls to the outside, away from the fence.

Rabbets can also be cut across the grain on the end of stock using a single blade, the fence, and the miter gauge. The fence is positioned the correct distance from the outside of the blade for the width of the rabbet, and the blade must be set to the correct height. Pass the stock across the blade using the miter gauge, Fig. 8-94. The stock is pulled away from the fence the width of the saw kerf for each additional cut until the waste is removed.

A dado can also be cut with a single blade. The layout for the dado is made on the edge of the stock. The blade is adjusted to a height equal to the depth of the dado. The stock is guided across the blade using the miter gauge. Several cuts are made to remove the waste stock, Fig. 8-95. If several dado cuts are to be made which

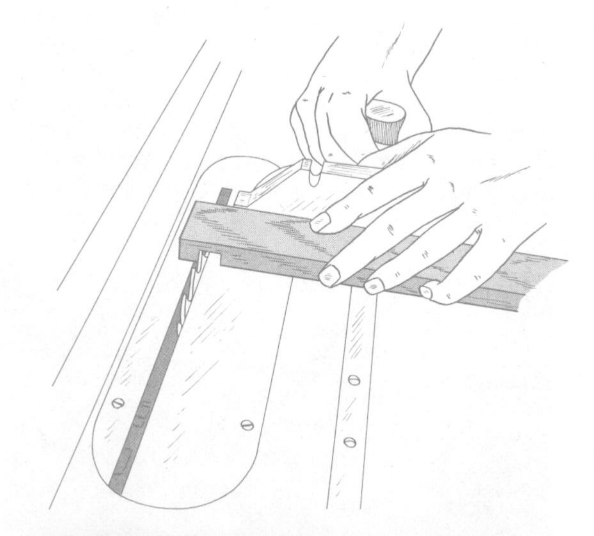

Fig. 8-94. Cutting a rabbet across the grain with a single blade using the fence as a stop.

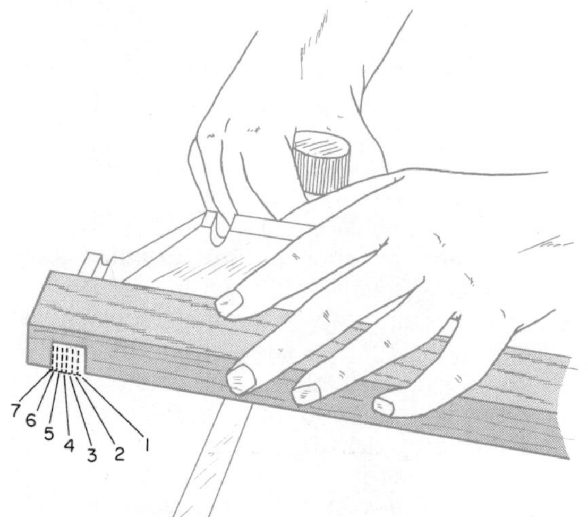

Fig. 8-95. Using a single blade to cut a dado on the circular saw.

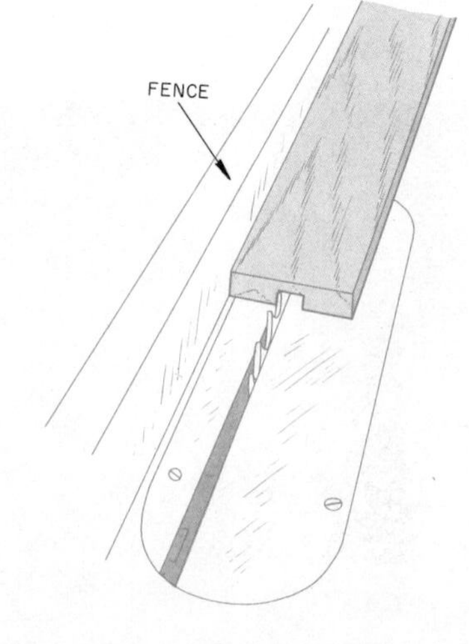

Fig. 8-96. Using a single blade and the fence to cut a groove on the circular saw.

are alike, the fence can be used for positioning the outside cuts.

A groove can be cut using a single blade and the fence, Fig. 8-96. The fence is set to match the layout for the first cut for one side of the groove. It is then moved the width of the kerf for each additional cut until the waste stock is removed. If several pieces are to be cut, time can be saved by feeding each piece through the setup before moving the fence.

The dado head is faster and safer for cutting rabbets, grooves, and dado cuts. There are several types of dado heads which can be used — Fig. 8-97 shows two common types.

The industrial dado head consists of two identical saw blades and a set of blades called chippers. The blades are 1/8″ thick and the chippers range in thickness from 1/6″ to 1/4″ thick. A standard assortment can be placed in combinations to cut grooves from 1/8″ to 13/16″. The quick-set dado head consists of one piece and is adjusted by revolving the outside cutters to "dial" the desired width.

A. Industrial dado heads. (H. K. Porter Co., Henry Disston Div.)

B. Quick-set dado head. (Rockwell Mfg. Co.)

Fig. 8-97. Dado heads.

The dado head is placed on the saw arbor like other blades. One blade of the dado head is placed first. The desired number of chippers are then placed, followed by the other blade. The chippers are swaged and should be staggered so they do not align directly with a tooth on the blade. A special throat plate which has a wide opening is used with the dado head.

Feed the stock into the blade more slowly when using a dado head than when using a single blade. Much more stock is removed with each cut with the dado head.

The fence is used to rabbet stock the length of the board, Fig. 8-98. Note that

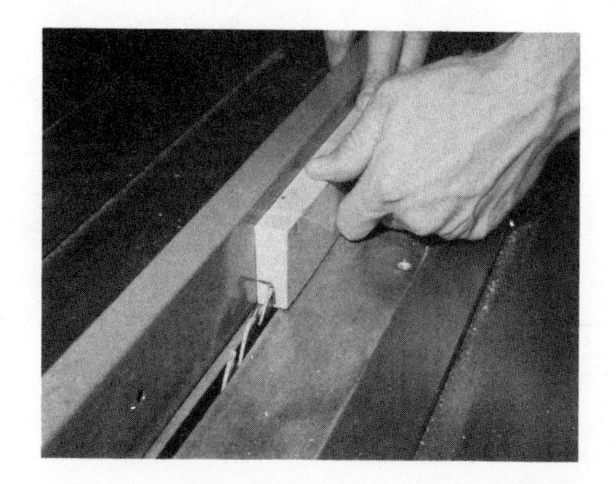

A. Edge cut.

B. Face cut.

Fig. 8-98. Rabbeting stock the length of the board.

an auxiliary facing fence or board has been attached to the fence. This makes it possible to place the fence tight against the blade without damage to the blade or fence.

A rabbet may be cut across the grain of a board with a dado head, Fig. 8-99.

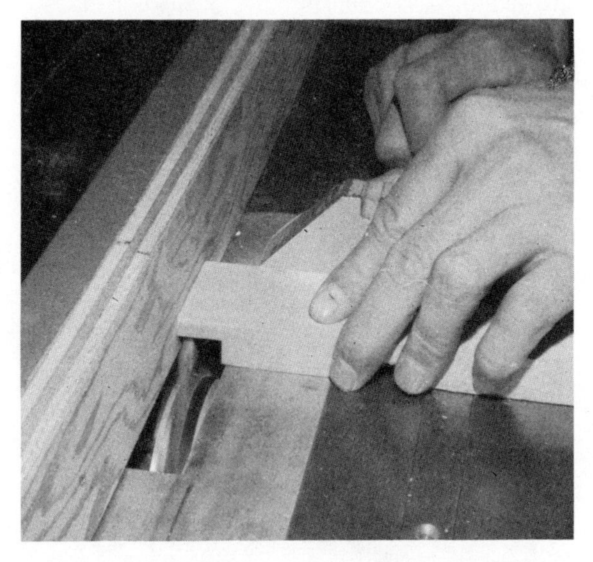

Fig. 8-99. Cutting a rabbet across the grain with a dado head on the circular saw.

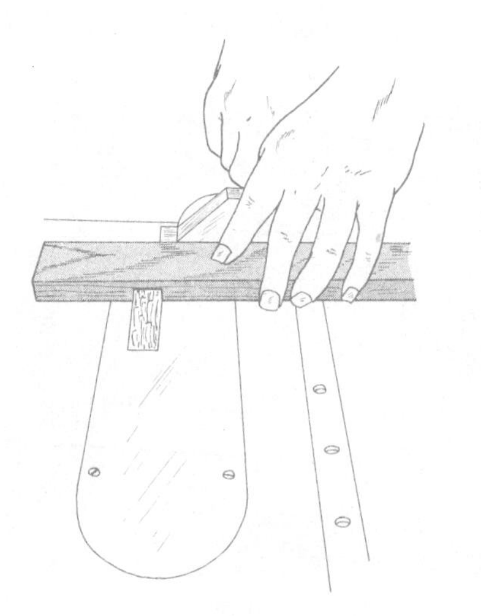

Fig. 8-100. Cutting a dado across the grain with a dado head on the circular saw.

The miter gauge is used to advance and guide the stock into the dado head.

If several pieces are to be rabbeted the same distance from the end, the fence can be used as a stop. A dado is cut across the grain using the same method as for rabbeting, Fig. 8-100.

Cutting tenons on the circular saw. The tenon is cut in the same manner as a rabbet. The dado head will speed the operation. However, it can also be performed with a single blade. The tenon is first laid out on the end of the stock. The fence is set

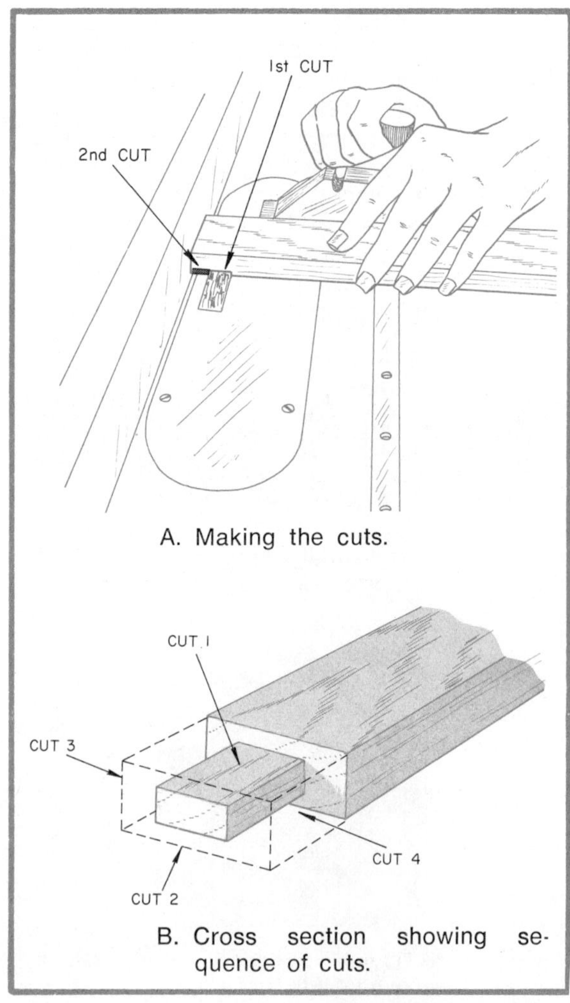

A. Making the cuts.

B. Cross section showing sequence of cuts.

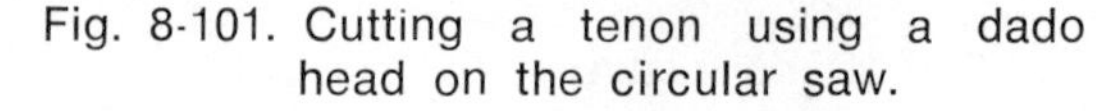

Fig. 8-101. Cutting a tenon using a dado head on the circular saw.

the correct distance from the blade to serve as a stop for cutting the four sides, Fig. 8-101.

Safety considerations for the circular saw.

1. Keep the guard, splitter, and anti-kickback attachments in place whenever possible.
2. Use the correct blade for the operation being performed.
3. Make certain the blade is sharp. A dull blade can cause a kickback.
4. Stand to one side of the blade when operating the saw.
5. Plan how to perform the operation before you start the machine.
6. Stock cut on the circular saw must have at least one flat surface and one straight edge. Place the flat side down and the straight edge against the fence or miter gauge.
7. Stock being cut must be guided by either the fence *or* the miter gauge. *The miter gauge and fence are used together only for special setups. NEVER CUT FREEHAND ON THE SAW.*
8. Avoid reaching across the blade.
9. Keep the saw table free of tools. Use a bench to stack stock which is to be cut.
10. Adjust the blade so that it extends ¼″ above the stock.
11. Check the stock for defects. It must be free of loose knots or splits which may be thrown out of the saw.
12. Make sure the teeth on the blade point toward you when in the operating position.
13. Feed the stock into the blade with smooth, even pressure.
14. Only the operator should be near the saw. When an assistant is used for supporting long stock, he should only support it. The operator controls the machine.
15. All adjustments are made with the power disconnected.
16. Use a push stick when the space between the fence and blade is less than 4″.
17. Check with your instructor before performing operations with which you are not familiar.
18. Stop the machine after completing the cut. Remain at the machine until the blade stops.
19. Clear the table and surrounding floor area of scraps of wood when finished with the saw.
20. Have the instructor check all special setups.

RADIAL ARM SAW

The radial arm saw is capable of performing many of the same operations commonly done on the circular saw. It is used for a wide range of work from rough cutting to finish cutting. The machine consists of a motor mounted on a yoke attached to a radial arm, Fig. 8-102. The motor travels on the radial arm track. The radial arm is mounted on a pivot and can be adjusted to any angle. The motor also pivots in the yoke so angle cuts can be made. Compound angles can be cut by pivoting both the radial arm and the motor. The blade is mounted directly on the motor shaft. The radial arm saw is convenient for cutting long stock to length. The overarm is raised

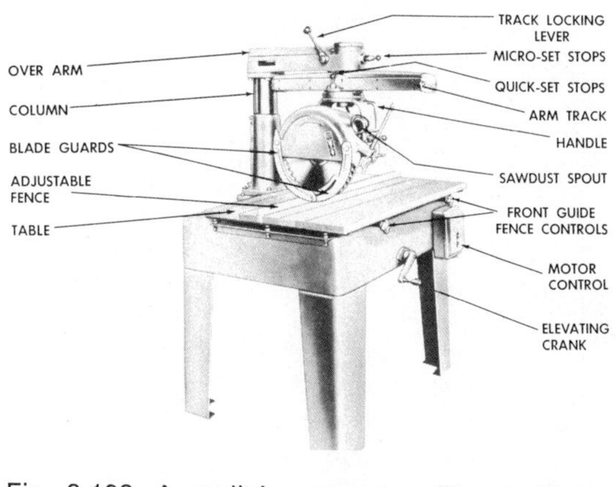

Fig. 8-102. A radial arm saw. (Power Tool Div., Rockwell Mfg. Co.)

Fig. 8-103. Crosscutting stock on the radial arm saw.

Fig. 8-104. Ripping stock on the radial arm saw.

or lowered to adjust the depth of cut. The stock is placed on the table and the blade is advanced into the stock for crosscutting, Fig. 8-103. This is easier than feeding large stock into the saw as is done with the circular saw. The blade is set parallel to the fence for ripping. The stock is advanced into the blade, Fig. 8-104.

The blade is installed on the radial arm saw in a manner similar to the circular saw. The size of the saw is given as the maximum diameter blade it will carry. The blades used on the radial arm saw are like those used on the circular saw. Blade selection is made according to the type of operation performed. A crosscut or combination blade is used for crosscutting; rip or combination blades are used for ripping.

Saw rotation. Special attention should be given to the direction of rotation of the blade. Crosscutting operations require that the blade be advanced into the stock in the same direction as the blade rotation. Figure 8-103 shows the relationship between blade rotation and direction of feed. Due to this direction of feed, the blade may advance into the stock too rapidly. To avoid this, the saw handle should be gripped firmly and pulled smoothly into the work. Ripping, on the other hand, requires that the stock be fed into the blade against its rotation direction.

Crosscutting on the radial arm saw. Crosscutting is the operation most often performed on the radial arm saw. The radial arm is positioned at a right angle to the guide fence. The work is held firmly against the fence. The layout line on the stock is placed in alignment with the blade. If several duplicate pieces are to be cut, a stop block may be clamped to the fence. If the longest portion of the stock extends to the left, the work is held against the fence with the left hand and the right hand is used to pull the saw across the work. Remember, the handle should be gripped firmly, as the saw may tend to advance forward too rapidly. After completing the cut, return the saw to its original position and turn off the power.

Ripping stock to width. To set the radial arm saw for ripping, the blade is turned parallel to the fence by turning the yoke assembly. The blade is positioned the desired distance from the fence and clamped in place. The stock is fed into the blade the opposite direction of the blade rotation, Fig. 8-104. Make certain that a push stick is used to complete the cut. The splitter and anti-kickback device are used for ripping, Fig. 8-105.

Beveling stock across the grain. The same procedure used for crosscutting is used for bevel cutting across the grain, Fig.

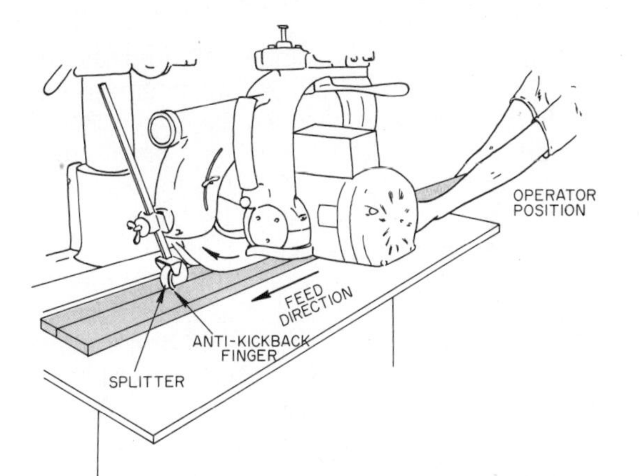

Fig. 8-105. The splitter and anti-kickback device must be used to rip stock on the radial arm saw.

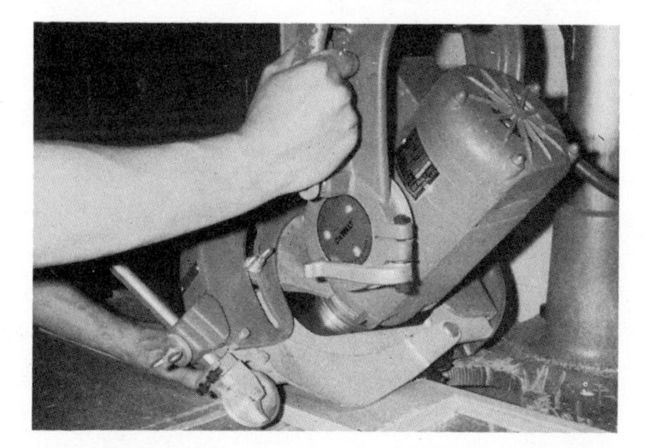

Fig. 8-106. Cutting a bevel across the grain on the radial arm saw.

8-106. The saw is pivoted in the yoke to the desired angle. To tilt the blade, the arm is elevated so the blade clears the table. The arm is lowered so the blade enters the surfaces of the table about 1/32″. The lower guard is removed for this operation.

Cutting miters and compound miters. To cut a simple miter the arm is rotated to the desired angle either right or left, Fig. 8-107. Miters are usually cut at a 45° angle. The stock is held against the fence.

A compound angle is a combination of a bevel and miter. The saw is pivoted in the yoke as for beveling, and the arm is turned in the desired angle as for mitering, Fig. 8-108.

Fig. 8-107. Cutting a miter on the radial arm saw.

Fig. 8-108. Cutting a compound angle on the radial arm saw.

Cutting dadoes and grooves. Dadoes and grooves can be cut with the radial arm saw using either a single blade or a dado head. For either method, the blade is raised to the desired height to give the correct depth of cut. If the dado head is used, it is mounted using the method similar to that used for the circular saw, Fig. 8-109. If a single blade is used, the dado is first laid out on the face and edge of the stock. The work is held against the fence and the required number of cuts are made to remove the waste stock. Stops can be setup if several duplicate pieces are being made for either type of blade. Stops will speed the operation and insure greater accuracy.

To cut a groove, set up the radial arm saw as in the procedure for ripping. Either the dado head or a single blade may be used. The saw is rotated to position the blade parallel with the fence. The yoke is locked on the radial arm the correct distance from the fence. The depth of cut is determined by the height of the blade above the table. The groove is cut using the ripping technique. If several duplicate pieces are cut using the single blade method, time can be saved by cutting each one before changing the setting. The anti-kickback device is used for the operation, Fig. 8-110.

Fig. 8-109. Cutting a dado on the radial arm saw. (Power Tool Div., Rockwell Mfg. Co.)

Fig. 8-110. Cutting a groove on the radial arm saw. (Power Tool Div., Rockwell Mfg. Co.)

INDUSTRIAL SEPARATION

Vast improvements have been made in the methods developed for separating wood. It is thought that prehistoric man developed stone implements used for separating wood thousands of years ago, Fig. 8-111. Through the centuries man has improved his methods of separating wood. Today much of the woodworking industry is equipped with automated equipment. All of the separating machines are either rotary, reciprocating, or linear, Fig. 8-112. The first sawing machine known to be used

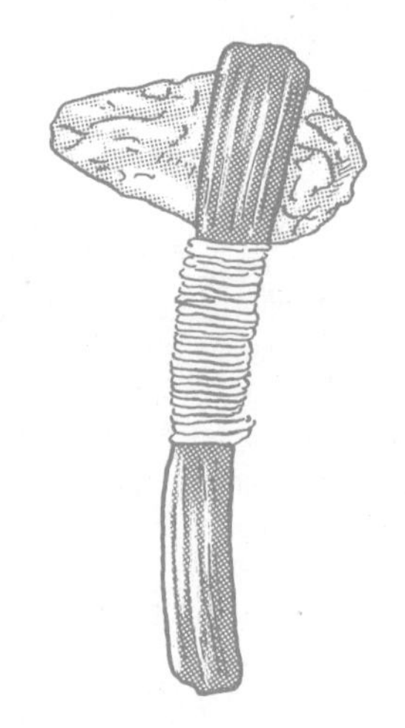

Fig. 8-111. A stone used by early man to separate fibers.

in America was a reciprocating saw. The jigsaw used in modern woodworking industries is an improvement upon this machine. The early reciprocating saws were used to cut boards from logs. The circular saw was the first continuous-cutting saw introduced. It used a cutting disc on which teeth were cut. The band saw was developed a few decades later and was also a continuous-cutting machine. These three basic machines were improved upon many times to increase their efficiency. The first machines were powered either by wind, water, or animal power.

The increased demand for goods at a competitive price has motivated industry to improve the separating methods. Most industrial separation of wood is still performed by the cutting processes described in this chapter. The greatest disadvantage of cutting with a blade is waste created by the saw kerf.

Extensive research has been performed to find new methods of separation. Jets of water under extreme pressure ranging up to 50,000 pounds per square inch have been tried, Fig. 8-113. LASER beams have

Fig. 8-113. Separation of wood with high-pressure water jets. (Forest Products Laboratories)

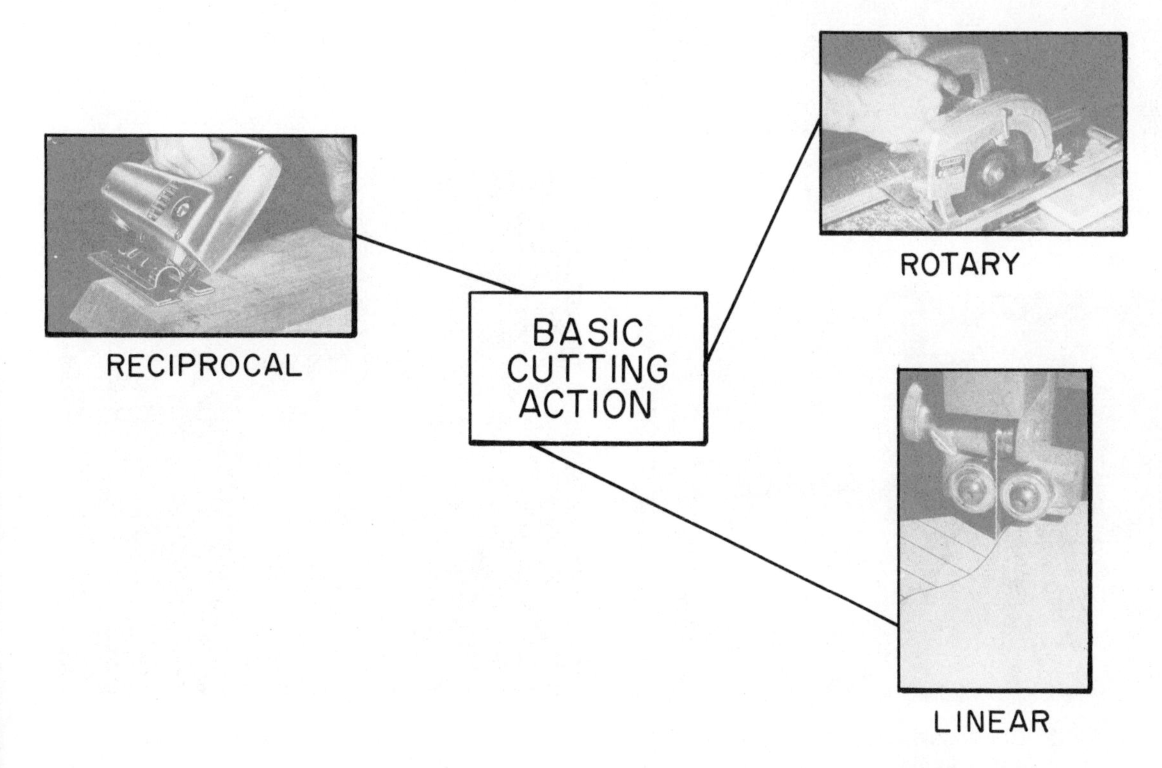

Fig. 8-112. Basic cutting action of separating machines.

also been tried as a means of separating wood. LASERS are based on the principle of amplification of light, Fig. 8-114. Researchers will no doubt continue to find improved techniques for separation. However,

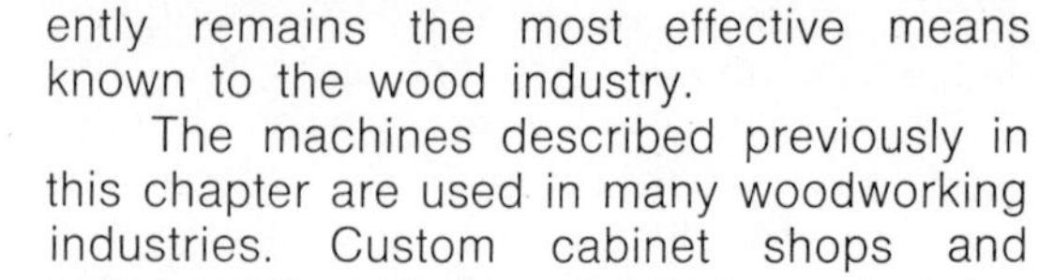

mechanical separation with a blade presently remains the most effective means known to the wood industry.

The machines described previously in this chapter are used in many woodworking industries. Custom cabinet shops and patternmaking shops use this type of equipment. Industries which produce large quantities of a product on a mass-production basis have machines designed for a specific purpose. In the lumbering industry where logs are separated into boards, giant band saws are used such as shown in Fig. 8-115. The blades may be in excess of 12″ wide and run on wheels of up to 10′ in diameter. Some machines may also use blades over 50′ long. The blades on band saws used to saw logs and resaw lumber often have teeth on both edges, Fig. 8-116. A band saw with this type of blade cuts with the log moving in either direction. The log travels on a platform that is advanced inward towards the blade after each cut. The platform carrying the log may be set so that it advances automatically at the end of each cut.

Fig. 8-114. Separation of wood with amplified pulses of light. (Forest Products Laboratories)

Fig. 8-115. Band saw used to separate logs into lumber. (American Forest Industries, Inc.)

Fig. 8-116. Band saw blade having teeth on both edges. (Southern Forest Products Assoc.)

Rotary saws using a circular blade are used extensively in the wood industry. Many of these machines are designed and equipped for a specific operation. Some are equipped with an automatic feed system. An operator presses a button and the saw travels at a uniform speed through the board. The distance the saw travels is preset for the particular operation.

Circular saws similar to the ones discussed previously in this chapter are widely used in the wood industry. Sometimes the saws have a table split down the center, making it easier to cut large sheets of stock. The sliding table can also be clamped in place and used with a miter gauge. Many of these machines are equipped with hydraulic or pneumatic systems which raise, lower, and tilt the blade when an operator pushes a control, Fig. 8-117.

High-production industrial separating machines are often equipped with feeding chains. The stock moves through the blade on a belt which also keeps the stock in alignment. Rollers on the overhead assembly keep the boards firmly in place against the endless feed belt. The rate of feed can be varied for different types of stock.

The gang ripsaw has an arbor equipped with a number of blades, Fig. 8-118. Stock can be ripped to the desired width by positioning the blade at intervals on the arbor. This saw also has an automatic feed — the stock is fed through the blade by means of pressure rollers. The arbor is powered by a large motor assembly. The stock is guided by the automatic feed system and provides an accurate cut.

Double - end cutoff saws are used to cut stock to length. They are equipped with two saw blades or separate arbors. The blades are spaced apart to the length of stock which is to be cut, Fig. 8-119. One blade is usually stationary and the other is adjustable to the length of stock to be cut. The stock is advanced into the blade on continuous feed chains. Rubber rollers exert downward pressure and hold the stock firmly in place for cutting. The arbors

Fig. 8-118. A gang saw. (American Forest Institute)

Fig. 8-117. A circular saw with an automatic adjustment system. (Greenlee Bros. & Co.)

Fig. 8-119. Double-end cutoff saw. (Greenlee Bros. & Co.)

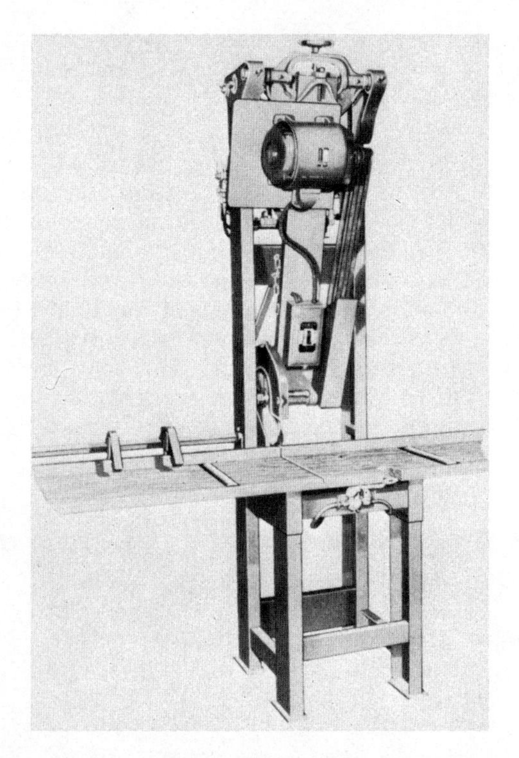

Fig. 8-120. Swing saw. (Fairfield Engineering and Manufacturing Co.)

can be tilted to an angle for bending and special cutting.

Swing saws such as shown in Fig. 8-120 are used a great deal for cutoff work. The blade is pulled into the stock by the operator.

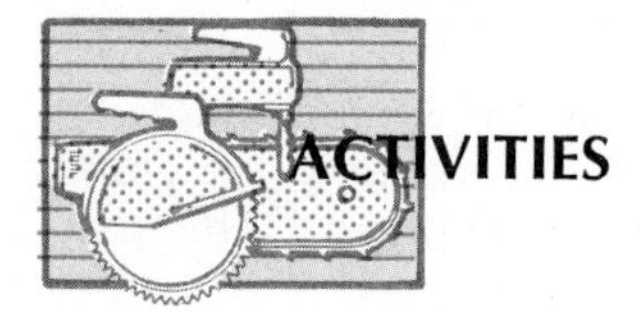

PROBLEM

Study the floor plan below and determine the number of windows and the number of cripples needed for those windows. Which of these two options would you choose: Option 1 — cut all cripples on the job with hand saw or portable power saw; Option 2 — cut all cripples in the shop with a radial arm saw or table saw and then transport them to the job.

Compute which standard length of 2×4 would result in less waste to cut 2′ 5″

cripples. How many of these standard lengths will you need for the entire floor plan of windows?

Lay out and cut one length for a cripple according to each of the specifications below, recording the time it takes for each operation in a table like the example shown here. (*Do not write in your textbook.*)

A. Lay out and cut one cripple length with a hand saw.
B. Lay out and cut one cripple with a portable electric saw.
C. Lay out and cut one cripple with a table saw set up to cut duplicate pieces.
D. Lay out and cut one cripple with a radial arm saw set up to cut duplicate pieces.

Determine which method should be used to cut three cripples.

Determine which method should be used to cut 100 cripples.

Example table:

SAW	SETUP AND LAYOUT TIME	PRODUCTION TIME
Hand		
Portable Electric		
Table		
Radial Arm		

EXAMPLE CHART

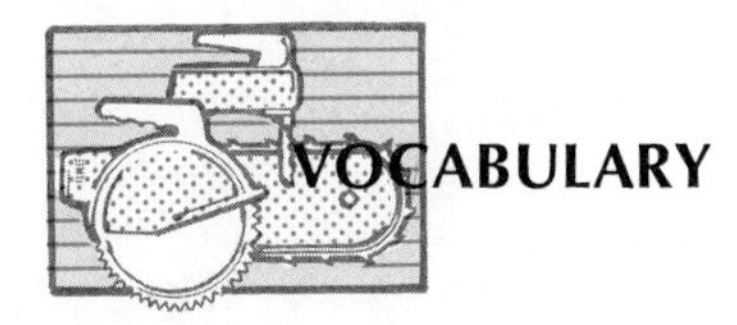

VOCABULARY

Crosscut
Kerf
Radial arm saw
Combination saw blade
Bevel
Chamfer
Jigsaw
Hand saw
Kickback
Resawing
Arbor
Splitter
Clearance Block
Dado head

CHAPTER 9
PRINCIPLE OF PLANING

INTRODUCTION

Planing (or surfacing) of stock is one of the most important production operations in the woodworking industry. Planing is the process of producing a smooth, flat surface by the removal of stock. The stock is removed in the form of thin shavings or chips by means of an *edge-cutting tool.* The production of a wood product usually requires that at least one edge or surface be flat and smooth. There are many types of machines and tools which may be used for this process. They range from simple hand tools to complex machines used in industry, Fig. 9-1. A combination of tools and machines is often used. Regardless of the tool or machine used, the cutting principle remains the same.

PRINCIPLE OF PLANING

When edge-cutting tools are used for making a surface smooth and flat, the wood fibers are sheared. Figure 9-2 shows the cutting action as the fibers are sheared by the sharp leading edge of the blade. This principle remains the same for the different types of edge-cutting tools and machines, Fig. 9-3.

Fig. 9-2. Shearing action of an edge cutting tool.

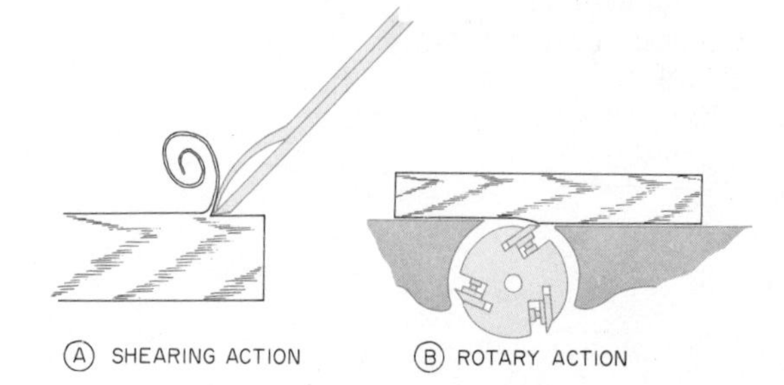

Fig. 9-3. Comparison of a rotary and straight line edge cutting tool and machine.

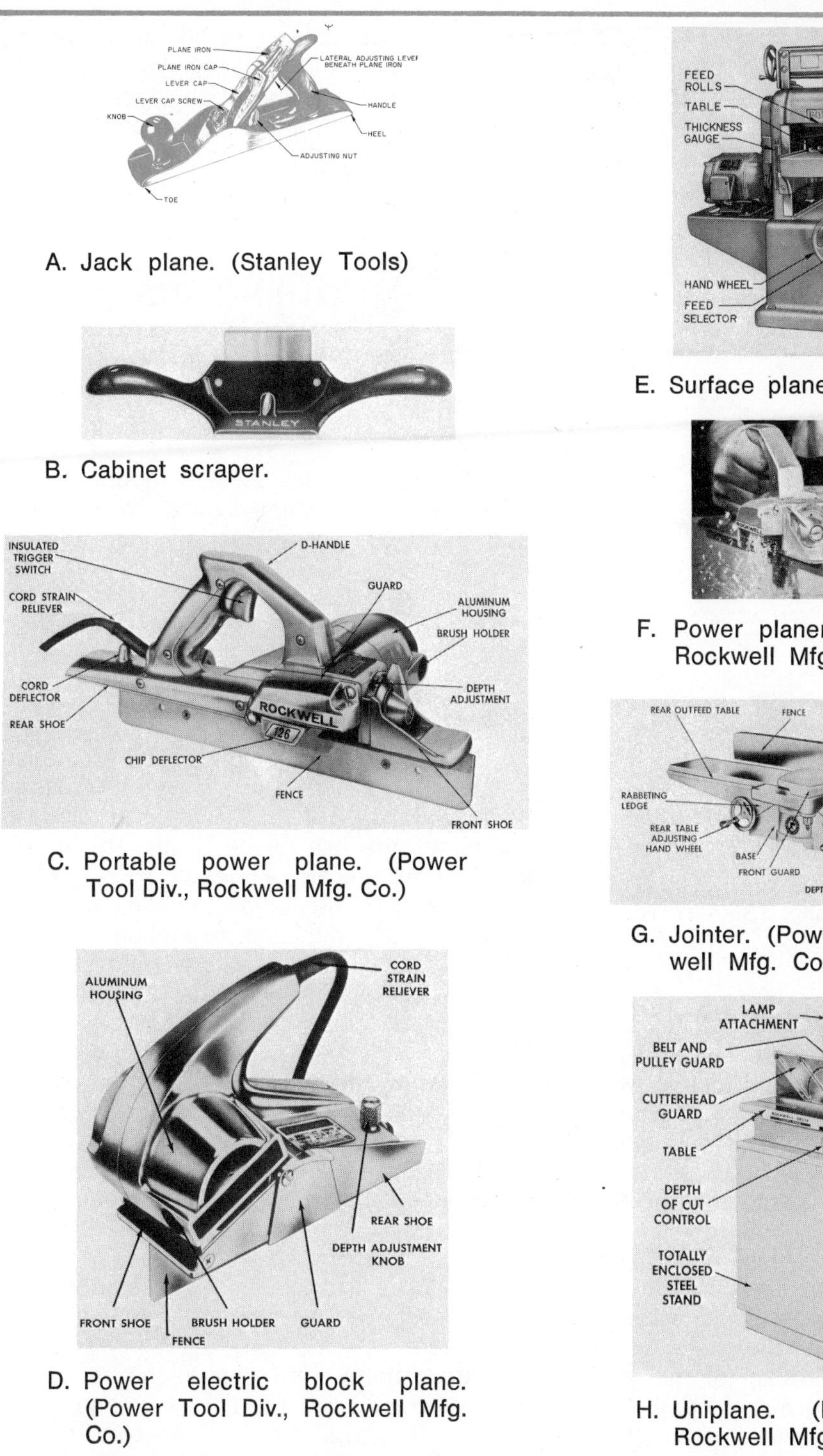

A. Jack plane. (Stanley Tools)

B. Cabinet scraper.

C. Portable power plane. (Power Tool Div., Rockwell Mfg. Co.)

D. Power electric block plane. (Power Tool Div., Rockwell Mfg. Co.)

E. Surface planer.

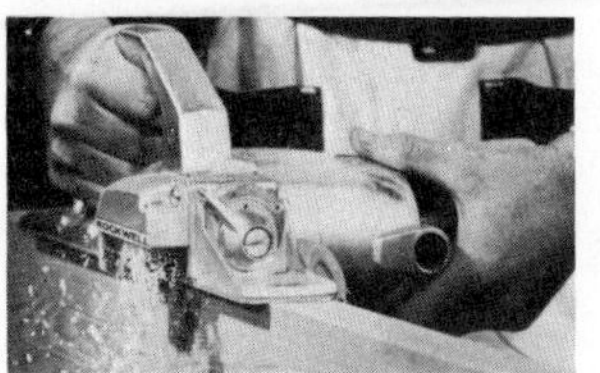

F. Power planer. (Power Tool Div., Rockwell Mfg. Co.)

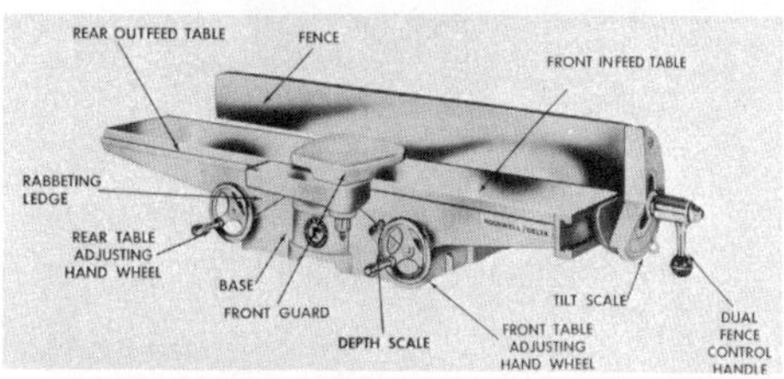

G. Jointer. (Power Tool Div., Rockwell Mfg. Co.)

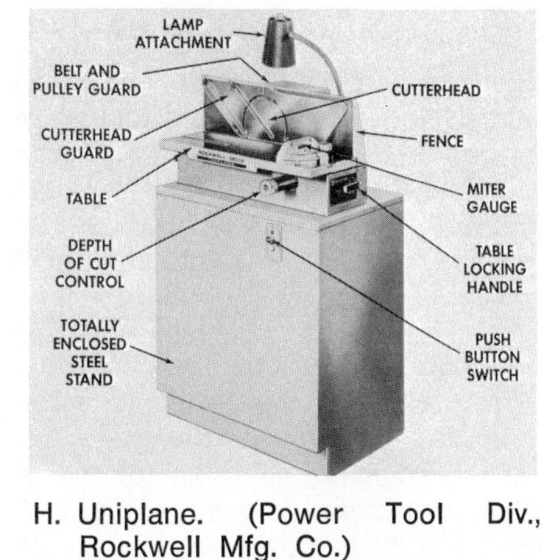

H. Uniplane. (Power Tool Div., Rockwell Mfg. Co.)

Fig. 9-1. Tools and machines used to produce flat surfaces.

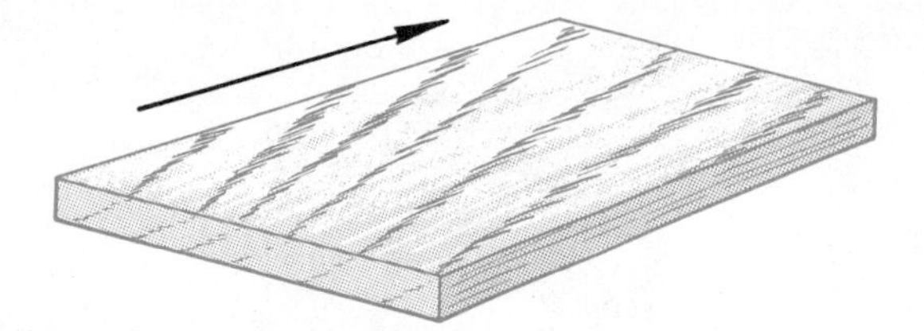

Fig. 9-4. Direction of the grain of wood.

Fig. 9-5. Defects may cause the grain to run in several directions.

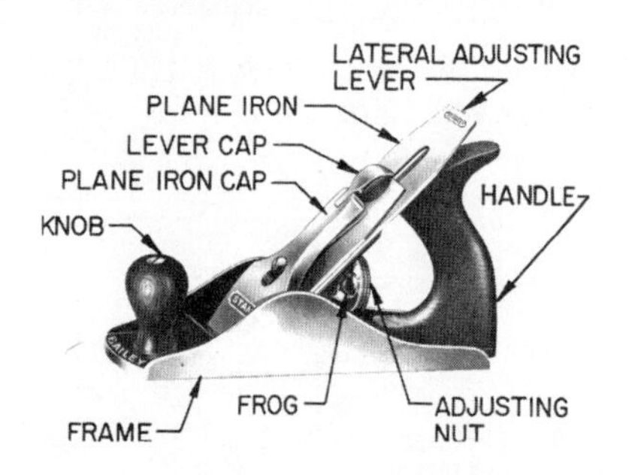

Fig. 9-6. A smoothing plane. (Stanley Tools)

DIRECTION OF THE GRAIN FOR PLANING

Special attention must be given to the direction of the wood grain. Stock should always be planed *in the direction of the grain,* or the result is that the wood fibers are torn and raised rather than sheared. This is like rubbing an animal's fur the wrong way.

The direction of the grain of a piece of wood can usually be determined by a careful inspection of the face and edge of the board. An example of the grain direction and the planing direction for the face and edges of a board is given in Fig. 9-4.

All wood does not have straight grain and cannot be planed with one continuous cut. Knots and similar defects may cause the grain to run in several directions within the same board, Fig. 9-5. Special care must be taken when planing such stock. It can either be planed in several directions, or a hand scraper may be used.

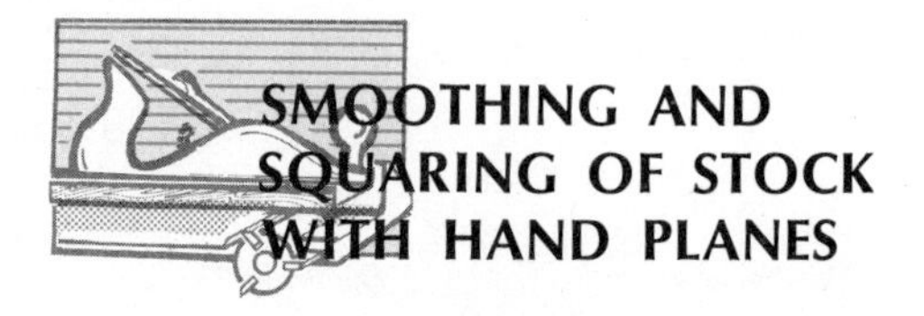

SMOOTHING AND SQUARING OF STOCK WITH HAND PLANES

Hand planes are used to reduce stock to the desired size and to smooth the surface. There are many types of planes used in woodworking, but regardless of the plane, the cuts are very similar. This chapter will cover the more commonly used planes often referred to as *bench planes.*

TYPES OF PLANES

The main difference in the various planes is in the size and the type of cutting blade. This difference in size and blades determines the use of the plane. The most universal planes are: (1) smooth, (2) jack, (3) jointer, (4) fore, and (5) block planes. These planes, with the exception of the block plane, differ only in size. The cutting blade or *plane iron* and other adjustments of the block plane are different from the others.

Smooth plane. The smooth plane ranges in length from 5½″ to 10″. It uses a plane iron ranging from 1¼″ to 2⅜″ wide, depending upon the length. The 8″ smoothing plane with 1¾″-wide plane iron is the most common size, Fig. 9-6. It is used for smoothing uneven surfaces, planing small stock, or to finish the smoothing of large surfaces after a larger planer has been used.

Jack plane. The jack plane is constructed exactly like the smooth plane, but is larger. It is available in a variety of sizes ranging in length from 11½″ to 15″ and widths of the plane iron may vary from

1¾″ to 2¼″. The most common size is the 14″ length with a 2″ plane iron, Fig. 9-7. The jack plane is regarded as an all-purpose tool for most planing operations.

Jointer plane. The jointer plane, Fig. 9-8, is the largest of the bench planes, ranging in size from 20″ to 24″ long with plane irons from 2⅜″ to 2⅝″ wide. It is used for planing long surfaces, such as the edge of a door or for preparing the edges of long boards for gluing.

Fore plane. The fore plane is slightly larger in both length and width than the jack plane. The most common size is 18″ long with a plane iron 2⅜″ wide. It is used for planing large surfaces such as table tops, edges of long boards for gluing, and furniture legs.

Block plane. The block plane, Fig. 9-9, is the smallest of the common types of planes used to smooth surfaces and remove stock. It is designed to be held in one hand and is generally used for planing end grain.

PARTS AND ADJUSTMENTS OF PLANES

Planes consist of many parts and adjustments, Fig. 9-7. The purpose of the planing process is to make a surface or edge true (level or square). Selection of the proper plane for a specific job is based upon the length of the *plane bottom.* A plane bottom that is too short for the stock will follow the contours of the piece and fail to true the surface or edge as shown in Fig. 9-10. The plane bottom must be long enough to remove the high points on the stock in the first cut.

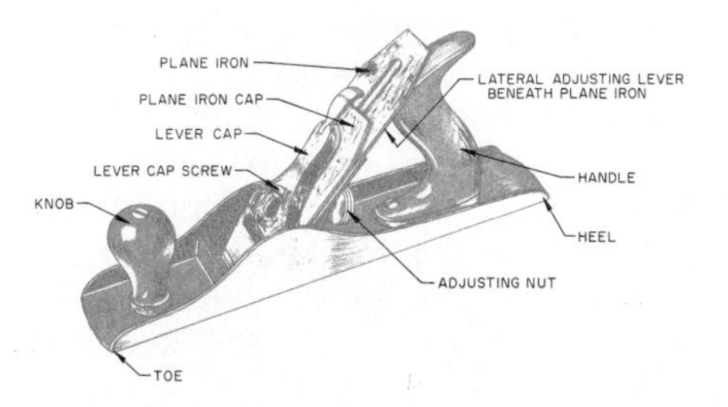

Fig. 9-7. A jack plane. (Stanley Tools)

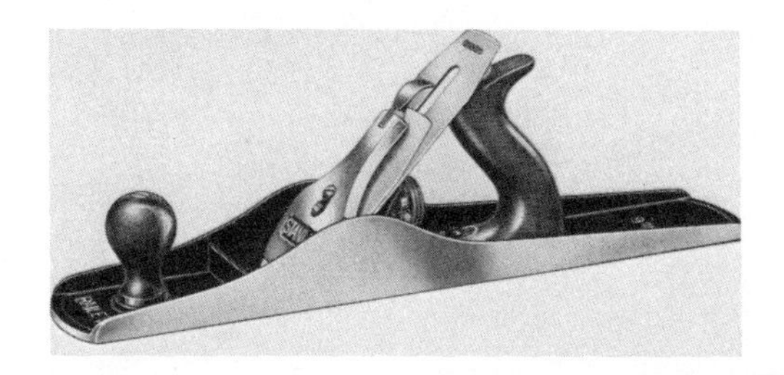

Fig. 9-8. A jointer plane. (Stanley Tools)

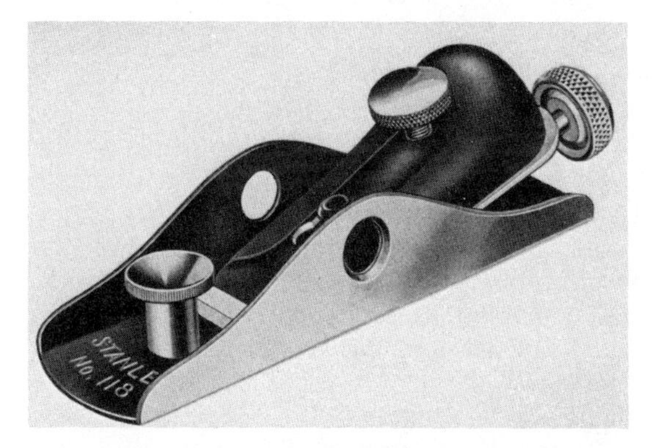

Fig. 9-9. A block plane. (Stanley Tools)

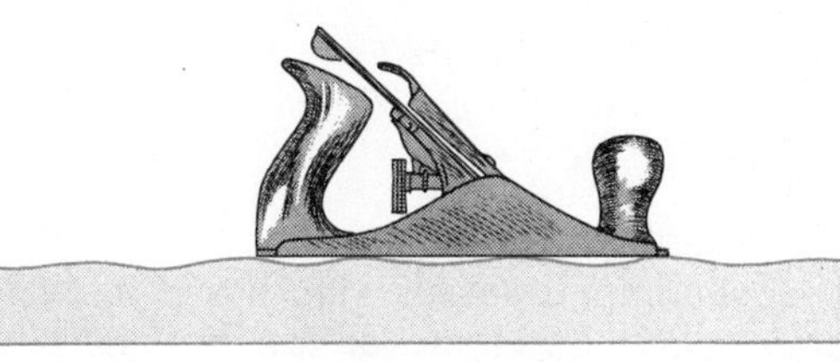

A. LENGTH OF PLANE WILL SERVE TO LEVEL THIS SURFACE

B. PLANE LENGTH IS TOO SHORT TO LEVEL THIS SURFACE

Fig. 9-10. The bottom of the plane must be long enough that it does not follow the contours of the board.

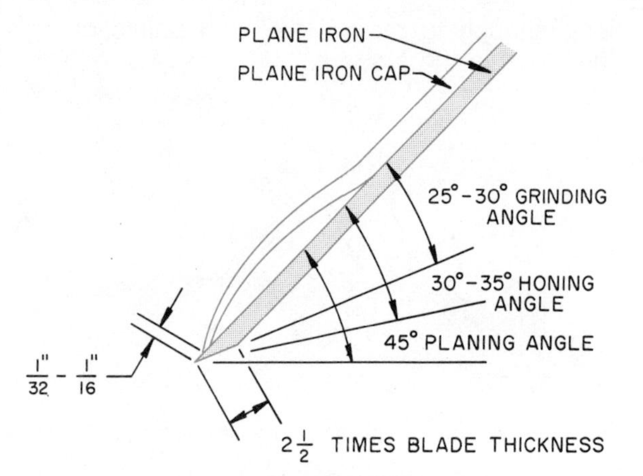

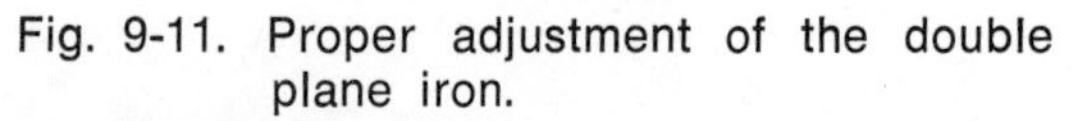

Fig. 9-11. Proper adjustment of the double plane iron.

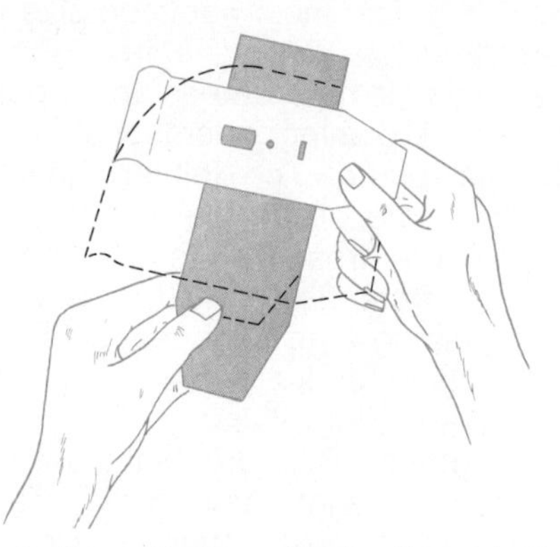

A. Drop the top blade through the slot at right angles.

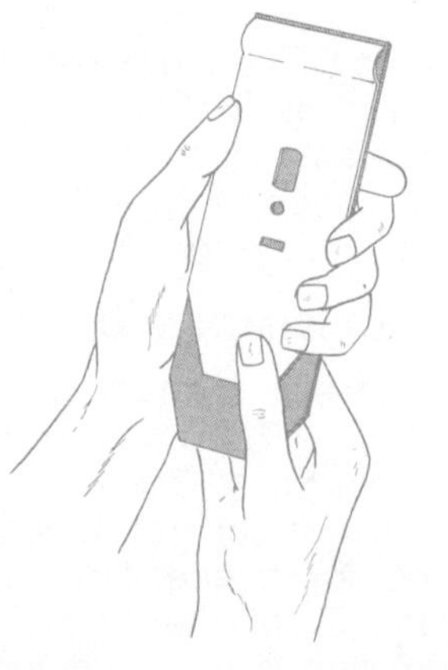

B. Move the top blade to the end of the slot, turn the blade into position, and tighten the screw.

Fig. 9-12. The assembly of the double plane iron.

The plane shown in Fig. 9-8 has a *double plane iron* assembly. This consists of a *plane iron* (cutting edge) and a *plane iron cap* held together by means of the *plane iron screw.* The plane iron cap causes the shavings to curl and break as they are cut, but keeps the wood from splitting before it is cut. It should be properly adjusted to the cutting edge, which is normally between 1/32″ and 1/16″ behind the cutting edge, Fig. 9-11. The double plane iron should be assembled with the plane iron cap fitting tightly against the plane iron. Chips will lodge between the plane iron and plane iron cap if they do not fit properly, and prevent the plane from cutting.

Care must be taken to avoid striking the sharpened edge with the plane iron cap during assembly of the double plane iron. The assembly is made as shown in Fig. 9-12. The plane iron screw is dropped through the hole with the plane iron and cap held at right angles to each other. The cap is placed on the side opposite the beveled edge, then is moved to the end of the slot, still at a right angle to the plane iron. The irons are then turned parallel to one another, the cap is moved to within 1/32″ or 1/16″ of the cutting edge, and the screw is tightened, Fig. 9-11.

The double plane iron is placed in the plane with the bevel side down. The slot in the plane iron is placed over the *lateral adjustment lever,* and is held in the plane by means of a *lever cap*, Fig. 9-7. Remember, care should be taken to avoid striking the sharpened edge with the plane iron cap, because this will dull the keen cutting edge. The lever cap screw should be adjusted so the cam of the lever cap snaps firmly in place.

The plane iron is adjusted parallel to the plane bottom by means of the lateral

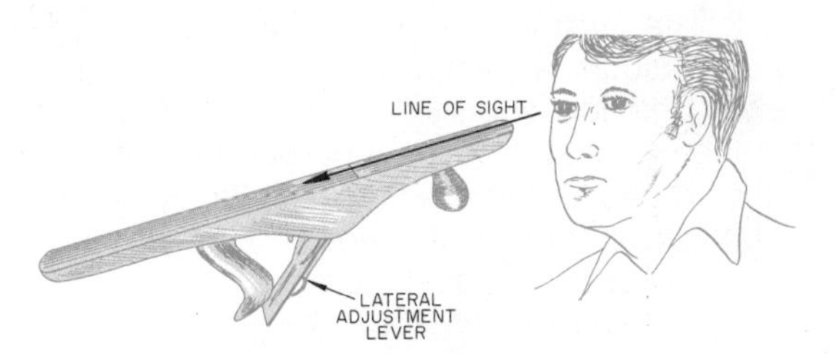

Fig. 9-13. Adjusting the plane iron with the lateral adjustment level and checking to see that it is parallel.

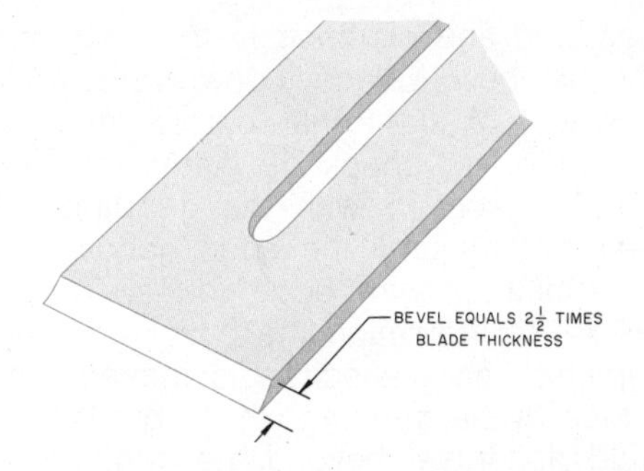

Fig. 9-15. The bevel angle of a double plane iron.

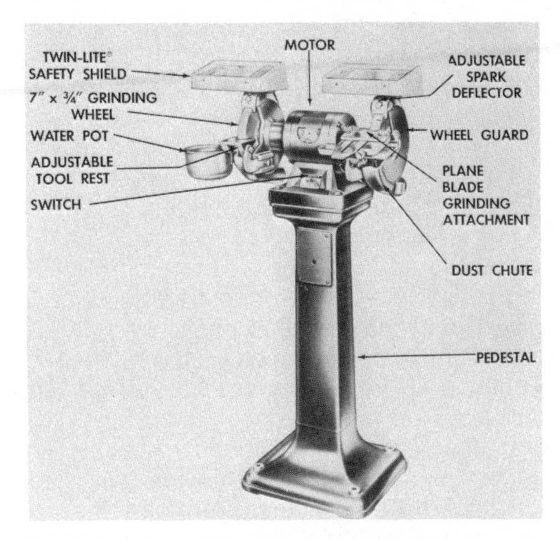

Fig. 9-14. A general-purpose tool grinder. (Power Tool Div., Rockwell Mfg. Co.)

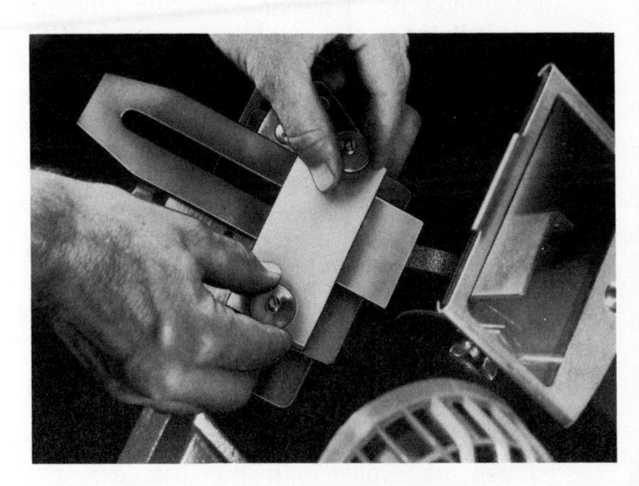

Fig. 9-16. A tool grinder equipped with a special attachment for plane irons.

adjustment lever, Fig. 9-13. It is adjusted so it projects about 1/16″ through the mouth of the plane. The *adjusting nut* raises and lowers the plane iron through the mouth. Turn the plane over and determine if the plane iron projects an equal distance on both sides of the mouth. Adjust the assembly until the plane iron is parallel to the bottom and projects equally on both sides of the mouth.

CONDITIONING THE PLANE IRON

The plane iron must have a very sharp edge to produce quality work. A plane iron is sharpened by *grinding* and *whetting* (sometimes called *honing*). The grinding operation shapes the plane iron and removes any major defects such as nicks. Only the cutting edge of the plane iron is whetted, making the edge extremely sharp.

Grinding plane irons. Grinding is done on a tool grinder such as shown in Fig. 9-14. It is not necessary to grind the plane iron each time it is used, but only when the iron is nicked or needs to be shaped. The adjustable tool rest is set at the proper angle to produce the required bevel, usually an angle of 25° to 30°, Fig. 9-15. This produces a beveled surface about two and one-half times the thickness of the plane iron. A more accurate bevel can be ground if the grinder is equipped with a grinding attachment, Fig. 9-16. The plane iron is disassembled and clamped in the attachment, with the bevel placed downward against the wheel. A light trial cut is made and the angle is checked. After the setup is made, proceed with the grinding. Continue grinding until the edge becomes thin and a butt or wire edge forms. The wire edge is an indication the bevel grinding is complete. The plane iron is moved across the face of the grinding wheel, but the edge should not travel beyond the center of the wheel. The outside corners can be ground slightly more than the center or this can be performed in the oilstone whetting process.

Oilstones for whetting edge-cutting tools. An oilstone is used to whet the final keen edge on the plane iron. There are several types of oilstones, classified as either *natural* or *artificial.* An *Arkansas stone* is a natural stone, whitish in color, and can be used to place a very keen cutting edge on a plane iron. Since the beginning of the 20th Century, many high-quality, man-made abrasive stones have been manufactured. The most common of these man-made stones are *aluminum oxide* and *silicon carbide.* The aluminum oxide stone is grayish in color, and the silicon carbide stone is brownish-red. The manufactured stones are available in three grades: fine, medium, and coarse.

The oilstone is used with a light lubricant such as kerosine. The oil washes away the loose particles of metal from the stone so that it does not become loaded and reduces the cutting efficiency of the plane iron. The oilstone should be kept moist with oil at all times.

The plane iron must be whetted after each grinding to restore the keen cutting edge. The plane iron is removed from the plane and placed on the oilstone with the bevel down, Fig. 9-17. The bevel is placed flat on the stone and then raised a few degrees. The iron is then moved back and forth *over the entire surface of the stone.* If only a small section of the stone is used, it will cause the stone to become dished (grooved). If an oilstone does become grooved, it can no longer serve the purpose of restoring keen cutting edges.

The whetting should continue until a wire edge appears across the entire cutting edge. The wired edge can be felt by carefully sliding the thumb over the edge of the iron. The wired edge can be removed by turning the iron with the bevel up and stroking it on the oilstone. The iron must be held flat against the stone. After a few strokes, turn the iron back to the bevel side and stroke it a few times. Alternate the iron back and forth until the wire edge breaks

Fig. 9-17. Whetting a plane iron on an oilstone.

Fig. 9-18. Removing the wired edge on a plane iron.

off, Fig. 9-18. The keen edge can be tested for sharpness by cutting on a scrap piece of wood. The sharpened iron should produce a smooth, feather-like shaving.

SMOOTHING AND LEVELING STOCK WITH A HAND PLANE

The parts of a product must be square if they are to fit together properly. A set procedure is used for squaring and smoothing the surfaces of stock. The procedure normally used is shown in Fig. 9-19. Steps used to square all of the surfaces are discussed below.

The best surface of the stock is selected to plane first. Hand planing will remove the imperfections in the stock left from machine planing. Remember, the direction of the grain can be determined by inspecting the edge of the board. The board is placed flat on the bench and secured between a bench stop and the vise dog so it can be planed with the grain, Fig. 9-20..

The plane is passed over the board in the direction of the grain with the depth adjusted until a feather-like shaving is produced. The first few strokes may not cut the entire length of the board, but will remove the high spots on the surface of the stock.

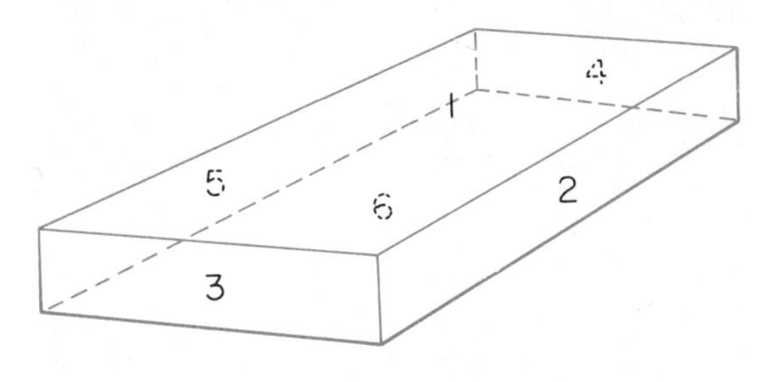

Fig. 9-19. Sequence for squaring stock by planing.

When these high spots are removed, the plane will produce a shaving on the entire length of the board. Only enough stock should be removed to produce a smooth, flat surface. The plane should be held with the knob in the left hand and the handle in the right hand. To start the cut, apply pressure only to the knob of the plane. Even pressure is then applied to the knob and handle when both are on the surface, and less pressure is placed on the knob as the toe leaves the surface. This procedure reduces the possibility of planing a convex surface, Fig. 9-21.

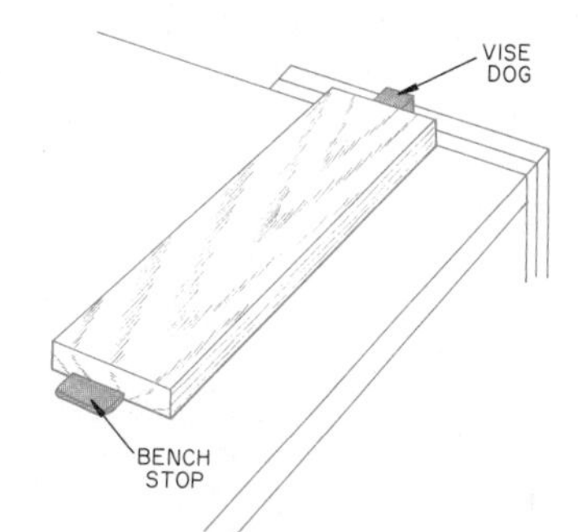

Fig. 9-20. Using a bench stop and vise dog to secure stock for planing.

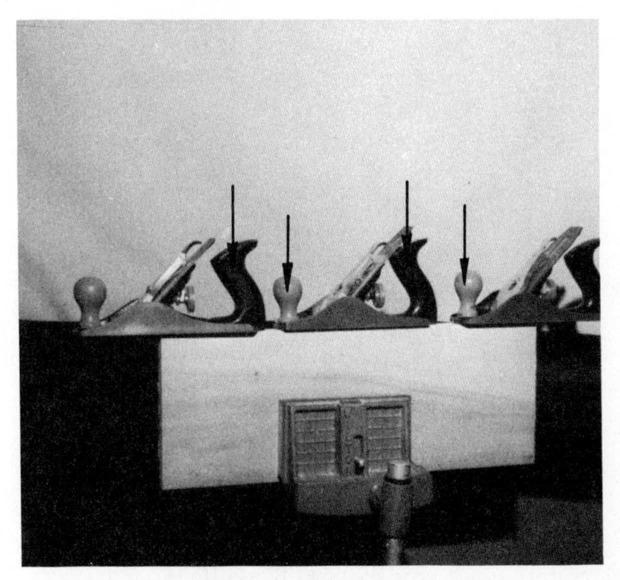

Fig. 9-21. Applying pressure to the plane to produce a smooth cut.

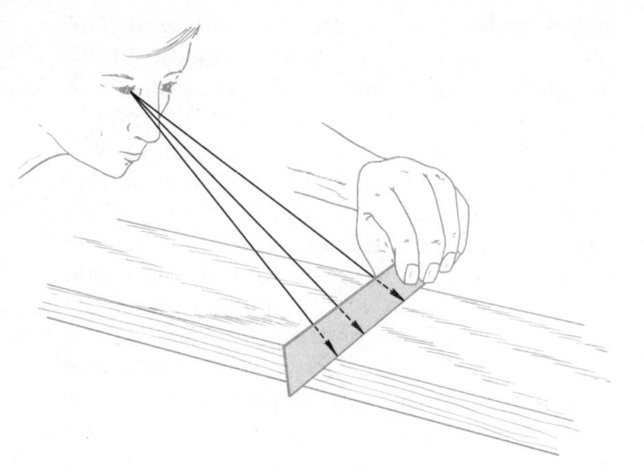

A. Across the grain.

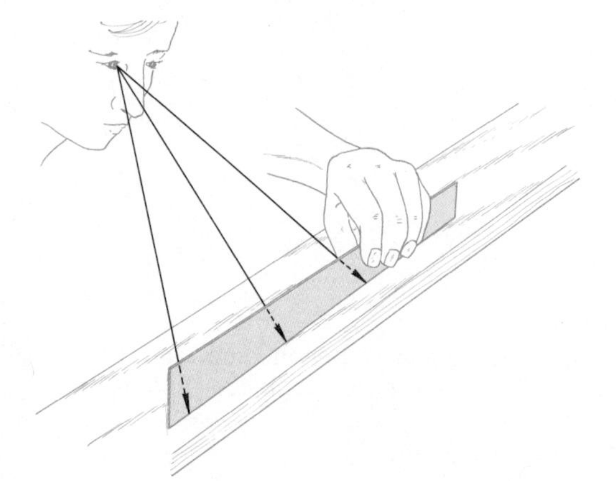

B. Along the length of the grain.

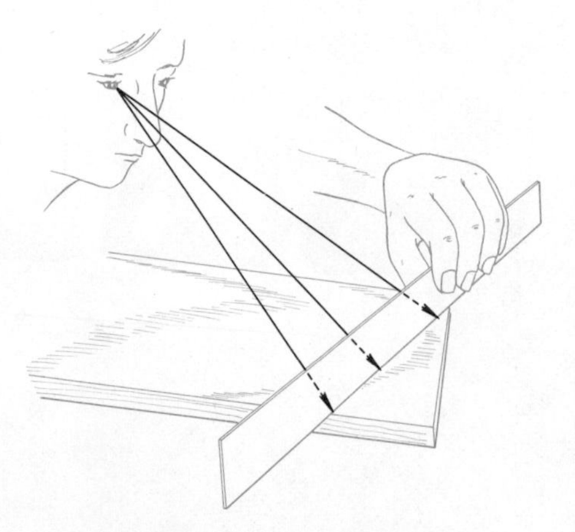

C. Across the corners.

Fig. 9-22. Checking a board for squareness.

Check the surface with a straightedge or a square. Hold the board up to the light and see if light passes between the board and the square, Fig. 9-22. If light shows, this indicates low spots. Slide the straightedge the full length and width of the board. Also check the board for flatness by laying the straightedge from corner to corner. Mark the high spots with a pencil and remove the excess material with the plane.

After smoothing and leveling the surface, the first edge is planed to form a right angle with the face. The board is placed in a vise as shown in Fig. 9-23. The best edge should be selected for planing, and should be planed until it is square with the first surface or *working face.* Pressure is applied to the plane as shown in Fig. 9-24. The edge of the board is tested for squareness to the working face, Fig. 9-25.

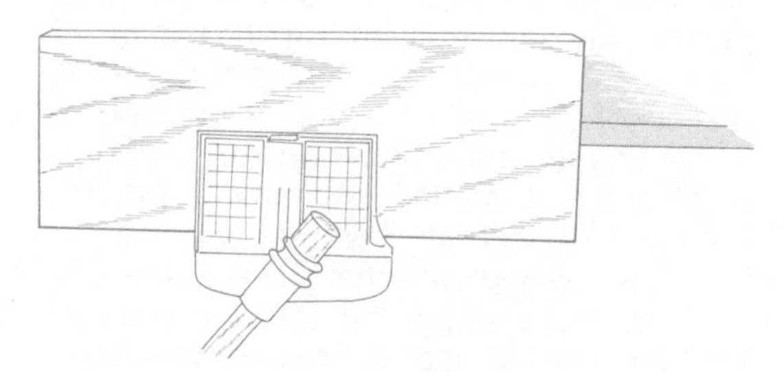

Fig. 9-23. Position of board in vise for edge planing.

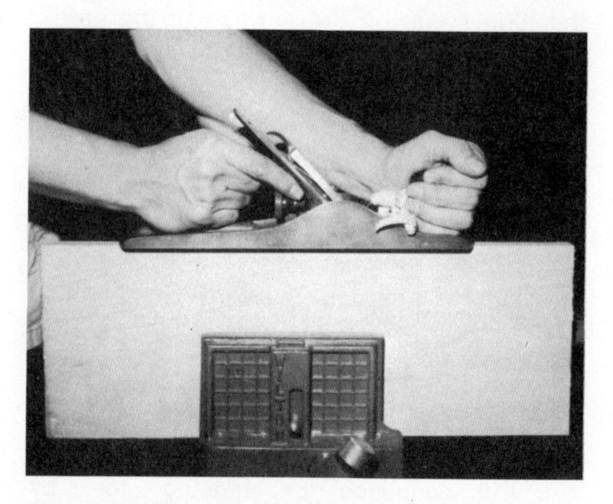

Fig. 9-24. Planing the edge of a board.

PLANING END GRAIN

The first end of the stock is planed square to the working face and working edge. The best end is selected and placed in the vise as shown in Fig. 9-26. The end grain must be planed differently from the edge or face, as it will tend to split if special care is not used.

1. Plane two-thirds of the width of the stock from one edge, then reverse the direction and complete the surface, Fig. 9-27. The end is tested for squareness with the working face and edge with a try square.
2. Splitting can be avoided by cutting a chamfer on one edge of the stock, Fig. 9-28. The end can then

Fig. 9-25. Testing the edge for squareness to the face of the board.

Fig. 9-26. Position of board for end planing.

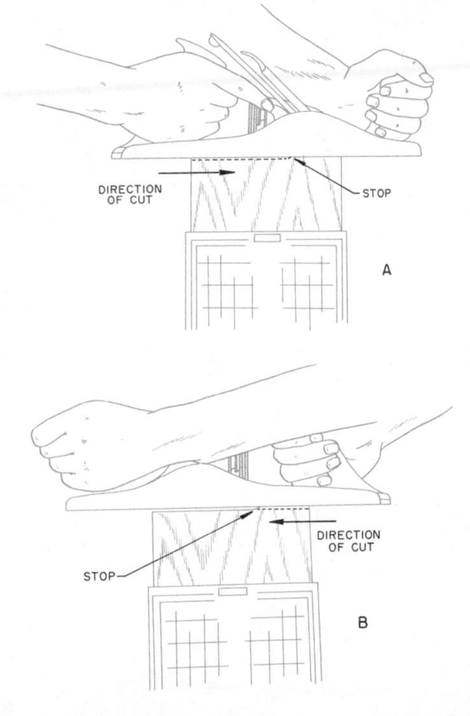

Fig. 9-27. Planing end grain from two directions.

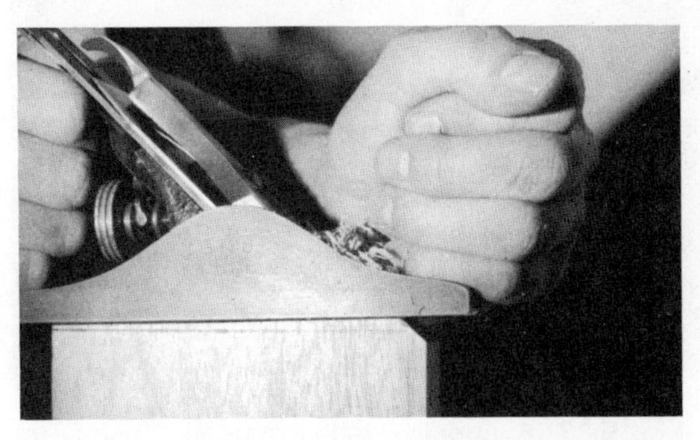

Fig. 9-28. Prevent splitting during end grain planing by chamfering one edge.

Fig. 9-29. Supporting end grain with a piece of scrap wood.

Fig. 9-30. A block plane may be used to plane the end grain of narrow boards.

be planed toward the chamfer without damage to the edge. The chamfer must be cut in the waste stock.

3. A piece of scrap stock can be clamped next to the edge as shown in Fig. 9-29. The scrap stock supports the edge and prevents splitting. This method makes it possible to plane the entire width of the board.

In all of the above methods it is helpful to place a layout line on the stock with a try square from the working face and edge. The end should also be tested with a square as the layout line is approached.

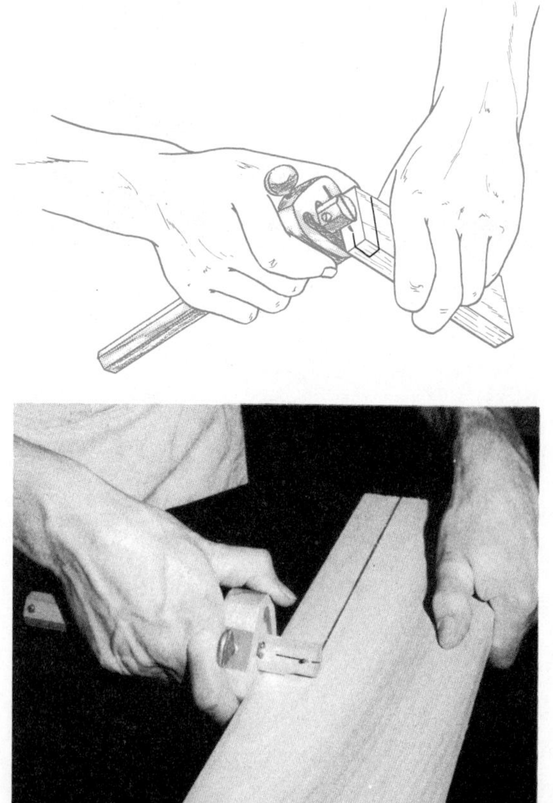

Fig. 9-31. Using a marking gauge to indicate thickness prior to planing.

USING THE BLOCK PLANE

The block plane is especially useful when planing the end grain on narrow boards and may be used to plane the edge of small pieces, Fig. 9-30. The single plane iron used in the block plane is placed in the plane at a much lower angle, resulting in an easier cut of end grain fibers. The block plane is held in one hand and guided with the fingers of the other hand.

After the working face, edge, and end are smoothed and squared, the stock is cut to length. The second end is then planed square using one of the methods previously described. The stock is ripped to width and planed square with the other surfaces. The final step is to plane the opposite face parallel to the working face to the desired thickness. A *marking gauge* is used to mark the ends and edges for the desired thickness, Fig. 9-31. The opposite face is planed to the layout lines, and a square is used to check the stock for squareness.

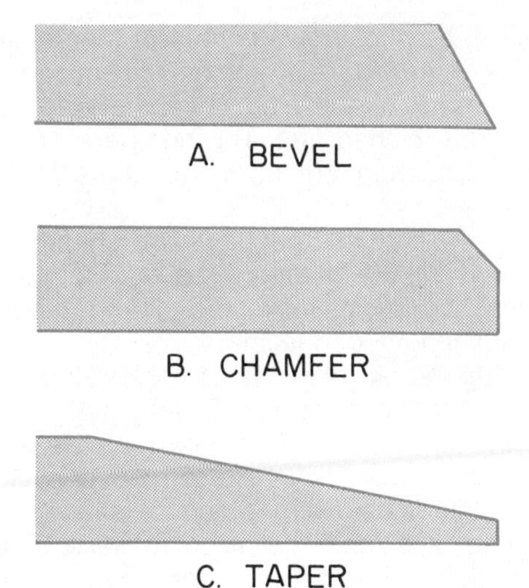

Fig. 9-32. Decorative cuts.

PLANING BEVELS, CHAMFERS, AND TAPERS

Bevels, chamfers, and tapers are angular cuts placed on the corner, edge, or surface of stock, Fig. 9-32. The angular cuts are often used as decorative cuts.

Bevels. The bevel is an angular cut on the entire length of the edge or end of a board, Fig. 9-32A. The bevel may be used on one or all edges and ends, and can be planed at any desired angle.

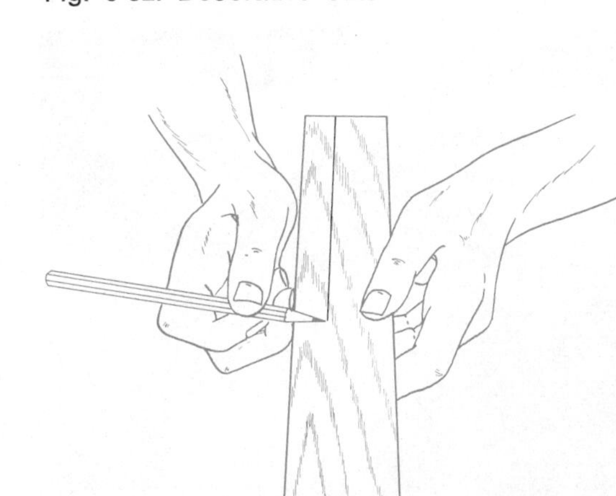

Fig. 9-33. Using a pencil and the fingers to gauge a layout line on the stock.

Chamfers. A chamfer goes only part way across the surface. It is usually planed on the stock at a 45° angle, Fig. 9-32B. The chamfer and bevel are usually laid out with a sliding T-bevel, or gauged with a pencil or a marking gauge. The pencil is more satisfactory than the marking gauge, as the spur point of the marking gauge will damage the wood. The layout is made by gauging the line with a pencil with the stock held in the fingers, Fig. 9-33. The lines are placed on the surface and adjacent edge to layout a chamfer. The bevel or chamfer is planed with a jack plane or block plane. It is easier to place small stock in a handscrew clamp and then fasten it in a vise, Fig. 9-34. When an end and a side are to be beveled or chamfered, the edges are planed first. The ends are planed with a shearing cut or at an angle to the stock.

The angle of the bevel or chamfer is checked with a sliding T-bevel, Fig. 9-35. To test the angle, move the sliding T-bevel the full length of the stock.

Tapers. A taper is a decorative cut which runs the direction of the grain. It may run the full length or any part of the total distance, and it may be used on one side or all four sides of a piece of stock. The purpose of a taper cut is to make a piece of stock smaller on one end than on the other. This cut is commonly used to improve the design of furniture legs. It makes the legs lighter in weight, improves the proportions, and gives them a more graceful appearance.

The stock to be tapered is first squared, then laid out according to the desired taper. The size of the small end is marked on the end of the stock, and a line is drawn on the edges of the stock. The stock is clamped in a vise and planed toward the small end of the taper, Fig. 9-36. Some of the excess stock can be removed with a saw and then planed smooth and true.

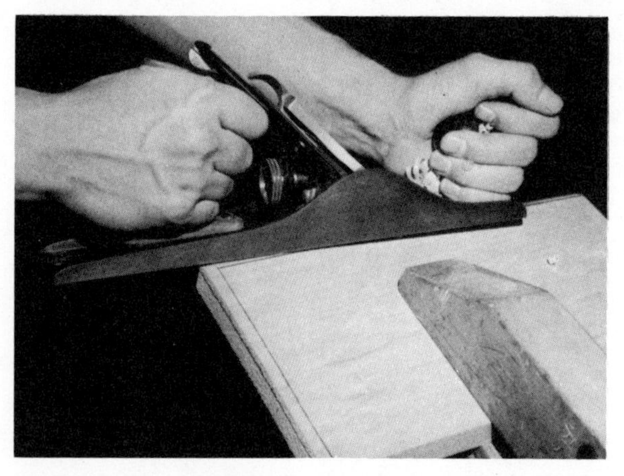

Fig. 9-34. Clamping and planing bevels and chamfers.

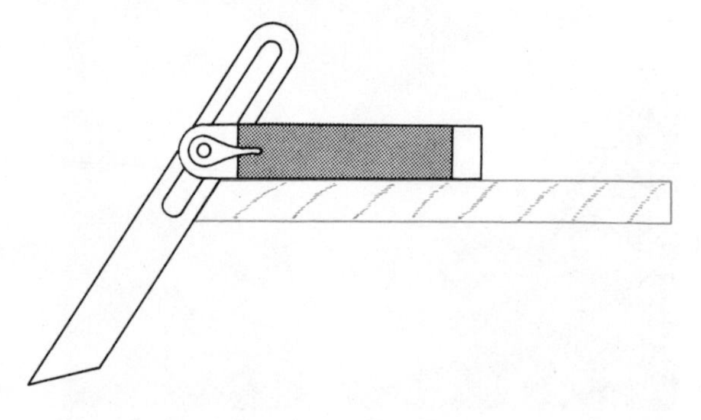

Fig. 9-35. Using a sliding T-bevel to test the angle of a bevel on a chamfer.

SPECIAL-PURPOSE PLANES

There are also many special-purpose planes which are designed to perform a particular job. A few of the most common include: (1) router plane, (2) rabbeting

A. Making the layout on the stock.

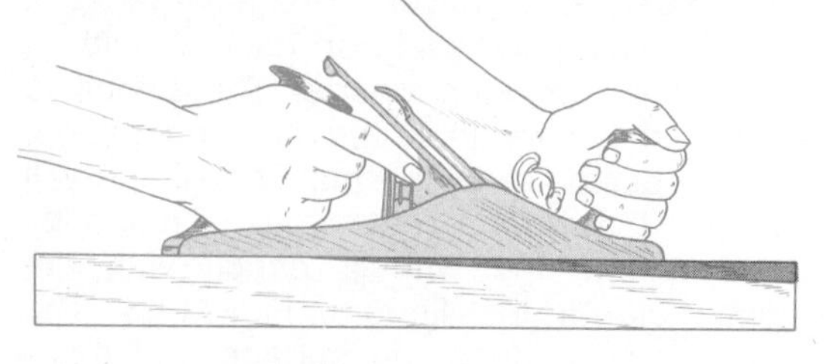

B. Planing toward the end of the taper.

Fig. 9-36. Planing tapers.

plane, (3) circular plane, (4) spokeshave, and (5) forming plane.

Router plane. The router plane, Fig. 9-37, is designed to remove stock between two parallel saw cuts. It is equipped with 1/4" or 1/2" square cutters or with V-shaped cutters. The depth of the cut can be adjusted by loosening the *collar clamp screw,* then raising or lowering the cutters by the *adjusting nut,* Fig. 9-37.

The router plane is used for cutting grooves and dadoes. It cuts a groove parallel with the sides of the stock as the base of the plane is supported on the surface being cut.

To cut a dado using the router plane, the stock is first laid out. The shoulder (side) cuts are made with a backsaw. A chisel is used to remove the stock to within 1/16" of the bottom of the dado. The router plane is then used to finish cutting the bottom, Figs. 9-38 and 9-39. The router plane can also be used to smooth the bottom of rabbet joints.

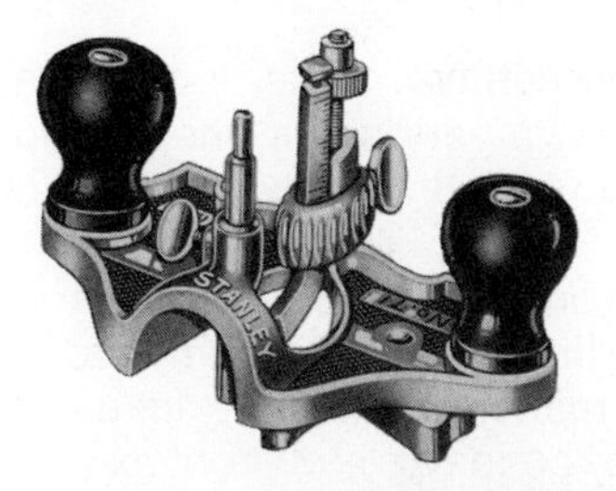

Fig. 9-37. A router plane. (Stanley Tools)

Fig. 9-38. Removing stock from the bottom of a dado using a router plane.

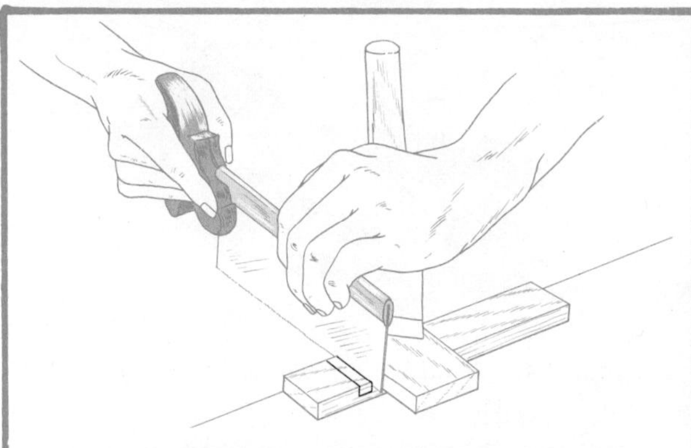

A. Follow layout lines and make shoulder cuts with a backsaw.

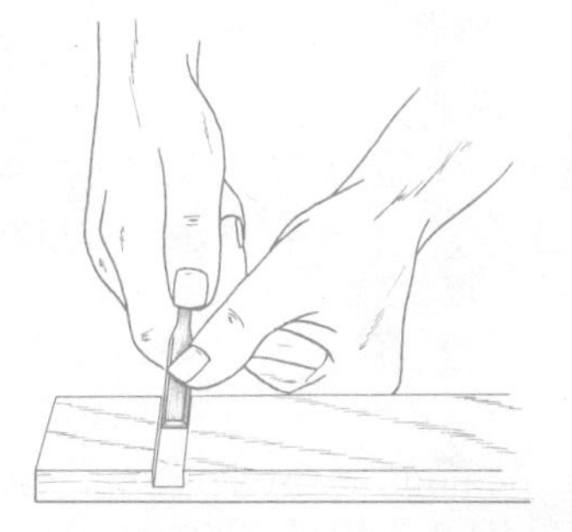

B. Remove stock in cut to within 1/16" of the bottom of the dado.

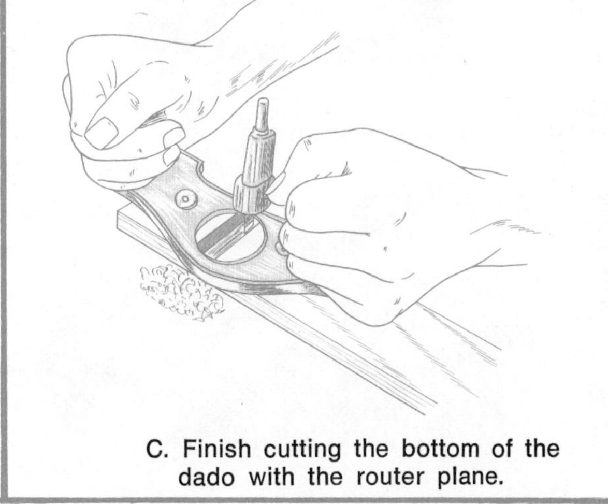

C. Finish cutting the bottom of the dado with the router plane.

Fig. 9-39. Cutting a dado with the router plane.

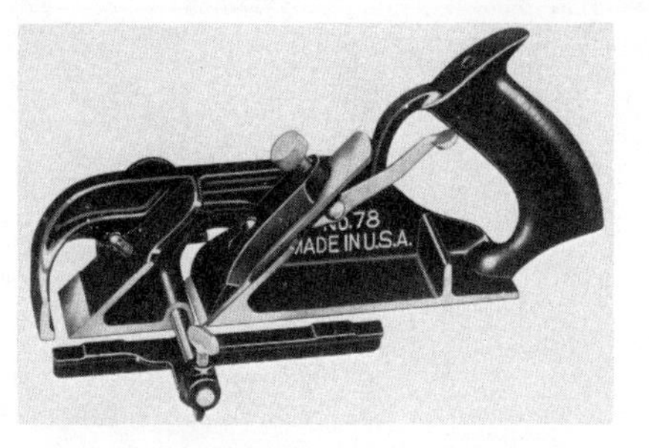

Fig. 9-40. A rabbeting plane. (Stanley Tools)

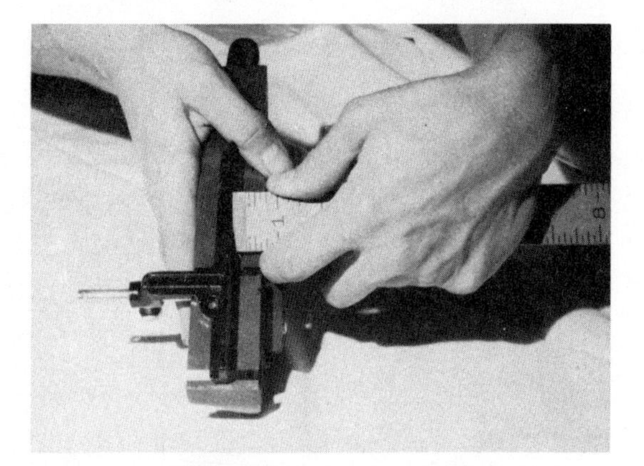

Fig. 9-41. Setting the fence on the rabbet plane with a rule.

Fig. 9-42. A circular plane.

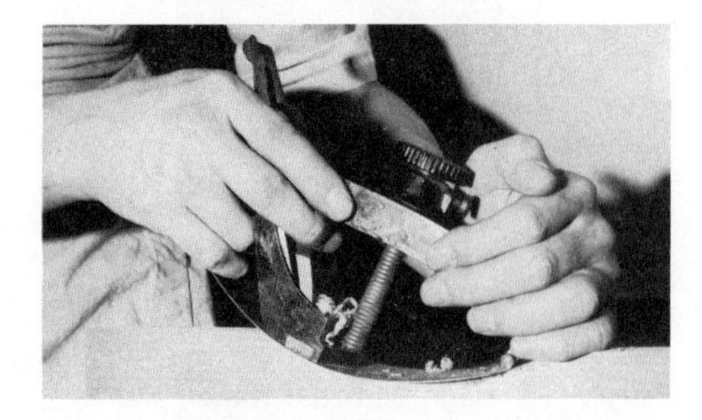

Fig. 9-43. Planing a concave surface with a circular plane.

Rabbet plane. The rabbet plane, Fig. 9-40, is used to cut a rabbet along the edge of stock. The rabbet plane can be adjusted so that both the width and depth of a rabbet can be controlled. The width is controlled by the position of the *fence,* set with a rule as shown in Fig. 9-41. The depth of cut is regulated by the *depth gauge.* The plane iron must be sharp and can be conditioned as discussed earlier in this chapter.

Circular plane. The circular plane is designed to plane and form convex and concave surfaces, Fig. 9-42. It has a double plane iron assembly. The frame of the plane has a flexible bottom which can be adjusted into a convex or concave position by regulating the *adjusting thumbnut* on the top of the plane. The adjusting nut is regulated until it fits the contour of the work, Fig. 9-43.

The circular plane is used similar to a jack or smoothing plane. The plane is grasped by the nose and the heel, and equal pressure is applied to both. A light cut is taken with each stroke. The circular plane must be used in the direction of the grain.

Spokeshave. The spokeshave is a plane with a very small base or bottom, Fig. 9-44. It cuts in the same manner as a plane, and the blade is shaped and cuts like a plane iron. The blade is inserted in the frame with the bevel downward. The short bottom makes it possible to use the plane for both concave and convex work.

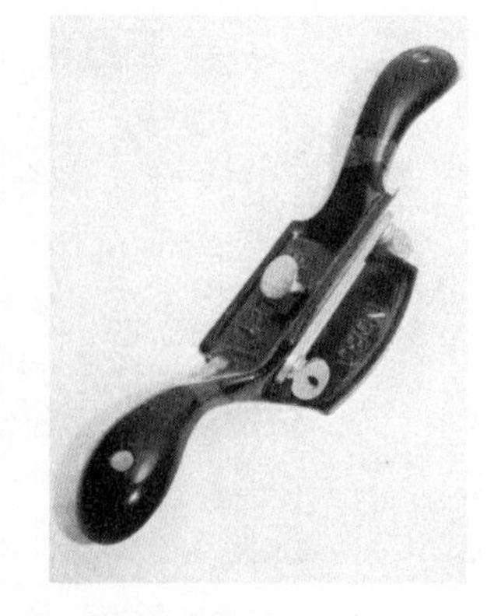

Fig. 9-44. A spokeshave.

The spokeshave is usually pulled as shown in Fig. 9-45. The blade is set for the depth of cut with the *adjusting nuts.* The adjusting nuts are also used to align the cutter with the bottom of the plane. The spokeshave is especially useful for shaping irregular work as shown in Fig. 9-45. The setting should produce a thin, uniform shaving. When the setting is too deep, the spokeshave will chatter and tear the wood.

Forming plane. A forming plane, as shown in Fig. 9-46, contains a serrated bottom. The serrated teeth are often compared to a mass of tiny block planes. There are holes between each of the teeth through which the shavings pass. The cutting edges cannot be restored when dulled, therefore the bottom or cutting edge is replaced when it no longer cuts properly. There is no adjustment on the forming plane for the depth of cut. It is used in the same manner as a jack plane. The forming plane is used for fast cutting and trimming.

SMOOTHING SCRAPERS

It is sometimes difficult or impossible to smooth some wood surfaces with a plane. Wood which is cross-, curly-, or wavy-grained must be smoothed with a scraper because a plane tears the grain. The excess material should first be removed with the plane and the surface should be leveled. Then the plane is used with a light shearing cut and the scraper is used to smooth the surface. There are several types of scrapers, but all cut in basically the same manner.

HAND SCRAPERS

The *hand scraper* shown in Fig. 9-47 is a piece of rectangular or curved tool steel. A burred edge is formed on the edge of the scraper and the cutting action is performed by this burred edge. The hand scraper is grasped between the thumb and

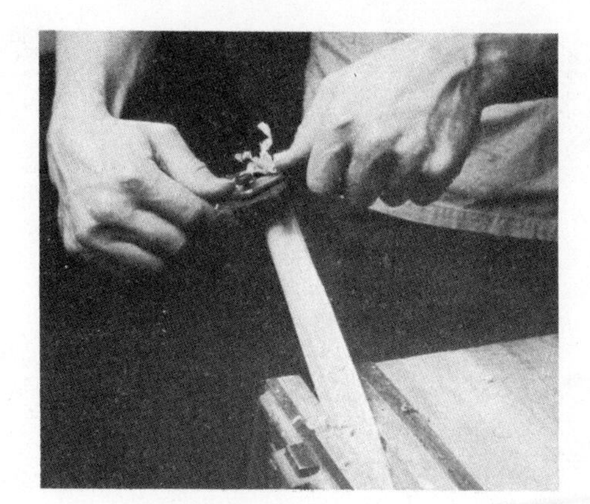

Fig. 9-45. Shaping irregular stock with a spokeshave.

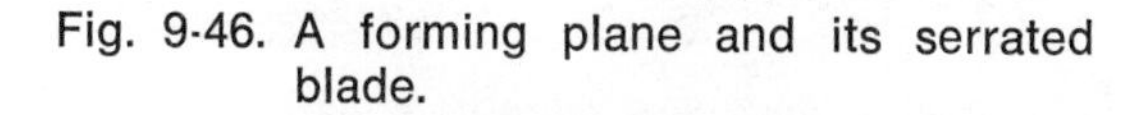

Fig. 9-46. A forming plane and its serrated blade.

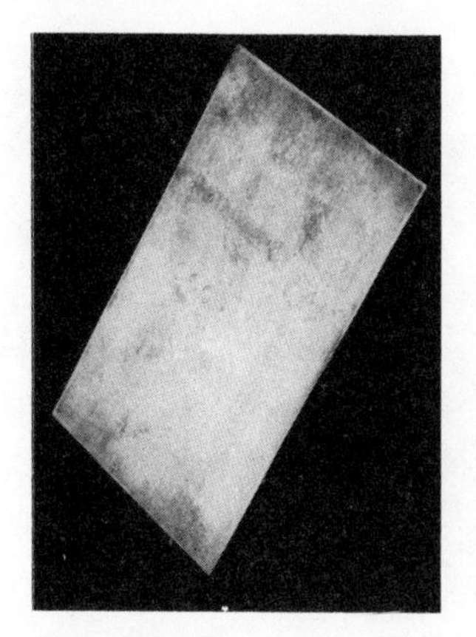

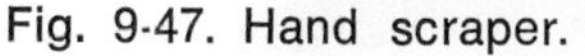

Fig. 9-47. Hand scraper.

Fig. 9-48. Smoothing stock with a hand scraper.

fingers on each hand. The thumbs are used to apply pressure for scraping, Fig. 9-48. The edge of the scraper is placed on the stock and tilted to an angle of about 75°. The scraper then can be either pushed or pulled; however, a heavier cut results when it is pushed

CONDITIONING HAND SCRAPERS

To condition and sharpen a hand scraper, place it in a bench vise and remove the worn edge with a mill file as shown in Fig. 9-49. Draw-file the cutting edge with the mill file held at a 90° angle to the side of the scraper. A few strokes will true the edge and form a wired edge.

To smooth this wired edge, place the filed edge on an oilstone (Fig. 9-50) and whet the edge. The scraper is then laid flat and whetted on each side, then placed in a vise and burnished on both edges. A burnisher is a hardened piece of oval-shaped tool steel used to turn the edge of the scraper into a hook (burred edge) which serves as the cutting edge, Fig. 9-51. The first stroke with the burnisher is performed with the burnisher at a 90° angle to the side of the scraper. The angle is then lowered to about 85° and the burnishing is continued until a hook appears on the edge. Firm pressure is applied to the burnisher.

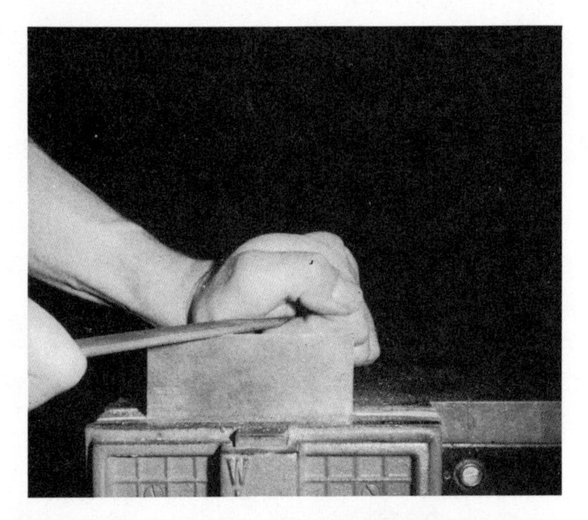

Fig. 9-49. Draw-filing edge of a hand scraper.

Fig. 9-50. Whetting a hand scraper.

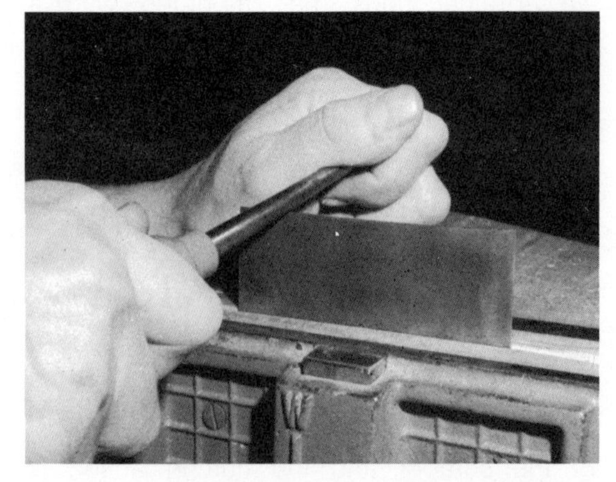

Fig. 9-51. Using a burnisher to form a hook on the edge of a hand scraper.

HAND SCRAPERS FOR SMOOTHING IRREGULAR SHAPES

A *swan-neck hand scraper* shown in Fig. 9-52 is used for scraping irregularly shaped pieces such as coves. It is reconditioned in a manner similar to the above method for hand scrapers.

The *cabinet scraper* is similar in appearance to a spokeshave, but removes stock like a hand scraper, Fig. 9-53. It is held by both handles like the spokeshave. These large handles make it easier to hold than the hand scraper, and more pressure can be exerted on it. The cabinet scraper is especially desirable for scraping large surfaces prior to sanding with abrasive paper. It is also used to smooth curly-grained wood that cannot be smoothed with a hand plane.

To adjust a cabinet scraper, loosen the adjusting thumbscrew and slip the blade in from the bottom. Place the scraper on a flat surface with the blade in position with the bevel side pointing toward the adjusting thumbscrew. Tighten the thumbscrew and clamp the blade in position. The blade is flexed by tightening the adjusting thumbscrew, causing the blade to protrude below the bottom of the frame. The scraper is now ready to cut the surface of the stock.

The blade of a cabinet scraper is reconditioned like the hand scraper; however, the cutting edge of the blade is ground to an angle of 45°.

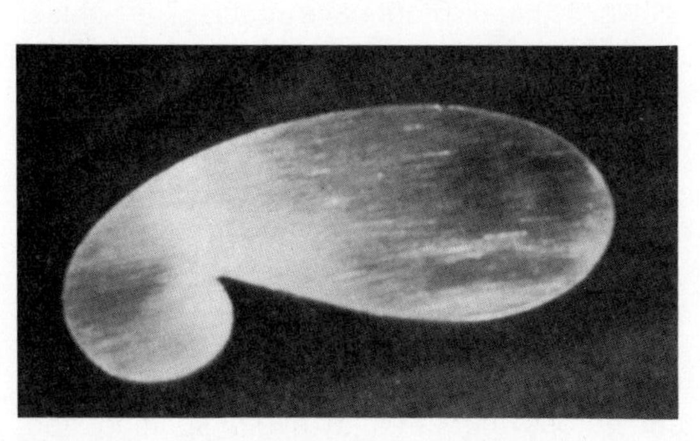

Fig. 9-52. Swan-neck hand scraper.

SHAPING OF STOCK WITH HAND CHISELS

A chisel is another form of edge-cutting woodworking tools. The principle of cutting with a chisel is essentially the same as that of a plane. There are two basic types of chisels: (1) tang, and (2) socket.

Tang chisels. The end of a tang chisel is tapered and fits into a handle of either wood or molded plastic, Fig. 9-54. It is designed for light work, and should not be pounded with a mallet. The paring and light cutting should be done only with hand pressure.

Socket chisels. The handle of the socket chisel fits into a hollow barrel on the end of the blade, Fig. 9-55. The socket

Fig. 9-53. Cabinet scraper.

Fig. 9-54. Tang chisel.

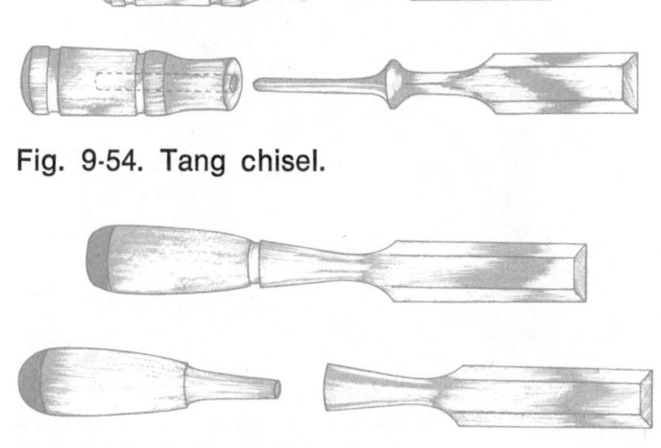

Fig. 9-55. Socket chisel.

Fig. 9-56. Whetting a chisel.

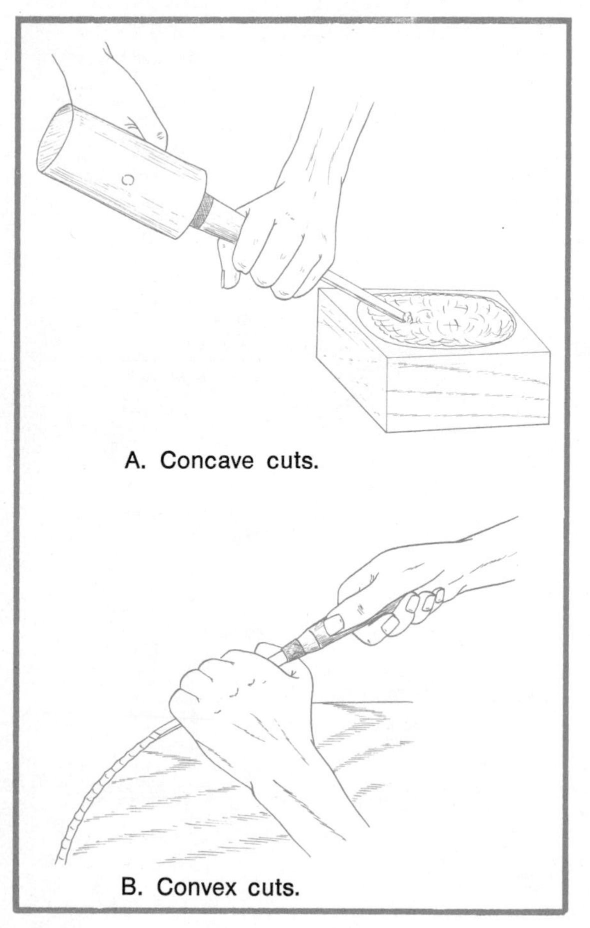

A. Concave cuts.

B. Convex cuts.

Fig. 9-57. Making cuts with a chisel.

chisel is normally a heavier-duty chisel, and can be struck with a wooden mallet to remove larger quantities of stock.

SIZES AND CONDITIONING OF CHISELS

The size of a chisel is given by its width. Most chisels are available in standard lengths. Shorter chisels (called butt chisels) may also be purchased.

Both tang and socket chisels are available in a wide range of sizes from 1/8″ to 2″. A set of chisels ranges in size intervals of 1/8″ from 1/8″ to 1″.

Chisels are widely used for joint construction and for the removal of excess material from stock. The blade of the chisel has a ground bevel at a 30° angle. To perform quality work, chisels must have a keen cutting edge. Chisels are ground on a tool grinder when they become badly nicked or when the angle must be reshaped. They should be whetted after each grinding or whenever they become dull to restore the cutting edge, Fig. 9-56. The same procedures are used for reconditioning a chisel as were described for a plane iron.

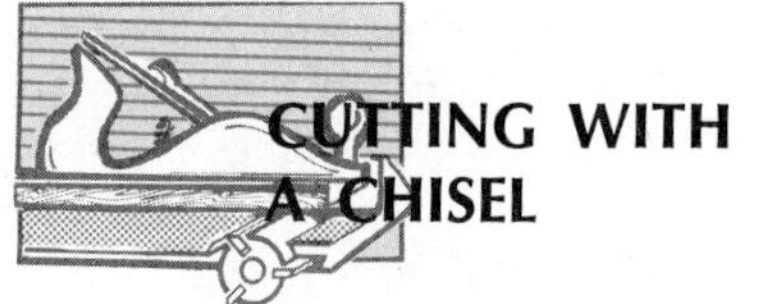

CUTTING WITH A CHISEL

A concave or convex contour can be formed with a chisel, Fig. 9-57. The stock must be secured in a vise or clamped to a bench. The chisel is turned with the bevel down to make concave cuts. The chisel handle is held in the right hand, and the fingers on the left hand are used to guide the chisel blade. Pressure is exerted on the chisel with the right hand and the chisel is moved slightly from side to side to create a shearing cut. If a large amount of stock is to be removed, a mallet may be used to strike the socket chisel, Fig. 9-58.

Always cut with the grain to cut the length of a board. Cut from both sides to cut across the grain to avoid splitting and chipping on the side of the stock, Fig. 9-59.

PORTABLE POWER PLANE

The portable power plane is capable of performing many of the planing operations otherwise performed with a hand plane, Fig. 9-60. However, the power plane reduces the time and labor required to perform these planing tasks. It is widely used by carpenters for planing doors and other large surfaces on the site. The knives in the power plane are *spiral cutters.* The cutter is mounted directly on the motor shaft. The cutter head is driven at a speed of about 20,000 RPM.

The plane is equipped with a *fence* which can be adjusted to any desired angle for planing bevels, Fig. 9-61. The fence can

Fig. 9-58. Striking a socket chisel with a mallet to remove stock.

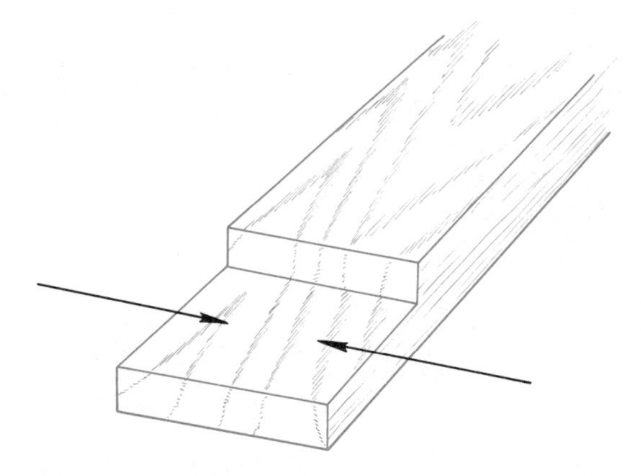

Fig. 9-59. To cut across the grain, chisel toward the center from both sides.

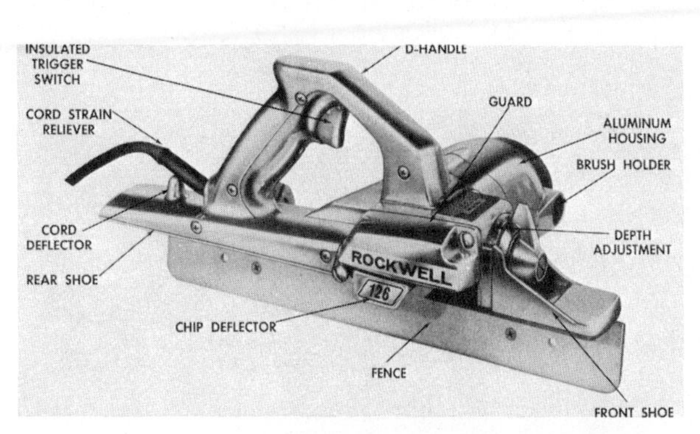

Fig. 9-60. Portable power plane. (Power Tool Div., Rockwell Mfg. Co.)

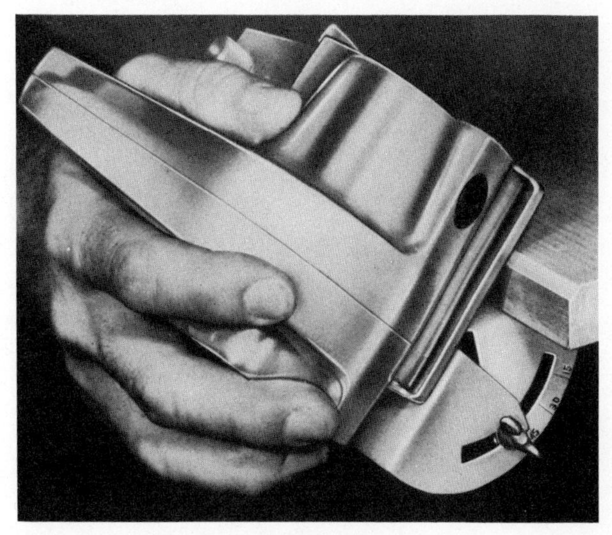

Fig. 9-61. Planing a bevel with the fence adjusted to an angle.

also be removed so the plane may be used to smooth large surfaces, Fig. 9-62.

The depth of cut is regulated by raising or lowering the *front shoe.* The adjustment of the *rear shoe* remains unchanged and is designed to remain level with the spiral cutter.

In addition to the standard power plane, smaller ones called block planes are also available, Fig. 9-63. The power block plane is also equipped with a guide fence and an adjustable shoe for depth adjustment. The shoe on the power block plane is of one piece which pivots on the front of the cast housing of the tool. The power block plane is used for truing short surfaces.

Fig. 9-62. Planing a surface with the fence removed.

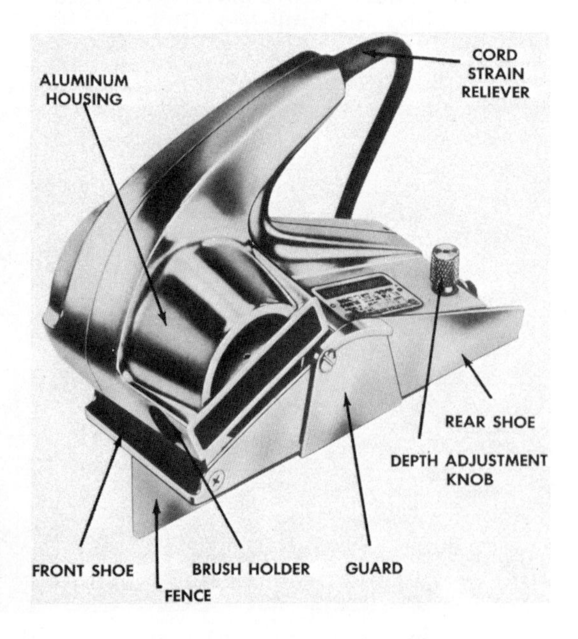

Fig. 9-63. Power block plane. (Power Tool Div., Rockwell Mfg. Co.)

OPERATING THE POWER PLANE

The power plane is held very similarly to a jack plane. The work must be secured in a vise or other holding device. The front shoe and fence are placed firmly against the board to surface an edge. The pressure is placed on front of the plane to start the cut. As the cutting proceeds, even pressure is placed on the front and rear of the plane, then the pressure is placed on the rear of the plane to finish the cut, Fig. 9-64.

STATIONARY POWER PLANES

JOINTER

The jointer is a stationary power planing machine. It is designed to perform work of the same type done with a hand plane or portable power plane. The jointer is primarily used for truing surfaces and edges of stock, cutting rabbets, bevels, chamfers, and tapers. Figure 9-65 shows a typical jointer and its parts.

Fig. 9-64. Planing an edge with a power plane. (Power Tool Div., Rockwell Mfg. Co.)

SIZE AND PARTS OF THE JOINTER

The size of a jointer is determined by the length of the cutter knives or maximum width of stock which can be surfaced. Jointers are available ranging in size from 4″ to 30″. Common sizes for all-purpose shop jointers are usually 6″ or 8″ models.

The parts of the machine consist of a heavy cast steel base, infeed table (front table), outfeed table (rear table), cutterhead, fence, and guard, Fig. 9-65. The *infeed* and *outfeed tables* are vertically adjustable (up and down) on most jointers, but on some models, only the infeed table can be adjusted. The tables are adjusted by the *adjusting hand wheels.* The cutting action of the jointer occurs when the stock is advanced into the revolving *cutterhead,* Fig. 9-66. The cutterhead is a solid steel cylinder in which slots are cut to hold the *cutting knives.* The knives are held in the head by means of a *gib* which also serves as a *chip breaker.* The gips are tightened into place by means of a bolt. The cutterhead is set in ball bearings and revolves at a speed of between 3000 to 4500 RPM. It may have from two to six knives. Most shop models have either three or four knives.

The *fence* is used to aid in holding the stock when performing all types of jointing operations. It can be moved back and forth across the width of the machine to distribute the wear on the knives. In addition, the fence on most machines can be turned so the stock is fed into the knives at an angle, Fig. 9-67. This setup produces a shearing cut and an improved cut on curly-grained wood.

The fence is usually set at a right angle to the cutterhead, but there are times when the fence will be tilted at more of an angle to produce special cuts. The jointer may be equipped with a protractor scale to show the angle of the fence. However, it is best to use a square to check the setting, Fig. 9-68. A sliding T-bevel is used to set the fence at an angle other than 90°.

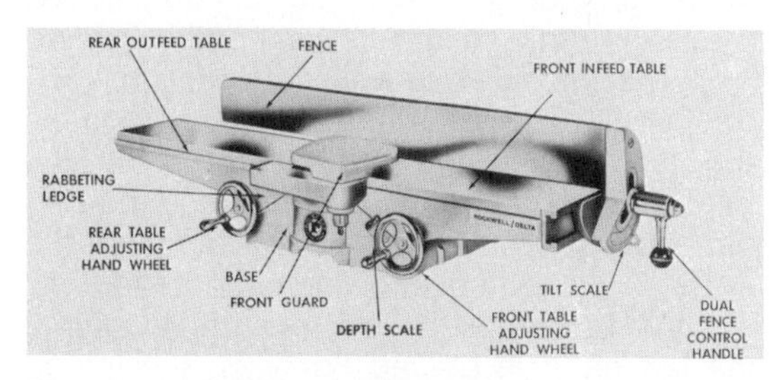

Fig. 9-65. Jointer. (Power Tool Div., Rockwell Mfg. Co.)

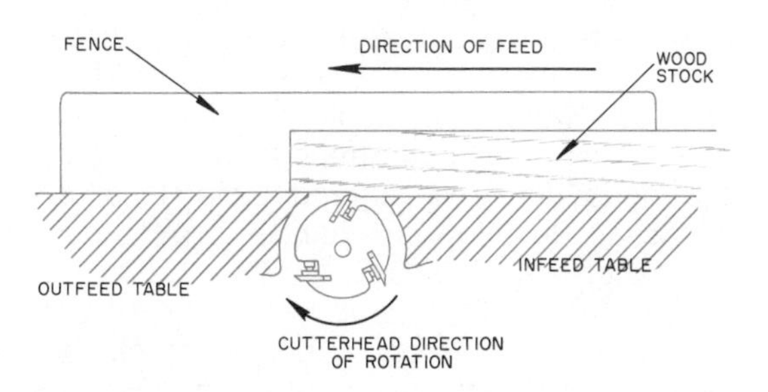

Fig. 9-66. Cutterhead assembly and cutting action of a jointer.

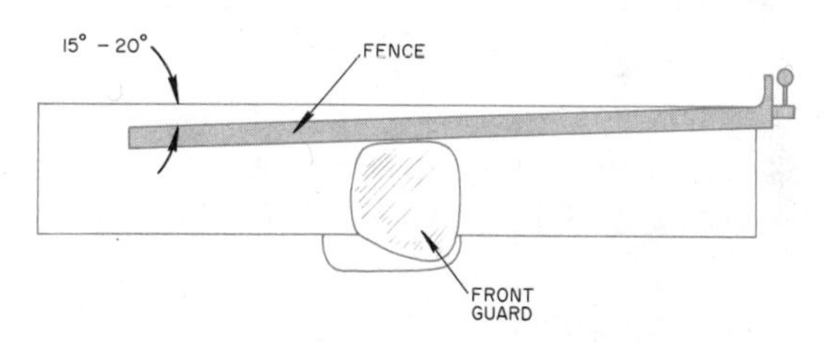

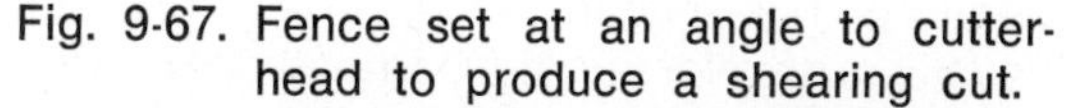

Fig. 9-67. Fence set at an angle to cutterhead to produce a shearing cut.

Fig. 9-68. Using a try-square to set the fence to a 90° angle.

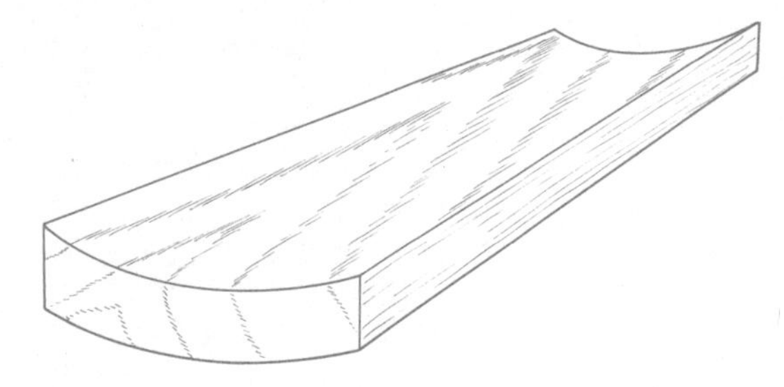

Fig. 9-69. A cupped board.

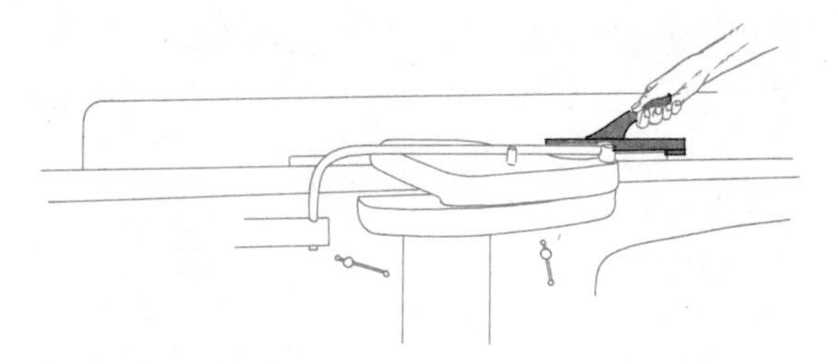

Fig. 9-71. Use a push block to surface a cupped board.

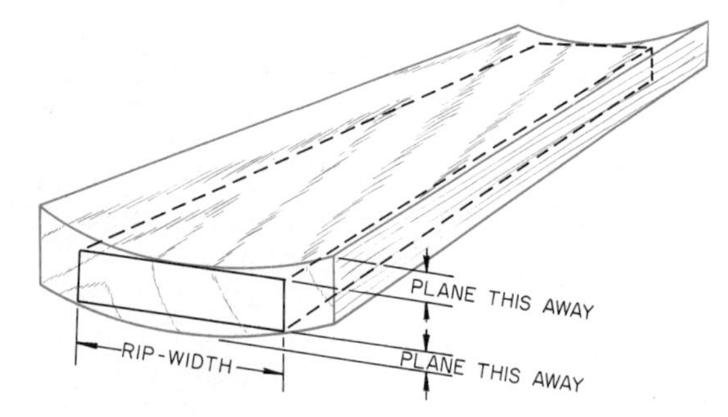

Fig. 9-70. Remove excessive cupping by ripping stock to rough width.

SMOOTHING AND LEVELING A SURFACE WITH PLANING MACHINES

The production of a piece of stock which is square on all four surfaces must begin by leveling one face surface. Boards usually are affected by some type of distortion. For example, they may be cupped, twisted or bowed, Fig. 9-69. One face surface must be planed true before the other surfaces can be cut and planed. If the board is wide and badly cupped, an excessive amount of the thickness will be removed if it is planed true. It is often better to roughcut the board to length and width before planing, Fig. 9-70. The stock can be cut to length with a radial arm saw, band saw, or handsaw. The board can be ripped with a handsaw or a band saw. *Do not attempt to rip or crosscut stock on the circular saw until one edge and surface is flat and level.*

To surface a board which is cupped, the concave side is placed against the table. The board is supported on the high points along the edges of the board, preventing the board from rocking as it passes over the cutterhead on the outfeed table. The left hand is raised and placed on the outfeed table. The cut is made by pushing the stock over the cutterhead with a *push block* held in the right hand, Fig. 9-71.

PLANING AN EDGE

Edge-jointing is the most common operation performed on the jointer. A board is edge-jointed to remove any bow in the stock, straighten the edge, and prepare stock for gluing. The fence is normally set at a 90° angle to the cutter head.

The infeed table is lowered for a 1/16″ depth cut. The flat surface is held against the fence, with the board placed on the infeed table with the grain in the correct direction. The machine is started and the stock is advanced into the cutterhead.

As the stock is fed into the machine, the left hand is placed firmly on the infeed table to hold the stock down. The right hand pushes the board into the cutterhead. The operator stands close to the machine with his feet toward the machine. He should be in a comfortable position and should not have to "walk" the stock. The left hand is lifted over the cutterhead as the board advances onto the outfeed table. It is unnecessary to allow the hands to pass near the cutterhead, Fig. 9-72.

A push stick is not necessary to start the cut for edge-planing narrow stock, but as the stock advances into the cutter, a push stick is used to finish the cut. After

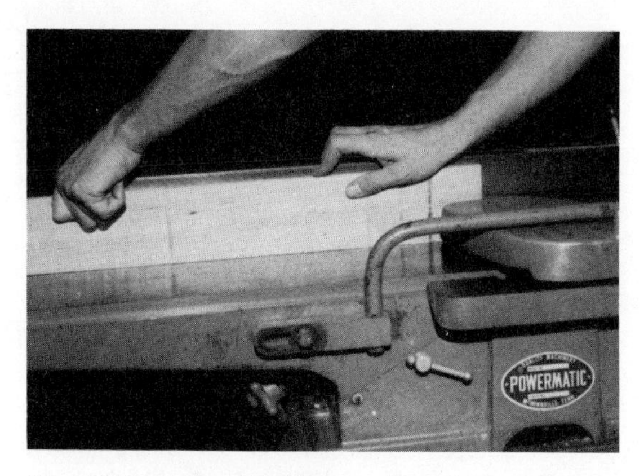

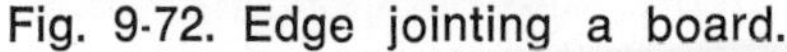
Fig. 9-72. Edge jointing a board.

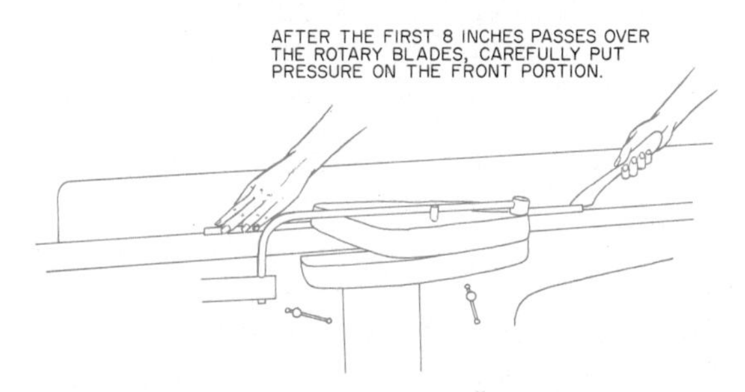

Fig. 9-73. Planing narrow stock using a push stick and the left hand.

the stock is well onto the outfeed table, place the left hand on the outfeed table, Fig. 9-73.

PLANING END GRAIN

A board less than 12" wide should not be planed on the jointer. Special precaution should be used during the jointing process to avoid splitting of the end grain. First joint the board about 1" from the edge using very light cuts, then turn and cut the other edge, Fig. 9-74.

PLANING BEVELS AND TAPERS

Bevels and tapers can be planed on stock using the jointer. The fence is adjusted to the desired angle and the stock is passed over the cutterhead as in edge-jointing, Fig. 9-75. The chamfer or bevel

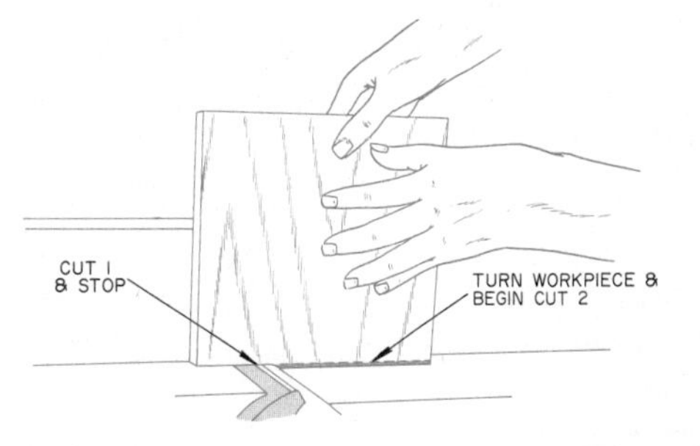

Fig. 9-74. End grain is planed by making two cuts.

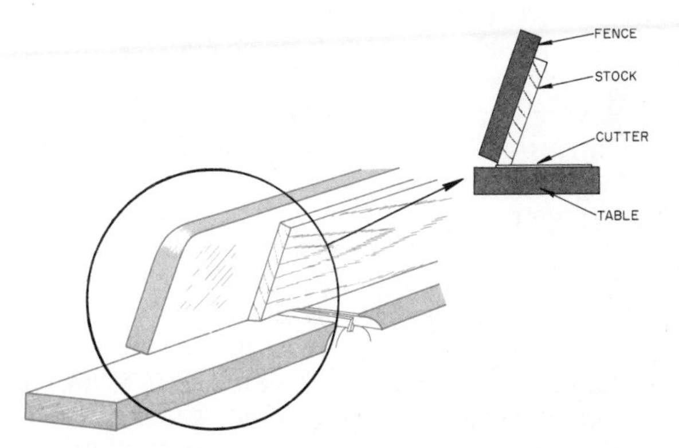

Fig. 9-75. Planing a bevel on the edge of stock.

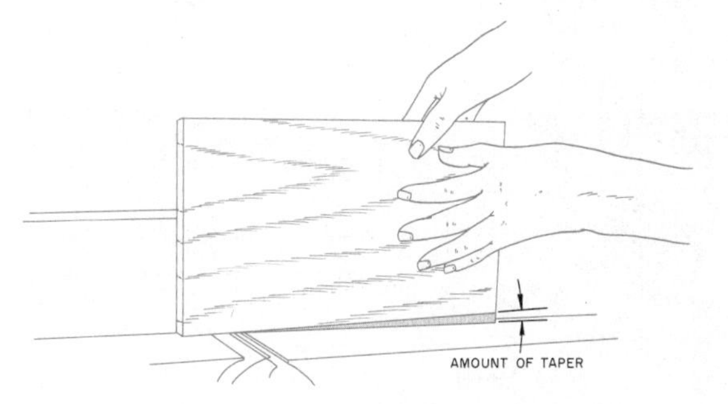

Fig. 9-76. Jointer setup for cutting a taper.

may be cut with the fence tilted either away from or toward the table. It is usually preferred to tilt the fence toward the table as this forms a "vee" to hold the stock. It is best to cut the bevel using several cuts. End grain cuts are made first to bevel wide stock on all sides. The desired amount of taper must first be marked on the stock to plane tapers on the jointer, Fig. 9-76. The method of jointing the taper will vary depending on the length of the taper. Tapers which are not longer than the infeed table can be made in one cut. The infeed table is lowered an amount equal to the total amount of the taper. A stop block is clamped to the jointer fence at the end of the stock, the stock is placed against the stop block, and the machine is started. The guard is carefully pulled back and the stock is slowly lowered into the cutterhead. The tapered end is held against the infeed table and advanced across the cutter. If duplicate parts are needed, they can be cut at the same setting. If the depth of the taper is very deep, it can be made in two passes, cutting one-half of the total depth each cut. Any irregularities at the beginning of the cut can be removed with a very light cut the total length of the work. This is the reason for setting the stop block slightly forward in the initial setup.

Two or more cuts can be made for tapers longer than the length of the infeed table. The total length of the taper is divided into portions and the above procedure is used. If the total length of the taper is 30″ for example, it would be cut in two passes.

Fig. 9-77. Cutting a rabbet on a jointer.

If the total depth of the taper is ½″, the depth of cut for each pass is ¼″.

CUTTING RABBETS ON THE JOINTER

The jointer can be set to produce a fine rabbet for joining stock. To adjust the jointer for rabbeting, the fence is moved toward the front of the machine. The portion of the knives which remains exposed will be the width of the rabbet. The guard must be removed for the rabbeting operation, therefore, special precaution must be taken to avoid an injury. The depth of the rabbet is determined by the amount the infeed table is lowered. The rabbet should be cut in two or more cuts, as a much smoother cut results if lighter cuts are used, Fig. 9-77. Use a push block or push stick to advance the stock over the jointer knives to rabbet narrow pieces.

INSTALLING AND ADJUSTING JOINTER KNIVES

To change or replace the jointer knives, first disconnect the power from the machine. Remove the safety guard and the fence, lower the infeed table to its lowest position, and with a piece of soft wood, wedge the cutterhead. The knife should be in an upward position. The wedge can be driven gently between the cutterhead and the base casting. Loosen the gib bolts and remove the knife. Rotate the cutterhead and remove all of the knives. Install newly-ground knives in the cutterhead. The gib screws are tightened only enough to hold the knives in place. All of the knives must be extended exactly the same height from the cutterhead. Make certain each knife is replaced with the bevel towards the outfeed table. The knives are adjusted to exactly the same height as the outfeed table.

A magnet may be used to insure the correct height for each knife, Fig. 9-78. A gauge block is attached to the table to insure the exact placement of the magnet for each knife. A mark is placed on the magnet so that the same spot is used each time, Fig. 9-78. The magnet is moved from one

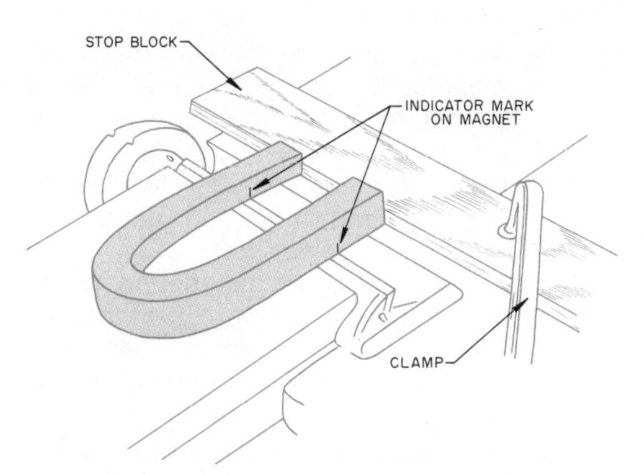

Fig. 9-78. Adjusting the height of jointer knives using a magnet.

end of the knife to the other to insure the exact height of the entire knife.

Tighten the bolts firmly after the knives are set to the right height. Tighten the center bolts first and work toward the outside edges.

Some jointers are equipped with special set screws for raising and lowering the knives. It is relatively easy to set the knives with a straightedge or special gauge tool, Fig. 9-79. A wooden straightedge is placed on the outfeed table, the cutterhead is revolved by hand and moves the straightedge a slight amount. If the knives are exactly the same height, the straightedge will move the same amount each time. A mark is made at the edge of the outfeed table. The distance that the straightedge moves when the knife revolves is then marked. The distance between each mark is equal when the knives are properly adjusted. The knives can be raised and lowered by turning the screw lifters. Special gauges can also be used to check and set the knife height, Fig. 9-80.

It is not always necessary to grind or replace the dulled jointer knives. Sometimes the knives can be restored by whetting with the knives still in the machine, Fig. 9-81. The jointer cutterhead must be rotated to a position where the front bevel of the knives just touch a straightedge. The cutterhead is then wedged in this position,

Fig. 9-79. Adjusting the height of jointer knives using a straightedge.

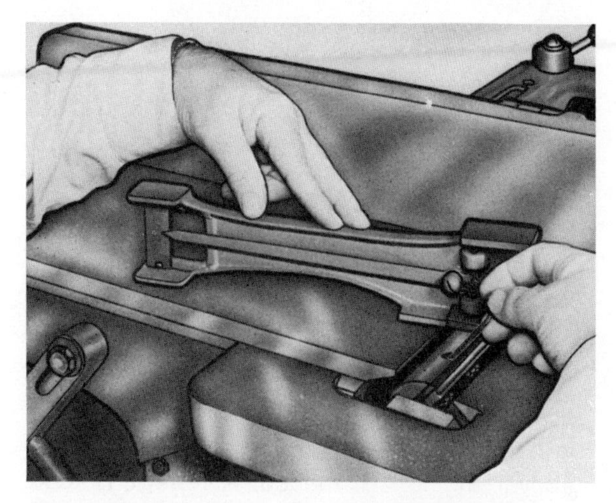

Fig. 9-80. Adjusting the height of jointer knives using a knife setting gauge.

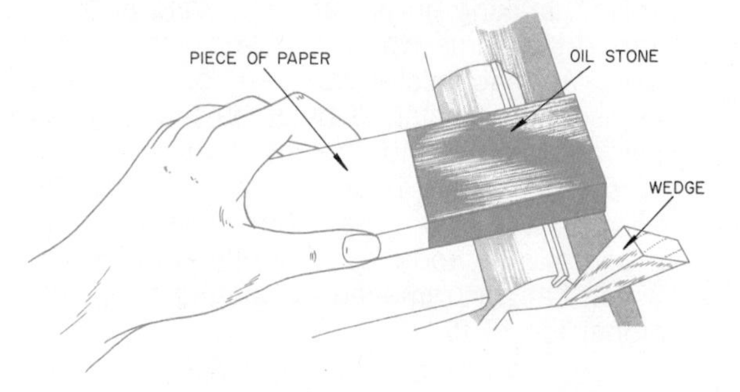

Fig. 9-81. Whetting jointer knives in the machine.

and the knives are whetted with a medium-coarse oilstone. The portion of the oilstone on the outfeed table is wrapped in paper to prevent damage to the table surface. The

Fig. 9-82. Removing the wired edge from whetted knives.

stone is moved back and forth across the length of the knives. Only the short portion of the cutting edge back slightly onto the bevel should be whetted. A renewed, keen edge can be produced with several strokes of the oilstone.

Repeat the whetting on each of the knives, making sure that each knife is positioned the same when it is wedged in position. A machinist rule can be used to measure the distance between the edge of the front table and the edge of the knife. This will insure exact positioning of the knives from one to another. The whetting process will produce a slightly wired edge which can be removed by means of a slipstone, Fig. 9-82.

SAFETY CONSIDERATIONS FOR THE JOINTER

Remember, a jointer is only as safe as its operator. Follow the hints below to prevent accidents:

1. Keep guards in place and use them when practical.
2. Avoid using machine tables as work areas or placing metal tools on them while the machine is running.
3. Do not wear loose sleeves or clothing that might be caught in moving machine parts.
4. Keep knives sharpened and properly edged to prevent kickbacks.
5. Stand clear and out of line with revolving parts when turning on the power or making heavy cuts.
6. Extra caution should be used when working on thin or short pieces. Only very light cuts should be made on such small pieces and they should be held with a push stick. Use push stick on all cuts when possible.
7. Check the depth of cut and the resistance of the knives carefully when making your first cut and after each adjustment of the table.
8. By all means, do not "play around" with a running machine. Nine accidents out of ten are the results of carelessness and playing with the machine as though it were a toy.
9. The jointer stand should be grounded to a water pipe or central grounding system.
10. Always disconnect power source when servicing machine.
11. Never make machine adjustment when unit is running.

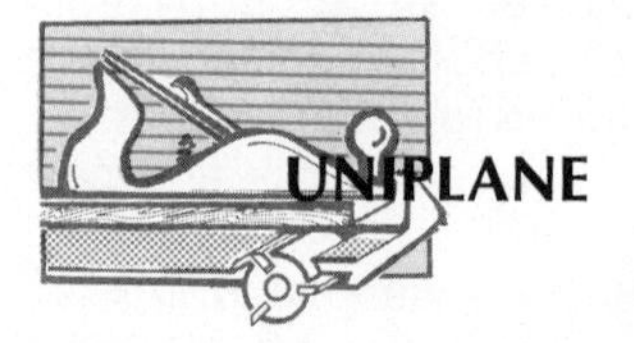

The uniplane is a relatively new machine which is capable of performing many of the same operations as a jointer, Fig. 9-83. Its appearance is also very similar to a jointer. The cutterhead, consisting of eight cutters, is circular in shape and rotates at a speed of 4000 RPM in a vertical plane, Fig. 9-84. Four of the cutters score the surface of the wood and the other four remove the stock by a shearing cut. The cutterhead is designed to surface a piece of stock 6″ wide. The maximum depth of cut is 1/8″.

The table is designed to tilt from 0-45° for planing chamfers and bevels, and is milled with a slot for using a miter gauge for squaring the end of narrow stock and similar cuts, Fig. 9-85. The miter gauge can be set at an angle for chamfering small stock, Fig. 9-86. The uniplane is equipped with a plastic shield which moves up and out of the way when the stock is advanced into the cutterhead.

Unlike the conventional jointer, the uniplane can be used for any size of stock, Fig. 9-87. Due to the downward rotary

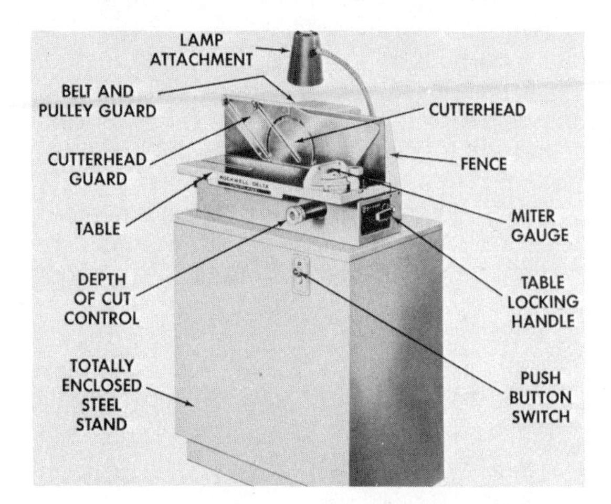

Fig. 9-83. Uniplane. (Power Tool Div., Rockwell Mfg. Co.)

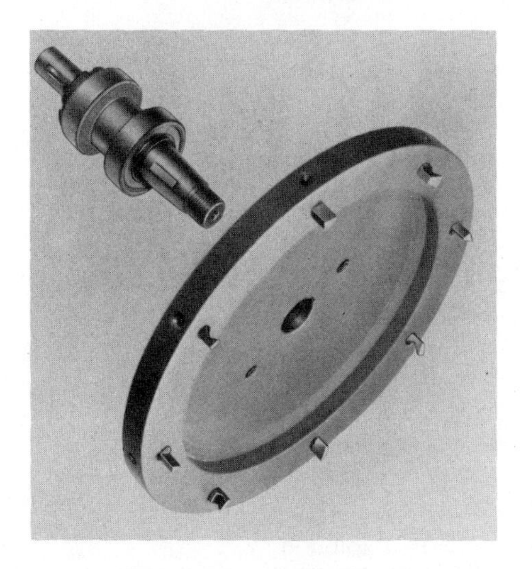

Fig. 9-84. Cutterhead assembly on a uniplane. (Power Tool Div., Rockwell Mfg. Co.)

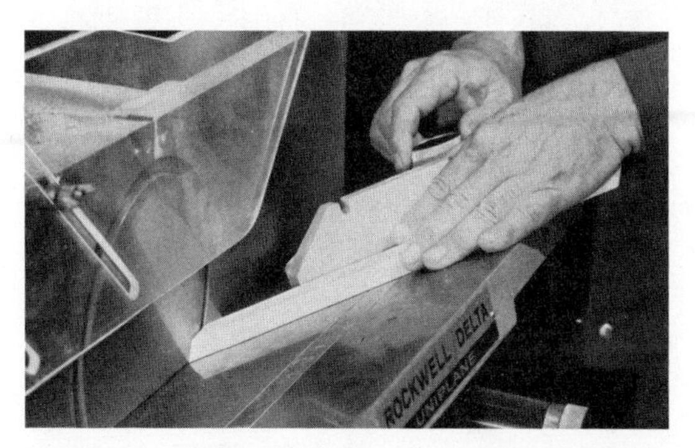

Fig. 9-85. Using a miter gauge attachment to guide narrow stock across the cutterhead.

Fig. 9-86. Using an angled miter gauge setting to chamfer small stock.

Fig. 9-87. Surfacing extremely small stock on the uniplane.

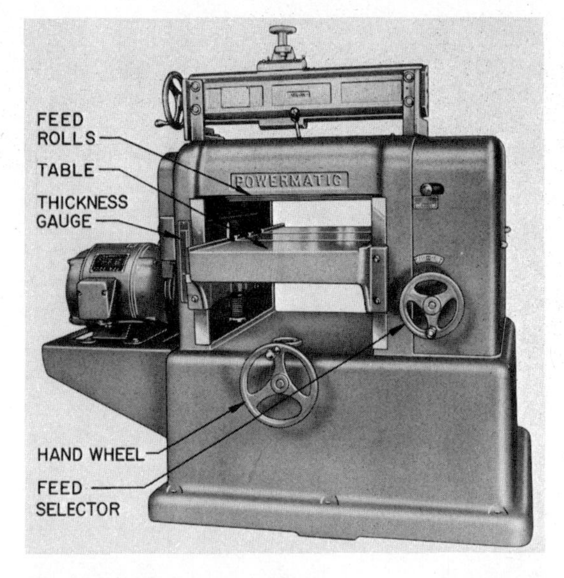

Fig. 9-88. Thickness planer.

cutting action, the potential of a kickback is eliminated. Like other surfacing machines, a push stick is used when the hands are passed close to the cutterhead.

THICKNESS PLANER

The thickness planer is sometimes referred to as a surfacer, Fig. 9-88. The basic purpose is to reduce the thickness of stock and to smooth a surface. The cutting principle of the planer is the same as that of the jointer, except the stock is fed automatically into the planer. The size of the planer is determined by the maximum width of stock which it will plane. Planes are available in sizes from 12″ to 52″. The most

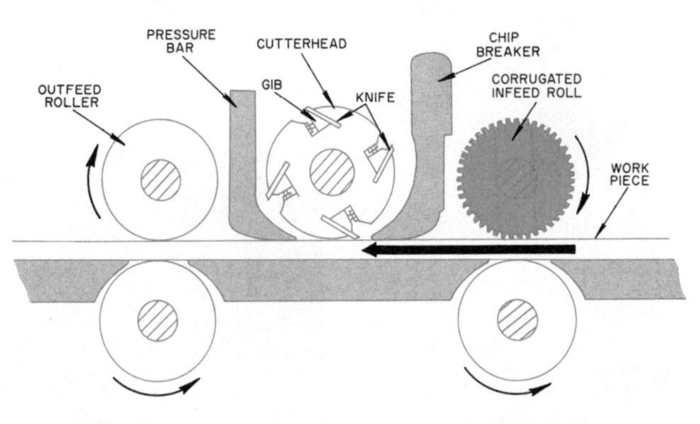

Fig. 9-89. Cross-section view of a thickness planer.

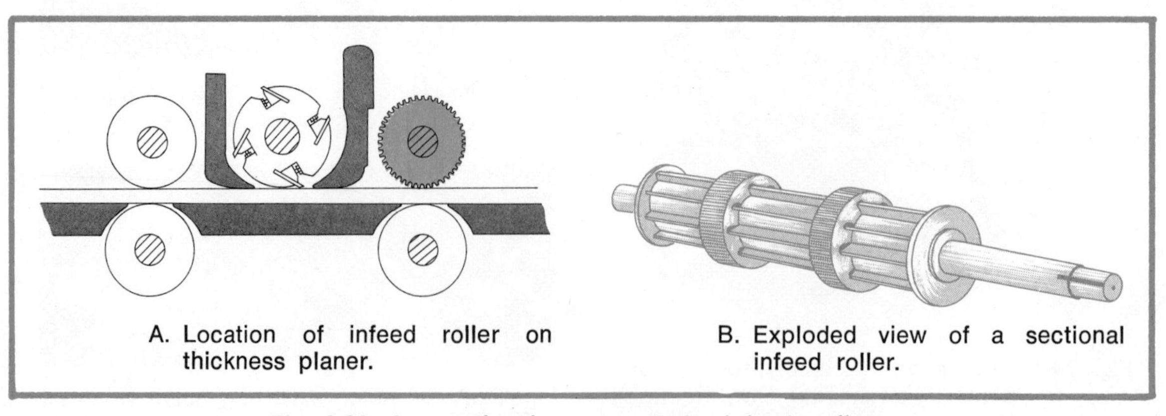

A. Location of infeed roller on thickness planer.

B. Exploded view of a sectional infeed roller.

Fig. 9-90. A sectional corrugated infeed roller.

common sizes for general use are the 12″, 18″, and 24″ models.

PARTS OF THE SURFACE PLANER

A cross-sectional view of a thickness planer will help to explain its operation, Fig. 9-89. The basic functional parts of the machine consist of the infeed rolls, cutterhead, chip breaker, pressure bar outfeed rolls, infeed table, and outfeed table. The stock is advanced into the cutterhead by means of the *infeed rolls.* The lower infeed roll is usually a solid, smooth roll. The upper infeed roll is of a sectional construction and is normally corrugated so it can grip the stock and force it into the cutterhead. It is sectional so that it will allow boards slightly different in thickness to feed in at the same time, Fig. 9-90. The lower infeed roll tends to reduce the friction of the board on the table and aids in carrying the board into the cutterhead.

The *chip breaker* is also sectional. Each section is designed to operate independently of the other sections. Pressure is applied to varying thicknesses of boards as they pass under the cutterhead. Each section of the chip breaker is forced against the wood spring tension, preventing the wood from being torn as the rotating knives reduce the thickness of the stock, Fig. 9-91.

The *cutterhead* consists of a steel cylinder with slots milled along its length to hold three or four cutter knives. Bolts apply pressure to gibs to hold the knives firmly in place. The cutterhead revolves above the stock as it is fed into the machine and surfaces the top of the material.

The *pressure bar* applies pressure to the board after it passes under the cutterhead and holds it firmly against the outfeed table. It prevents the stock from chattering as the cut is being made. The *outfeed rolls* are smooth and carry the planed stock out of the machine.

The depth of cut is controlled by raising and lowering the table assembly. The *infeed* and *outfeed tables* are aligned and move up and down together. The table is moved by means of the *elevating handwheel.* A *scale* mounted on the frame of the machine indicates the thickness at which the machine is set.

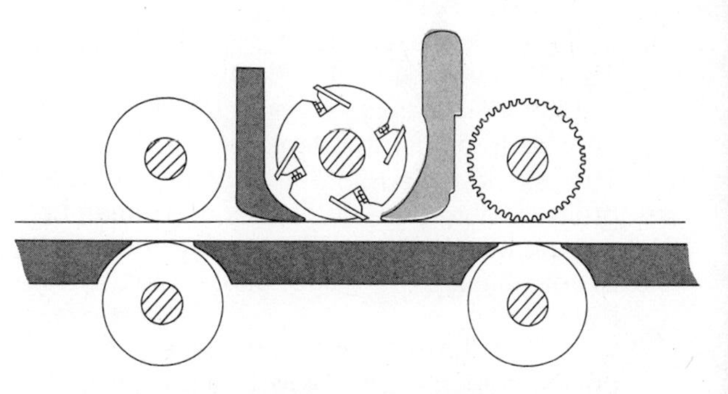

Fig. 9-91. Even pressure is applied to the entire width of the stock by the sectional chip breaker to prevent the wood from splitting.

CONTROLS AND ADJUSTMENTS ON THE SURFACE PLANER

The number of controls on the machine are determined by the size and model of the planer. Most machines have a control to change the rate of feed of the stock into the cutterhead. This is given as surface feet per minute (SFM). The cutterhead speed operates at a constant speed of about 3600 RPM.

The rate of feed or SFM will be determined by the type of stock. Extremely hard wood is fed into the machine at slower speeds. The rate of feed will offset the quality of the planed surface. If wide boards are planed, the SFM should be reduced.

SURFACING STOCK ON THE PLANER

The stock to be planed on the thickness planer must have one smooth, flat surface. The first side can be prepared either on a jointer or with a hand plane. If warped stock is surfaced, the pressure of the rollers will flatten the stock temporarily, but when the pressure is reduced, the board will resume its warped shape. The planer is not designed to straighten warped stock.

The first cut on the planer is made with the machine adjusted for a cut depth of 1/16″. For example, if a board measures 13/16″ thick, adjust the machine so the graduated scale is on ¾″. The handwheel is used to adjust the depth of cut, Fig. 9-92. The stock is fed into the planer with the grain in the correct direction. Remember, the direction of the grain can be determined by examining the edge of the board, Fig. 9-93.

Start the machine and place the stock on the infeed table. Slowly push the board into the machine until it is engaged by the infeed rollers. Walk to the rear of the machine and support the board as it comes out of the outfeed rollers. If a second cut is needed, adjust the depth of cut and pass the board through the planer again. A cut of 1/16″ is an average cut for most types of wood. If several pieces are to be planed to the same thickness, plane all of them at the same setting before readjusting the depth. Start with the thickest board. If considerable stock must be removed, plane both sides of the board to prevent cupping due to unequal moisture content of the wood. The outer portion of the board may have a slightly higher content than the center.

Be sure to support long boards as they come out of the outfeed rollers. Otherwise, the weight of the board forces the end into the cutterhead after it passes under the chip breaker. This causes a slight irregularity called a "clipmark" on the end of the stock, Fig. 9-94.

Stock shorter than the distance between the infeed and outfeed rollers requires special consideration. The stock must be

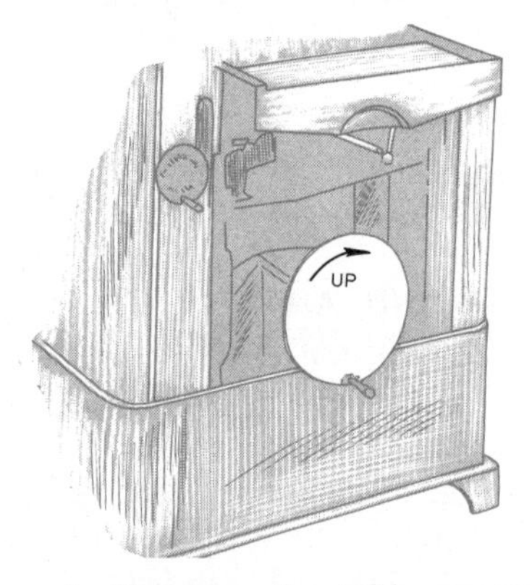

A. To move the table assembly up to remove less stock, turn the handwheel clockwise.

B. To move the table assembly down for a deeper cut, turn the handwheel counterclockwise.

Fig. 9-92. The handwheel adjusts the depth of cut.

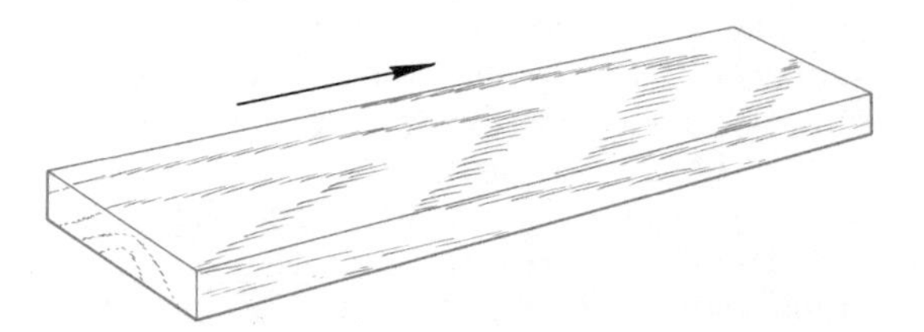

Fig. 9-93. Determine the direction of the grain before feeding stock into the planer.

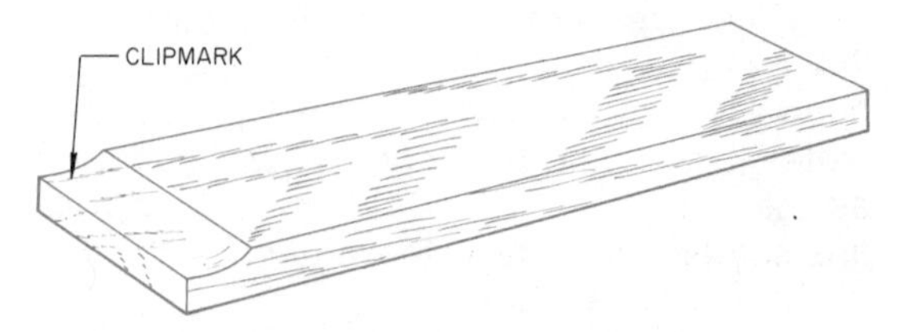

Fig. 9-94. An irregularity in the cutting of stock caused by failure to support long stock as it comes out of the outfeed rollers.

in contact with either the infeed or outfeed rollers at all times. If it is too short, it will lodge between the rollers. To cut these shorter pieces, butt the ends of several pieces together so they push each other through. However, the last piece must be long enough to run between the rollers, Fig. 9-95.

Stock less than 1/4" thick will vibrate excessively unless it is placed on top of a heavier board. This is sometimes referred to as "piggybacking". The stock used as a carrier board must be planed true before it is used. A stop block glued to the end of the carrier board will keep the stock in place.

The planer is often used to plane stock square such as pieces to be used for furniture legs. To square the stock, first plane two adjacent surfaces true on the jointer. The sides opposite the jointed surface can be planed square on the surface planer. The same setting is used for both surfaces. Make sure you stand to one side as you feed the stock into the planer, Fig. 9-96.

RECONDITIONING THE SURFACE PLANER

Some surface planers have special grinding attachments so the knives can be reconditioned without removing them from the cutterhead, Fig. 9-97. The grinder is mounted on the special *ways* (steel rails) and can be moved back and forth over the cutting knives.

The manufacturer's manual should be consulted before servicing and adjusting the surface planer. A routine lubrication and preventive maintenance schedule should be established for all machines.

SAFETY CONSIDERATIONS FOR THE PLANER

1. This machine has been designed with as many safety features as humanly possible; however, always remember that a planer is only as safe as its operator.
2. BEFORE starting the planer, be sure to check the following:
   (a) Table must be completely free of all foreign matter.
   (b) Cutterhead knives MUST be inspected before each operation. Check for tightness in cutterhead and make certain knives are not fractured in any place. Flying knives are DANGEROUS.
   (c) The knives must be sharp.
3. Check material thickness and depth of cut desired. Never overload

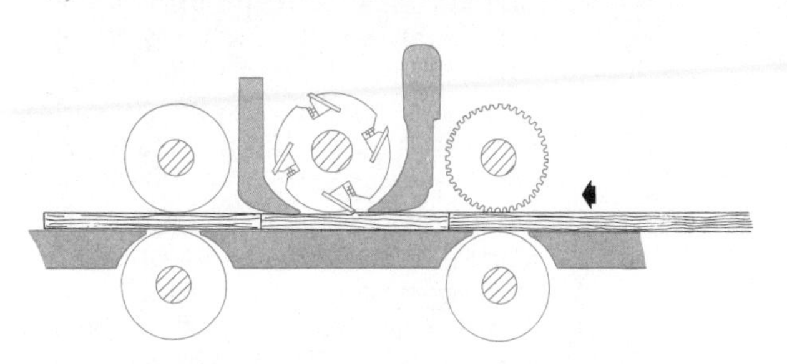

Fig. 9-95. Butting ends of short stock for planing.

Fig. 9-96. Stand to one side when feeding the planer.

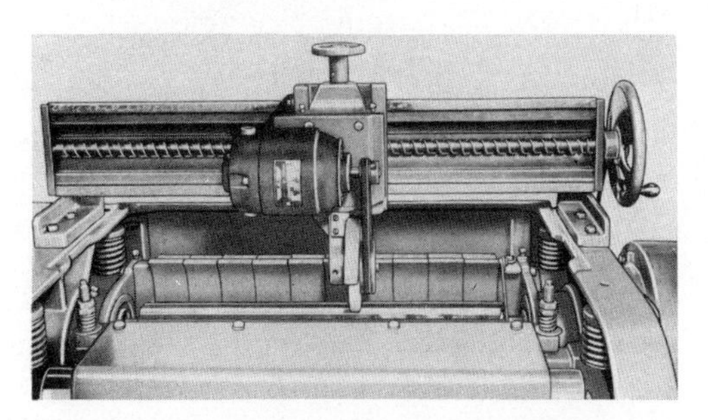

Fig. 9-97. Sharpening cutter knives with a special grinding attachment.

planer, or try to cut beyond its capacity.

4. As material is fed into the machine, stand to one side of the board (never directly behind) and near a switch. Kickback caused by improper gripping of lumber by infeed roll and chip breaker can cause serious injury.
5. Never stand directly behind or walk behind machine when it is running. Direction of cutterhead rotation usually throws chips or any foreign material from rear of machine.
6. In case it is necessary to stop material as it is feeding through machine, disengage feed clutch and turn machine off. Wait until cutterhead has completely stopped before lowering table to remove material. Attempted removal while cutterhead is turning may cause kickback.
7. Never "horse around" a running planer. "Play" should absolutely be forbidden as nine out of ten accidents are the results of carelessness and playing with machines as though they were toys.
8. Always stop machine for adjustment or when leaving immediate area. Disconnect power source when working on or around any moving parts.
9. Never feed two boards through a planer with solid infeed rolls at once (side by side or stacked). Kickback can result and the board may fly from machine with the velocity of a bullet. When sectional infeed rolls and chip breakers are installed, it is possible to feed several narrow boards through machine safely.
10. Keep all guards properly positioned for protection of the operator.
11. Do not wear long or loose sleeves or neckties when operating planer.
12. Extra care should be taken when running short pieces. Butt with another piece of material of equal thickness and stand ASIDE.
13. Base of machine should be grounded to a water pipe or neutral ground system.

INDUSTRIAL PLANING MACHINES

Planers are available with either single or double cutterheads. When only one cutterhead is used, it is normally mounted on the top. High-production surface planers are equipped with both *upper* and *lower cutterheads,* which are staggered rather than directly above and below one another, Fig. 9-98. These industrial machines are capable of planing stock up to 6′ in width and up to 2′ in thickness. The double surfacer has two sets of *chip breakers* and *pressure bars*, one set for each cutterhead.

High-production surface planers are often equipped with a *hopper feeding system.* The stock is stacked on the conveyor and carried into the machine, one piece at a time. A conveyor then carries the surfaced stock to the next machine for further processing.

Fig. 9-98. A double surface planer equipped with a conveyor system to move heavy stock to the feed rollers. (Greenlee Tool Co.)

Some single cutterhead surface planers are also equipped with conveyor belts instead of feed rollers. This facing planer is designed for truing one surface of the stock, and is actually a form of power-fed jointer.

The conveyor belt is designed with spring-loaded *fingers* which conform to the shape of the warped board. The fingers support the work while the top surface is being trued. The stock is then sent through a conventional single cutterhead planer and surfaced to a uniform thickness.

The facers and planers are typically setup together in production operations. The stock is first cut to rough length with the cutoff saws. From the cutoff operation, the stock proceeds by conveyor to the facer which planes the bottom surface of the stock true. The stock then goes to the single cutterhead planer where the top surface is planed parallel to the bottom surface and the stock is planed to the desired thickness.

Other production machines called *planer molders* are capable of surfacing stock on four sides simultaneously to produce various types of mill work, Fig. 9-99. The cutterheads located on the top, bottom, and both sides are equipped with custom-designed knives ground to produce the desired shapes in molding. Some of the larger molding machines are capable of handling stock 4″ to 6″ thick and up to 20″ wide.

Fig. 9-99. Combination planer-molder for planing and shaping four sides simultaneously. (J. A. Fay and Egan Co.)

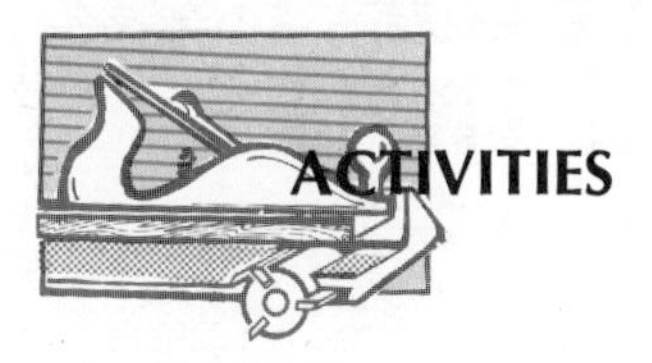

ACTIVITIES

PROBLEM 1

Describe the procedure for squaring a piece of solid lumber stock using power tools and machines. Identify all of the factors that must be considered to create the first square face, edge, and end surface. After you have planned the procedure carefully, select a scrap piece of wood and request teacher approval to gain experience squaring lumber.

PROBLEM 2

Assume that you need ten pieces of squared stock $\frac{5}{8}'' \times 3'' \times 9''$. What would be the most efficient method of producing them using hand tools? Also, select the most efficient method of producing them using hand tools. Also, select the most efficient method using power tools. Can the rate of production be increased if you square both edges and faces before cutting the stock to length?

VOCABULARY

Edge cutting tool
Plane iron
Oilstone
Whetting
Bevel
Chamfer
Tang
Shearing cut
Plane iron cap
Lateral adjustment lever

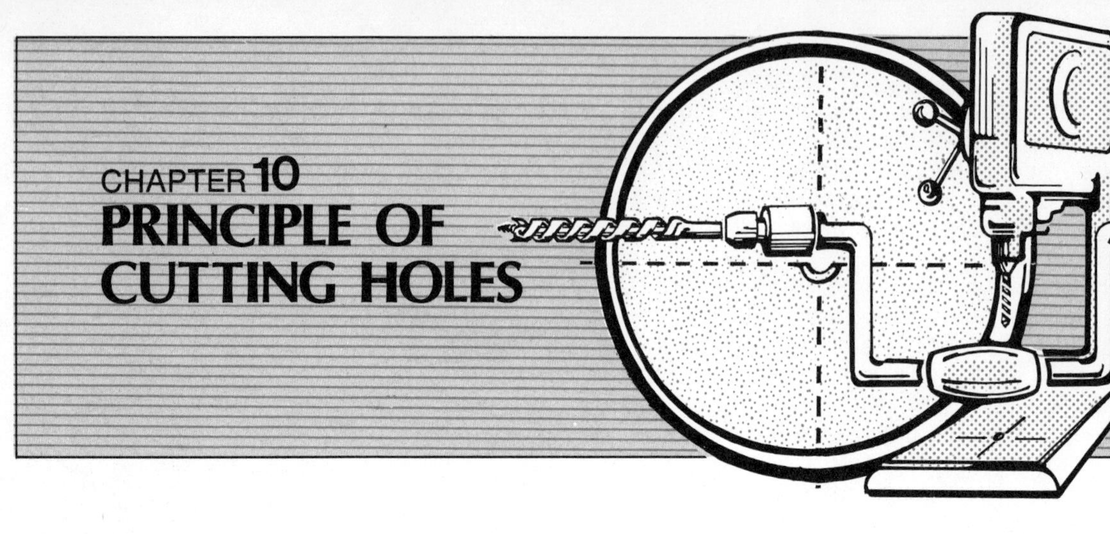

CHAPTER 10 PRINCIPLE OF CUTTING HOLES

INTRODUCTION

Cutting holes is one of the most fundamental processes for producing a wood product. The shape and size of the hole may vary from round to square or from tiny to large. Regardless of the shape or size of the hole, the method for cutting remains basically the same. The hole may be cut

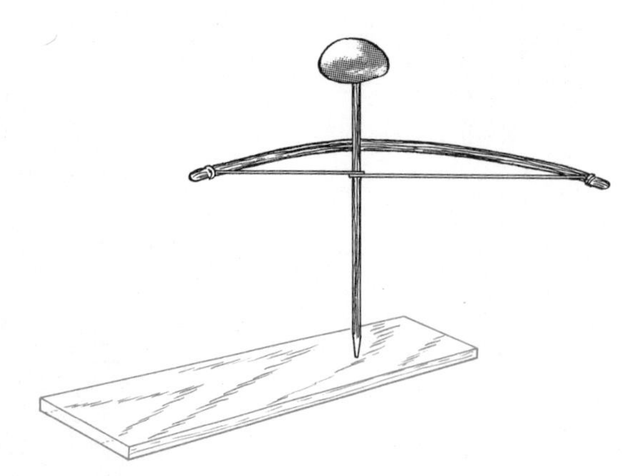

Fig. 10-1. A bow drill.

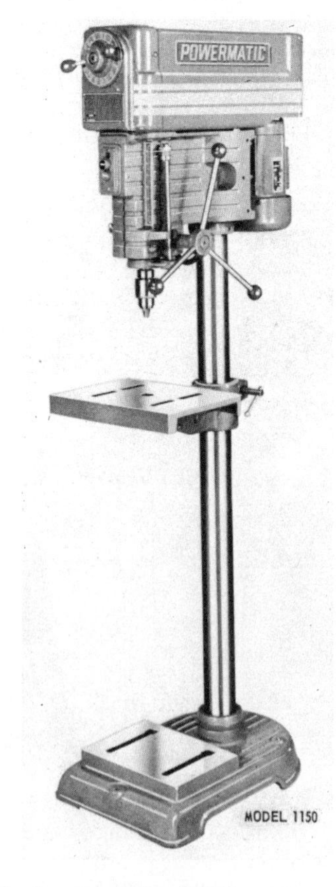

Fig. 10-2. A modern drill press. (Powermatic Machine Co.)

with a sophisticated boring machine or a simple handtool. Figure 10-1 shows a bow drill which can be used to cut a circular hole in wood. Even though the method was developed hundreds of years ago, the same principle of cutting is still used.

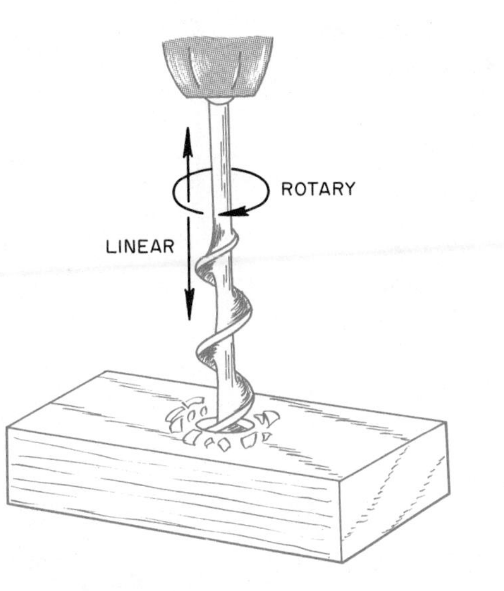

Fig. 10-3. Principle of cutting holes.

Hole cutting is regarded as *linear control of a rotary tool.* The principle applies to the bow drill or the modern drill press shown in Fig. 10-2. Most modern drilling and boring tools utilize the mechanical advantage of leverage. The cutting of holes utilizes pressure combined with a cutting edge revolved against the stock. The cutting edge is of hardened steel capable of cutting a softer material. The pressure exerted on the tool causes the cutting edge to advance into the material. Figure 10-3 illustrates the principle of cutting using linear pressure and a rotating cutting edge. The pressure and speed varies with the size and type of cutting tool as well as the type of material being cut. Generally, the larger the cutting tool, the slower the rate of rotation.

There are numerous types of cutting tools called bits, drills, and hole cutters available for cutting holes in wood, Fig. 10-4. The cutting edges are especially shaped to perform a particular job. Holes larger than 1/4″ are usually referred to as *bored,* and are cut by tools called *bits.* Holes 1/4″ or smaller are referred to as *drilled,* and are cut by *drills.*

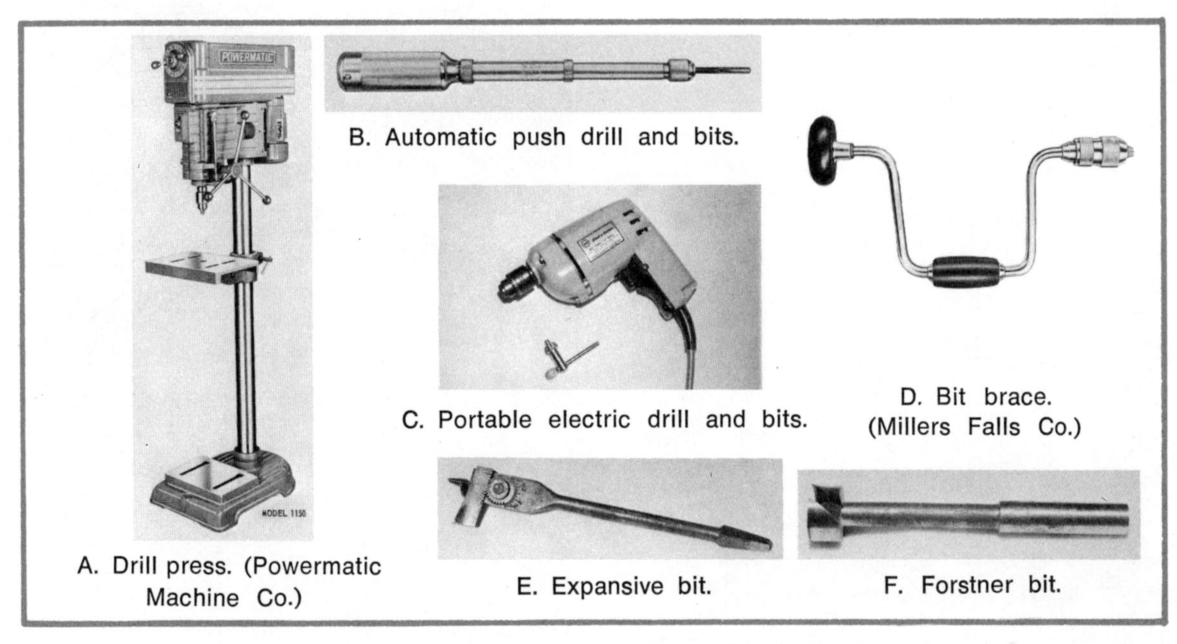

A. Drill press. (Powermatic Machine Co.)

B. Automatic push drill and bits.

C. Portable electric drill and bits.

D. Bit brace. (Millers Falls Co.)

E. Expansive bit.

F. Forstner bit.

Fig. 10-4. Common tools for cutting holes.

Holes are often cut in wood for assembling two or more parts in product fabrication. The holes are used for installing mechanical fasteners or for joint construction.

CUTTING HOLES WITH A HAND DRILL

A hand drill or automatic drill is used to cut holes less than ¼″ in diameter, Fig. 10-5. The drill bit is held in the chuck. The size of the hand drill is determined according to the chuck capacity of either ¼″ or ⅜″. The ¼″ size is the most common. Hand drills are equipped with a three-jaw chuck designed for use with twist drill bits, Fig. 10-6.

Twist bits are available in sizes ranging from 1/16″ to ¼″ and increase in size by increments of 1/64″.

The drill bit is tightened in the chuck by holding the chuck and turning the handle clockwise. The chuck will center the drill between the three jaws. The stock to be drilled should be clamped in a vise or secured to a bench. A piece of scrap stock

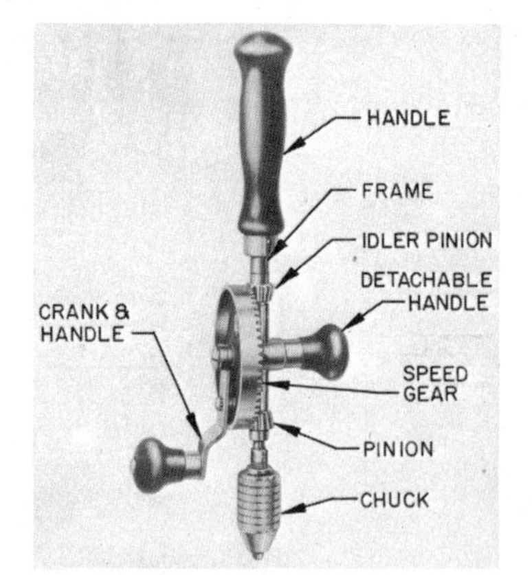

Fig. 10-5. Basic parts of a hand drill. (Stanley Tools)

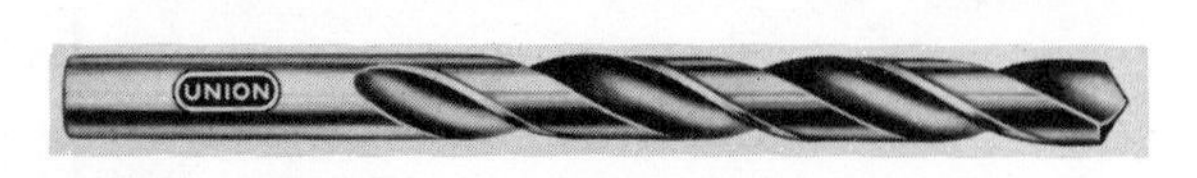

Fig. 10-6. A twist bit. (Union Twist Drill Co.)

Fig. 10-7. A back-up board keeps the back of the wood from splitting.

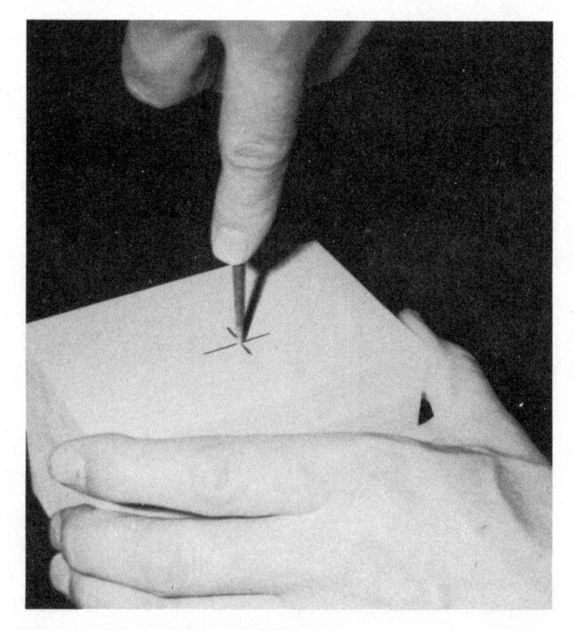

Fig. 10-8. Using a scratch awl to mark the location of the hole on the stock.

is clamped to the back of the board if the hole is drilled completely through to prevent the stock from splitting, Fig. 10-7.

The hole should be carefully located with the proper layout tools. The exact location for the hole should then be started with a scratch awl, Fig. 10-8, to insure exact location of the hole.

The hand drill is held by the handle and the crank is turned. Care must be taken to keep the bit perpendicular to the surface of the stock, as shown in Fig. 10-9.

Slight pressure is applied to the handle as the crank is turned. To remove the bit, continue turning the crank and withdraw it slowly. When deep holes are drilled, it may be necessary to withdraw the drill and clean the chips several times.

It is often necessary to drill numerous holes to a uniform depth. A depth gauge is helpful for this operation. A piece of wood or dowel rod can be used as a depth gauge. A piece of stock is cut to the desired length and a hole is drilled in the center through the length of the stock. The drill projects through the stock the depth of the desired hole, Fig. 10-10. The bit can be moved in and out of the drill chuck to make slight adjustments.

Fig. 10-9. Drilling a hole with a hand drill.

A piece of masking tape will also serve to indicate the depth of the drill to drill holes, Fig. 10-11. However, tape serves only as an indicator, not a stop. The tape is wrapped around the bit the desired distance from the end of the bit.

Fig. 10-10. A stop gauge for a twist bit made from a block of wood.

Fig. 10-11. Using masking tape as an indication of the depth of the hole.

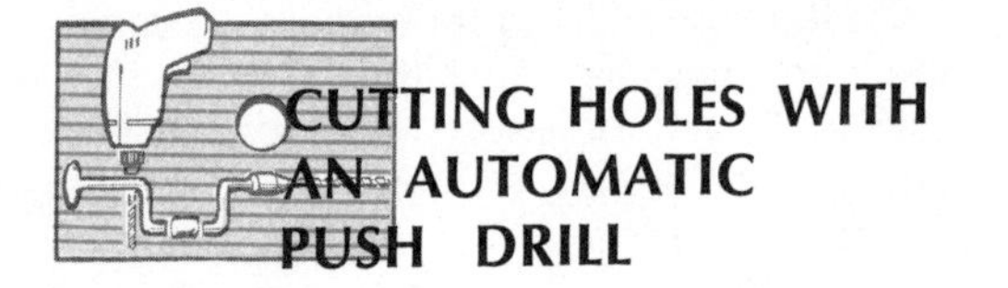

CUTTING HOLES WITH AN AUTOMATIC PUSH DRILL

The automatic push drill is sometimes used for cutting small holes, Fig. 10-12. The push drill is equipped with a set of eight bits, ranging in size from 1/16″ to 11/64″ increasing by 1/64's. The bits are specially notched to fit the chuck. The bit in the automatic push drill is rotated by pushing down on the handle. The drill handle returns to its original position when the pressure is released. It can then be pushed again. Each stroke of the handle revolves the bit several times, Fig. 10-13.

A small jig can be used to drill holes at a desired angle, as shown in Fig. 10-14. The jig is placed on the desired location and clamped in place.

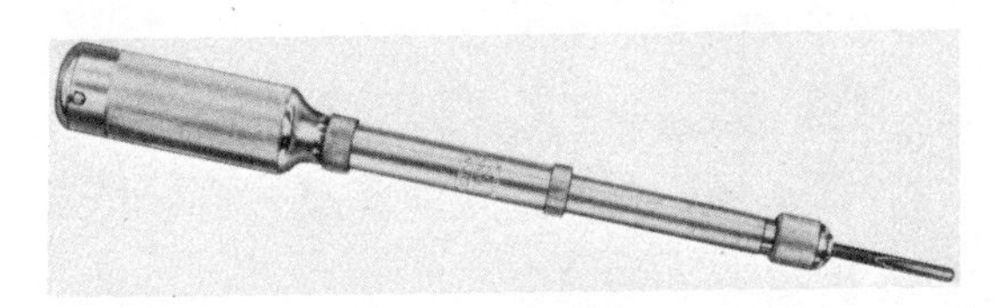

Fig. 10-12. An automatic push drill and bits.

Fig. 10-13. Drilling small holes using an automatic push drill.

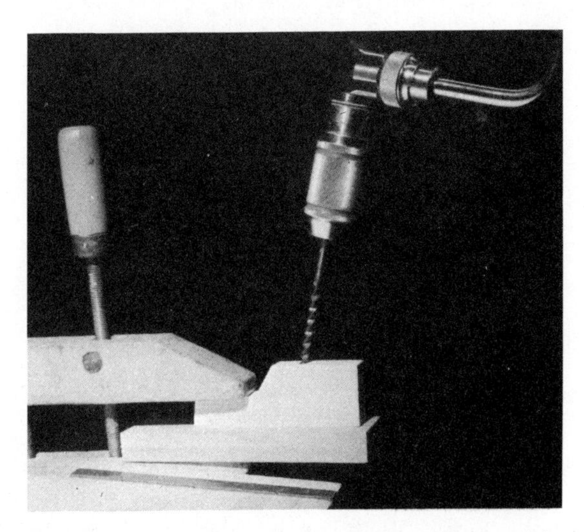

Fig. 10-14. Drilling holes at an angle with stock held in a jig.

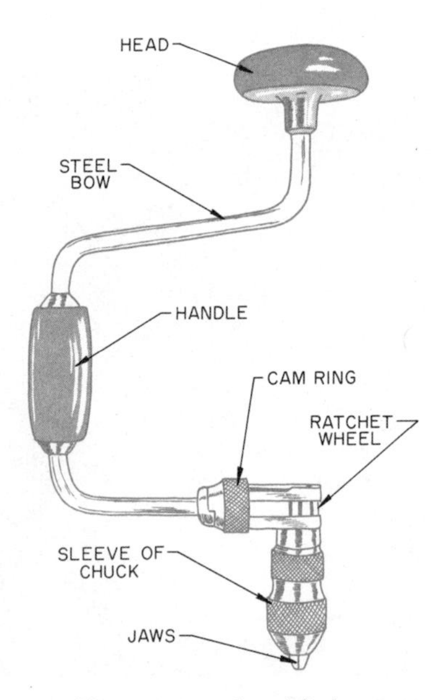

Fig. 10-15. The parts of a bit brace.

CUTTING HOLES WITH A BRACE AND BIT

A bit brace is a tool used to hold the bit to cut larger holes, Fig. 10-15. The holes are usually larger than ¼″, so it is most often referred to as boring. The size of a brace is determined by the sweep of the handle, Fig. 10-16. The sweep is the diameter of the circle made when the handle is revolved one complete turn. The sweep provides the mechanical advantage or leverage for turning the bit. Braces are available ranging in size from 6″ to 16″. A 10″ brace is common for general use. Most braces are equipped with a ratchet device to make it possible to turn the bit without making complete turns of the handle. This is an advantage when holes are being drilled in small pieces and accurate depth cuts are being made.

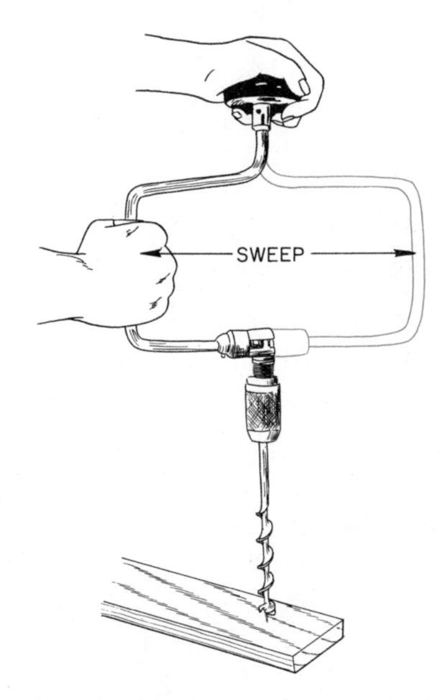

Fig. 10-16. The size of the brace is determined by its sweep as it is revolved one turn.

The brace chuck has two jaws which grip the tang of the bit as it is tightened. The chuck is tightened by turning the handle clockwise.

CUTTING HOLES WITH AN AUGER BIT

There are several types of auger bits which are used in the bit brace, Fig. 10-17. The tang fits into the chuck. The feed screw pulls the bit into the stock as it turns and centers the hole. The rate of feed is determined by the pitch of the threads on the feed screw. The outside of the hole is scored by the spur (sometimes called nibs) as the bit is pulled into the stock. Chips are then cut by the cutting lips and forced through the throat. The separated chip is removed from the hole by means of the twist.

Auger bits range in size from 3/16″ to 2″. Standard sets usually consist of sizes ranging from ¼″ to 1″ by 1/16″ increments. The size of the bit is stamped on the tang, and is given as a number such as a 7. This would mean that the bit would cut a hole with a diameter of 7/16″; a number 8 would cut a hole of 8/16″ or ½″; etc. Standard auger bits are from 7″ to 10″ in length.

The layout for locating a hole for boring with the bit and brace is the same as described for hand drills. The stock to be bored should be secured in a vise or clamped to the bench. A piece of scrap stock is clamped behind the stock to provide support to prevent chipping on the back side as the bit cuts through. The feed screw will also enter the scrap stock and continue to pull the bit completely through the stock.

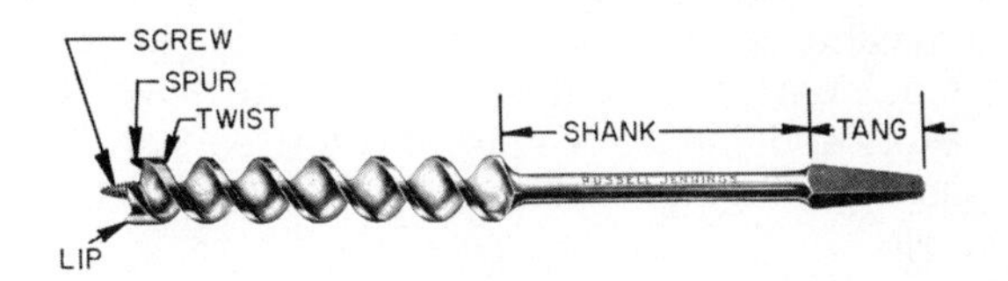

Fig. 10-17. Parts of an auger bit.

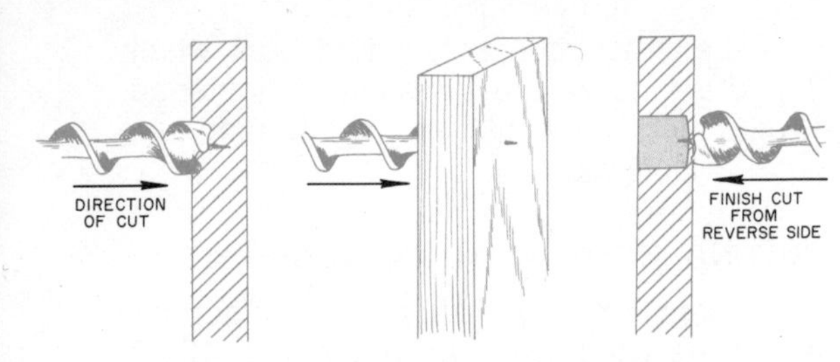

Fig. 10-18. Boring a hole without a back-up board requires boring from both sides.

Fig. 10-19. Checking vertical alignment of the bit with a try square.

If a backup board is not used, the boring should be done from both sides. Just as the feed screw breaks through the board, stop and remove the bit. Turn the stock over and finish boring the hole from the back side, Fig. 10-18.

Only light pressure is applied to the brace to bore holes with the auger bit. The feed screw pulls the bit into the stock, causing it to cut. The brace is turned clockwise for cutting. It is necessary to exert additional pressure to bore end grain, as the feed screw tends to "strip" out in the hole.

Care must be taken to hold the brace perpendicular to the surface of the work.

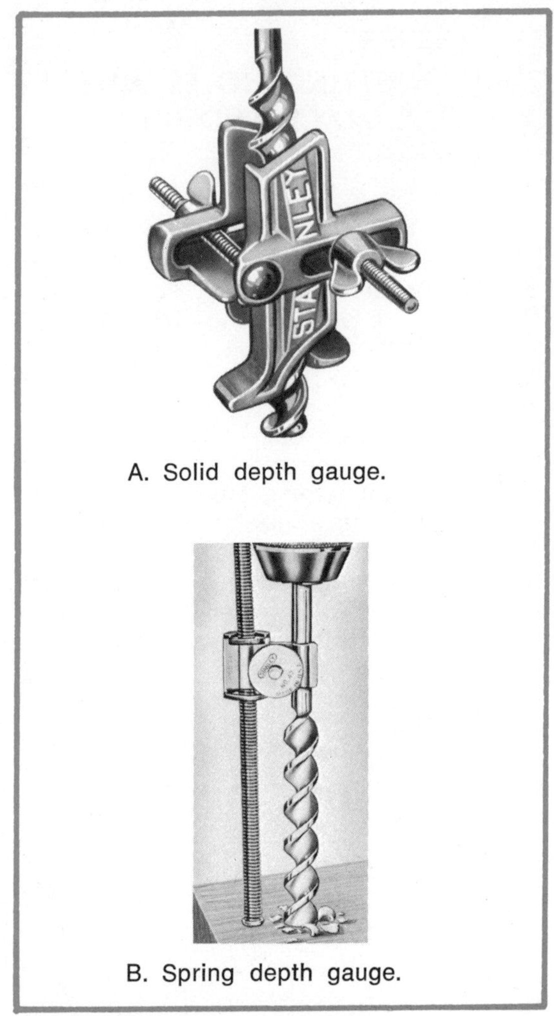

A. Solid depth gauge.

B. Spring depth gauge.

Fig. 10-20. Depth gauges. (Stanley Tools)

To insure an accurate hole, a try square is used to check the vertical alignment of the bit, Fig. 10-19.

GAUGING THE DEPTH OF A HOLE AS IT IS BORED

If a hole only goes part way through the stock, a depth gauge is used to regulate the depth, Fig. 10-20. In addition, a depth gauge can be made by drilling a hole through a block of wood as shown in Fig. 10-21.

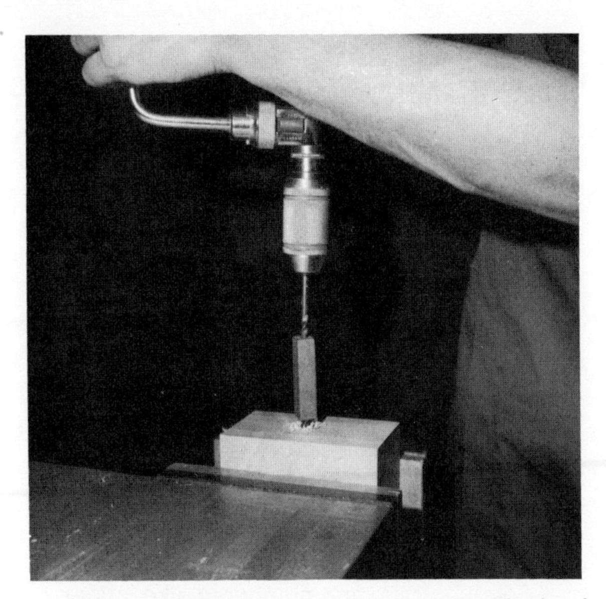

Fig. 10-21. Depth gauge made of a block of wood.

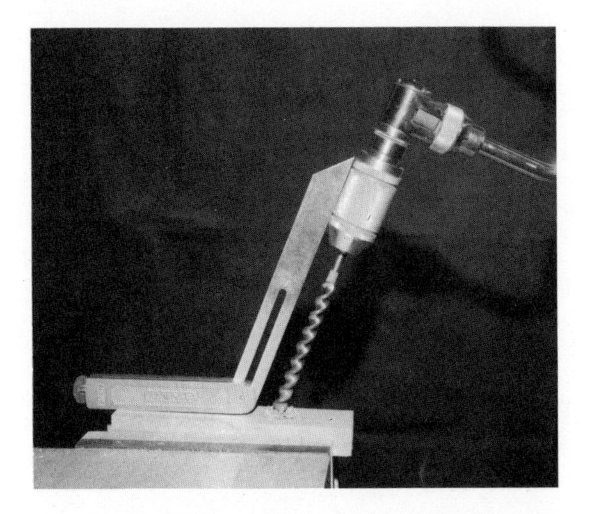

Fig. 10-22. Using a sliding T-bevel as a guide for boring holes at an angle.

BORING HOLES AT AN ANGLE

A sliding T-bevel is used as a guide to bore holes at angle. After the sliding T-bevel is set to the desired angle, it is placed on the stock and the bit placed parallel to the blade, Fig. 10-22.

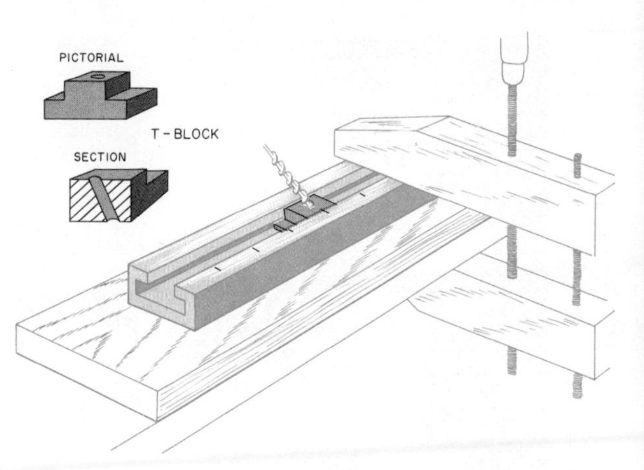

Fig. 10-23. A jig for boring duplicate holes at an angle.

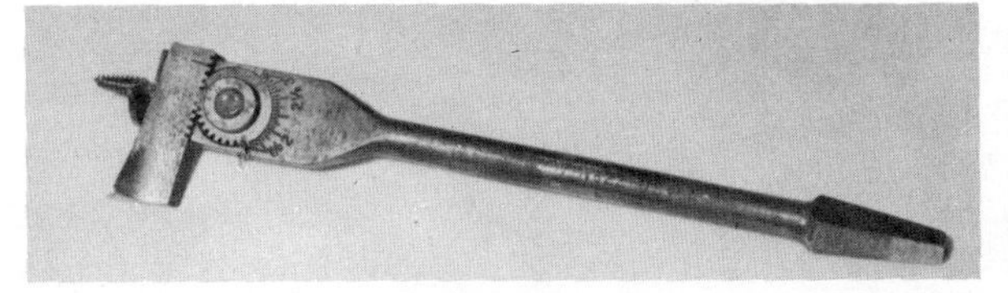

Fig. 10-24. An expansive bit.

If several holes are bored at the same angle, a jig can be made to simplify the job. A hole is bored at the desired angle in a piece of scrap stock. The stock is then clamped over the layout with the feed screw striking the center of the hole, Fig. 10-23.

CUTTING HOLES WITH AN EXPANSIVE BIT

Holes over 1″ to 3″ in diameter are usually cut with an expansive bit. This bit is equipped with two adjustable cutters, Fig. 10-24. The cutter is adjusted so the distance from the outside of the cutter to the feed screw is the radius of the hole to be cut. Some expansive bits have a scale on the side of the bit for setting the desired hole size.

Again, a piece of backup stock is clamped to the back of the board to prevent splitting and to help the feed screw pull the cutter through the stock, Fig. 10-25.

Fig. 10-25. Cutting a hole with an expansive bit.

CUTTING HOLES WITH A FORSTNER BIT

A forstner bit is used when it is necessary to drill a hole only part way through the stock. The forstner does not have a feed screw, Fig. 10-26. A forstner bit is also good for boring holes in end grain.

If the stock is thick enough so the feed screw of a regular auger bit does not break through the opposite side, it will work well for starting the hole for the forstner bit. Another method is to bore a hole through a scrap block the size of the forstner bit to be used. The block is clamped in place where the hole is to be cut and serves as a guide to start the forstner bit, Fig. 10-27.

Forstner bits are numbered like auger bits to show the size. They range in size from 1/4″ to 2″ in diameter.

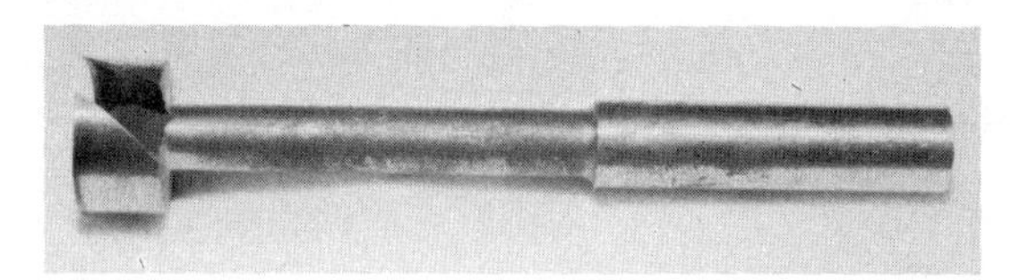

Fig. 10-26. A forstner bit.

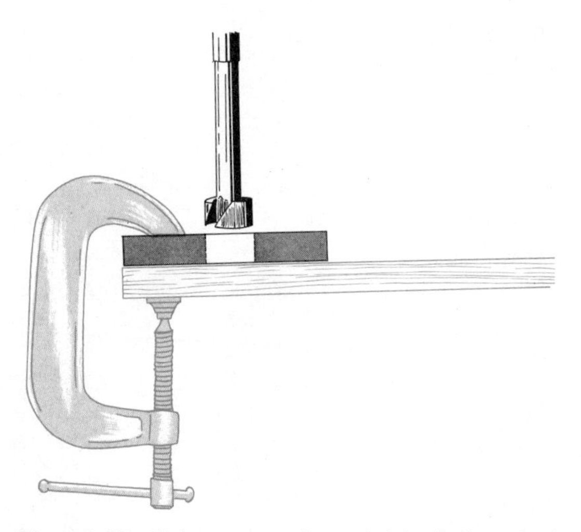

Fig. 10-27. Using a pre-bored block to start a forstner bit.

RECONDITIONING DRILLING AND BORING TOOLS

If a hole-cutting tool is to function properly, it must be properly maintained. It must be kept sharp and free of rust. Whenever a tool is exposed to moisture, it should be wiped dry prior to storing. Auger bits must be stored in individual compartments so the spurs are not dulled.

SHARPENING AN AUGER BIT

A dulled auger bit can be restored by sharpening it with a special auger file or a small flat file. The bit is supported on a bench with the screw pointed upward and the spur is filed with forward strokes, Fig. 10-28. To sharpen the cutters support the bit on a board with the screw pointed downward, using forward strokes to file the cutter edges, Fig. 10-28. The spurs and cutters should be filed to a keen, razor-like edge.

The screw and outside portions of the auger bit should never be filed.

GRINDING A TWIST DRILL

A twist drill can be reconditioned by grinding on a tool grinder. The tool rest is set on a horizontal plane. The drill lip is held at a 59° angle to the face of the wheel. The cutting edge is raised and the opposite end is lowered at the same time. The drill is also rotated to produce the correct angle, Fig. 10-29.

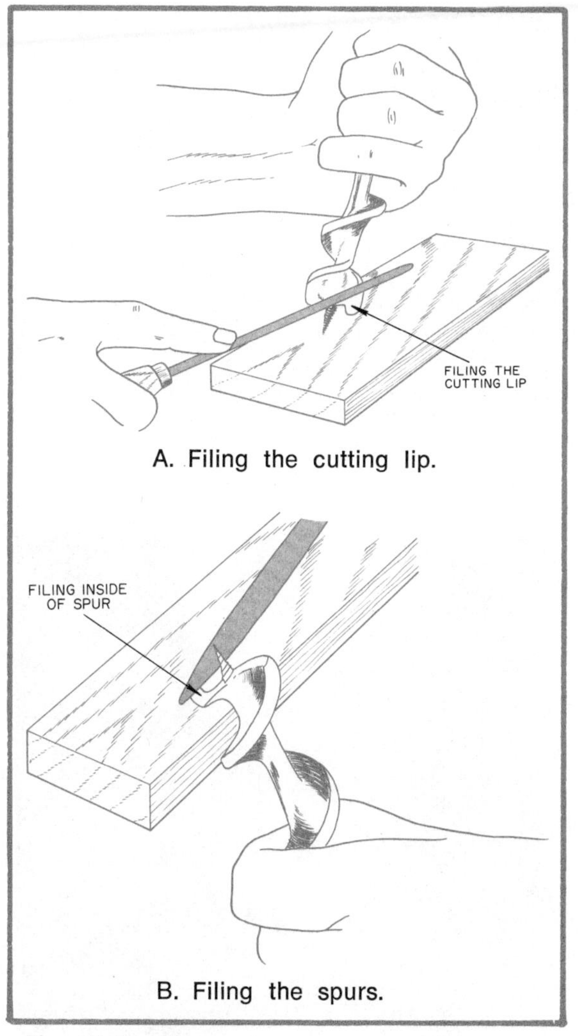

A. Filing the cutting lip.

B. Filing the spurs.

Fig. 10-28. Sharpening an auger bit.

PORTABLE POWER DRILLS

Portable power drills such as the one in Fig. 10-30 are used to perform many of the same operations performed by hand tools and stationary drilling and boring machines. They are usually powered by small electric motors; however, an increasing

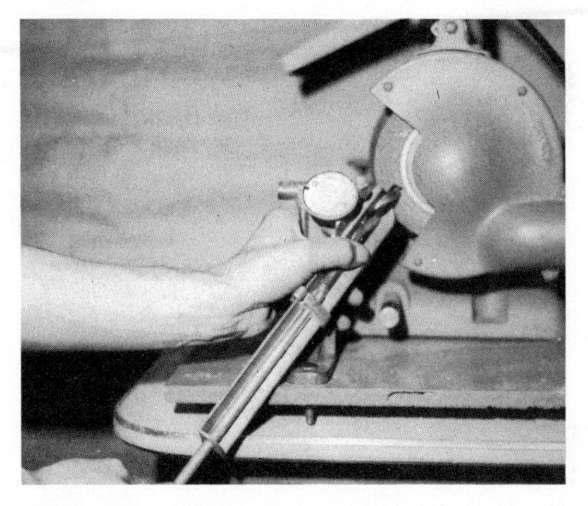

Fig. 10-29. Grinding a twist bit.

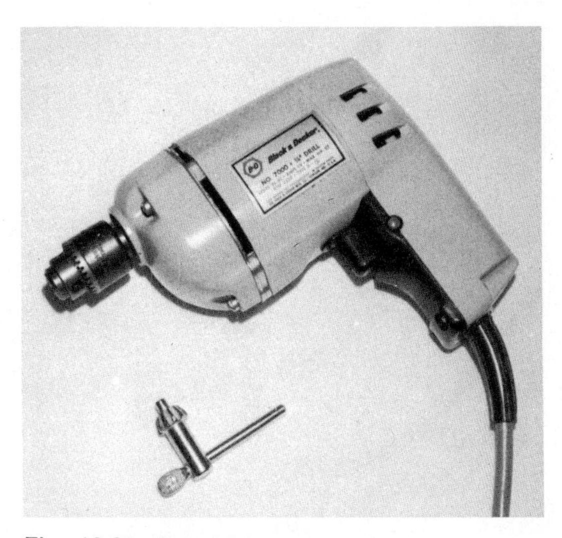

Fig. 10-30. Portable power drill.

number used in industry are pneumatic or air-driven. Many of the electric-powered drills contain a rechargeable power pack. This makes it possible to use the drill on the job site without an electrical power source.

The size of the portable power drill is determined by the maximum drill diameter which can be fitted into the chuck. The most common sizes are the 1/4″, 3/8″, and 1/2″. In most cases, the drill is equipped with a key chuck, Fig. 10-30. The chuck consists of three hardened steel jaws in which the bit is centered when it is tightened in place.

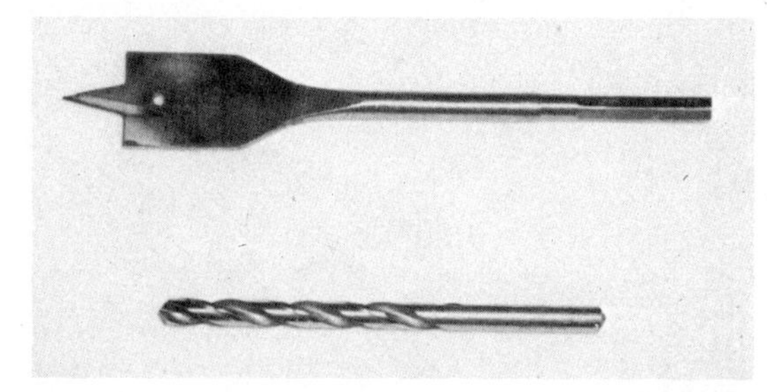

Fig. 10-31. Bits used with a portable power drill.

These drills are available with either a fixed chuck speed or a variable speed control device. If the drill has a fixed speed, the chuck speed decreases with the size of the drill. That is, a 1/4″ portable power drill develops more RPM's at the chuck than a 3/8″, and so forth.

Larger bits are used at lower speeds than small bits. This is important, as excessive speed on large bits may cause the bit to overheat and be destroyed.

TYPES OF BITS USED WITH PORTABLE POWER DRILLS

The two types of bits most commonly used for drilling wood with a portable power drill are the *twist drill* and the *spade bits,* Fig. 10-31. The spade bit, sometimes also called a speed bit, is available in a wide range of sizes from 1/4″ to 1 1/2″ by 1/16″ increments. It has become increasingly popular for boring wood because the shank is 1/4″ in diameter on all sizes. Notice the shank sizes on the various-sized bits in the set shown in Fig. 10-32. This makes it possible to fit bits into the 1/4″ chuck and drill holes up to 1 1/2″ in diameter.

Fig. 10-32. A set of spade bits.

Fig. 10-33. Using the portable power drill.

USING THE PORTABLE POWER DRILL

The portable power drill is light in weight and easy to use. The hole is laid out and centered with a scratch awl in the same manner as described for drilling with other hole-cutting tools. When starting the hole, the drill is supported in both hands, Fig. 10-33. One hand is placed on the handle positioned over the trigger starting switch. The drill pilot is placed on the layout mark and positioned at the proper angle to the surface being drilled. Light pressure is applied to the drill as it is started. The amount of pressure required to complete the boring will depend upon the size and type of bit and the material being drilled. The pressure is relieved as the bit breaks through the opposite side to avoid excessive chipping. A block of scrap stock can be clamped to the back side of the stock to eliminate excessive splitting.

Portable drills can be placed in drill stands as shown in Fig. 10-34. The stand provides many of the advantages of the stationary drill press. The table can be tilted for drilling holes at an angle.

SAFETY CONSIDERATIONS FOR THE PORTABLE POWER DRILL

1. Make certain the electrical plug is properly grounded.
2. Select the proper bit and secure it in the chuck.
3. Small stock must be held secure in a vise or with clamps.

STATIONARY POWER DRILL PRESS AND MORTISING MACHINES

The drill press is a very versatile machine and can be used for boring, drilling, mortising, routing, sanding, and shaping operations, Fig. 10-35. Drill presses are

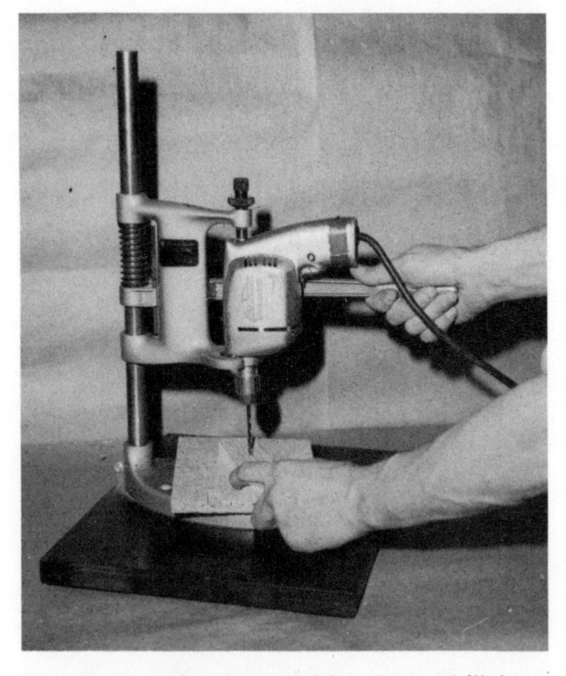

Fig. 10-34. Using a portable power drill in a drill stand.

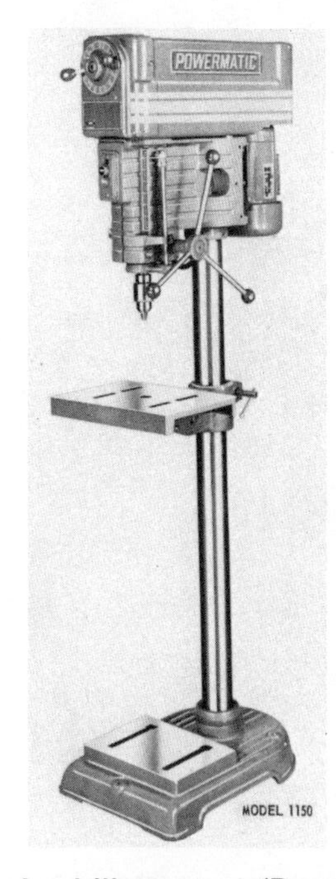

Fig. 10-35. A drill press. (Powermatic Machine Co.)

available in many different models and sizes. In addition to bench models, there are also floor models. The size of a drill press is determined by the distance from the center of the chuck to the column multiplied by two. They range in size from 14″ to 20″. The 15″ size is a common general purpose machine, Fig. 10-36. The distance from the column to the center of the drill chuck of a 15″ drill press measures 7½″.

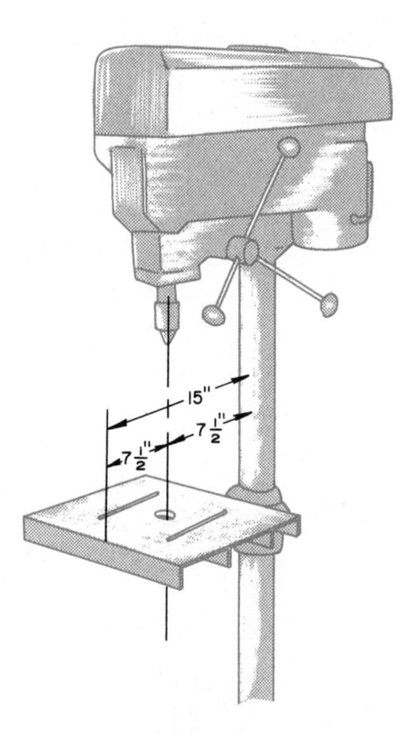

Fig. 10-36. Sizing of a 15″ drill press.

DRILL PRESS SPEED CONTROL

The speed is adjusted by means of moving V-belts on step pulleys or a variable speed mechanism, such as shown in Fig. 10-37. The adjustable variable speed provides a wide range of speeds from 500 to about 4500 RPM. Figure 10-38 shows a stepped pulley assembly for changing the spindle speed of the drill press. Various speeds are needed for different size bits and other operations. The routing and shaping operations require high speeds, and speeds of between 500 to 2000 RPM are generally used for cutting holes in wood.

PARTS OF THE DRILL PRESS

The basic parts of the drill press consist of the *head, column, base* and *table*. The head, base and table are supported by the column. The table can be moved up and down along the column, and can also be tilted for drilling holes at an angle. The head assembly consists of a spindle on which the chuck is attached. The spindle revolves inside a sleeve called a *quill.* A gear assembly moves the quill up and down when the feed lever is turned. A coil spring returns the quill to the up position when the

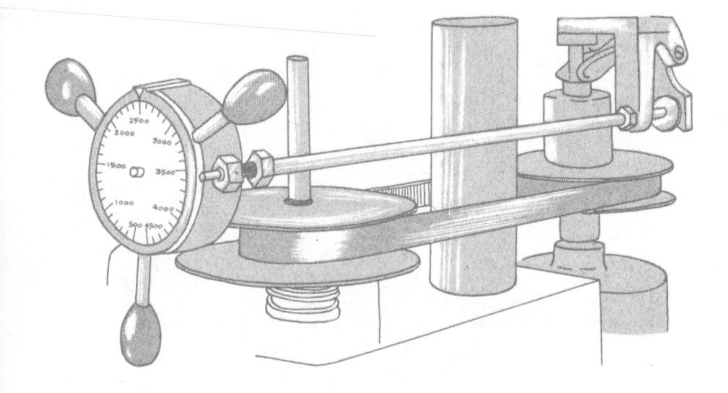

Fig. 10-37. Variable speed change mechanism on a floor model drill press.

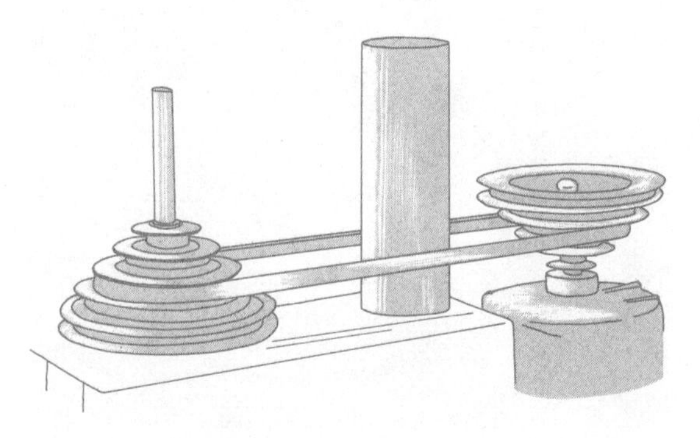

Fig. 10-38. Step pulley assembly for speed changes.

feed lever is released. Most drill spindles are equipped with a three-jaw chuck, usually ½″ in size. The bit is tightened in the chuck by means of a chuck key. The base serves as the support for the machine and can also be used as a table for drilling large stock. For this purpose, the table is turned to the side and the work is placed on the base.

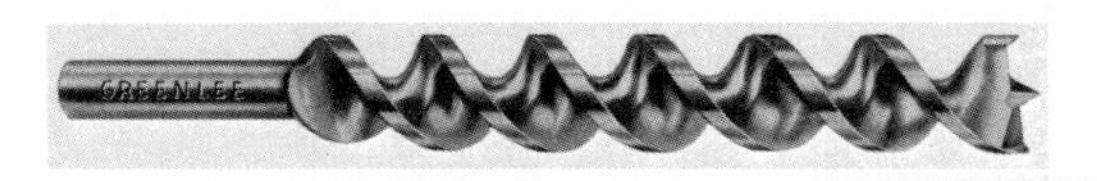

A. Machine spur bit. (Greenlee Tool Co.)

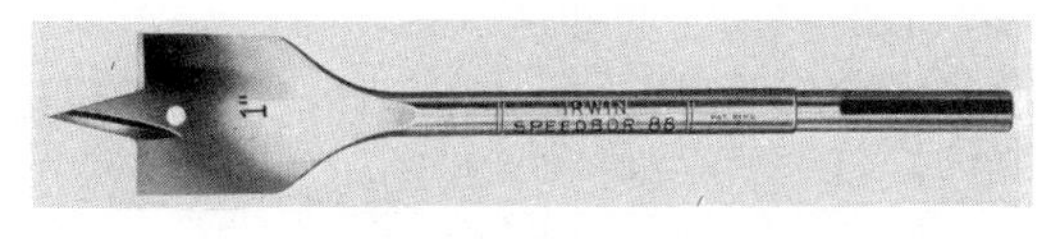

B. Spade bit. (Irwin Auger Bit Co.)

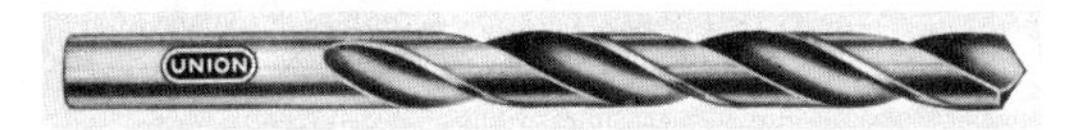

C. Twist bit. (Union Twist Drill Co.)

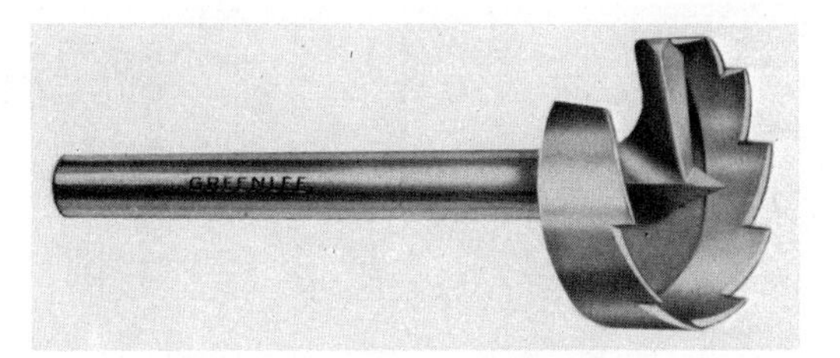

D. Multi-spur bit. (Greenlee Tool Co.)

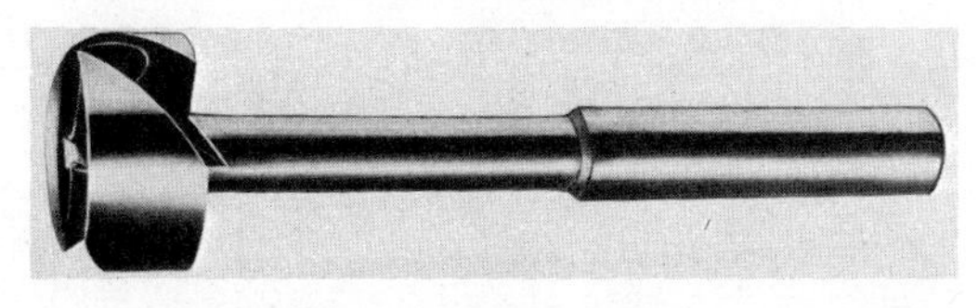

E. Forstner bit. (The Conn. Valley Mfg. Co.)

Fig. 10-39. Common bits for the drill press.

COMMON BITS FOR CUTTING HOLES WITH THE DRILL PRESS

There are numerous types of bits used in the drill press, Fig. 10-39.

Machine spur bits. Machine spur bits with brad points are used for drilling holes larger than 3/16″ in diameter. There are two spurs on the outside which score and cut the stock, producing smooth clean holes. These bits are available in sizes from 3/16″ to 1¼″ in diameter.

Twist bit. The twist bit is used for drilling holes ranging in size from 1/64″ to ½″ in diameter. It is a good, all-purpose bit and especially useful for drilling holes less than ¼″ in diameter. Twist bits used for drilling in wood are ground with a sharper point, which provides faster cutting. The *included angle* (total lip to lip angle) should be between 60° and 80° for drilling in wood. The cutting edge of twist bits for drilling all other materials (including metal) have an included angle of 118°.

Multi-spur bit. A multi-spur bit is commonly used for boring large holes. It is available in sizes ranging from ½″ to 4″ in diameter, and has a brad point, one cutting edge, and a number of spurs. The spurs score and cut the outer edge of the hole, and the cutting edge removes the center portion of the hole from the spurs to the brad center. The spurs on the outer edge make it an ideal cutter for cutting holes in plywood and veneer, as it causes very little splitting and chipping of the veneered surface.

Forstner bit. The forstner bit is used when a hole is cut only part way through the stock. It does not have a center screw. The outer rim of the bit is sharp and both cuts and guides the tool.

CUTTING HOLES ON THE DRILL PRESS

Select the proper bit for the job to be performed, and insert it into the chuck. The bit must be centered in the chuck and run true. Tighten the bit firmly in the chuck, using the chuck key. The full depth of the shank of the bit should extend into the chuck, Fig. 10-40. Make certain the chuck key is removed after mounting the bit. Adjust the height of the table, allowing about an inch of clearance between the bit and the stock. A piece of scrap stock should be placed between the table and the piece being drilled. The waste stock will prevent chipping out on back side of the stock and protect the table surface.

The speed of the drill press will vary according to the size and type of bit and kind of stock. In general, the spindle speed should not exceed 1000 RPM for bits over ½″ in diameter. Excessive speed will cause the bits to overheat. Smaller bits can be operated at higher speeds, up to 3000 RPM.

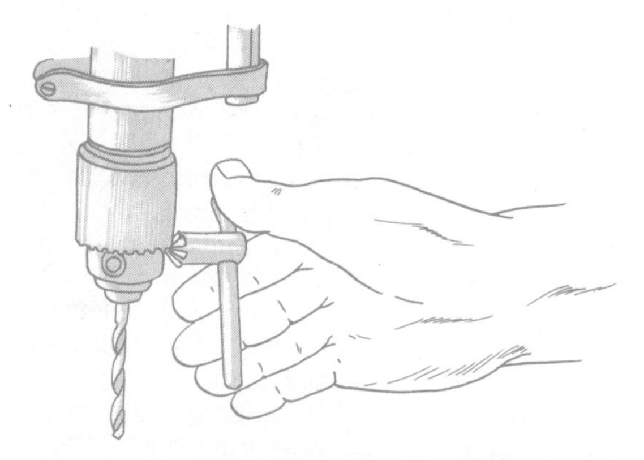

Fig. 10-40. Placing a bit into the drill chuck with the entire shank extended into the chuck.

The holes are laid out using the same procedure as previously discussed for cutting holes with hand tools. The stock is held flat on the table, the drill press is started, the bit is brought downward, and the layout mark is placed directly beneath the center of the drill bit, Fig. 10-41. The bit is then advanced into the stock the desired depth by turning feed lever.

HOLDING SMALL STOCK ON THE DRILL PRESS

To drill small pieces on the drill press, the stock should always be clamped in place with a hand screw, Fig. 10-42. Sometimes larger stock can be placed in a hand screw and held firmly against the table. However, if small stock is not clamped, it may be jerked out of your hand and cause an injury.

Fig. 10-41. Drill press and stock set up for drilling holes in flat stock.

Fig. 10-42. Small stock is held in a hand screw for drilling holes.

The depth stop may be used to bore a hole to specified depth. The stop rod nuts are locked for the desired depth on the stop rods, Fig. 10-43.

DRILLING HOLES IN ROUND STOCK

A V-block is used when drilling holes in the side of round stock to center the stock and prevent it from turning, Fig. 10-44. The V-block can be clamped to the table which is turned to a vertical position for drilling in the end of round stock, Fig. 10-45.

DRILLING HOLES AT AN ANGLE

The table is tilted when a hole is drilled at an angle. A sliding T-bevel is used to adjust the table to the correct angle, Fig. 10-46. A more accurate hole can be drilled if the stock is clamped to the table.

Fig. 10-43. Setting the depth of the hole with stop rod nuts screwed into position.

Fig. 10-45. Using a V-block to drill holes in the end of round stock.

Fig. 10-44. Using a V-block to drill holes in the side of round stock.

Fig. 10-46. Using a sliding T-bevel to set the angle of the table for angled holes.

DRILLING DUPLICATE HOLES

A number of holes drilled an equal distance apart can be drilled using a simple jig such as the hinged jig shown in Fig. 10-47. A dowel pin jig may be used by placing a locating pin in each hole as it is drilled. The distance from the center of the locating pin to the center of the drill bit must equal the desired spacing of the holes.

Many duplicate pieces can be drilled with a hole in the same location by means of a simple jig such as the pocket jig in Fig. 10-48. The jig eliminates the need for the layout of each hole and insures exact location. If the piece has a hole located in the same location on both ends, the same jig serves both purposes.

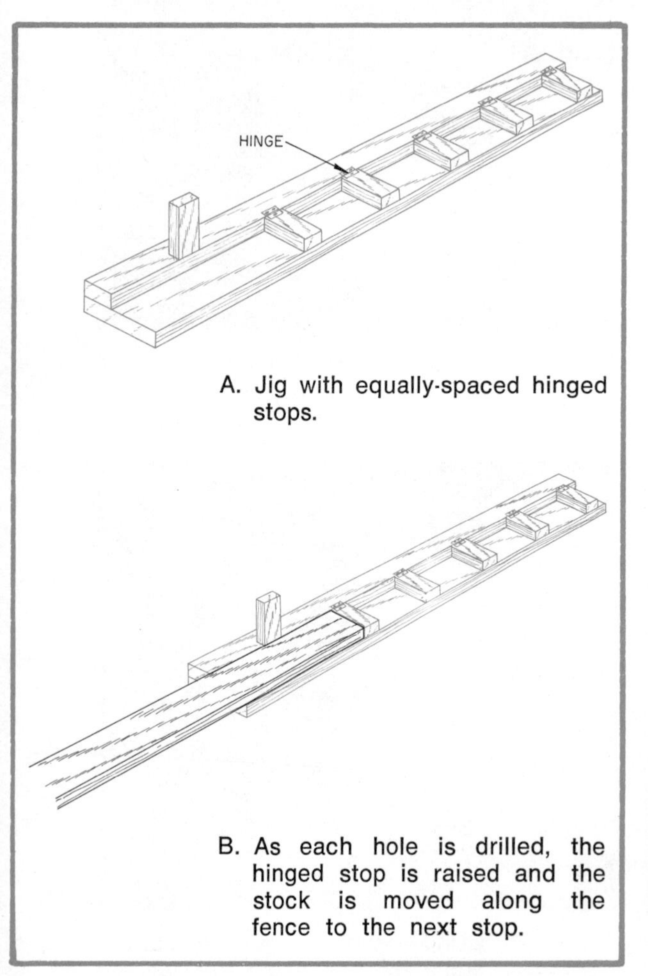

A. Jig with equally-spaced hinged stops.

B. As each hole is drilled, the hinged stop is raised and the stock is moved along the fence to the next stop.

Fig. 10-47. Equal spacing of drilled holes.

DRILLING HOLES IN IRREGULAR SHAPES

Irregular shapes can be drilled by clamping the stock in position with a hand

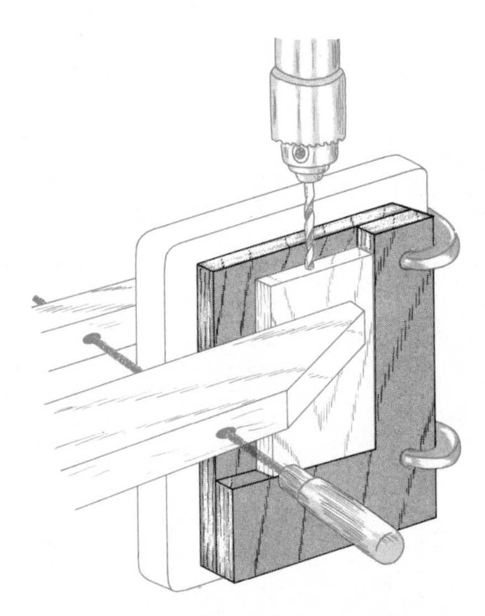

Fig. 10.48. Pocket jig for locating holes in duplicate pieces of stock.

Fig. 10-49. Set-up for drilling holes in irregular stock.

screw as shown in Fig. 10-49. A small try square is used to align the hole.

CUTTING LARGE HOLES ON THE DRILL PRESS

Special types of tools are used for cutting large holes on the drill press. The work must be clamped securely to the drill press table, and a fly-cutter or hole saw is used, Fig. 10-50.

A board must be placed under the work to prevent damage to the drill press table and the cutting tool. The hands must be held out of the danger zone of the revolving beam when the fly cutter is used. It is especially useful as it can be adjusted for any size hole. Both the fly cutter and the hole saw are operated at low speeds of 400 to 500 RPM.

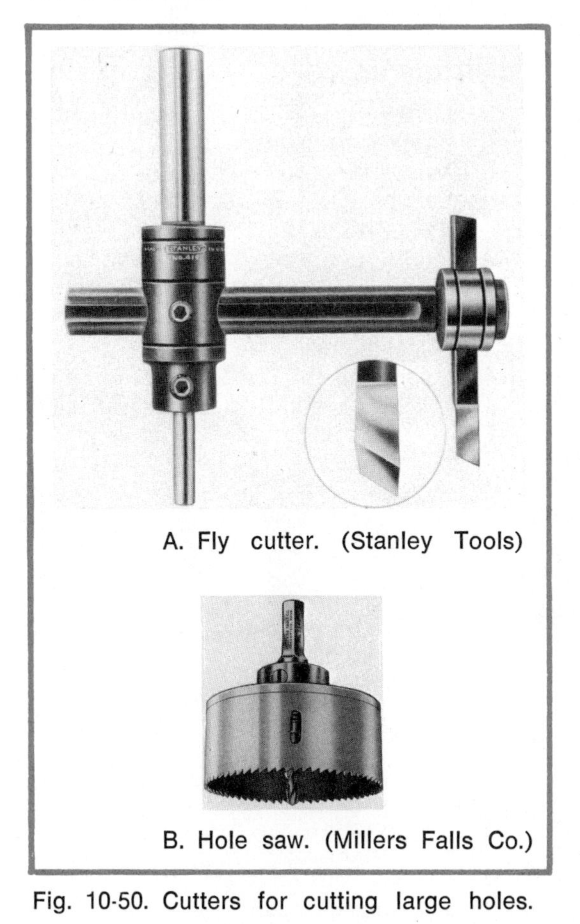

A. Fly cutter. (Stanley Tools)

B. Hole saw. (Millers Falls Co.)

Fig. 10-50. Cutters for cutting large holes.

CUTTING PLUGS AND DOWELS

It is often necessary to cut dowels and plugs to match the particular wood being used. A plug cutter will produce round wooden plugs to fit into counterbored holes to cover screw heads and other types of fasteners. The spiral plug cutter is used to cut plugs up to 2″ long from end grain or across the grain, Fig. 10-51. The cutter is designed so that each plug cut pushes the previous plug out of the cutter. The spiral plug cutter is available in a variety of sizes.

MORTISING ATTACHMENT ON DRILL PRESS

A mortise is a square or rectangular hole cut in wood. The purpose of the mortise is to receive the tenon when joining stock with the mortise and tenon joint, Fig. 10-52. The mortise can be cut in numerous ways, ranging from hand tools to sophisticated mortising machines. The most efficient

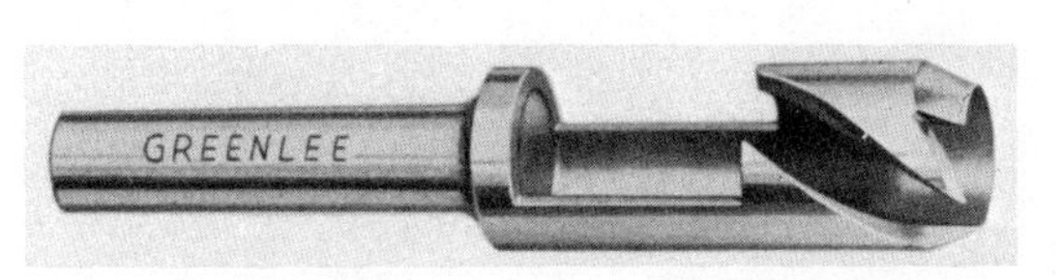

Fig. 10-51. A spiral plug cutter. (Greenlee Tool Co.)

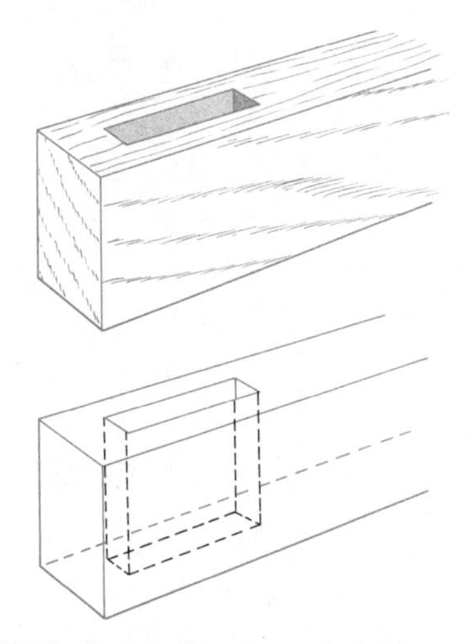

Fig. 10-52. A mortise for receiving a tenon.

method is with a mortising machine or with an attachment for the drill press. The cutting portion of the mortising attachment consists of a hollow chisel and a bit, Fig. 10-53. The chisel revolves inside the hollow chisel and removes most of the stock. The "corners" of the hole are sheared by the hollow chisel as it is forced into the hole. The inside of the chisel is beveled on the cutting edge to make a shearing cut.

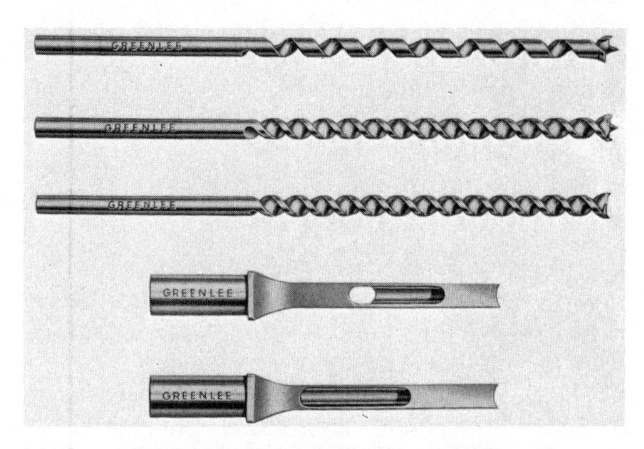

Fig. 10-53. Mortising chisels and bits. (Greenlee Tool Co.)

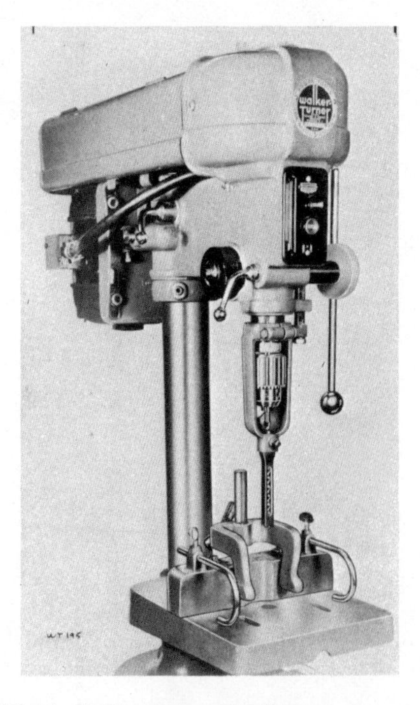

Fig. 10-54. Drill press attachment for mortising. (Walker-Turner Div., Rockwell Mfg. Co.)

The mortising attachment for the drill press consists of a chisel, bit, fence, chisel holder, and hold-down, Fig. 10-54. The fence to which the hold-down is attached holds the works securely to the table in alignment with the chisel. The chisel holder is clamped to the quill of the drill press, the depth stop bracket is removed from the quill to install the chisel holder, and the bit is held by the drill chuck.

Hollow mortising chisels and bits are available in sizes from 1/4″ to 3/4″ by 1/16″ graduations. Common sizes are 1/4″, 3/8″, and 1/2″. The width of the mortise is determined by the size of the chisel. The length of the mortise is determined by moving the stock for desired number of cuts.

INSTALLING THE MORTISING ATTACHMENT

1. Remove the feed depth stop bracket from the quill.
2. Mount the hollow chisel holder and attach the depth stop rod in the chisel bracket. (The stop rod regulates the mortise depth and prevents the bracket from turning.)
3. Install the hollow chisel in the chisel holder.
4. Insert the bit in the chuck. The spurs of the bit extend from 1/32″

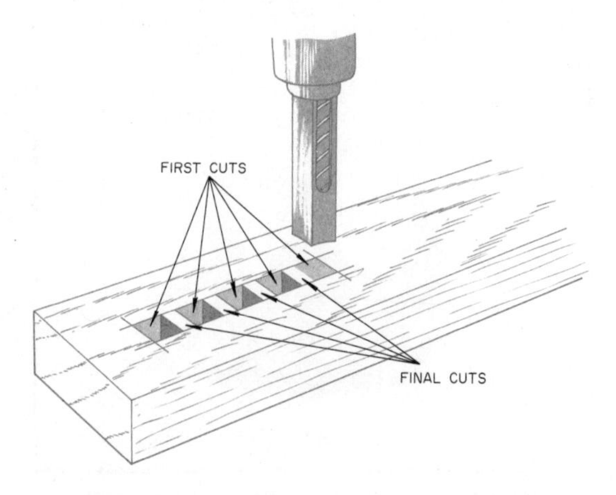

Fig. 10-55. Sequence of cuts for a mortise.

to 1/16″ below the chisel. (If there is not proper clearance between the bit and chisel, overheating will result.) Tighten bit firmly in the chuck.

5. Attach the fence to the table. Set the fence parallel with the front surface of the hollow chisel.
6. Layout the mortise of the stock to be cut and make the final location of the fence to the chisel.
7. Set the depth stop for the depth desired.
8. Check the clearance of the bit in the chisel by revolving the chuck by hand.

OPERATING THE MORTISER

After adjustment of the machine using the procedure previously described, the mortiser is ready to operate. The first set of cuts should be made as shown in Fig. 10-55. This method of making alternate cuts eliminates stress otherwise exerted from pressure on three sides of the chisel. Bending of the chisel also causes additional friction from the rotating bit.

To joint legs and rails with a mortise and tenon joint as shown in Fig. 10-56, the mortise is cut from two adjacent sides. The first cut should be made with the depth only to where it will meet the second mortise. This method reduces the internal splintering which may occur.

The time required to layout the mortise can be eliminated if stops are used to locate the mortise. It is sometimes easier to use an auxiliary fence with stops to locate the top and bottom cuts such as shown in Fig. 10-57.

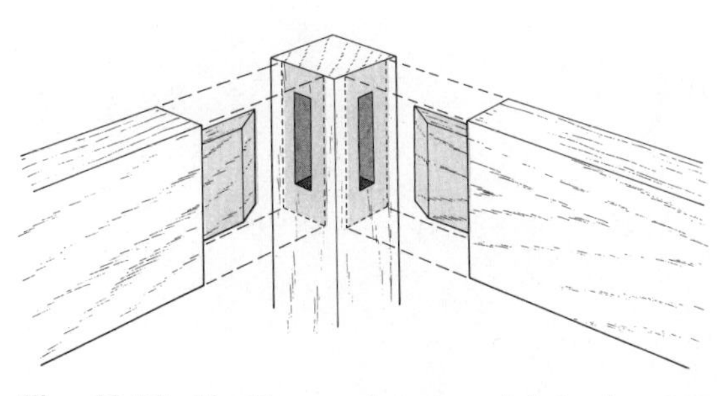

Fig. 10-56. Mortise and tenon joints for leg and rail construction.

ROUTING ON THE DRILL PRESS

A special attachment for routing can be mounted on most drill presses. The drill press should not be used without the special adaptor unless the chuck is a fixed part of the drill spindle. The drill press can be setup with a fence and table for cutting a groove the length of the stock or a rabbet along the edge, Fig. 10-58.

OPERATING AND SAFETY SUGGESTIONS FOR DRILLING AND BORING MACHINES

1. Remove or fasten loose articles of clothing, such as necktie, sleeves,

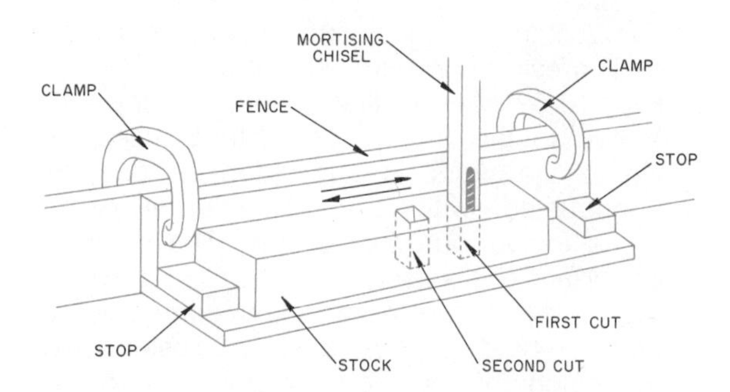

Fig. 10-57. Cutting duplicate pieces using auxiliary fence and stops.

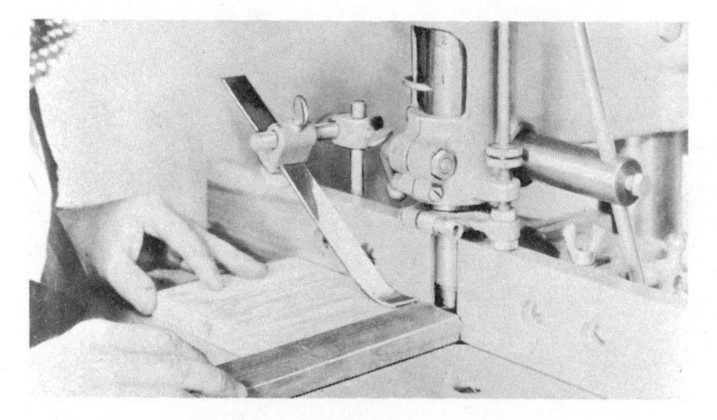

Fig. 10-58. Routing attachment on a drill press. (Delta Div., Rockwell Mfg. Co.)

coat or jacket, etc. before operating power drilling machine.
2. Remove jewelry from hands and arms before operating power drills.
3. Wear a face shield or goggles to protect the face and eyes from flying chips.
4. Always clamp work securely — do NOT hold material with hand while drilling.
5. Disconnect the power source to the machine when servicing and adjusting.
6. Use correct size drill and drill speed.
7. Properly ground the machine to a water pipe, radiator or general grounding system to safeguard against electrical short.

SANDING ON THE DRILL PRESS

A sanding drum can be placed in the drill press chuck for sanding operations, Fig. 10-59. An auxiliary table clamped or bolted to the drill press table is used on which the work is supported. It can be constructed with several different sized holes for various size drums. The sanding drum can be raised and lowered through the hole as the abrasive becomes worn and is held at the desired position by tightening the quill clamp.

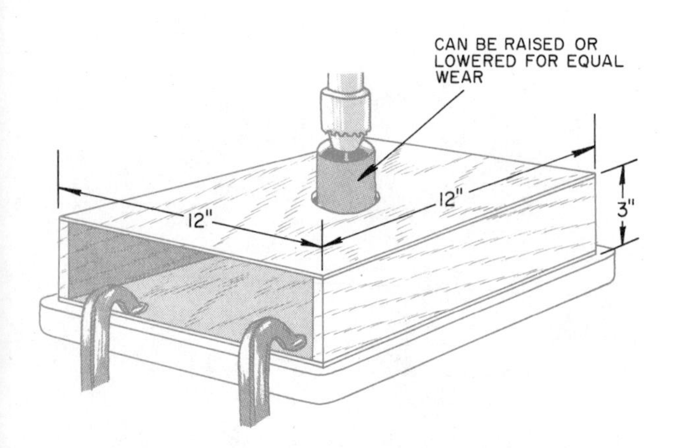

Fig. 10-59. Sanding drum attachment and an auxiliary table on a drill press.

INDUSTRIAL HOLE-CUTTING MACHINES

Industrial hole-cutting machines are designed to produce high-quality work in the minimum amount of time. To accomplish this, machines are used which will drill several holes at one time. Figure 10-60 shows a universal spindle boring machine in which hydraulic power moves the spindles to the work. The length of the stroke can be adjusted from ½″ to 12″ for various types of work. The position of the boring spindle can be changed to accommodate different types of work.

Figures 10-61 and 10-62 show various types of automatic boring machines commonly used in the wood industry. Each ma-

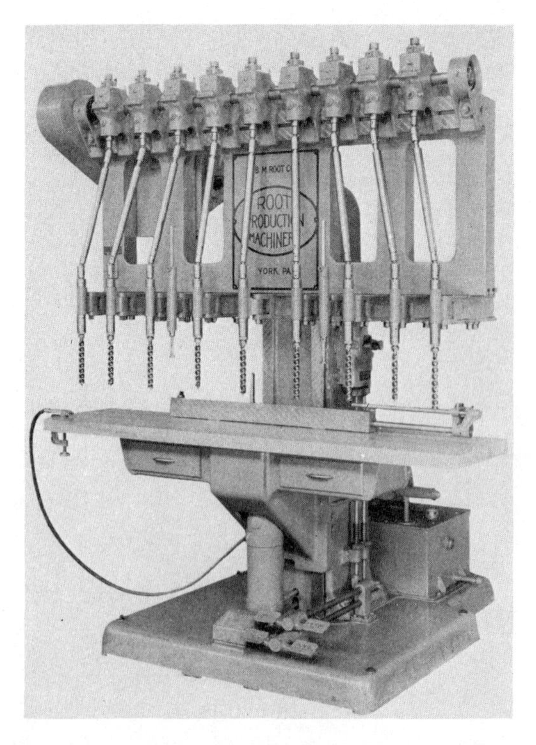

Fig. 10-60. Multiple spindle boring machine. (B. M. Root Co.)

chine can be setup for a particular operation for cutting holes on various products.

Fig. 10-61. Automatic multiple spindle boring machines. (Culley Engineering & Mfg. Co.)

Fig. 10-62. Setups for automatic boring machines. (Culley Engineering & Mfg. Co.)

The machines are designed to bore a number of holes at one setting. Machines such as these reduce production costs by reducing labor costs.

The mortising machine shown in Fig. 10-63 employs a hollow mortising chisel. The machines are often equipped with a power feed mechanism to raise and lower the chisel into the work, and have tables which can be moved in and out, back and forth, and up and down.

The chain-saw mortiser has replaced the hollow-chisel mortiser for high production runs of some products. Figure 10-64

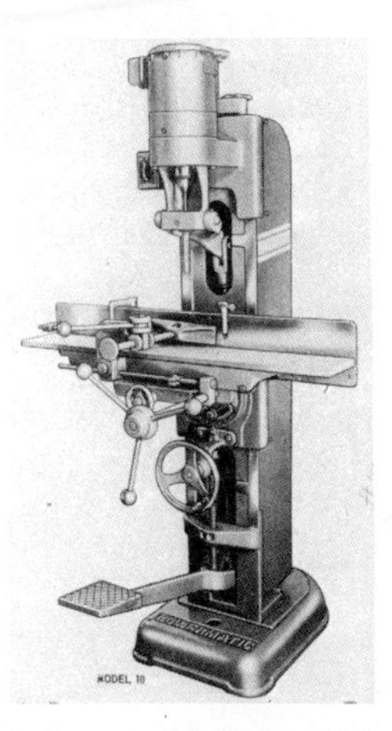

Fig. 10-63. Hollow chisel mortiser. (Powermatic Machine Co.)

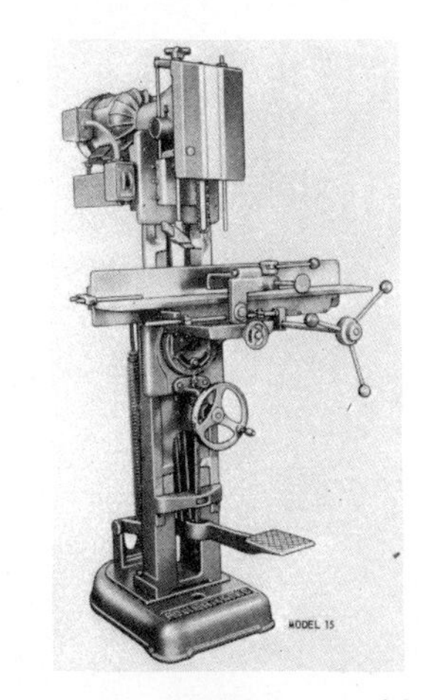

Fig. 10-64. Chain mortising machine. (Powermatic Machine Co.)

shows a continuous chain mortising machine. With this type of machine, the chain is lowered into the work and the stock is advanced to the desired distance. The chain mortiser produces more work in a shorter time than a hollow chisel machine.

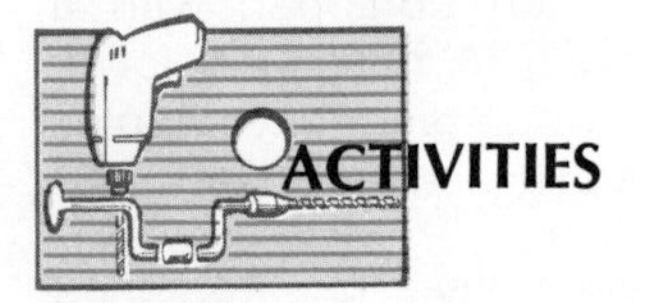

ACTIVITIES

PROBLEM

Assume that you are the supervisor of production. The product which you manufacture requires twenty ¼″ holes on the face surface of a ½″ × 4″ × 12″ piece of stock. The holes are premarked with a template. The calculated cost for a work station and a man is $0.20 per minute. How much would it cost your company to drill the twenty holes in each piece by hand and by machine?

Use a timing device to record how long it takes to produce a hole with:

A. Hand drill

B. Portable electric drill

Compute the total cost of this operation for both methods and the total length of time for each.

Using the above problem, what would the operating cost be if you had five operators at five work stations at the same time?

VOCABULARY

Chuck
Depth gauge
Expansive bit
Machine spur bit
Forstner bit
V-block
Drilling jig
Mortise
Spade bit

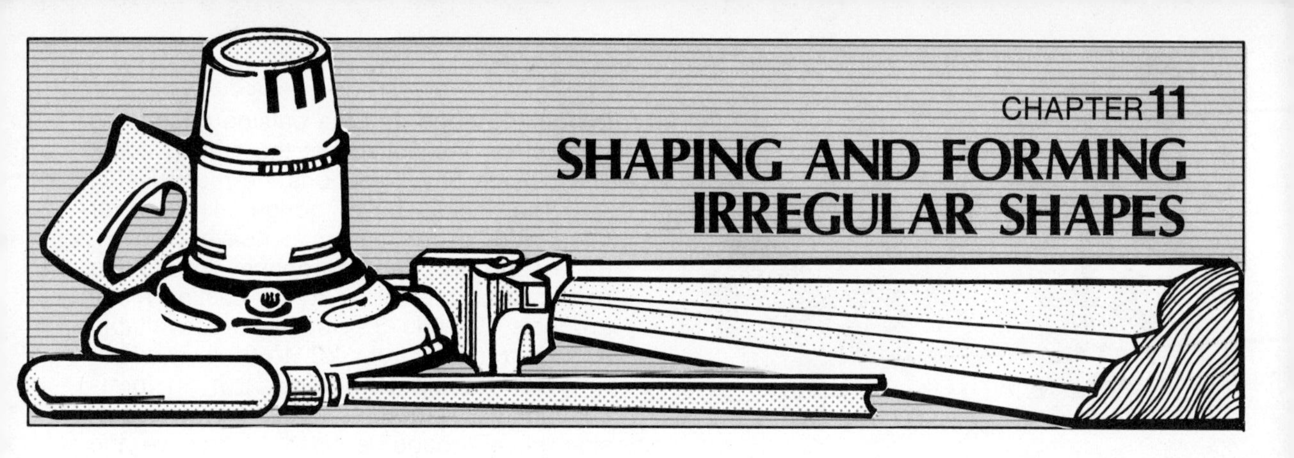

CHAPTER 11

SHAPING AND FORMING IRREGULAR SHAPES

INTRODUCTION

An irregular shape is generally more difficult to form than a square or straight shape. The irregularity in shape may be on the inside or outside of a product. In some cases, both the inside and outside may be irregular, Fig. 11-1. Unless automatic machines can be used, the forming of irregular shapes is an expensive process.

HAND TOOL PROCESSES FOR SHAPING IRREGULAR SURFACES

Irregularly shaped work is difficult to form and often involves considerable work with hand tools. The most common hand tools for carving irregular shapes are *gouges* and *carving sets*; however, files, rasps, and spokeshaves are also commonly used.

GOUGES

A gouge is similar to a chisel but is half round. There are two types of gouges: (1) *inside,* and (2) *outside,* Fig. 11-2. An inside gouge has a bevel on the inside of

Fig. 11-1. Objects requiring the forming of irregular shapes.

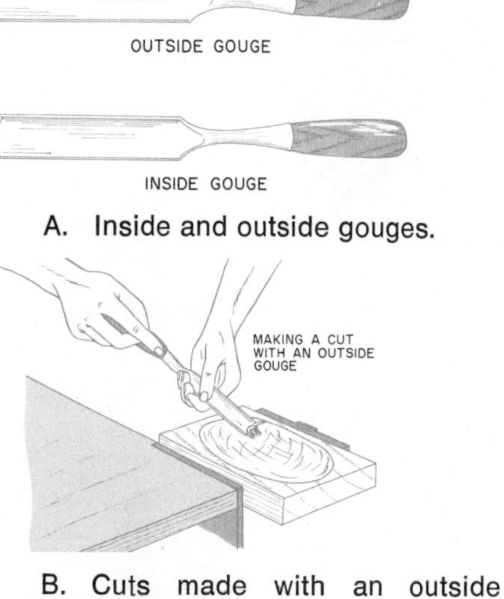

A. Inside and outside gouges.

B. Cuts made with an outside gouge.

Fig. 11-2. Gouges.

Fig. 11-3. Foundry patterns which consist of irregular shapes.

Fig. 11-4. A free form bowl which has been sculptured with gouges.

Fig. 11-5. Removing large quantities of stock with a gouge and mallet.

the cutting edge and the outside gouge has a bevel on the outside of the cutting edge. The outside bevel gouge is the most useful.

The inside bevel gouge is used for deep work or for making concave grooves. Patternmakers often use gouges for making patterns used in a foundry for casting metal products, Fig. 11-3. Gouges are available in sizes ranging from 1/8″ to 2″ in width.

Wood sculpturers who carve art forms commonly use gouges, Fig. 11-4. The outside bevel gouge is used to remove the stock from the center. The gouge is forced into the work with a forward and circular movement. When large amounts of stock are to be removed, a wood mallet may be used to strike the gouge, Fig. 11-5. The work must be secured in a vise or clamped to a bench.

CARVING TOOLS

Detail carving is done with a *carving tool.* Carving tools are available in sets of various sizes or can be purchased separately. Figure 11-6 shows a typical set of carving tools, which includes small gouges, skew chisels, fluters, and corner chisels. The size of the tools in a carving set varies from 1/16″ to 1″ in width.

CONDITIONING HAND GOUGES

Like all other edge-cutting tools, the gouges must be sharp. The outside gouge may be ground on the face of a grinding wheel when it is nicked or when the angle must be reshaped. It must be rolled as it is ground to produce an even bevel, then it is

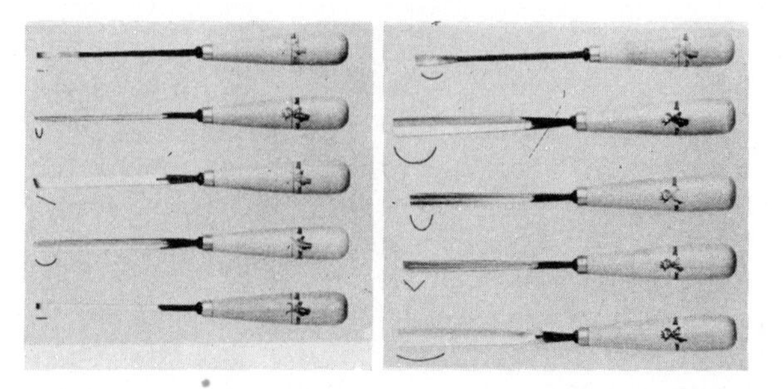

Fig. 11-6. A typical set of carving tools. (Buck Brothers)

honed with an inside slipstone, Fig. 11-7. The inside gouge is ground on a stone with a special shape, then is whetted on the outside edge with a slipstone.

FORMING IRREGULAR SHAPES WITH A SPOKESHAVE

The spokeshave is very similar to a plane except it has a short bottom which allows the spokeshave to follow the contour of the work. The handles are mounted on the side. The cutting action is like that of a plane. The cutting edge is conditioned like that of any other plane iron. The operator can either push or pull the spokeshave into the work, Fig. 11-8.

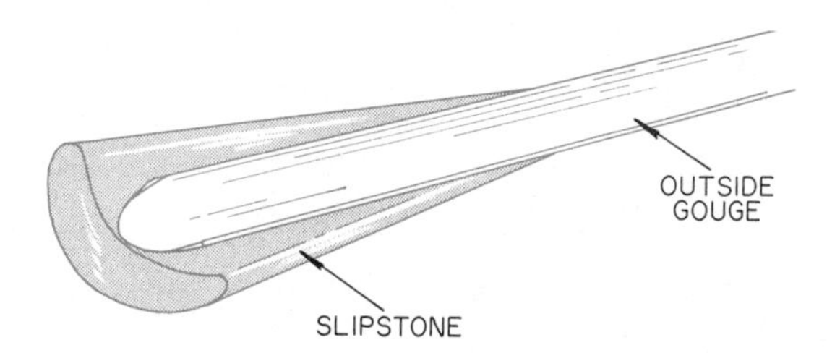

Fig. 11-7. Honing an outside gouge with a slipstone.

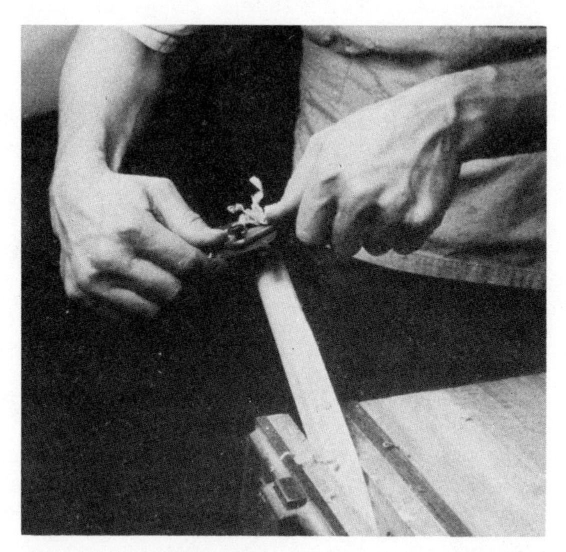

Fig. 11-8. A spokeshave forming a rounded shape.

FORMING IRREGULAR SHAPES WITH MULTI-BLADE FORMING TOOLS (SURFORMS)

There are several types of multi-blade forming tools which are used to shape wood, Fig. 11-9. The blades are not

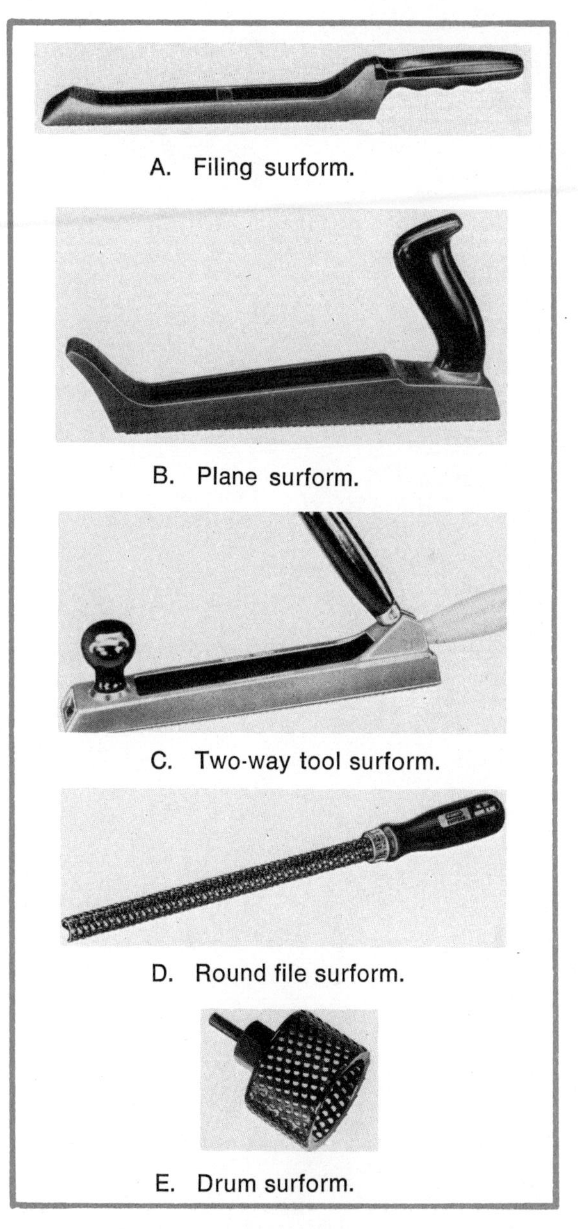

A. Filing surform.

B. Plane surform.

C. Two-way tool surform.

D. Round file surform.

E. Drum surform.

Fig. 11-9. Various multi-blade surform forming tools. (Stanley Tools)

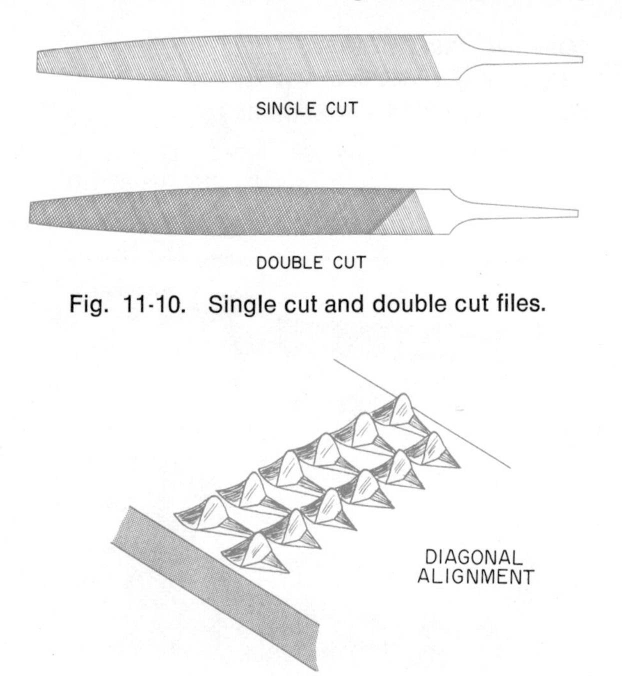

Fig. 11-10. Single cut and double cut files.

Fig. 11-11. The triangular teeth of a rasp cut faster and remove more stock than a file.

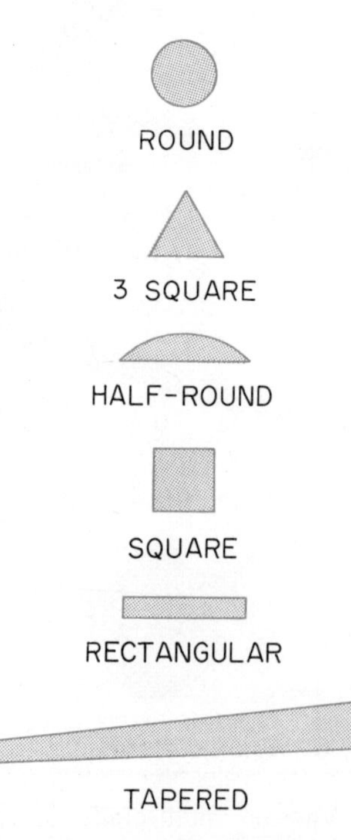

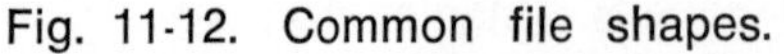

Fig. 11-12. Common file shapes.

reconditioned, but instead are replaced when dull. The cutting edges are made from extremely hard steel and cannot be easily reconditioned. These tools are used in the same manner as planes and files, depending upon their shape.

FILES AND RASPS FOR FORMING IRREGULAR SHAPES

Wood files and rasps are used to smooth and shape surfaces on which edge-cutting tools cannot be used. Files are of varying shapes, sizes, and types of cut. They may make either single cuts or double cuts, Fig. 11-10. The teeth on a file form groups of cutting edges.

Rasps have rows of triangular-shaped projections (also called teeth), Fig. 11-11. The rasp cuts much faster than a file due to the larger teeth. The surface which has been worked with a rasp is rough as it is used to remove excess stock and the file is used to smooth the surface.

COMMON FILE SHAPES

Files and rasps are available in many shapes such as round, flat, rectangular, half round, and triangular, Fig. 11-12. Various shapes of files allow the tool to follow the contour of the stock being shaped. Some files have teeth on the edges and others are smooth on the sides.

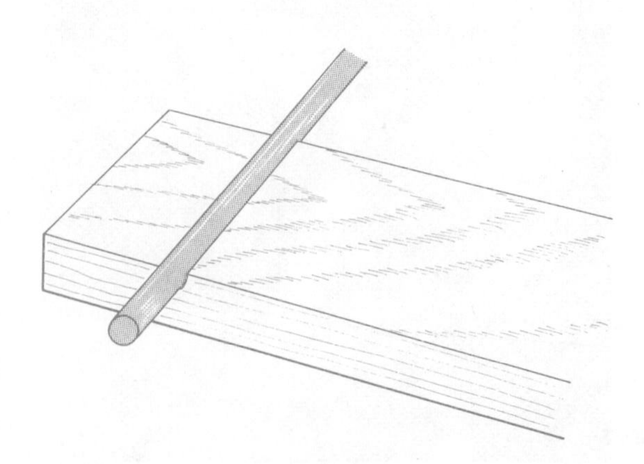

Fig. 11-13. A round file is used to smooth a round surface.

The pointed end of a file is called the tang. A handle is placed on the tang end of the file. The tang is not considered as part of the file when the length is given. Files range in size from 3″ to 14″ for general use.

The file and rasp are used to form irregular shapes on either the inside or outside of the stock, Fig. 11-13. Remember, a rasp is used to remove large quantities of stock, and a file is then used to smooth the surface.

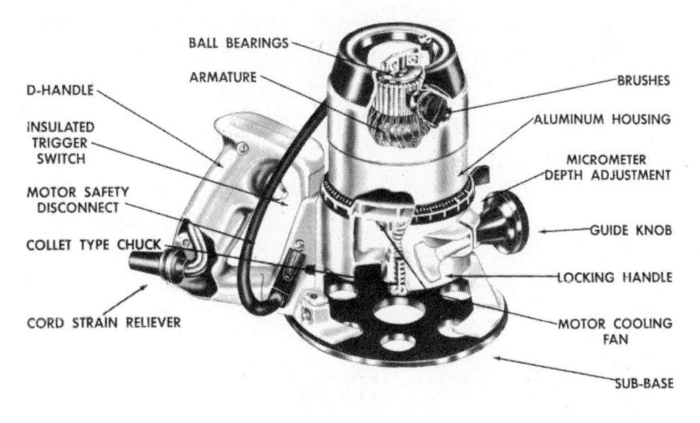

Fig. 11-14. Portable power router. (Power Tool Div., Rockwell Mfg. Co.)

PORTABLE POWER TOOLS FOR SHAPING IRREGULAR SURFACES

The portable router is rapidly becoming a widely accepted and versatile power tool, Fig. 11-14. It is used for all types of work, ranging from industrial production to hobby work in the home workshop. The tool consists of a motor and base. A collet chuck is attached to the hollow motor spindle to grip the tool bit, Fig. 11-15. Standard routers have a collet which holds a 1/4″ diameter tool bit. Heavy-duty industrial routers have collets for tools with 1/2″ diameter shanks. The spindle is held by a locking device on some routers. To change bits, the collet can be loosened with one wrench on this type, whereas the collet is loosened with two wrenches on models without a locking device, Fig. 11-16. To install a bit, the shank should extend into the collet not less than 1/2″. Wrenches for changing bits are furnished with the router, and the proper wrench should always be used to avoid damaging the collet chuck.

The size of the router is given as the horsepower (hp) rating of the motor. Standard routers range in size from 1/4 h.p. to 2 1/2 h.p. Heavy-duty routers may have ratings up to 8 h.p. The operating speed of the motor ranges from 20,000 to 24,000 RPM. The router bit revolves in a clockwise

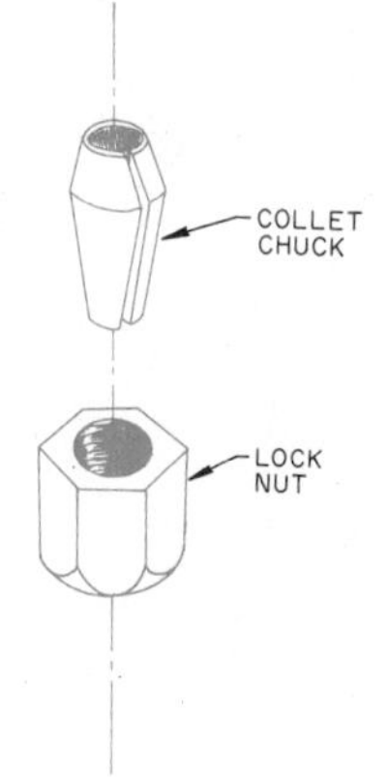

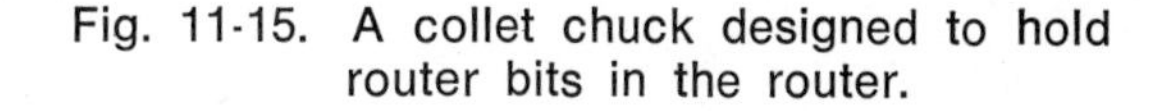

Fig. 11-15. A collet chuck designed to hold router bits in the router.

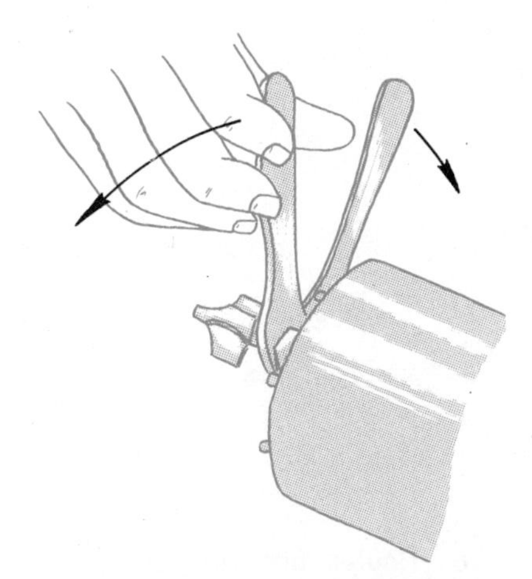

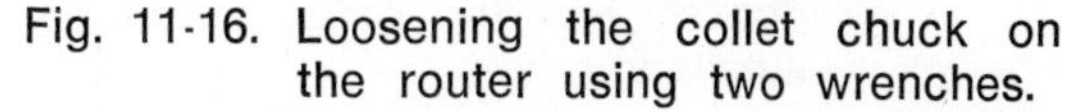

Fig. 11-16. Loosening the collet chuck on the router using two wrenches.

direction when used in an upright position. The router is moved from left to right into the stock. The router is moved counterclockwise around the stock when routing around a circle, Fig. 11-17.

The motor unit is adjusted in the base unit to determine the depth of cut. The housing of the motor on the inside of the base unit is threaded. The exact depth of cut can be adjusted by turning the router in the base, then the motor unit is clamped at the desired depth with a locking screw.

ROUTER BITS AND CUTTERS

Router bits are made from tool steel. Tool steel bits with carbide tips are used for cutting plastic laminates and for production work. Router bits are of one piece or multi-pieces and are called shaper cutters, Fig. 11-18. The one-piece cutter is the most

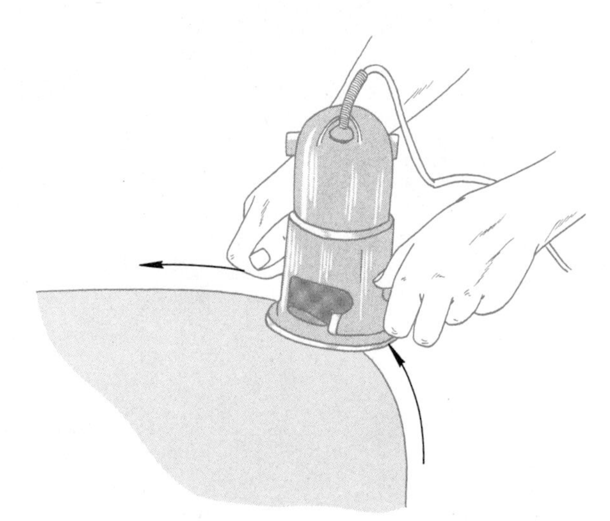

Fig. 11-17. Cutting direction of the router.

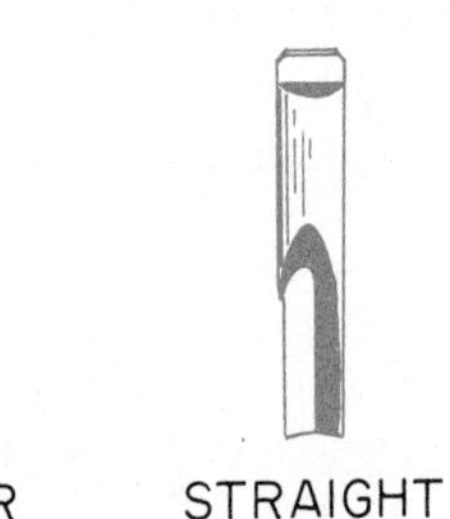

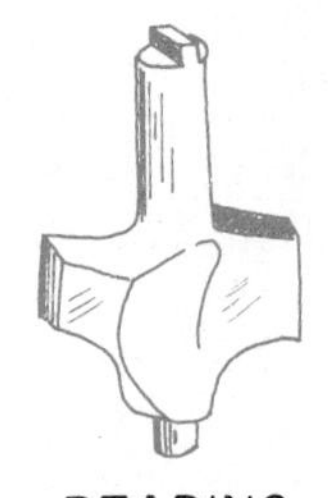

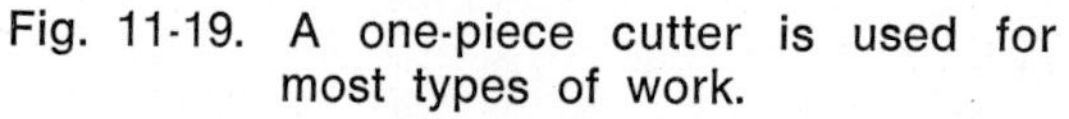

Fig. 11-19. A one-piece cutter is used for most types of work.

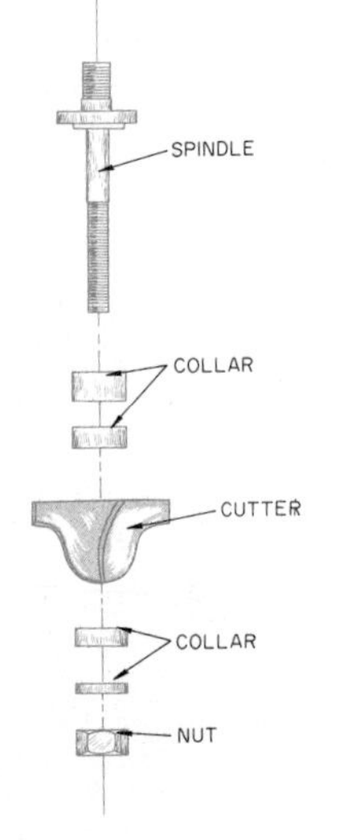

Fig. 11-18. Router bits are designed to fit directly into the collet or on an auxiliary arbor.

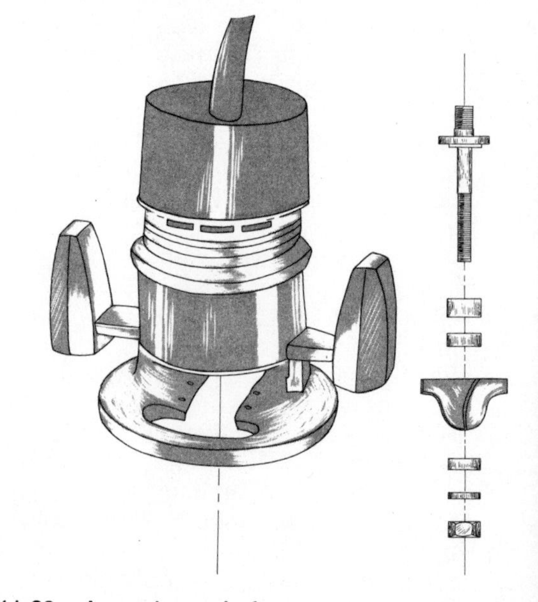

Fig. 11-20. An arbor shaft mounts on the router shaft.

common type of router bit and is called one-piece because the shank and the cutter are machined from the same piece of steel, Fig 11-19. A shaper cutter for the router consists of a separate shaft or arbor on which the cutters are placed. The arbor mounts on the router shaft, Fig. 11-20, and the router is then mounted in a special table for use with shaper cutters. The cutters are used in combination with spacer collars, which control the depth of cut. Various shapes of cutters are available.

Some one-piece router bits have a smooth section called a *pilot* on the bottom of the bit. This pilot makes the bit follow the contour of the stock, and controls the horizontal depth of cut, Fig. 11-21. The pilot bits are used for making decorative cuts on the edge of stock.

Rabbeting bits. A rabbeting bit is used for cutting rabbet joints on the edge of stock. It has a pilot which controls the width of the rabbet, but the depth is controlled by the motor and base adjustment. Rabbet bits are available for cutting different widths, Fig. 11-22.

Straight bits. Straight bits serve for cutting grooves, dadoes, rabbets, and for routing stock from background areas for recesses. The straight bit has cutting surfaces on the ends and sides, Fig. 11-23.

Chamfer bits. Chamfer bits are designed to cut a 45° angle on the edge of stock. They also have a pilot which controls the horizontal depth of cut, Fig. 11-24.

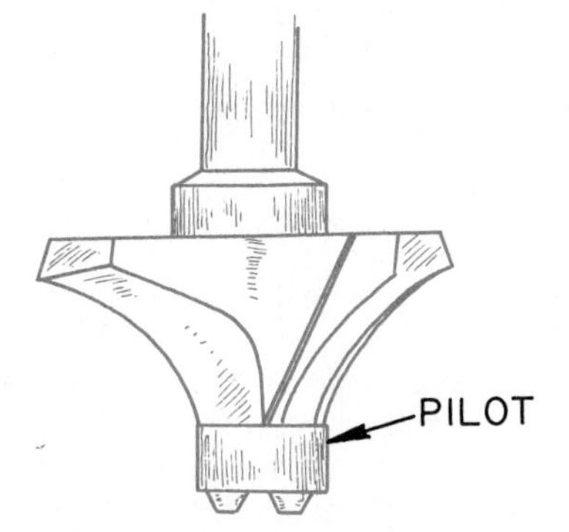

Fig. 11-21. The pilot on a one-piece cutter controls the horizontal depth of the cut.

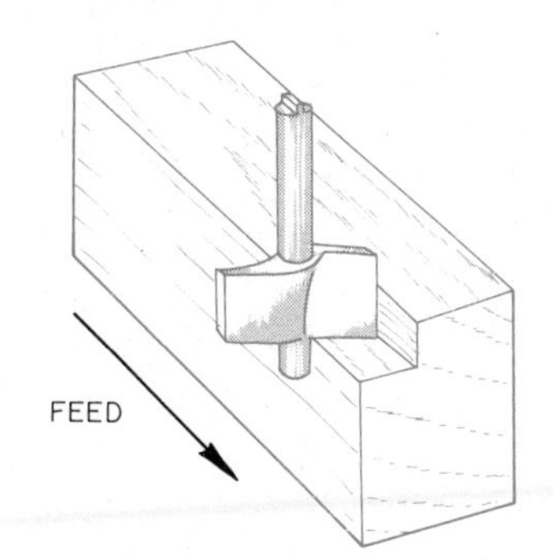

Fig. 11-22. A rabbeting bit on which the pilot controls the width of the rabbet.

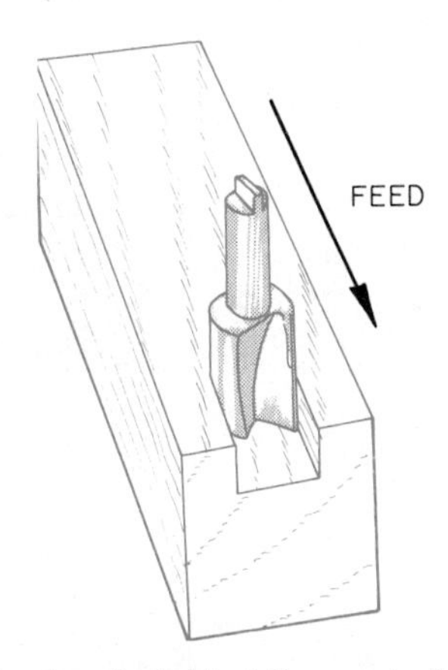

Fig. 11-23. A straight bit used for cutting many types of joints.

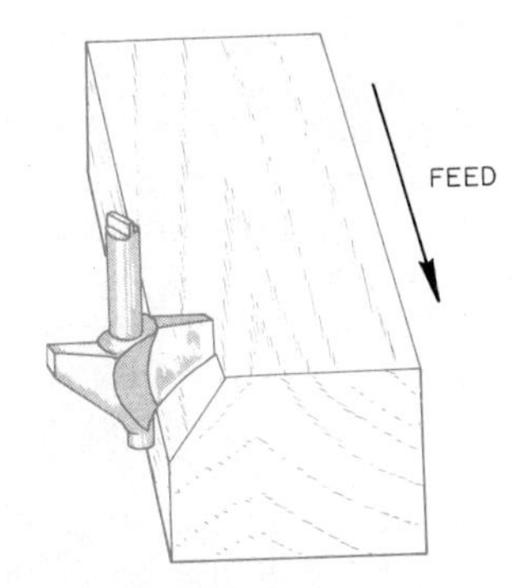

Fig. 11-24. A chamfer bit.

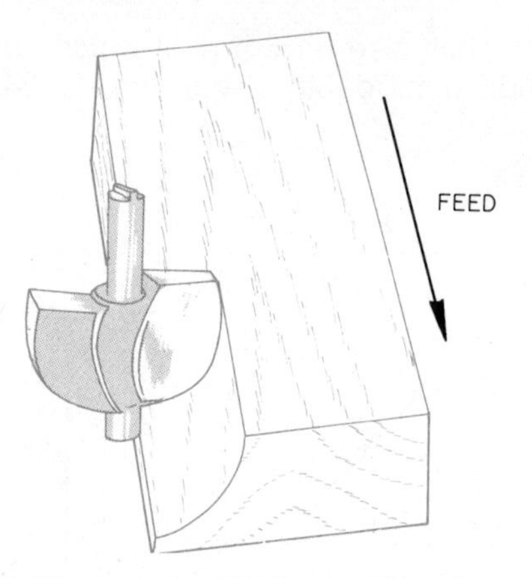

Fig. 11-25. A cove bit for cutting decorative concave cut.

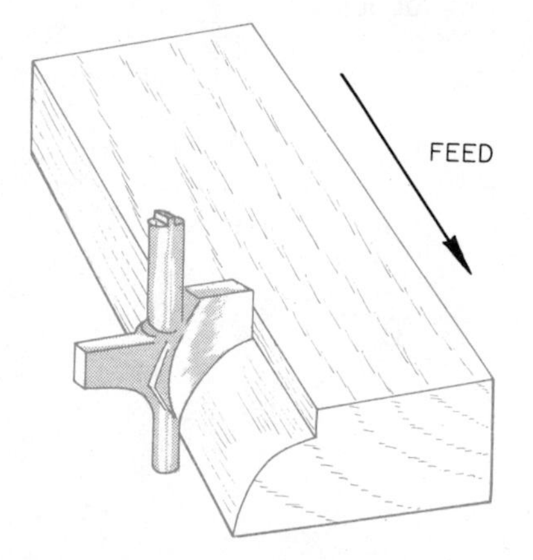

Fig. 11-26. Round over bits produce a convex decorative cut.

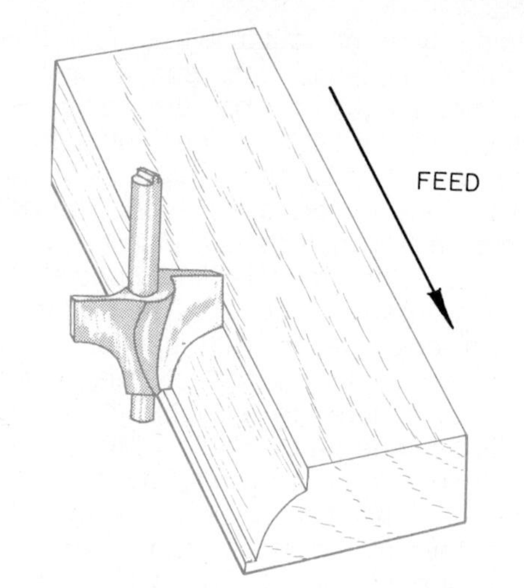

Fig. 11-27. A beading bit produces a decorative cut.

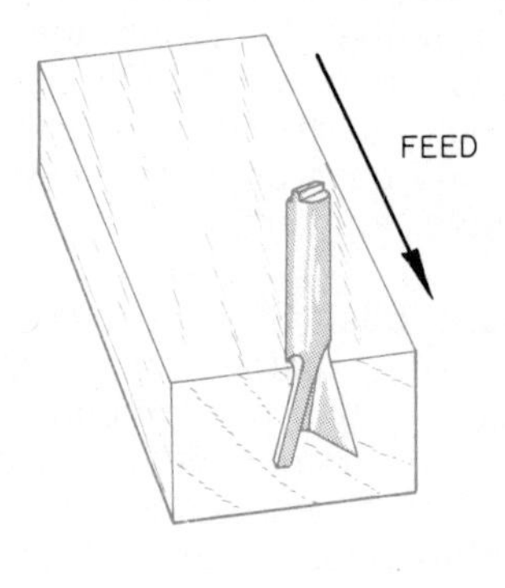

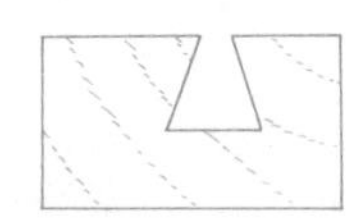

Fig. 11-28. A dovetail bit.

Cove bits. Cove bits produce a concave cut along the edge of the stock, guided by the pilot, Fig. 11-25.

Round over bits. Round over bits make a convex radius on the edge of the stock. This cut is made on the corners of furniture tops, Fig. 11-26.

Beading bits. Beading bits are used to make decorative cuts on the edges of furniture tops. The beading bit is guided by means of a pilot, Fig. 11-27.

Dovetail bits. Dovetail bits are used for cutting dovetail joints (locking joints commonly used for drawer construction). The bit is used with a template guide, Fig. 11-28.

Core box bits. Core box bits are often used for making patterns for foundry work. They produce concave cuts in the stock. The bit does not have a pilot, Fig. 11-29.

Veining bits. Veining bits are much like core box bits, except somewhat smaller. The bit is also similar in appear-

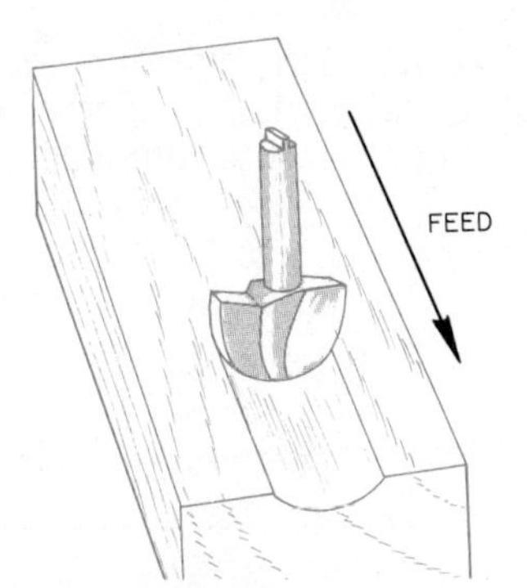

Fig. 11-29. A core box bit.

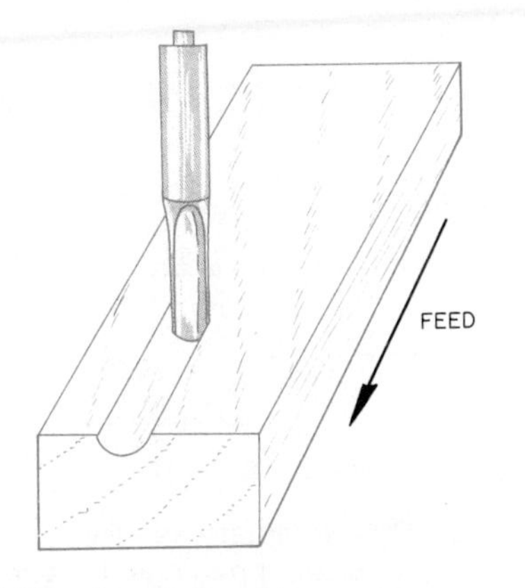

Fig. 11-30. A veining bit.

ance to the straight bit, except it has a rounded end, Fig. 11-30.

METHODS OF GUIDING THE ROUTER

There are several methods by which the router is guided in the work. The simplest guiding method is the use of a router bit with a pilot, as the pilot is smooth and follows the contour of the work, Fig. 11-31.

Straight, dado, and groove cuts can be made by clamping a straightedge on the surface of the stock. The base of the router runs along the straightedge, Fig. 11-32. When several cuts are to be made at a right angle to the edge of the stock, a special T-square can be clamped to the stock, Fig. 11-33. Most router bases are designed to

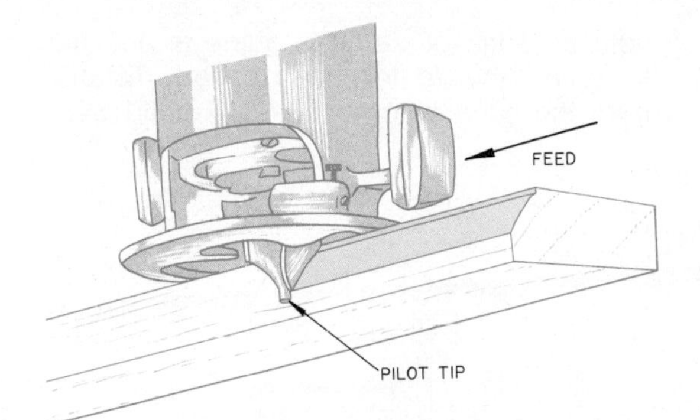

Fig. 11-31. A router guided by a bit with a pilot tip.

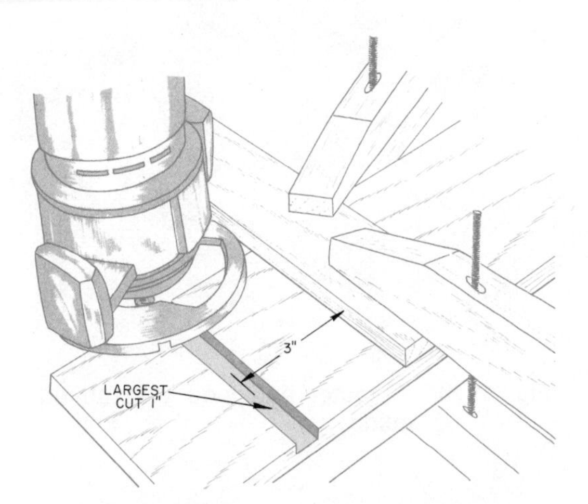

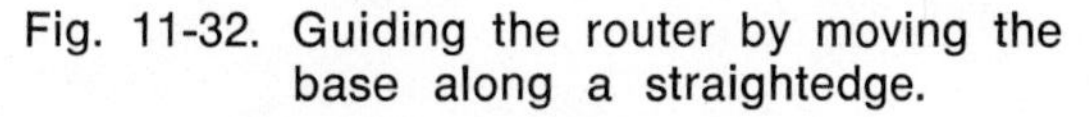

Fig. 11-32. Guiding the router by moving the base along a straightedge.

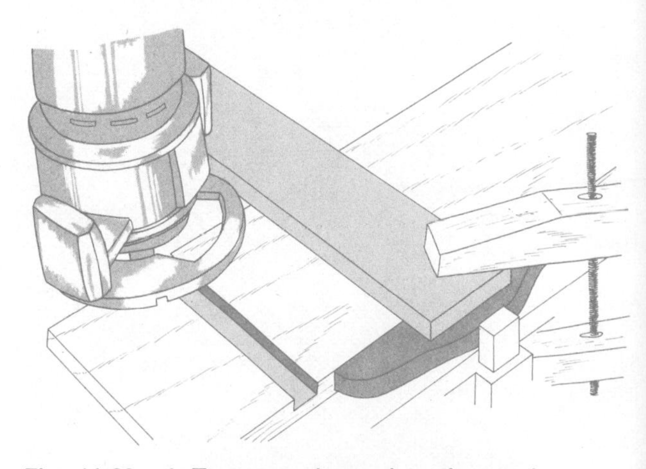

Fig. 11-33. A T-square clamped to the stock to guide the base of the router.

hold an edge guide. This guide is adjusted to guide the router the desired distance from the edge along a board, Fig. 11-34.

The router can also be guided by means of a template for some types of work, Fig. 11-35. The template guide is constructed from 1/4″ plywood or hardboard. The template is tacked or clamped to the stock. The straight bit is most often

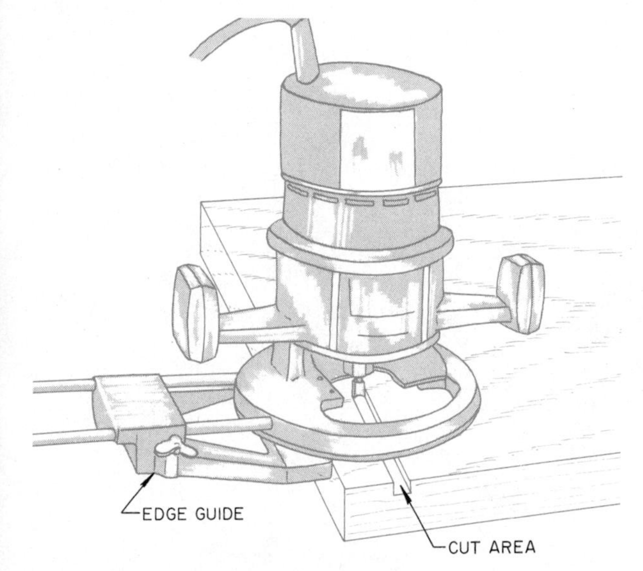

Fig. 11-34. A router equipped with an edge guide.

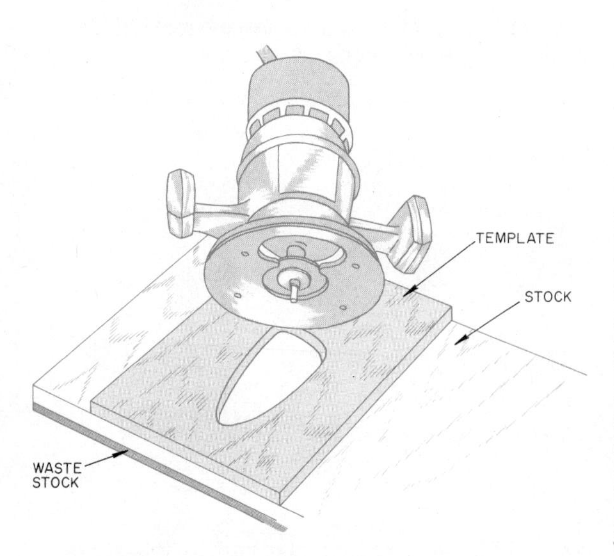

Fig. 11-35. A template guide for the router.

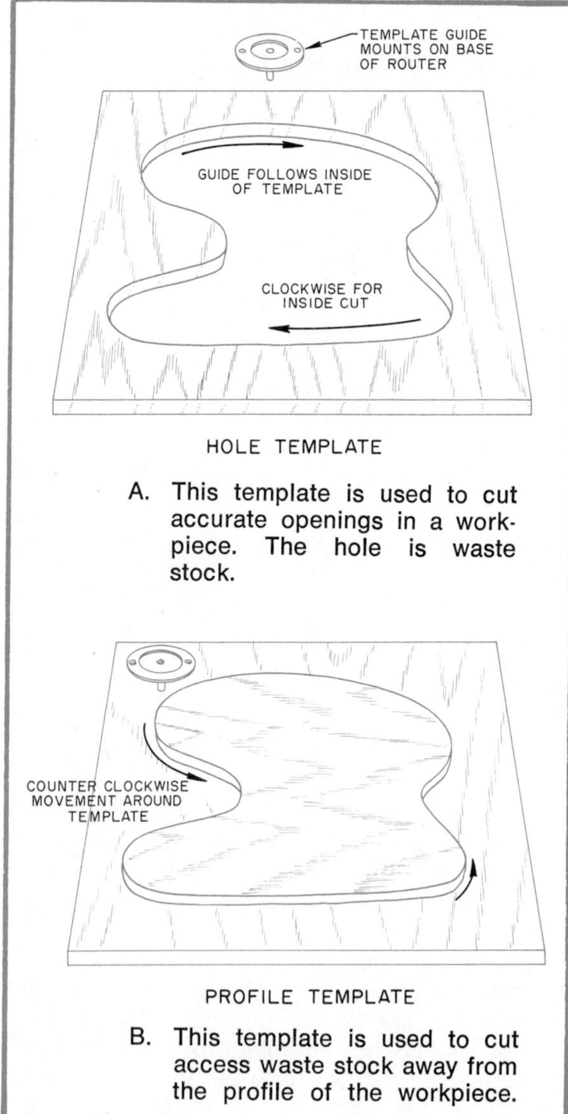

A. This template is used to cut accurate openings in a workpiece. The hole is waste stock.

B. This template is used to cut access waste stock away from the profile of the workpiece.

Fig. 11-36. A template and template guide.

used to cut a recess using a template. A template guide is attached to the base of the router. The outside of the template guide follows the template, Fig. 11-36.

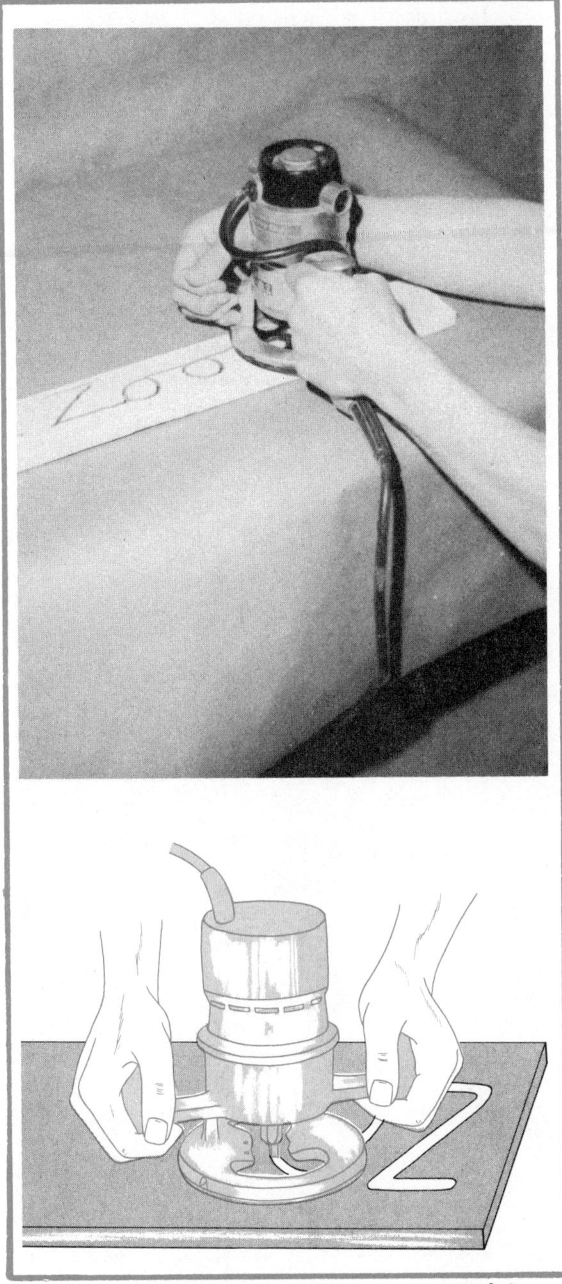
Fig. 11-37. Freehand routing a nameplate.

The router can also be used for freehand routing with no special guides or attachments, Fig. 11-37. A veining bit or straight bit is generally used. Freehand routing requires practice. The depth of the cuts should not exceed 3/8″. Hold the router securely and follow the layout lines for freehand work.

CUTTING A DOVETAIL JOINT

The router is commonly used for cutting dovetail joints for drawer construction. A special attachment is used with the portable router for making dovetail joints, Fig. 11-38. The dovetail attachment is secured to a bench with the front of the base projecting slightly. The template guide tip is mounted to the base of the router. A dovetail bit is placed in the router collet. The bit is set so it extends exactly 9/32″ below the router base. Stock which has been squared on the ends is clamped in

Fig. 11-38. A template guide for cutting dovetail joints. (Power Tool Div., Rockwell Mfg. Co.)

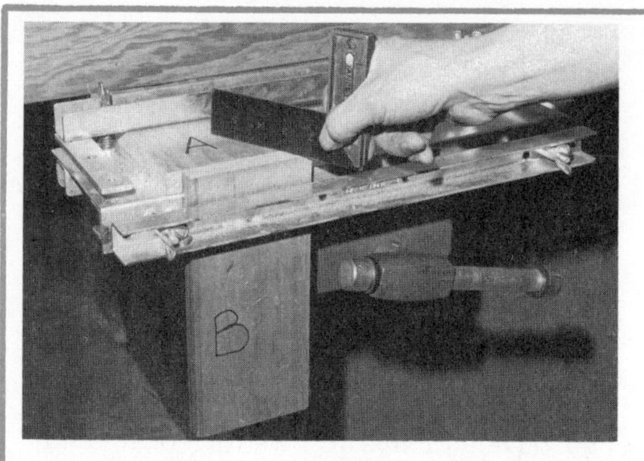

A. Checking squareness of the ends of the stock and the clamped position.

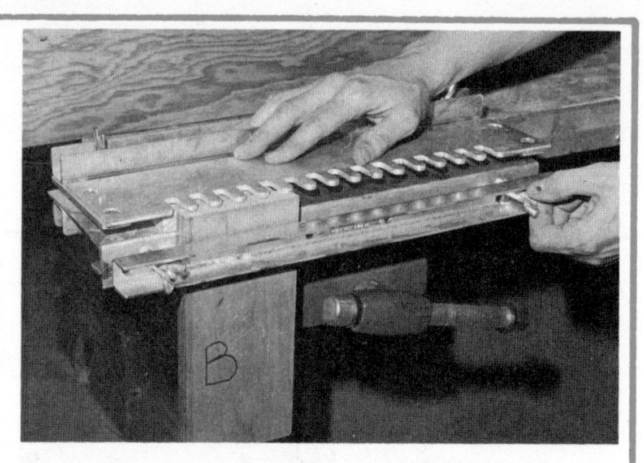

B. Positioning the template guide accurately.

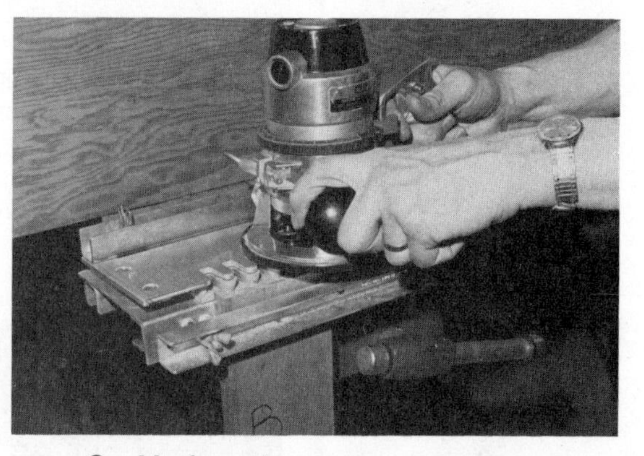

C. Moving the router following the template guide.

D. View from beneath router shows the cutting action of the bit.

E. Completed dovetailed pieces fitted together to form two sides of a drawer.

Fig. 11-39. Procedure for cutting dovetail joints with a router.

the attachment, Fig. 11-39A. Care must be taken to place the stock squarely in the clamp. The parts are placed against the alignment pins on the side of the attachment. The part marked "A" is the drawer front. The inside of the drawer face is placed up. The part marked "B" is the drawer side. The inside surface of the drawer side faces out on the attachment.

The template guide is positioned after the two pieces of stock are secured in the clamp, Fig. 11-39B. The router is placed on the template guide. You must always make certain the bit is not in contact with the fingers of the template. The router is started and moved to follow the fingers of the template, Fig. 11-39C. After the cut is complete, shut off the router and allow it to stop before removing it from the template guide.

NOTE: *Do not raise the router from the template with the motor running.*

The first cut should be made on scrap stock. After completing the cut, check the fit of the joint. If the fit is too loose, lower the dovetail bit. If the fit is too tight, raise the bit slightly. To adjust the depth of the fit, move the template in or out by use of the adjusting nuts.

The stock is positioned on the left end of the attachment for cutting the right front corner and the left rear corner of the drawer. The right end of the attachment is used for cutting the left front corner and right rear corner of the drawer.

Drawers with an overlapping front or rabbeted front are dovetailed one piece at a time. A gauge block is used to position the drawer front in the attachment, Fig. 11-40, and to locate the stock in the clamps. The gauge block is removed to cut the dovetail on the front. A piece of scrap the same thickness as the drawer front is used to cut the side and is left in the clamp when the side is cut. The scrap is placed in the top position and the side is in the vertical position. The completed drawer is shown in Fig. 11-41.

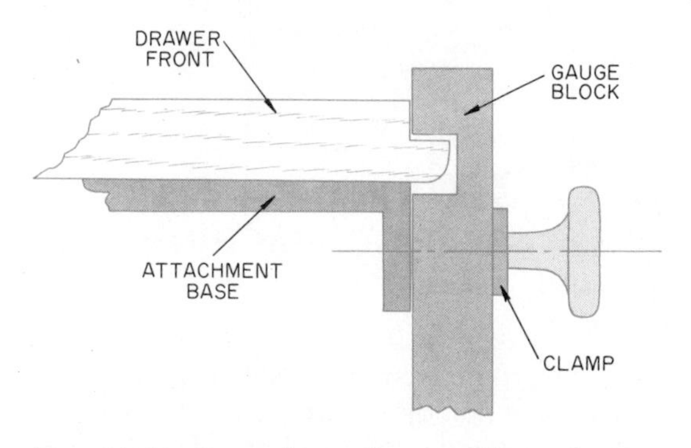

Fig. 11-40. Procedure for routing drawers with overlapped fronts.

Fig. 11-41. A drawer assembled with dovetail joints.

A. Router positioned in table.

B. Mounting which holds router in position.

C. A modern overarm router.

Fig. 11-42. Mounting a router in a stationary base.

MOUNTING THE ROUTER IN A STATIONARY TABLE

The router can be mounted in a special table to convert it to a bench shaper, Fig. 11-42. The table is generally designed so the router can be tilted to produce various types of cuts. An arbor spindle attached to the router motor replaces the collet chuck for this purpose. A fence guides the stock and controls the depth of the cut. Collars can also be mounted on the spindle to control the depth of cut on irregularly shaped stock.

The router can be used as a stationary shaper in a shop-built table using a collect chuck and pilot-tipped bits, Fig. 11-43. A fence can also be attached for cutting grooves and dadoes. A straight bit is used for this type of cutting.

SAFETY CONSIDERATIONS IN OPERATING A ROUTER

1. Disconnect the router from the power source when making adjustments or changing bits.
2. Wear goggles or a protective face shield.
3. Hold the base of the router firmly against the work before starting the machine.
4. Keep the hands and fingers clear of the cutters.
5. Dress appropriately for operating machines. Do not wear loose jewelry or clothing which may become entangled in the machine.

STATIONARY SHAPING MACHINES

The *spindle shaper* consists of a vertical hollow spindle on which various shaped cutters are mounted, Fig. 11-44. The spindle revolves at a speed of about 10,000

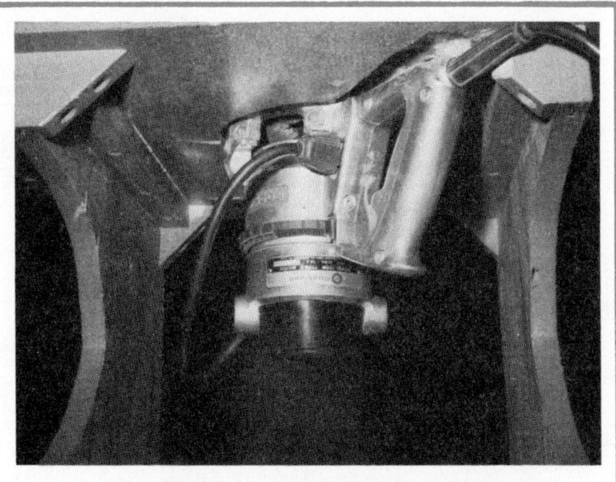

A. Router positioned in table and secured by mounting bracket.

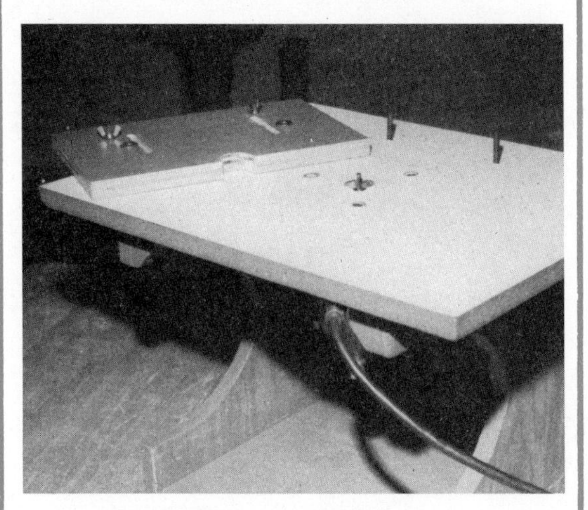

B. Detail of guide fence.

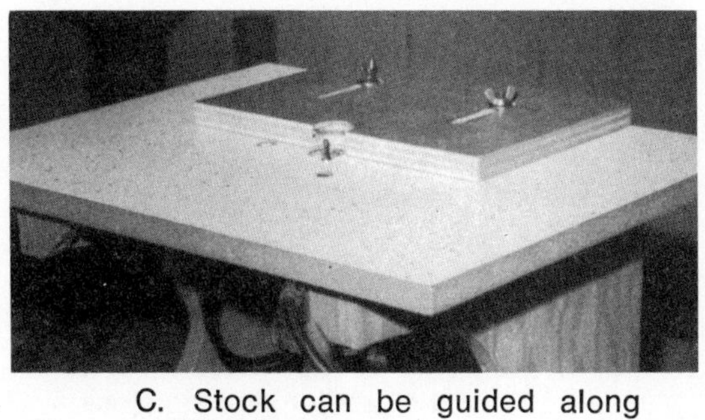

C. Stock can be guided along the fence.

Fig. 11-43. Mounting a router in a shop-built table.

RPM. The direction of rotation of the cutter is reversible on most shapers.

Shapers are used to cut molding, grooves, and rabbets, as well as decorative edge cuts. The size of the shaper is given by the size of the spindle and table size. The most common general-purpose shapers have 5/16″ to 3/4″ spindles.

The spindle can be raised and lowered by means of a hand wheel to adjust the depth of cut. The fence is adjustable and is sectioned in two front and rear parts, which move independently of each other. The table top is machined and has a miter gauge slot in which a miter gauge can be placed for grinding the stock.

TYPES OF SHAPER CUTTERS AND MOUNTING CUTTERS

The solid-wing cutter is most commonly used on the shaper, Fig. 11-45.

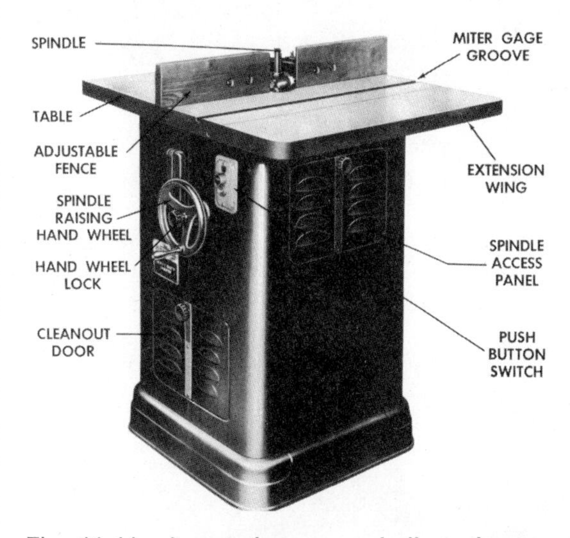

Fig. 11-44. A stationary spindle shaper. (Power Tool Div., Rockwell Mfg. Co.)

Fig. 11-45. A solid spindle shaper cutter.

However, the cutters are available in a vast array of shapes, Fig. 11-46. The solid cutter is generally regarded as the safest type of cutter for most operations.

The cutters are mounted on the spindle in combination with spacer collars for many types of work. A lock nut is placed under the spindle nut. *Never operate the shaper without the lock nut in place.* The manufacturer's operating manual should be consulted to change spindles and make special setups. *Disconnect the power to make adjustments or change cutters.*

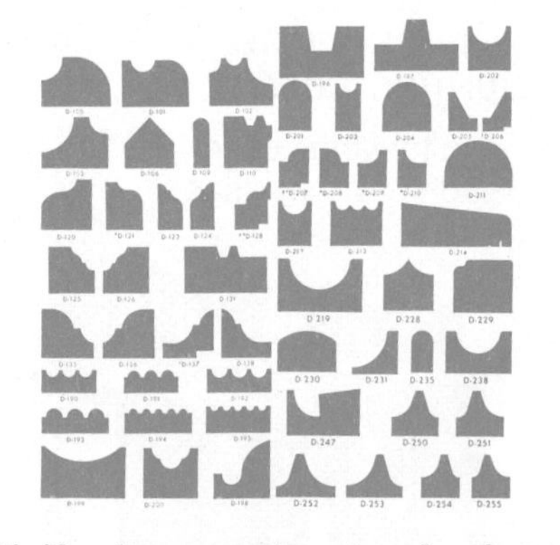

Fig. 11-46. An assortment of shapes of spindle shaper cutters. (Rockwell Mfg. Co.)

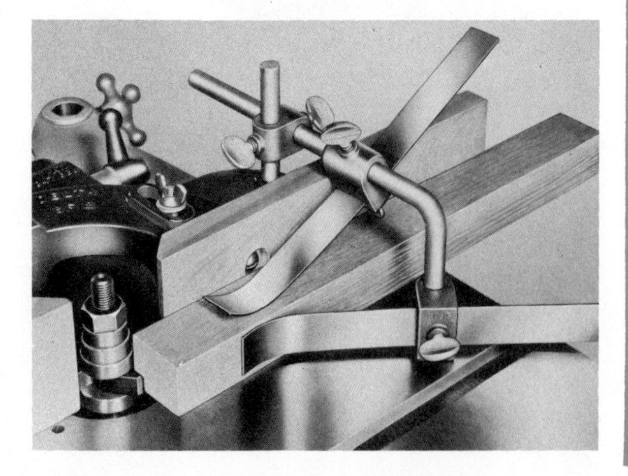

Fig. 11-47. Use of the fence and holddowns for shaping stock safely.

OPERATING THE SPINDLE SHAPER

There are several basic methods of guiding and supporting stock as it is passed through the shaper knives. The two simplest methods include using the fence and the miter gauge.

Shaping stock along a fence. The fence is used for shaping straight stock. The position of the cutter is adjusted by raising or lowering the spindle and moving the fence in or out. The spindle should turn opposite the direction of the feed. Hold-

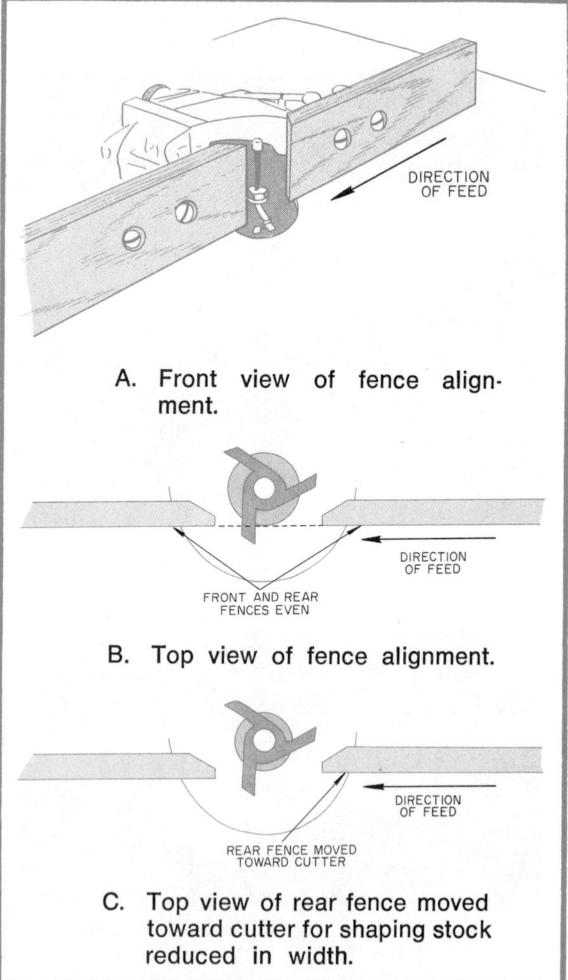

Fig. 11-48. Alignment of fence for shaping operations.

downs are placed against the stock to hold it against the fence, and are also placed on top of the stock to hold it against the table, Fig. 11-47. It is much safer if the cutters are positioned to cut on the bottom rather than the top of the stock.

If the stock is reduced in width, the rear fence is moved toward the cutters a distance equal to the amount of stock removed. When stock is not removed across the entire thickness of the board, the front and rear fence remain in alignment, Fig. 11-48.

Shaping using a miter gauge. End cuts can be made in stock using a miter gauge to which the stock is clamped. The fence is used to determine the horizontal depth of cut. The miter gauge should be used for all end cuts if the stock is less than 12″ wide, Fig. 11-49.

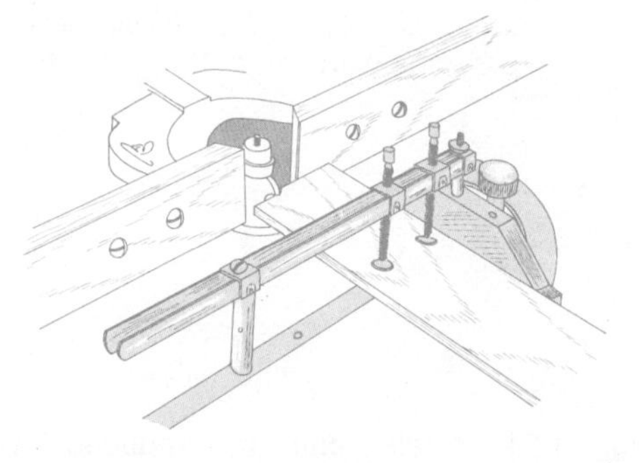

Fig. 11-49. Stock clamped in a miter gauge for making end cuts.

CUTTING WITH COLLARS

Irregular edges can be shaped using collars to control the horizontal depth. The diameter of both the collar and the cutter determines the depth of cut. A collar may be placed above or below a single cutter, between two cutters, or two collars may be used, placing one above and one below a cutter, Fig. 11-50. The placement of a collar above the cutter provides more safety during the shaping operation. A tapered pin which fits into a tapered hole in the table top serves as a guide pin to use with the depth collar. *Make certain the ring guard is in place before using the collar and pin method.* To start the cut, hold the stock against the guide pin and gradually feed the stock against the collar into the cutters, Fig. 11-51. If deep cuts are needed, it is best to make two cuts.

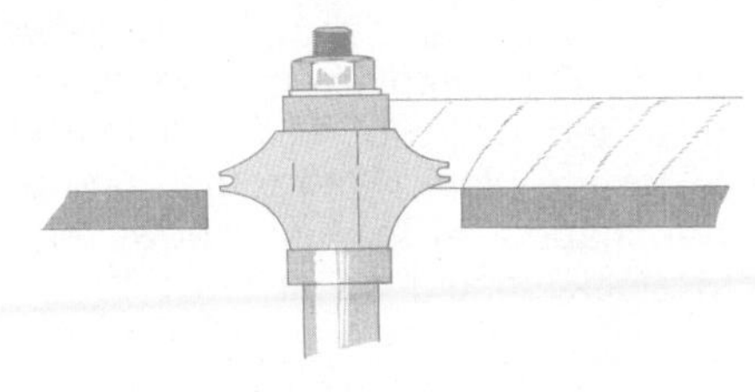

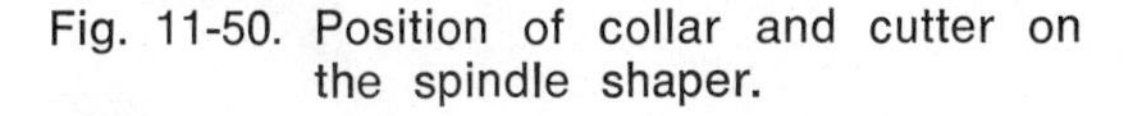

Fig. 11-50. Position of collar and cutter on the spindle shaper.

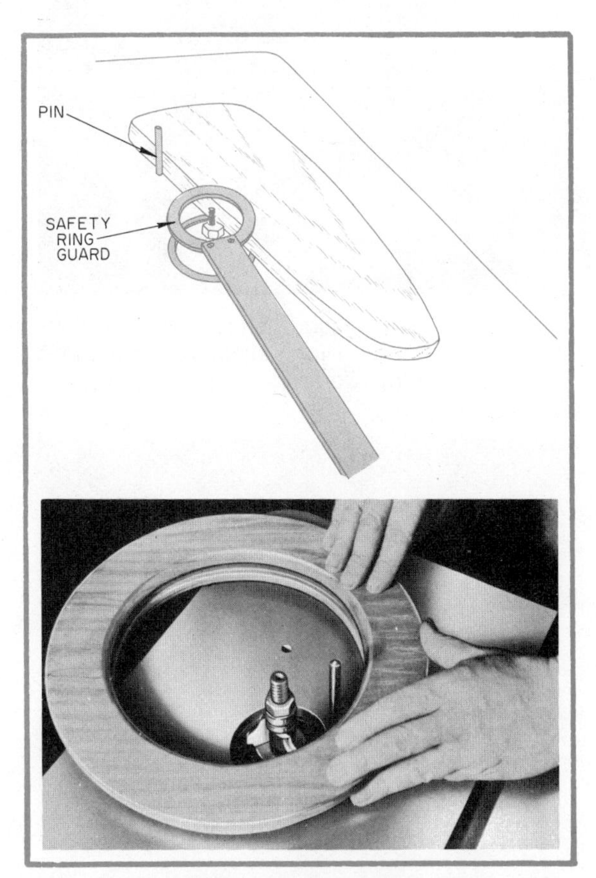

Fig. 11-51. Shaping stock using a pin and collar and the safety guard in position.

FORMING DUPLICATE SHAPES

Duplicate pieces can be formed on the shaper using a template and pin. A template of the desired shape is first constructed, the exact size determined by the size of the collar and type of cutter. Two identical pattern templates are made and the blank stock is sandwiched between, Fig. 11-52. The template is guided against the collar and pin, Fig. 11-53. If thin stock is used for the blanks, more than one piece can be shaped at a time, however, the pattern stock should be at least ½″ thick.

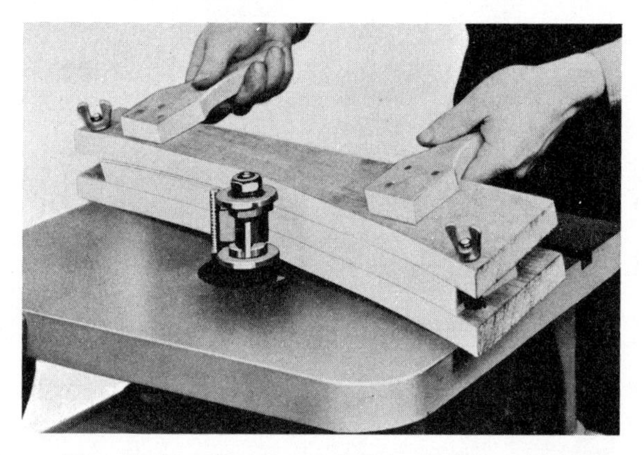

Fig. 11-52. Stock sandwiched between templates for shaping. (Rockwell Mfg. Co.)

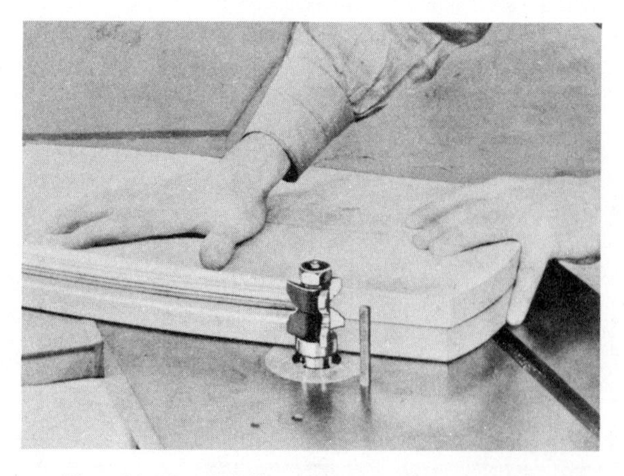

Fig. 11-53. Guiding the template against the collar and pin. (Rockwell Mfg. Co.)

SAFETY CONSIDERATIONS FOR OPERATING THE SHAPER

1. Before starting machine be sure to check the following:
   a. table for foreign matter.
   b. quill taper and spindle taper for chips, dust, and alignment.
   c. spindle for possible damage.
   d. spindle nut for possible damage.
   e. security of draw.
   f. collars for smooth surfaces.
   g. cutters for sharpness.
   h. installation of retaining washer.
   i. cutterhead for tightness under nut.
   j. spindle rotation.
2. Make certain spindle threads and spindle nut are in good condition. Rounded corners prevent locking of head securely.
3. Use only safety type cutterheads for maximum safety.
4. As material is fed into knives, stand well to right front of the machine (unless reverse rotation of spindle is being used).
5. Since a shaper operates with cutterhead exposed, keep hands free of danger area by using a push stick when cutting small pieces.
6. Always wear goggles or a safety shield to protect your face and eyes from flying chips.
7. Never "horse around" a running machine. Play should be absolutely forbidden as 9 out of 10 accidents are the result of carelessness and playing with machines as though

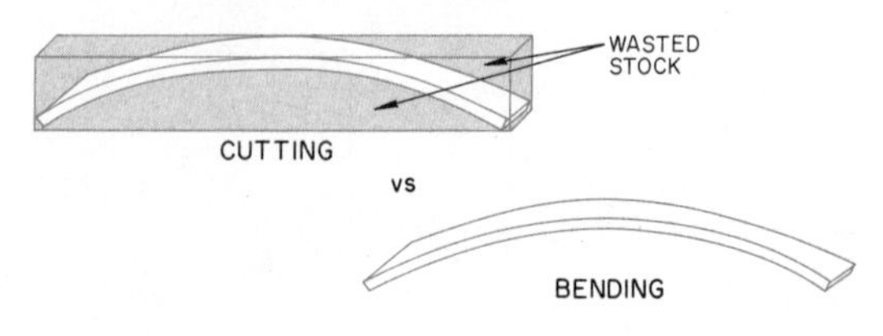

Fig. 11-54. Waste can be reduced when stock is bent rather than formed by cutting.

they were toys. Respect the machine and make it work for you.

8. Keep all guards in place and use them. Safety ring guards and spring-loaded hold-downs should be used whenever the work permits.
9. Never wear long or loose sleeves, neckties, or jewelry when operating shaper.
10. "Good housekeeping" is essential for safety. Keep working area free and clean at all times.
11. Always stop machine for adjustment or when leaving immediate area. Disconnect power source when working on or around moving parts.
12. Use starting pins for free hand shaping.

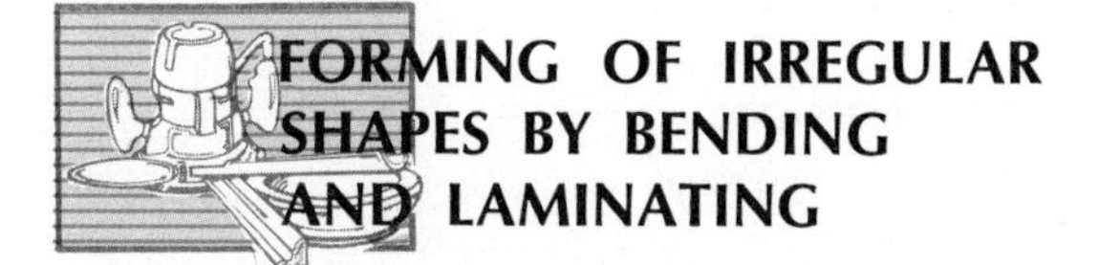

FORMING OF IRREGULAR SHAPES BY BENDING AND LAMINATING

When irregular shapes are formed by separating with saws and edge-cutting tools, there is often an excessive amount of waste. Therefore, wood is often formed into an irregular shape by bending and laminating, Fig. 11-54. Lamination is the process of gluing many layers of thin stock together. The layers of stock are clamped in a form of the desired shape and the glue is allowed to dry, Fig. 11-55. The glued stock will retain the shape of the form after the glue has dried. Thin pieces of wood are used because they can be bent easily.

Sometimes it is not possible to retain the necessary strength qualities of an irregular shape formed by cutting the parts. Often cross-grained parts will break easily if they are subjected to a force. Bending or laminating eliminates this type of problem.

The designs for many products require that bent or laminated parts are used. Examples of such parts are shown in Fig. 11-56.

FORMING WOOD

Wood can be formed if it is steamed or soaked before bending. Soaked or steamed wood can be compared to a living tree which sways in the wind without breaking. In other words, the high moisture content of the soaked or steamed wood allows the wood to bend without breaking.

When wood bends, part of it is compressed and part of it is stretched, Fig. 11-57. The water-softened wood becomes

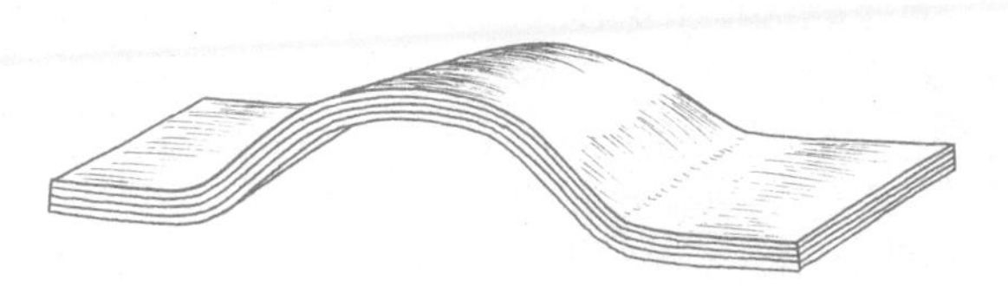

Fig. 11-55. Irregular shapes can be formed by lamination.

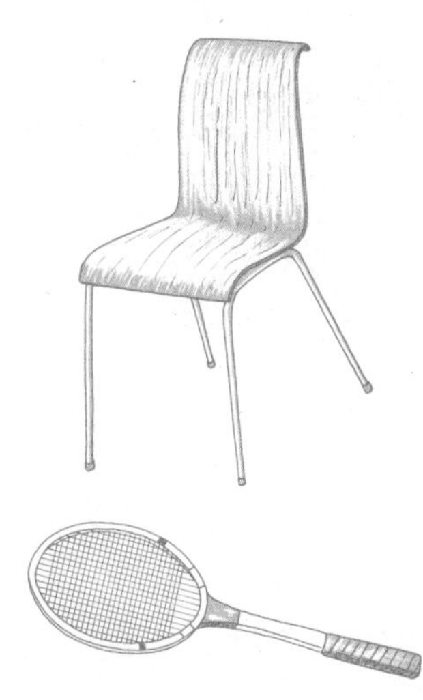

Fig. 11-56. Irregular shapes formed by laminating or bending processes.

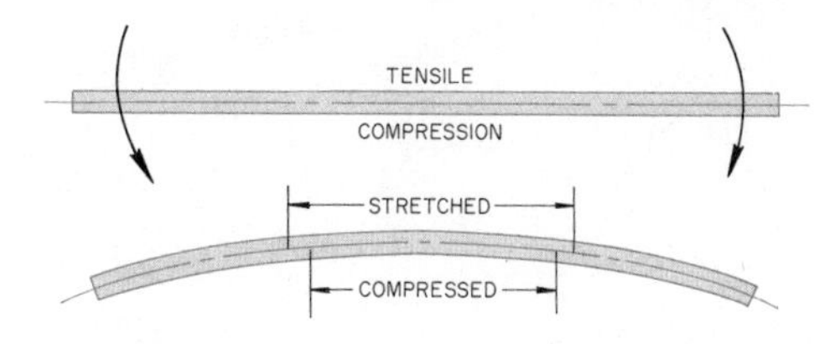

Fig. 11-57. Stress placed on wood as it is bent to shape.

much more pliable, elastic, and compressible.

SELECTING WOOD FOR BENDING

The stock for bending should be straight-grained and free of defects. Air-dried lumber with a moisture content of about 12% to 15% is the most satisfactory. Hardwoods including oak, maple, elm, ash, walnut, and sweet gum are among the best for bending.

PREPARING SOLID WOOD FOR BENDING

As stated, the wood fibers can be made pliable by either soaking or steaming the wood in water. Steaming is even more effective than soaking. The wood should be supported above the water from which the steam is produced. Time needed to soften wood for bending depends on the *thickness, species of wood,* and the *original moisture content.* The amount of softening required depends upon the radius of the bend. It is usually necessary to experiment to determine the exact conditions for specific types of work.

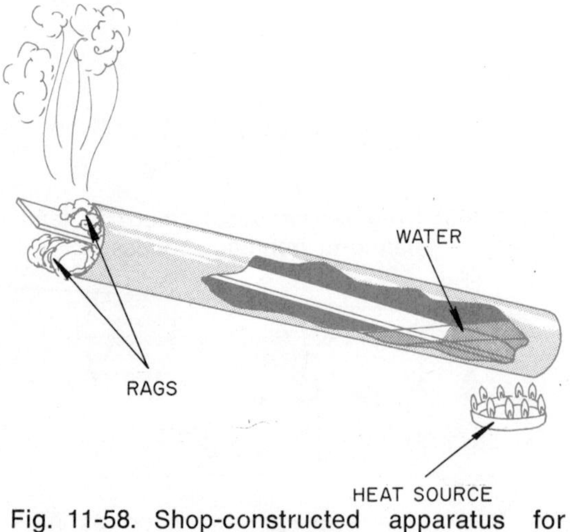

Fig. 11-58. Shop-constructed apparatus for bending wood.

STEAM BENDING

Wood can be steamed for bending in highly-controlled steam chests; however, for small pieces, a shop-built apparatus will provide ample steam. A pipe sealed on one end and placed in a heat source will provide steam for small pieces. Care must be used to prevent the apparatus from boiling dry. A rag stuffed in the top end of the pipe helps to hold the steam, Fig. 11-58. A covered pan with a support screen above the water level also works well for small pieces.

After the stock has been softened or plasticized, it is removed from the steam and immediately placed in a form. Figure 11-59 gives an example of a simple shop-made form for clamping wood, in which the wood is left until it dries. Various types of clamps can be constructed for different shapes.

KERF BENDING

Thick pieces of stock can also be bent by kerfing. For this method, saw kerfs are cut across the grain on the inside of the stock to be bent. The kerfs reduce the amount the stock has to compress to bend. Stock bent this way is often used for table aprons and similar purposes. It has less strength, and the inside of the curve does not have a desirable appearance, Fig. 11-60.

The number and depth of kerfs needed depend on the type and thickness of the wood. For a sharp radius, the kerfs are cut to within 1/6″ of the outside surface of the bend, Fig. 11-61. The number of kerfs

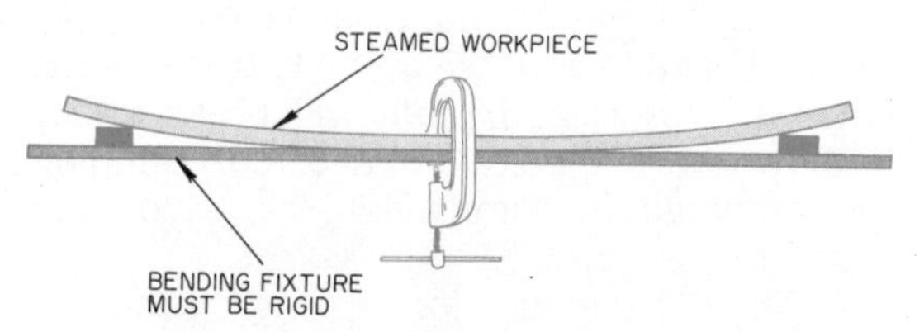

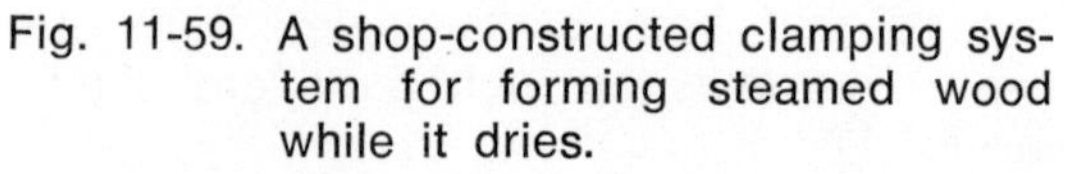

Fig. 11-59. A shop-constructed clamping system for forming steamed wood while it dries.

needed can be determined by experimenting on scrap stock of the same kind which is to be bent. The kerfed wood is placed in a form to produce the desired shape. Glue, combined with fine saw dust, placed in the open kerfs, will help hold the shape after it is formed and the glue has dried.

LAMINATING IRREGULAR SHAPES

Lamination is the process of gluing several thin layers of stock together parallel to the grain to form larger members. This method is widely used in the construction industry to form structural members, Fig. 11-62. The laminated structural members offer many different designs not possible with other materials. Due to the great strength of the members, vast buildings can be spanned without center supports. This is one reason laminated beams are often used in large gymnasiums.

Laminated beams are also favored because of their resistance to fire. Wood resists the transmission of heat and chars very slowly. The char has high insulating value and provides additional protection for the inner wood from fire temperatures, Fig. 11-63. The heavy wood timbers remain standing, whereas the heavy steel members collapsed from the high temperatures of the fire.

Heavy laminated timbers are generally produced in factories which specialize in

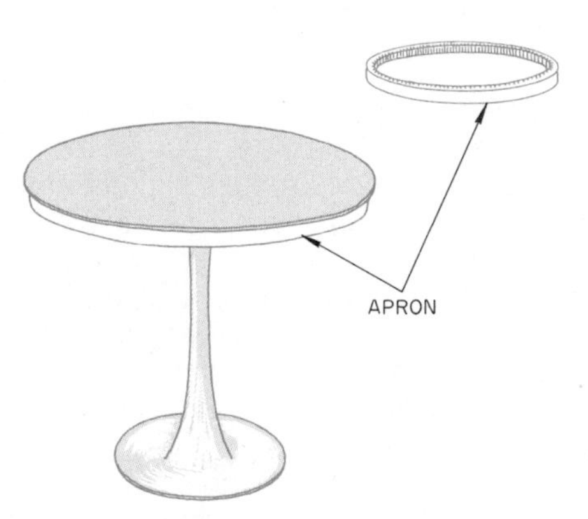

Fig. 11-60. A table apron formed by kerf bending.

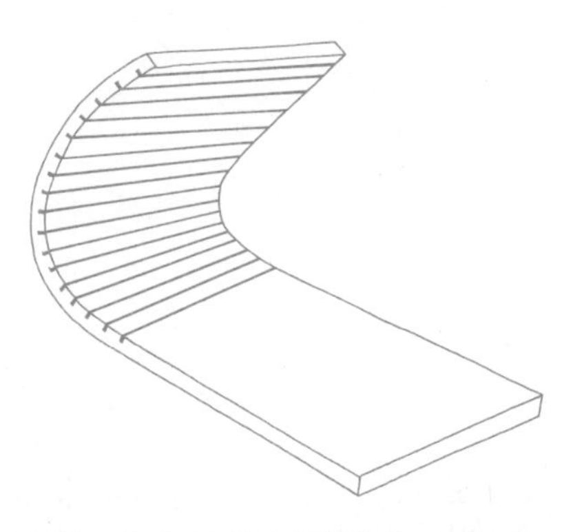

Fig. 11-61. A board kerfed for bending.

Fig. 11-62. Laminated structural supports were used in the construction of this building.

Fig. 11-63. Remains of a building after a fire shows strength of laminated wood beams as compared to structural steel.

laminated structural members, Fig. 11-64. Huge presses form and hold them while the glue dries.

Please note that this method of lamination is different than that used for plywood construction. In the manufacture of plywood, the grain of each ply adjacent to the next one is at right angles, Fig. 11-65.

Fig. 11-64. Laminated timber used in construction.

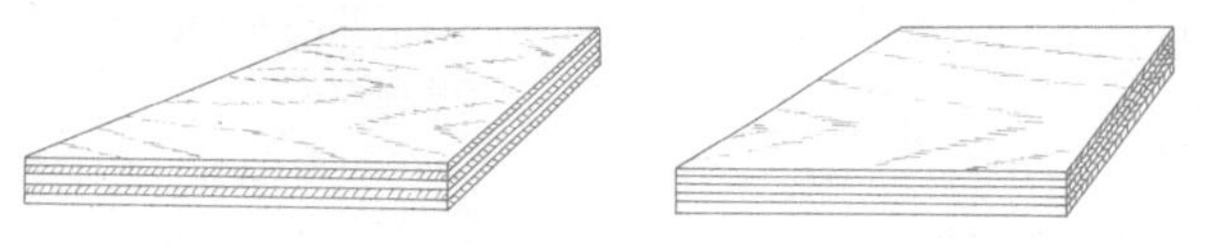

Fig. 11-65. Comparison of the direction of the grain for plywood (left) and structural laminates (right).

The same principle of lamination is used for other types of laminated parts. Lamination is widely used in the sporting goods industry, Fig. 11-56, and the furniture industry, Fig. 11-66. Veneer is used to produce the type of laminated products shown in these figures. Veneer is usually less than ¼″ in thickness, and is available in thickness of 1/28″, 1/20″ and 1/16″. The standard veneer thickness for hardwoods is 1/28″.

Veneer stock is rough cut to the approximate length and width for laminating. Veneer can be cut to the approximate size using scissors, jigsaw, band saw, or a metal squaring shear. If the veneer tends to split as it is cut, it can be dampened with water. The addition of moisture reduces the problem of splittage.

FORMS FOR LAMINATION

A form of the desired shape must be constructed to hold the lamination in the desired shape while the glue dries. The form should be constructed slightly smaller than the finished product size as the laminated member will tend to spring open slightly. Most laminations are clamped in a two-part form, Fig. 11-67. The open space between the two parts must be exactly the same width as the thickness of the lamination.

Convex shapes can be shaped with a one-piece form. A flexible metal strap ap-

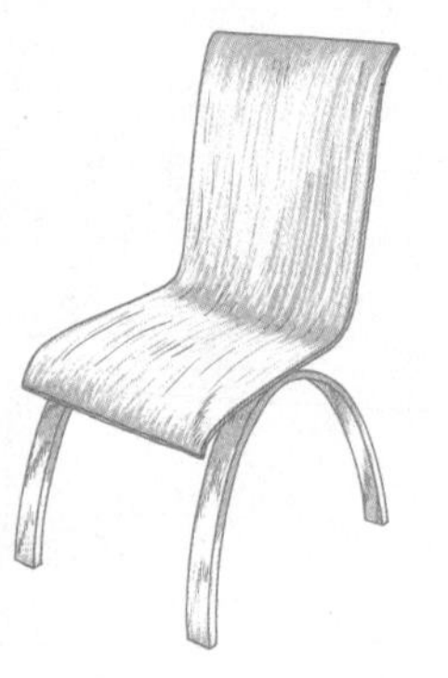

Fig. 11-66. A chair designed for and produced by the lamination process.

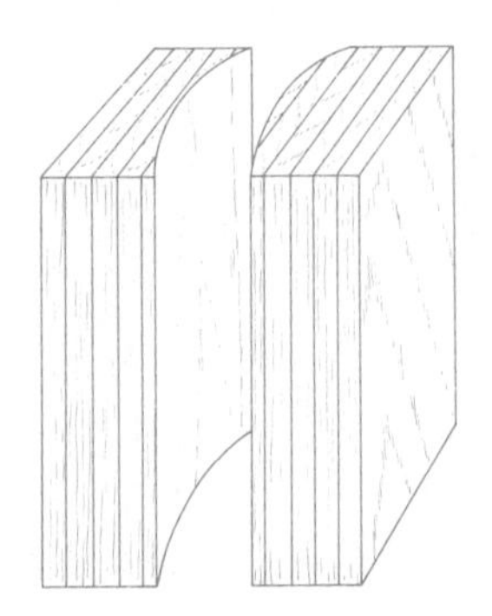

Fig. 11-67. A two-piece lamination form for clamping and drying.

plies firm pressure to the layers of veneer, Fig. 11-68. Pressure is applied by tightening the bolt attached to the end of the band.

Pressure can be applied to a two-piece form by either hand clamps or various types of presses. The forms should be covered with either thin metal or plastic. The glue tends to squeeze from the veneer and will dry on the form.

A lamination press for clamping small laminated parts uses a rubber bladder inflated with air which applies pressure to a one-piece form, Figs. 11-69 and 11-70. Another machine press for laminating uses molds to form the desired shapes, Fig. 11-71.

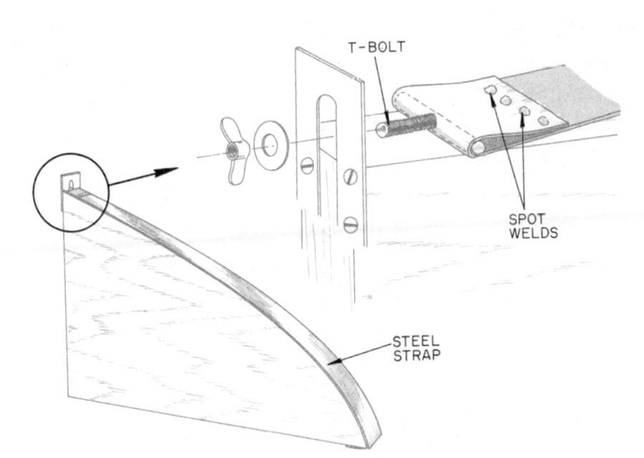

Fig. 11-68. A steel strap laminating form.

Fig. 11-70. A chair back formed by the bladder press shown in Fig. 11-69. (Vega Industries)

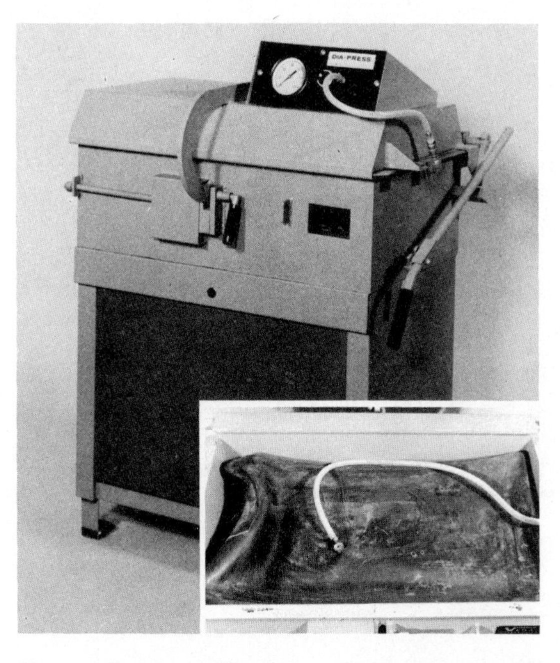

Fig. 11-69. A lamination press uses an inflatable bladder to apply pressure to the stock to form the desired shape. (Vega Industries, Inc.)

Fig. 11-71. Forming press to produce curved stock. (Hardwood Plywood Manufacturers Assn.)

The mold in this press is tapered and rounded to protect the rubber diaphragm. A polyethylene film is used over and under the veneer to prevent sticking of glue which squeezes out as the form is shaped.

An adhesive should be used which will allow adequate assembly time. Plastic resin is a good all-purpose glue for lamination work. Resorcinol resin glue should be used for water-skies and similar sporting goods equipment. The glue can be applied using a brush or small roller. The layers should be stacked in the proper order as they are coated with glue.

The laminate should remain in the forms until it is completely dry, normally requiring overnight clamping. The curing time can be greatly reduced for thermosetting adhesives by applying heat.

After the laminate is removed from the form, it must be cut and trimmed to finish size. This can be done with hand tools or on a jigsaw or band saw. Special care must be taken when cutting irregular shapes on machines. The final product is sanded and a finish is applied.

Fig. 11-72. An automatic stair stringer router. (RUVO Automation Corp.)

INDUSTRIAL MACHINES FOR SHAPING AND FORMING IRREGULAR SHAPES

Many production woodworking industries are dependent on various types of routers and shapers to produce a particular product. Many of the machines are multipurpose machines and others are special-purpose. The machine in Fig. 11-72 is an automatic stair stringer router which, when set up for a specific size step, routes the recess for the treads and risers in one operation on both stringers. Special industrial machines are used in the wood industry to form irregular shapes, Fig. 11-73.

Fig. 11-73. Smoothing irregular shapes which have been formed on shaping machines. (Norton Coated Abrasive Division)

ACTIVITIES

PROBLEM 1

You have learned from studying the previous chapter that there are several ways to fabricate an irregularly shaped part. One method may be used instead of another because of the product demands for strength, appearance, and product costs. There are three products shown here consisting of irregularly shaped parts. You already know that irregularly shaped parts can be formed by steam bending, kerf bending, lamination and carving operations. Evaluate the products and determine the process which would be most suitable for shaping the parts for the products. Explain your reasons for selecting a particular forming technique.

PROBLEM 2

If you have permission from your teacher, try forming an irregular shape by steam bending, sawing kerfs in the stock, and by laminating. After forming the irregular shape determine which is the strongest, has the best appearance, and was the easiest to construct.

PROBLEM 3

You have learned from Chapter 2 that many operations can be performed using different tools and machines. If you were the owner or manager of a manufacturing industry you would make *decisions* about which processes to use for manufacturing a product. The factors which will influence your decision are (1) the types of equipment available to your company, (2) the number of products to be produced, and (3) the type of product you manufacture. As a student in a woodworking class, you must also make decisions. Many of the same factors influencing industry will influence your decisions.

For example, you are constructing book ends such as shown below. It is obvious that you must perform many operations. List the tools and machines which you could use to form each of the component parts. It may be necessary to review the chapter to complete the list.

Lamination	Laminating form
Collet chuck	Kerfing
Template	Dovetail bits
Template guide	Pilot bits
Plasticized	Spacer collars

CHAPTER 12 FORMING CYLINDRICAL SHAPES

INTRODUCTION

The theory of cylindrical shaping is based on the principle of stock rotation combined with the presence of a cutting or scraping tool. Woodturning on a lathe is an example of cylindrical shaping. The stock can be held in position for rotation in a variety of ways; however, the principle of shaping remains the same, Fig. 12-1.

The shape of the cylindrical form created is determined by two factors: (1) the tool applied to the revolving stock, and (2) the method in which this tool is applied to the stock.

Turning and forming cylindrical shapes was done in ancient times, Fig. 12-2. The principle of shaping remains the same today, even though highly-sophisticated automated machines are now used in industry.

As stated, the principle of forming cylindrical shapes involves the rotation of the stock against a cutting or scraping tool. The rate of forming depends upon the rate of rotation, size of the tool, and amount of force applied to the tool.

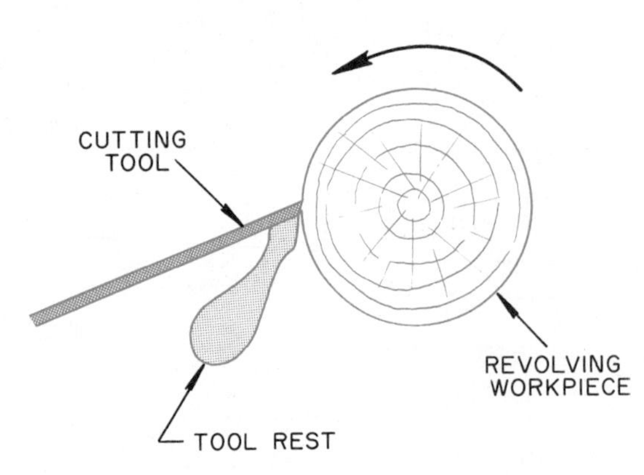

Fig. 12-1. Principle of cylindrical shaping.

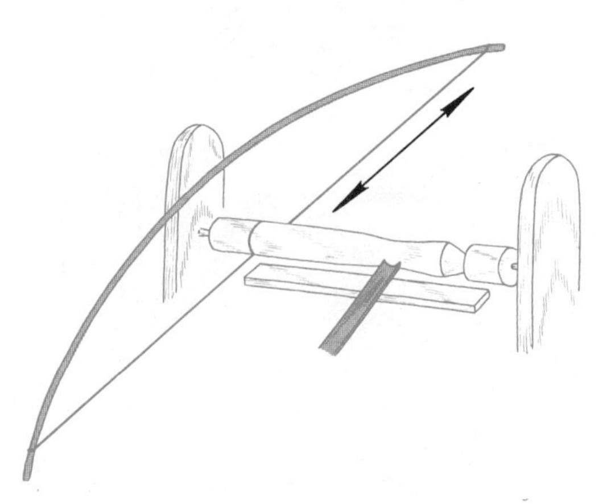

Fig. 12-2. An ancient bow lathe which was used to form cylindrical shapes.

FORMING CYLINDRICAL SHAPES ON THE LATHE

The wood lathe differs from any other woodworking machine in that a finished product can be produced on the one machine. To understand the operation of the woodturning lathe you will need to become familiar with the operational parts of the machine, Fig. 12-3. The basic parts of the lathe consist of the *headstock, tailstock, bed,* and *tool rest.*

Headstock. The headstock is attached to the left end of the lathe bed. A hollow spindle threaded on both ends runs the length of the headstock. The right-hand end of the shaft has right-hand threads and the opposite end has left-hand threads. The faceplate is turned on the spindle for faceplate trimming. The inside of the spindle is tapered to receive the *live center* (sometimes called *spur center).* Since the spindle is hollow, the live center is removed by means of a *knockout bar* inserted through the hollow headstock spindle. The power for revolving the stock is produced in the headstock assembly.

HEADSTOCK SPINDLE
SWITCH
HEADSTOCK
INDEXING PIN
THREAD PROTECTOR
HAND WHEEL AND INDEX
SPEED CONTROL HAND WHEEL
CALIBRATED TOOL SUPPORT
TOOL SUPPORT BASE
BED GAP
LOCKING HANDLE FOR TOOL SUPPORT BASE
VARIABLE SPEED DIAL
TAILSTOCK LOCKING CLAMP
SPINDLE
SPINDLE LOCK
HAND WHEEL
TAILSTOCK
BED
STEEL CABINET

Fig. 12-3. Woodturning lathe. (Power Tool Div., Rockwell Mfg. Co.)

Tailstock. The tailstock also consists of a tapered hollow spindle in which the *dead center* (*cup center*) is placed. The tailstock spindle can be moved in and out by means of the hand wheel. When the hand wheel is turned counterclockwise, the tailstock spindle is pulled into the tailstock casting and forces the dead center out of the spindle.

Tool rest. The tool rest is mounted in a tool support base. The tool support base can be secured anywhere along the bed of the lathe. There are several different types and sizes of tool rests used on the lathe. The most common size is 12″ long. The top surface of the tool rest is smooth to provide for easy movement of the tools. When the tool support becomes rough, it can be reconditioned by draw filing with a mill file. It should be filed only enough to remove the nicks.

Bed. The size of a wood lathe is designated by the largest diameter of work which can be turned over the bed. A 12″ lathe would be capable of swinging a 12″ diameter disk. The disk is attached to a faceplate and mounted on the headstock spindle. A 12″ lathe measures 6″ from the center of the live center to the top of the bed. The length of the lathe bed may be 36″, 48″, or 60″ long, but a 48″ lathe bed is a good all-purpose length. The maximum size disk which can be turned is increased on lathes of the gap-bed type. This increases the size of faceplate work which can be turned on the inside of the headstock, Fig. 12-4.

Fig. 12-4. The gap bed allows larger stock to be turned on the lathe.

LATHE DRIVE SYSTEMS

Lathes are either belt-driven or have a direct-drive motor. The speed can be changed by means of a step pulley arrangement or variable-speed pulley on belt-driven machines. The direct-drive lathe is equipped with an automatic speed control. A direct-drive and variable-speed pulley assembly provides a speed range from 350 to 4000 RPM. A standard lathe equipped with step pulleys can be operated at four different speeds, Fig. 12-5. Common speeds are about 900, 1350, 2150, and 3300 RPM.

A. Speed change mechanism in slowest operating position.

B. Speed change mechanism in fastest operating position.

Fig. 12-5. Speed change mechanisms on a lathe.
The spring-loaded pulley on the drive motor expands or contracts opposite the top pulley.

KINDS OF TURNING TOOLS

A standard set of lathe turning tools usually consists of six different shapes. The standard set consists of a parting tool, gouge, skew, spear point, flat nose tool, and round nose tool.

Parting tools. Parting tools are designed for cutting recesses and cutoff work. The parting tool is commonly used for making sizing cuts on cylindrical turnings. It is ground with a double angle, that is, either cutting edge can be used for removing stock.

Gouges. Gouges are used for roughing square stock to a round shape or for forming cove cuts. The gouge is ground with a bevel on the outside.

Skews. Skews are used for smoothing cylinders, cutting shoulders, forming beads, and vee grooves. The skew is ground with a double angle and can be used as either a cutting or scraping tool.

Spear point. Spear point tools (also called diamond point tools) are used where the shape fits the contour of the work.

Flat nose. Flat nose tools are scraping tools and are used extensively for removing stock in faceplate turning. They can also be used for some types of spindle turning.

Round nose. Round nose tools are also scraping tools and are used for both faceplate and spindle turning.

MEASUREMENT TOOLS USED FOR WOODTURNING

In addition to the lathe cutting tools, measuring tools such as the inside caliper, outside caliper, rule, and dividers are used extensively, Figs. 12-6 and 12-7.

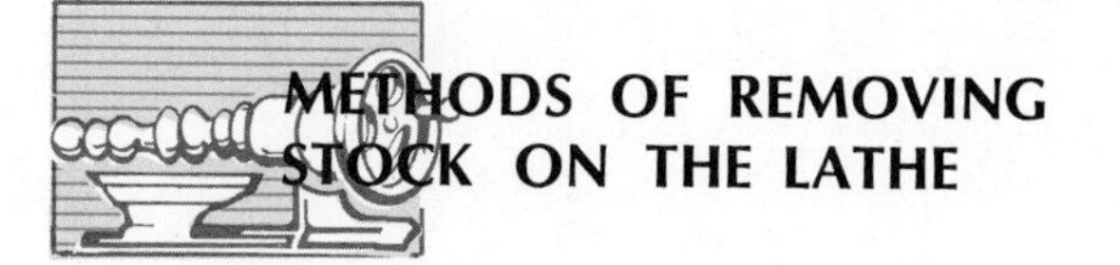

METHODS OF REMOVING STOCK ON THE LATHE

There are two basic methods of removing stock on the lathe: (1) *cutting*, and (2) *scraping*. The cutting techniques produce a very smooth surface. Scraping tends to tear the wood fibers and creates a rough surface. Tools designed for *scraping* are the round nose, flat nose, spear point, and parting tools. Tools designed for *cutting* are the gouge and skew. However, the gouge and skew can be used for scraping if held in the correct position. Figure 12-8 shows stock being removed by the scraping method on a cylindrical workpiece with a skew held flat on the tool rest and forced into the workpiece. Cutting with the skew, on the other hand, is performed by holding it at an angle and applying the center portion of the cutting edge to the stock. This requires practice, and beginners are often more successful in using the scraping method.

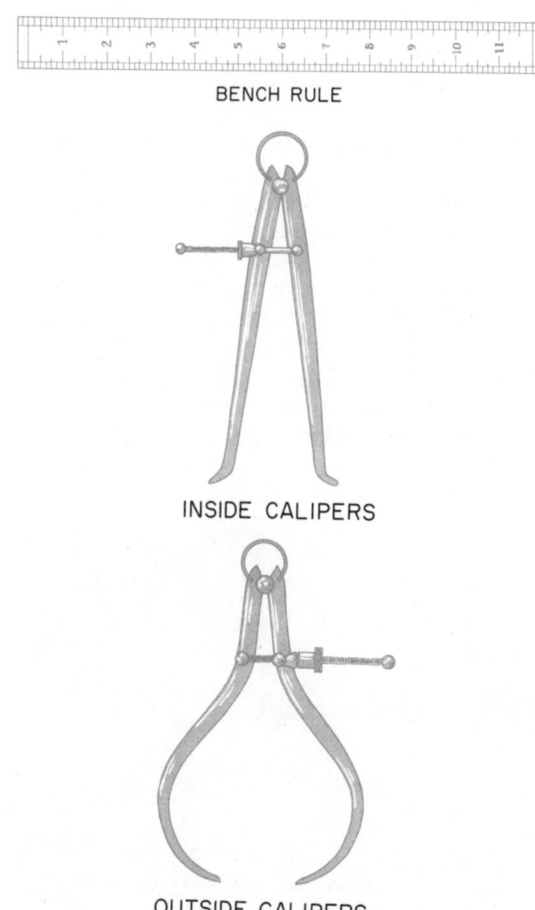

Fig. 12-6. Useful measurement tools for turning.

RECONDITIONING LATHE TOOLS

The lathe tools must be sharp to produce quality work. If the tools are not

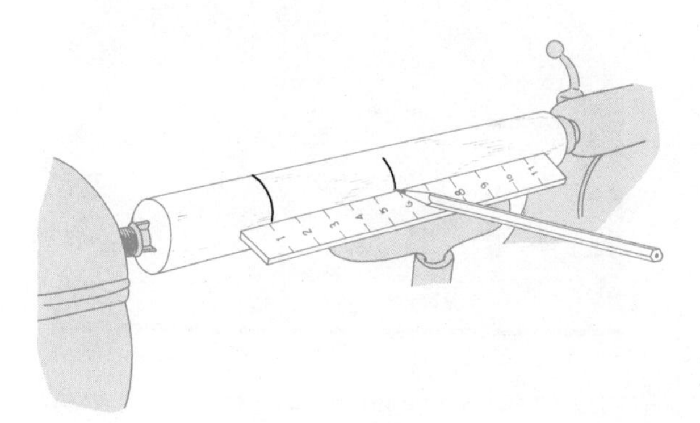

Fig. 12-7. Using a bench rule to lay out the length for a cut.

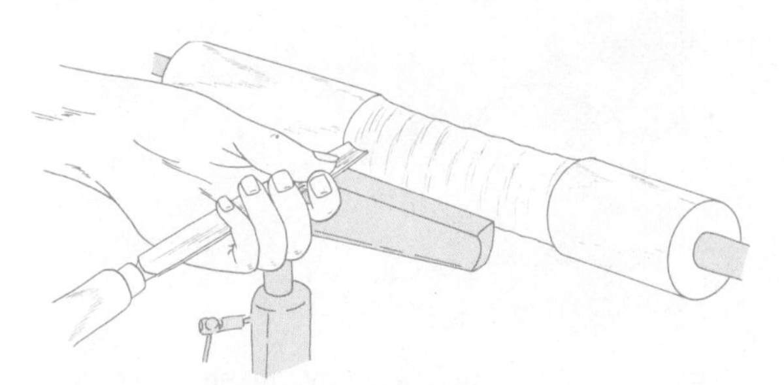

Fig. 12-8. Removing stock by the scraping method.

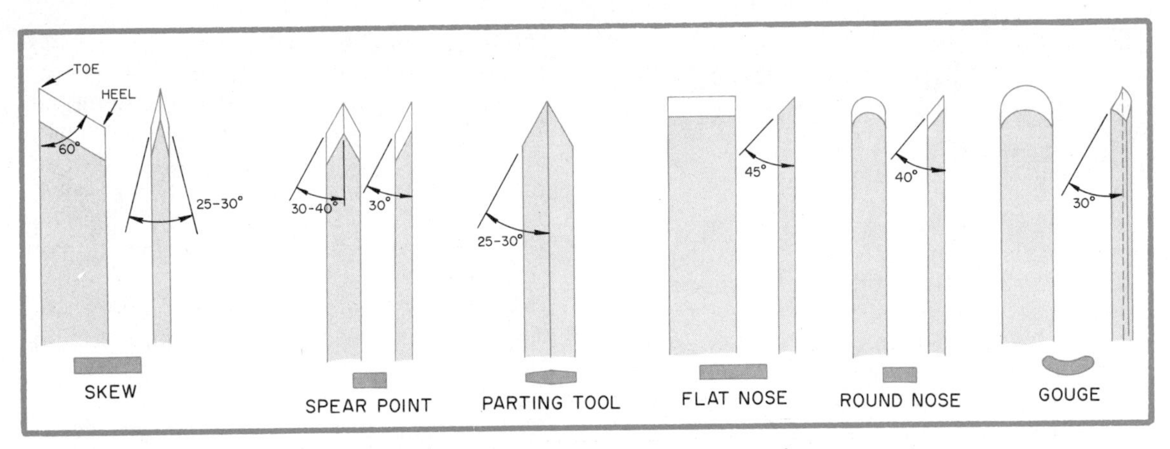

Fig. 12-9. Angles of common lathe tools.

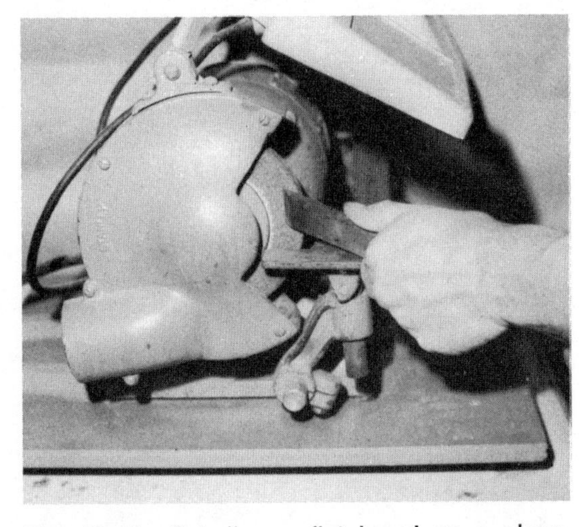

Fig. 12-10. Grinding a flat bevel on a skew.

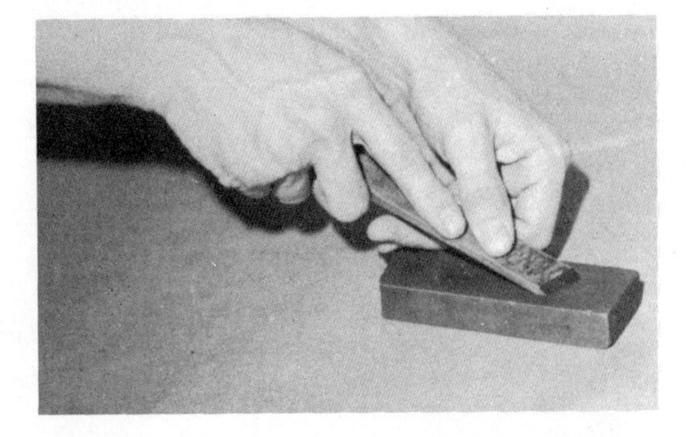

Fig. 12-11. Honing a skew on an oilstone to remove the wired edge formed by grinding.

nicked and the angle has not changed, the cutting edge can be restored by whetting to the angles shown in Fig. 12-9.

Most lathe tools are flat-ground on their bevel side on the side of a grinding wheel until a burr or wired edge forms, Figs. 12-10 and 12-12. The tools are then whetted on an oilstone to remove the wired edge and form a keen edge, Fig. 12-11. The gouge, however, is whetted on the inside surface with a slipstone, Fig. 12-13. The parting tool and flat nose tool should be ground on the edge of the oilstone to produce a hollow grind, as this will result in higher quality work, Fig. 12-14.

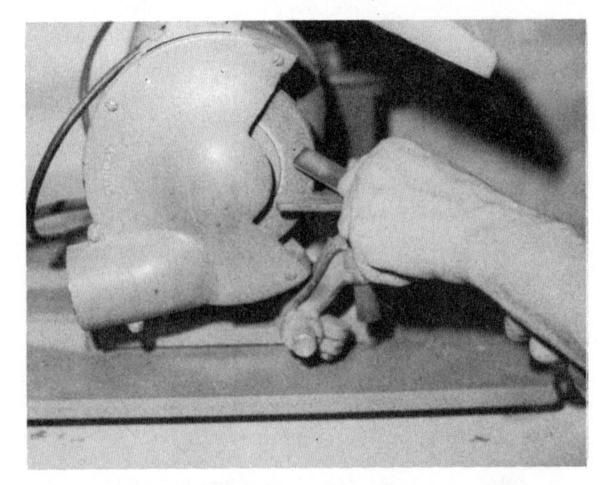

Fig. 12-12. Grinding a gouge on the grinder.

METHODS OF SECURING STOCK IN THE LATHE FOR TURNING

There are two basic types of woodturning. They are: (1) *faceplate turning,* and (2) *spindle turning,* Fig. 12-15. The difference is in the method by which the stock is held in the lathe. For spindle turning, the stock is held between the live center and the dead center. Faceplate turning is performed with the stock attached to a faceplate with screws. The faceplate is threaded on the headstock spindle.

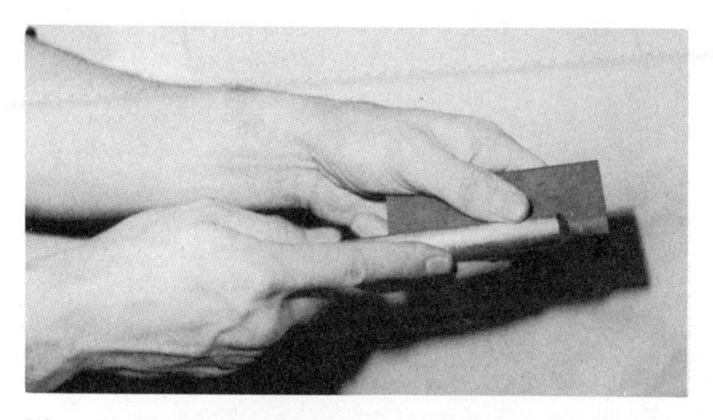

Fig. 12-13. Whetting the gouge on the inside surface of a slipstone.

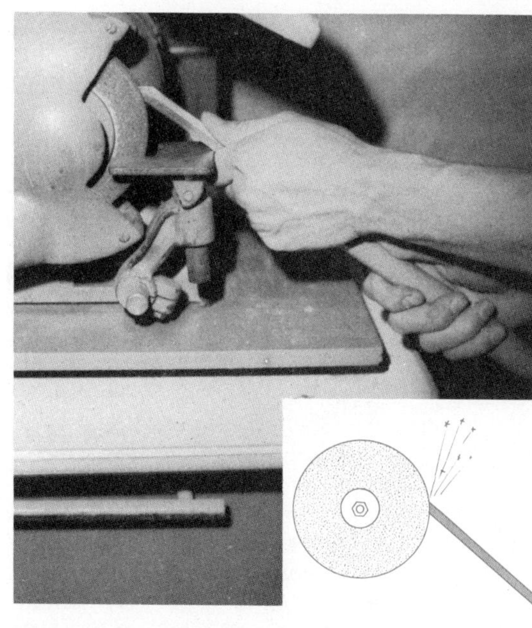

Fig. 12-14. Grinding a parting tool on the edge of the grinder.

SPINDLE TURNING BETWEEN CENTERS

Stock which is turned must be solid and free of defects such as knots and checks. The stock is cut square and the ends are squared before it is mounted in the lathe. The workpiece should be about 1/4" larger in cross-section than the finish turning.

The center is located on the end of the stock to receive the live and dead center. Diagonal lines are drawn across the corners using a straightedge to mark the center on both ends of the work, Fig. 12-16.

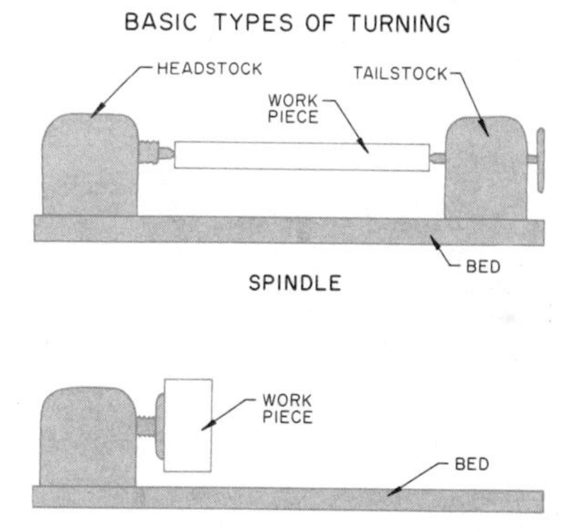

Fig. 12-15. Basic methods of woodturning.

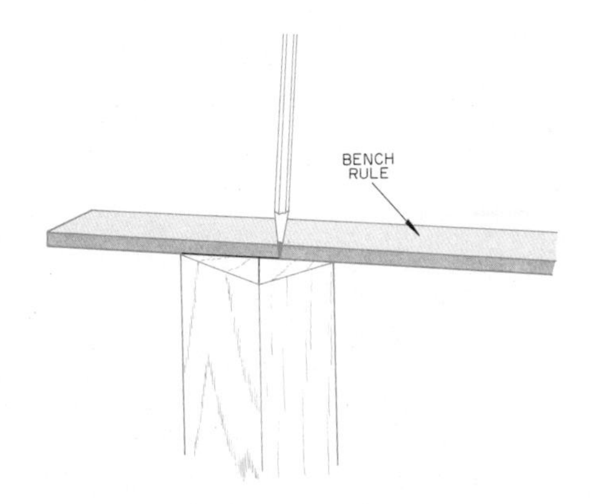

Fig. 12-16. Marking the center in the end of the stock by drawing diagonal lines.

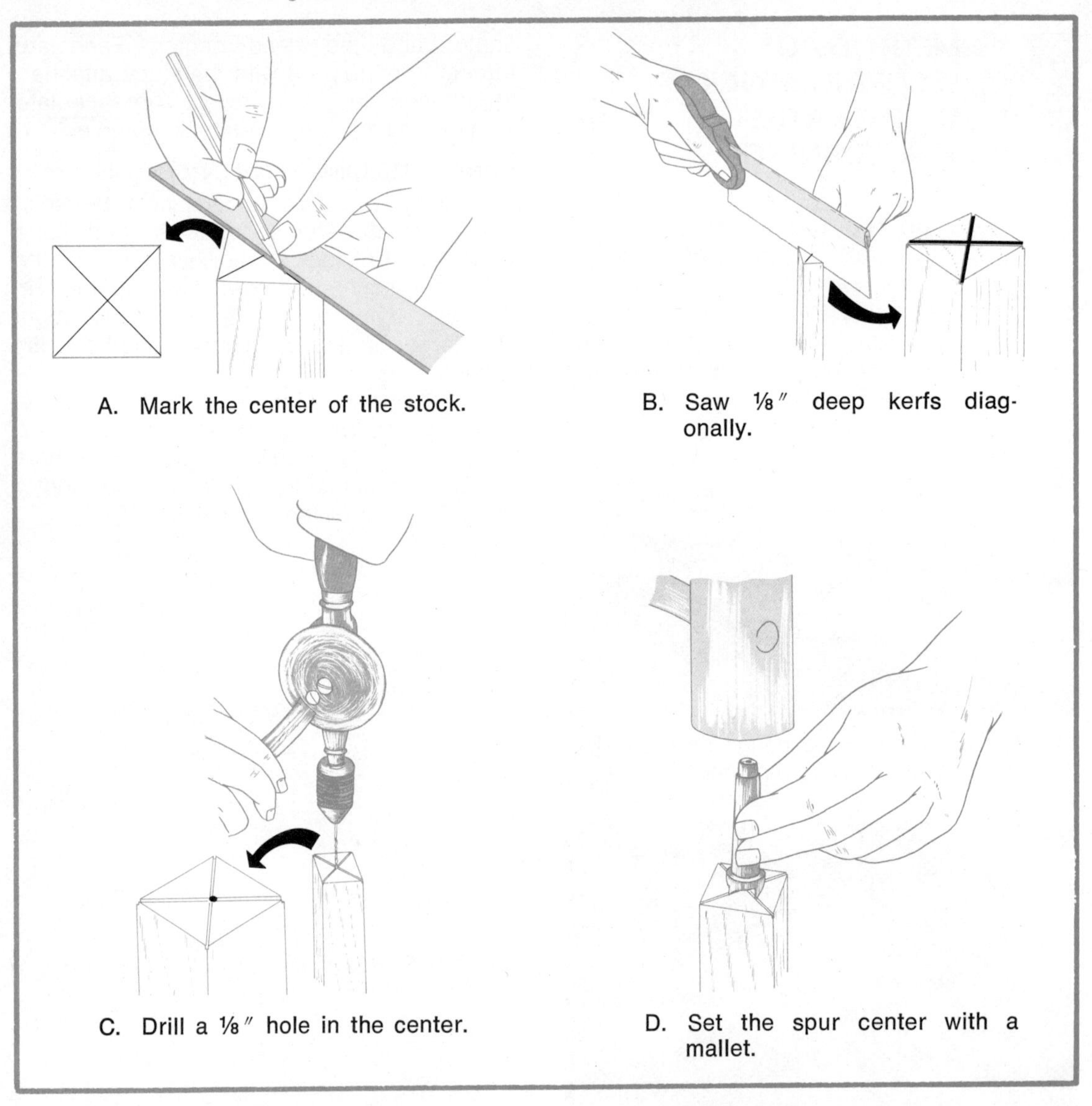

Fig. 12-17. Procedure for installing live centers.

A 1/8″ deep saw kerf is cut on the diagonal lines on one end of hardwood stock for the spur live center. In softwoods it is not always necessary to make the kerfs. A 1/8″ diameter hole, 1/4″ deep, is drilled to receive the center pin of the spur center for mounting hardwood to prevent splitting, Fig. 12-17. Striking the live center with a mallet sets the spurs in the saw kerfs. The spurs drive the work.

The stock for spindle turning is placed in the lathe as shown in Fig. 12-18. A drop of oil or paraffin wax is applied to the cup center to reduce friction and overheating. The tool support is positioned 1/8″ away from the workpiece and 1/8″ above the center. *Always revolve the stock by hand before starting the lathe.* Also, be certain the lathe speed is set at the lowest possible setting before engaging the power.

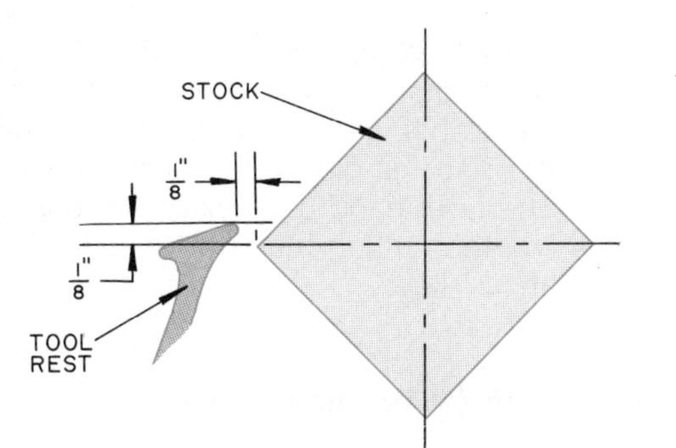

Fig. 12-18. Positioning stock for spindle turning.

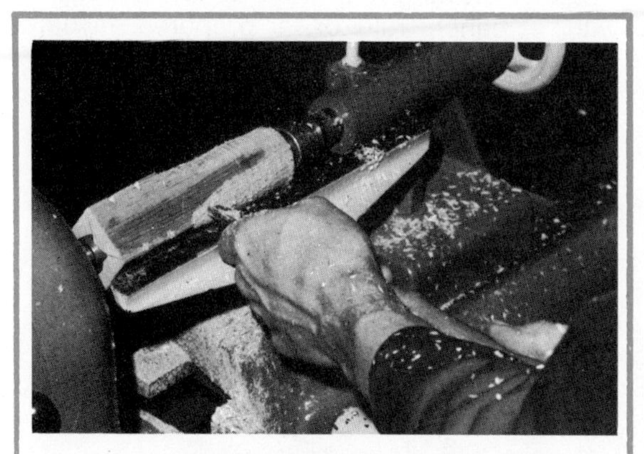

A. Rough turn by making short series of cuts.

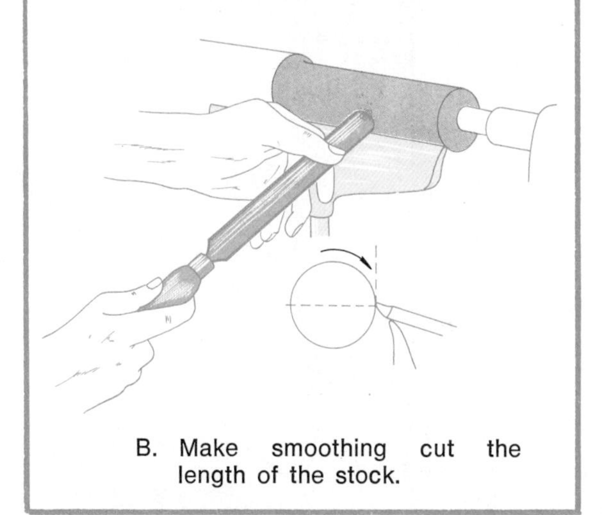

B. Make smoothing cut the length of the stock.

Fig. 12-19. Using a gouge to form cylindrical shapes.

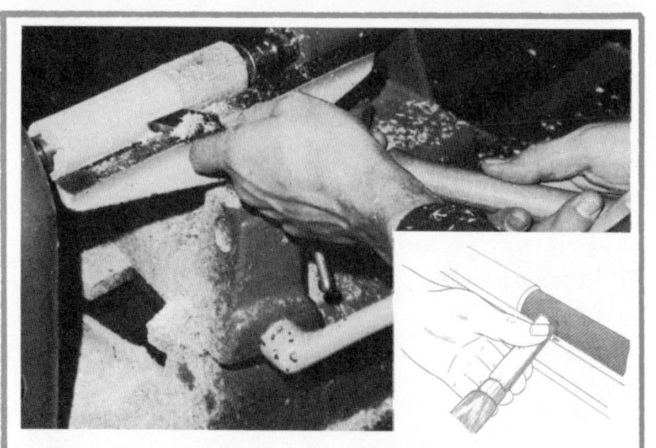

A. Scraping to smooth the cylinder.

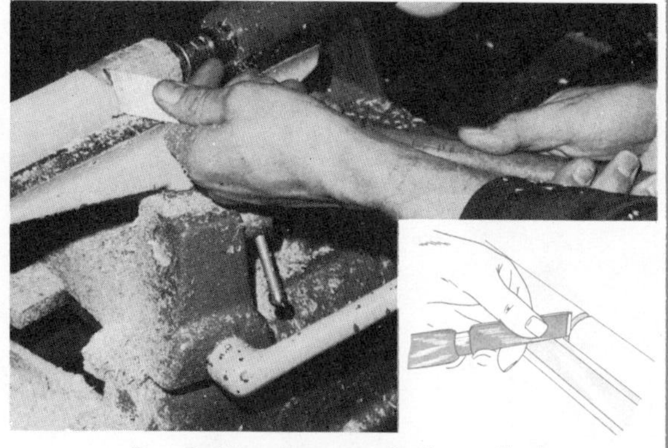

B. Cutting to smooth the cylinder.

Fig. 12-20. Smoothing a cylinder with a skew.

ROUGH TURNING STOCK

A large gouge is used for rough-turning the stock to a cylindrical shape, Fig. 12-19. The lathe is operated at the lowest speed until the stock is turned round. A series of short cuts are made rather than long, continuous cuts. The tool is moved from left to right or toward the tailstock. After the stock is rough-turned, a smoothing cut the length of the stock can be made.

SMOOTHING CYLINDRICAL SHAPES

The cylinder is finished smooth and cut to size with a large skew. The skew can be used as either a scraping or edge-cutting tool, Fig. 12-20. Remember, for beginners it is best to use the scraping method. The skew is held flat on the tool rest and parallel with the floor for cutting. The tool rest is adjusted so the cutting edge strikes on or slightly above the center of the stock. Shearing cuts are made with a large skew. The center portion of the tool is placed in contact with the workpiece and the bevel of the skew lays on the round workpiece. The skew is turned slightly and advanced along the work, the fingers serving as a guide along the tool rest to control the depth and position of the tool, Fig. 12-21. The handle of the skew is held firmly in the right hand. Use extreme caution when using the cutting method for removing stock with a skew.

LAYOUT FOR CYLINDRICAL WORK

After the workpiece is turned to a cylindrical shape the diameter is tested with an outside caliper. A layout is made from the working drawing on the cylindrical shape with a bench rule or a prepared layout template, Fig. 12-22. The layout lines will be visible when the stock is revolving. Figure 12-33 is an example working drawing which could be used for turning stock. Sometimes it is best to make the drawing full-scale, as the measurements can be taken directly from the full-scale drawing

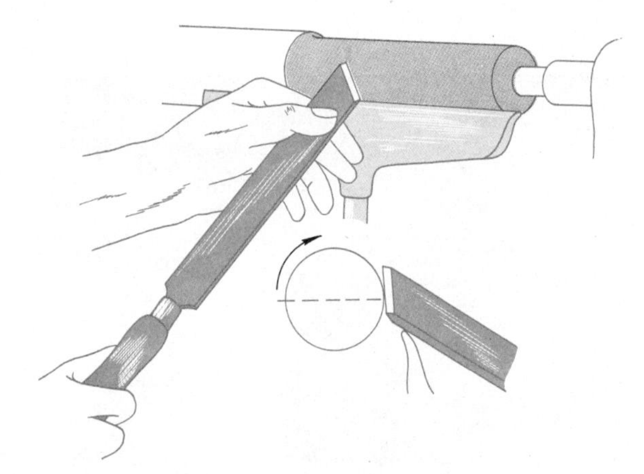

Fig. 12-21. Cutting with a skew using the fingers along the tool rest to control depth of cut.

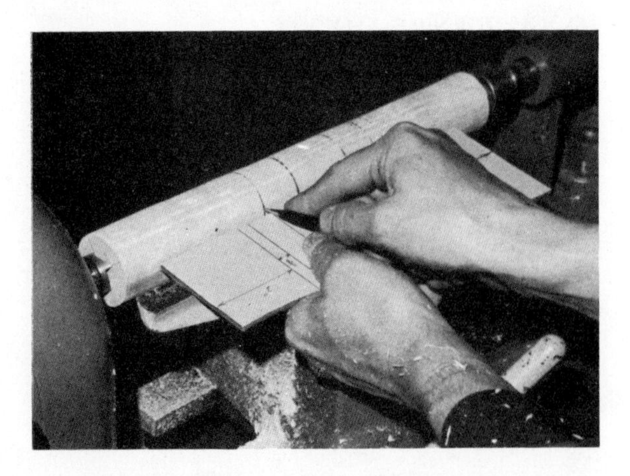

Fig. 12-22. Marking layout lines on the stock with a prepared template.

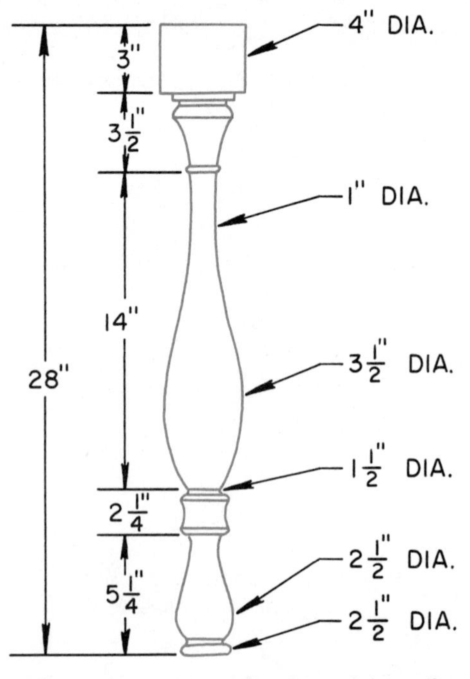

Fig. 12-23. A working drawing giving dimensions for turning.

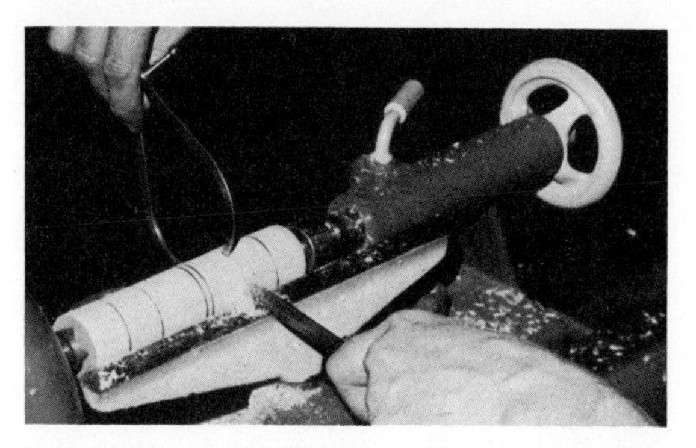

Fig. 12-24. Using outside caliper to check the size of stock as it is turned.

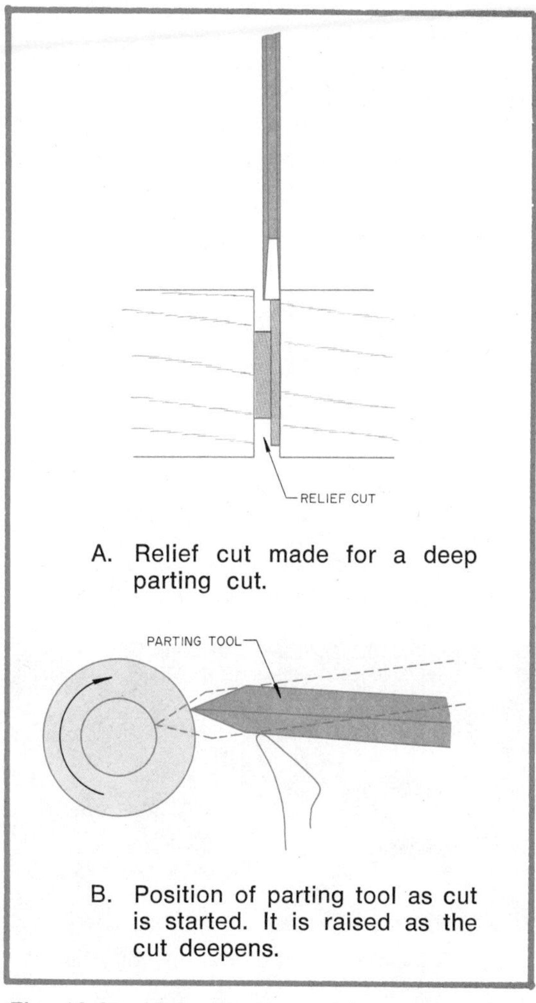

Fig. 12-25. Make a relief cut for deep parting cuts.

and transferred to the workpiece, Fig. 12-22. The diameter at various places along the stock must be shown on the drawing, as well as the distances from one diameter to the next.

MAKING SIZING CUTS WITH THE PARTING TOOL

The parting tool is used for making the sizing cuts. It is held with the narrow edge supported on the tool rest. The work can be cut to size holding the parting tool in one hand and a caliper in the other, Fig. 12-24. Care must be used when using this method. Excessive pressure on the caliper will force it over the stock. When the parting or sizing cut is more than 3/8″ deep, a relief cut must be made as shown in Fig. 12-25. Deep cuts may cause the parting tool to overheat.

TURNING TAPERS

The stock is rough-turned to slightly over the finished size. The layout is then made from the working drawing. With a parting tool, cut both ends of the taper to the desired size. On long tapers, sizing cuts can also be made along the length of the stock, Fig. 12-26. The gouge is used to remove the excess stock. Finish turning is then done with a skew using either the scraping or cutting method.

The stock layout should be about 1/8″ longer than the finished length. The extra length is required to smooth the ends. The ends are finished using a skew and the

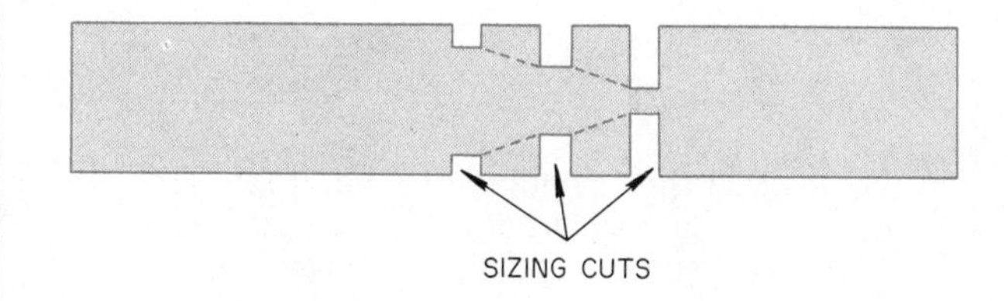

Fig. 12-26. Sizing cuts can be made prior to turning tapers.

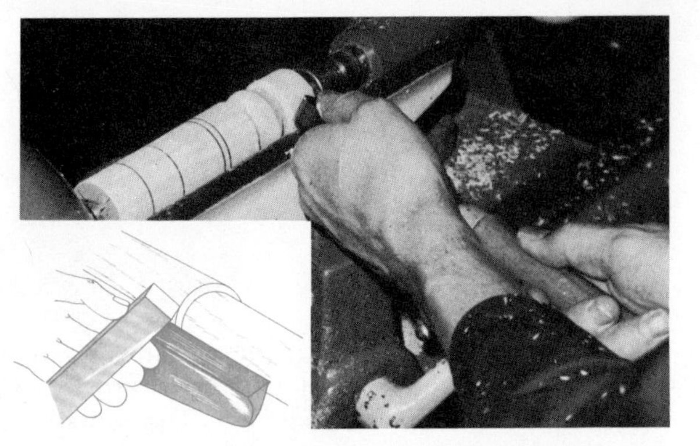

Fig. 12-27. Making a vertical shoulder cut with the toe of the skew.

vertical cut is made with the toe of the skew, Fig. 12-27.

FORMING COVES

A cove is a concave cut which can be formed on the lathe. The layout for the cove is the same as for other types of lathe work such as the taper. A round nose tool can be used to scrape the cove shape. The tool is forced into the work and the handle is moved from side to side, Fig. 12-28.

Coves can also be cut using the shearing method and a gouge. After the layout is completed, the gouge is forced straight into the work to remove the excess stock. The gouge is then used as a cutting tool by placing it nearly on edge with the handle held high. The handle is gradually

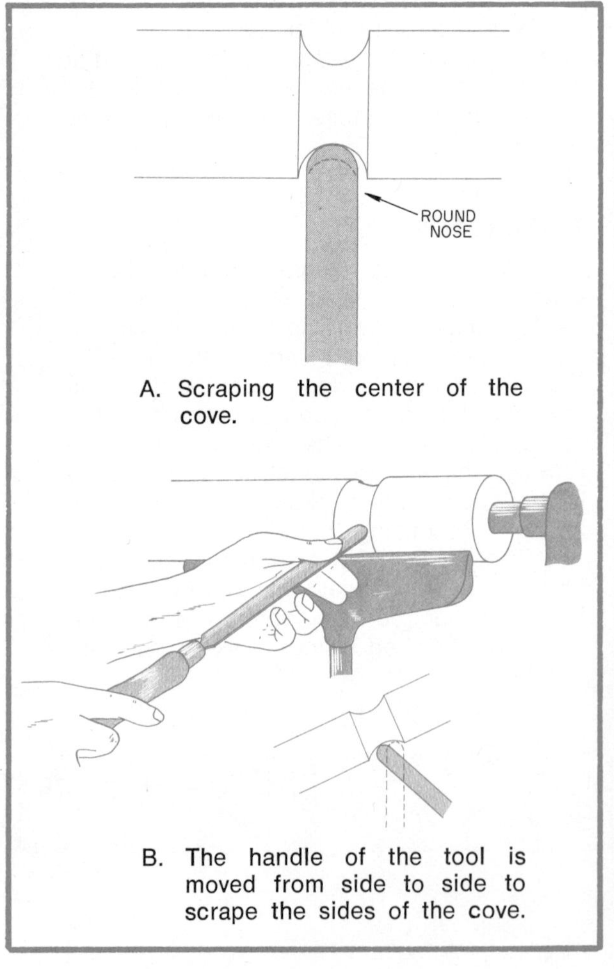

A. Scraping the center of the cove.

B. The handle of the tool is moved from side to side to scrape the sides of the cove.

Fig. 12-28. Scraping a cove with a round nose tool.

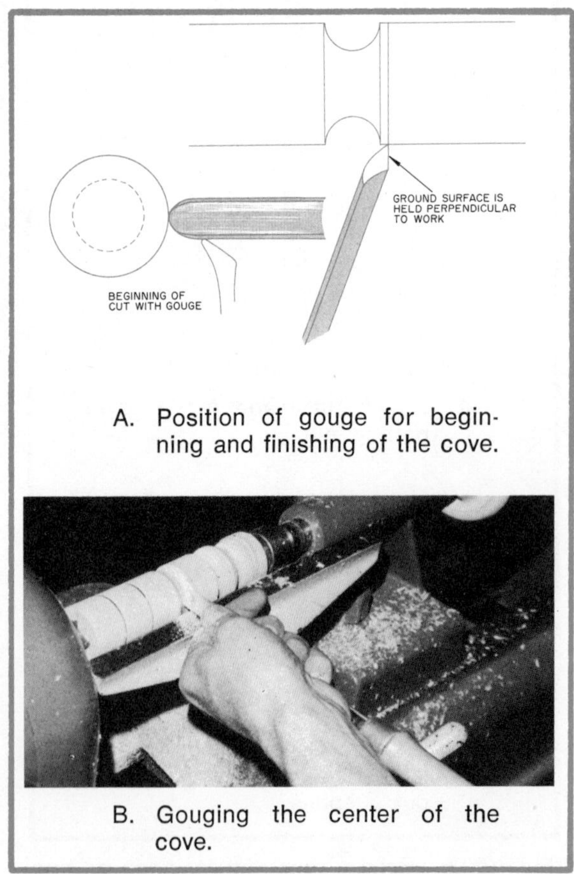

A. Position of gouge for beginning and finishing of the cove.

B. Gouging the center of the cove.

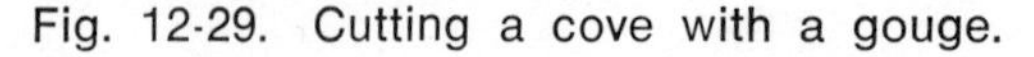

Fig. 12-29. Cutting a cove with a gouge.

lowered and the cutting edge is rolled to the bottom of the cove cut, Fig. 12-29. The same procedure is used to cut the opposite side of the cove, Fig. 12-30.

TURNING VEES ON THE LATHE

A vee cut can be scraped with a diamond or spear point tool. The tool is forced into the stock at the desired angle, Fig. 12-31. Each side of the vee is scraped to the correct depth.

A vee can also be cut using a skew. The vee is first laid out with a pencil to locate each edge of the cut. The skew is supported on edge with the heel down, and is advanced into the stock at the center of the vee cut. The cut is gradually completed working from one shoulder of the vee to the center, Fig. 12-32. The heel of the tool is used for making the cut as it is lowered into the stock toward the center. A vee formed with a shearing cut of the skew will be very smooth.

FORMING BEADS

The position of each bead is laid out on the stock. The bead can be formed by either the cutting or scraping method, but the scraping method is the easiest for the beginner. The parting tool and spear point are used for scraping beads. The beads are first separated by sizing cuts with the parting tool. It is forced to the desired depth of the bead. The spear point is then forced into the stock and rotated to form the bead, Fig. 12-33.

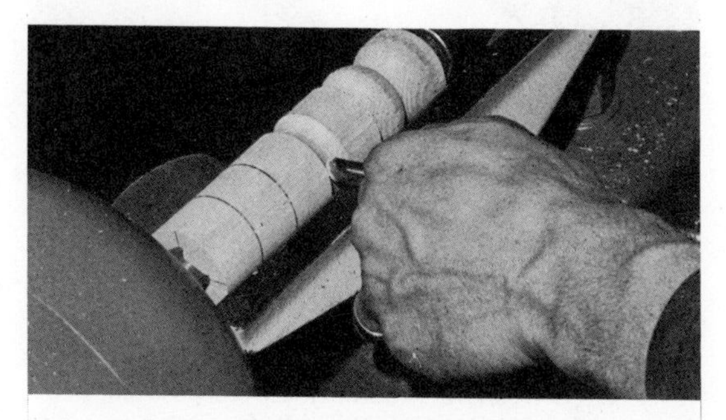

Fig. 12-30. Finishing the left side of the cove cut with the gouge.

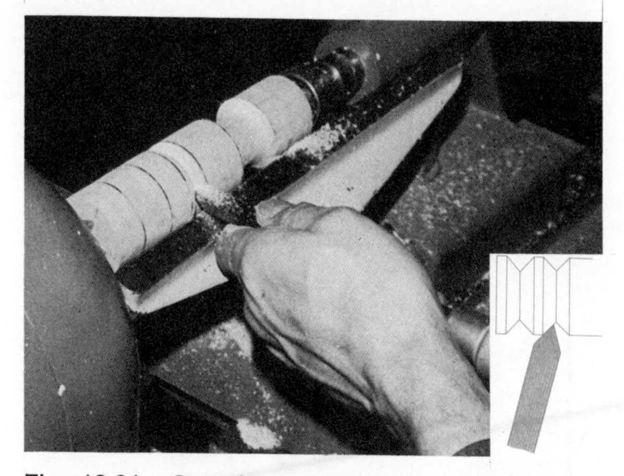

Fig. 12-31. Scraping a vee with a diamond or spear point tool.

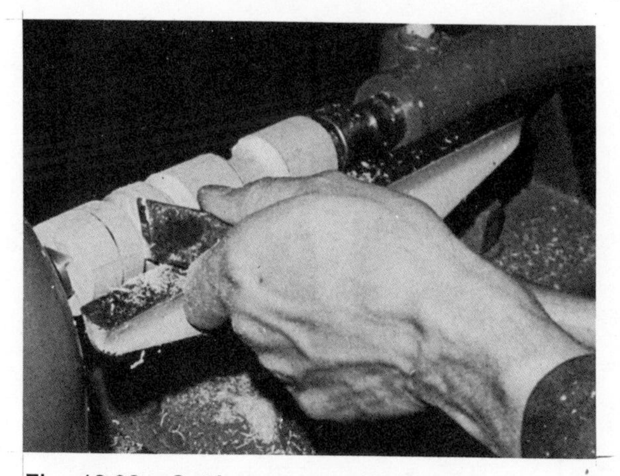

Fig. 12-32. Cutting a vee with a skew.

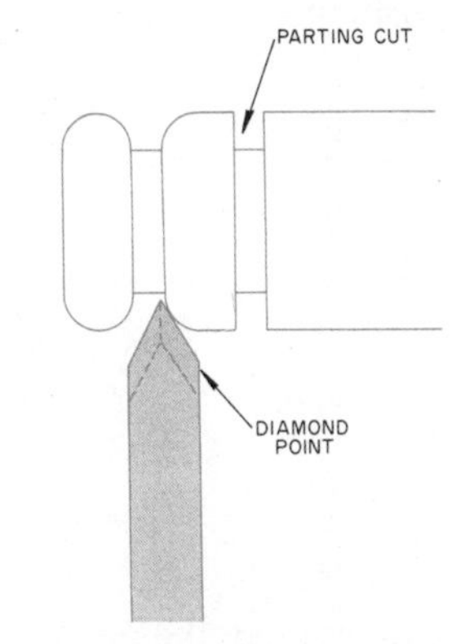

Fig. 12-33. Scraping beads with a diamond or spear point tool.

When beads are formed by the cutting method, the parting tool is used for sizing the beads. The center of each bead is marked. The skew is placed on the center of the bead and the cut is started with the heel. The tool is gradually moved from the horizontal to a vertical position as the handle is raised. The tool is rotated 90° from the start to finish, Fig. 12-34.

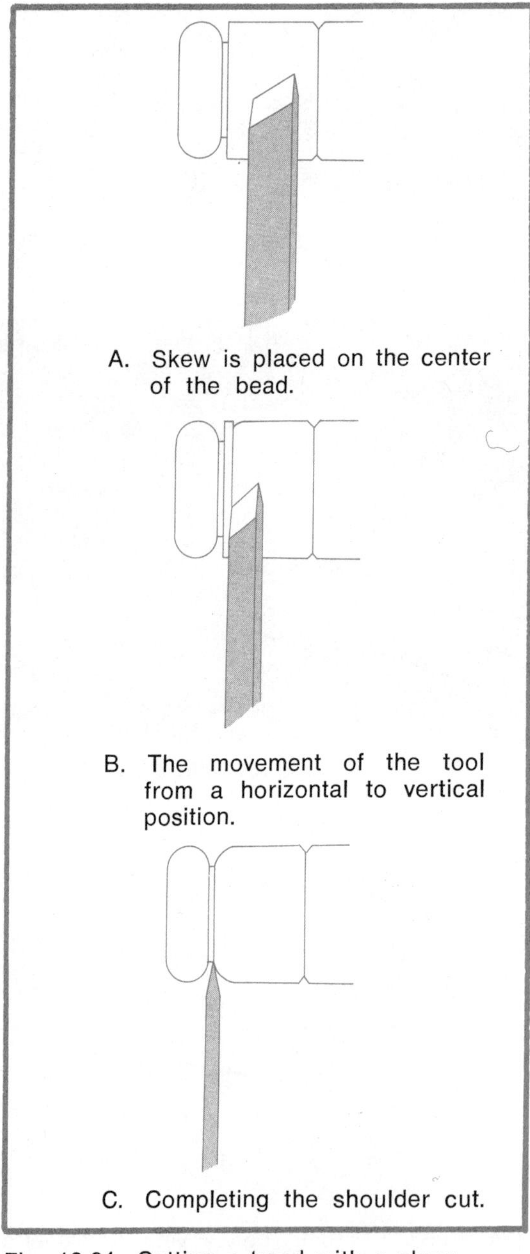

Fig. 12-34. Cutting a bead with a skew.

FORMING STOCK WITH SQUARE SECTIONS

Sometimes it is desirable to leave a portion of the stock square as is often done for legs in table construction. The stock must be perfectly square before it is turned. Special care must also be taken when centering the stock. To prevent splitting, the stock is "nicked" with a sharp skew at the point where the stock changes from square to round, Fig. 12-35. It is sometimes helpful to wrap the portion of the leg where the square starts with masking tape to safeguard against chipping.

After the stock is nicked, a parting tool is used to size the stock. The spindle is then turned the same as other types of turning. The square shoulder is finished using either a skew or diamond point tool.

When a hole is necessary the entire length of a turning, it can be made before the stock is turned. Two pieces of stock may be glued together for turning after a dado of the desired size is cut in each piece, Fig. 12-36. A square plug is placed in each end. The plug is drilled out after the turning has been completed. This

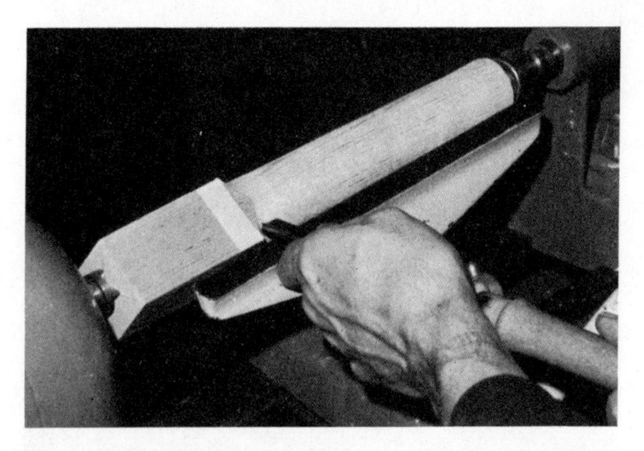

Fig. 12-35. Turning stock to leave a square portion.
The masking tape helps to prevent chipping of the square portion.

method insures you of a straight hole the entire length of the turning.

SANDING SPINDLE TURNINGS

The stock can be sanded on the lathe after it has been turned to shape. The lathe should be run at a speed of about 1500 RPM for general sanding. *Always remove the tool rest before sanding on the lathe.*

A piece of abrasive paper is held by each end against the revolving stock. A narrower piece of abrasive paper should be used for small, intricate turnings, and the width may increase with the size of the turning, Fig. 12-37. The paper can be forced into small concave areas using a piece of wood cut to the shape of the workpiece, Fig. 12-38. Abrasive papers for this type of sanding range from 80 (medium coarse) to 400 (extra fine). The rough sanding can be done with grit No. 100 and then No. 150. The final sanding can be done with the No. 220 abrasive paper.

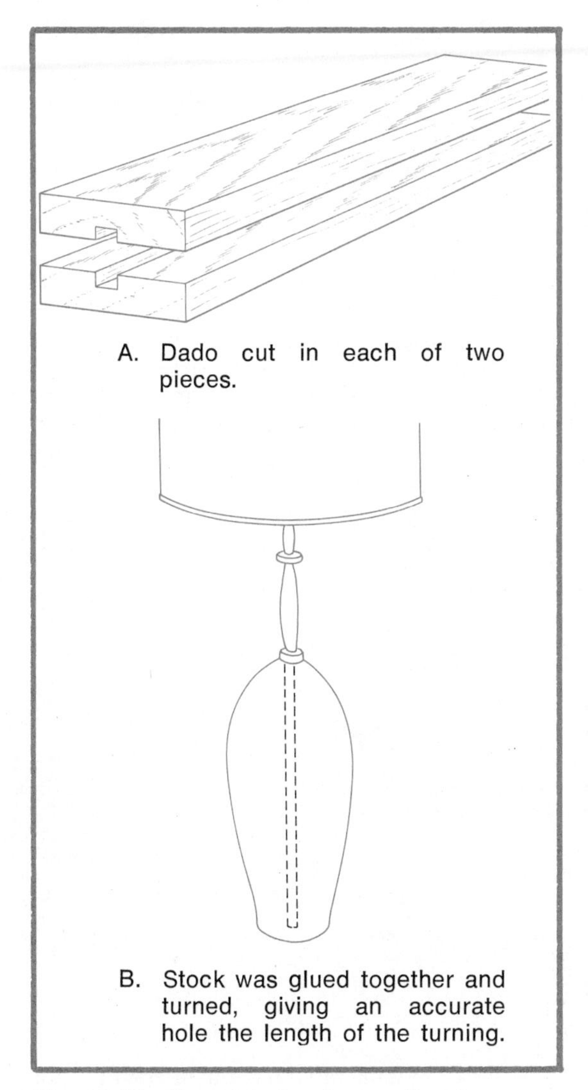

Fig. 12-36. Precut holes in woodturning.

REMOVING THE LATHE CENTERS

The live center is removed from the headstock spindle using a *knockout bar*,

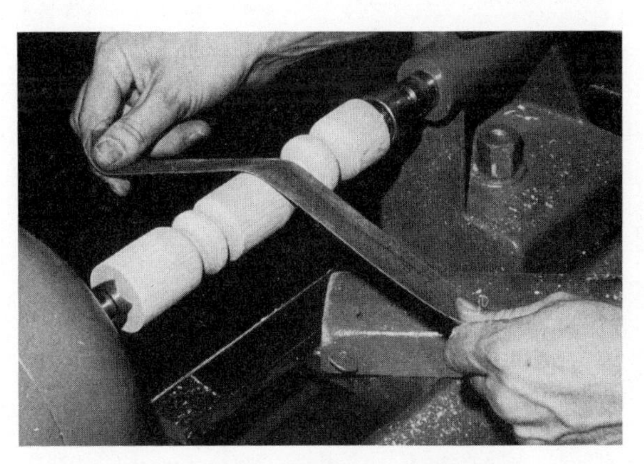

Fig. 12-37. Sanding turned spindles with narrow strip of abrasive.

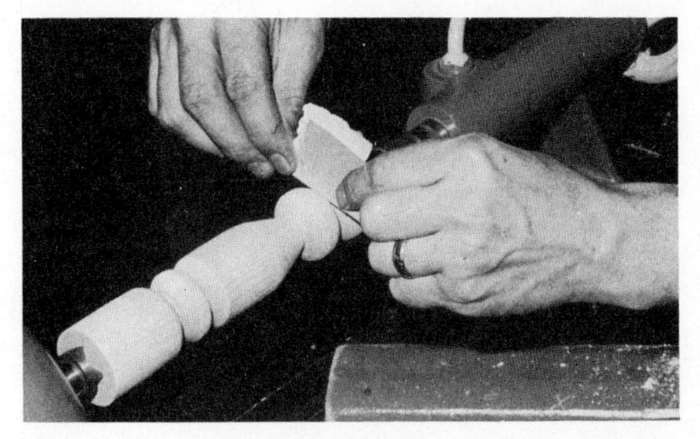

Fig. 12-38. Sanding the small concave areas of a turned spindle.

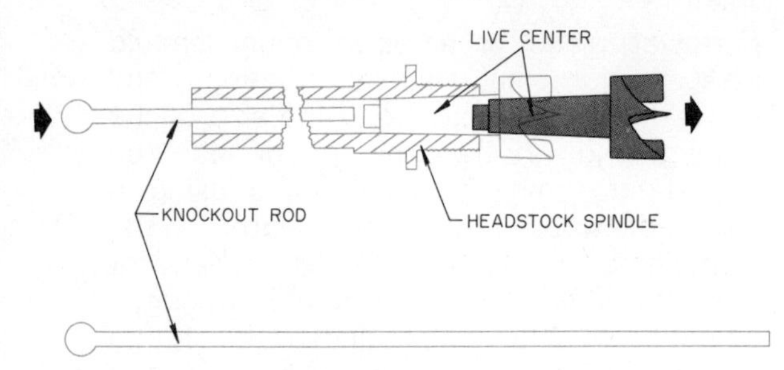

Fig. 12-39. Remove the live center from a headstock spindle with a knock-out rod.

Fig. 12-39. The bar is inserted through the outside hole of the headstock spindle. A gentle tap with the bar will loosen the live center. You should be prepared to catch the loosened center with your free hand, as the center can be damaged if it is allowed to drop on the floor. Always remember to remove the knockout bar from the headstock spindle.

The dead center is removed from the tailstock by turning the hand wheel in a counterclockwise direction. This will force the center to drop out.

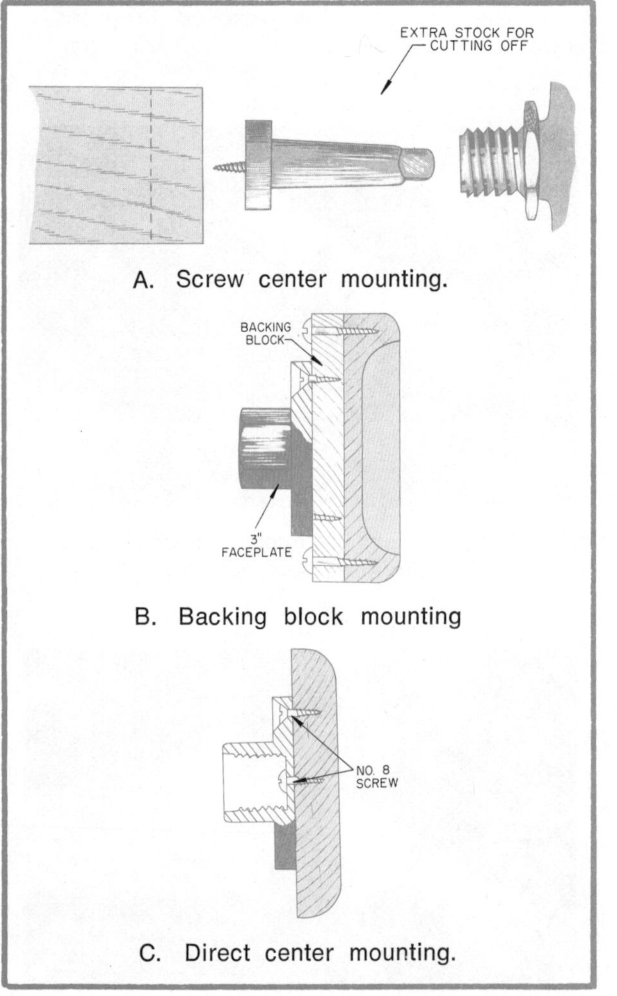

Fig. 12-40. Methods of mounting stock for faceplate turning.

FACEPLATE TURNING

The faceplate method of turning is used to form shapes such as bowls. The stock is removed by scraping rather than cutting. Before the stock for faceplate turning is mounted in the lathe it is rough cut to a round shape on the band saw.

The stock is mounted directly to the faceplate, or sometimes a backing block is used when the screws will interfere with the inside turning, Fig. 12-40. Small faceplate turning less than 3″ in diameter can be mounted on a screw center which mounts in the hollow spindle of the head-

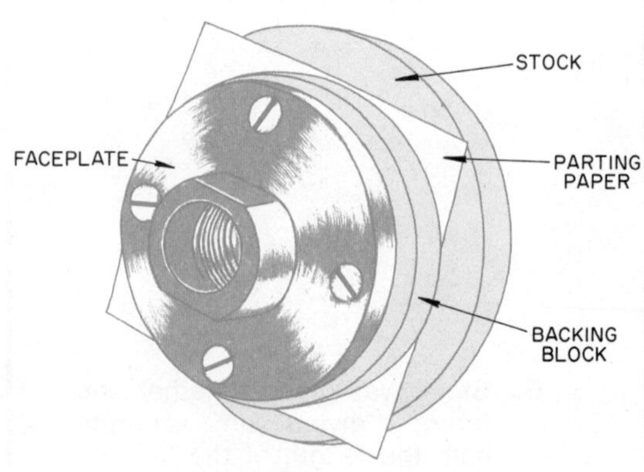

Fig. 12-41. Stock mounted on faceplate with a backing block.

stock. The work is attached to the faceplate with #8 flathead screws.

When a backing block is used, it is glued to the turning. A piece of paper is placed in the joint between the turning stock and the backing block, Fig. 12-41. The paper makes it possible to remove the backing block from the completed turning without damaging it. The glue should dry overnight before the workpiece is mounted on the lathe.

MOUNTING THE FACEPLATE

The faceplate is turned firmly onto the threads of the headstock spindle with hand pressure. The live center is removed from the spindle using a knockout bar before the faceplate is mounted. The lathe is operated at the lowest speed until the outside is turned round. On large faceplate turnings, the tailstock and dead center can be used to support the work, Fig. 12-42.

The tool rest is set slightly above center for rough turning the outside cylinder. A flat nose or spear point tool is used to turn the diameter of the disk, Fig. 12-43. All stock is removed by scraping on faceplate work.

The tool rest is turned parallel with the face of the work for shaping the inside, using a flat nose, round nose, or spear point tool. The cut is started on the center and moved toward the outside.

Fig. 12-42. Supporting faceplate stock with a dead center for roughing to a round shape.

MAKING DUPLICATE PARTS

A lathe can be equipped with a duplicator such as the one shown in Fig. 12-44 which makes it possible to produce a number of parts of exactly the same size. It is useful for turning items such as furniture legs. The tool is mounted on a carriage which moves in and out along the bed of the lathe. The design is cut in a template

Fig. 12-43. Turning the outside of a cylinder using a flat nose tool.

Fig. 12-44. A lathe equipped with a duplicator to produce identical turnings.

and a guide follows the contour of the template, moving the cutting tool in and out of the stock. The template or pattern differs for various models of duplicators. A turning of the desired shape is used as the pattern or a template is cut from thin material such as hardboard, Fig. 12-45.

FINISHING ON THE WOOD LATHE

It was previously stated that a product can be produced from start to finish on the lathe, including the application of the protective finish. A paste wax finish can easily be applied by placing wax in a folded cloth. It is then held against the turning as it revolves. The heat melts the wax and it is absorbed by the wood. After it has dried for a few minutes, it is polished with a soft cloth. The lathe is always run at a slow speed for applying finish.

A French polish finish is very satisfactory for many types of lathe turnings. It consists of white shellac, boiled linseed oil, and denatured alcohol. A few drops of shellac are applied to a soft cotton cloth folded into a pad; next, about one-half as much alcohol is applied to the pad; then, about three drops of linseed oil are added. The mixture is held lightly against the turning until the pad becomes dry. Shellac and oil are added as the pad dries. The procedure is repeated until the entire turning is coated. Several coats are applied to produce a fine finish.

Bowls and food trays can be finished with mineral oil. The tool rest must be removed for applying finish to prevent injury to the operator.

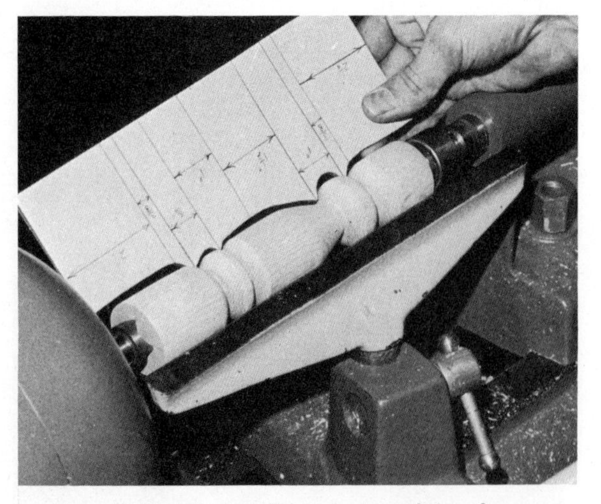

Fig. 12-45. A hardboard template for producing duplicate turnings on the lathe.

SAFETY CONSIDERATIONS FOR THE LATHE

1. Wear eye protection when operating the lathe.
2. Remove loose clothing such as ties. Shirts should be tucked in, and long sleeves should be tightly rolled up above the elbows.
3. Always start the lathe at low speeds.
4. *Remove the tool rest completely to sand on the lathe.*
5. Always keep the tools sharp. Dull tools "grab" in the work and could be thrown from your hands.
6. Adjust the tool rest so the tools are always on or above the center of the work.
7. Use only scraping cuts on the inside of faceplate turnings.
8. When holding sandpaper, place it against the work so it cannot become wrapped around the workpiece.
9. Make certain the tailstock spindle is locked tight before starting the lathe.
10. Lubricate the dead center when placing stock in the lathe for spindle turning.
11. Always remove the knockout bar from the headstock spindle.

INDUSTRIAL MACHINES FOR FORMING CYLINDRICAL SHAPES

Woodturning, as it has been described in previous sections in this chapter, is not used much for mass production in the wood industry. The lathe is used for pattern-making, furniture repair, and as a hobby. It is not used industrially because the production rate is too slow. Hand turning also requires highly-skilled craftsmen. To meet the industrial production demands, automated machines have been developed to form cylindrical shapes. The automated machines can produce products at a much lower cost, Fig. 12-46.

The cutting tools used on automated woodturning machines are usually sets of knives ground to the shape of the finished turning. These machines are sometimes called back-knife lathes. The blank stock automatically feeds in the machine and is chucked in place. This machine can turn stock up to 3″ in diameter and 30″ long. It is often used for such work as chair legs and chair back spindles.

Fig. 12-46. An automated carving machine which carves eight products at one time. (Winchester - Western Div., Olin - Mathieson Chemical Corp.)

ACTIVITIES

PROBLEM

Many products such as the chair shown below consist of several identical parts such as the legs, rungs, and spindles. These parts were produced in a factory on an automatically controlled machine. When duplicate parts are produced on a standard woodturning lathe each part must be carefully measured and turned to produce identical parts. However, the processes can be simplified by using a template.

Construct a working drawing for a turned component part such as a candle holder or table leg. Make a full size template on a piece of thin hardboard or heavy construction paper to check the product profile.

VOCABULARY

Live center
Dead center
Calipers
Spindle turning
Faceplate turning
Skew
Gouge

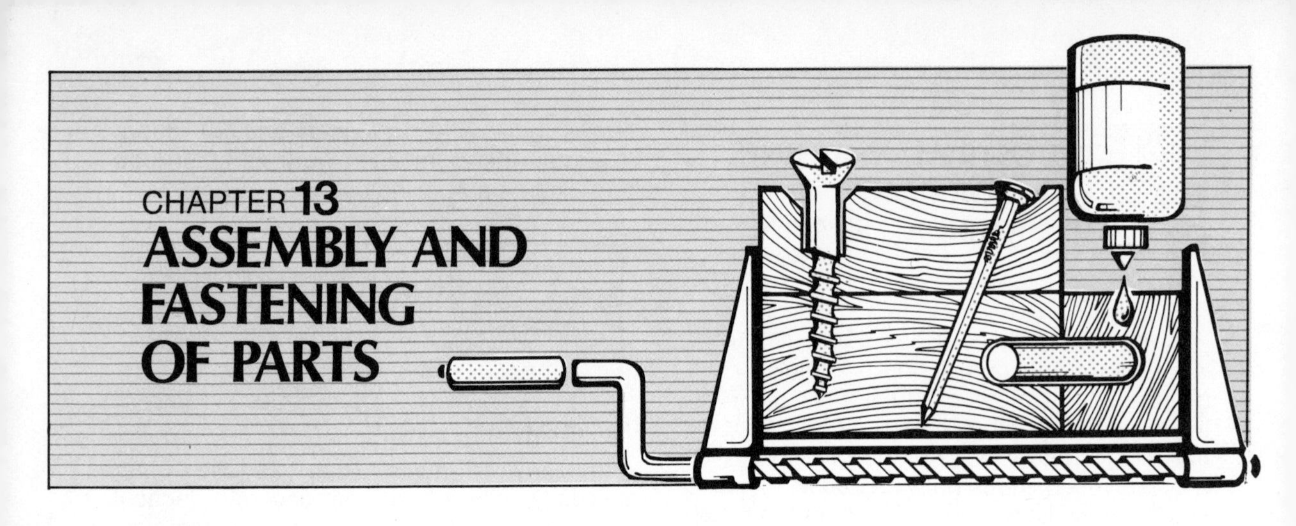

CHAPTER 13 ASSEMBLY AND FASTENING OF PARTS

INTRODUCTION

Fasteners used in the woodworking industry can be divided into two categories: (1) *mechanical,* and *(2) chemical.* The mechanical fasteners include such items as nails, screws, bolts, staples, and numerous special-purpose fasteners. The chemical fasteners include a vast array of adhesives and glues. Both chemical and mechanical fasteners are sometimes used in the production of a product.

MECHANICAL FASTENING

The most common type of mechanical fasteners used is nails. Nails are used because they are easy to install without special-purpose tools and result in a satisfactory joint for many uses. In industrial production nails are rapidly being replaced by staples. The holding power of staples is nearly the same as nails. However, staples can be installed with automatic guns much faster than nails, thus reducing production costs. Screws have the greatest holding power of any of the mechanical fasteners.

NAILS

There are many different kinds of nails, each for a different purpose. The nails most commonly used for fastening wood are shown in Fig. 13-1.

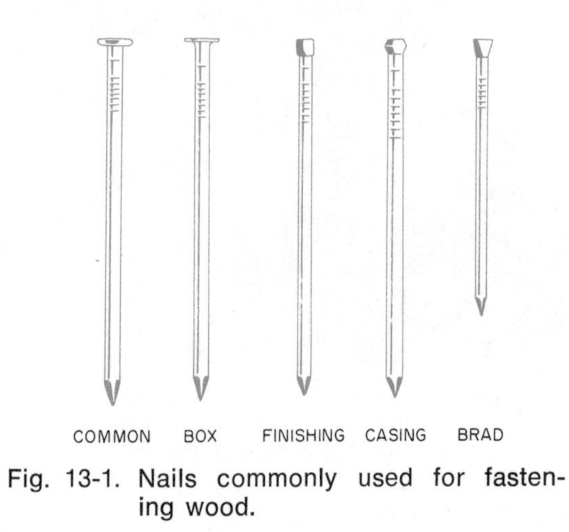

Fig. 13-1. Nails commonly used for fastening wood.

Common nails. Common nails have flat heads and have the largest diameter of any of the nails. They are used in carpentry construction where appearance is not important. Common nails have greater holding power than most other nails because of the larger diameter.

Box nails. Box nails are smaller in diameter and have a thinner head than common nails. They are used when it is likely that common nails will split the wood. They are often used in the manufacture and construction of wooden boxes and crates and are also being used extensively for lightweight framing purposes.

Casing nails. Casing nails have a head shaped like a cone, providing good holding power. These nails are used when it is necessary to set the head of the nail below the surface of the wood, such as installation of doors and windows and other trim in house construction.

Finishing nails. Finishing nails are smaller than casing nails in cross-section and have a rounder, smaller head, making them easier to set without splitting the wood. Finish nails are usually set below the surface of the wood and are also used for installing trim in house construction.

Brads. Brads are shaped like finishing nails and are used when a small fastener is needed.They are available in smaller sizes than the other nails.

NAIL SIZES

The unit denoting the size of a nail is called *penny*. Penny is often abbreviated with the lower case "d", referring to the length and diameter of the nail, Fig. 13-2. The term is thought to be derived from the cost per 1,000 nails. For example, 1,000-6d nails would have cost six pennies. The smallest nail is a 2d or two penny. A 2d nail is 1″ long. As nails increase in penny size, the length of the nail increases by ¼″. For example, a 3d nail is 1¼″ long and a 4d nail is 1½″ long.

The penny system of measurement is used for common, box, and finish nails. Brads and small wire nails are sized by their actual length and gauge size. Brads are available in gauge sizes from 12 to 20; the higher the gauge number, the smaller the diameter of the brad. Regardless of the size of the nail, they are all sold by the pound.

SPECIAL-PURPOSE NAILS

In addition to the nails previously described, there are many special-purpose nails, Fig. 13-3. For example, colored nails

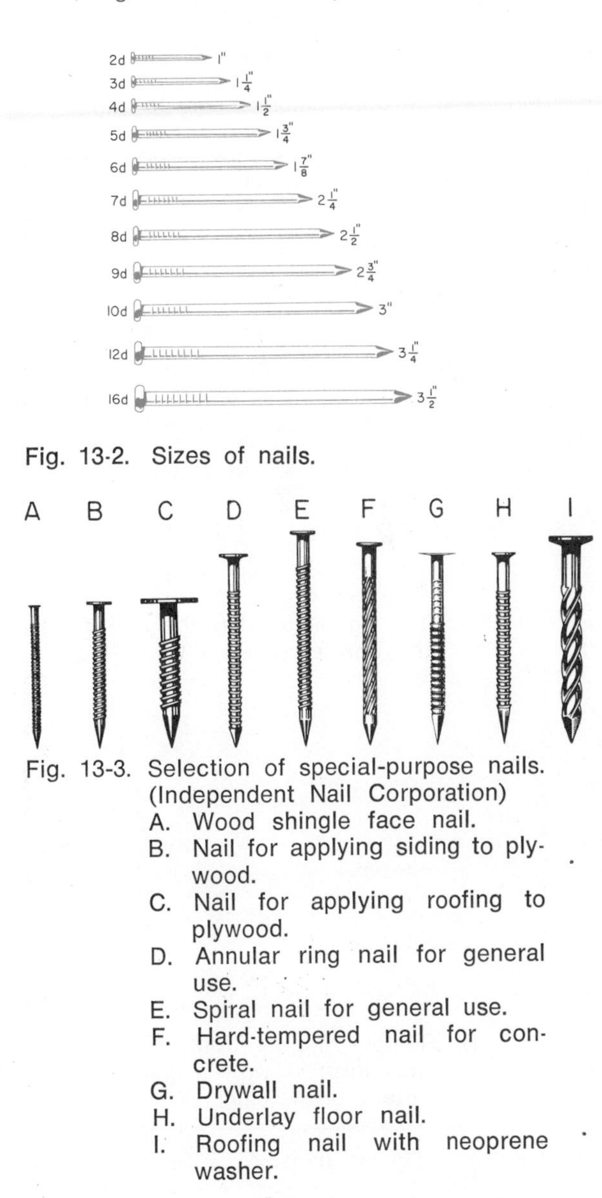

Fig. 13-2. Sizes of nails.

Fig. 13-3. Selection of special-purpose nails. (Independent Nail Corporation)
A. Wood shingle face nail.
B. Nail for applying siding to plywood.
C. Nail for applying roofing to plywood.
D. Annular ring nail for general use.
E. Spiral nail for general use.
F. Hard-tempered nail for concrete.
G. Drywall nail.
H. Underlay floor nail.
I. Roofing nail with neoprene washer.

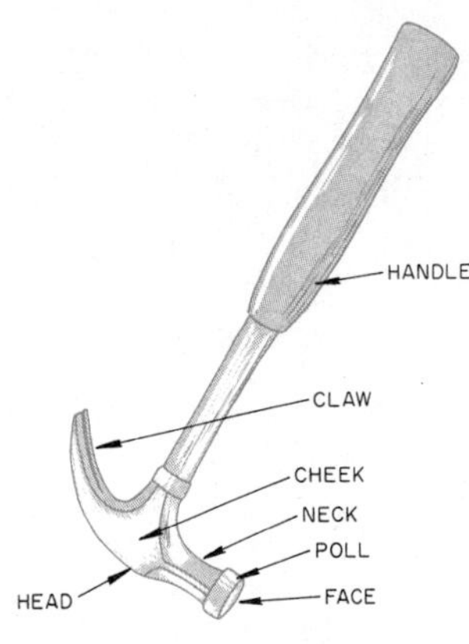

Fig. 13-4. A claw hammer and its parts.

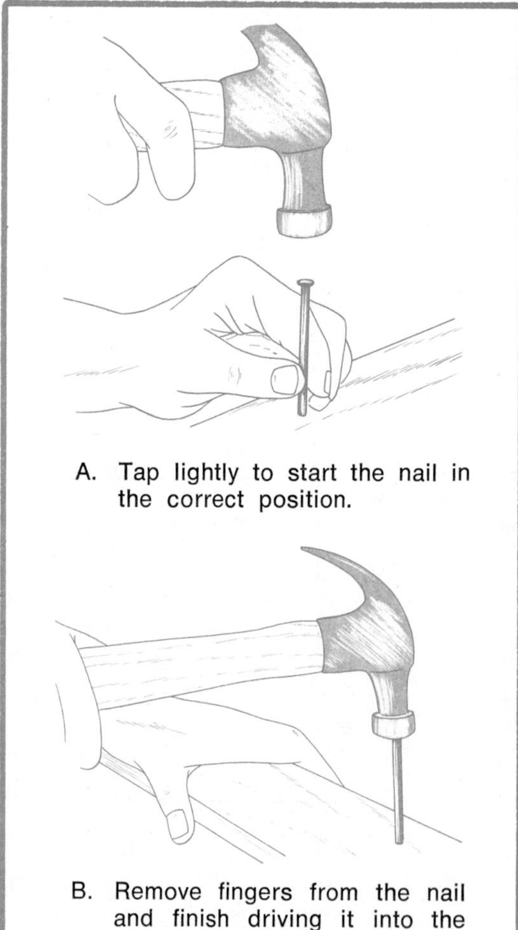

A. Tap lightly to start the nail in the correct position.

B. Remove fingers from the nail and finish driving it into the surface of the stock.

Fig. 13-5. Driving nails.

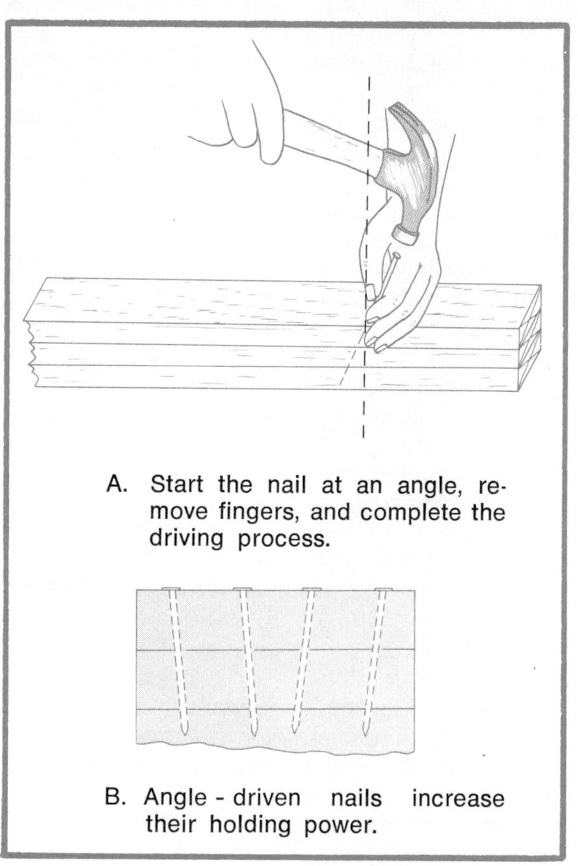

A. Start the nail at an angle, remove fingers, and complete the driving process.

B. Angle - driven nails increase their holding power.

Fig. 13-6. Installing nails to increase holding power.

are available to match the color of the wood panels in which they are installed. Large nails with grooves or rings on them are used for attaching boards to posts. Ring shank nails are also for nailing flooring because they will not loosen easily.

Most nails are made of mild steel; however, some are made of copper, aluminum, brass, and special alloys. Nails made of mild steel are coated with a surface material to keep them from rusting or to increase their holding power. Nails used outside are often coated with zinc and are called galvanized nails which will not rust. Some nails are cement-coated to improve their holding power.

INSTALLING NAILS

Nails are driven with a claw hammer. There are two common types of claw hammers, one type has a curved claw and the other a straight or ripping claw.

The curved claw is the most suitable for general use, Fig. 13-4. The size of the hammer is determined by the weight of the head, varying from 5 oz. to 20 oz. A 13 oz. or 16 oz. hammer is good for general work. The straight claw hammer (also called a ripping claw) is used for rough construction such as framing carpentry. The straight claw makes it easy to pull nails from temporary forms.

Driving nails. Start the nail with very light blows, holding the nail in one hand and the hammer in a firm grip with the other hand. After the nail is started, remove your hand from the nail and finish driving it level with the surface of the stock, Fig. 13-5. Keep your eye on the nail. Use light blows as the nail nears the surface and be very careful not to make hammer marks around the nail. Nails will hold more if they are driven into the wood at a slight angle instead of straight into the wood, Fig. 13-6.

Setting nails. Finish, casing, and brad nails are usually set below the surface of the wood. The hole made by the nail is filled so that it does not show. A nail set is used to sink the nail below the surface, Fig. 13-7. Nail sets are available in different sizes, so use a nail set that is the correct size for the nail.

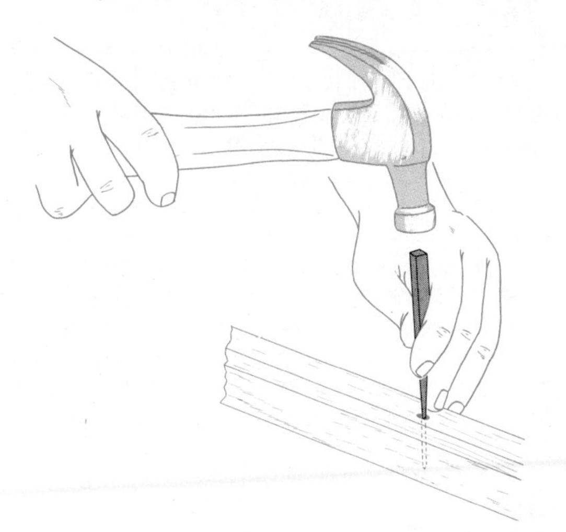

Fig. 13-7. Using a nail set to sink the nail below the surface.

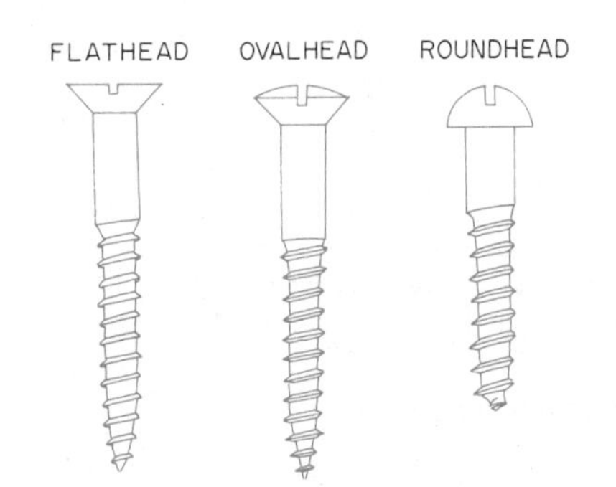

Fig. 13-8. Common types of screw heads.

SCREWS

Screws have two definite advantages over most other mechanical fasteners for wood: (1) screws provide greater holding power, and (2) screws can be removed to dismantle a structure.

Screws require more time to install than nails and are more expensive. For this reason, screws are usually used for high-quality furniture construction. Screws are often used when a product is dismantled for shipping and put together after it is purchased.

SCREW SIZES

Screws are available in a wide variety of sizes, with common sizes ranging from 1/4″ to 6″ in length. The diameter of the screw is referred to as the gauge size, which varies from 0-20.

TYPES OF SCREW HEADS

There are three basic types of screw heads: (1) *round*, (2) *flat*, or (3) *oval*, Fig. 13-8.

Flat head and oval head screws are used if the head is to be countersunk below the surface of the wood. A round head is installed if the head remains visible on the surface. Screws are made with either a slotted head or a phillips head as shown in Fig. 13-9.

Screws are made from a number of different kinds of metals; however, mild steel is the kind most often used. For some purposes screws made of mild steel without a coating would be referred to as *bright*. When the screws are exposed, the screws are coated. Common coatings are chrome and nickel. Brass screws are used if the product will be exposed to moisture such as in building boats.

WRITING SCREW SPECIFICATIONS

To order screws it is necessary to state what type of screw is desired. This would include the *length*, *diameter*, *type of head*, and the *finish*. For example, a specification may be written 1″ No. 7 F. H. chromium. This would mean the screw would be 1″ long with a gauge size of 7, the head would be flat, and it would have a chromium finish.

INSTALLING SCREWS

To install a screw properly, two holes of different sizes must be drilled. One hole is called the *pilot hole* and the other is called the *shank hole.* The shank hole is drilled the size of the screw. The second hole is drilled the size of the root diameter of the screw. The shank and root diameter is shown in Fig. 13-10 and the relationship between the pilot hole and the shank hole is shown in Fig. 13-11.

To obtain the maximum holding power for screws, it is important that the pilot holes and shank holes be the correct size. If the holes are not of the proper size, the wood may split when the screw is installed. Table 4 gives the correct size holes to drill for common-sized screws.

In some soft woods such as pine and willow, it may be possible to install the screws without drilling pilot holes. The pilot hole can sometimes be made with an awl. However, in the very hard woods such as oak and maple, the pilot hole must be drilled, as the screw will likely twist off in the hard woods if the pilot hole is not drilled.

The length of the screw selected will be determined by the thickness of the wood being joined. Care should be taken

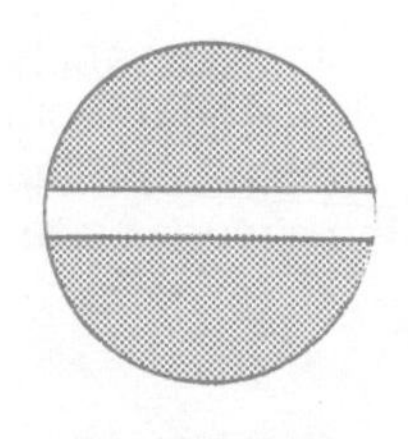

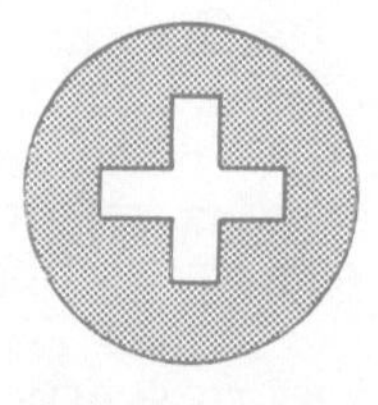

SLOTTED PHILLIPS

Fig. 13-9. Slotted head and phillips head of screws.

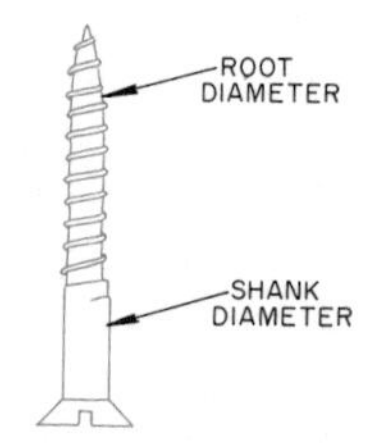

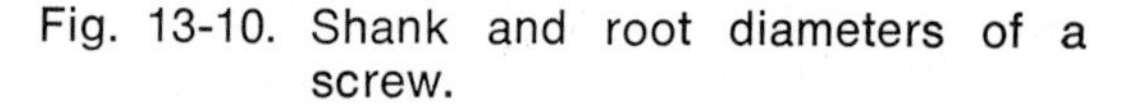
Fig. 13-10. Shank and root diameters of a screw.

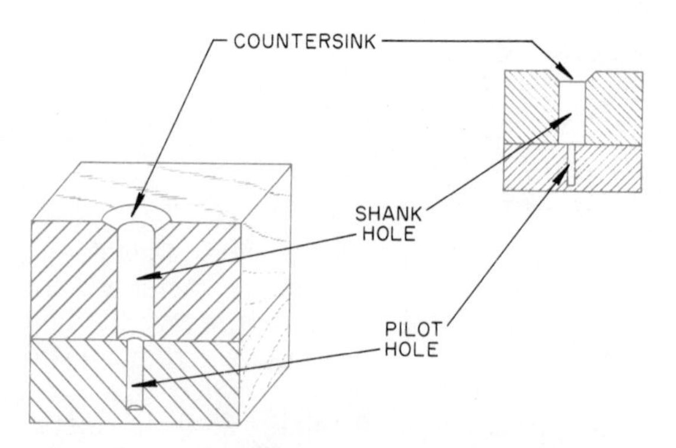

Fig. 13-11. Relationship of the pilot hole, shank hole, and countersink for a screw.

Table 4
Screw and Screw Hole Chart

Gauge No. Wire Size	Counter-bore for head Drill Size	Shank Hole Drill Size	Pilot or Lead Hole Drill Size	
			Hard Wood	Soft Wood
0	.119(1/8)	1/16	3/64	
1	.146(9/64)	5/64	1/16	
2	1/4	3/32	1/16	3/64
3	1/4	7/64	1/16	3/64
4	1/4	1/8	3/32	5/64
5	5/16	1/8	3/32	5/64
6	5/16	9/64	3/32	5/64
7	3/8	5/32	1/8	3/32
8	3/8	11/64	1/8	3/32
9	3/8	3/16	1/8	9/64
10	1/2	3/16	5/32	9/64
12	1/2	7/32	3/16	9/64
14	1/2	1/4	3/16	11/64
16	9/16	9/32	15/64	13/64
18	5/8	5/16	17/64	15/64
20	.650(11/16)	11/32	19/64	17/64
24	.756(3/4)	3/8	21/64	19/64

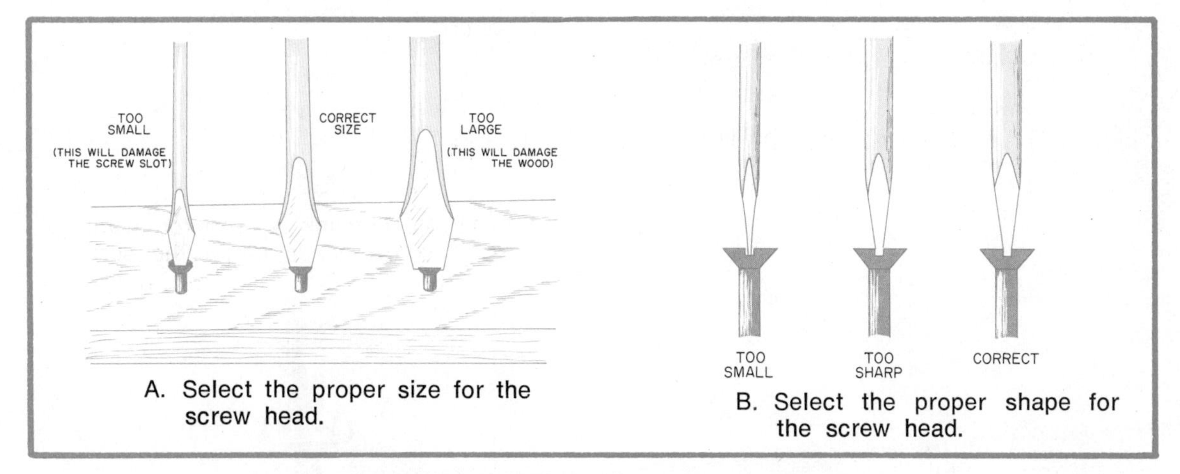

Fig. 13-12. Selecting the proper screwdriver.

to select a screw of the proper size so that all of the threaded part will be in the second piece. Sometimes this may be impossible when joining thin pieces. The holding power of end grain is considerably less than face or edge grain, so a longer screw should be selected when it is installed in end grain.

Screws can be installed more easily if a lubricant such as bees-wax or wax is coated on the threads prior to installation.

SELECTING AND USING SCREWDRIVERS

A screwdriver should be selected which properly fits the screw slot, Fig. 13-12. If the screwdriver is too large, it will damage the screw head, or if the driver is too small, it will damage the screwdriver. The screwdriver should be ground and shaped so that it is flat and fits all the way into the slot. A phillips screwdriver is used for installing screws with phillips slots, Fig. 13-13.

Screwdrivers are available in a variety of sizes and lengths. The most common sizes have blades from 2½″ to 6″ long, Fig. 13-14. Screwdriver bits held in a hand brace are used in many industrial situations.

Quick-return ratchet screwdrivers are used to save time and labor when a product requires the installation of several screws, Fig. 13-15. These screwdrivers operate in a reciprocal (pushing up and down) action rather than a rotary (turning) action. Ratchet screwdrivers are slightly longer to accommodate a quick-return spring which holds the blade in the slot and automatically returns the handle to the driving position after each stroke.

COUNTERSINKING SCREW HEADS

When it is necessary to set a screw head below the surface, it is called *countersinking*, Fig. 13-16. Special bits called screw-mates are available for drilling the

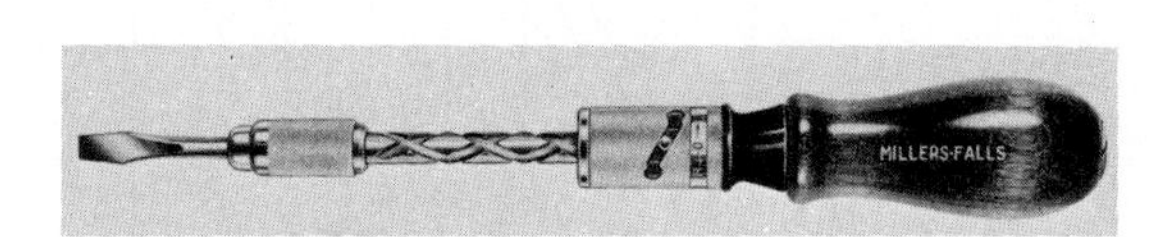

Fig. 13-15. Quick-return ratchet screwdriver. (Millers Falls Co.)

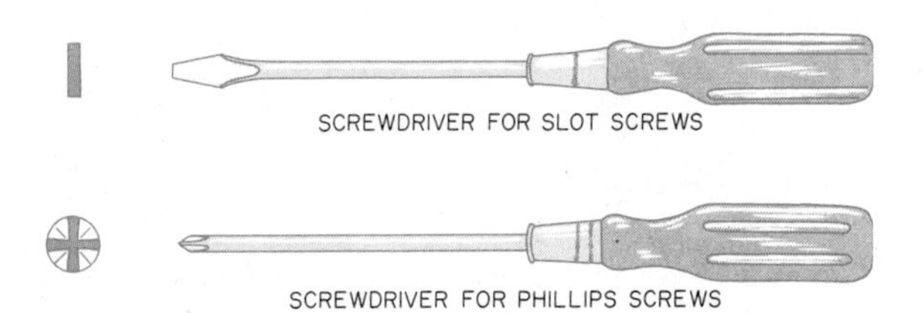

Fig. 13-13. A standard and phillips head screwdriver.

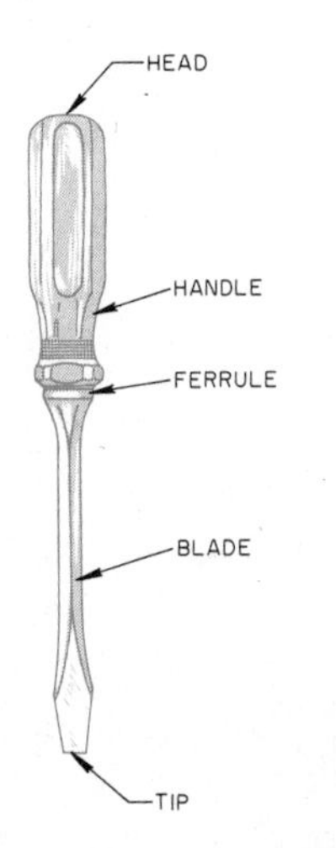

Fig. 13-14. Parts of a screwdriver.

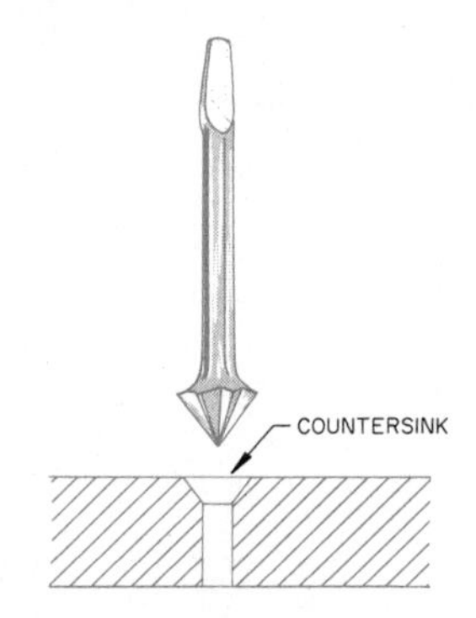

Fig. 13-16. Countersink is used for installing screw heads below the surface.

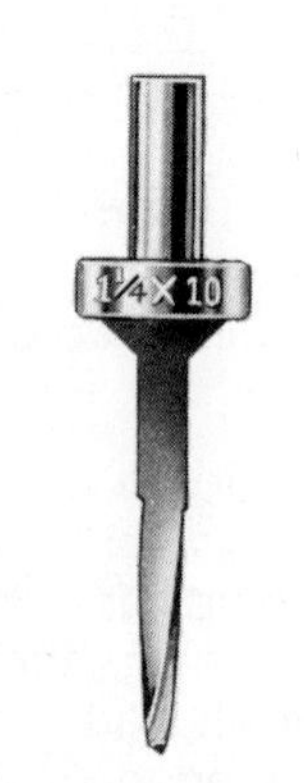

Fig. 13-17. Screw-mate countersink. (Stanley Tools)

holes for installing screws. This tool saves time and insures proper installation. It drills the hole the correct depth, the proper

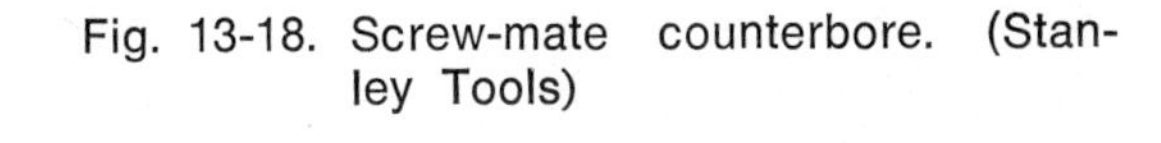
Fig. 13-18. Screw-mate counterbore. (Stanley Tools)

size for both the clearance and pilot hole, and countersinks the head of a flat screw, Fig. 13-17. The proper size screw-mate must be used. For example, a screw-mate with 1¼″ × 8″ stamped on it would drill the correct size hole for an 8-gauge screw, 1¼″ long. Screw-mates are used for installing screws with the head countersunk or counterbored, Fig. 13-18.

SPECIAL-PURPOSE MECHANICAL FASTENERS

Among other fasteners sometimes used are corrugated fasteners, chevrons, wood joiners, carriage bolts, hanger bolts, lag screws, and mounting plates for legs. Figure 13-19 gives examples of some of

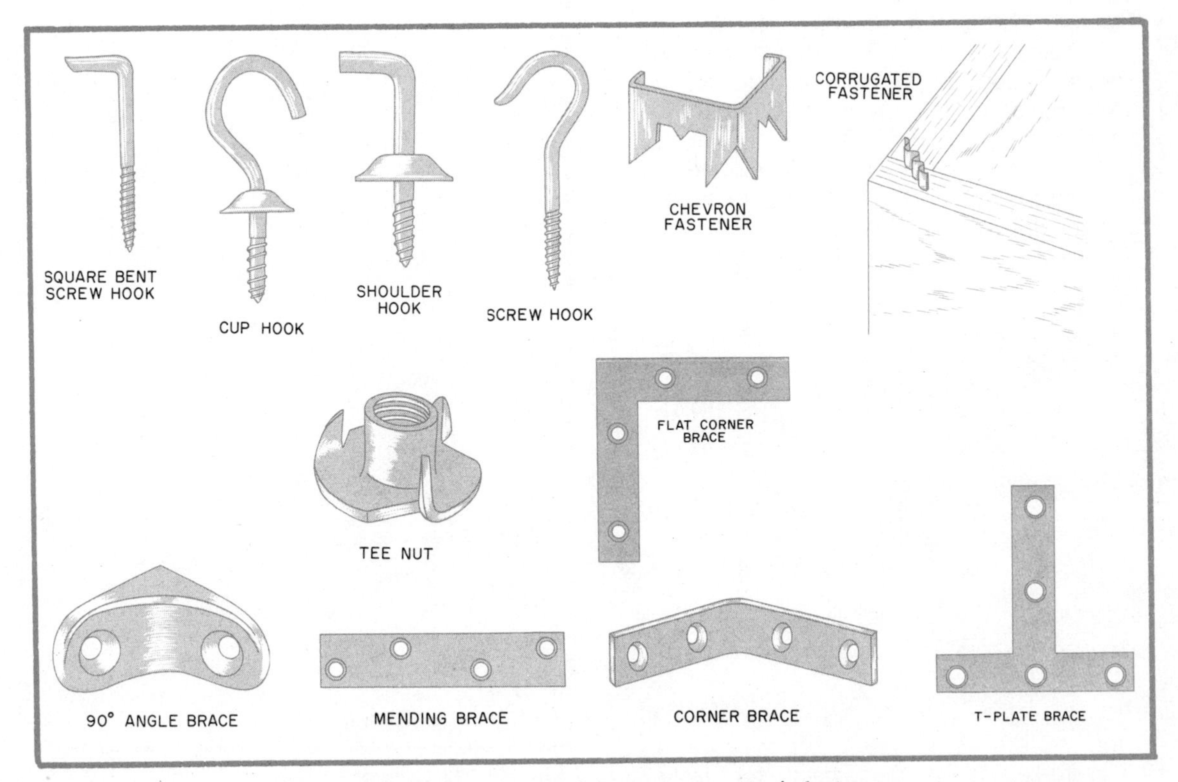

Fig. 13-19. Common special-purpose wood fasteners.

these common special-purpose fasteners used with wood.

Dowel pins combined with adhesive or chemical fasteners are a type of mechanical fastener commonly used to join two pieces of wood. They are available in a variety of designs, such as the common ones shown in Fig. 13-20.

CHEMICAL FASTENING

Chemical fastening is the process of joining two or more pieces together by means of glues and adhesives. An adhesive is a material which has an attraction for material and causes two or more like or unlike materials to bond together. Adhesives have become an essential element of the wood industry. They are used for gluing stock together, for making larger sizes, and for assembling parts, as well as being widely used in the manufacture of sheet material such as plywood, particle board, and hardboard, Fig. 13-21.

Adhesives have been developed which are stronger than the base materials joined. Many adhesives have been developed for

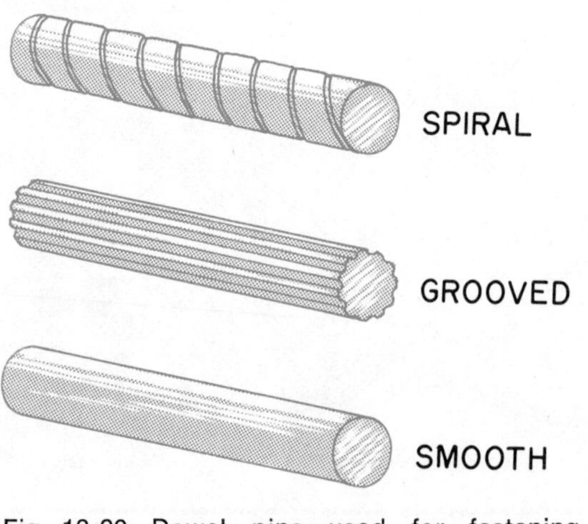

Fig. 13-20. Dowel pins used for fastening wood.

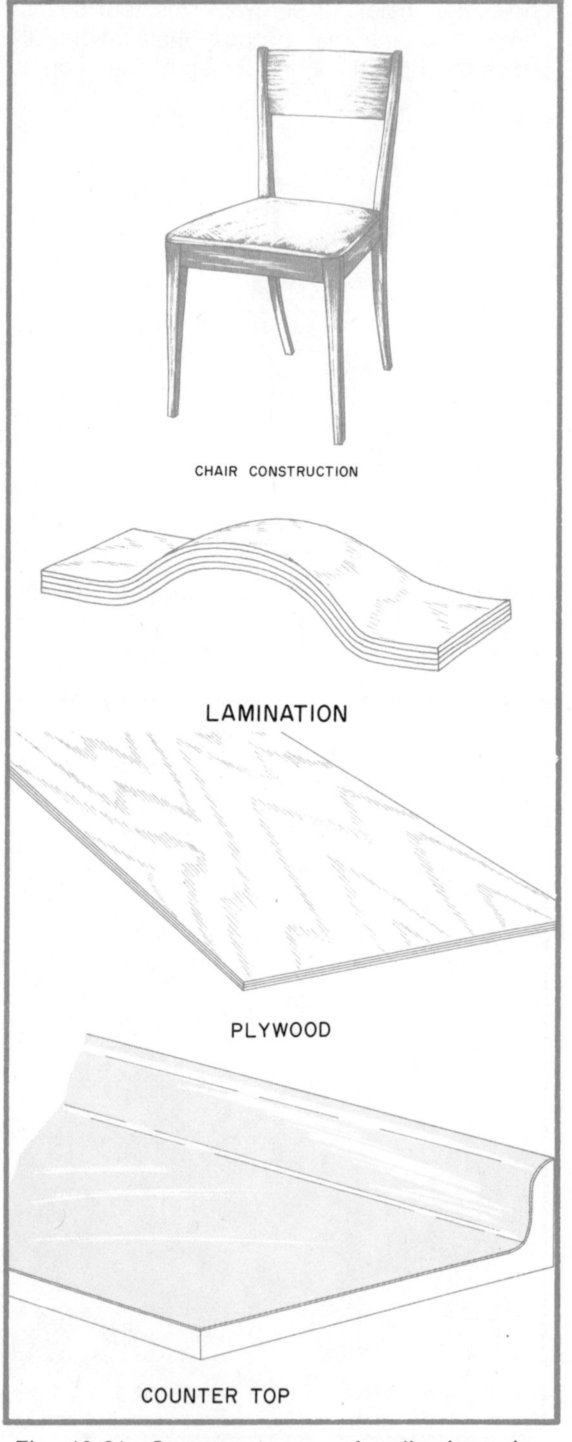

Fig. 13-21. Common uses of adhesives in wood industries.

special purposes. The process of bonding two or more materials together is often referred to as gluing.

Chemical fastening has several advantages over mechanical fastening for some types of assemblies. The appearance of a chemically-fastened joint is usually more pleasing and, in general, a glued joint is as strong or stronger than a mechanically-fastened joint. Additional strength can be produced by combining mechanical and chemical fastening.

Fig. 13-22. Common adhesives used for fastening wood products.

COMMON ADHESIVES

Joint preparation and proper selection of adhesives are the most important factors in the gluing process. The main consideration in glue selection is the use for which the finished product is intended, that is, whether the product will be exposed to water or high moisture. Adhesives are categorized as being waterproof, water-resistant, or non-waterproof. Products such as boats must be assembled and fastened with a waterproof glue. Furniture and similar products can be fastened with a non-waterproof adhesive.

Many glues have been produced for specified gluing jobs other than moisture conditions. The kind of wood is also a factor for glue selection, as well as preparation of the joint. Some types of glue are more satisfactory than others for loose fitting joints. Several types of glue have a tendency to stain the wood. These and other factors will be discussed for each of the types of glue in this chapter.

TYPES OF ADHESIVES

Glues can be classified as either natural or synthetic, according to their origins. Of course, natural adhesives are products of natural materials and synthetic adhesives are products of our modern chemistry. Constant research is conducted in chemical laboratories to produce adhesives to bond wood and other fibrous materials more successfully, Fig. 13-22.

The synthetic resins or adhesives are also classified as *thermosetting* or *thermoplastic* resins. Thermosetting resins cure or "dry" by means of a catalyst and cannot be softened by heat or the addition of water or a solvent because they are more resistant to moisture and heat.

Thermoplastic resins cure by means of absorption and evaporation of the moisture, thus can be reactivated by means of heat and water or a solvent.

SELECTION OF AN ADHESIVE

There is really no adhesive which can be called an all-purpose adhesive. The demands of the job must be considered when fastening with an adhesive. The following description of common adhesives will serve as a selection guide.

NATURAL ADHESIVES

Natural adhesives are produced from a variety of natural substances. The following are common natural adhesives.

Animal glue. Animal glue (sometimes called hideglue) is made from blood, hides, hide fleshings, and sinews of cattle, horses, and sheep. Most animal glues must be soaked and heated before using. It is also available in a ready-to-use liquid form.

Animal glue has good holding power but low water resistance, and dries or cures by absorption and evaporation. Animal glue was widely used until the development of synthetic resins.

Casein glue. Casein glue is of natural origin and is derived from milk curd combined with lime and other chemicals. Casein glue is available in powder form to which water is added prior to use. It is mixed to the consistency of thick cream and allowed to set for about 10 to 15 minutes before being applied to the wood surface. Casein glue may have a tendency to stain some types of wood. It is water resistant, and in addition, cures rather slowly and allows about 15 to 20 minutes assembly time before the bonding strength is impaired. Casein glue is used extensively for lamination work, veneering, and large carpentry structures.

Numerous other glues of natural origin have been developed over the years, including vegetable glues, soybean glues, and blood albumen. In most cases, the synthetic adhesives have proved superior to those of natural origin. For this reason, the natural glues are of less commercial importance.

SYNTHETIC ADHESIVES

Synthetic adhesives include those developed by modern chemistry and derived from other than natural substances. As stated, the synthetic adhesives are either thermosetting or thermoplastic.

Thermoplastic adhesives. The largest group of synthetic adhesives is the thermoplastic classification. A thermoplastic adhesive cures by means of evaporation and absorption and does not undergo a chemical change. The thermoplastic materials will soften when heat and/or water is added, and as a result, general thermoplastic adhesives are not satisfactory when exposed to high moisture or elevated temperatures. There is a vast array of synthetic thermoplastic adhesives. Only the most common will be discussed.

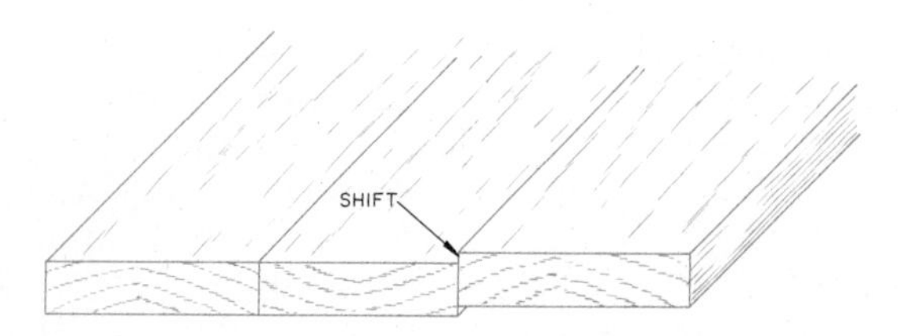

Fig. 13-23. An edge joint shift caused by inadequate gluing.

Polyvinyl resin emulsion glue has become widely known as white glue. It is available in plastic squeeze bottles in ready-to-use form, has good holding power, and can be removed from the clamps in about 1 to 2 hours. Polyvinyl is non-staining, economical to use, and produces a clear-colored glue line. It is commonly used as an adhesive when a non-water resistant glue is satisfactory. The glue line with polyvinyl glue tends to remain elastic which may cause slight movement of the glued joint when stressed, Fig. 13-23. This condition can be overcome by reinforcing the joint with splines or dowel pins. The "shifting" of an edge joint adhered with polyvinyl may be caused by either stress or expansion and contraction of the wood.

Polyvinyl glue cannot withstand high heat or excessive moisture. Excessive glue must be removed with a chisel or scraper before sanding, as it tends to soften due to the friction of sanding and "load" the abrasive paper. It also has limited assembly time, allowing only a few minutes.

Aliphatic resin glue has working characteristics similar to polyvinyl. The necessary clamp time is in the 1-hour range. The chief advantage of aliphatic resin is its resistance to heat, as it can be used in heat ranges up to 250°F. without joint failure. It also has resistance to shifting. Its disadvantage is that it is not water-resistant. This glue can be softened by the application of heat and addition of moisture.

Contact cement is widely used in the wood industry for the application of plastic laminates and veneers. It comes in a ready-to-use liquid form and is yellowish in color. Its working characteristics are unlike those of any of the other adhesives. Both surfaces to be bonded together with contact cement are coated, the adhesive is allowed to dry until it is no longer tacky when touched with a piece of paper, and then the coated surfaces are placed in contact with one another, bonding immediately upon contact. Extreme care must be used when positioning the two parts, as they cannot be moved into place once contact has been made.

The alignment can be insured if strips of paper are placed between the two surfaces until the two pieces are properly aligned. A permanent bond results from pressing the two surfaces together with the hands — clamps are not required. Contact cement has high resistance to water and will withstand considerable heat.

Thermosetting adhesives. A thermosetting adhesive is one which undergoes a chemical change during the drying or curing stage. Once a thermosetting adhesive has cured, it cannot be softened by applying heat or moisture. Various types of thermosetting adhesives are discussed in the following paragraphs.

Plastic resin adhesive is available in powdered form and is mixed with water to prepare for use. It is widely used for cabinetmaking and veneering work. The bond is highly water-resistant. One of the chief advantages of plastic resin is that it allows 10 to 15 minutes assembly time for complicated assemblies before it must be clamped. It is non-staining and produces a light brown glue line. Plastic resin can be cured using an electronic glue welder.

Resorcinal resin glue is a waterproof adhesive ideal for marine and exterior use. It consists of two parts packaged in separate containers. The resin portion is liquid and the catalyst may be either liquid or powder. The mixture is dark red and produces a very dark glue line. The resorcinal resin glues are generally more costly than most others, but they allow ample time for assembling and clamping of complicated shapes.

Urea resin glue is available in powder and dry form. The powdered form contains a catalyst which is activated when water is added, and mixes to a creamy consistency. Urea resin glue produces a light tan glue line when cured. It is one of the most successful glues to use with high-frequency glue welders.

This glue is highly water-resistant, and is widely used for hot-press curing such as veneering and plywood fabrication.

Urea resin sets up slowly and works well when longer assembly times are required. The lid must be replaced tightly on the can, as it tends to absorb moisture from the air which shortens its shelf life.

Epoxy resin is often used in woodworking to bond wood to a dissimilar material. Epoxy can be used to bond wood to metal, plastic, glass or nearly any type of material. It is especially useful for attaching hinges or other hardware when mechanical fasteners cannot be used. Epoxy resin is waterproof. It is available in two parts in liquid form. One part is the resin and the other part is the catalyst. It is not satisfactory for bonding large wood surfaces to other wood surfaces due to the cost.

Table 5 on page 321 will be helpful in selecting adhesives. The major uses and characteristics for common adhesives are given.

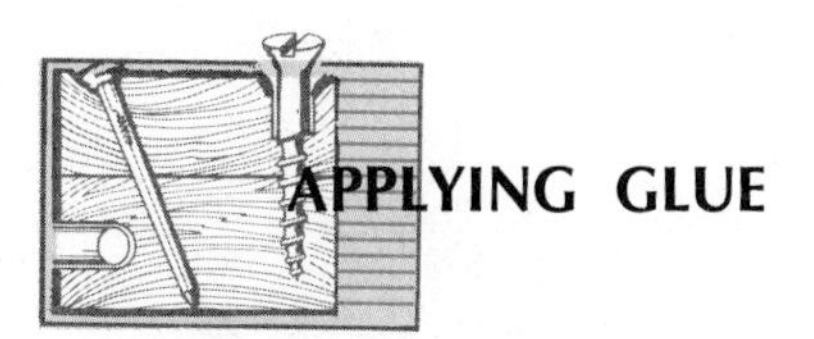

APPLYING GLUE

Ample glue must be applied to the stock. It is best to put a light coating on both surfaces. A brush can be used for applying most glues on small surfaces; however, a small rubber roller is used to coat large surface areas. The roller is especially helpful for applying adhesive to veneer.

Excessive glue will be forced out when the two surfaces are clamped. It is important not to over-apply the glue, but surfaces which do not have ample glue will result in what is called *starved joints.* Excessive clamping pressure may force all the glue out and result in a starved joint.

ELECTRONIC GLUE CURING

Certain types of thermosetting adhesives can be cured with a high-frequency glue welder, Fig. 13-24. The electronic welder creates or generates high-frequency energy which passes through the electrodes in the hand gun. The passage of the energy causes the material near the electrodes to become warm, and causes the thermosetting glue to cure more rapidly. The hand gun is equipped for interchanging electrodes, Fig. 13-25.

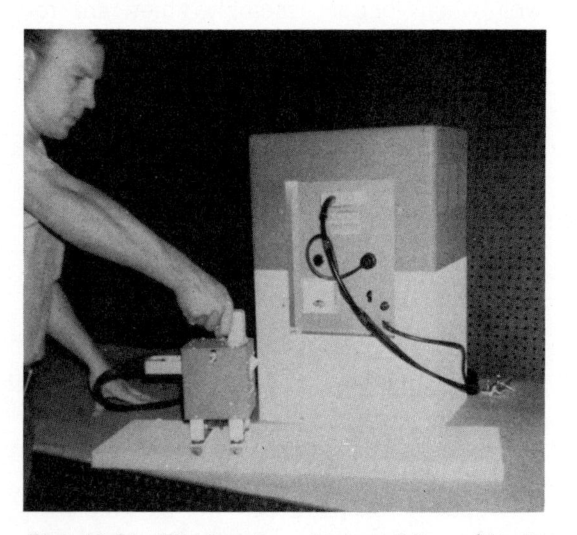

Fig. 13-24. Electronic glue welder. (Work-rite Products Co.)

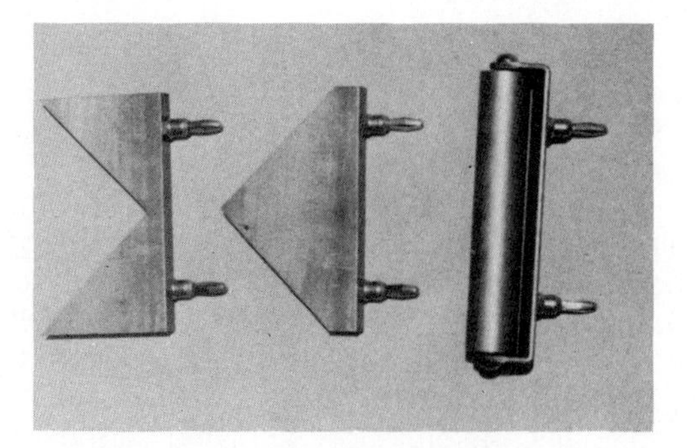

Fig. 13-25. Electrodes of different shapes which conform to the work. (Work-rite Products Co.)

WOOD JOINTS

The effectiveness by which two or three pieces of wood are attached together is dependent upon the type of joinery construction. There are many different types of joints, and some are more suitable than others for a particular product. Even though many types of mechanical and chemical fasteners are available, the selection of the proper joint is of utmost importance.

There are three basic factors which will affect the selection of the type of joint: (1) *strength,* (2) *appearance,* and (3) *simplicity of construction.* Each of these factors must be considered when selecting a joint for fabricating a product. The design of a joint may be very simple or complex. Some joints are interlocking and do not require mechanical or chemical fasteners to hold them together. The strength of a joint is directly dependent upon the amount of wood in contact, and the type of joint design can increase the amount of contact.

LEVEL OF RESTRICTION OF MOVEMENT OF WOOD JOINTS

Joints can be classified as to the amount of restriction. The simplest joint has restriction in only one direction, whereas others may have restriction in all

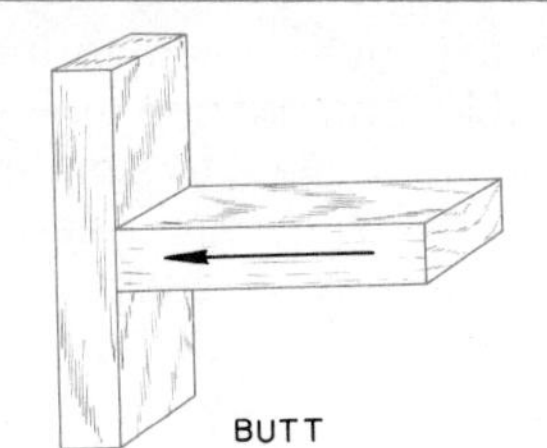

A. One level restriction of a butt joint.

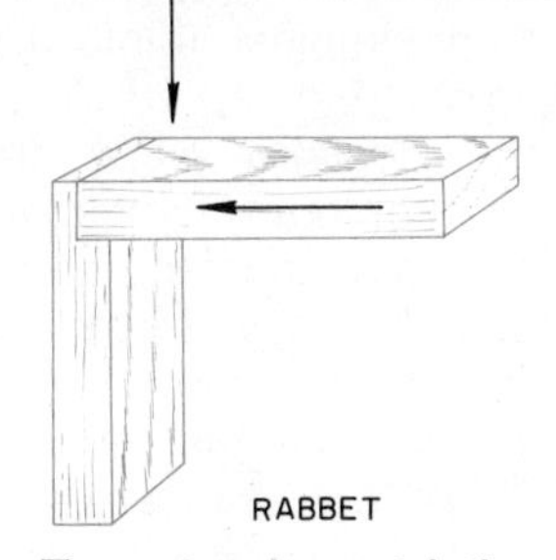

B. Two level restriction of a rabbet joint.

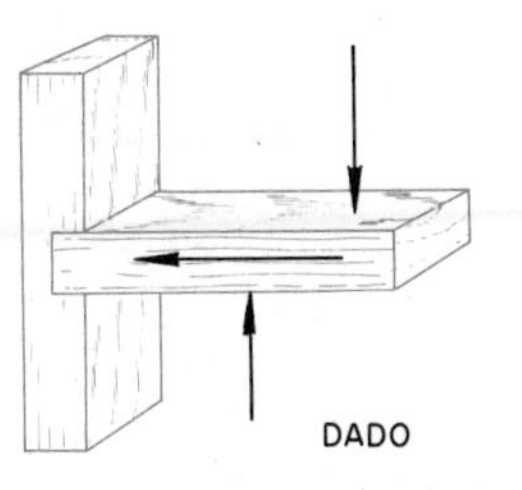

C. Three level restriction of a dado joint.

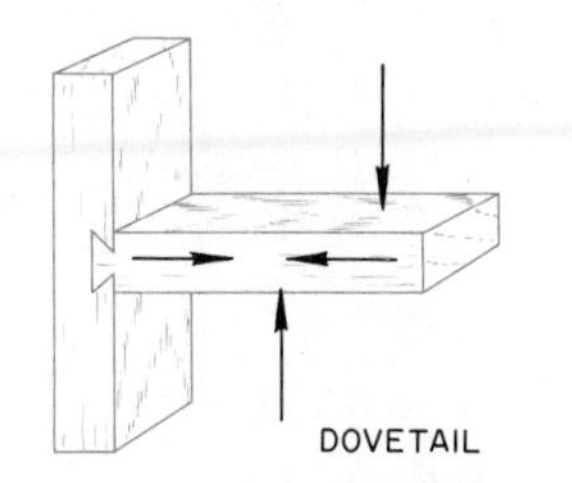

D. Four level restriction of a dovetail joint.

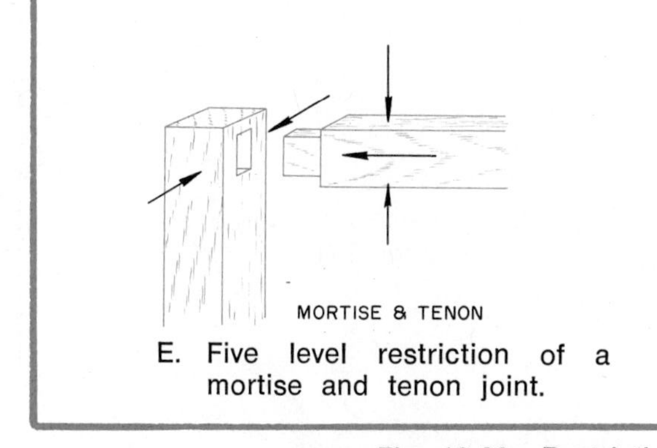

E. Five level restriction of a mortise and tenon joint.

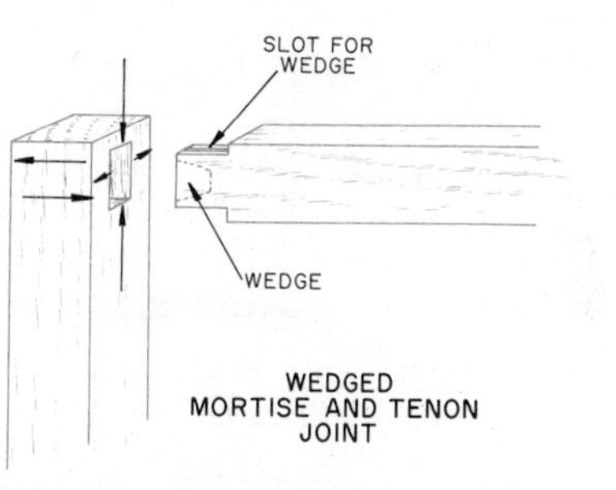

F. Six level restriction of a wedged mortise and tenon joint.

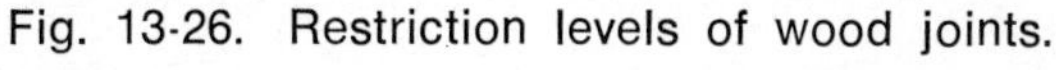

Fig. 13-26. Restriction levels of wood joints.

six directions. Generally, the greater the restriction, the greater the strength, Fig. 13-26.

A one-level restriction on movement of a joint has resistance to movement in only one direction, Fig. 13-27. The restriction is in the direction of the joining surface. As the restriction on the movement increases, the strength generally increases. The more restrictive joints are usually also more difficult to construct.

The woodworking industry must be concerned with all three factors of joint

Fig. 13-27. A joint with one level restriction has resistance to movement in only one direction.

selection. The simplest joints are the *butt* and *edge* joints, Fig. 13-28. The butt joint is used when a relatively low-strength joint can be used. For many types of work, the appearance is not satisfactory as the end grain is exposed. The butt joint is not strong because end grain does not hold mechanical or chemical fasteners well.

An edge joint is used to increase the width of stock, and is commonly used for box, table top, or similar simple construction, Fig. 13-29. They are often reinforced with dowel pins or a spline, Fig. 13-30. Wood which is used for edge joining to form a larger piece should not be over 4″ wide. If the stock is wider than 4″, it should be ripped into narrow strips. When the stock is reglued, the annual rings should

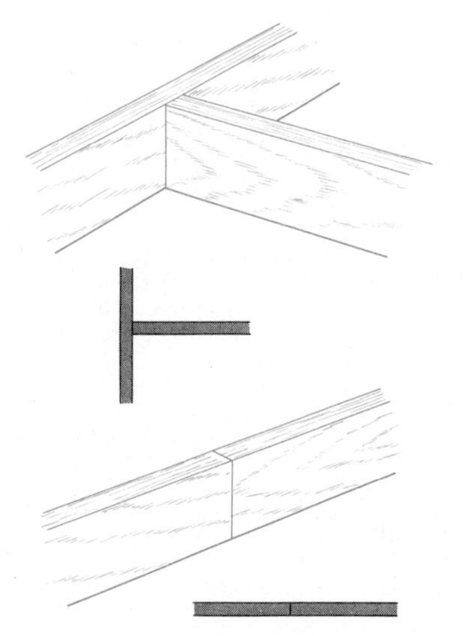

Fig. 13-28. Edge and butt joints have one level restriction.

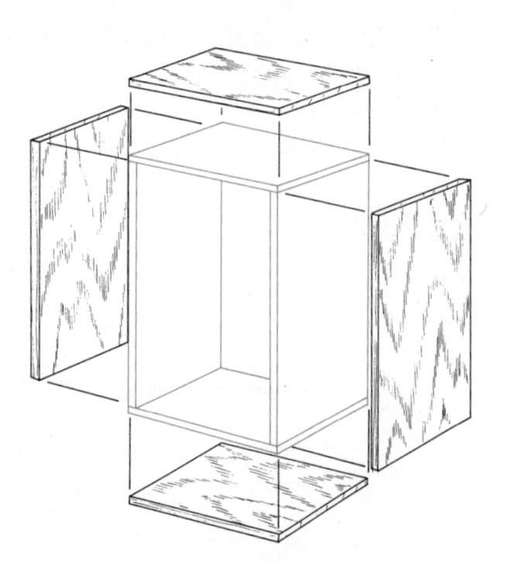

Fig. 13-29. Edge joints used to glue the tops and sides of simple construction.

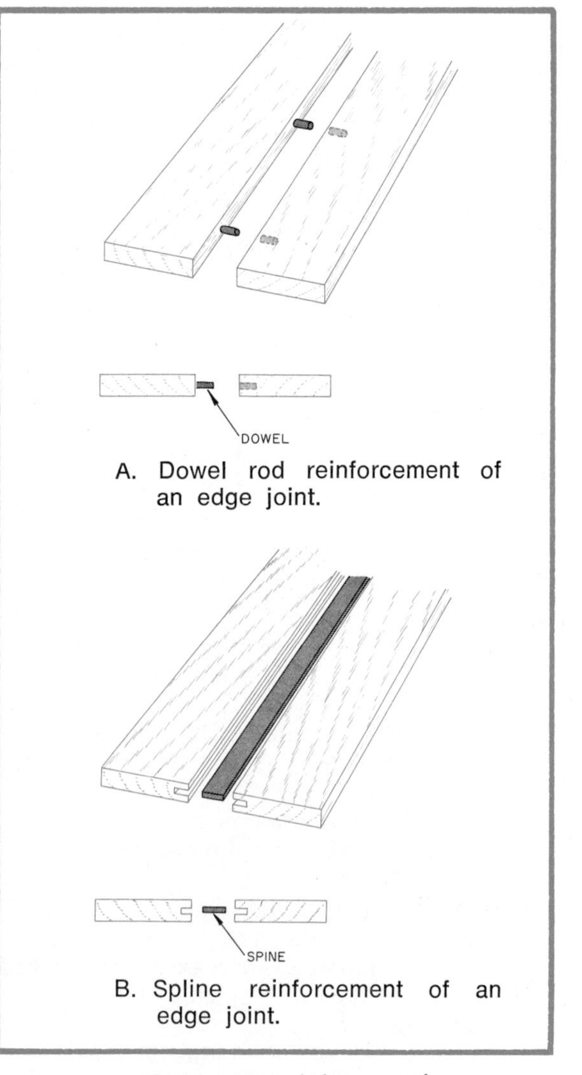

A. Dowel rod reinforcement of an edge joint.

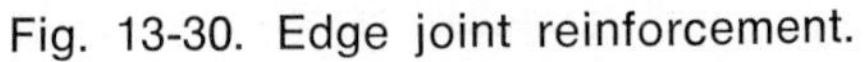

B. Spline reinforcement of an edge joint.

Fig. 13-30. Edge joint reinforcement.

be alternated, to help prevent warpage and cupping of the surface, (Fig. 13-31).

A *rabbet* joint is a typical class two joint based on the level of restriction, Figs. 13-32 and 13-33. It is one of the most commonly used in wood construction. The surface area for attachment is increased in comparison to a butt joint. The rabbet joint is commonly used in furniture construction, is a relatively simple joint to construct with hand tools or machines, and the rabbet may be cut on either the end or edge of the stock, minimizing the amount of exposed end grain.

Dado joints are restricted in movement in three directions, Fig. 13-34. The dado joint is a strong joint and is not difficult to construct. It is used in cabinet construction. A plow is also like a dado but is placed on the face of the stock, Fig. 13-35.

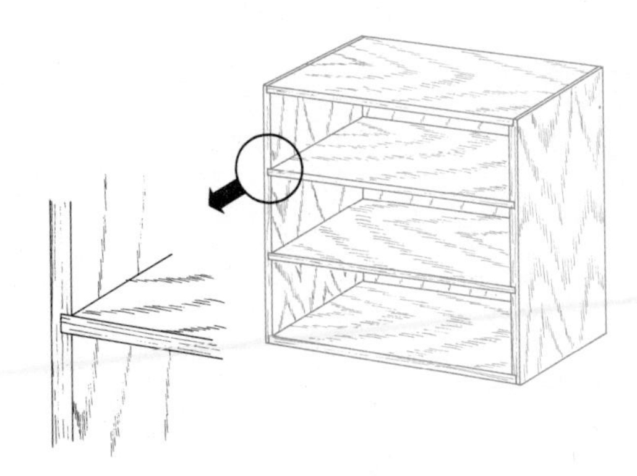

Fig. 13-34. A dado joint and its use in cabinet construction.

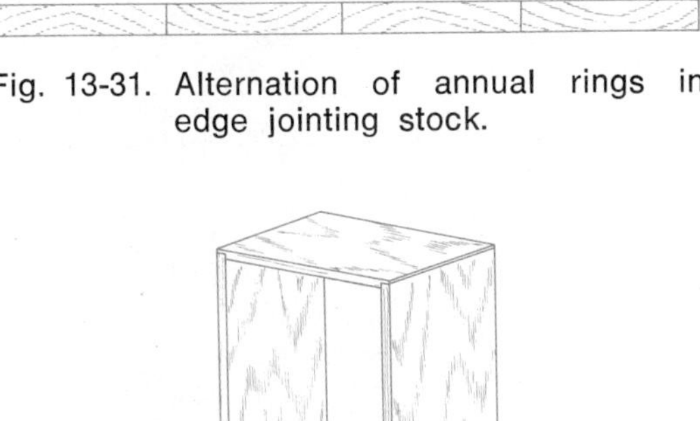

Fig. 13-31. Alternation of annual rings in edge jointing stock.

Fig. 13-32. A rabbet joint is a typical two level restriction joint.

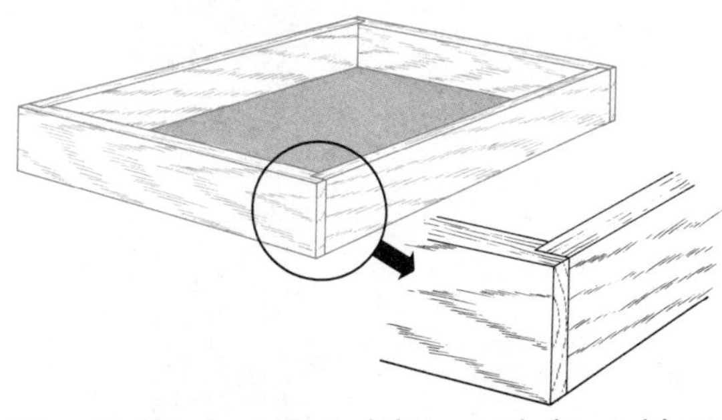

Fig. 13-33. A rabbet joint used in cabinet or drawer construction.

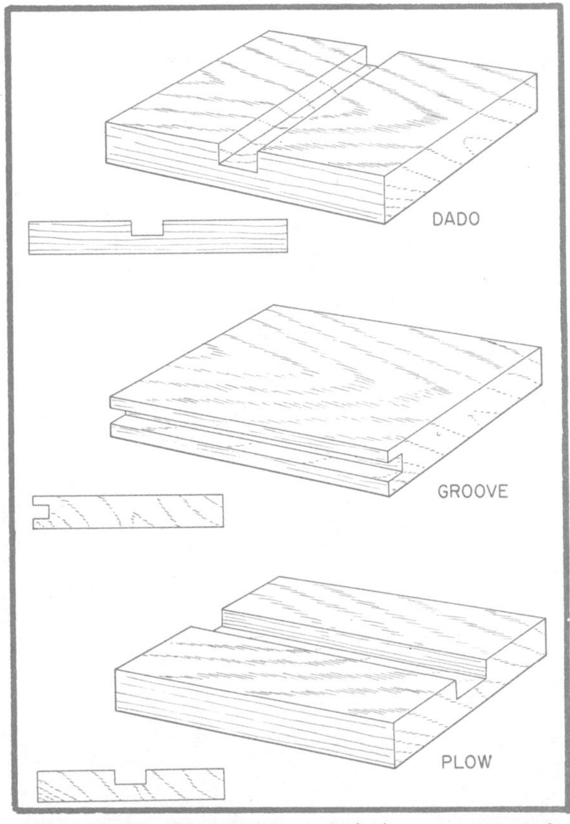

Fig. 13-35. Comparison of dado, groove, and plow cuts.

A mortise and tenon joint, Fig. 13-36, is used in the construction of fine furniture. It can be classed as a level five joint, as the only direction in which it can move is outward.

Fig. 13-36. A mortise and tenon joint and its use in table construction.

There are numerous modifications of the basic joints, many of them designed for a specific purpose, Fig. 13-37.

CONSTRUCTION OF COMMON JOINTS WITH HAND TOOLS

Butt joints. The butt joint is the simplest to construct. It can be fastened with nails, screws, glue, or a combination of fasteners. The first step in the construction of a butt joint is cutting the stock square. The layout should be placed on the face and edge as shown in Fig. 13-38. A try square is used to lay out the cutting lines. The end of the stock may be squared with either a handsaw or a backsaw, however, a backsaw is best for cutting small pieces, Fig. 13-39. A straightedge clamped

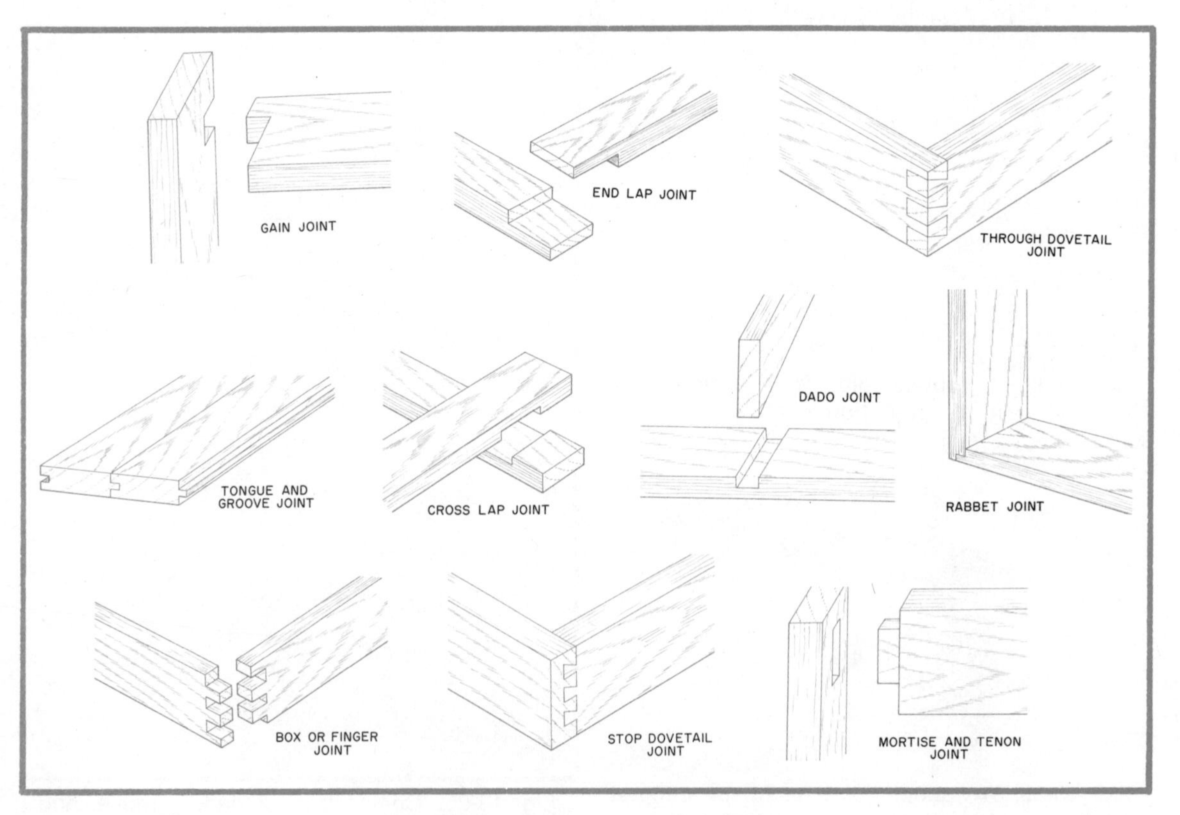

Fig. 13-37. Common wood joints.

along the layout lines will help insure a square cut. Hold the saw firmly along the straightedge and follow the vertical line on the edge of the stock.

Edge joints. The edges of the two pieces being joined must be placed square and straight, Fig. 13-40. Both pieces may be placed in a bench vise and planed at the same time. The joints can be fastened by means of adhesive or mechanical fasteners or a combination of the two. The most common mechanical fastener used is the dowel pin, which is available in numerous types and sizes, Fig. 13-41.

Dowel joints. The dowel joint is often used to strengthen many kinds of common joints. The insertion of the dowel pin insures proper alignment and adds strength to the joint.

A doweling jig, Fig. 13-42, is used to align holes in the stock. The positions of

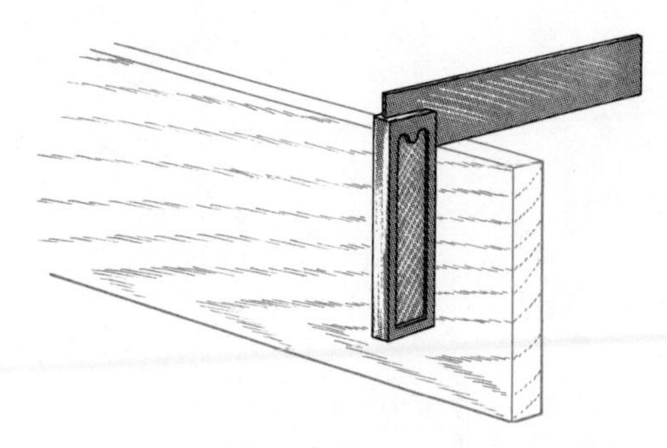

Fig. 13-40. Checking the edge of stock for squareness for an edge joint.

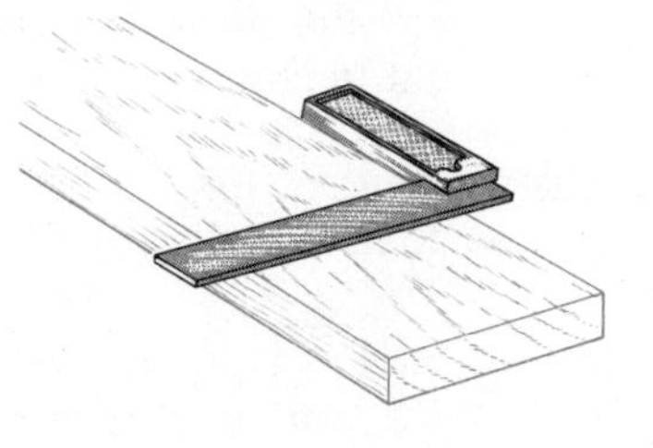

Fig. 13-38. A try square is used to mark the layout for squaring the end of stock.

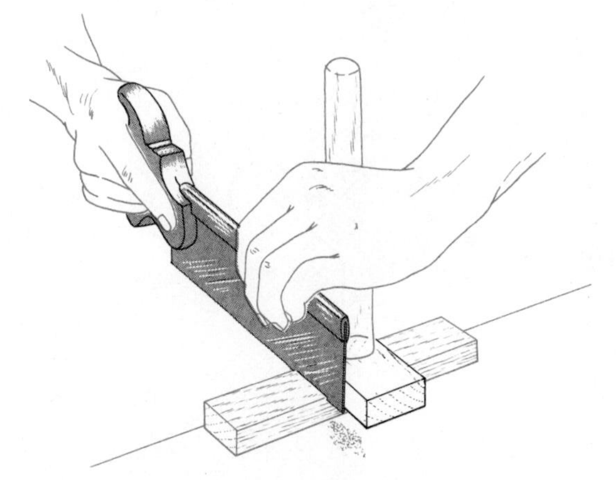

Fig. 13-39. Squaring stock with a backsaw and a straightedge.

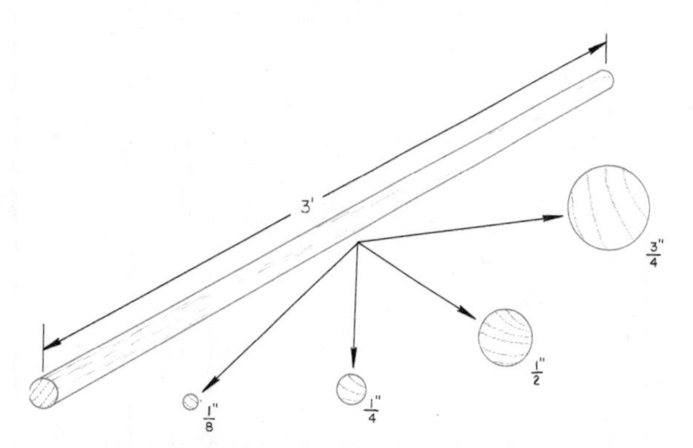

Fig. 13-41. Dowel pin sizes commonly used as fasteners.

Fig. 13-42. Doweling jig. (Stanley Tools)

the dowels are first located on the stock. The holes are bored into each piece an equal distance, Fig. 13-43. The size of dowel pin used is generally one half of the diameter of the smallest piece being joined. Dowels 3/8″ are often used for doweling stock 3/4″ to 1″ in thickness. The boards to be edge joined with dowels are placed in a vise. A try square is used to square lines across the edge of both boards. The doweling jig is adjusted to center the dowel on the layout line. The holes are drilled to a depth of about 1″ to 1 1/4″ in each piece. A depth gauge attached to the bit controls the depth of the hole, Fig. 13-44.

The dowel joint is often used in leg and rail and frame construction, Fig. 13-45. Procedure for placing dowels in all construction joints is very similar. When a doweling jig cannot be used, the location of the holes can be transferred from the working drawing. Two lines which intersect are used to locate the dowel pin position, Fig. 13-46. A slight indention with a scratch awl on the center of the layout lines will insure greater accuracy for boring; special care must be taken to bore the holes straight.

Regardless of the type of dowel joint, the stock must be square. Any irregularity in squareness will not produce a tight-fitting joint.

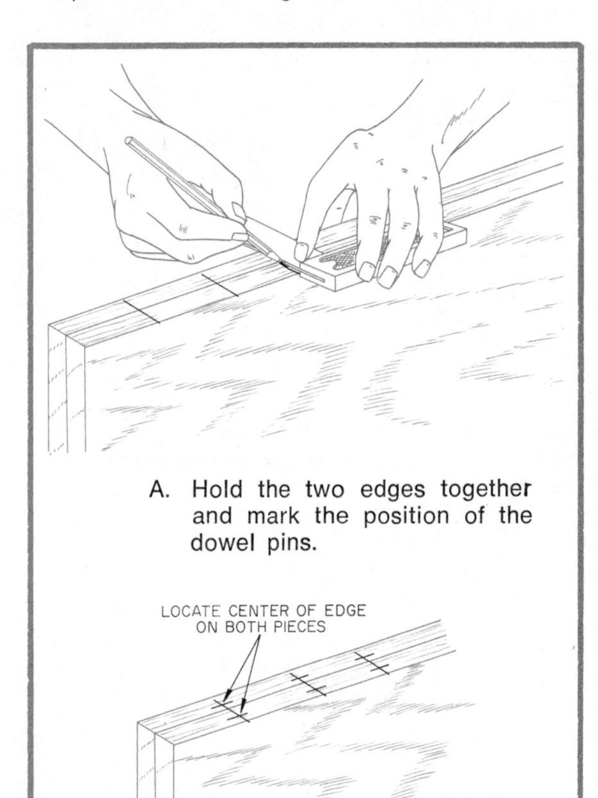

A. Hold the two edges together and mark the position of the dowel pins.

B. Locate the center of the edge of both pieces at each dowel position.

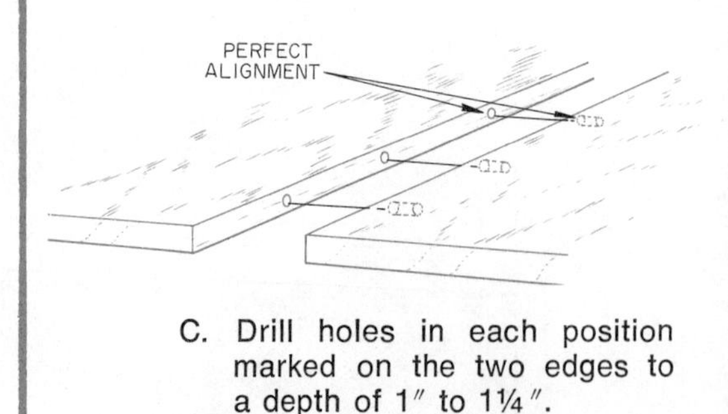

C. Drill holes in each position marked on the two edges to a depth of 1″ to 1 1/4″.

Fig. 13-43. Layout and placement of dowel pins for edge joints.

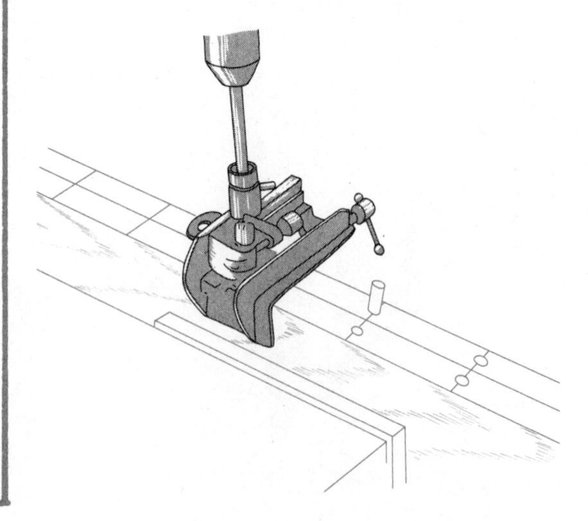

Fig. 13-44. Cutting dowel pin holes using a doweling jig and a depth gauge.

Dowels are often used to reinforce miter joints. The dowels can be placed in either of two ways as shown in Fig. 13-47. If the appearance of the joint is not important on one side, the dowels can be drilled from that side. Otherwise, a doweling jig can be used to align the holes which are bored only part way through the stock. The location of the dowel pins is marked on

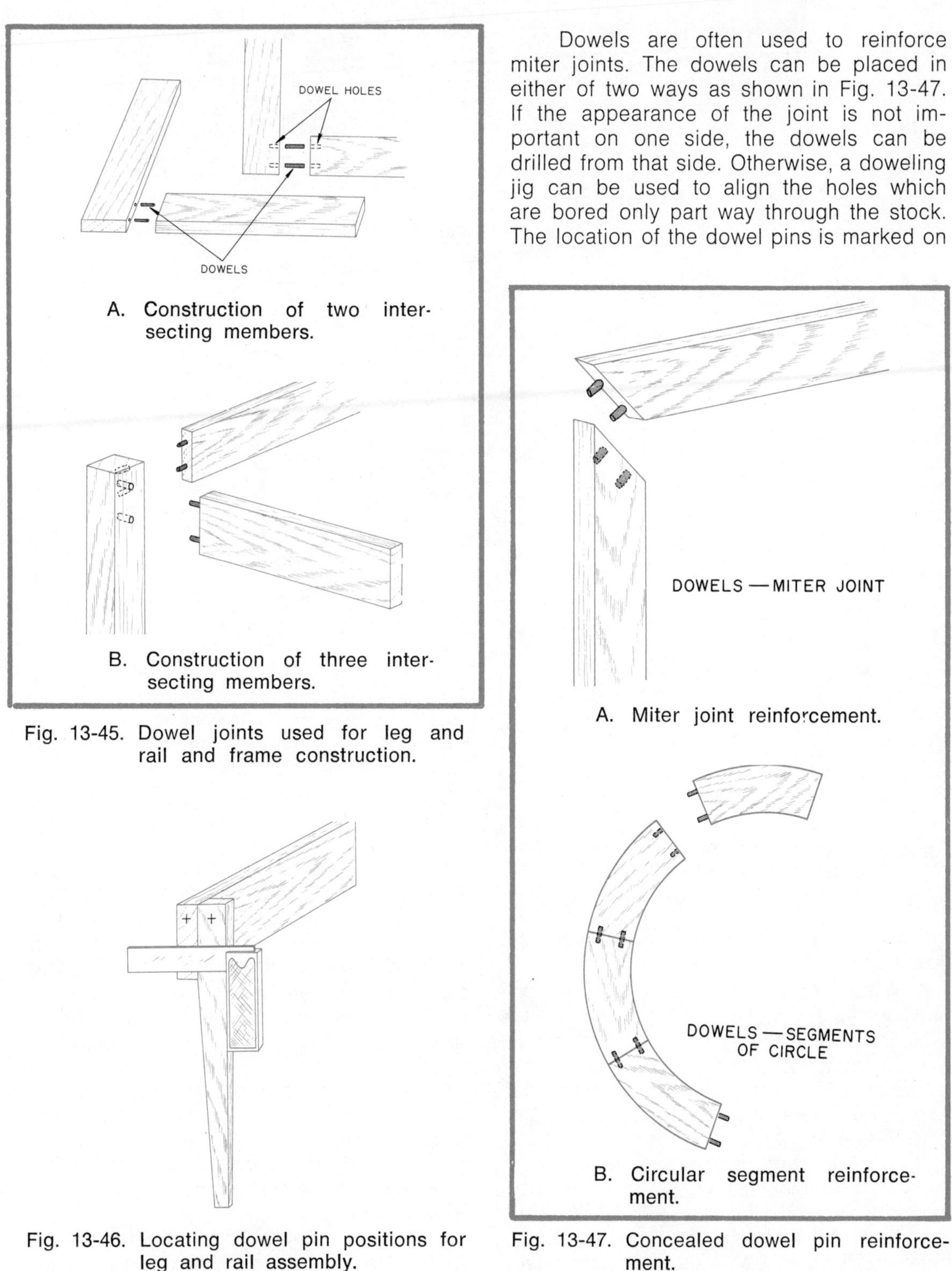

A. Construction of two intersecting members.

B. Construction of three intersecting members.

Fig. 13-45. Dowel joints used for leg and rail and frame construction.

Fig. 13-46. Locating dowel pin positions for leg and rail assembly.

A. Miter joint reinforcement.

B. Circular segment reinforcement.

Fig. 13-47. Concealed dowel pin reinforcement.

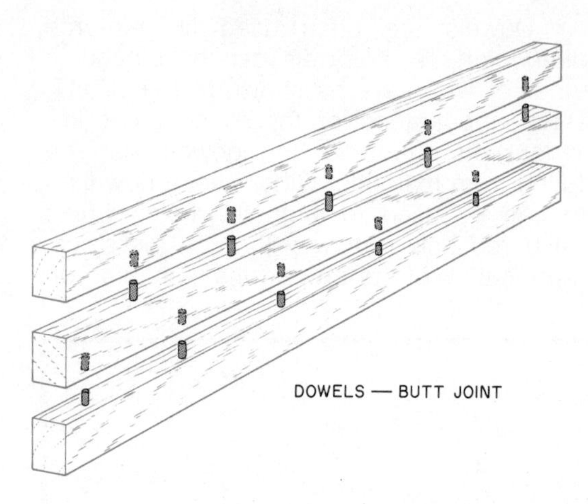

Fig. 13-48. Locating dowel pin holes for concealed reinforcement of a butt joint.

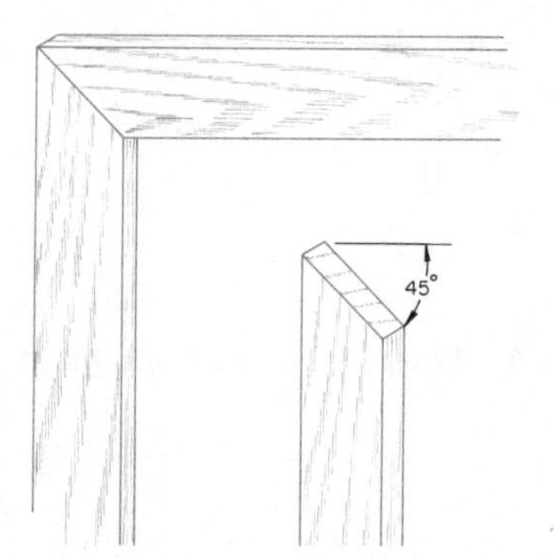

Fig. 13-49. Simple 45° miter cuts joined together to form a 90° corner.

the stock. The holes are bored with a doweling jig, Fig. 13-48.

Miter joints. The miter joint is often used when it is not desirable to have end grain exposed. A typical example is for a picture frame. A simple miter joint is one in which two pieces of stock cut at 45° angles join together and form a right angle or 90° corner, Fig. 13-49.

A miter joint can be cut with hand tools in a number of ways. The simplest means is with a miter box, Fig. 13-50. The miter box can be set for any desired angle between 30° and 90°. Automatic stops are

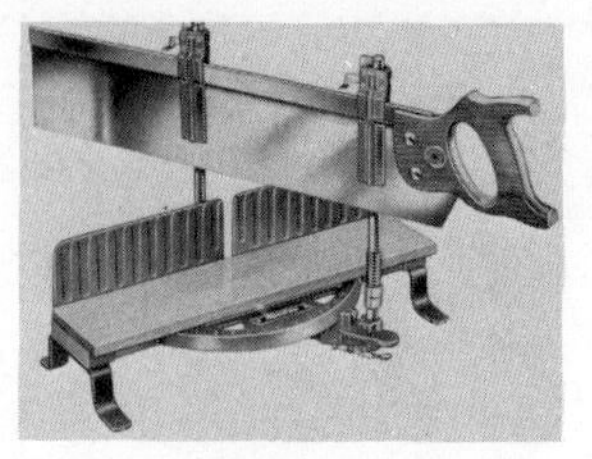

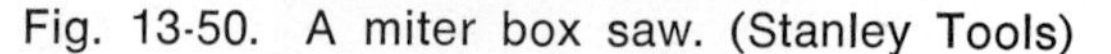

Fig. 13-50. A miter box saw. (Stanley Tools)

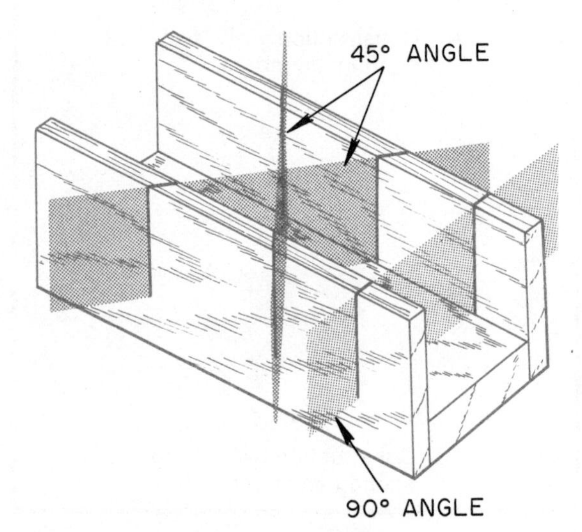

Fig. 13-51. A shop-made miter box for cutting 45° angles.

placed every 5° on the adjustment. A simple miter box for cutting 45° angles can also be made in the shop, Fig. 13-51.

The 45° angle can also be marked on the stock using a combination square. The line can be carefully cut using a backsaw. A board clamped along the line for a straightedge helps to insure an accurate cut, Fig. 13-52.

Miter joints are not strong joints because they involve the joining of end grain; thus, mitered joints often require reinforcement. There are several methods of reinforcing miter joints, some of which are shown in Fig. 13-53.

Rabbet joints. The rabbet joint is among the most-used in the wood industry.

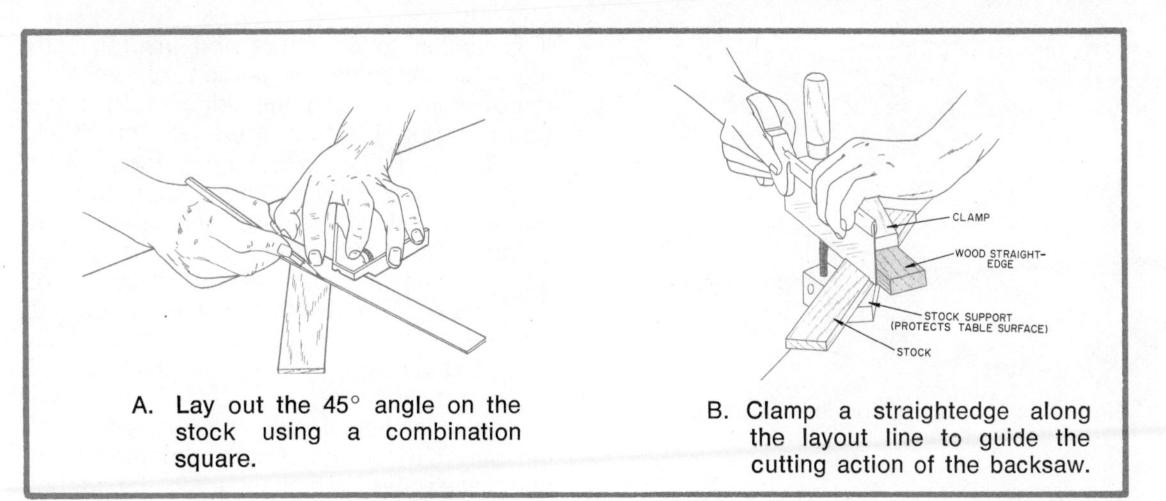

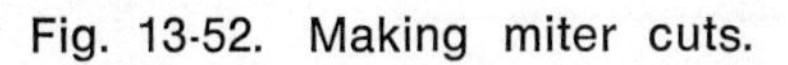

Fig. 13-52. Making miter cuts.

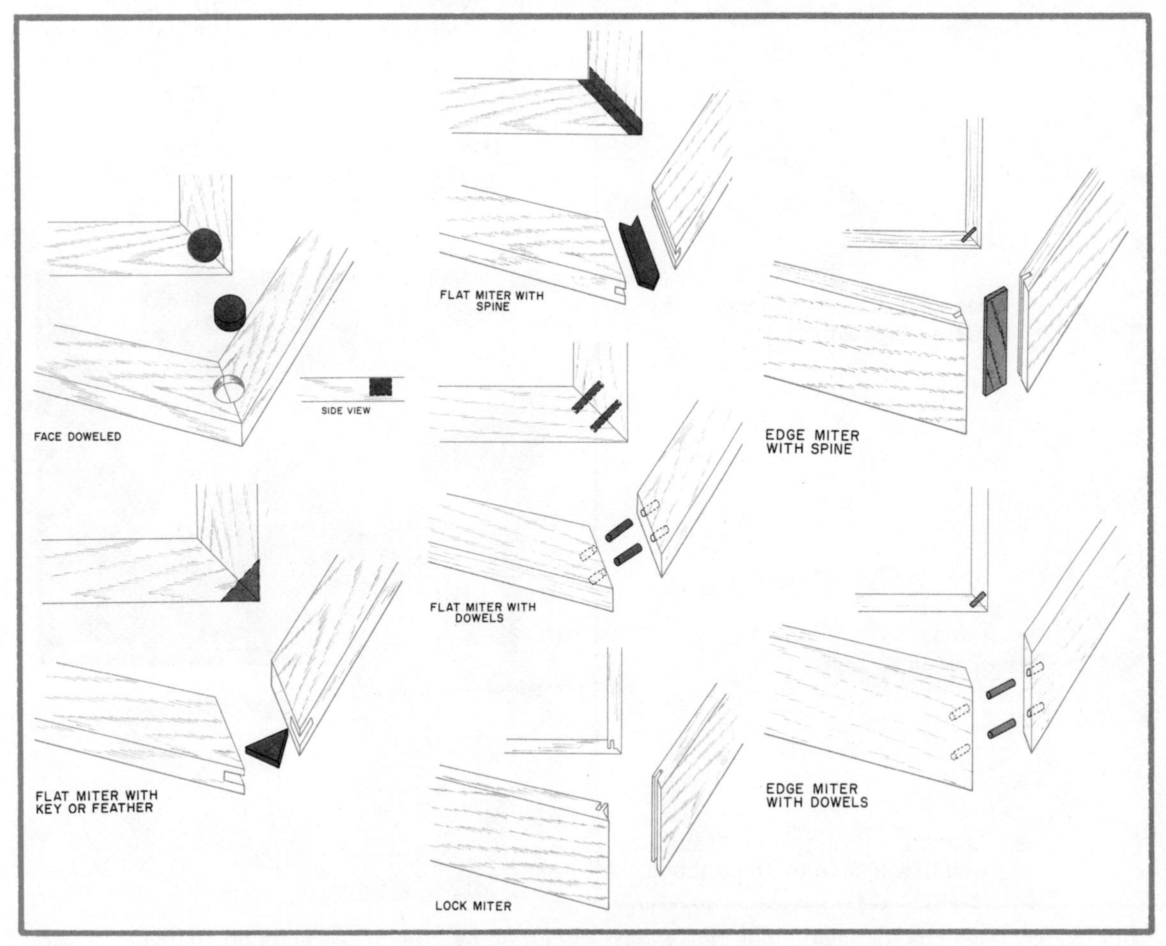

Fig. 13-53. Common methods of reinforcing miter joints.

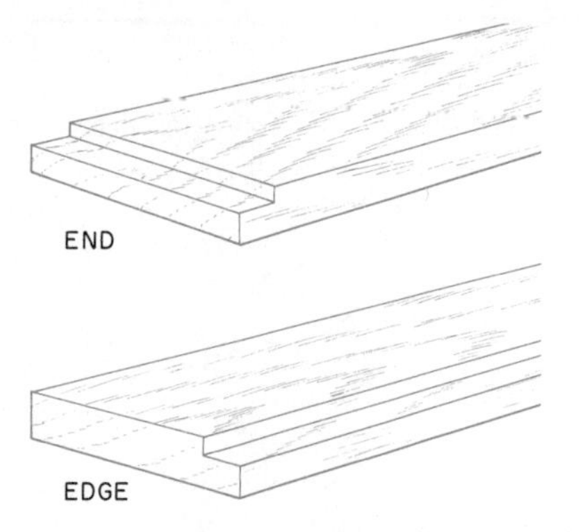

Fig. 13-54. A rabbet formed on the end and edge of stock.

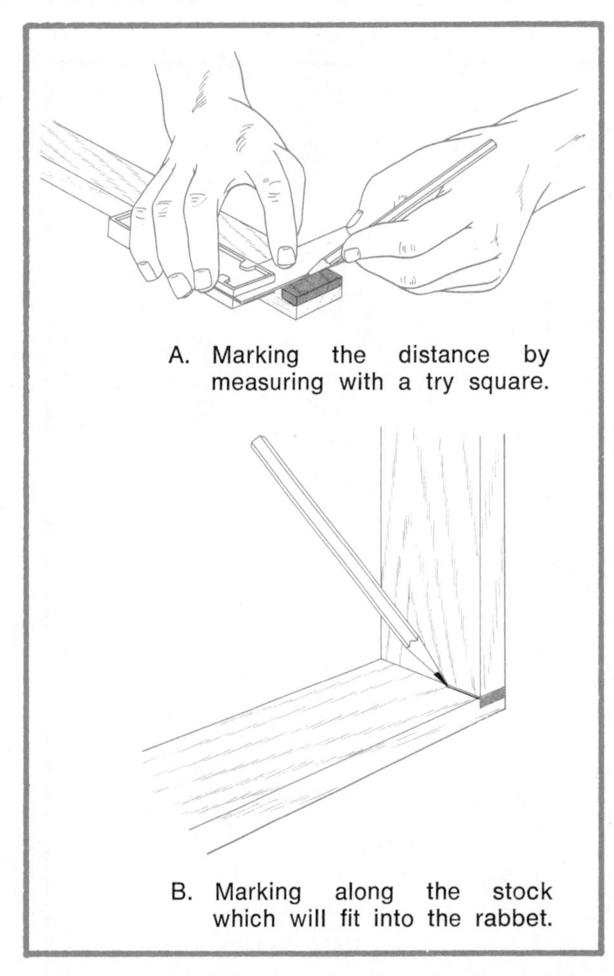

Fig. 13-55. Methods of laying out the width of a rabbet.

It is simple to construct and insures considerable strength. A portion of stock is removed from along the edge or from the end of a board to form a rabbet, Fig. 13-54.

The first step in cutting a rabbet joint is the proper layout. The layout can be made with a try square and a marking gauge. The depth of the rabbet is generally one-half to two-thirds the thickness of the stock. The width of the rabbet is marked on the face side of the stock. This can be done by measuring the thickness of the board which fits the rabbet. The measurement is then transferred to the face of the board to be cut. Another method is to place the board which is to fit the rabbet on the face surface and mark the desired width with a knife or sharp pencil, Fig. 13-55. The depth of the cut is laid out with a marking gauge, Fig. 13-56. The same procedure is used for marking a rabbet on the edge of the stock.

The stock which forms the recess in the rabbet cut can be removed by any

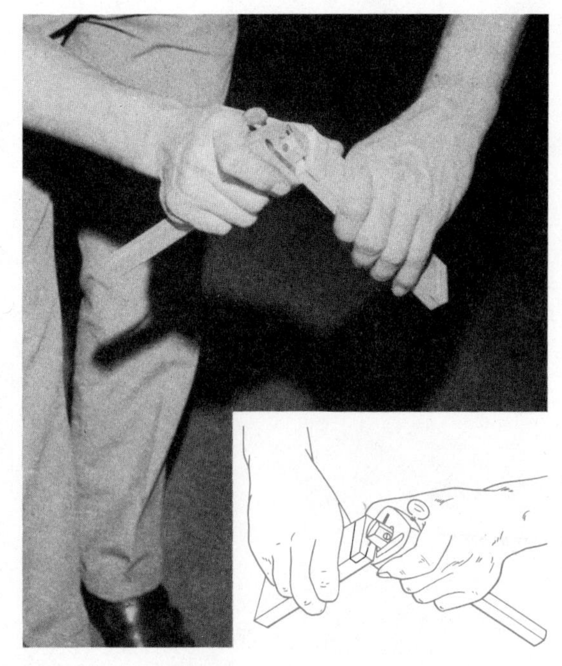

Fig. 13-56. Using a marking gauge to lay out the depth of the rabbet.

of several methods. A backsaw is generally used to cut the depth of rabbet. The board is held firmly to the bench using a clamp. The cut can be made with greater accuracy if a straightedge is clamped along the layout line.

Several saw kerfs made in the waste stock to the layout line will be helpful for removing the excess stock with a chisel, Fig. 13-57. A chisel the same width as the rabbet is best for removing the stock, Fig. 13-58.

The second cut of a rabbet can also be finished with a backsaw. When a saw is used, care must be taken to make the kerf in the waste stock.

A rabbet which is parallel to the grain along the edge of the stock can be cut using one of the methods previously described. However, it is often cut using a rabbeting plane. A rabbeting plane such as the one shown in Fig. 13-59 can be adjusted for width of cut by means of the adjustable fence and can also be adjusted for the desired depth. The plane iron is adjusted in the rabbet plane for a thin shaving. The depth gauge is set for the desired depth, Fig. 13-60. The fence is also

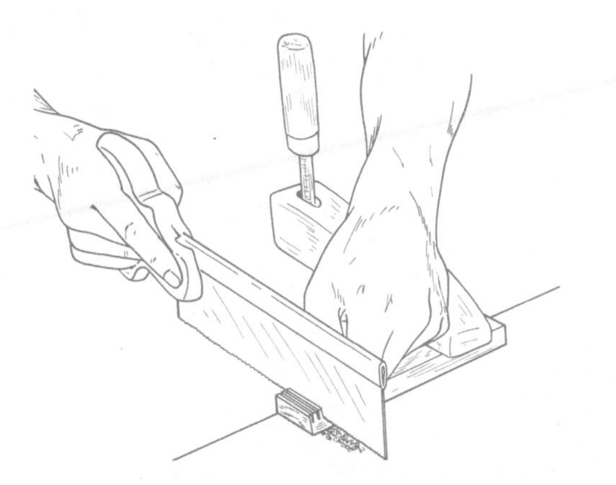

Fig. 13-57. Saw kerfs made in the waste stock to the layout line prior to cutting the rabbet.

Fig. 13-59. A rabbeting plane with an adjustable blade.

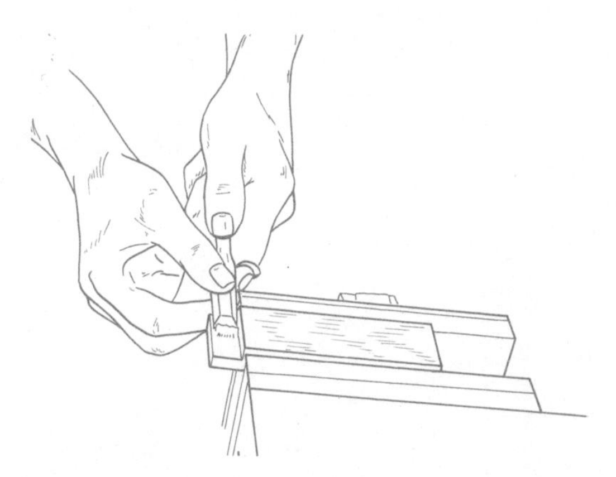

Fig. 13-58. Finish removing stock by using a chisel the width of the rabbet.

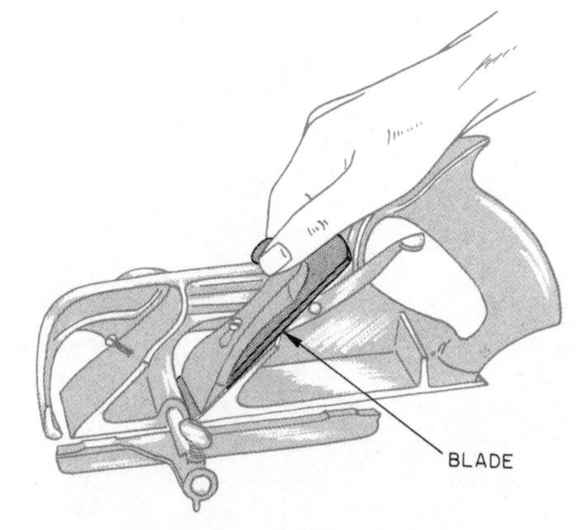

Fig. 13-60. Setting the depth of the rabbet cut on the plane.

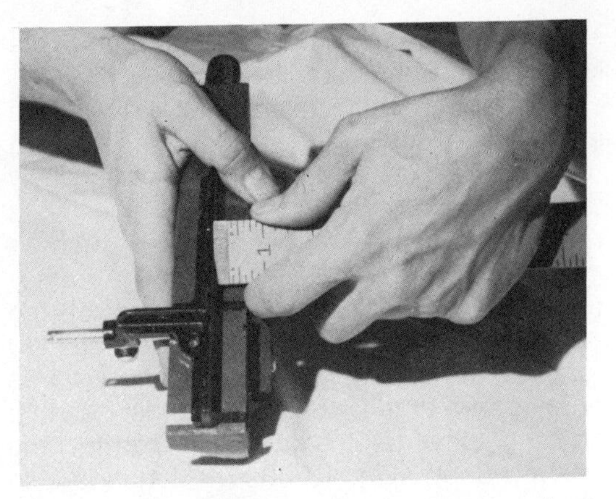

Fig. 13-61. Setting the fence of the plane for the correct width of the rabbet.

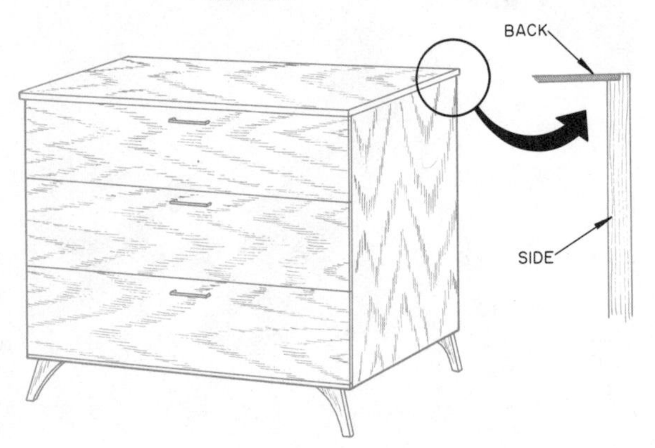

Fig. 13-62. The rabbet joint is used to install the back in many types of furniture.

set with a rule for the correct distance, Fig. 13-61.

The rabbet joint is widely used for construction and assembling of furniture, Fig. 13-62.

Dado joints. The dado joint is one of the most useful joints in the wood industry. It provides sufficient strength for most types of construction. The dado joint is cut across the grain and never touches the end of the stock.

A dado is quite simple to construct. It can be cut using a variety of methods and tools. The dado is cut to a depth of one-half to two-thirds the thickness of the stock. After the position for the dado is located, the layout is made with a try square. The layout lines are placed on both edges and the face on which it is cut. The width is cut so that the second piece fits into it snugly. The end of the second piece must be cut square to insert into the dado, Fig. 13-64. The outside shoulder cuts are made with a backsaw to the bottom layout lines. The kerf must be made in the waste stock. Extra cuts made in the center of the dado help in the removal of the excess stock, Fig. 13-65. The sawing operation may be more accurate if a straightedge is clamped along the layout line.

The excess stock is removed with a chisel, the cuts being made with the chisel from each side, Fig. 13-66. The last portion of the stock may be removed with a router plane to produce a flat, level bottom. The

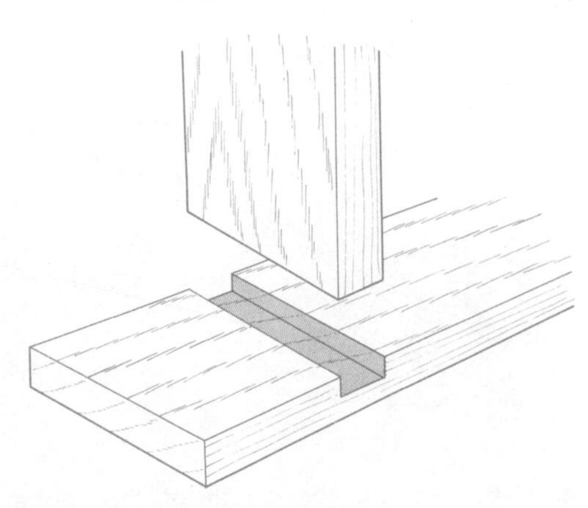

Fig. 13-63. A dado joint.

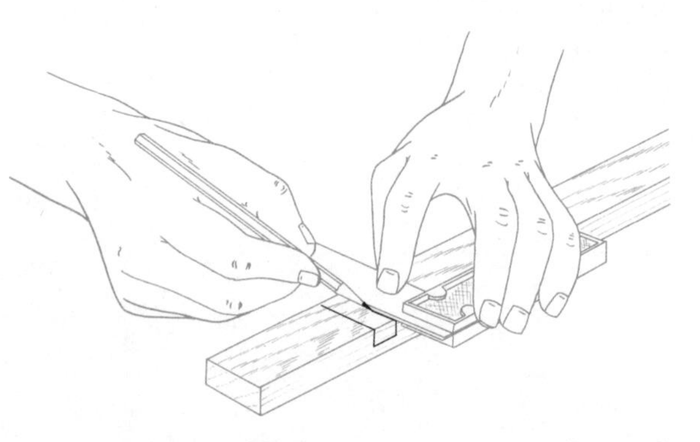

Fig. 13-64. Using the try square to lay out a dado.

cut should be made from both sides to avoid splitting out the edge of the board, Fig. 13-67.

An alternate method of laying out the dado is to use the second piece as a gauge. The dado is located as previously described and the first line is laid out with a try square. A straightedge is clamped along the first layout line. The second piece is placed against the straightedge and the other layout line is made. The depth of the dado is marked on both edges of the stock, Fig. 13-68.

When the layout is in the same position on two opposite pieces, they should

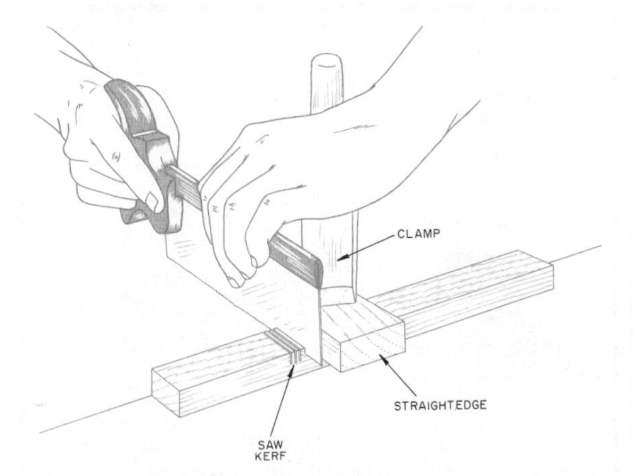

Fig. 13-65. Saw kerfs are cut in the waste stock with a backsaw guided along a straightedge clamped to the stock.

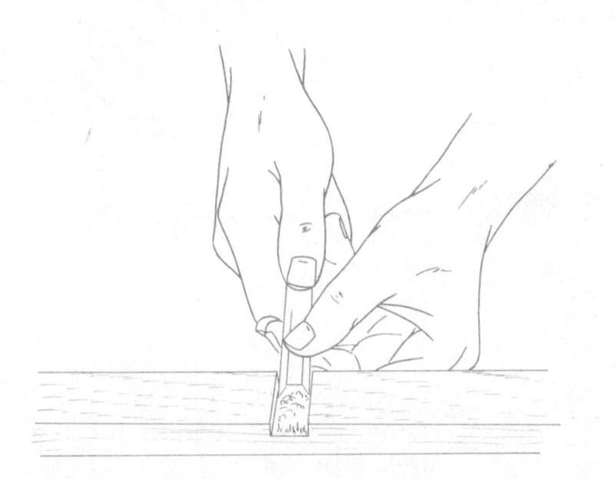

Fig. 13-66. Removing excess stock from the dado using a chisel.

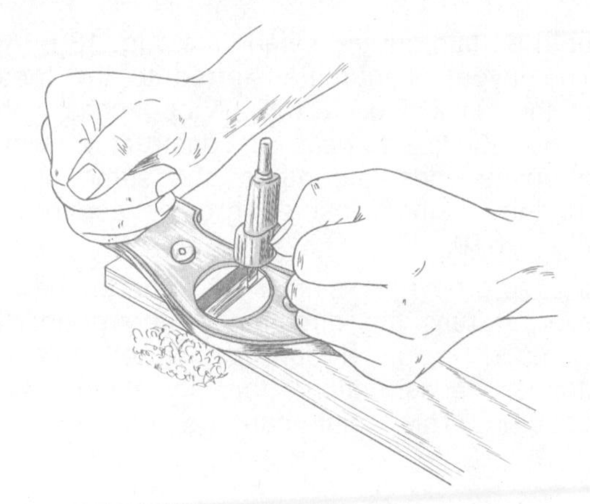

Fig. 13-67. Completing the dado joint with a router plane.

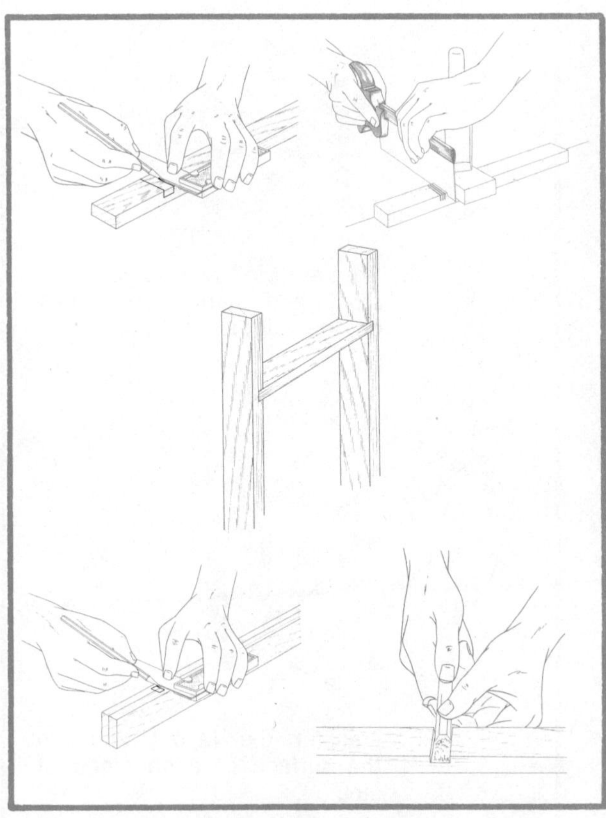

Fig. 13-68. Sequence for laying out and cutting a dado joint.

both be laid out the same time, Fig. 13-69A. The layout is then transferred to the face of the stock, Fig. 13-69B. A bookcase with dado joints is usually constructed with opposite ends laid out at the same time, as this method insures greater accuracy, Fig. 13-70.

A groove is similar to a dado; however, it runs the direction of the grain. It is used in panel door construction and drawer construction for installing the bottoms. This joint can be constructed

Fig. 13-70. Bookcase constructed with dado joints.

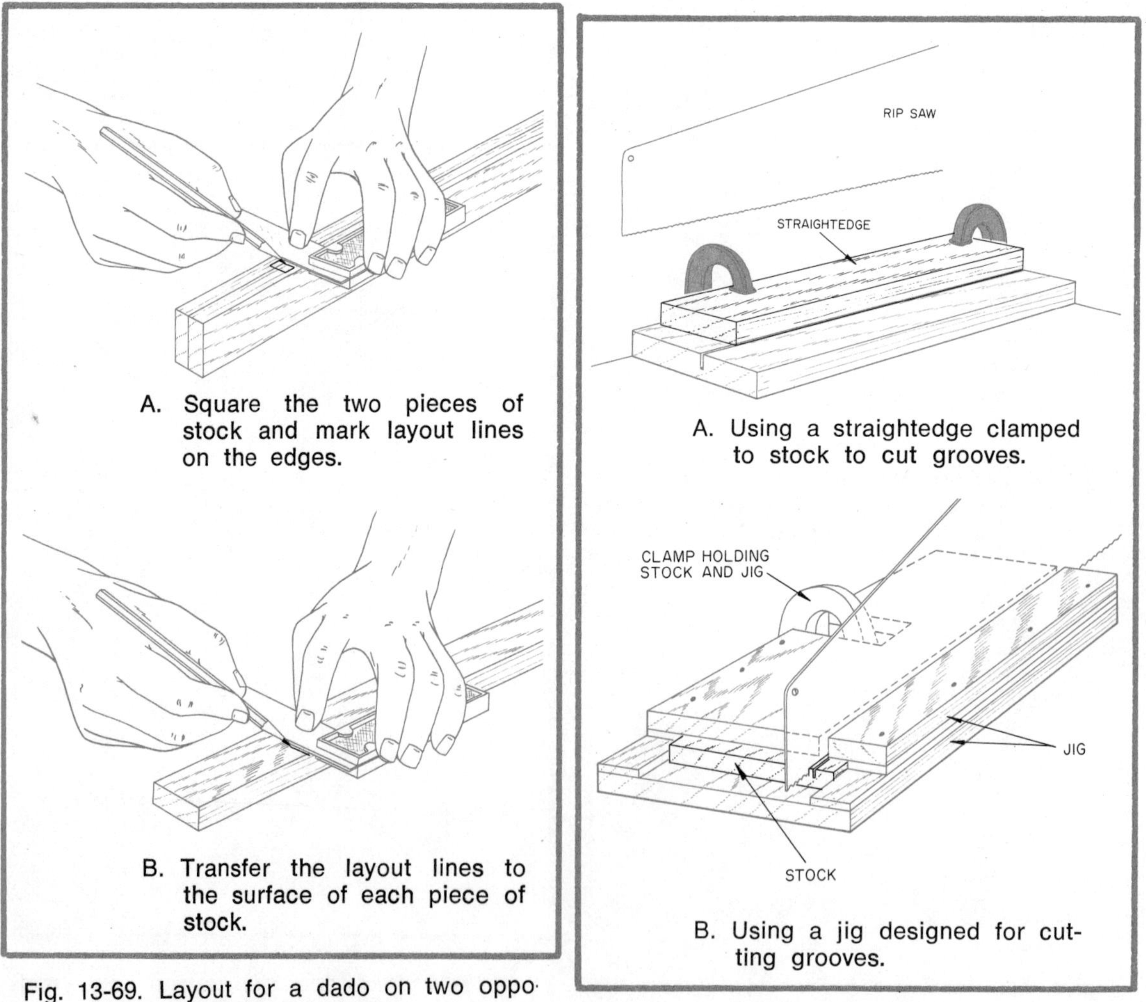

A. Square the two pieces of stock and mark layout lines on the edges.

B. Transfer the layout lines to the surface of each piece of stock.

Fig. 13-69. Layout for a dado on two opposite pieces of stock.

A. Using a straightedge clamped to stock to cut grooves.

B. Using a jig designed for cutting grooves.

Fig. 13-71. Cutting grooves with hand tools.

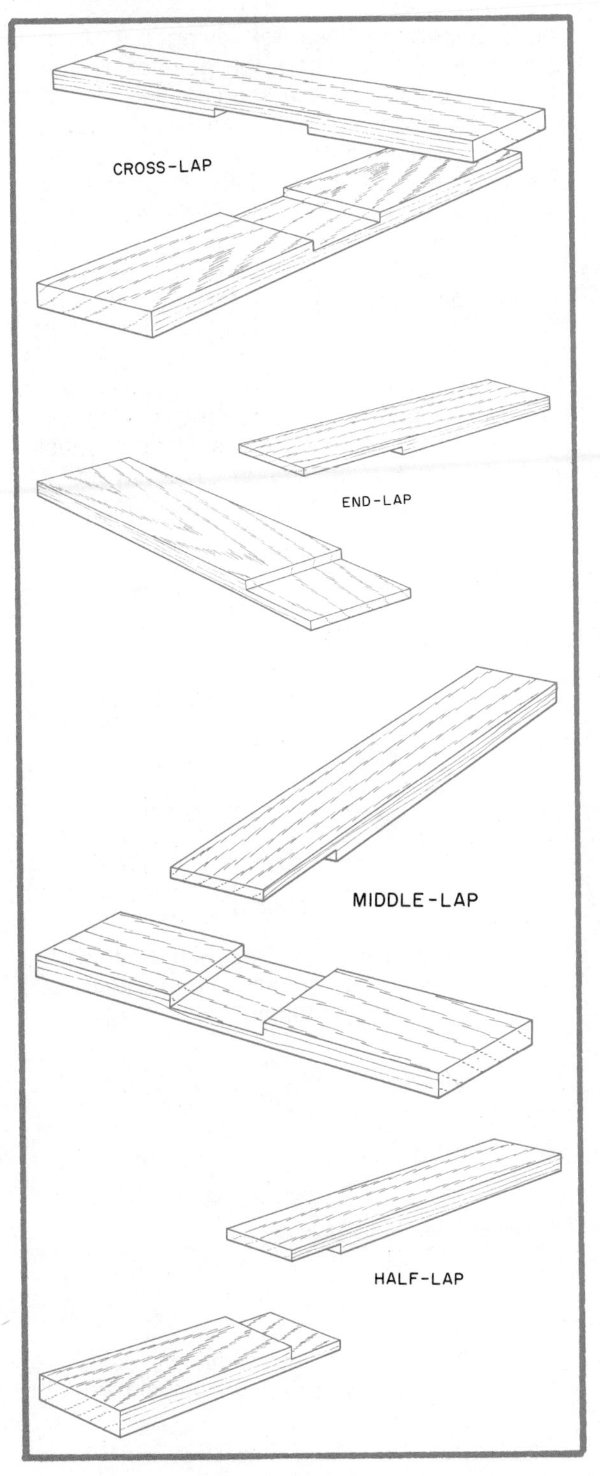

Fig. 13-72. Common types of lap joints.

with hand tools, although it is easier to construct using machines. The grooves are cut using much the same procedure as for dadoes. The layout parallel to the grain is made with a marking gauge. A straightedge or special jig such as shown in Fig. 13-71 can be helpful when grooves of a standard width such as 1/8″ or 1/4″ are cut. The excess stock is removed with a chisel or router plane, Fig. 13-74.

Lap joints. The lap joint is similar to a combination of a rabbet and dado joint. There are several types of lap joints such as shown in Fig. 13-72. Regardless of the type, one-half of the stock is removed from each piece. The end lap consists of two rabbet joints, the cross lap is made up of two dado joints, and a middle lap consists of a dado and a rabbet.

The procedure for cutting all laps is the same as for cutting a dado and rabbet. The layout may be done with a try square or the stock can be put in place and marked, Fig. 13-73. When this method is used, one line must be placed on each piece. This line locates the position and squares the two pieces to one another.

Mortise-and-tenon joints. The mortise-and-tenon joint is commonly used in the furniture industry. It is a very strong joint and is used in leg and rail types of table

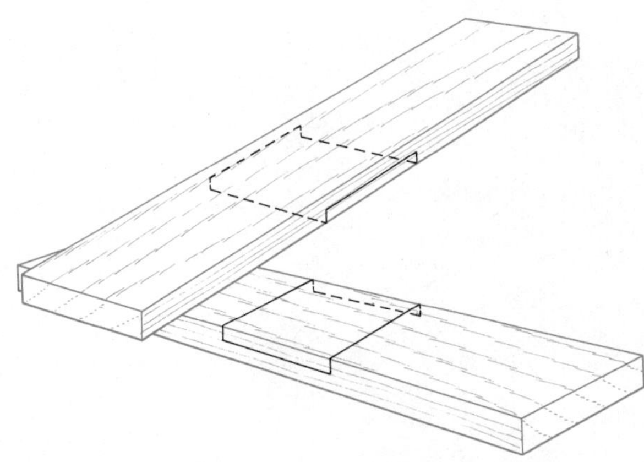

Fig. 13-73. Layout for a cross lap joint.

construction, Fig. 13-74. This joint consists of two parts which are naturally the mortise and the tenon, Fig. 13-75.

The mortise is cut before the tenon, and the tenon is cut to fit tightly in the mortise. In most types of construction such as the example shown in Fig. 13-74, several identical mortise-and-tenon joints are constructed. Therefore, the tenons and the mortises can be laid out on the stock at the same time.

After the stock in which the mortises are to be cut is squared and sized, it can be clamped together for laying out. A try square is used to mark the position for the top and bottom of the mortise, Fig. 13-76.

The excess stock can be removed from the mortise with an auger bit. The inside of the mortise is then squared using a chisel. An exact center line scratched with an awl will help to locate the bit for accurate boring. A depth gauge attached to the bit will insure uniform depth, Fig. 13-77. Sometimes the entire mortise is cut with a chisel.

The stock for the rails on which tenons are to be cut is squared and cut to the desired size first. Additional length must be allowed for the tenon. The layout is made with all of the pieces the same size

Fig. 13-74. A table constructed with mortise and tenon joints.

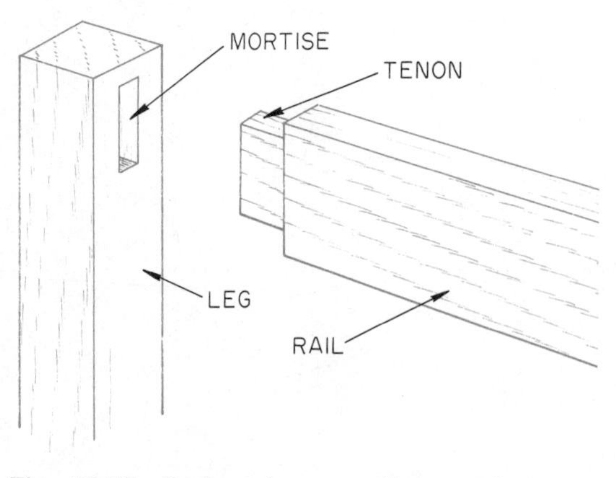

Fig. 13-75. Parts of a mortise and tenon joint.

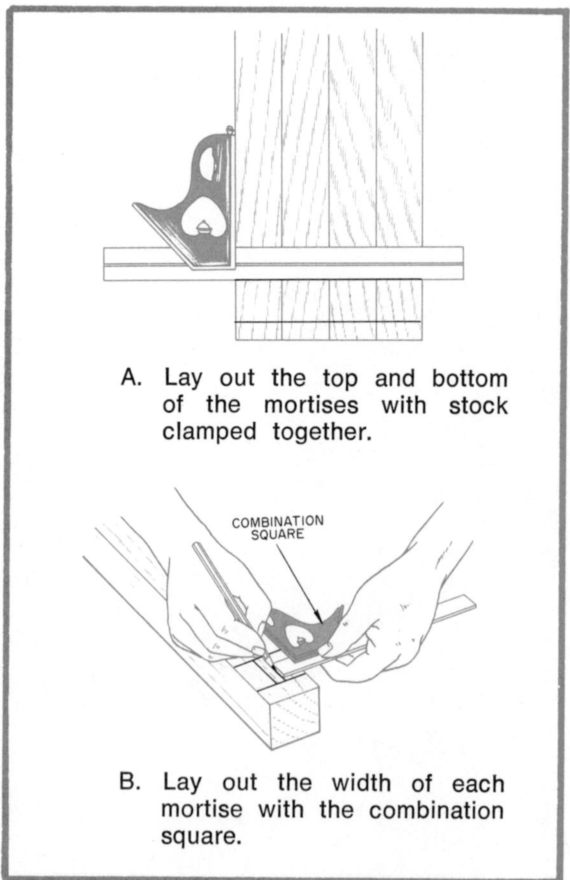

A. Lay out the top and bottom of the mortises with stock clamped together.

B. Lay out the width of each mortise with the combination square.

Fig. 13-76. Layout for duplicate mortises.

clamped together as previously described for laying out a mortise. The exact size of the tenon is determined by the size of the mortise.

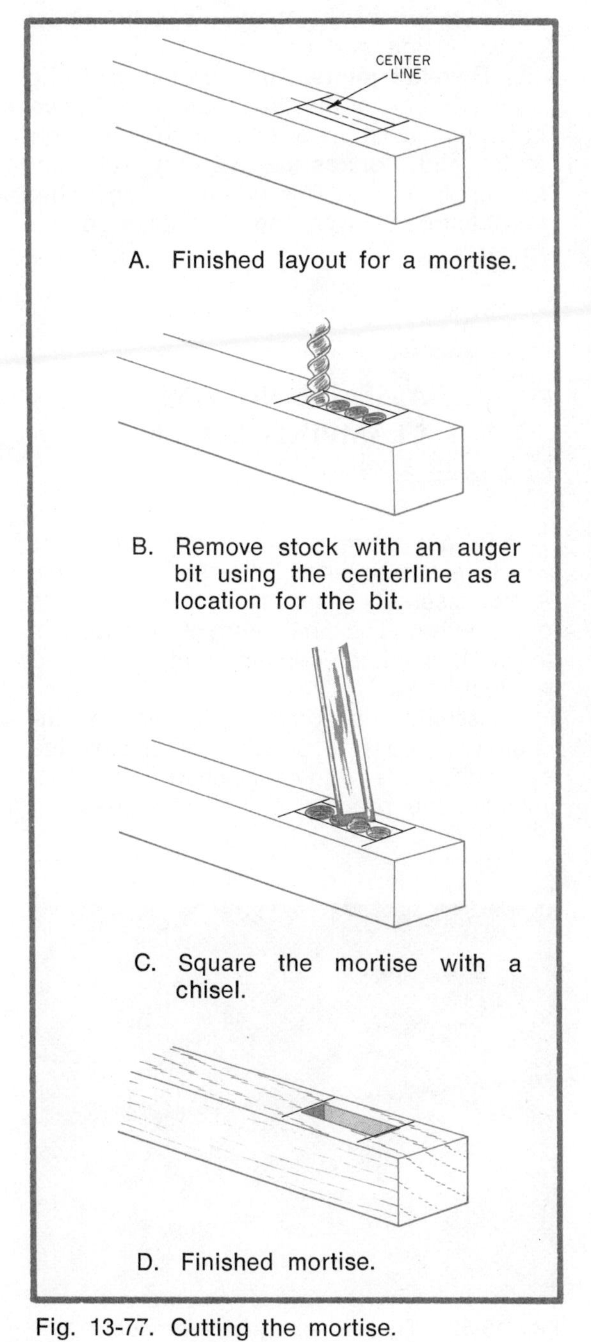

A. Finished layout for a mortise.

B. Remove stock with an auger bit using the centerline as a location for the bit.

C. Square the mortise with a chisel.

D. Finished mortise.

Fig. 13-77. Cutting the mortise.

The tenon consists of the cheek and the shoulder, Fig. 13-75. The layout is made with a mortise gauge or a marking gauge and try square. Figure 13-78 shows the completed layout for a tenon.

When cutting the tenon, the shoulder cuts are made first, using a backsaw. A piece of stock clamped along the line will help insure a straight cut, Fig. 13-79. Care must be taken not to make the shoulder cuts too deep.

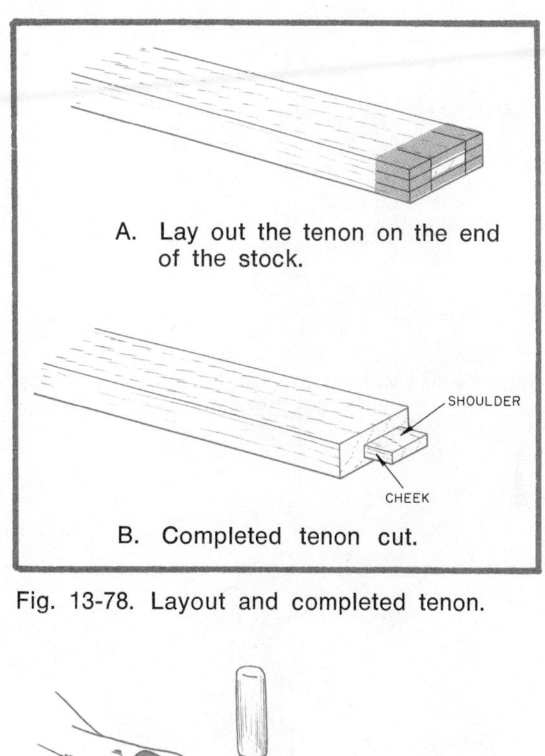

A. Lay out the tenon on the end of the stock.

B. Completed tenon cut.

Fig. 13-78. Layout and completed tenon.

Fig. 13-79. Making the shoulder cut on a tenon with a backsaw.

The cheek cuts are also made with a backsaw. A cutting jig makes it easier to make the cheek cut, Fig. 13-80. The stock on which the saw is supported is the thickness of the waste stock minus the width of the saw kerf.

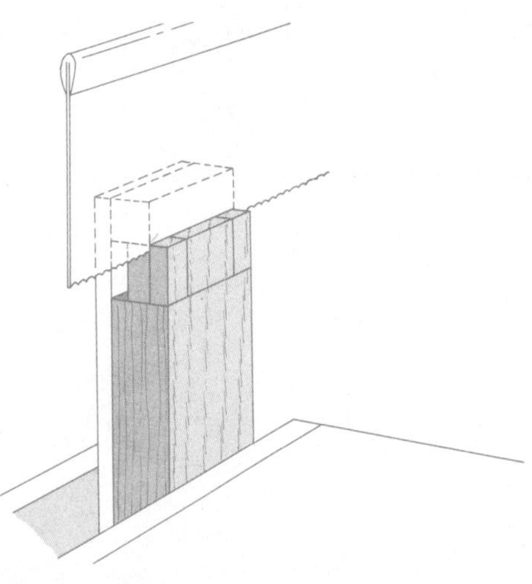

Fig. 13-80. Making the cheek cut on a tenon using the backsaw and a simple jig.

Fig. 13-81. A dovetail joint.

Each individual tenon is fitted to a mortise. It is best to number each joint and make the final fitting. A chisel is used to fit the joints. If the number is placed on the end of the tenon, it will not be cut away. The number for the mortise can be placed on an inside surface.

Dovetail joints. The dovetail joint, Fig. 13-81, is one of the finest joints for drawer construction. It is a difficult joint to construct, and requires the use of a router and a template guide. This process is described in Chapter 11 with the discussion of the router.

ASSEMBLING AND CLAMPING STOCK

Products are manufactured from parts which are combined into subassemblies. A subassembly consists of several parts put together. The final product is a total of the subassemblies combined into a finished product.

Assembly is commonly a separate industrial process. It consists of placing the separate parts into a finished unit. This includes the process of applying chemical

Fig. 13-82. Trial assembly of a product.

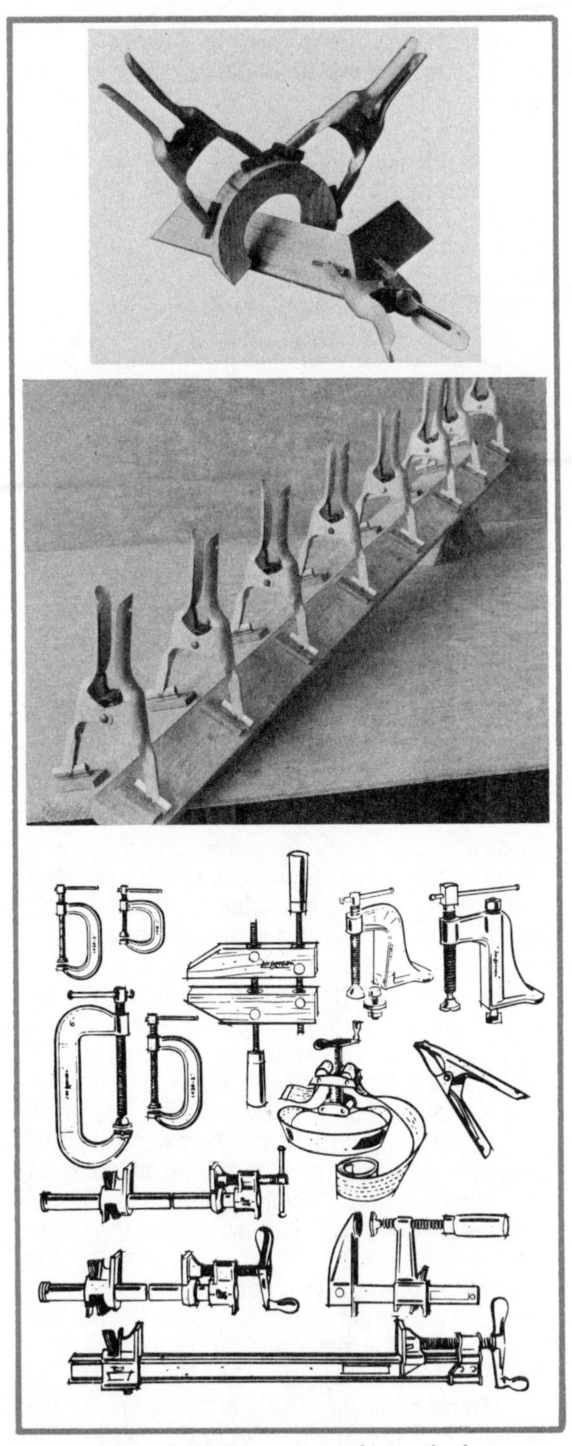

Fig. 13-83. Common types of wood clamps. (Adjustable Clamp Company and Arvids Iraids Multi Clamps)

fastening materials and the installation of mechanical fasteners. It also involves placing the work in clamps until the adhesive has cured if this type of fastening is used.

Before a product is assembled and fastened, a trial assembly should be made. Large, complicated products normally consist of many sub-assemblies. The trial assembly is used without adhesive and other fasteners. The purpose is to determine how the parts and subassemblies fit, Fig. 13-82. If a problem occurs, or any parts do not fit properly, the product can be dismantled. The necessary fitting changes can be done without damaging the parts.

CLAMPING PARTS AND SUBASSEMBLIES

Stock is commonly glued together to form larger pieces. Often the edge joint is used for this purpose. After the stock has been prepared, it must have a layer of glue applied to it and be placed in clamps while the glue cures. There are numerous clamps which can be used to hold the glued stock, as shown in Fig. 13-83.

Bar clamps. The bar clamp is available in a variety of sizes which range from 2′ to 10′ in length. It is used for assembling products and edge gluing as shown in Fig. 13-84. The clamps are placed alternately

Fig. 13-84. Clamping edge-glued stock with bar clamps.

on the top and bottom to equalize the pressure and prevent bowing of the stock. A piece of scrap stock should be placed between the clamp and the stock being glued to prevent damage to the finished edge.

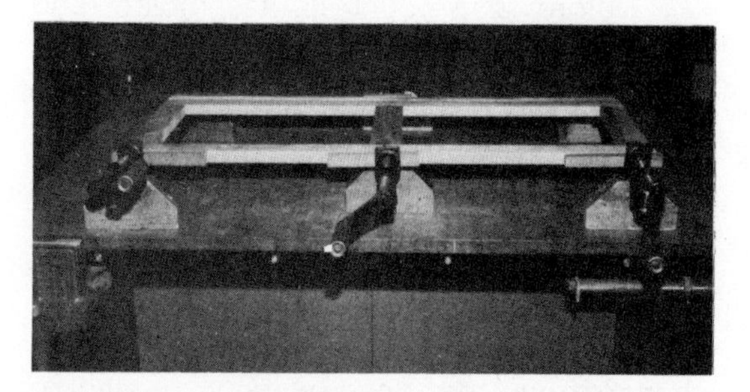

Fig. 13-85. Clamping a frame assembly with bar clamps.

Fig. 13-86. Using a hand screw to clamp stock while the glue cures.

Fig. 13-87. Using a C-clamp to apply pressure to a splintered section.

When edge gluing, the adhesive should be applied to both surfaces. Only moderate pressure is required to hold the stock in firm contact while the glue cures. Remember, excessive pressure will force the adhesive out, causing a condition called *starved joint*.

Bar clamps are also used for other assemblies such as shown in Fig. 13-85 where a paneled frame is being clamped with bar clamps. The corners are joined with mortise-and-tenon joints.

Hand screws. Hand screws are used for clamping small pieces or several layers into a pad, Fig. 13-86. In addition, the hand screw is helpful for holding stock when cutting and shaping. The jaws should be adjusted so they are parallel and exert even pressure. The outside spindle can be tightened to increase the leverage. Hand screws are available with jaws from 4″ to 18″ long.

C-clamps. C-clamps (sometimes called carriage clamps) are used extensively in woodworking. They are manufactured in a wide variety of sizes with openings ranging from 1″ to 12″. The C-clamp is sometimes used for applying pressure to small surfaces, such as in Fig. 13-87 where a C-clamp is being used to hold a chipped edge while the glue dries. A small block of wood is placed between the clamp pad and the stock to prevent surface damage, and a small piece of waxed paper is placed between the stock and the block of wood to prevent the block of wood from sticking to the surface. The C-clamp is also used for clamping "setups" on woodworking machines.

Miter clamps. A miter clamp is used for clamping products such as picture frames, Fig. 13-88. It is designed to apply pressure on two parts which are cut at a 45° angle.

Band clamps. The band clamp is designed for clamping irregular shapes for gluing. It consists of a canvas band and a mechanical ratchet device, Fig. 13-89.

Frame clamps. Frame clamps are used to clamp mitered joints or square or rectangular objects, Fig. 13-90.

Spring clamps. Spring clamps can be used when light clamping pressure is needed on small objects, Fig. 13-91. They are available in a variety of sizes.

Spring clamps which have pivoting jaws are also used. Each jaw has two rows of serrated teeth which grip the surface of the stock being glued. This type of spring clamp can be used for clamping irregularly shaped objects. In addition, they can be used for clamping mitered corners, Fig. 13-92.

Fig. 13-90. A mitered corner of a frame clamped in a frame clamp.

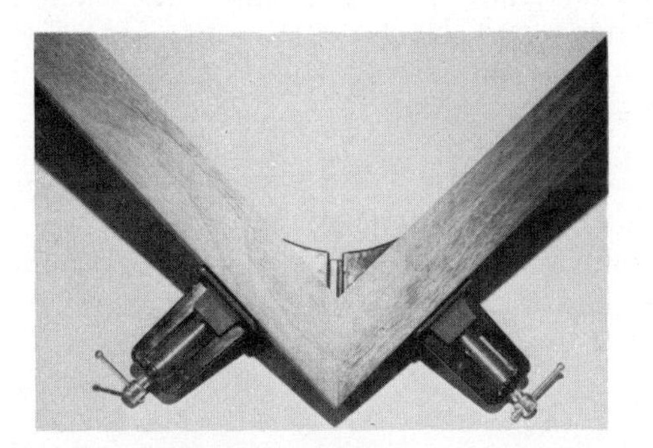

Fig. 13-88. Clamping a glued miter joint with a miter clamp.

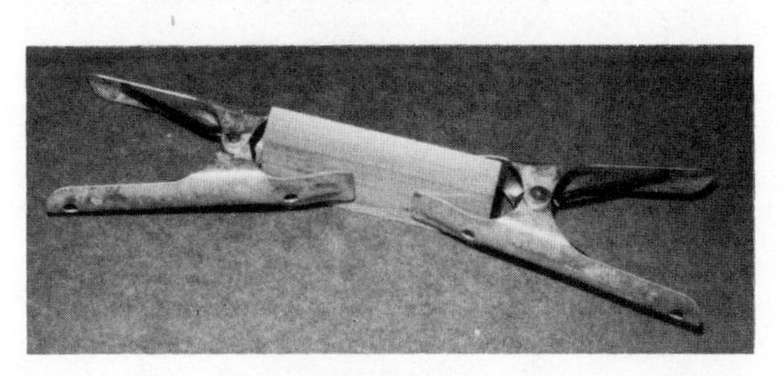

Fig. 13-91. A spring clamp holds narrow strips in place.

Fig. 13-89. Band clamp used to clamp irregular shape.

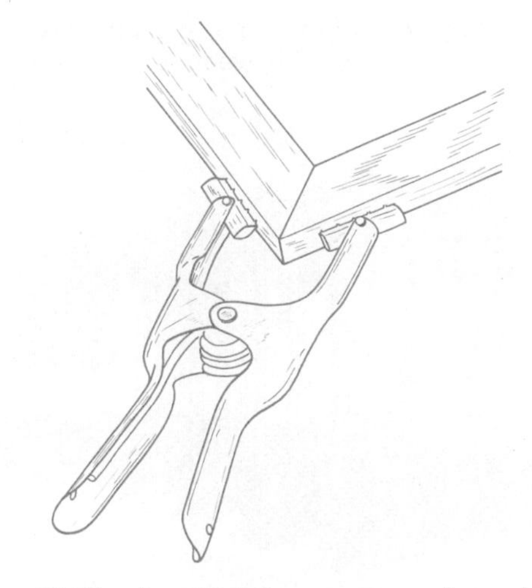

Fig. 13-92. A special-purpose spring clamp with pivoting jaws.

Fig. 13-93. Adjustable shop-constructed frame clamp.

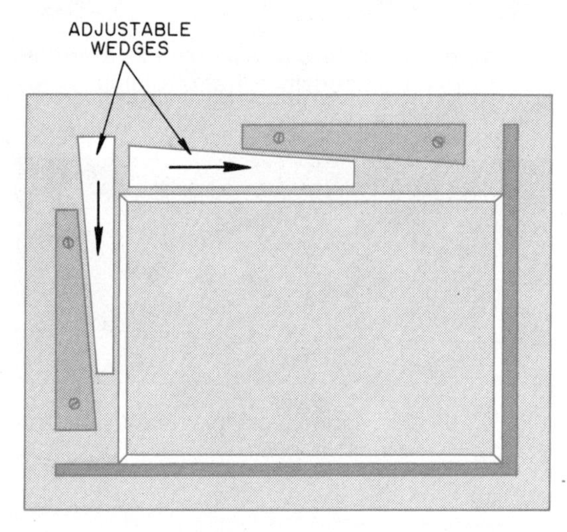

Fig. 13-94. Adjustable shop-constructed clamp with wedges.

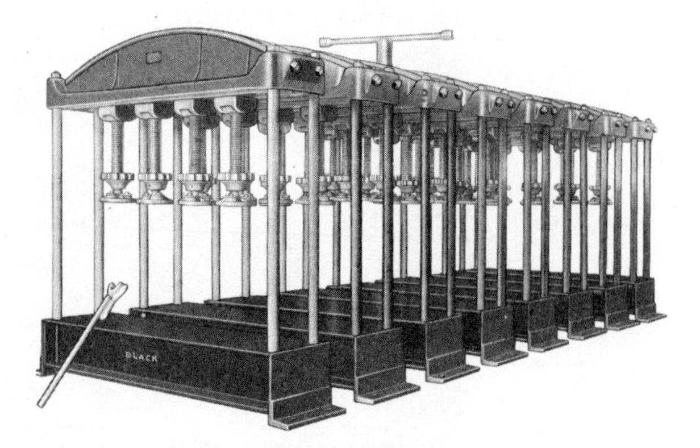

Fig. 13-95. Laminating press. (Black Bros.)

Special clamping devices. Often clamping devices can be constructed which perform better than commercial clamps, such as the clamping jig for applying pressure to mitered joints for frame construction, shown in Fig. 13-93. Figure 13-94 shows another type of shop-constructed clamping device. Proper pressure against the stationary frame insures squareness.

GLUE PRESSES

In addition to various types of hand clamps, there are also gluing presses. A press is designed to apply pressure to large areas. This type of press, often called a laminating press, is used for clamping veneer and plastic laminates, Fig. 13-95. Force is exerted on sheet stock by turning the screw thread.

Pneumatic or air pressure lamination presses are also used. The exact pressure can be regulated by the volume of air which is allowed to enter the inflatable rubber bladder. The inflatable bladder is

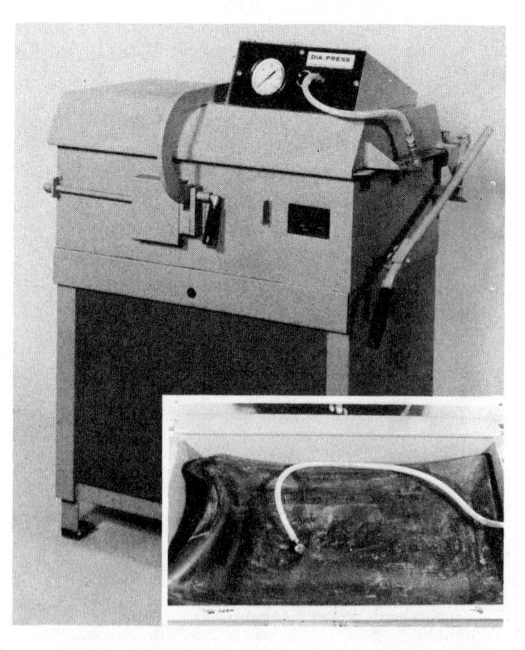

Fig. 13-96. Inflatable bladder press. (Vega Industries)

also designed to conform to irregularly shaped work, Fig. 13-96.

HARDWARE FOR ASSEMBLY AND FASTENING

As you have learned, wood can be fastened by a wide variety of mechanical fasteners. All types of mechanical fasteners may be called *hardware.* Many special-purpose hardware fasteners have been developed. The following section will describe various types and methods of installation.

ADJUSTABLE SHELF SUPPORTS

A cabinet is more functional if the spacing between the shelves can be easily changed through use of an adjustable shelf support, Fig. 13-97. This type of support is installed extending from the top to the bottom of the cabinet. It can be screwed to the inside of the cabinet or mounted flush by cutting a groove. Four assemblies are required in each cabinet. This support allows for the shelves to be moved at one inch increments. The standards are available in lengths up to 6′ long.

Shelf supports such as shown in Fig. 13-98 are also used for adjustable shelving. This type requires only the bracket which fits into a ¼″ hole drilled in the side of the cabinet. It is inexpensive and can be placed where the shelf is most likely to be moved. If shelf brackets are not available, short sections of dowel rods cut flat on one side can be cut and inserted in pre-drilled holes, Fig. 13-99.

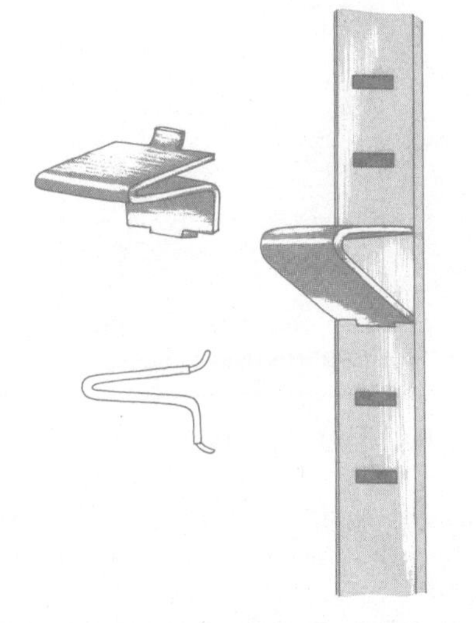

Fig. 13-97. Adjustable shelf standard and support.

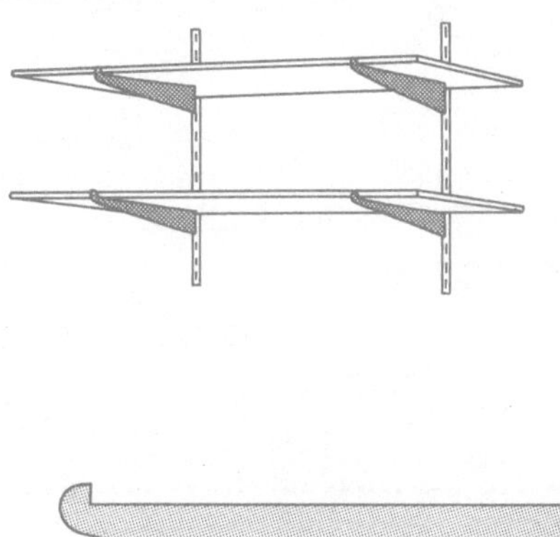

Fig. 13-98. Adjustable shelf bracket.

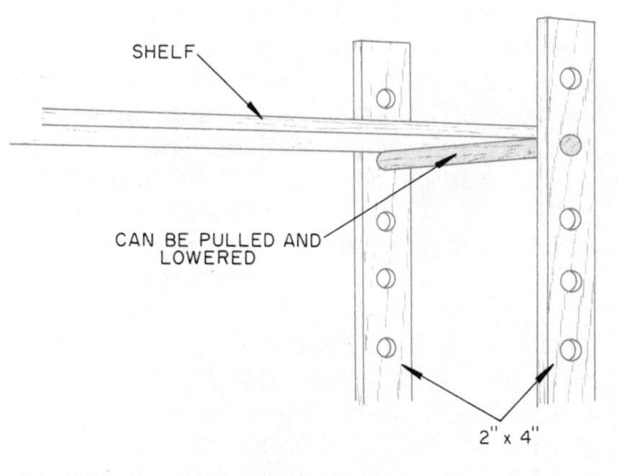

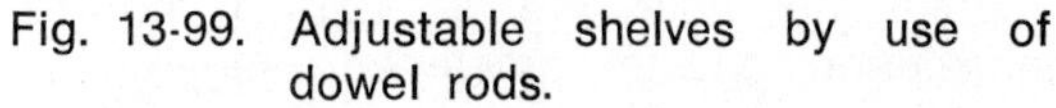

Fig. 13-99. Adjustable shelves by use of dowel rods.

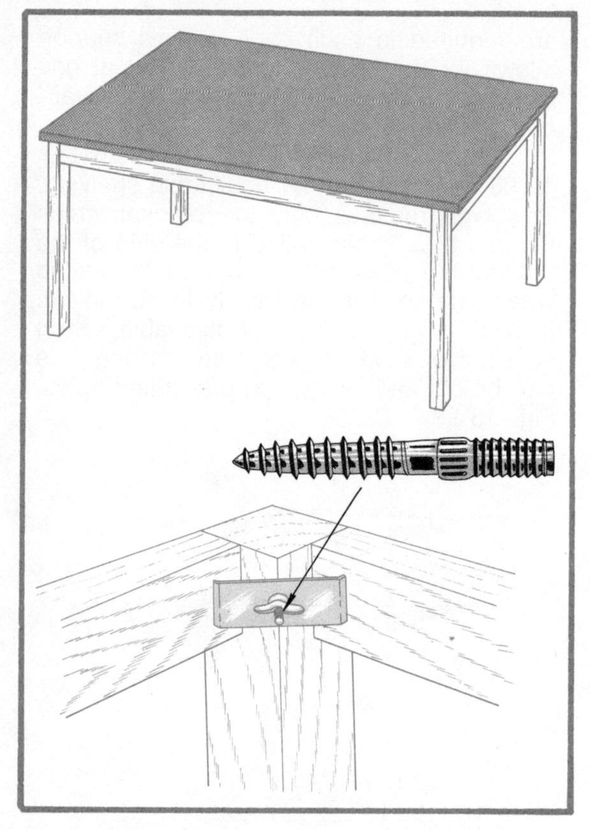

Fig. 13-100. Table brace used for mounting legs on a simple table.

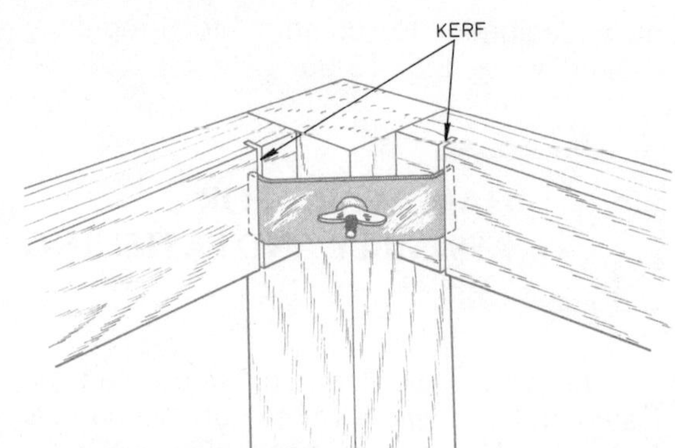

Fig. 13-101. Method of installing table brace.

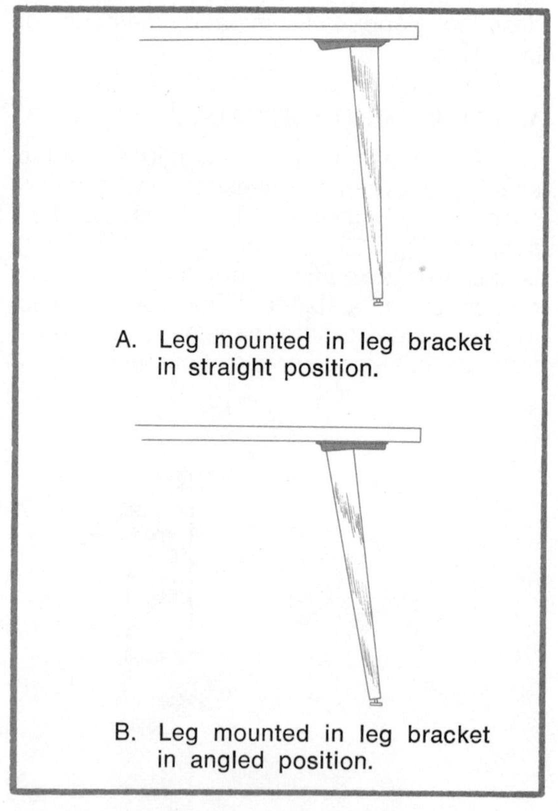

A. Leg mounted in leg bracket in straight position.

B. Leg mounted in leg bracket in angled position.

Fig. 13-102. Leg mounting bracket.

LEG MOUNTINGS

Many methods can be used to mount furniture legs. Often they are mounted with permanent joints such as mortise-and-tenon or doweled joints. Special hardware is also available for mounting legs. Table corner braces can be used to hold the legs to the rails, Fig. 13-100. This method is often used on manufactured furniture so the legs can be dismantled for shipment.

The corner table brace is installed by cutting a dado in the end of each rail which joins the leg. A hanger bolt is placed in the leg. The nut can be removed from the projecting end of the hanger bolt to dismantle the table, Fig. 13-101. A mortise-and-tenon joint is sometimes used, but no adhesive is used on the joint. This insures positive alignment and increases the strength.

Leg mounting brackets are available which allow the leg to be mounted straight

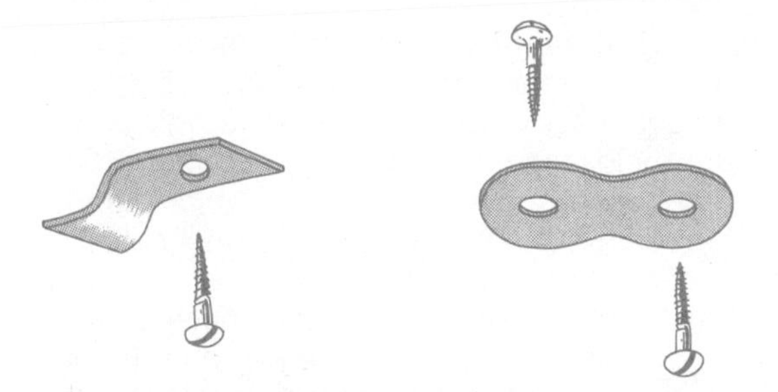

Fig. 13-103. Hardware for fastening table tops.

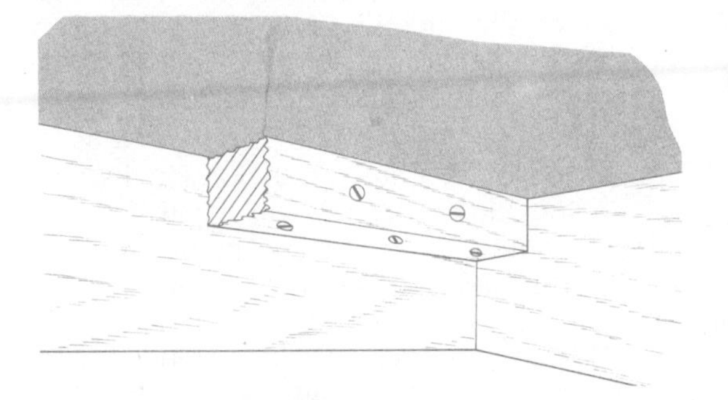

Fig. 13-104. Fastening table top with slotted wood strips.

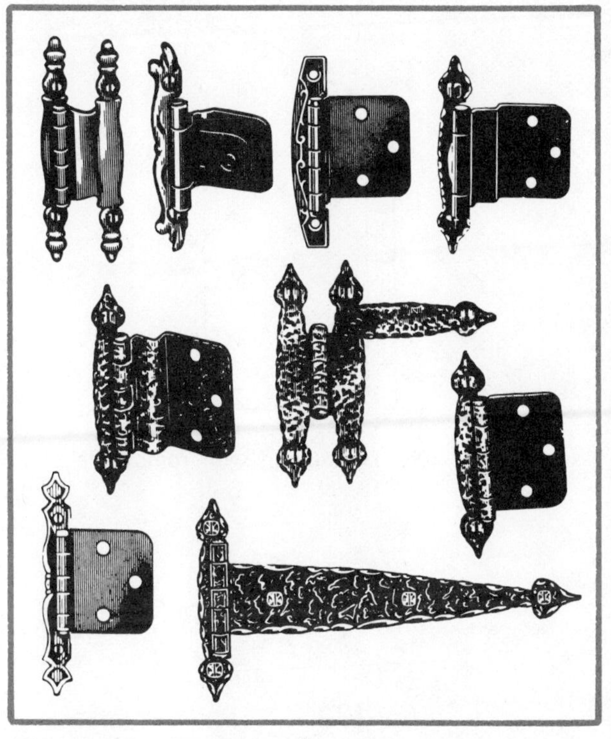

Fig. 13-105. An assortment of decorative hinges.

or at an angle, Fig. 13-102. The bracket is fastened to the table top with screws, and a lag bolt is installed in the leg. The opposite end of the lag bolt has threads which turn into the bracket. This assembly is easy and can be dismantled.

HARDWARE FOR ATTACHING TABLE TOPS

Large table tops must be attached so they are free to expand and contract as the moisture content changes. The most common type of hardware for this purpose is shown in Fig. 13-103. This type fastener is installed by cutting a saw kerf about ⅜″ deep in the table rail. The kerf is cut down from the top edge to place slight tension on the top against the rail. The opposite end is screwed to the top. The fastener allows the top to move slightly in the saw kerf.

If commercial hardware is not available, wood strips with slotted holes can be used, Fig. 13-104. The holes which are against the top are slotted to allow for movement.

HINGES

There are numerous types of hinges used for installing doors on cabinets. All hinges have about the same efficiency if the correct size is selected. Concealed hinges are designed so they are not seen when installed and the door is closed. Other types of hinges are decorative and are intended to enhance appearance, or are designed for various periods of furniture. Such furniture periods would be Contemporary, French Provincial, Early American, Mediterranean, as well as many others. The selection of hardware should harmonize with the particular piece of furniture, Fig. 13-105.

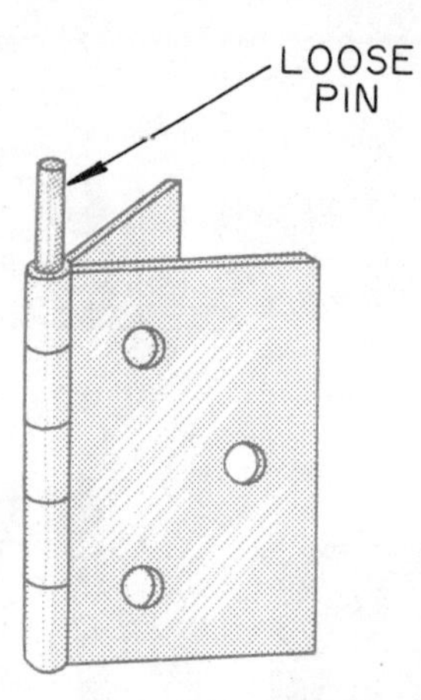

Fig. 13-106. Butt hinge with removable pin.

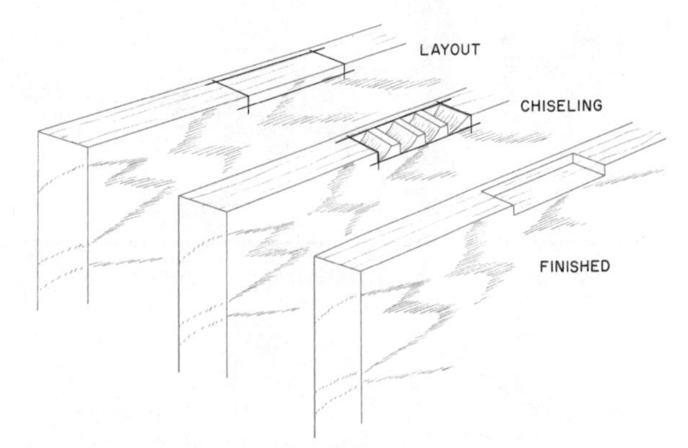

Fig. 13-107. Chiseling a gain for a butt hinge.

Butt hinges. Butt hinges are used for hanging many types of doors. The butt hinges have two equal size leaves which are held together with a steel pin. Some butt hinges have a fixed pin which cannot be removed. Others have a removable pin called a loose pin, Fig. 13-106. Larger butt hinges for doors in a house have removable loose pins. The size of butt hinges is given by the width and length of the individual leaves. Common butt hinges range in size from ½″ to 4″. The thickness of the leaf increases with the size and quality of the hinge.

Butt hinges are normally installed so only the pin and knuckle is exposed when the door is closed. This requires that a recess called a *gain* is cut for the leaf in which the hinge will fit. The layout for the hinge should be carefully made with a knife. The gain can be cut with a hand chisel or power tools, Fig. 13-107.

Surface hinges. Surface hinges are mounted on the door frame and door. This type of hinge does not require that a gain be cut. They are normally decorative and are designed to complement the design, Fig. 13-108.

Semiconcealed hinges. Semiconcealed cabinet hinges have only the leaf mounted on door jamb exposed. This hinge is often referred to as ⅜″ offset or inset hinges, Fig. 13-109. A semiconcealed hinge requires that a ⅜″ rabbet be cut on all sides

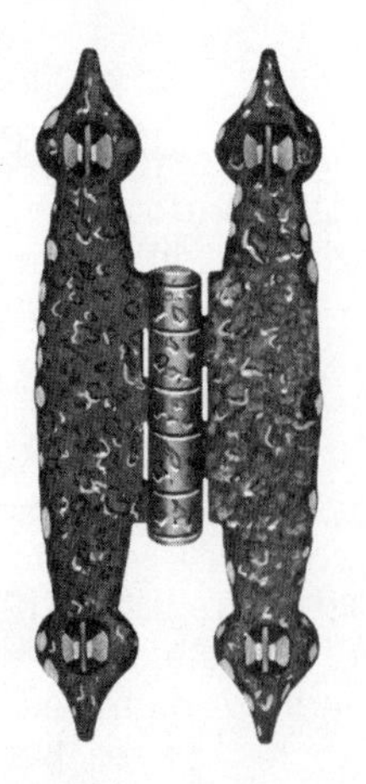

Fig. 13-108. Decorative surface hinge. (Amerock Corp.)

Fig. 13-109. Semi-concealed hinge. (Amerock Corp.

of the door. The rabbeted portion of the door fits inside of the cabinet frame, Fig. 13-110.

Concealed hinges. Concealed hinges, Fig. 13-111, are installed when it is desirable not to have the hinge exposed. They generally come with templates and instructions for installation.

DRAWER AND DOOR PULLS

Drawer or door pulls should be both functional and enhance the appearance of the cabinet. There is a wide assortment available, so the pulls should match the design of the product, Fig. 13-112.

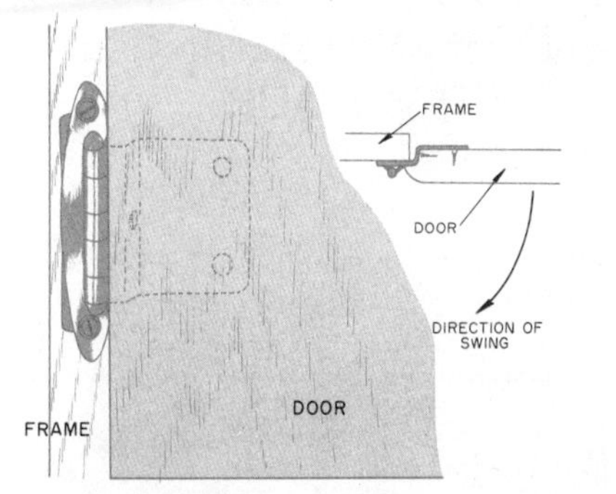

Fig. 13-110. Installation of a semi-concealed hinge.

Fig. 13-111. A concealed hinge. (Shelby Furniture Hardware Co.)

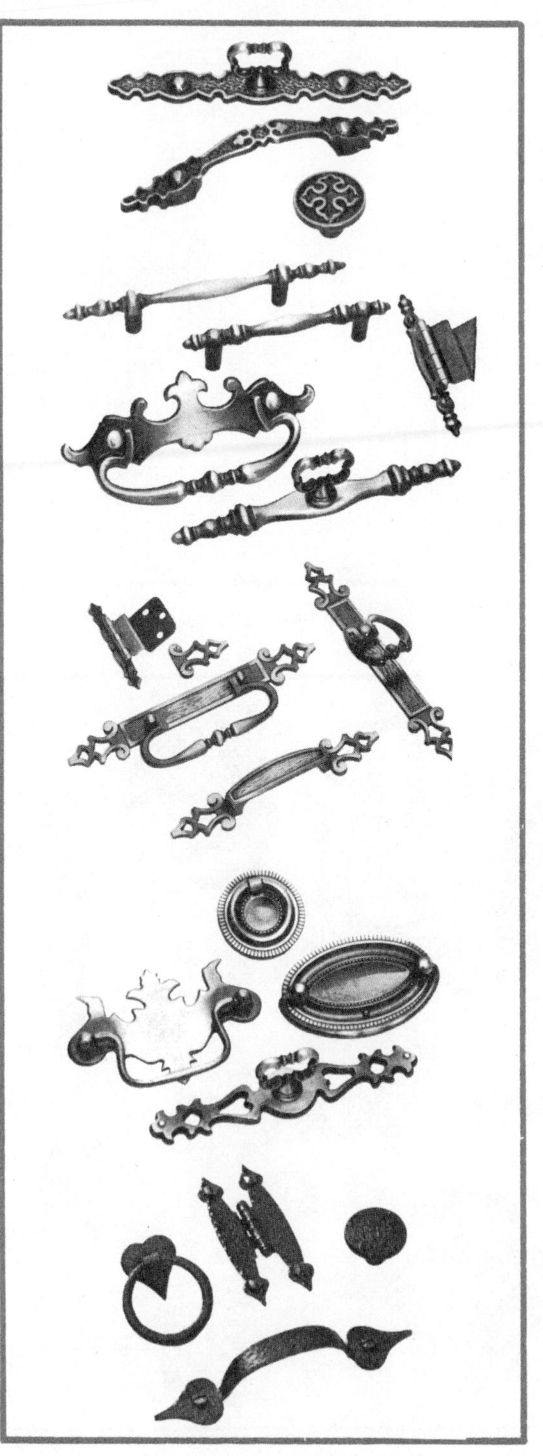

Fig. 13-112. An assortment of knobs and pulls. (Amerock Corp.)

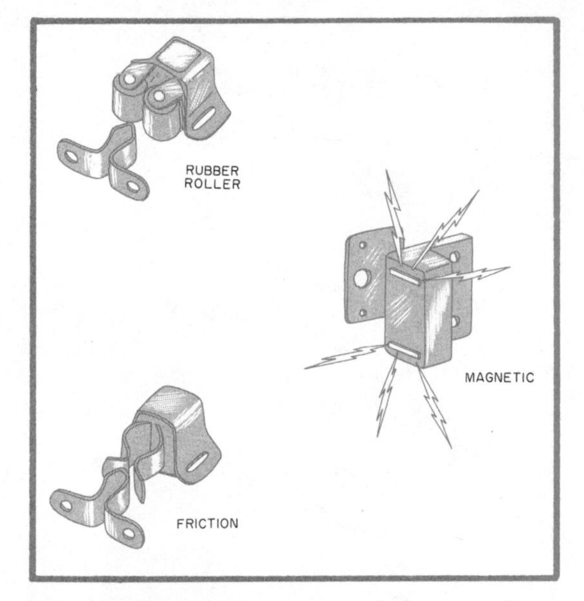

Fig. 13-113. Common types of door catches.

Fig. 13-114. Special-purpose hardware. (Amerock Corp.)

A door catch should hold the door tightly closed; however, the door should open with only slight pressure. The most common types of catches include magnetic, friction, and rollers which are shown in Fig. 13-113.

SPECIAL-PURPOSE HARDWARE

Figure 13-114 shows some of the available types of special-purpose hardware commonly used in the wood industry.

INDUSTRIAL ASSEMBLY AND FASTENING TECHNIQUES

As stated earlier, assembly is one of the major production processes in pro-

Fig. 13-115. Small portable glue spreaders. (Black Bros. Co., Inc.)

ducing a product. To reduce production costs, industry has developed numerous methods for speeding the operation. Jigs are developed for holding various shaped parts in position for assembly.

INDUSTRIAL GLUE APPLICATION

Many types of glue applicators have been developed, such as the small portable glue spreaders shown in Figs. 13-115, 13-116, and 13-117. These not only speed the job, but also apply the correct amount of glue. Many glue applicators are motor driven. Large production models have conveyors which move the stock under spread rollers. This type of machine is commonly used for applying glue to veneer. The stock often passes through the glue machine into automatic clamps which cure the glue by means of high frequency heat.

Fig. 13-116. A drum glue applicator. (Gulf Oil Co.)

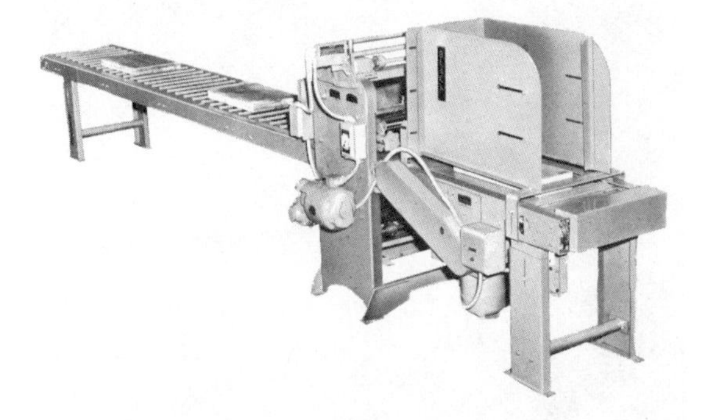

Fig. 13-117. A glue applicator with an automatic conveyor. (Black Bros. Co., Inc.)

INDUSTRIAL ASSEMBLY AND CLAMPING MACHINES

Clamping operations are carefully sequenced in industry. The clamping machines and devices are often designed for a particular purpose or product. The stock to be glued is machined and prepared before it is moved to the gluing area. A prime concern in industry is reduced curing time to reduce the need for storage. The large revolving clamping machine shown in Fig. 13-118 is equipped with rows of bar clamps attached to an endless conveyor. The machine is used for gluing edge-to-edge joints for table tops and furniture requiring large surfaces. It is loaded with the stock, and the glue is cured by the

Fig. 13-118. A revolving clamping machine. (Richmond Cedar Works)

time the machine makes a complete revolution and the stock is removed.

Large frame clamps operated by air pressure are used for clamping many large flat frames, Fig. 13-119. The frame is adjustable so various sized products can be clamped. The clamp not only applies pressure, but also clamps the product square.

Fig. 13-119. An air-controlled frame clamp. (Biltbest Windows)

Furniture such as cabinets and dressers are clamped in a large case clamp. It consists of two flat surfaces which move together and apply even pressure. The case clamps may operate either manually or with air pressure, Fig. 13-120.

INDUSTRIAL FASTENING MACHINES FOR ASSEMBLING

Industry has replaced the nail and hammer in most cases with automatic nailing guns. The nailing guns can install nail fasteners much faster, resulting in less labor cost. Most automatic nailing machines are operated with compressed air. Figure 13-121 shows studs for a manufactured house being fastened in place with an automatic nailing gun. The nails for automatic nailers are in clip form, Fig. 13-122. This provides for rapid filling of the gun.

Automatic nailers are used in factories as well as on the job site. They are commonly used for installing roof sheathing and sub-floors in house construction, Fig. 13-123. A portable air compressor is used for power for an on-site job.

Nail driving devices are also used which do not use air pressure. Flooring

Fig. 13-120. Air-controlled case clamp for gluing large furniture. (Black Bros. Co., Inc.)

Fig. 13-121. Fastening studs with an automatic nailer. (Senco Products, Inc.)

may be installed with a automatic stapler, Fig. 13-124. The nail is installed by striking the ram with a heavy rubber mallet. Clips of nails are installed in the gun and fed into place as each nail is driven.

There are numerous types of automatic nailers and staplers available. They are becoming widely accepted because of the labor savings. In most cases, the holding power of nails installed with automatic nailers is as great or greater than for hand-driven nails.

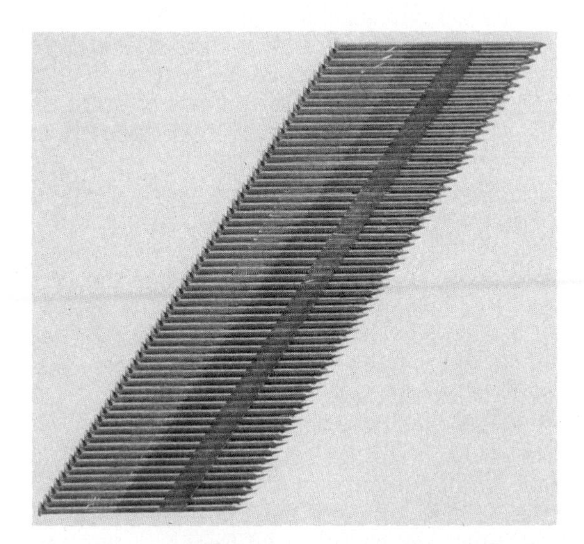

Fig. 13-122. Nails for an automatic gun. (Senco Products, Inc.)

TESTING THE HOLDING POWER OF JOINTS

The comparative strength of various types of fasteners can be tested to determine the selection of the best one for a particular job. They should be selected on the basis of (1) holding power, (2) ease of construction, and (3) appearance. It is impractical and costly to construct joints which are exceedingly stronger than the job requires.

Fig. 13-123. Automatic nailers used to apply sheathing. (Senco Products, Inc.)

Fig. 13-124. An automatic stapler used to apply subflooring.

There are several variables or factors which will influence the strength of a wood joint. The following factors directly affect the strength of a joint:

1. Kind of joint
2. Species of wood
3. Moisture content of the wood
4. Type of adhesive
5. Type of mechanical fastener
6. Technique for applying and clamping the adhesive
7. Method of installing mechanical fasteners
8. Conditions to which the complete joint is exposed (for example — high heat, high moisture, etc.)

A joint should be subjected to a condition similar to that which it will be subjected to in actual use since all joints are not subjected to the same kind of force. You should evaluate the way the product will be used and set up tests which will be similar to the actual forces. For instance, a chair should be constructed so the joints would withstand the force of abused usage like a child jumping in it even though it is not intended to be used in that way.

Numerous types of testing machines can be used to evaluate the strength of a joint. If testing equipment is not available, shop-constructed testing equipment can easily be designed, such as that used in the activities at the end of this chapter. Sand is added to a pail until the joint fails. The weight required to fracture the joint is recorded and compared with other test specimens.

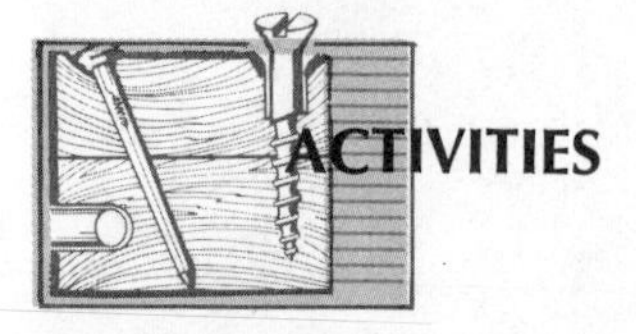

ACTIVITIES

PROBLEM

The types of joints which are used for assembling a wood product will depend upon the *strength required, appearance,* and *ease of construction.* The strength of the joint is usually the most important single factor for making a decision as to the type of joint which will be used.

The strength of a joint is in relationship to the *level of restriction.* As you recall from reading the chapter, the level of restriction varies from one (1) through six (6). For example, a butt joint is a *level one* joint because the movement is restricted in only *one direction.* A rabbet joint is representative of a level two restriction, and so forth.

You can compare the relative strength of joints by means of simple experimentation. First, construct a joint representative of each level of restriction from one through six. Use two pieces of wood ¾″ × 2″ × 6″ for each joint. The wood must be of the same species for each of the joints such as pine or maple.

The joints can then be tested by clamping them in a bench vise one at a time and attaching a container such as a pail to one side of the joint as shown in the figure below. Then add sand to the pail until the joint fails (breaks). After the joint fails, weigh the sand to determine the amount of force required to break the joint.

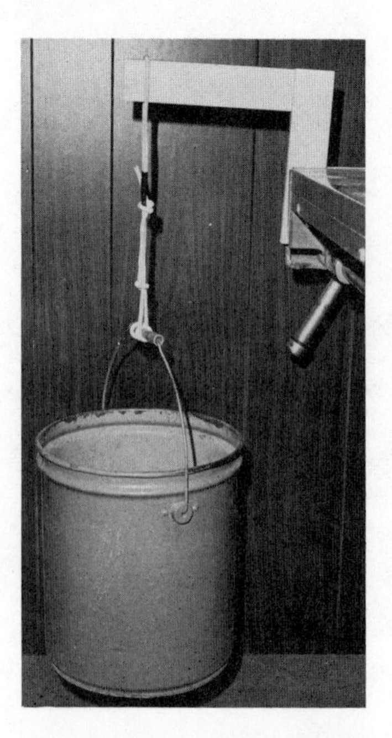

As the samples break, record the weight required to break the joint. After all of the samples have been broken, plot the weights on a graph and compare the strength of the joints in relationship to the level of restriction.

NOTE: The sand container should be suspended not over three or four inches above the floor. Keep your feet away from the container as you add the sand to avoid injury as the container drops to the floor.

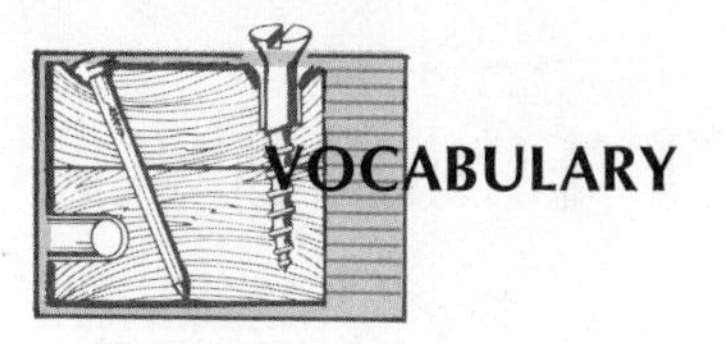

Mechanical fastener
Chemical fastener
Penny "d"
Shank diameter
Root diameter
Pilot hole
Shank hold
Thermosetting
Thermoplastic

Table 5
SELECTION GUIDE FOR COMMON ADHESIVES

TYPE	ORIGIN	MOISTURE RESISTANCE	PREPARATION FOR USE	APPLICATION PROCEDURE	ADVANTAGES AND MAJOR USES	DIS-ADVANTAGES
Liquid Animal Glue	Animal bones and hides	Non-water Resistant	Ready to use	Both surfaces	Widely used for furniture construction. Has good crack filling characteristics	Lack of water resistance
Casein Glue	Milk Curd	Water Resistant	Mix with water	Both surfaces	Good for woods with high oil content. Used in carpentry construction, lamination, and cabinet making	Stains some types of woods
Polyvinyl resin (White Glue)	Chemical	Non-water Resistant	Ready to use	Both surfaces	Inexpensive, dries clear, non-staining. Used for furniture construction	Non-waterproof joints may creep. Non-heat resistant, fouls abrasive paper
Aliphatic resin	Chemical	Non-water Resistant	Ready to use	Both surfaces	Fast drying, strong, resists moderate heat, clear glue line, non-staining. Used in furniture construction	Non-moisture resistant
Plastic resin	Chemical	Moisture Resistant	Mix with water	Both surfaces	Provides strong joint, non-staining, cures with electronic welder. Used in general woodworking where moisture resistance is important	Must be used with tight fitting joints. Must be clamped for several hours
Urea resin	Chemical	Moisture Resistant	Mix with water	Both surfaces	Best suited of most adhesives to electronic drying. Provides strong joint. Used in woodworking where strong moisture resistant joints are necessary	Must be used on low moisture content wood. Must be used with tight fitting joints
Resorsinal resin Glue	Chemical	Water Proof	Mix Catalyst and resin	Both surfaces	Waterproof and is used in boat-building and products exposed to high moisture	Expensive, tends to stain wood
Epoxy	Chemical	Water Proof	Mix Catalyst and resin	Both surfaces	Use for joining non-similar materials. Strong, high gap filling qualities	Expensive
Contact Cement	Chemical	Highly Moisture Resistant	Ready to use	Apply to both surfaces and allow to dry then place surfaces together	No clamps required, ideal for veneering and plastic laminates	Must have ample ventilation. Parts cannot be moved after contact is made
ALWAYS READ AND FOLLOW THE MANUFACTURERS INSTRUCTIONS						

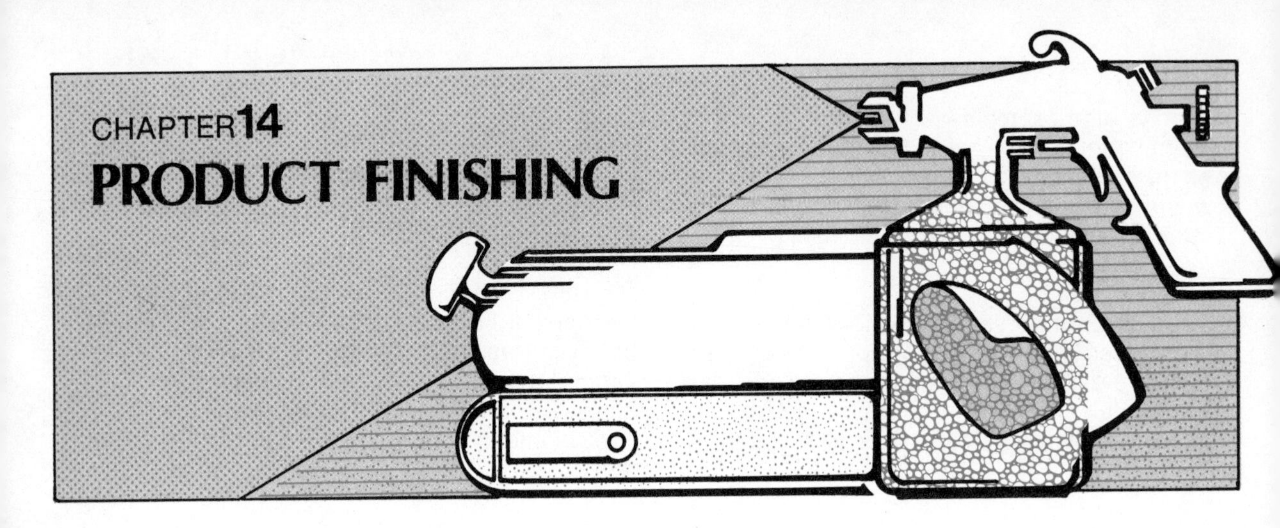

CHAPTER 14 PRODUCT FINISHING

INTRODUCTION

Product finishing is begun only after a product is well-constructed with all joints fitting together properly. The surface is prepared for finishing by means of sanding. This surface preparation is one of the most important phases of product finishing, as the quality of the finish can be no better than the condition of the surface prior to the application of the finishing material.

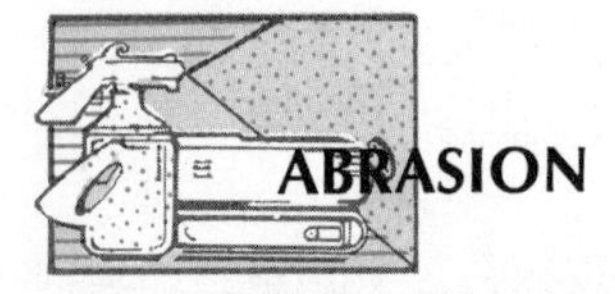

ABRASION

Surface preparation requires an understanding of the principle of *abrasion* (sanding). Abrasion can be defined as the process of wearing away a soft material with a hard material. Various types of abrasive papers (often called sandpaper) serve this purpose. Figure 14-1 shows the principle of abrasion, in which the sharp points of the abrasive material actually cut the fibrous wood structure.

PURPOSES OF SANDING

Sanding serves two purposes: (1) the first sanding step removes excess wood, and (2) the second step smooths the surface. A course grit abrasive is used for the first sanding, leaving deep scratches in the surface. These scratches are then removed through the use of finer grits of abrasive. The finer grits are used until a smooth, polished surface appears. The sanding process removes knife marks made by machines, dents, chips, and other imperfections. It is also necessary to sand the wood immediately prior to finishing to remove any

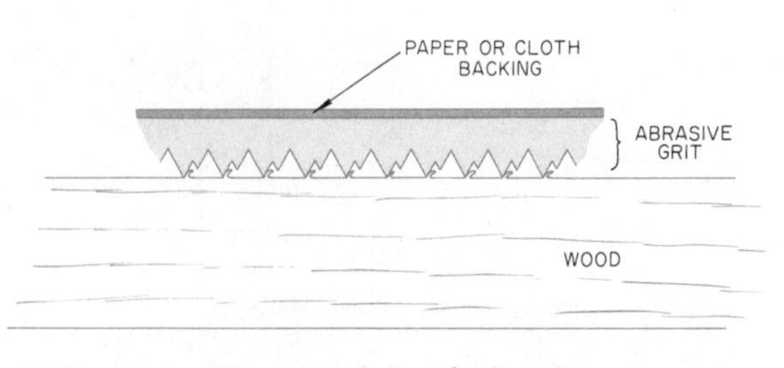

Fig. 14-1. The principle of abrasion.

raised grain caused by the absorption of moisture from the atmosphere.

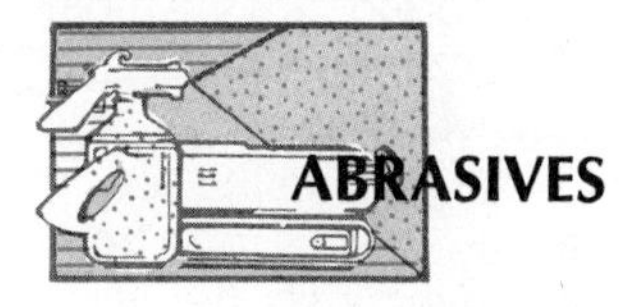

Abrasives are classified as either natural or man-made. A natural abrasive is composed of minerals extracted from the earth while man-made abrasives are manufactured. The most common natural abrasives are *flint* and *garnet,* and man-made abrasives are *aluminum oxide* and *silicon carbide.*

Flint. Flint abrasive can be recognized as quartz extracted from the earth and it is whitish in color. It is used very little in the woodworking industry today as it lacks durability and is limited to hand sanding.

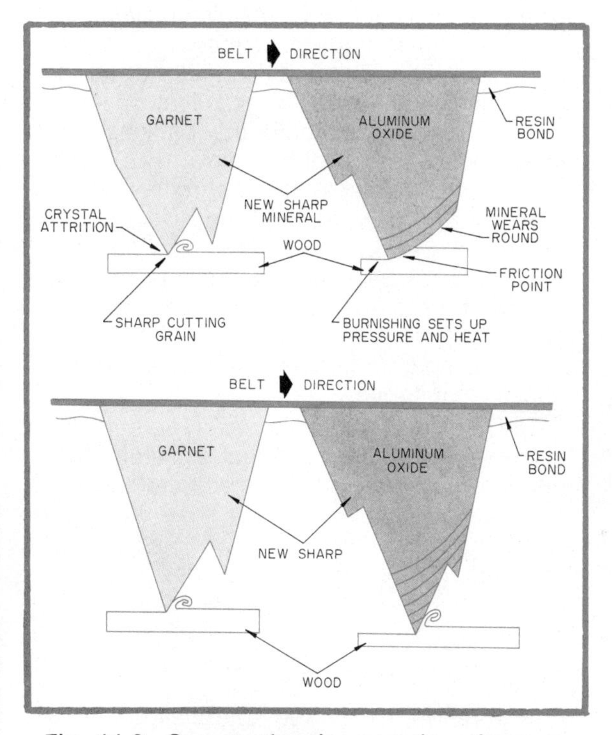

Fig. 14-2. Garnet abrasive remains sharp as the crystals fracture and expose new sharp cutting edges. (H. H. Connally)

Garnet. Garnet is reddish-brown in color, and is widely used in the wood industry. It is much harder than flint and remains sharp. Some types of abrasives wear round and smooth, losing the cutting effectiveness, but as garnet wears, the crystals break and leave sharp jagged edges, Fig. 14-2.

Aluminum oxide. This abrasive is manufactured from bauxite, iron filings, and coke. Bauxite is an ore from which aluminum is produced. In its finished form, aluminum oxide abrasive is brown in color, is very hard, and wears longer than most abrasive materials.

Silicon carbide. Silicon carbide is blue-black in color and extremely hard, nearly as hard as industrial diamonds. The grains fracture and remain sharp due to their hardness. This type of abrasive is well-suited to machine and high speed sanding operations.

GRADING OF ABRASIVES

Abrasives are produced in a wide range of grit sizes. As stated, the course grits are used for removing excess stock and deep scratches, and the finer grits are for producing smooth surfaces. The grit size is determined by the silk screen mesh through which the particles were sifted, or in other words, in accordance to the number of openings to the linear inch in the screen. An example is 100 mesh-screen which has 100 openings along the side in each of two directions, or 10,000 openings in each square inch. Grit size may range from 12 (very coarse) to 600 (extra fine). The grit size of abrasive paper was given as oughts and numbers, but the hundreds numbering system is most often used for modern abrasives. Table 6 shows the grit sizes and the appropriate types for various jobs.

ABRASIVE BACKING MATERIALS

There are different types of backing to which the abrasive grits are attached. They include paper, cloth, combination paper-

Table 6

Comparison of Abrasive Grit Sizes and Common Uses

Coarseness	Mesh or Grit No.	Ought Symbol*	General Purposes
Very Fine	400 360 320 280 240	10/0 9/0 8/0 7/0	Used for rubbing and polishing finish after it has been applied
Fine	220 180 150 120	6/0 5/0 4/0 3/0	Last sanding prior to the application of stain and final finish
Medium	100 80 60	2/0 1/0 1/2	Rough sanding before final sanding before applying finish
Coarse	50 40 36	(1) (1½) (1½)	For removing deep dents and imperfections and removing excess stock
Very coarse	30 24 20 16 12	(2½) (3) (3½) (4) —	Rough shaping and forming of stock.

*The ought numbers are given for comparison only, since the grit numbering system was revised in April, 1966, to use the hundred series.

cloth, fiber, and combination cloth-fiber. The various types of backings are also available in different weights and grades. The weight of paper backings is given as A, C, D, and E. The A grade is the lightest in weight and is used for the finer grits. E weight paper backing is a heavy production grade.

There are several types of cloth backings for abrasives. Cloth is generally used on production abrasives where flexibility and strength are factors. Jean and drill cloth are common cloth backing materials. Jean cloth is designated by a "J" and drill cloth by an "X". Jean cloth is more flexible and is used for sanding belts. Abrasive grits for heavy sanding disks are usually adhered to the rigid drill cloth.

Cloth and paper are used in combination when a strong but flexible abrasive backing is needed. Fiber backing is very stiff and is limited to the manufacture of heavy sanding disks. The strongest combination is cloth and fiber.

The abrasive is attached to the backing using numerous types of adhesives. Hide glue is used for adhering grit to most paper when it is used for dry sanding operations. When the abrasive is used for wet sanding, waterproof synthetic resins bond the grit. If high heat is generated from friction, synthetic resins may also be used on paper backings intended for dry sanding.

MANUFACTURE OF ABRASIVES

Abrasives are also classified as either *open coat* or *closed coat.* The entire surface of the backing material is covered for closed coat abrasive, but only about 70% of the backing material is covered with grit on open coat abrasives. Closed coat abrasives are used on heavy production sanding machines because they provide longer abrasive life. The open coat abrasive material does not "load up" as quickly when sanding soft and resinious materials.

The grit is applied to the backing material by either the gravity method or the electrostatic method. In the first method, the grit is dropped onto the backing material by means of gravity flow. The backing material passes through rollers which apply the adhesive, then grit is dropped onto the backing and dries to the adhesive. The backing material is printed before it passes through the coating process. The number of grit, weight of the backing, and trademark are printed on the backing material.

The electrostatic process for applying abrasive grits uses an electrical field produced by passing an electrical charge between two electrodes. The electrical charge causes the abrasive grits to be attracted to the backing material in a vertical position.

PURCHASING ABRASIVES

Abrasives may be purchased in a variety of forms. This includes sheets, belts, disks, and cylinders. It is available in sizes to fit standard sanding machines.

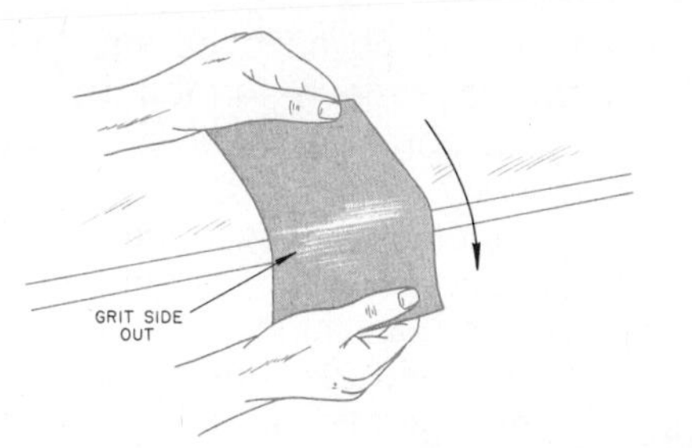

Fig. 14-3. Breaking the backing material on sheet abrasive.

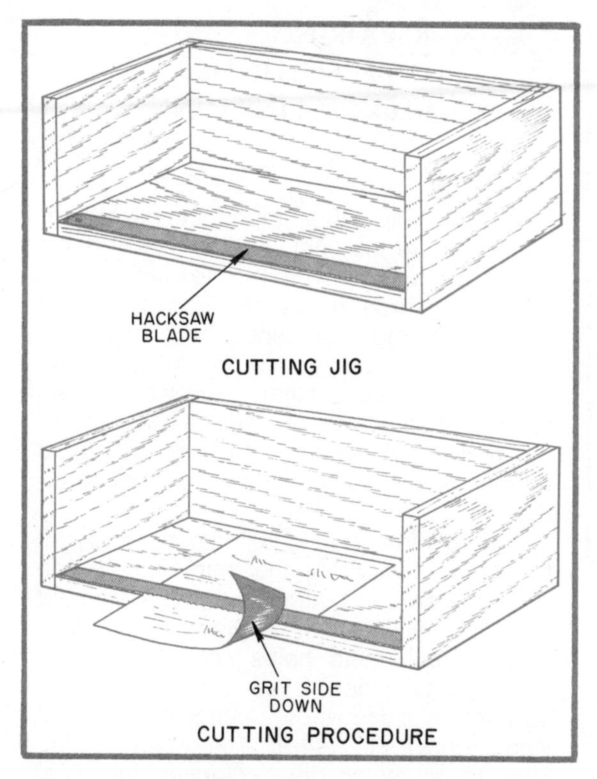

Fig. 14-4. Cutting jig and procedure for tearing abrasive sheets.

Sheet abrasives may be purchased in quantity, and it is more economical to purchase packages of 50 to 100 sheets.

USING ABRASIVES

The life of the abrasive sheet can be extended by breaking the backing over a contoured edge prior to hand sanding, Fig. 14-3. This prevents the backing from tearing when used with a sanding block.

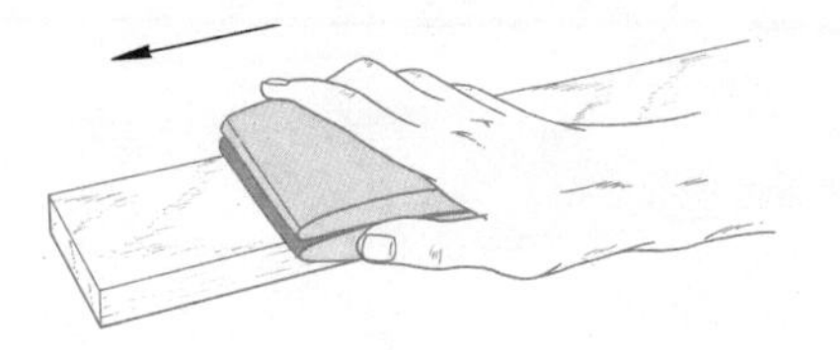

Fig. 14-5. Sanding with the grain using a sanding block.

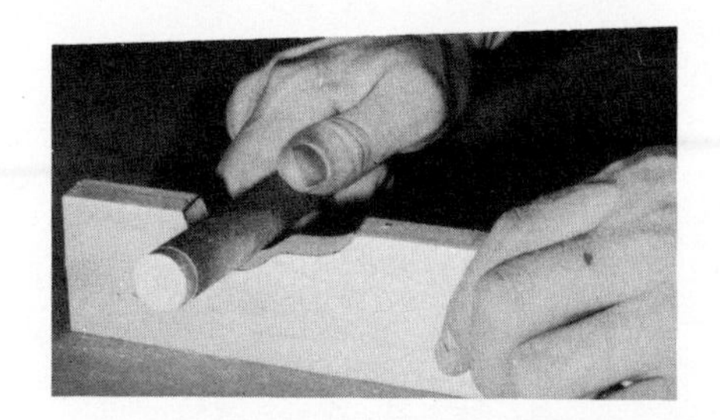

Fig. 14-6. Sanding irregular shapes with abrasives wrapped around a dowel rod.

Most sheet abrasive must be divided into smaller pieces before being used. This can be done by tearing along a straightedge, or a shop-built tearing jig shown in Fig. 14-4. The abrasive grits are placed opposite the cutting edge for tearing.

HAND SANDING

Hand sanding is necessary even if machines are used to produce a perfect surface for finishing. Flat surfaces are sanded with the abrasive paper attached to a sanding block, Fig. 14-5. The abrading action should be in the direction of the grain, as cross grain sanding leaves deep scratches which are hard to remove. Firm pressure is applied to the sanding block as it is moved back and forth. Stock is easier to sand when fastened in a bench vise or with clamps. Always hold the sanding block flat against the surface, sanding the entire surface evenly.

Irregular shapes can be sanded by wrapping strips of abrasive around a dowel rod, Fig. 14-6. Small pieces can be sanded

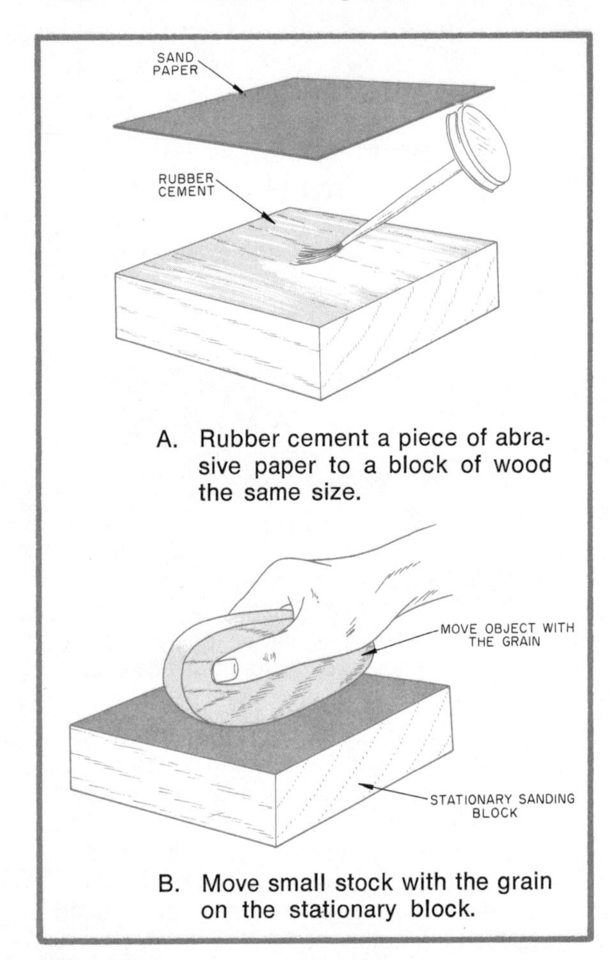

A. Rubber cement a piece of abrasive paper to a block of wood the same size.

B. Move small stock with the grain on the stationary block.

Fig. 14-7. Sanding small pieces of stock.

Fig. 14-8. Removing a small dent with steam.

faster if they are moved back and forth on a flat abrasive surface formed by rubber cementing an abrasive sheet to a block of wood, Fig. 14-7.

Hand sanding can be done with three grits of abrasive, always starting with the coarse grit and finishing with a fine grit. Grits 100, 150, and 220 are satisfactory for most sanding operations.

REPAIRING IMPERFECTIONS BEFORE FINISHING

Dents and other imperfections are repaired before the final sanding. Small dents can often be removed by steaming, using a soldering iron or clothes iron as a heat source. A rag dampened with water is placed on the dent and a hot iron is put on the rag to produce steam. The wood fibers tend to absorb moisture and swell, causing the collapsed wood fibers to return to normal size and the dent to disappear, Fig. 14-8.

Some imperfections must be repaired by filling. Sometimes the chip or hole can be repaired by cutting out the damaged area, then filling it with a piece of matching wood. Cracks and holes can also be repaired with wood fillers, including stick shellac, plastic wood, putty sticks, and wood putty, Fig. 14-9. The color of the filler must match the color of the final finish. Stick shellac is available in a variety of colors which will match most finishes. It is applied with a hot knife. Plastic wood can be mixed with oil stain before it is placed in the hole. Two applications of plastic wood are used on large holes to allow for shrinkage as it dries. Wood putty (sometimes called water putty) is mixed with water to form a paste. It can be colored to match the final finish with dry, powdered water stain. Putty sticks are used after the final finish has been applied.

All parts should be presanded before the product is assembled to prevent the need to sand into tight corners. Avoid sanding the joints as they will become rounded and will not fit properly. After the product is assembled, it should be finish-sanded. At this time, carefully inspect the surface. *All glue must be removed before finishing materials are applied.* Areas covered with glue will not absorb the finish evenly, and a poor appearance will result. During the finish sanding, all *arrises* (corners) are rounded slightly to remove the sharp edges. After final assembly and sanding, the surface is wiped with a lightly dampened sponge to raise the grain fibers which were not completely cut during sanding, Fig. 14-10. After the wood dries, the surface is lightly sanded with fine abrasives to remove the raised fibers. This process is called "raising the grain."

SURFACE PREPARATION WITH SANDING MACHINES

The sanding action of abrasive machines is either stroke, rotary, or a combination of the two. Of the three basic types of portable power sanders, belt sanders operate on the principle of the stroke motion, disk sanders are rotary, and orbital sanders are a combination of the stroke and rotary action, Fig. 14-11.

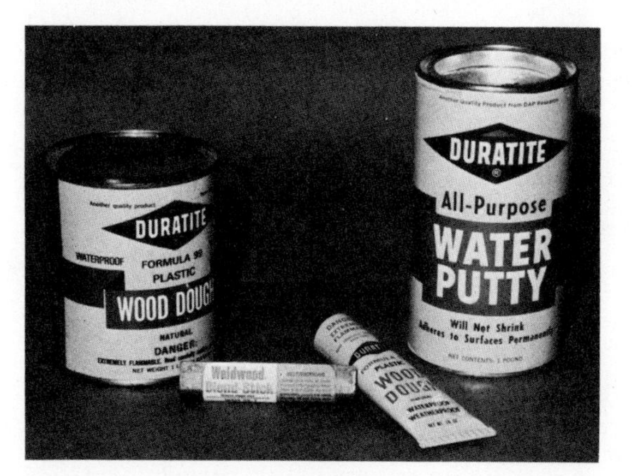

Fig. 14-9. Repair materials for wood surfaces.

PORTABLE POWER SANDERS

Portable belt sanders. Belt sanders such as the one shown in Fig. 14-12 are

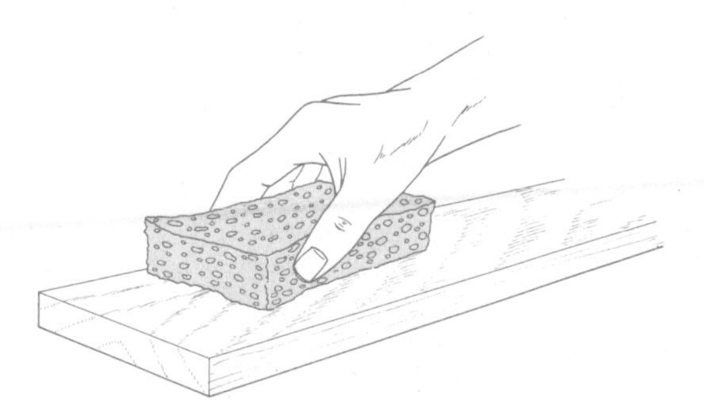

Fig. 14-10. Raising the grain fibers with moisture.

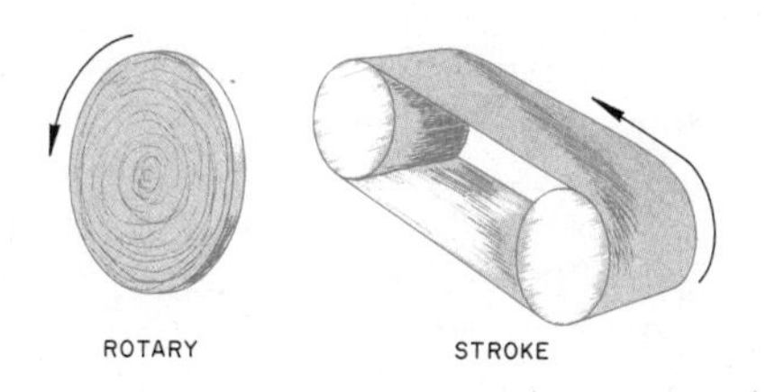

Fig. 14-11. Cutting action of abrasive machines.

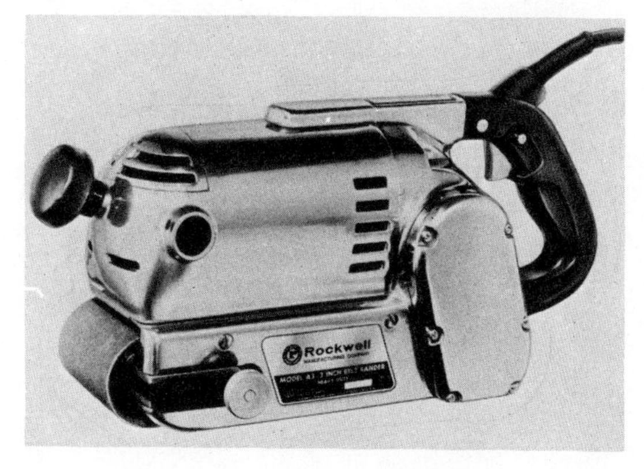

Fig. 14-12. Portable belt sander. (Power Tool Div., Rockwell Mfg. Co.)

used for smoothing large, flat surfaces. The mechanism consists of two rollers, one of which is powered while the other is an idler used to tighten and track the belt. The belt runs over a platen or shoe and provides a flat surface. The sander is powered by a small electric motor which is either belt- or gear-driven. The size of the belt sander is determined by the width of the belt. Common belt sizes range from 2″ to 4″ in width. Cloth belts available in various grit sizes are used on the portable belt sander.

Always disconnect the sander prior to installing a belt. A belt is then installed by releasing the tension on the front roller and placing the belt on the rollers with the arrow on the inside of the belt pointing in the direction of travel. Tension is applied to the belt by releasing the idler roller spring mechanism. Track the new belt by turning the tracking adjustment.

To sand stock with a portable belt sander, clamp the stock to a bench or a vise. The sander is started and slowly lowered onto the work. It should be moved evenly from end to end on the workpiece. It is moved toward the opposite edge of the stock with each stroke. Care must be taken not to allow the sander to "cut" too long in one place — the entire surface should be equally sanded. The weight of the sander provides ample pressure, therefore, additional pressure should be avoided. It is important that the sander be held flat against the workpiece and not allowed to rock on the rollers, Fig. 14-13. The power cord must be held back away from the abrasive belt.

Portable disk sanders. Disk sanders, Fig. 14-14, are used less in the wood industry than other types of sanders. The disk on which the abrasive is attached rotates on a motor shaft. The disk-like pad on which the abrasive is mounted is usually rigid rubber. Pressure is exerted on the sander during use until the pad flexes slightly. Kits are available to convert portable power drills into disk sanders. The rotary action of the disk leaves swirls in the work which must be removed by straight-line sanding.

Portable finishing sanders. There are two types of finishing sanders, both with oscillating action, Fig. 14-15. These are the *orbital* and *in-line* sanders.

The finishing sanders are limited in use to final sanding before finish application, and are not designed to remove large amounts of stock. The orbital finish sander cuts more rapidly, but its greatest disadvantage is that it leaves small swirls in the wood which must be removed with a straight-line sander or by means of hand sanding. Finishing sanders are powered

Fig. 14-13. Smoothing a flat surface with the portable belt sander.

Fig. 14-14. A portable disk sander.

by either air or electric motors, Fig. 14-16. The straight-line sander moves back and forth in the direction of the grain. This type sander can be used in the final surface preparation stage with very fine abrasives.

Various models of finishing sanders have different mechanisms for attaching the abrasive paper. The operator's manual must be consulted for a particular model. Standard sheet abrasive can be cut to the required sizes and fitted on finishing sanders. When using the finishing sander, the operator should refrain from exerting pressure, as the weight of the sander is adequate for proper cutting action.

STATIONARY SANDING MACHINES

Stationary disk sanders. The disk sander is used to smooth and shape straight and convex surfaces, Fig. 14-17. The abrasive disk is usually paper-backed and is cemented with a special cement or adhesive to the metal disk. The size of the disk sander is determined by the size of the disk. The table on most disk sanders can be tilted to support the work at the desired angle, and usually has a miter gauge slot in which guides can be placed to support the work. The miter gauge can be set and used with the tilted table to form compound angles, Fig. 14-18. The

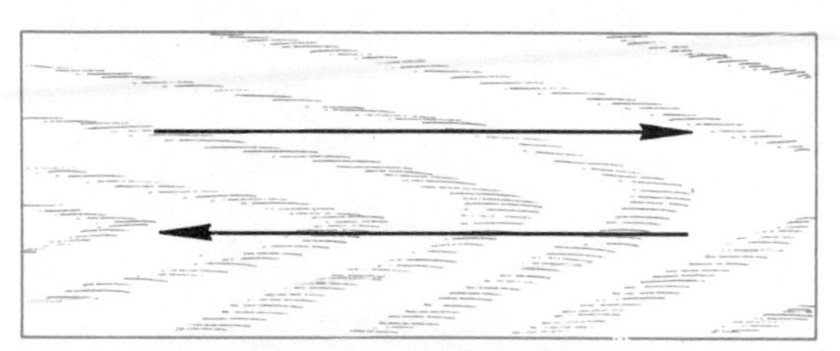

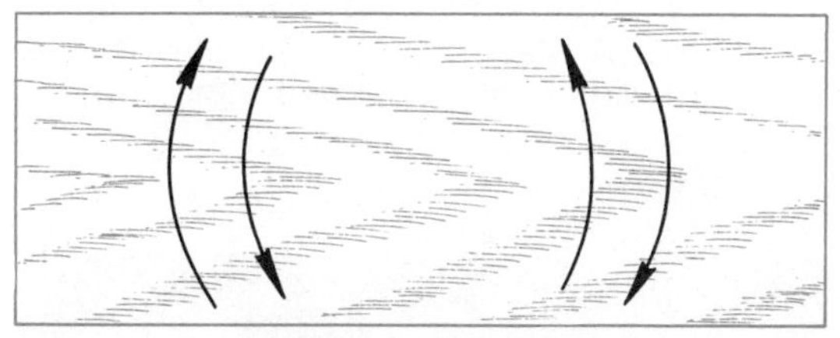

Fig. 14-15. Sanding action of in-line and orbital finishing sanders.

Fig. 14-16. An air powered, straight-line finishing sander.

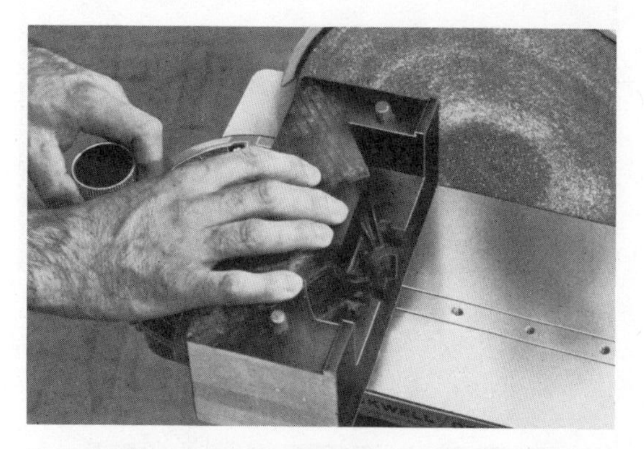

Fig. 14-17. A stationary disk sander. (Power Tool Div., Rockwell Mfg. Co.)

Fig. 14-18. Forming a compound angle on the disk sander.

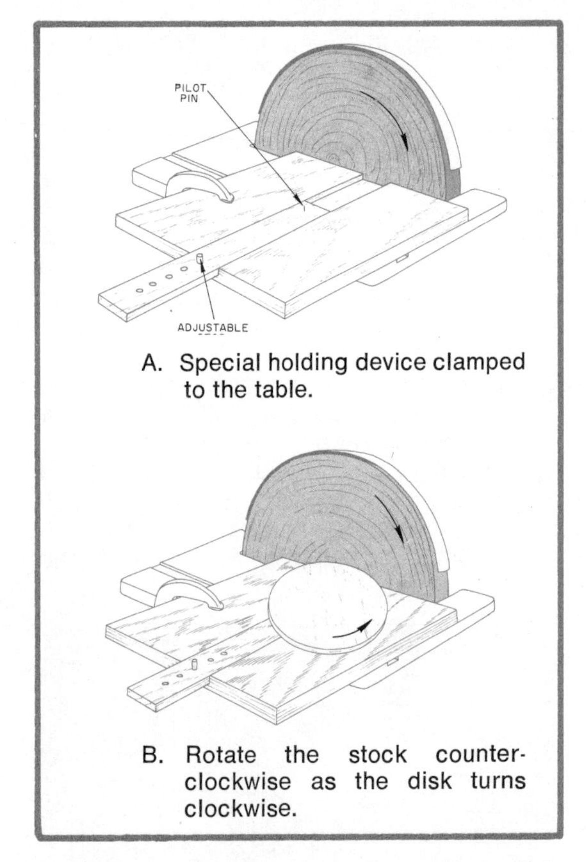

A. Special holding device clamped to the table.

B. Rotate the stock counter-clockwise as the disk turns clockwise.

Fig. 14-19. Sanding circles on the disk sander.

Fig. 14-20. Smoothing and shaping a convex edge on the disk sander.

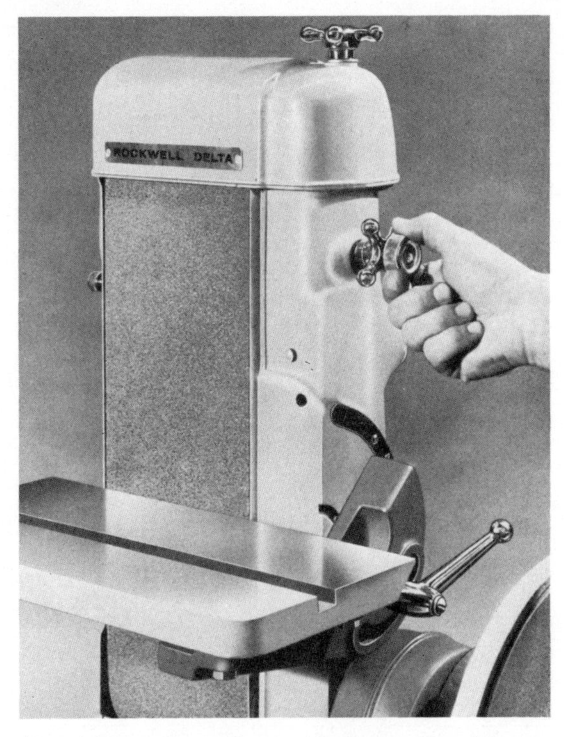

Fig. 14-21. A stationary belt sander. (Power Tool Div., Rockwell Mfg. Co.)

Fig. 14-22. Squaring small stock using a miter gauge and the belt sander.

disk sander can also be equipped with a shop-built holding device for sanding circles, Fig. 14-19. To smooth irregularly shaped convex curves on the disk sander, a layout line is drawn on the stock after the part has first been rough-sawed to shape, Fig. 14-20.

Stationary belt sanders. Belt sanders such as shown in Fig. 14-21 are used to smooth flat surfaces. They are available in a wide variety of sizes, determined by the width and length of the abrasive belt. Regardless of the size or type, the principle of operation remains basically the same. The machine consists of two drum-like pulleys, one powered and the other an idler. A cloth abrasive belt runs on a platen which is fastened between the two pulleys. One pulley moves to tighten and track the belt in a straight line.

The table on stationary belts such as shown in Fig. 14-21 can be tilted at various angles, and has a miter gauge slot in which a guide fits to support the work. The miter gauge is used to sand stock square, Fig. 14-22.

Most stationary belt sanders can be used in either a vertical or horizontal position. The table acts as a stop for sanding in a horizontal position, Fig. 14-23.

The top guard can be removed from the belt sander to smooth concave curves, Fig. 14-24. It is helpful to attach a shop-built fence to help support the work at a right angle to the idler drum. Special care must be taken when using the machine with the guard removed. Always replace the guard after sanding concave curves.

Spindle sanders. The spindle sander as shown in Fig. 14-25 consists of a ver-

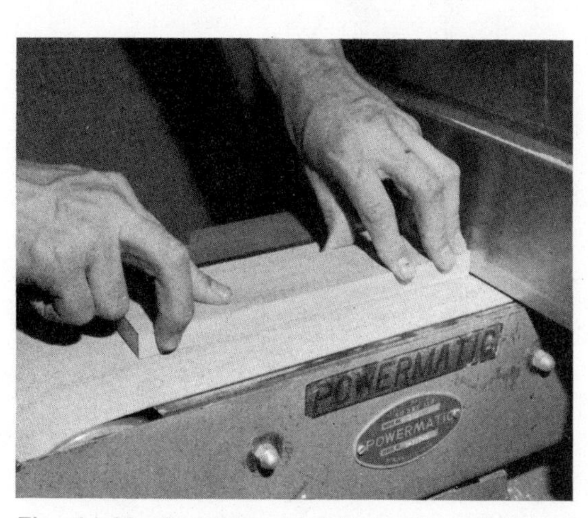

Fig. 14-23. Sanding stock flat with belt in the horizontal position.

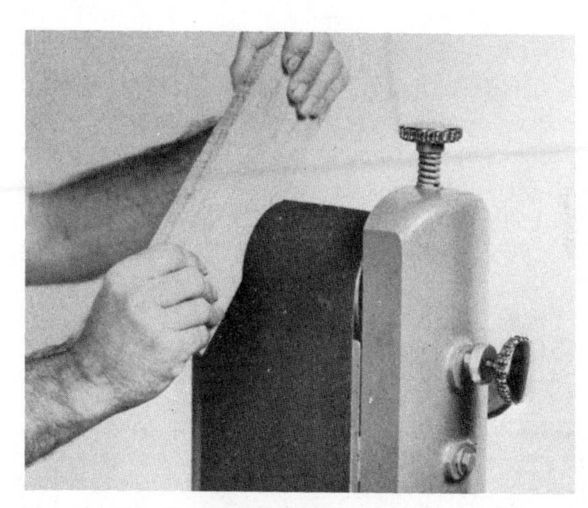

Fig. 14-24. Smoothing a concave edge on the idler pulley of a belt sander with the guard removed.

Fig. 14-25. A spindle sander. (Boice Crane)

tical spindle which projects through a hole in the table. The table can be tilted to the desired angle. The spindle is attached to an oscillating mechanism which moves the spindle up and down as it rotates. An abrasive-coated sleeve slips over the rubber spindle, and the rubber expands when the spindle nut is tightened to hold the sleeve in place. The spindle can be changed to a size suited to the contour of the work, Fig. 14-26.

Hand-stroke belt sander. This sander is similar in design to other belt sanders except the work is supported on a table which moves from side to side below the belt, Fig. 14-27. The belt mounted on the revolving drums can be lowered to the work below. On some machines, the table moves up and down to make contact with the belt. A paddle block is used to force the belt against the stock, moving back and forth the length of the work. This machine is used to smooth large surfaces such as table tops.

Fig. 14-27. Hand stroke belt sander. (Boice Crane)

Fig. 14-26. Smoothing irregular shapes on the spindle sander. (Boice Crane)

Fig. 14-28. Pneumatic sanding drum.

SPECIAL-PURPOSE SANDING MACHINES

Pneumatic sanding drum. The pneumatic sanding drum consists of an inflatable tube mounted on a shaft, with the abrasive sleeve mounted on the tube. When the drum is slightly inflated, the pliable tube causes the abrasive sleeve to conform to the contour of the workpiece, Fig. 14-28.

Brush-backed abrasive wheels. Figure 14-29 shows brush-backed abrasive wheels consisting of tabs of abrasive paper which project out from slots in the wheel. The tabs are backed with small brushes which force the abrasive against the work as it is pressed against the face of the wheel. As the abrasive wears, new strips stored on rolls inside the wheel are pulled out. The brush-backed sander is used for final smoothing and polishing of intricate irregular shapes.

Combination belt sander/grinder. This machine has a narrow abrasive belt which makes it possible to sand intricate shapes, Fig. 14-30. The table can be tilted at an angle for sanding compound angles.

Special belt sanders are designed for edge sanding. The drums move up and down as the belt revolves, giving equal wear to the entire surface of the abrasive. Many edge belt sanders are equipped with tables on the ends for sanding concave surfaces.

INDUSTRIAL SANDING MACHINES

Wide belt sanders are used in the wood industry for smoothing large sheets,

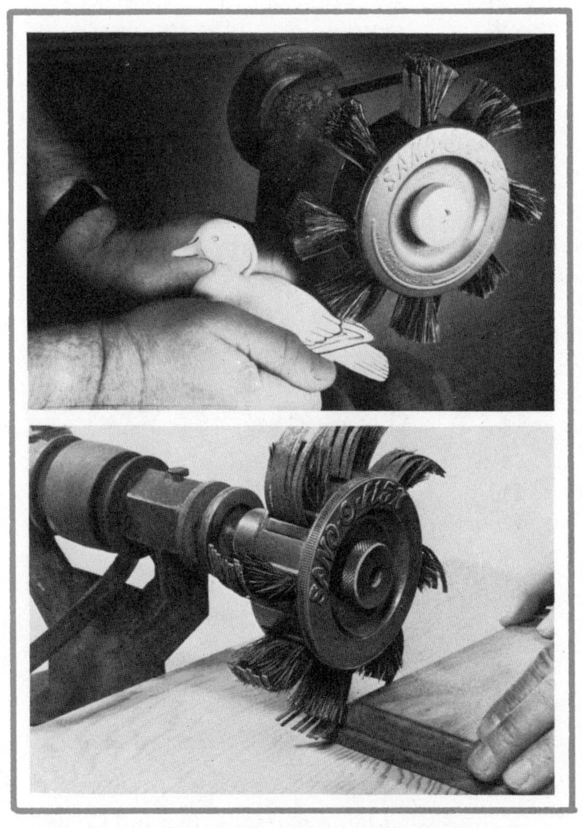

Fig. 14-29. Sanding with brush-backed sanding wheels. (Sand-O-Flex)

Fig. 14-30. An edge belt sander. (Norton Coated Abrasive Division)

such as the sander shown in Fig. 14-31, used for sanding plywood panels as they are manufactured. They are designed for sanding to a close tolerance, as the stock is passed through the machine on a conveyor belt.

Wide belt sanders are available in a variety of widths, ranging from 2′ to 6′. The rate of feed of the stock through the machine can also be varied, Fig. 14-32. The belt is automatically tracked on the drums.

Some belt sanders have two abrasive belts. The first belt rough sands the material to the approximate thickness, and the second belt is equipped with a fine grit belt and finish sands the stock smooth.

The wood industry also uses drum sanders, Fig. 14-33. The abrasive is mounted on a drum which revolves and smooths the surface of the stock as it is automatically fed through the machine by means of a conveyor system. There are also many special-purpose industrial sanding machines, Fig. 14-34.

Fig. 14-31. Wide belt sander smooth plywood panels. (Norton Coated Abrasive Division)

Fig. 14-32. The rate of feed of the stock can be varied on some wide belt sanders. (Norton Coated Abrasive Division)

WOOD COATINGS AND TREATMENTS

Finish is applied to wood for two basic purposes: (1) to preserve the wood, and (2) to beautify or enhance the appearance. Wood is a porous material and tends to absorb moisture. When it is used for exterior purposes, the moisture content may become high enough to cause the wood to rot. A surface finish applied to the wood

Fig. 14-33. A drum sander used for smoothing large surfaces. (Norton Coated Abrasive Division)

Fig. 14-34. Special-purpose industrial sanding machines. (Norton Coated Abrasive Division)

tends to repel moisture, Fig. 14-35. The paint protects the wood from moisture, fumes, and ultra-violet sun rays which also cause the wood to deteriorate. At the same time, the paint improves the appearance of the house. The life of a house which has been properly painted can exceed a hundred years.

Finishing is a major process in the woodworking industry. It is the final process in the manufacture of a product. The finish is applied as either a means of protecting or decorating the surface, and often serves both purposes. Finishing consists of covering the surface with a thin film of liquid material. The liquid dries or cures by means of oxidation or evaporation. Some finishes cure due to a chemical change.

CLASSIFICATION OF WOOD FINISHES

Finishes can be classified as either *opaque* or *transparent.* Opaque finishes which include such materials as paint, enamel, and pigmented lacquer are commonly used on wood when the grain does not have a naturally pleasing appearance because they cover up the surface. They are popular for color addition and variety in decorating schemes, Fig. 14-36.

Fig. 14-35. A house protected and made more attractive by paint. (Hedrick-Blessing)

Properly-selected opaque finishes are also effective for illumination purposes. Light-colored opaque finishes tend to reflect light, while dark opaque finishes absorb light. Various lighting effects can be created by the proper selection of finishes.

Fig. 14-36. Effective use of opaque finish for beautifying and protecting wood surfaces. (American Plywood Association)

Fig. 14-37. The natural beauty of wood remains visible through a protective transparent coating. (Brandt)

Transparent finishes, on the other hand, are applied to wood to accent the natural beauty of the wood. The grain figuration and the color of the wood is visible through the transparent finish although it still protects the wood from moisture and dirt, Fig. 14-37. Common types of transparent finishes include varnish, stain, lacquer, shellac, wax, and penetrating oils.

Finishes can also be classified as either *penetrating* or *build-up.* Some quantities of all finishes are absorbed into the pores of wood. There are some woods which tend to absorb more finish than others because they are more porous. The penetrating finishes include Danish oils, linseed oils, and some wax finishes. The chief advantage of the penetrating finish is the ease of application, as it is usually wiped on the stock with a clean cloth. The penetrating finish is also easy to repair as a surface is damaged.

Build-up finishes protect the wood by a film which is built up on the surface with each application. The build-up finish usually provides a higher gloss than a penetrating finish, and it provides better protection from moisture, ultra-violet rays, and dirt. A build-up finish can be wiped clean when it becomes covered with finger prints and grime. Paint, varnish, shellac, enamel, and lacquer are only a few of the many types of build-up finishes.

A build-up finish must be elastic as nonelastic finishes crack and check. That is, the film must expand with the wood when it increases in moisture content.

COMPOSITION OF FINISHING MATERIALS

Finishes are composed of vehicles, pigments, driers, and thinners. The vehicle is the liquid portion of a finishing material consisting of binders and thinners. It may be linseed oil combined with various types of resins for oil base finishes, or it may be water for latex finishes. The greatest percentage of volume in finishing material is the vehicle. Pigments are added to opaque finishes to produce a decorative color; driers are added to accelerate the drying and curing process; and thinners are added to produce proper consistency for application, Fig. 14-38.

STAINS

Wood may be stained to assimilate another type of wood, that is, less expensive wood may be stained to appear similar to expensive wood. Staining accents the grain and enhances its natural beauty. The staining process may be omitted in the finishing process, as some kinds of wood are more attractive if they are not stained. However, if staining is used, it is the first step in the application of the finish. Stain is used to cover sap streaks and give a uni-

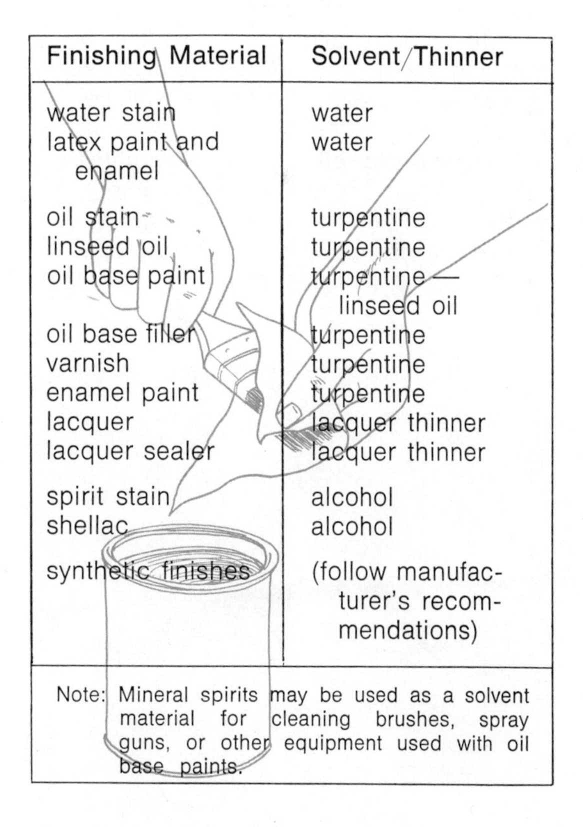

Finishing Material	Solvent/Thinner
water stain	water
latex paint and enamel	water
oil stain	turpentine
linseed oil	turpentine
oil base paint	turpentine—linseed oil
oil base filler	turpentine
varnish	turpentine
enamel paint	turpentine
lacquer	lacquer thinner
lacquer sealer	lacquer thinner
spirit stain	alcohol
shellac	alcohol
synthetic finishes	(follow manufacturer's recommendations)

Note: Mineral spirits may be used as a solvent material for cleaning brushes, spray guns, or other equipment used with oil base paints.

Fig. 14-38. Thinners and solvents for use with common finishing materials.

form color to an entire surface, Fig. 14-39. It consists of a vehicle combined with a coloring material, and is absorbed into the porous wood material.

There are four basic types of stains used in the wood industry: (1) oil, (2) water, (3) non-grain raising, and (4) spirit. The classification is based on the type of vehicle or solvent used. Oil stains are of two basic types: (1) *penetrating*, and (2) *pigmented*.

Penetrating oil stains. Penetrating oil stains are easy to apply by wipe-on or brush methods. They also have the advantage of eliminating the need to dampen the wood fibers and to sand prior to application because they do not raise the wood grain. However, their tendency to bleed (combine with top coats applied later) and to fade when exposed to direct sunlight are definite disadvantages. To minimize bleeding, penetrating oil stains may be followed by a washcoat of shellac which serves as a sealer. Washcoats will be discussed later in the chapter.

Pigmented oil stains. Pigmented oil stains (often called "wiping stains") are widely used in the furniture industry. Unlike penetrating oil stains, they do not bleed or fade. It is possible to produce an even color tone over the entire product with this stain.

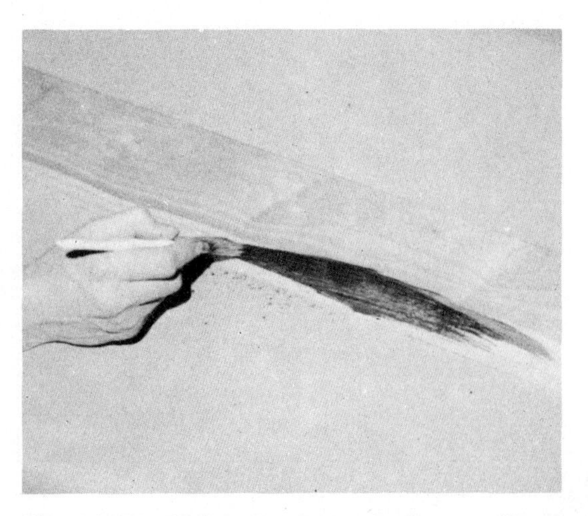

Fig. 14-39. Staining sapwood gives uniform color to surface.

Oil stains are applied by spray, brush, dipping, or swabbing with a cloth. The stain is allowed to be absorbed into the wood, but excess stain is wiped off the surface before it completely dries. The grain of the wood may not show through if excessive stain is left on the surface.

Water stains. Water stains consist of colored powdered dyes sold in dry form then mixed with water. They provide a transparent (clear) surface, allowing the natural grain of the wood to remain visible. The stain is absorbed deep into the wood, and does not fade as much as other stains. The major disadvantage of water stain is that it tends to raise the grain of the wood, requiring one additional sanding operation. The surface of the wood is dampened with a sponge before staining which raises the grain. The raised grain is removed with 180 abrasive paper. The stain is then applied with a brush or sponge. Additional water should be applied to end grain to avoid excessive darkening of these areas. Care must be taken when applying stain to avoid streaks. The stain dries slowly and should dry overnight before additional finish is applied. Sand the stained surfaces lightly to remove the remaining raised grain.

Non-grain raising stains. Non-grain raising stains (NGR) have many of the fine characteristics of water-stains; however, they do not raise the grain. The NGR stain dyes are combined with glycol, and alcohol solvents and are applied by spraying, brushing, or dipping. They are widely accepted in the furniture industry.

Spirit stains. Spirit stains are not widely used in the furniture industry because they tend to fade. They consist of dye mixed with alcohol.

WASHCOATS

Stains are usually coated with a sealer often called a washcoat. A washcoat of shellac prevents the stain from bleeding into the finish coats. A good mixture is one part of 4 lb. cut shellac combined with seven parts of alcohol. Lacquer sealers are used as a seal coat for lacquer finishes. Wash or seal coats are sanded lightly before top coats are applied.

FILLERS

The pores of open-grained woods are usually filled with a paste filler to provide a smooth final finish. Common woods requiring filler are oak, walnut, hickory, ash, and butternut. A washcoat should be applied to the surface before filling, which also makes it easier to remove excess filler as it is applied.

Fillers can be purchased as either paste or liquid. The paste filler can be thinned to a creamy consistency with mineral spirits or turpentine. Thinner consistencies are used on wood with small pores. Filler may be purchased either natural in color or stained with oil stains to the desired tone.

Filler is applied to the properly-prepared wood with a stiff bristled brush, in both directions of the grain to fill all the pores, Fig. 14-40. The filler is left on the surface until it no longer appears wet. At this point, the filler on the surface is rubbed off with a burlap rag rubbed *across* the grain. Wiping in the direction of the grain removes the filler from the pores. Care must be taken to remove all excess filler before it dries completely, Fig. 14-41. Allow the filler to dry 24 hours before applying top coats.

GLAZING AND SPECIAL EFFECTS

Shading and tinting stains are sometimes used after staining and/or filling to produce an antique look. Shading stains consist of lacquer colored with dye. This is also a method of adding additional stain to give a uniform color to an entire surface, in which case it is applied before the last top coat is applied. When a shading stain is used for antiquing, it is applied before the top coats. Glazes available in many colors can be applied to the edges and in depressions to give the furniture an "aged" look.

BLEACHING

Bleaching is seldom necessary in the finishing operation. It is the process of removing color from dark wood when a lighter colored wood is desired. Special commercial bleaches are the most satis-

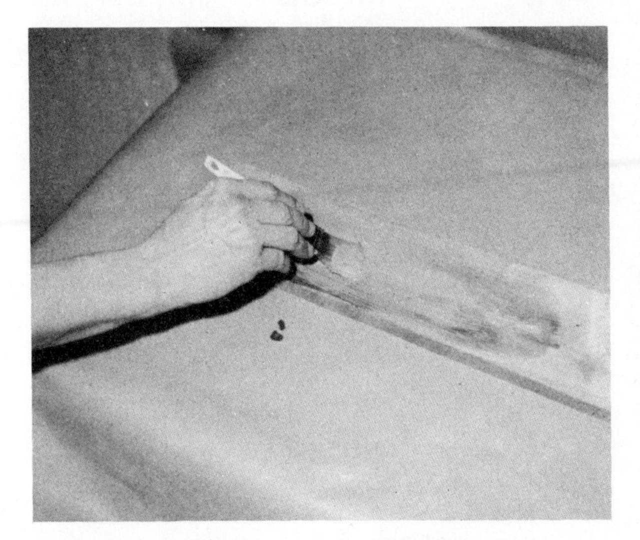

Fig. 14-40. Applying paste wood filler to fill pores.

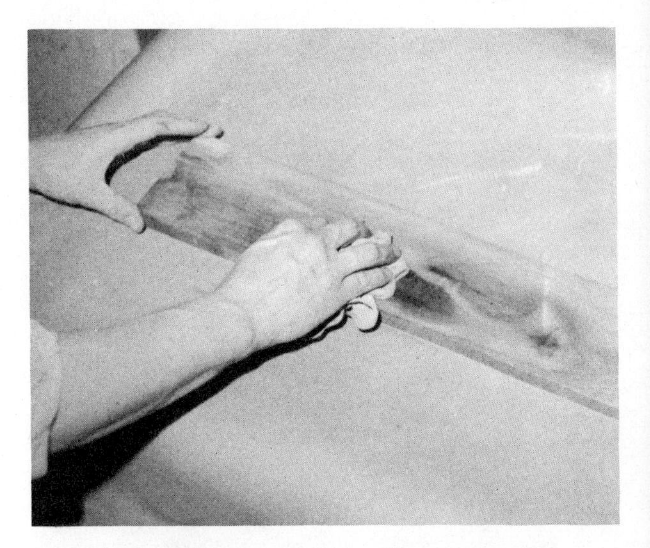

Fig. 14-41. Removing paste filler from the surface with a coarse textured cloth.

factory. Instructions for use are furnished with the product. The solution, usually applied with a sponge, can be harmful to the skin, so always wear rubber gloves to bleach wood.

SCHEDULING FINISH

Applying a fine finish requires a complex operation. The process can be simplified by means of a proper plan called a finishing schedule. The type of schedule will vary depending on the size of the product and type of finish. Always inspect the product after the final sanding operation. Any dents should be removed at this time. Steaming as shown in Fig. 14-42 will remove most dents. Special care should be taken to remove any glue from the surface, and nail holes and any imperfections in construction should be filled with plastic wood, stick shellac, or other special filler materials. All surfaces are sanded for the final time with 220 abrasive paper, then dust must be removed with a brush or air hose before final finishing can be started.

An example schedule for applying varnish finish on walnut oil stained mahogany might be as follows:

1. Prepare and inspect the surfaces.
2. Select and apply a walnut-colored oil stain. Allow to dry overnight.
3. Apply washcoat of shellac and allow to dry. Sand lightly when dry.
4. Apply paste wood filler to which walnut stain has been added to fill pores. Wipe off excess filler and allow to dry.
5. Cover paste filler with light washcoat to seal and prevent bleeding. Allow to dry and sand lightly.
6. Apply first top coat of clear varnish.
7. Sand first coat lightly with 180 abrasive paper after it has dried 24 hours.
8. Remove dust using a tack cloth (a rag dipped in turpentine).
9. Repeat steps 6, 7, and 8 until three coats of varnish have been applied.
10. Lightly sand and polish final coat.
11. Apply furniture polish and rub to desired luster.

Similar finishing schedules can be established for other types of finishes. Finishing is an important operation and should be a part of the total plan of procedure.

Fig. 14-42. Removing dents with steam.

SELECTING A TOP COAT

Many types of final finishes are available to the wood industry. The finish selected will depend on the particular purpose of the product and the desired appearance. The following common finishes are available. The special characteristics of each makes it more desirable for a specific purpose.

OPAQUE FINISHES

Oil base paint. Oil base paint has commonly been used for coating interior and exterior wood surfaces. It serves the functions of decorating and preserving the wood. The vehicle or base is, for the most part, linseed oil combined with resins.

Paint is selected as a finish for products produced from less expensive softwoods. A primer is recommended as an undercoat for oil base paints.

Enamel. Enamel is a varnish to which a pigment has been added. It dries to produce a much harder surface than paint, and also has a higher gloss than most paints. Enamel is very durable and is used for wood trim, cabinets, and bathroom and kitchen walls. It can be washed repeatedly when it becomes spotted. An undercoat is used before coating bare wood with enamel, and two coats are normally required for proper coverage. It can be applied by means of a brush, roller, or spray, and is available in gloss or a less shiny semi-gloss.

Latex emulsion paints. Latex emulsion paints are relatively new as a finishing material. They are easy to use because the painting tools can be cleaned with soap and water. Latex paints have water as the vehicle or base, and are available for both interior and exterior purposes. Latex paints are nearly odorless and dry to the touch in about one-half hour. The paint film is not as hard as oil base paint or enamel.

Pigmented lacquer. This opaque finish is limited in use to furniture finishing. It is the same as transparent lacquer only pigment has been added. The lacquer finish is very hard and dries very quickly. There are many types of lacquer available, each having special qualities. It cannot be used over old finishes which may already be on a product, as it softens and lifts other types of finishes. It may be either brushed or sprayed on wood products.

TRANSPARENT FINISHES

Shellac. The trachardia lacca bug of India converts the sap of certain trees to a substance which is processed and purified, and the end result is a transparent finishing material called shellac. Various "thicknesses" of shellac are referred to as *cut.* A common cut available commercially is the 4 lb. cut. This means that a mixture of 4 lbs. of shellac gum is combined with 1 gal. of denatured alcohol. Such a mixture would equal about 1.4 gal. of liquid shellac. A gallon container of shellac contains slightly less than 3 lbs. of dry shellac gum.

Shellac is naturally yellow but is bleached white for most purposes. When used as a top coat, 4 lb. cut shellac is mixed about half-and-half with alcohol; when used as a wash or sealer coat it is mixed about 1 part 4 lb. cut to 7 parts alcohol.

Shellac is a very hard finish; however, it has low resistance to moisture and today has limited use as a top coat. It is an excellent material for use as a washcoat.

Varnish. Varnish is a term including many transparent finishing materials. It includes both natural and synthetic resins, vehicles, driers, and solvents. In general, varnishes provide a hard, clear finish with a high luster.

SYNTHETIC RESINS

Synthetic resins are the result of vast research in the chemical industry. The most popular synthetic resins include the polyurethanes, polyesters, and epoxies.

Polyurethane. Polyurethane is extremely hard and has high resistance to wear and abrasion. It is available for use as either an interior or exterior finish, and is highly recommended for surfaces receiving heavy use such as table tops, desks, and floors. The major disadvantage of polyurethane is its tendency to yellow. It may be applied by spray or brush.

Polyester resin. Polyester resin consists of two materials which must be mixed before application. One part is the resin and the other is the catalyst. The two parts are commonly sprayed with two nozzles in industry, but they combine before touching the work. The material may also be combined and then brushed on the surface. Polyester finish is extremely hard and thick coatings can be applied.

Epoxy resin. Epoxy resin is similar to polyester resin in that it consists of two parts. Epoxy resin is extremely hard and is resistant to nearly all hazards that affect finishes. It is quite expensive and rather difficult to use, but it is very likely to be a prime finish of the future.

Research is constantly being conducted in an effort to develop improved finishes. Pigmented finishes are tested to develop materials which will not fade, Fig. 14-43.

PREPARATION AND STORAGE OF FINISHING MATERIALS

Finishing materials must be properly stored to prevent waste and to insure quality finishes. When using a finish, it is best to pour it from a large storage container into a small can to prevent contamination of the large quantity. The storage can must be sealed tightly to prevent skim layers from forming on the finishes. Clean the lip of the storage can to obtain a tight seal.

Some finishes which are used regularly can be stored in a service container, or a rubber stopper through which the brush extends is good for others, Fig. 14-44. Washcoats of shellac and sealers may be stored this way. The brush does not require cleaning each time.

Fig. 14-43. Painted surfaces are exposed to the sun rays and other elements of weather to test their resistance. (Rohm and Haas Co.)

METHODS OF APPLYING FINISHES

Finishing materials may be applied by various methods such as brush, roller, aerosal spray can, cloth pad, and air spray.

BRUSHES

Brushes are used for applying most types of finishes. There are many types, sizes, and qualities of brushes. A good

Fig. 14-44. A stopped container saves cleaning some types of brushes each time they are used.

brush is essential for high-quality finishing. It is helpful to select a brush suited to the type of work.

The parts of a brush include the handle, ferrule, plug, bristles, and the setting, Fig. 14-45. The bristles may be either natural or synthetic. High-quality brushes are generally pure bristle. The best bristle comes from the Chinese hog, although hair from other animals may also be used. Chinese hog hair is especially good due to its shape, which is split or flagged on the end and oval in the cross-section. The split ends tend to carry more paint, and the oval cross-section prevents tangling of the bristles. In addition, the bristles are tapered which makes them spring back into shape, and the length of the bristles vary, causing the brush to be tapered.

Bristles are held in the brush by means of the setting inside the ferrule, and are spaced apart by means of a plug. The plug forms a hollow area to carry paint.

Many brushes are being manufactured today with nylon bristles which have proven superior for use with latex emulsion paints. The nylon wears longer than animal hair.

Brushes are available in wide range of sizes. Generally it is best to select the smallest brush which can be conveniently used for the particular surface being finished.

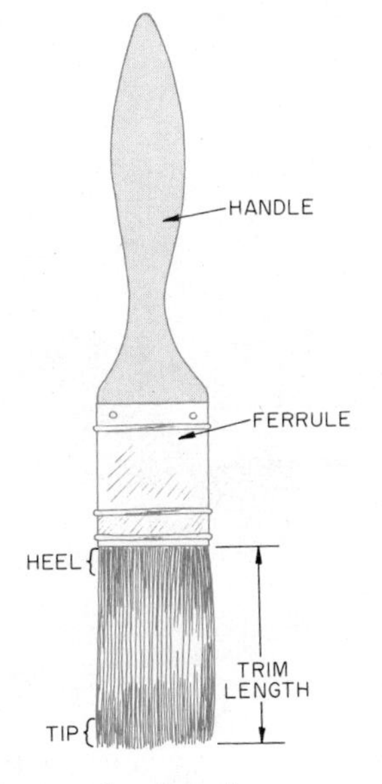

Fig. 14-45. Parts of a brush.

Applying finish with a brush. Special care must be used to prolong the life of a brush. If properly cared for, they will last for years. The following method of using a new brush will insure extended life and good results.

1. A brush often has a few loose bristles and may be dusty. To correct this, briskly but gently brush the bristles across the palm of your hand.
2. Dip the brush in solvent all the way to the ferrule. This prevents finishing materials from collecting in the heel of the brush.
3. Brush the material using the full width of the brush and not the edge.
4. Avoid "dabbing" with the brush as this tends to break and deform bristles.
5. Clean the brush immediately after finishing the job. Brushes should not be allowed to stand on the bristles in the bottom of a can.

Proper use of a brush. Select a brush which is best-suited to the material and the product. The brush should not be dipped in the finishing material over one-third to one-half the length of the bristles, Fig. 14-46. Remove the excess finishing material

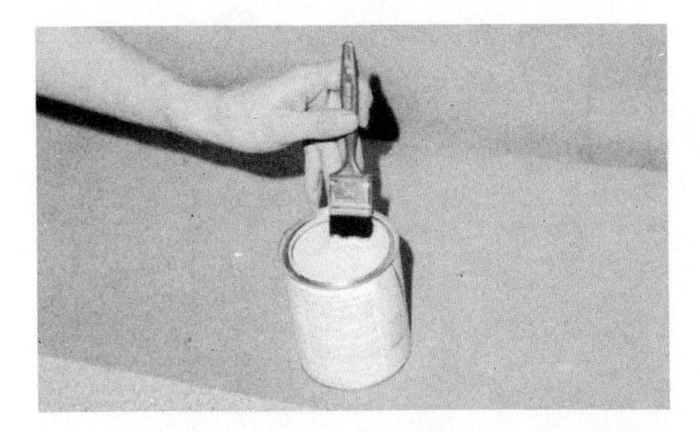

Fig. 14-46. Dipping the brush one-third to one-half the length of the bristles.

from the bristles as it is removed from the container, Fig. 14-47.

Apply the finish, brushing the direction of the grain as shown in Fig. 14-48. Brush from the center out in both directions. It is better to apply several coats than to apply excessive amounts at one time.

Fig. 14-47. Excess paint is removed from the bristles.

Fig. 14-48. Brushing finishing materials onto surface from center to outside edges.

Cleaning brushes. Proper cleaning and storage of brushes will increase the life of a brush. More brushes are ruined from improper cleaning than from excessive wear. The cleaning process is very simple and should become a habit. The following procedure will serve for cleaning most brushes:

1. Brush the excess material on the edge of the finish container. What remains can be brushed out on a paper towel or newspaper.
2. Wash the brush in the proper solvent, Fig. 14-49.
3. Wash the solvent from the brush with soap and warm water. Con-

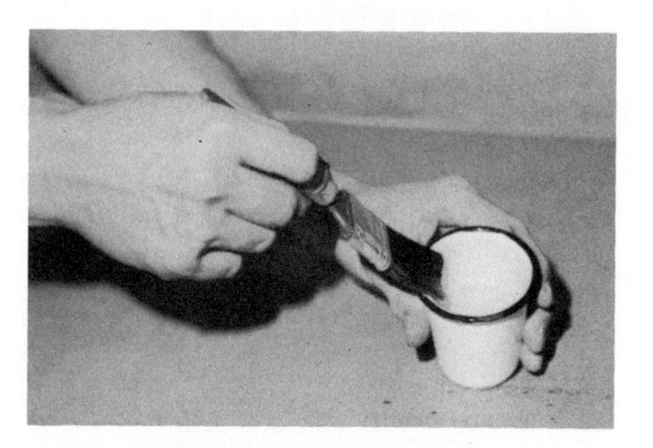

Fig. 14-49. Washing a small brush.

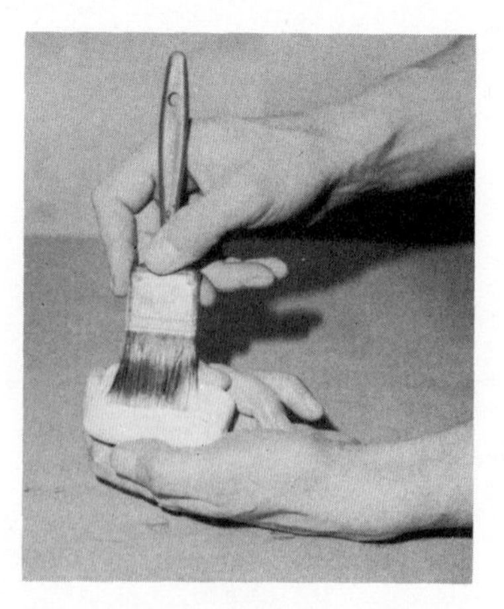

Fig. 14-50. Using soap and water to clean a brush.

tinue until no traces of finish or solvent appear, Fig. 14-50.

4. Rinse the brush in clear water and pat dry with a paper towel.
5. Wrap the bristles and prepare for storage. Wrapping holds the bristles straight and keeps them free of dust, Fig. 14-51.

ROLLERS

The hand roller has become an accepted method of applying paint by both professional painters and homeowners who do their own painting, Fig. 14-52. The roller method is an effective means of coating large surfaces.

The covering material on the roller must be matched to the surface and type of finish. A typical assortment of roller tools is shown in Fig. 14-53. Special rollers are available for corners; however a finish brush is generally used for inside corners. Extension handles can be attached to the roller to reach high surface areas.

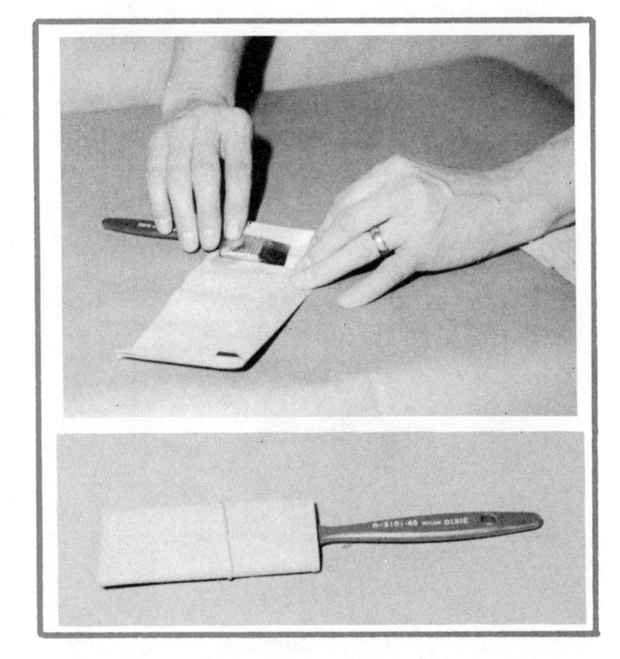

Fig. 14-51. Wrapping a brush for storage.

The roller covers may be washed and reused. The proper solvent must be used to remove the finishing material.

Fig. 14-52. Applying paint with a roller.

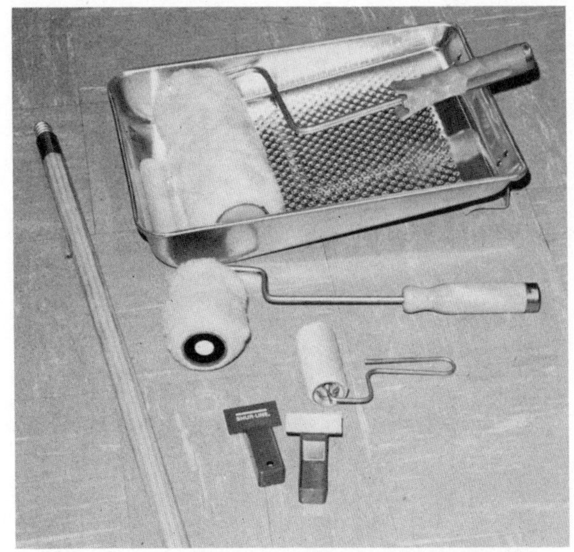

Fig. 14-53. Assortment of tools and equipment for applying finish with a roller.

AEROSOL SPRAY CANS

The aerosol spray method is ideal for finishing small products. A large assortment of finishing materials is available in aerosol spray cans, Fig. 14-54. It is an easy method of coating small areas.

The aerosol spray can contains the finishing material and Freon gas. The gas serves as the propellent for the liquid finishing materials. The Freon is in the form of liquid when placed in the can with the paint. The liquid Freon changes to gas at about 70° F. and provides spraying pressure of about 40 lbs. per square inch. A metal ball is placed in the bottom of the can to stir the paint when the can is shaken. The finish flows through a tube extending from the bottom of the can up through the spray valve. Shake the container well before spraying with the aerosol spray. The nozzle of the can is held about 8″ to 12″ from the surface. The valve is easily cleaned by turning the can upside down and spraying, as only gases escape when the can is inverted.

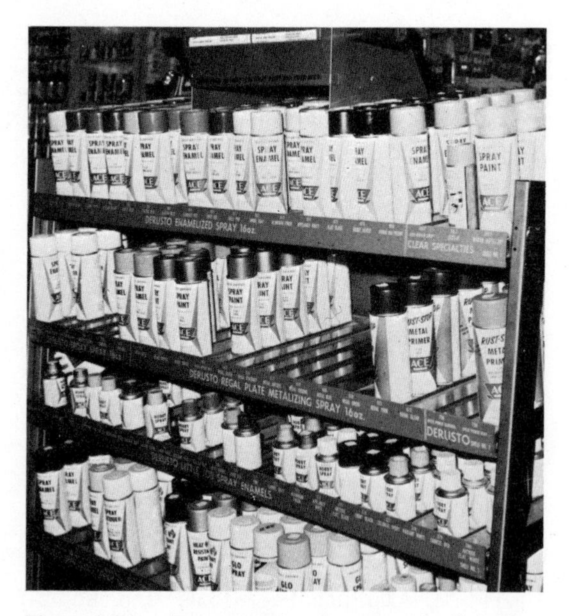

Fig. 14-54. Assortment of aerosal finishing materials.

AIR SPRAY

The air spray method of applying finish has become widely accepted in most industries. Many types of finishes with rapid drying characteristics have been designed for spray application. Spray application provides an excellent finish in a short period of time. The main equipment for spray application consists of an *air compressor, spray gun,* and a *pressure regulator.* The compressor must be provided with a power source such as an electric motor, Figs. 14-55 and 14-56. Various hoses and spray booth are also needed. A compressor must be large enough to provide about 40 psi of constant air pressure.

An efficient spray system requires the use of a pressure regulator to which supply

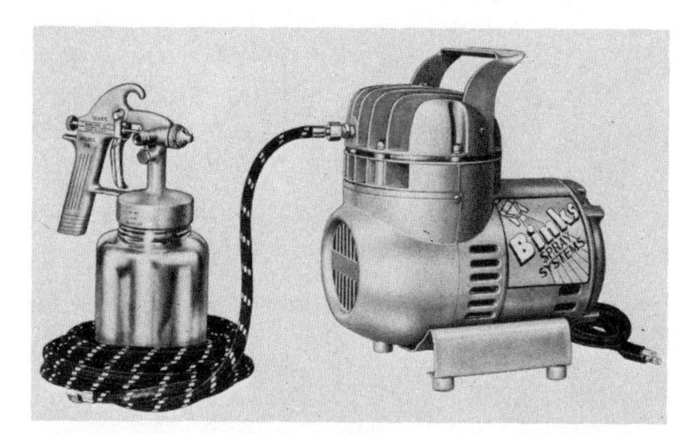

Fig. 14-55. A portable air compressor. (Binks Spray Systems)

Fig. 14-56. A general-purpose spray gun. (DeVilbiss Co.)

lines are attached, Fig. 14-57. The regulator is placed between the compressed air source and the spray gun. It provides a constant air pressure to the spray gun without fluctuation.

Most spray coating is performed in an enclosure called a spray booth. The booth is designed to serve several basic purposes. It filters the fumes from the air which could cause combustion or illness to the workers. The booth also eliminates the spray mist from landing on surfaces where it is not intended by the use of filters, which are designed to remove the spray particles from the air before they are emitted to the outside, Fig. 14-58.

Spray guns. Spray guns are of two different types depending upon where the air and fluid are combined. The guns are of the *internal* or *external* mix, Fig. 14-59.

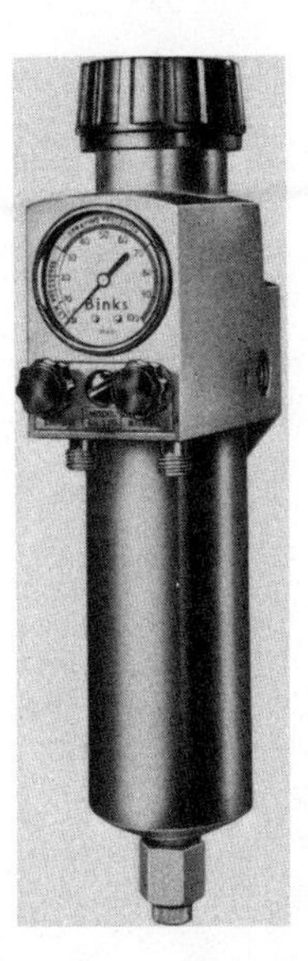

Fig. 14-57. Regulator to control air pressure to spray gun. (Binks Spray System)

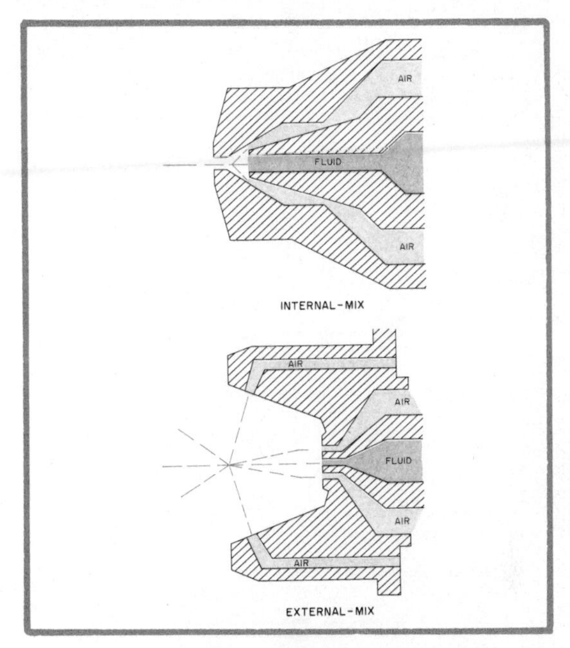

Fig. 14-59. Internal and external mix principles of spray guns.

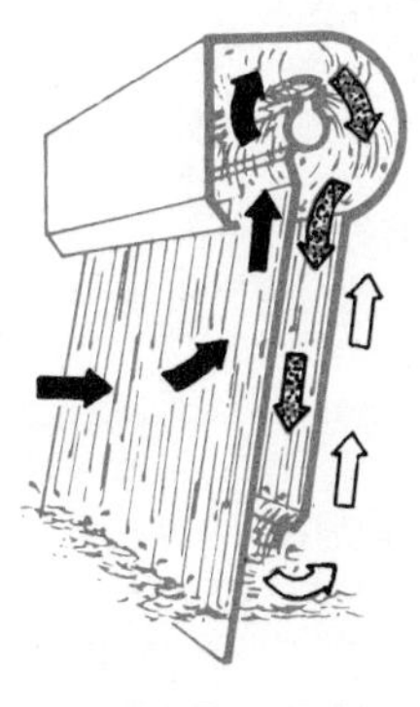

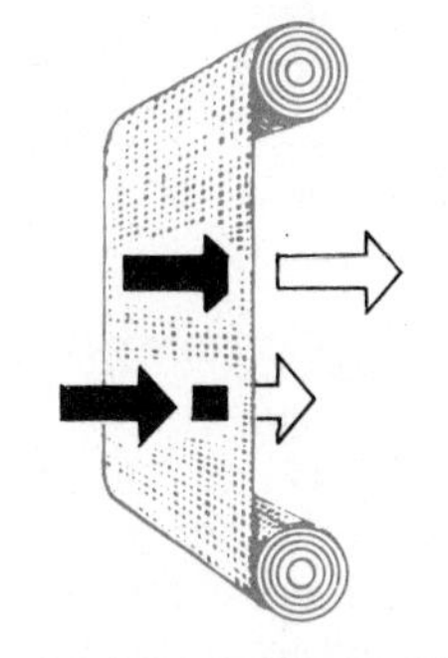

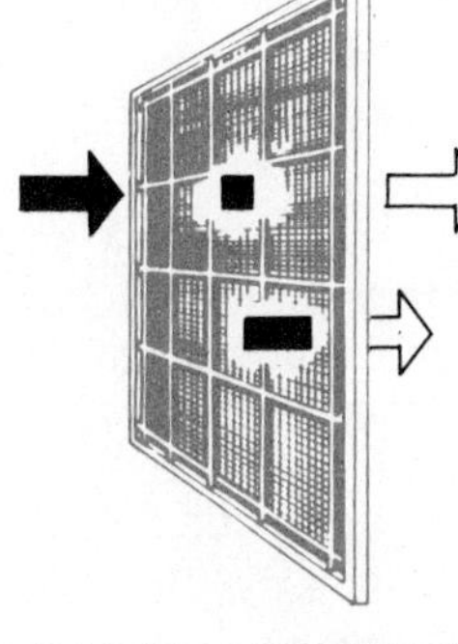

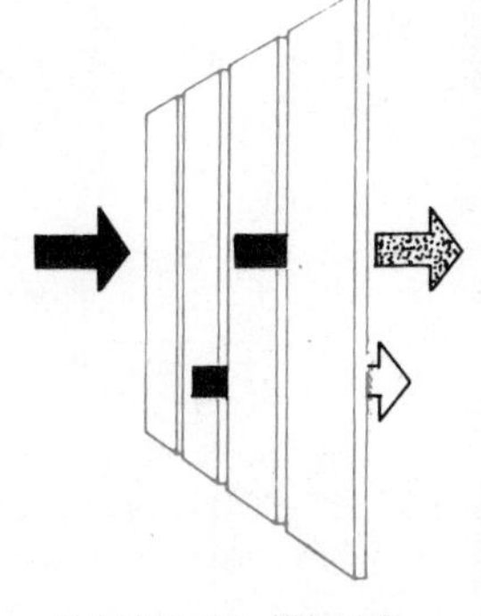

Fig. 14-58. Filtering systems in spray booths.

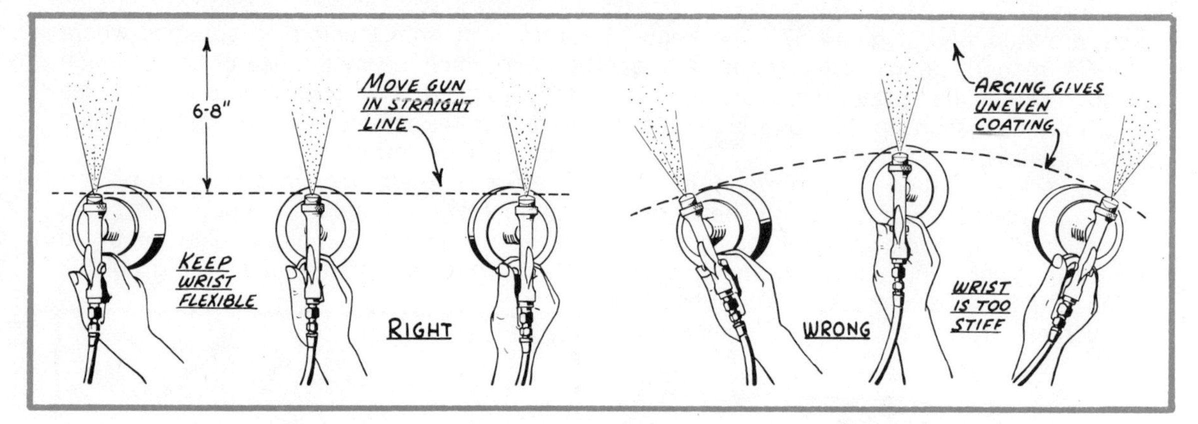

Fig. 14-61. Correct use of spray gun position and movement. (DeVilbiss Co.)

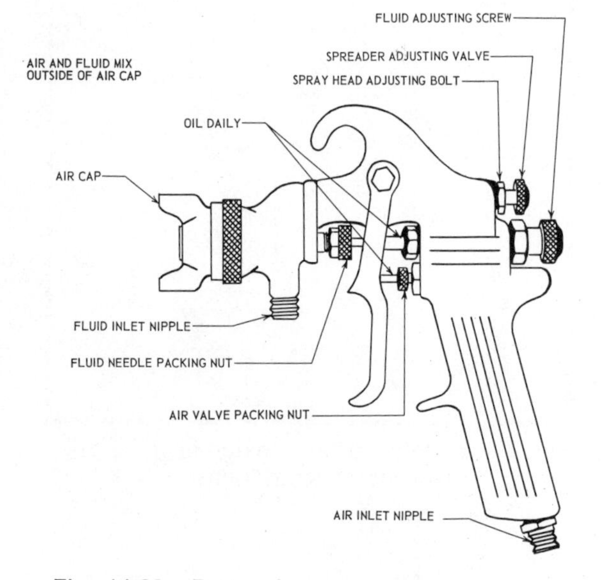

Fig. 14-60. Parts of a spray gun.

A spray is formed as compressed air and fluid mix. This mixture takes place inside the air cap on internal mix guns, then the mixture passes through the nozzle to the work surface. The air and fluid leave the nozzle separately and mix outside of the air cap on external mix guns. Figure 14-60 shows the parts of a typical spray gun.

Operating the spray gun. Successful spray coating requires that certain fundamentals be observed. The position of the spray gun must always be at a right angle to the work surface, Fig. 14-61. The gun is moved evenly across the surface as uneven coating results when the gun is tipped or rotated.

The wings of the spray nozzle can be turned to alter the spray pattern. The wings are set in a horizontal position for most work. The pattern is vertical when the gun is held horizontally. This position gives the widest possible pattern for maximum coverage. The spray gun is moved along about 6″ to 12″ from the surface.

The amount of finish applied to the surface is determined by several factors. The two most important factors are: (1) the speed at which the gun is moved, and (2) the distance it is held from the surface. The gun should be positioned and moved at a rate so that an even, wet coat is applied. Each pass should overlap the previous pass by one-half the width of the pattern, Fig. 14-62. This method will apply an even coat on the entire surface.

Excessive material spray on a surface will cause runs and sags. It is better to apply several coats after drying time has elapsed than to apply too heavy a coating at one time. Proper lighting is essential for good spray finishing.

The work should be placed in a spray booth so that it is between the spray gun and the exhaust system. Work placed on a

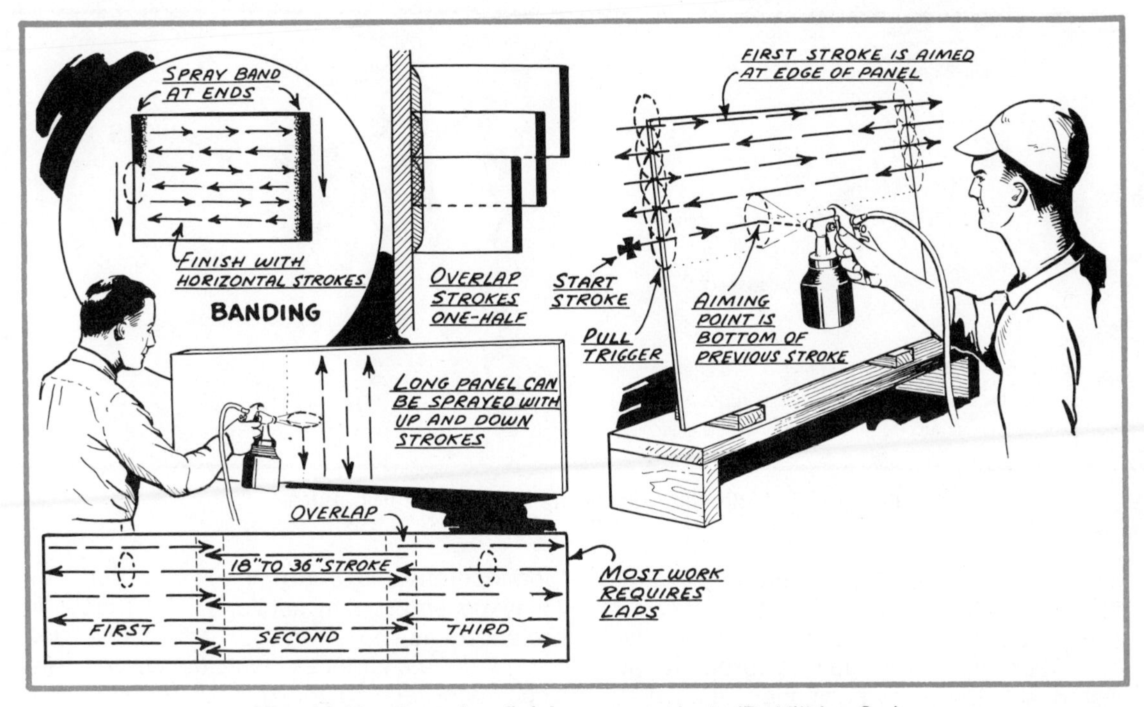

Fig. 14-62. Spraying finish on a product. (DeVilbiss Co.)

special turntable makes it easy to revolve the work, Fig. 14-63.

Preparing the finishing material for spray application. It is necessary to thin or reduce most finishing materials by about 25% for spray application. The proper thinner or solvent must be used for each type of material. Cleanliness is important to the finishing process. After thinning, the material is strained as it enters the cup. Nylon mesh fabric is satisfactory for most straining. This removes any solid particles which may become lodged in the nozzle of the gun.

Preparing to spray. The cup is attached to the nozzle assembly and the pressure line is connected. Between 25 and 35 lbs. of line pressure is used with most spray equipment and finishing materials. Practice strokes should be made on a test surface other than the product. The spray pattern is set by turning the adjustment valve. As stated, the shape of the pattern is altered by changing the position of the wing on the

Fig. 14-63. A special turntable on which a product may be rotated for spraying.

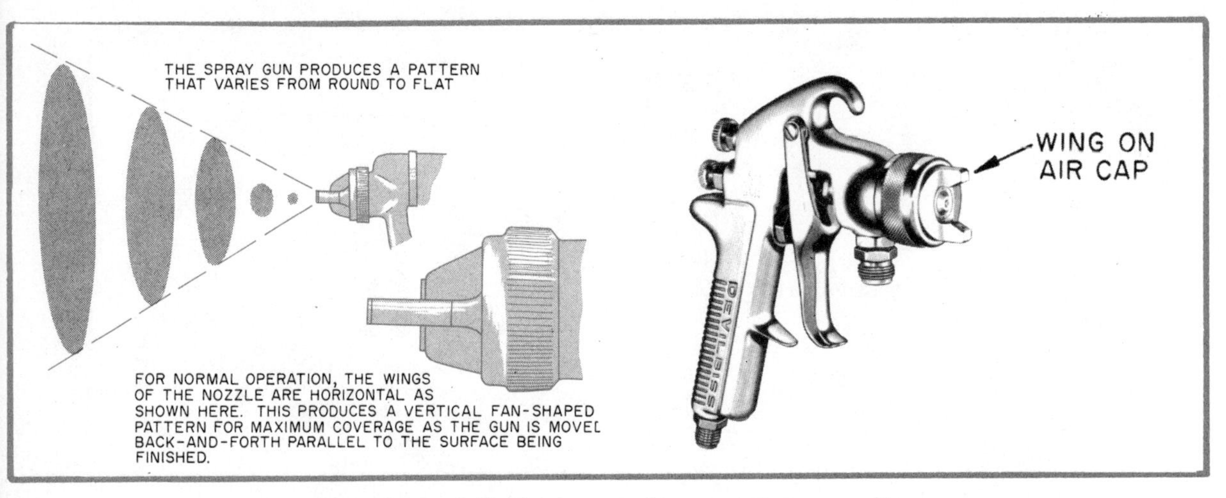

Fig. 14-64. Adjusting nozzle for correct spray pattern.

air cap, Fig. 14-64. The operator's manual furnished with a particular model spray gun should be consulted for specific details.

After the spray gun is properly adjusted and the work positioned, the spraying operation is performed. Large surfaces may be sprayed in sections. The trigger is released as the spray gun approaches the edge on all surfaces. Outside corners are coated by aiming the gun at the edge and allowing the spray to strike both right angle surfaces evenly, while inside corners are sprayed as shown in Fig. 14-65. After the corners have been coated, a normal spraying technique is used for the remaining surfaces.

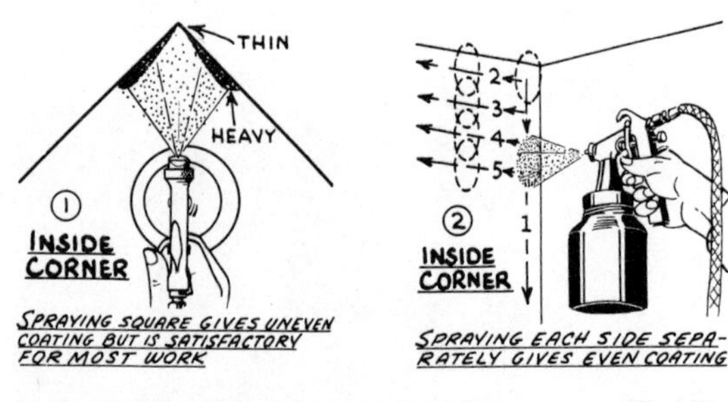

Fig. 14-65. Spraying an inside corner. (DeVilbiss Co.)

Cleaning the spray gun. The spray gun must be cleaned immediately after each job is completed, Fig. 14-66. This is a simple procedure and should become routine. The procedure for cleaning the spray equipment is as follows:

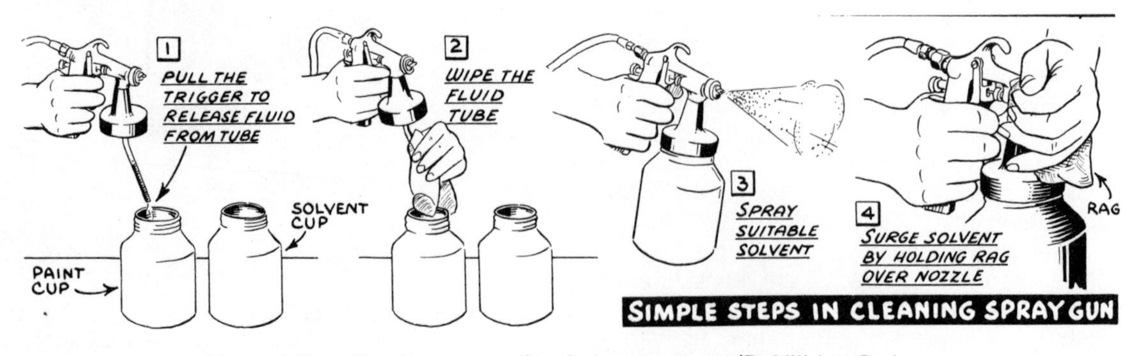

Fig. 14-66. Cleaning a suction-fed spray gun. (DeVilbiss Co.)

1. Remove the fluid cup from the cap.
2. Place a cloth over the air cap and trigger the gun. The excess spray material left in the gun will be forced back into the fluid cup.
3. Empty the fluid cup and wipe it clean, using the proper solvent for the material sprayed.
4. Fill the fluid cup one-half full of the proper solvent and reassemble the gun.
5. Spray the solvent through the gun to remove the finishing material remaining in the gun.
6. Remove and disassemble the air cap and place it in solvent to soak.
7. Scrub all remaining traces of finish from the air cap using a brittle brush. Dry the air cap with a lint-free cloth.
8. Empty the solvent from the fluid cup and dry the cup with a clean lint-free cloth.
9. The holes in the air cap can be cleaned with a toothpick. Never use a metal object to clean the air cap.
10. Reassemble the gun and store for future use.

Determining spray gun problems. There are several common types of problems which may develop, but the problem can be easily remedied if the cause can be determined. The following methods will resolve most spray gun problems. The type of pattern which the gun produces is the first key to determining malfunction, Fig. 14-67.

Other common spray application problems are "orange peel" and "blushing," and "run."

1. An *orange peel* condition is when the sprayed surface is rough and has a pebbled texture. This is often caused by inadequate thinning of the finishing material, or by the material being too dry before it strikes the surface. This problem

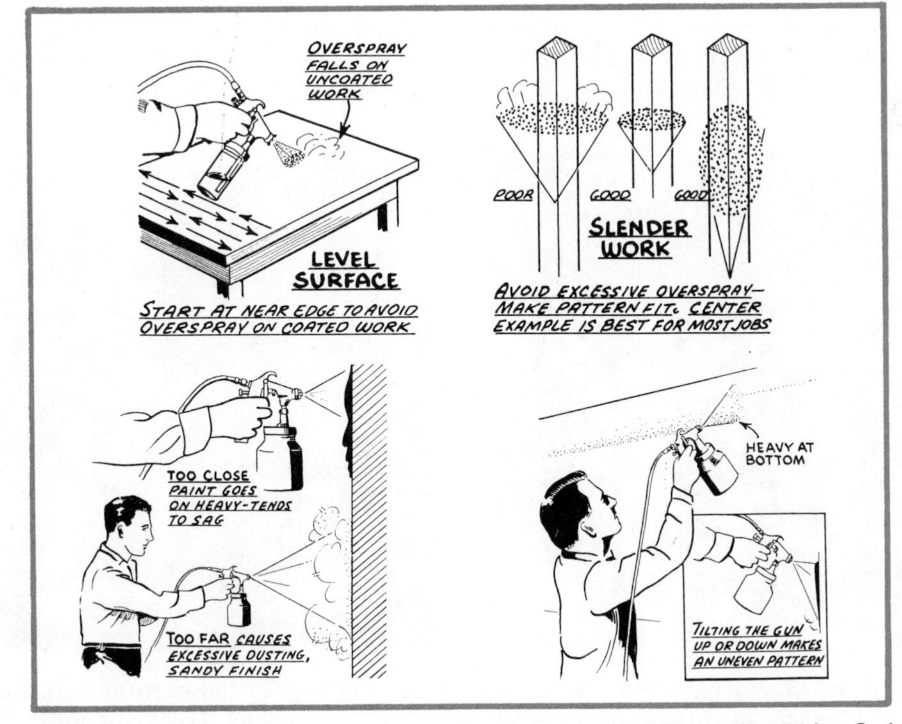

Fig. 14-67. Spray gun problems and suggested solutions. (DeVilbiss Co.)

can be corrected by thinning the material and holding the gun closer to the surface.

2. *Blushing* can be detected by a cloudy, milky appearing surface. This can be caused by excessive moisture due to high-humidity conditions, or by excessive thinning causing evaporation of the solvent.
3. *Runs* are a common problem for inexpensive spray finishes, and are caused by applying too much material to the surface. Sometimes the finish may be too thin or the gun is held too close to the surface and not moved quickly enough.

INDUSTRIAL FINISHING

Finishing is a major production process in the wood industry. Every attempt is made to speed the process as much as possible and still produce a quality product. Elaborate systems are often used to move the product along the finishing line, Fig. 14-68. The entire finishing process is usually planned so that a product is completely finished and ready for shipment when it reaches the end of the line.

The curing time of many types of finishes is accelerated by passing the freshly coated product through a heat source. This source may be electrical lamps or an oven, Fig. 14-69. The speed at which the product travels is closely controlled.

Several different methods are used in the wood industry for applying finishes. The method of application depends on the type of finish, size of the product, and the purpose of the product.

TUMBLING

Small products such as golf tees, toys, and knobs are finished by the tumbling method. The tumbling machine is very similar in appearance and operation to a small cement mixer, Fig. 14-70. The objects to be finished are placed in the barrel tumbler. About a quart of finish and a pint of thinner are placed in each drum as it is

Fig. 14-68. Product moves along the finishing line on a conveyor system. (Jasper Cabinet Co.)

Fig. 14-69. A drying oven to accelerate finish drying time. (Kroehler Furniture Mfg. Co.)

loaded or changed. The drum rotates at about 25 rpm during the tumbling process. The articles are removed from the tumbler after about 15 minutes. The freshly painted products are placed on a screen and allowed to dry. Sometimes the product is placed in an oven to speed the drying process. The tumbling process is sometimes repeated several times to give maximum paint coverage.

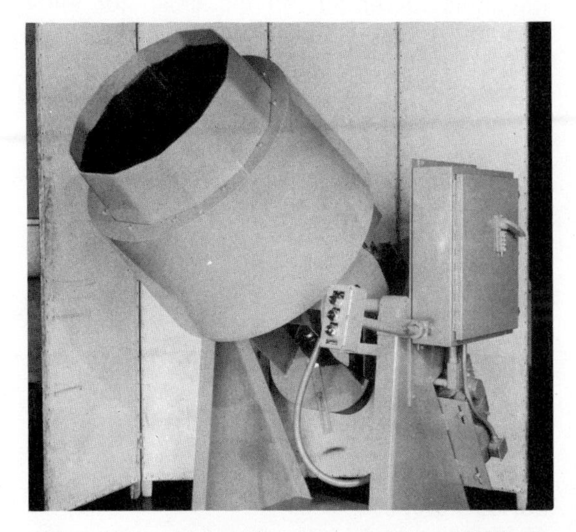

Fig. 14-70. A tumbler for applying finish to small products. (Baird Manufacturing Co.)

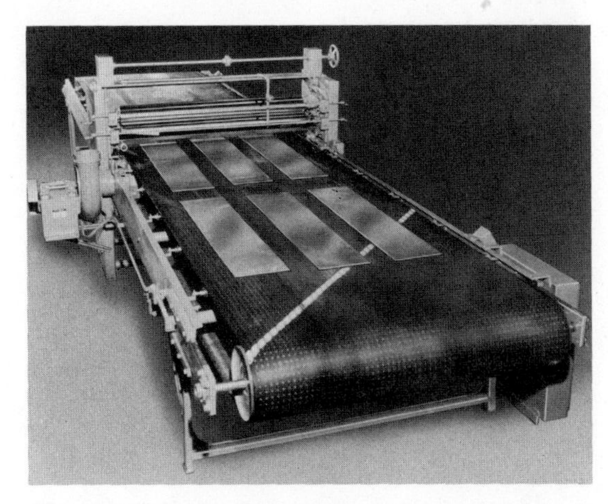

Fig. 14-71. A roller coating machine. (Westinghouse)

ROLLER COATING

The roller coating process is commonly used for coating large flat panels such as plywood. A conveyor system in which the speed can be varied carries the stock through the rollers. The system consists of three rollers, two rollers above the surface which is to be coated and one drive roller below. The upper rollers revolve in a bath of finishing material. The doctor roll is the metering roll and controls the film thickness on the applicator or transfer roll, Fig. 14-71.

CURTAIN COATING

The surface to be coated by curtain coating is passed under a continuous flow of the finishing material. The continuous stream of falling material is called a curtain. Curtain coating is one of the most rapid methods of applying finish. There are several methods by which the finish may be dropped. One of the simplest methods is by gravity flow, Fig. 14-72. The finish is stored in a reservoir and flows through elongated holes onto the surface. Thick finishing materials cannot be applied by this method. The thickness of the coating is controlled by the rate of flow and the

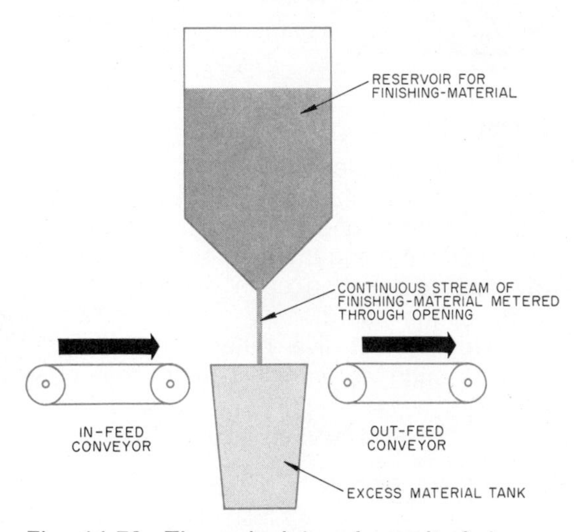

Fig. 14-72. The principle of gravity-fed curtain coating.

speed of the stock passing under the coater.

The curtain coating system may employ various feeding systems. The major variations are in how the finishing material is released from the storage hopper. It may include the use of pressure, vacuums or a combination of gravity and pressure.

ELECTROSTATIC SPRAY COATING

Electrostatic spray painting is based on the principle of unlike electrical charges attracting one another. The objects to be coated are given a positive charge while the coating material is given a negative charge. When the coating material is emitted into the air it is attracted to the positively-charged object. The waste of coating materials which results from most air spray operations is cut to a minimum, and the finish which results is very uniform and smooth. The process is generally restricted to metal objects because they can receive an electrical charge; however, wood objects can be coated with a material which will conduct an electrical charge for electrostatic finishing.

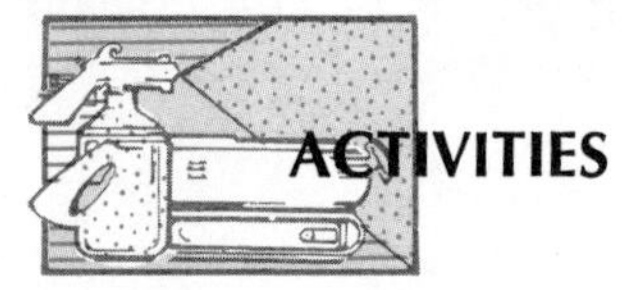

ACTIVITIES

PROBLEM

Select a piece of stock about one-fourth inch thick, three inches wide and fifteen inches long. Divide the board into five sections along the length. Place a piece of masking tape on each division line the width of the board.

Now select five different transparent finishes such as varnish, Danish oil, polyurethane, etc. Apply a different finish to each of the sections carefully following the manufacturer's recommendations for application.

After the finish has dried, devise a systematic method of subjecting the finish to common household dangers such as catsup, mustard, custard, alcohol, finger nail polish remover and ink. Use a dropper and place several drops of the household material on the finished samples. Allow the materials to remain on the finish for thirty minutes and then wipe off.

Inspect each of the finishes to determine how it has reacted to the various materials. Assume you are planning to finish an end table and must decide what finish you will apply.

The results of this experiment should help you decide which type of finish you will want to use.

VOCABULARY

Surface preparation
Abrasion
Open coat
Closed coat
Arrises
Washcoat
Orbital
NGR
Vehicle
Pigment
Top coat
4 lb. cut
Aerosal spray
Pressure regulator
"Orange peel"
Electrostatic spray

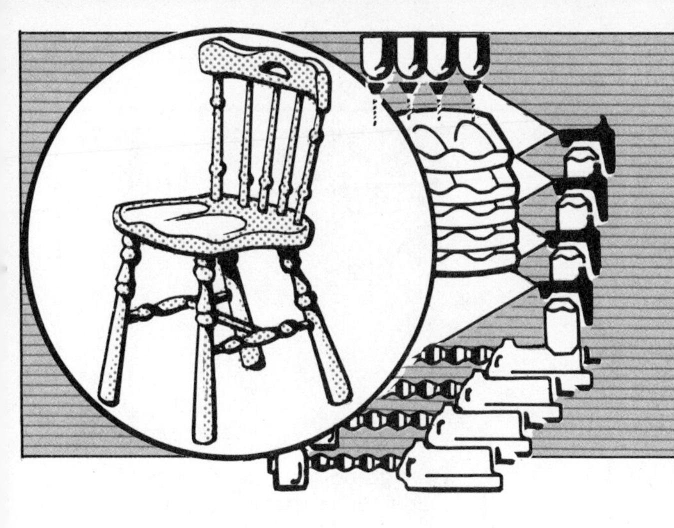

CHAPTER 15 ELEMENTS OF MASS PRODUCTION

INTRODUCTION

After a market study has been conducted and a product has been designed to satisfy the consumers, the next step is to identify *production* procedures. In large companies, the production planning is performed by the industrial engineering department. In small plants, the designers may also be responsible for planning production. The production planning team determines the most efficient methods of producing the best quality product at the lowest possible cost. They develop the *manufacturing plan.*

The preliminary manufacturing plan is developed from the designer's working drawings. A prototype is constructed to represent the production model of the product. Planners study the prototype and determine the sequence of machines to be used. Each part is studied and the total assembly is planned carefully. It is not uncommon for the production planners to suggest design alterations to the design team. A slight change in design may reduce the amount of material, machine time, or labor required during production. Every one of these reductions results in a cost savings.

As the manufacturing plan for production is developed, cost estimates are prepared for the approval of plant management. Management will study the design and production plan to determine if the product will be produced, Fig. 15-1. If high production and material costs will not produce a profit, the product will be redesigned or it will not be produced. The

Fig. 15-1. Management planning session discusses production and cost of products.

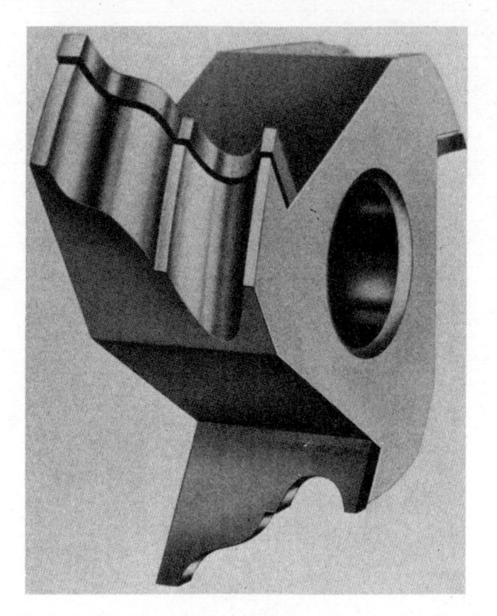

Fig. 15-2. A special shaper cutter designed and ground for producing a specific product.

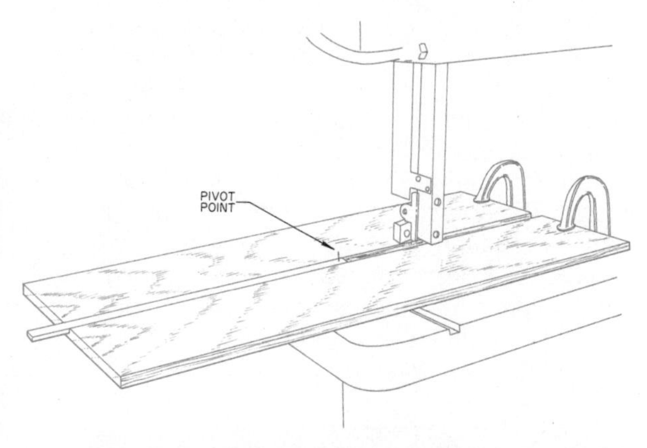

Fig. 15-3. A fixture used to hold a special piece in a machine as it is processed.

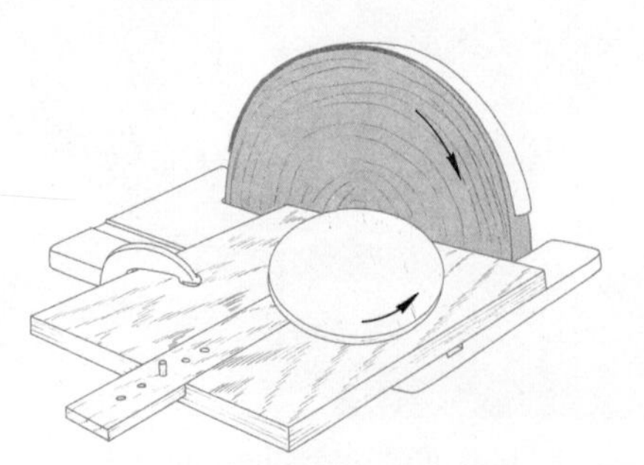

Fig. 15-4. A jig designed for sanding round disks to shape on the disk sander.

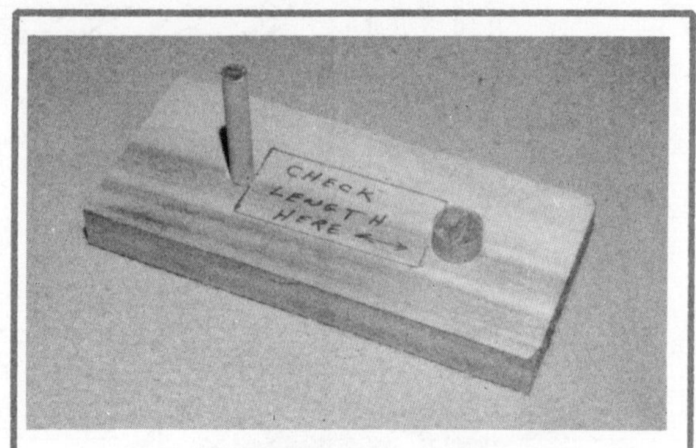

A. The go-no-go gauge may be made for a specific product.

B. The product piece is placed on the plug to check the size of the hole in one end.

C. The product is placed between the plug and the pin to determine if length is accurate.

Fig. 15-5. Go-no-go gauges.

design changes must not alter the function or consumer appeal of the product. Management's decisions are usually controlled by their sensitivity to product success with the consumer.

After the necessary design changes have been made and the estimated costs reveal that a profit will result from the investment of time, materials, and labor; management will approve the product for production. The production planner then proceeds with the actual manufacturing plan. The production planner decides which machines will be used for final production. He determines what special tools, jigs, and fixtures will be required. The special jigs and fixtures are designed and purchased or produced. Most large industries have specialists who design and construct the special tools, jigs, and fixtures.

Special machine cutters are designed and ground for a specific shape, Fig. 15-2. Special tools are costly to produce; however, they are designed to save time and materials. It is difficult to amortize (spread) the initial cost of special tools on small production runs. Therefore, the high volume of a production run helps to reduce special tooling cost.

A *fixture* such as shown in Fig. 15-3 is designed to hold stock on a standard machine. The machine operation is performed while the workpiece is held in the fixture. Standard machines such as saws, drill presses, and shapers can be equipped with fixtures to perform special operations. A *jig* is a device used for guiding a tool or holding the workpiece in place during production, Fig. 15-4.

When possible, repeated measurements of standardized parts on the assembly line is avoided. Instead, gauges sometimes called "go-no-go" gauges are used to determine if the part meets specifications. Gauges are used because less time is required to make specification judgments, Fig. 15-5.

PLANNING FOR PRODUCTION

The production planning department prepares a detailed listing of materials called a *bill sheet.* This bill sheet consists of all of the required stock to produce one unit. The bill sheet lists each part by (1) name, (2) number, (3) finish and rough size, and (4) kind of material. Any special notes about any part of the product are also included on the bill sheet, Fig. 15-6. The bill sheet is prepared from the working drawing.

IGLOO MANUFACTURING COMPANY
(Wood Products are our Specialty)

Bill Sheet for one Roll type Desk   Date 12 - 2 - 1970   Page 1 of 5

PART NO.	PARTS PER ARTICLE	NAME OF PART	FINISH thickness	FINISH width	FINISH length	MATERIAL thickness	MATERIAL kind	ROUGH SIZE thickness	ROUGH SIZE width	ROUGH SIZE length	ROUGH PIECES per unit article	SPECIAL NOTES
1	2	Drawer Side	3/8"	4"	12"	2/4	Popler	2/4	4 1/4"	12 1/2"	2	
2	1	Drawer Back	3/8"	3 3/4"	10"	3/4"	Poplar	2/4	4"	10 1/2"	1	
3	1	Drawer Front	3/4"	4 1/2"	10 3/4"	4/4"	Birch	4/4	5'	11"	1	
4	1	Drawer Bottom	1/4"	9 1/2"	11 1/2"	1/4"	Birch Plywood	1/4"	9 1/2"	11 1/2"	1	
5	2	Leg	Turn as shown in Drawing		28"	12/4 sq.	Birch	12/4	12/4	30"	4	
6	2	Front and rear rails	3/4"	3	40"	4/4	Birch	4/4	3 1/2"	42"	2	
7	2	Side rails	3/4"	3	17"	4/4	Birch	4/4	3 1/2"	19"	2	

Fig. 15-6. A product bill sheet.

A more complete materials bill including all specifications is developed from the bill sheet. This materials list is sent from the production planning department to the purchasing department to order the materials from suppliers.

ROUTE SHEET
IGLOO MANUFACTURING COMPANY

Product Name Roll Top Desk　　Part Name Drawer Front
Product Number 83-12'1　　Part Number 83-10
Material Birch　　Rough Size 4/4 x 5" x 12" (thickness width length)
Finish Size: Thickness; 3/4"　Width; 4 1/2　Length 10 1/4"
Rough Ft. per Article .35

OPERATION NUMBER	MACHINE STATION	DESCRIPTION OF OPERATION	OPERATOR	TIME	SPECIAL INSTRUCTIONS FOR OPERATOR
1	1	Rough cut to size	J.W.	2.0	
2	2	PLANE	GC	1.5	
3	4	Rip	AR	1.10	
4	5	joint edges to size	BZ	2.0	
5	7	cut to finished length	RS.	1.0	
6	9	rough sand	S.R.	2.10	
7	14	cut dovetail joints	J.A.	4.0	
8	15	route edge	J.S.	3.0	cutter #17
9	9	final sand	HB	2.5	

Fig. 15-7. A process route sheet.

Fig. 15-8. A production plan is developed prior to the start of production. (Winnebago Industries)

A *process route sheet* is prepared by the production planning department. The route sheet is prepared for each individual part of the product, Fig. 15-7. The route sheet accompanies the part as the operations are performed. The rough stock is selected according to specifications given on the process route sheet. The process sheet gives the operation number and machine or station number. The time required to perform the operation is also recorded. *Time and motion* studies are made on various operations. The purpose of time and motion studies is to determine better methods of performing the operations. Industry must always strive to reduce cost. The production taking place in Fig. 15-8 required a complete production plan before anyone actually worked on the product.

MASS PRODUCTION IN THE SCHOOL SHOP

The mass production of a product in the school shop requires the same planning as in an industry. The product must undergo the research and development stage. During the R and D stage, a product is selected which has student appeal. This is similar to determining the consumer demand for an industrial product. The product must be designed with class members playing the role of a design team. Numerous design alternatives are presented for the product, Fig. 15-9. The design is modified until it is appealing to the class members. Class members can act as management in the decision-making process. The design team, which may also include the engineering draftsmen, prepare the working drawings for the final product, Fig. 15-10.

PRODUCTION PLANNING

The next stage is production planning. A team determines the most efficient means of producing the product. The manufacturing plan is developed from the working drawings. The method of producing each part and the method of assembly is considered. Cost estimates are made for the product. This estimate includes the time and material cost. Additional changes may have to be made in the design to accommodate production at a reasonable cost.

The production planning team determines what types of jigs and fixtures will be required for production. The special jigs are designed, constructed, and tested. The jig must be designed to produce a part quickly, accurately, and safely. If the jig does not meet these production requirements, it must be redesigned. The bill sheet listing all materials is prepared for the product, Fig. 15-11. A complete set of materials specifications is developed from the bill sheet. This is used for purchasing materials from the suppliers.

A process route sheet is prepared for each part of the product. The route sheet moves along the production line with each part. An exact description of the operations necessary to each part is specified on this route sheet. The actual time for performing the operations is recorded by

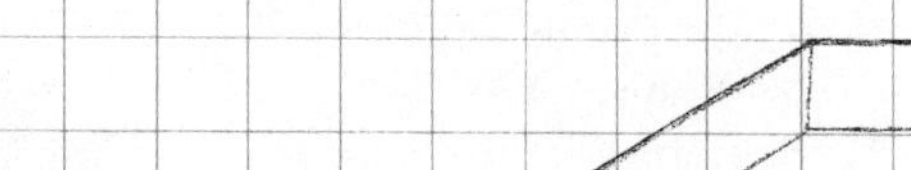

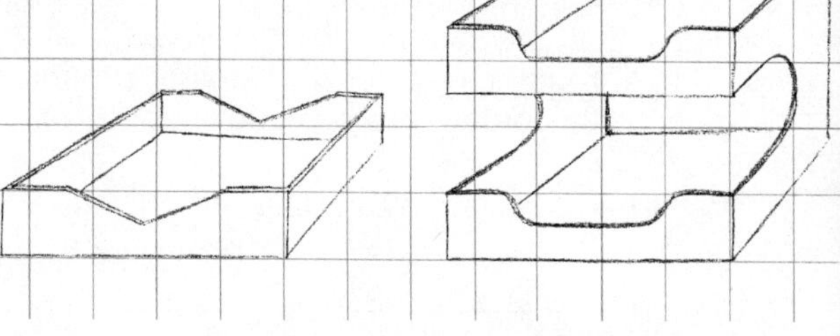

Fig. 15-9. Alternative design sketches for a product.

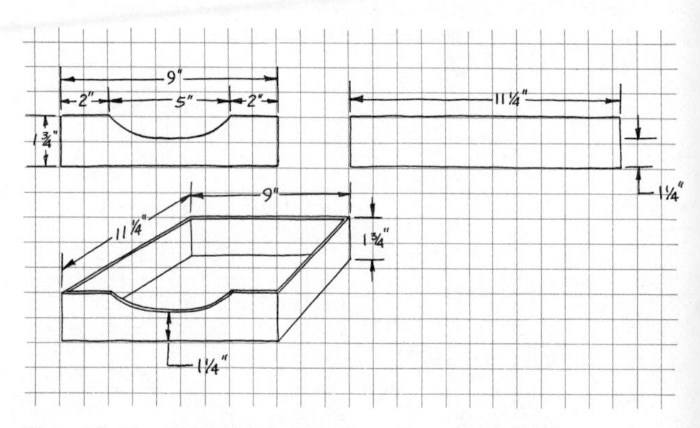

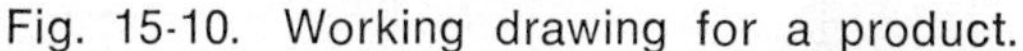

Fig. 15-10. Working drawing for a product.

SOUTH HIGH PRODUCTS DIVISION

Bill Sheet for one Desk Organizer   Date 10-1-1971

Page 1 of 1

PART NO.	PARTS PER ARTICLE	NAME OF PART	FINISH SIZE thickness	width	length	MATERIAL thickness	kind	ROUGH SIZE thickness	width	length	ROUGH PIECES per unit article	SPECIAL NOTES
1	1	Front	3/8"	2"	9"	2/4	willow	2/4	2 1/4"	9 1/4"	1	rabbet ends 3/16" x 3/8"
2	2	Sides	3/8"	2"	12"	2/4	willow	2/4	2 1/4"	12 1/4"	2	rabbet one end
3	1	backs	3/8"	2"	8 5/8"	2/4	willow	2/4	2 1/4	9"	1	
4	1	bottom	1/8"	8 3/4"	11 3/4"	1/8"	Standard Hardboard	1/8"	9"	12"	1	
												Note: All wood must be matched grain

Fig. 15-11. A complete bill sheet for a product is made before production begins in the school laboratory.

each worker. Figure 15-12 shows the route sheet which would accompany the fronts of the product shown. A sketch may be used to show special setups and operations that are difficult to describe.

A route sheet for the entire production of the product is also developed. This shows the sequencial operations beginning with the rough stock through the final finish. Remember, the efficiency of the production activity is dependent upon thorough planning. The production plan is illustrated using the standard symbols recommended by the American Society of Mechanical Engineers (ASME), Fig. 15-13. A triangle shows a point at which materials are either *stored or procured*; a circle stands for *operations and procedures;* a square is an *inspection point;* a rectangle shown connected to an operation circle indicates a *special* jig or fixture is used for the operation; and a large "D" indicates that the process is *delayed* temporarily at a given point. Figure 15-14 shows the steps of production as a product moves through the shop.

Inspection points are essential as the operations progress. The parts must be exactly the same within the predetermined tolerances. The jigs and fixtures, if properly designed, will reduce the likelihood of errors. The advantages of mass production are lost if the parts are not interchangeable.

ROUTE SHEET
SOUTH HIGH PRODUCTS DIVISION

Product Name Desk Organizer — Part Name Front
Product Number 1 — Part Number 1
Material Willow — Rough Size 3/4 x 2 1/4" x 9 1/4" (thickness width length)
Finish Size: Thickness; 3/8" — Width; 2" Length 9"
Rough Ft. per Article .13

OPERATION NUMBER	MACHINE STATION	DESCRIPTION OF OPERATION	OPERATOR	TIME	SPECIAL INSTRUCTIONS FOR OPERATOR
1	1	Rough cut			4" radius; 2"; 2"; 1"
2	2	Plane			
3	3	Rip			
4	8	Joint edges to size			
5	11	Cut to finish length			
6	14	Contour front			
7	15	Finish sand contour			
8	19	Groove edge for Bottom			1/8" x 1/8"
9	16	rabbet ends			
10	18	finish sand			
11	20	Assemble			3/8"; 3/16"
12	17	Finish			
13	18	Inspect			

Fig. 15-12. A completed route sheet for a desk organizer.

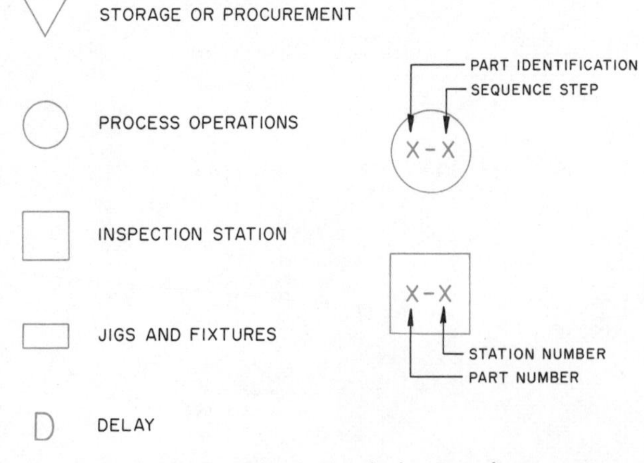

Fig. 15-13. Standard symbols to show production flow.

PROBLEM

You know that products can be produced at a lower cost if they are mass produced in large quantities. However, before a product is produced in a factory, much consideration must be given to how the product can be manufactured in the most efficient manner. This phase is called *production planning*. The first requirement is the working drawings. The drawings give the specifications for the final product. A bill of materials is then developed including the materials for the production of the product. Whenever possible, the bill of materials includes standard materials because this will reduce the product cost. A list of suppliers must also be developed

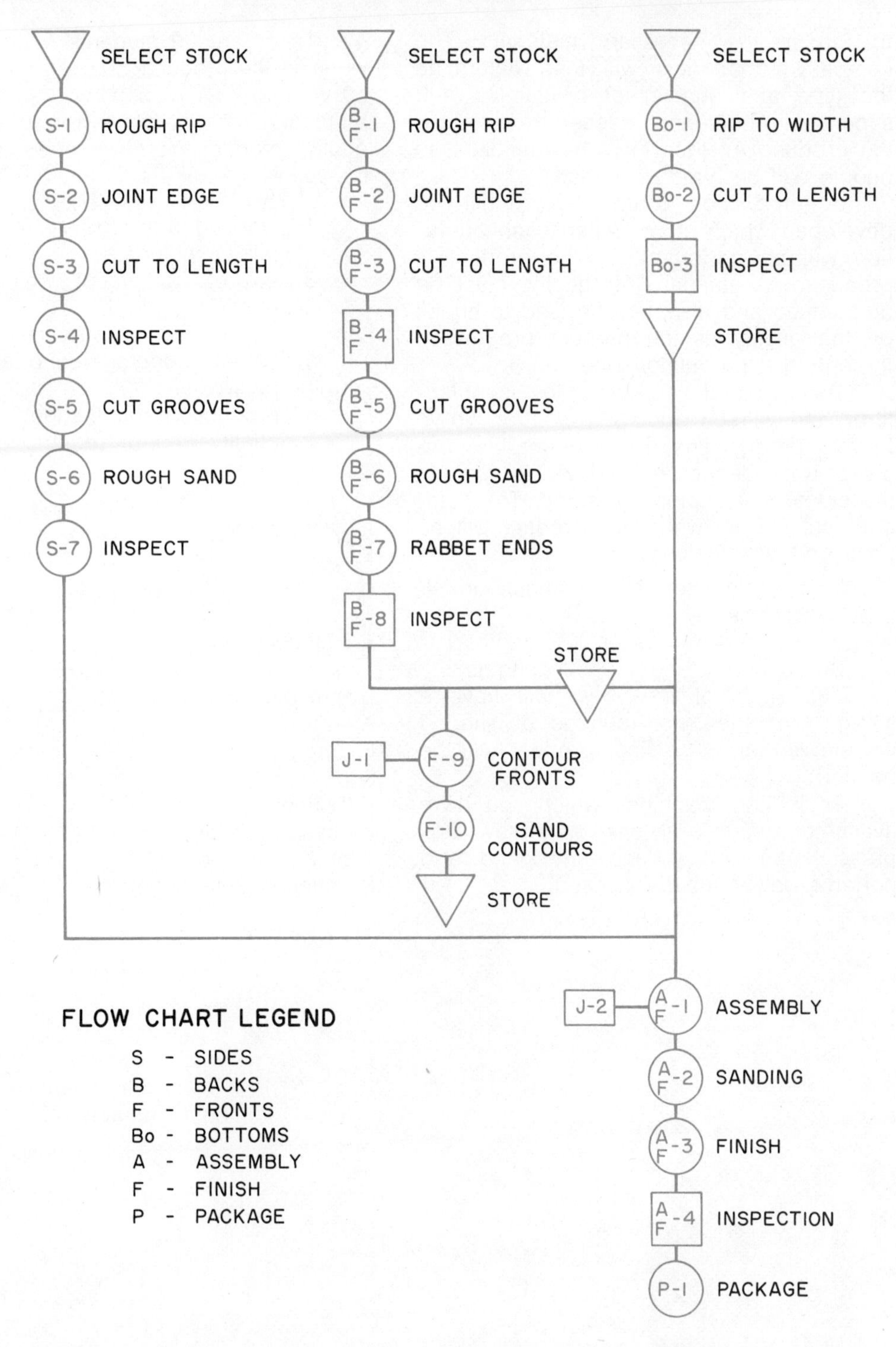

Fig. 15-14. Process chart showing production flow of parts which combine into a finished product.

to furnish the necessary materials. The company's managers will then determine the type and number of production and supervisory personnel needed to produce the product. All jobs must be identified so people can be hired or trained.

A production route sheet must be developed which shows when each operation will be performed as the product is produced. A standard for quality must be determined and a system devised to check on the quality as the product progresses through the production line.

Assume that you are a toy manufacturer and will be adding a wooden yo-yo to your product line. The product will consist of two sides, a center dowel to connect the sides, and a piece of string. The standard stock from which the product will be produced will include:

1. 2″ diameter maple stock in 4′ lengths.
2. ¼″ dowel rods in 36″ lengths.
3. suitable string in 500′ lengths.

The sides of the yo-yo will have a straight shoulder (not rounded or shaped on the corner). The final product will be painted.

Tools and machines which you have available include a jig saw, band saw, drill press, circular saw, and any hand and portable power tools you need.

There are 20 students in the class to work on the product. You will be producing 50 yo-yos. Your responsibility is to develop a management system for producing a quality product. You will need to develop the following:

1. working drawings
2. bill of materials
3. list of equipment which will be needed
4. list of possible supplies
5. list of production and supervisory personnel
6. production route sheet
7. quality control requirements.

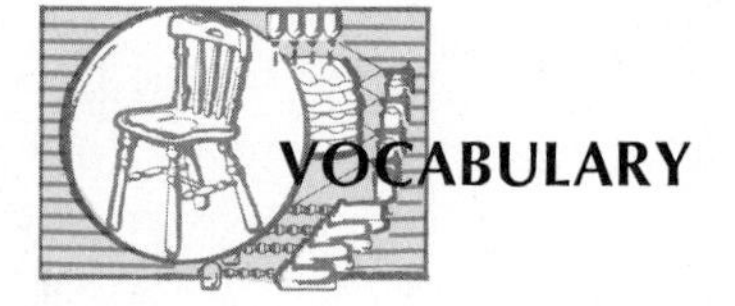

Manufacturing plan
Amortize
Fixture
Jig
Bill sheet
Process route sheet
Time and motion studies
Production route sheet

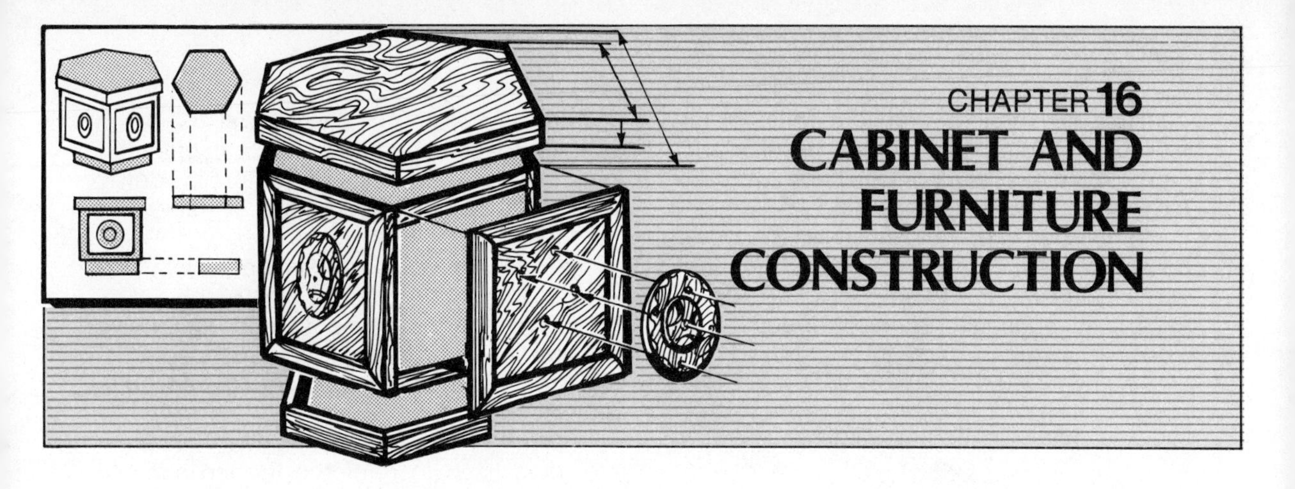

CHAPTER 16 CABINET AND FURNITURE CONSTRUCTION

INTRODUCTION

Furniture and cabinetmaking require working to close tolerances. A cabinetmaker is a highly-skilled craftsman. Care must be given to accurate measuring and cutting to construct cabinets. High-quality, professional cabinet work must have tight-fitting, well-constructed joints. Furniture and cabinets which will withstand long years of service must be well-constructed. These five basic types of construction may be used, depending upon the product and material used: (1) leg and rail; (2) box; (3) case; (4) frame and cover; and (5) carcass. Large furniture may involve more than one of the basic methods.

LEG AND RAIL CONSTRUCTION

Leg and rail construction is commonly used in the construction of furniture. It is used in the construction of many types of chairs, tables, benches, and similar furniture, Fig. 16-1. This type of construction

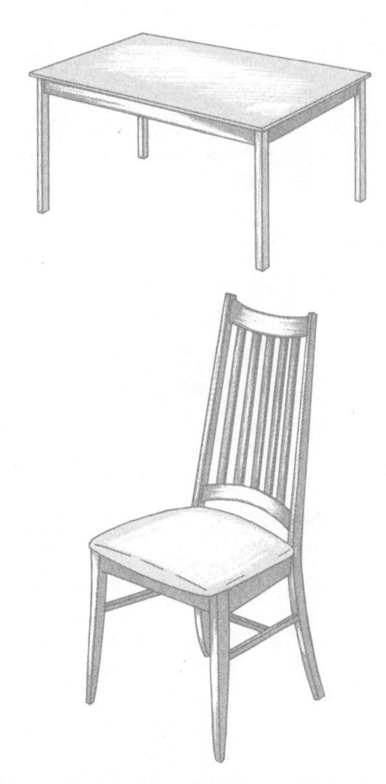

Fig. 16-1. These products were produced using leg and rail construction.

consists of four legs and four rails, generally using solid stock. The mortise-and-tenon joint has traditionally been used for leg and rail construction, Fig. 16-2. Since the advent of modern adhesives, the dowel joint has also been used extensively, Fig. 16-3. Combined with adhesive, the dowel joint is of about the same strength as the mortise and tenon. The dowel is much simpler to construct if power equipment is not available. The main factor in dowel joint construction is positive alignment of the holes.

A drawer is often combined with this method of construction in small tables, Fig. 16-4. One rail must be altered to receive the drawer. The drawer is suspended from the top as one method of guiding the drawer, Fig. 16-5.

Glue blocks or mechanical fasteners are often used to reinforce the leg and rail construction. Wood blocks can be cut and attached in the corner with glue and screws, Fig. 16-6. Special metal fasteners are also available to provide additional strength for the corner joints, Fig. 16-7.

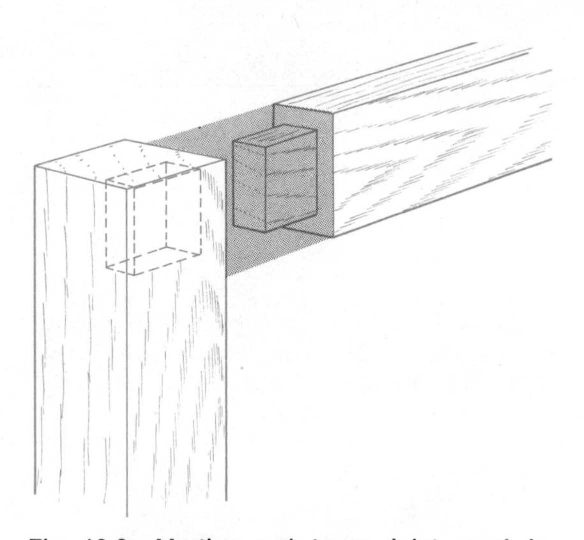

Fig. 16-2. Mortise-and-tenon joint used in leg and rail construction.

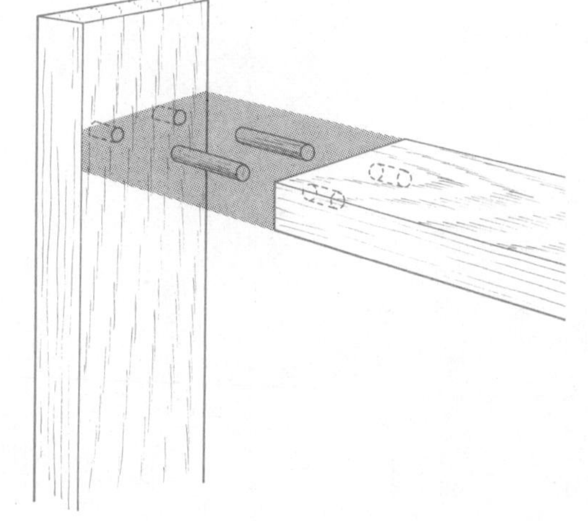

Fig. 16-3. Dowel joint used in leg and rail construction.

Fig. 16-4. A small table with a drawer constructed in the leg and rail method.

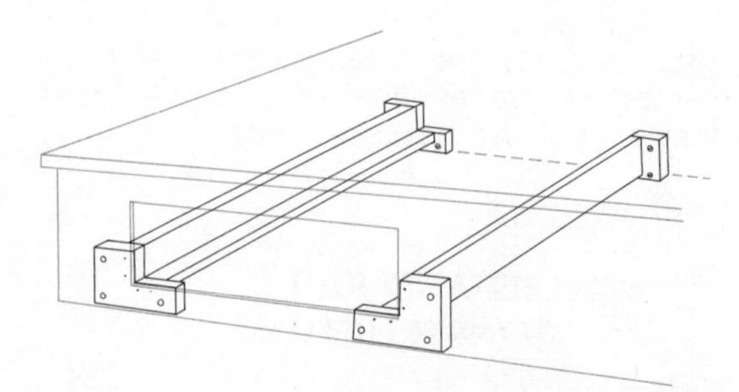

Fig. 16-5. Drawer guide assembly for leg and rail construction when drawer is placed under the table top.

Tops which are attached to leg and rail assembly are fastened so they can move because the top will expand and contract as the moisture content of the wood changes, Fig. 16-8.

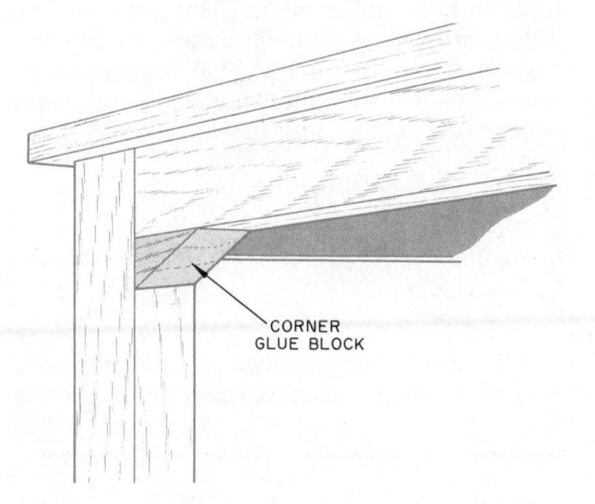

Fig. 16-6. Corner glue block for strengthening leg and rail joints.

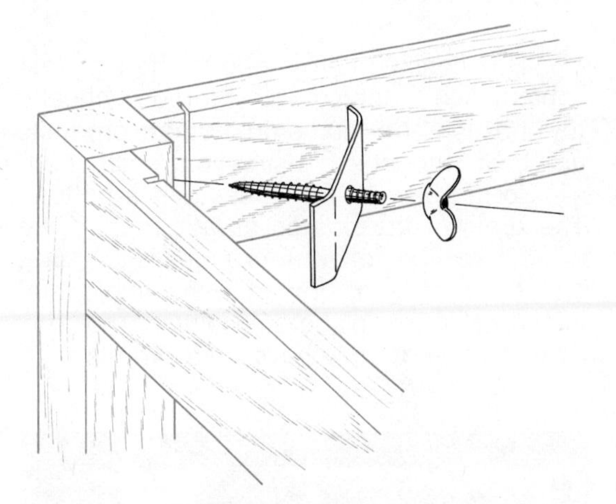

Fig. 16-7. Metal hardware designed to reinforce corner joint.

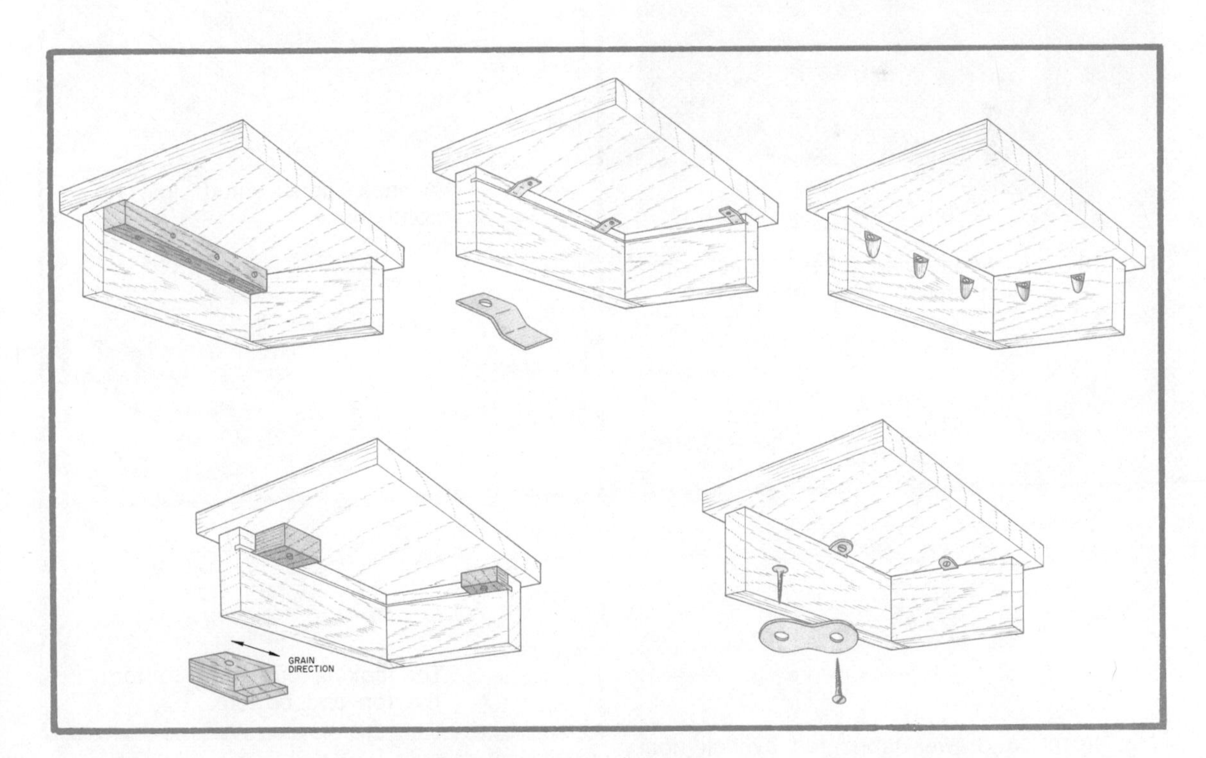

Fig. 16-8. Methods of reinforcing the joints of table tops in leg and rail construction.

BOX CONSTRUCTION

Box construction is one of the simplest construction techniques. Storage chests, toy boxes, and numerous other products are built with box construction which consists of a box with four sides and a bottom. The grain of the boards always runs in the same direction for all four sides, Fig. 16-9. The sides may be joined with simple butt joints or more elaborate joints. The box does not necessarily have a lid. The bottom is usually attached with either a rabbet or groove joint. A drawer is also a typical type of box construction, Fig. 16-10.

A box constructed with a lid can be built in one solid piece. The box is then cut apart with a circular saw and the top portion serves as a lid. This technique insures positive alignment between the lower portion of the box and the lid, Fig. 16-11.

CASE CONSTRUCTION

Built-in units such as kitchen cabinets and other small cabinets often use the case

Fig. 16-9. A cedar-lined chest constructed by means of box construction.

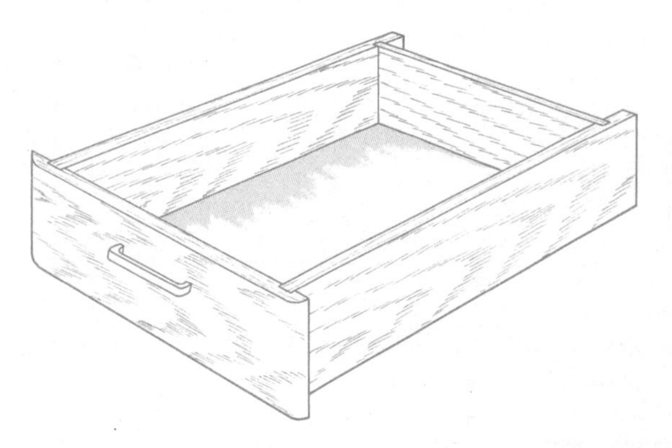

Fig. 16-10. A drawer fabricated by box construction.

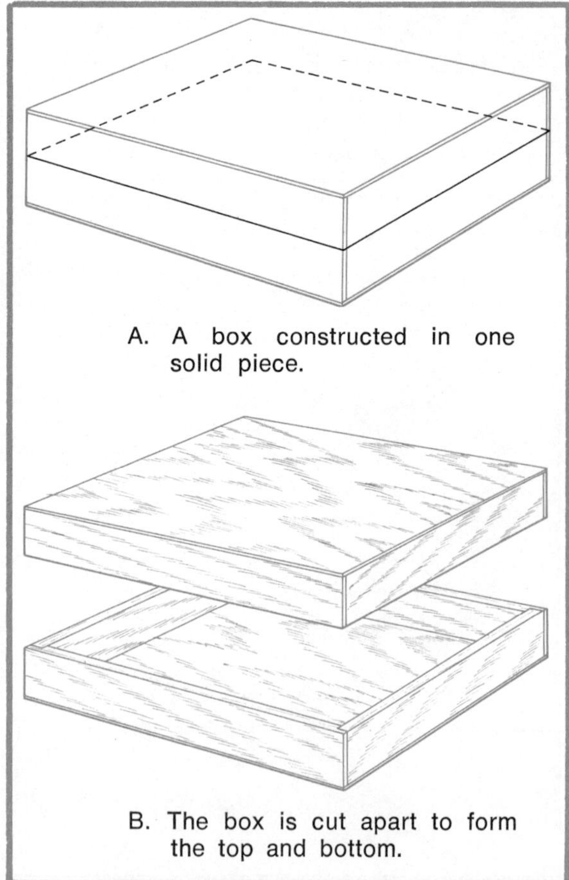

A. A box constructed in one solid piece.

B. The box is cut apart to form the top and bottom.

Fig. 16-11. Constructing boxes with lids.

construction method. This method is similar to box construction because it is actually a box turned on its side. The same material serves as both the exterior and interior of the structure. Figure 16-12 shows a drawing of a work bench storage cabinet designed for case construction. Plywood and other types of standard sheet materials are good for use in case construction. Additional structural support is not required with sheet material in case construction. Solid stock is used to construct the shelf glide assembly and frame, Fig. 16-13.

The bottom of the base cabinet is recessed, Fig. 16-14, making it possible to stand close to the cabinet. The recess is called a toe strip or toe kick. It can be built as part of the cabinet as shown in Fig. 16-15. Another method of constructing a toe strip is to add a *plinth* which is a separate base to which the cabinet is attached, Fig. 16-16.

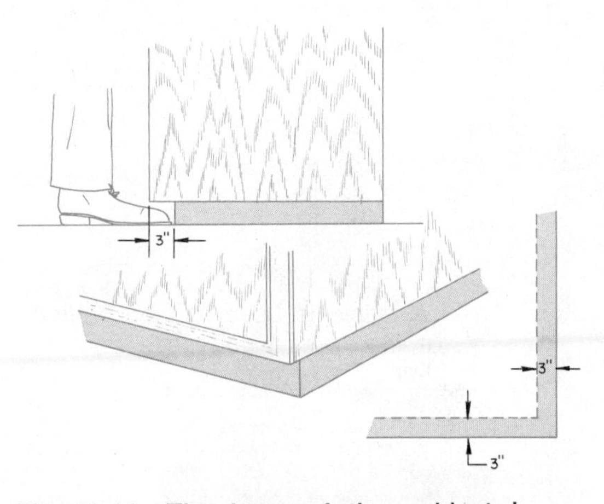

Fig. 16-14. The base of the cabinet is recessed for a toe kick.

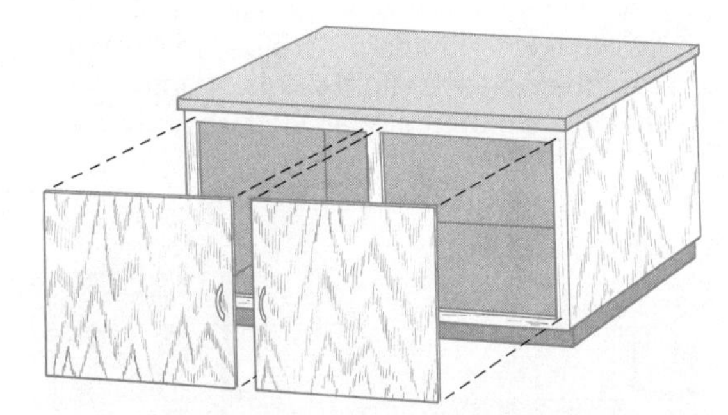

Fig. 16-12. Exploded view of work bench constructed by case construction.

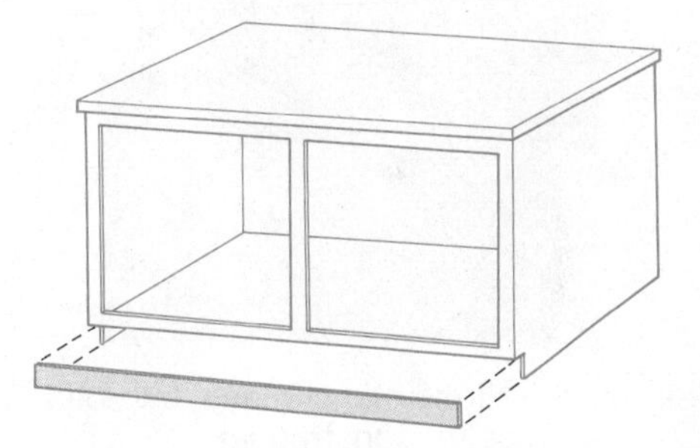

Fig. 16-15. The toe kick can be constructed as part of the cabinet.

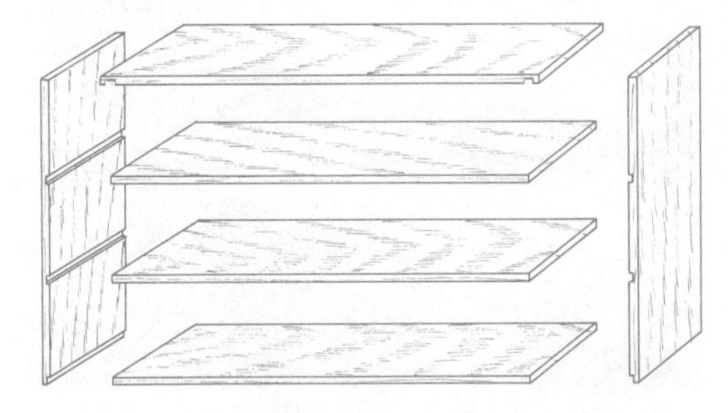

Fig. 16-13. Properly assembled heavy sheet material does not require additional structural support.

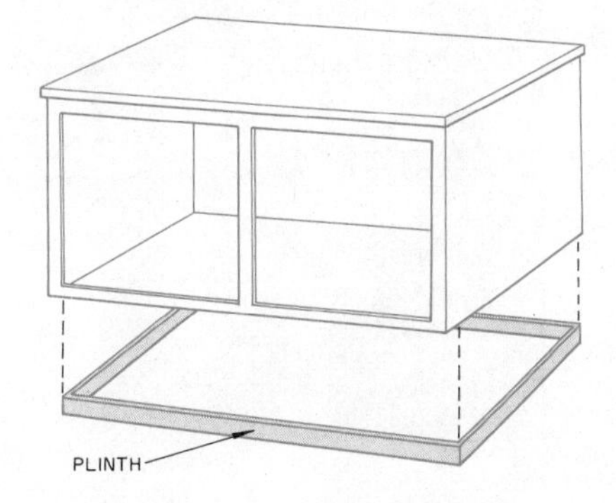

Fig. 16-16. The toe kick may be constructed separately.

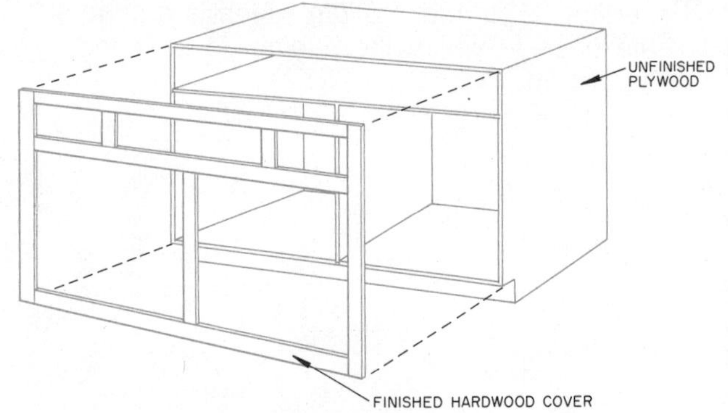

Fig. 16-17. A frame is used to face cabinets made of plywood case construction.

Cabinets which are constructed from plywood using case construction are normally faced on the front edge, Fig. 16-17. The drawers and doors are constructed to fit the facing frame.

Solid stock can also be used for case construction. The small nightstand shown in Fig. 16-18 was built with case construction. The facing is not necessary when solid stock is used instead of plywood, Fig. 16-19.

Fig. 16-18. Night stand made by solid wood case construction.

Fig. 16-19. Cabinet dry-sink constructed of solid case construction.

FRAME WITH COVER CONSTRUCTION

Cabinets built by means of *frame with cover construction* appear similar to case construction cabinets. This construction technique uses a lightweight frame cover with thin sheet material such as plywood.

Fig. 16-20. Kitchen cabinets constructed with frame and cover construction. (Universal Chief)

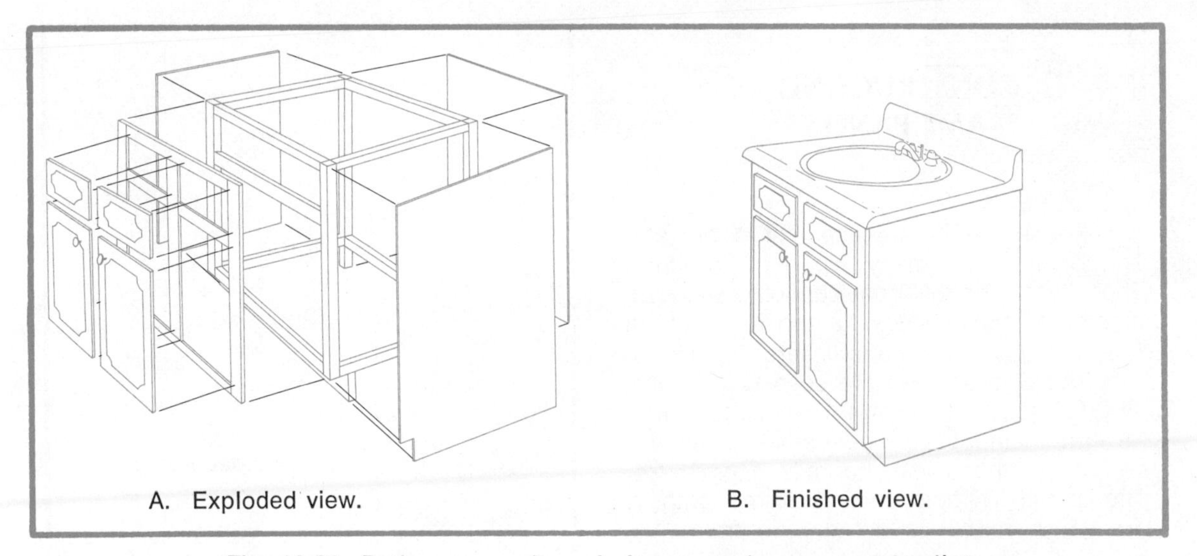

Fig. 16-21. Bathroom vanity of frame and cover construction.

Factory-manufactured kitchen cabinets and bathroom vanities are often constructed this way, Fig. 16-20.

The frames are constructed of solid wood about 3/4″ thick and 1 1/2″ wide. Figure 16-21 shows a detail drawing of a bathroom vanity unit. The front vertical members of the frame are extended beyond the sides. The raw edges of the cover material are concealed by the solid stock of the frame.

CARCASS CONSTRUCTION

Carcass construction is generally used for larger cabinets which may have a combination of doors and drawers, Fig. 16-22. The ends are often constructed with framed panels. This method is similar to case construction, but usually has more internal components. Furniture such as large desks, dressers, and buffets will often be constructed using the carcass method.

Fig. 16-22. A piece of furniture constructed by carcass construction. (Brandt)

CONSTRUCTING FRAME PANELS

Frame panels are used for doors and end sections of many types of furniture. The frame panel consists of two side members (called *stiles*), a top and bottom (called *rails)*, and a panel insert, Fig. 16-23.

The stile and rail are solid wood, usually 3/4″ to 1″ thick and 1 1/2″ to 2″ wide. Several methods can be used to join the corners of the stiles and rails, Fig. 16-24. The panel is held in the frame with a groove. A rabbet is cut on the frame in-

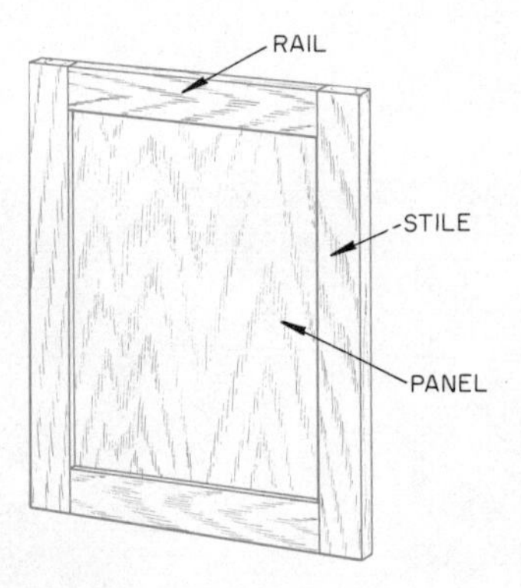

Fig. 16-23. Parts of a panel frame.

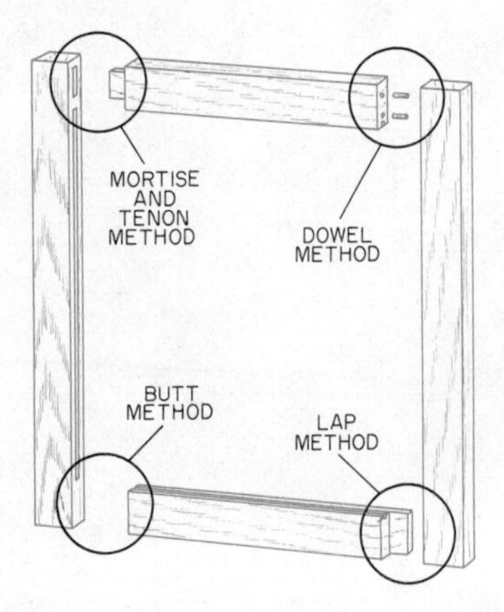

Fig. 16-24. Methods of joining frame panels.

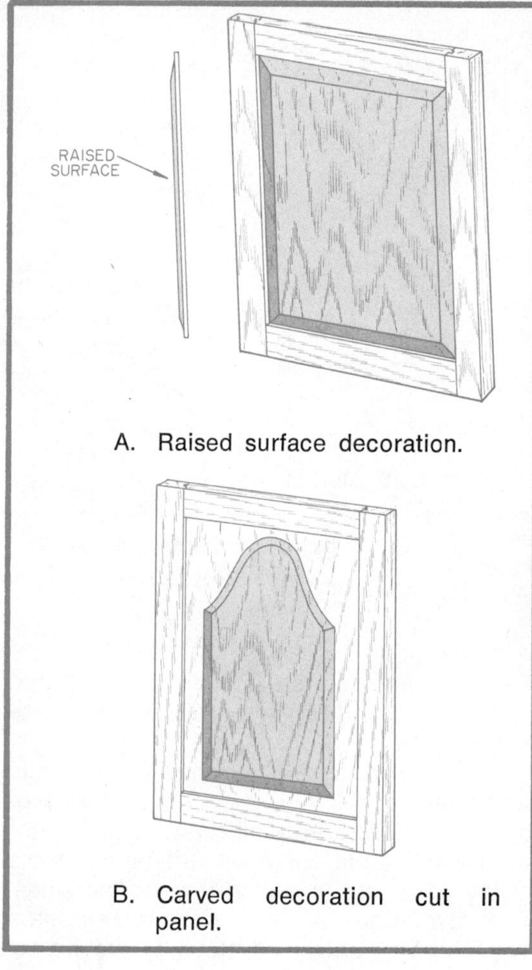

A. Raised surface decoration.

B. Carved decoration cut in panel.

Fig. 16-25. Door panels on furniture.

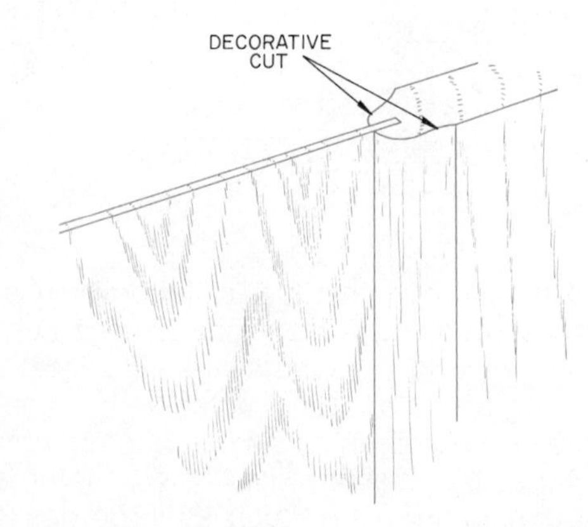

Fig. 16-26. The inside of the stiles and rails may have decorative cuts.

stead of a groove if a glass panel is placed in the frame.

The panel can be standard sheet material such as plywood or hardboard. Fine furniture generally has a raised decorative panel such as shown in Fig. 16-25A or may be carved with decorative cuts as shown in Fig. 16-25B. The stiles and rails shown in Fig. 16-26 have a decorative cut on the inside edge.

The greatest disadvantage of using plywood for frame panels is that special treatment is required to finish the exposed edges. There are several acceptable methods of concealing the exposed edges, Fig. 16-27.

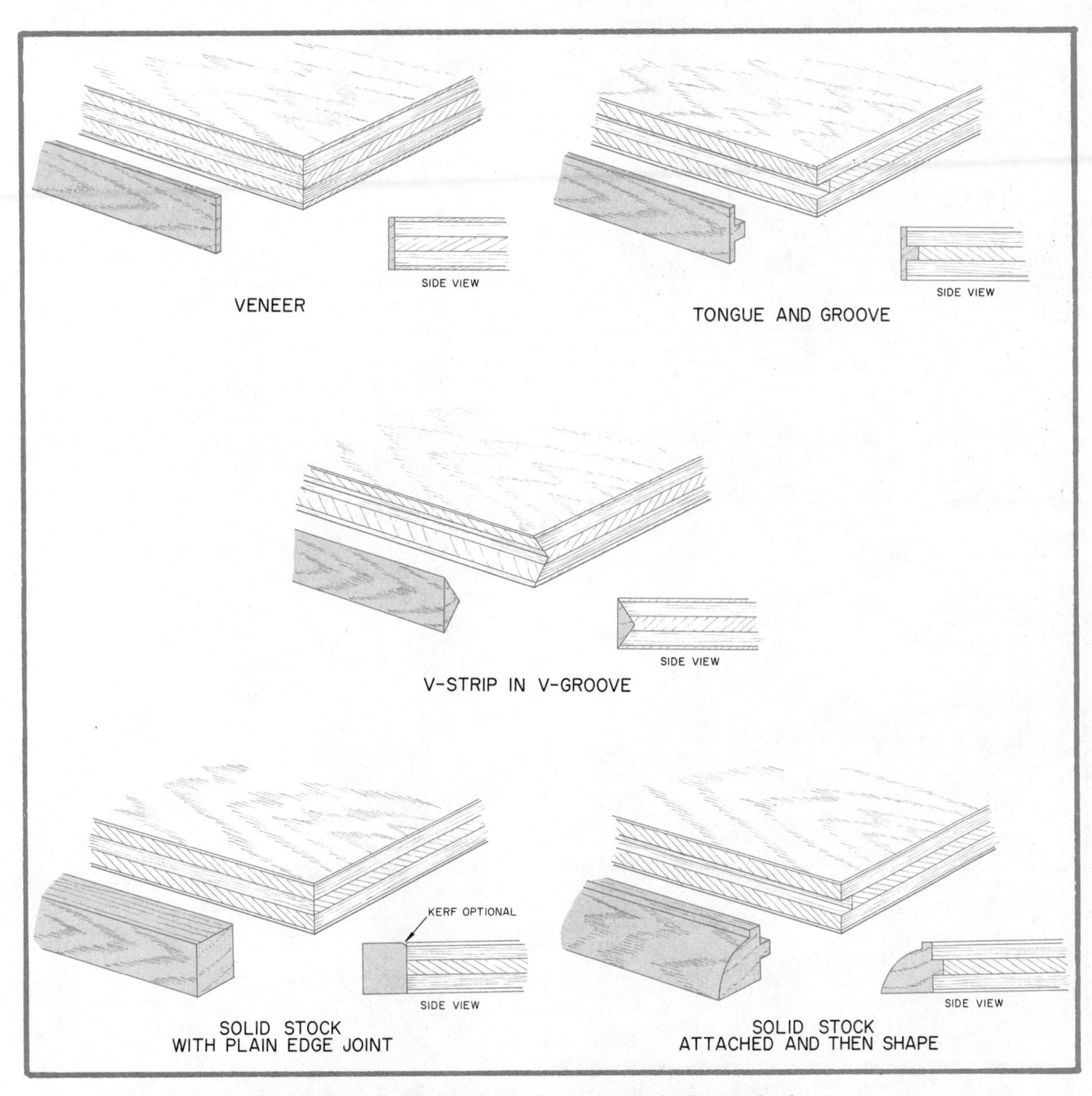

Fig. 16-27. Methods of treating exposed plywood edges.

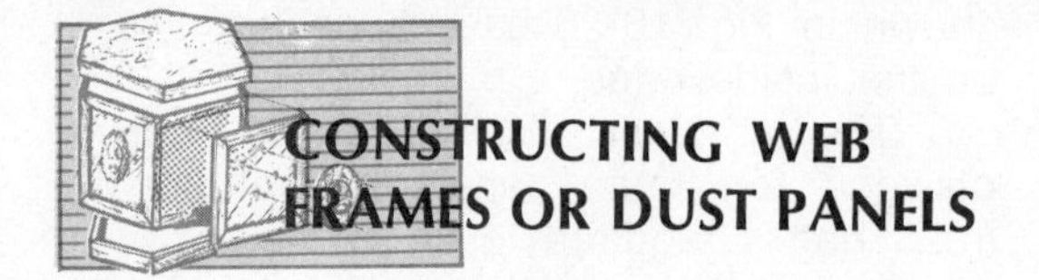

CONSTRUCTING WEB FRAMES OR DUST PANELS

The web frame or dust panel supports the drawers in carcass and case construction, Fig. 16-28. The panel can be constructed from 3/4″ × 1 1/2″ solid stock. The kind of wood selected for the front members depends on the type of drawer construction. If the drawers and doors are flush, the front member of the web must match the exterior of the cabinet. Overlapping drawers and doors conceal the web frame. Several types of joints including dowel, mortise and tenon, and end lap joints can be used to construct the web frame, Fig. 16-29.

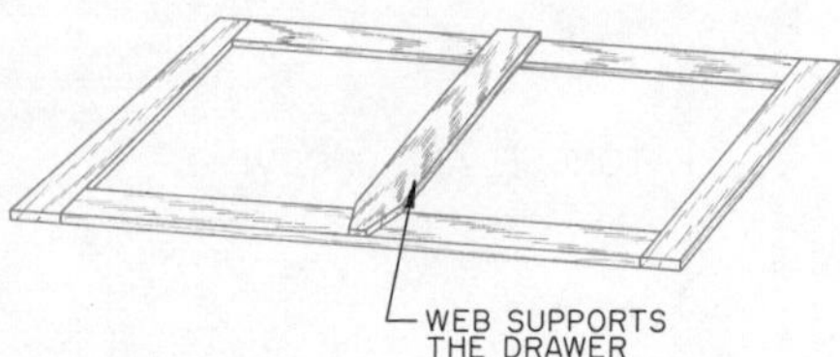

Fig. 16-28. A web panel for case construction.

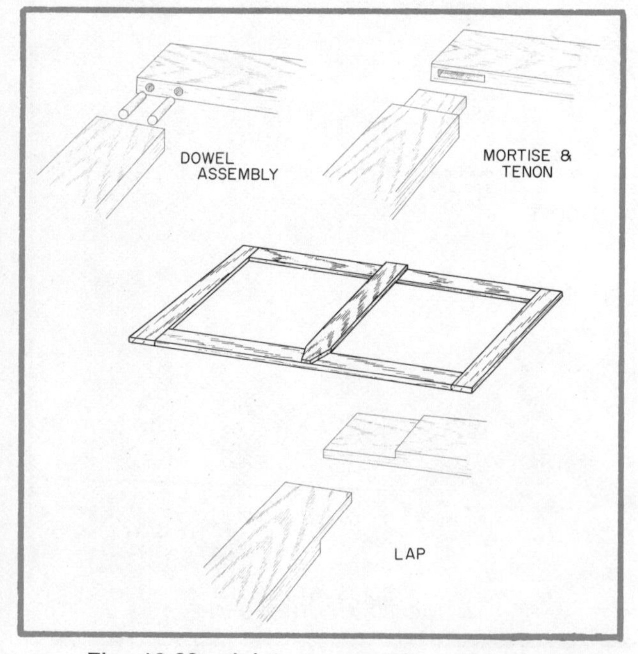

Fig. 16-29. Joints commonly used for web panel assembly.

DRAWER CONSTRUCTION

Two types of drawers are commonly used in cabinet and furniture construction: (1) flush drawers and (2) lip drawers. The difference is the method used for constructing the drawer front.

The web panel is referred to as a dust panel when the inside carries a panel, Fig. 16-30. This panel is a thin sheet of plywood or hardboard. Its purpose is to prevent dust from entering the drawers. Dust panels are generally included in high-quality furniture construction. The sheet material is placed in a groove on the inside of the frame.

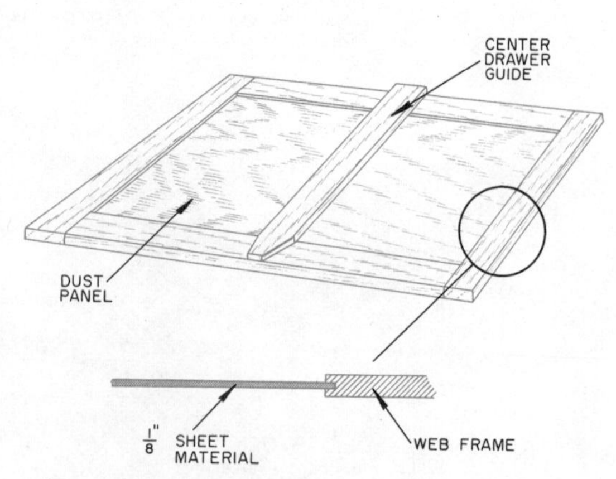

Fig. 16-30. A web panel with a solid dust panel center.

This house will provide many years of carefree living. The vertical Western Red Cedar siding will resist weather conditions. (Western Red Cedar Association)

Wood is used to construct the entire frame for this modern home. All construction is being done on the site. (Southern Forest Products Assn.)

Laminated beams provide structural support and an attractive interior ceiling. (Southern Forest Products Assn.)

IDENTIFY WOOD BY ITS COLOR AND GRAIN PATTERN

Black Walnut

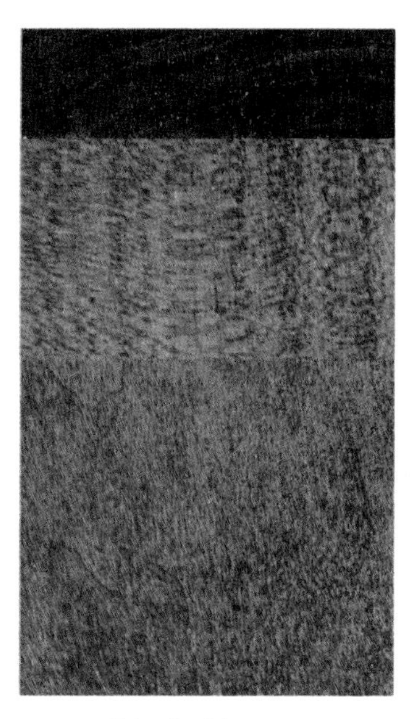
Black Cherry

Philippine Mahogany

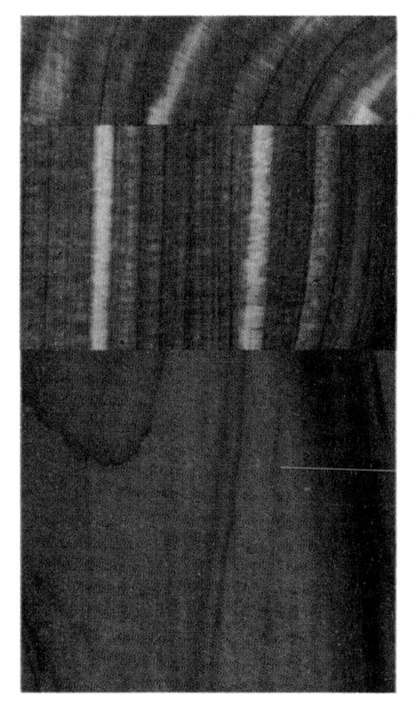
Eastern Red Cedar

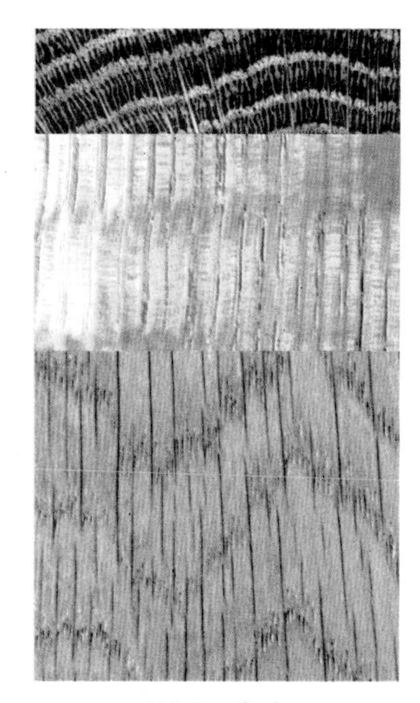
White Oak

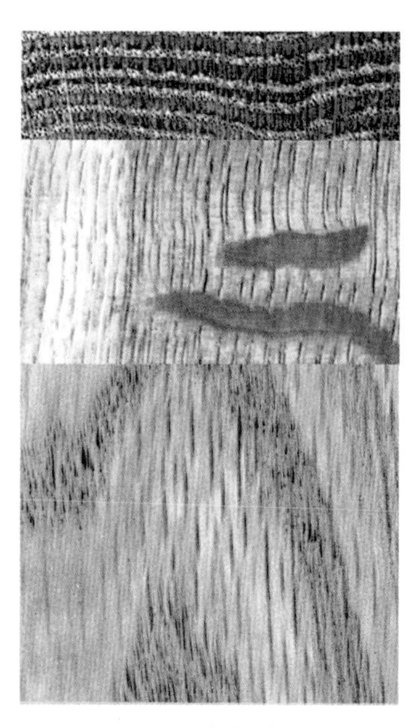
Red Oak

Ponderosa Pine*

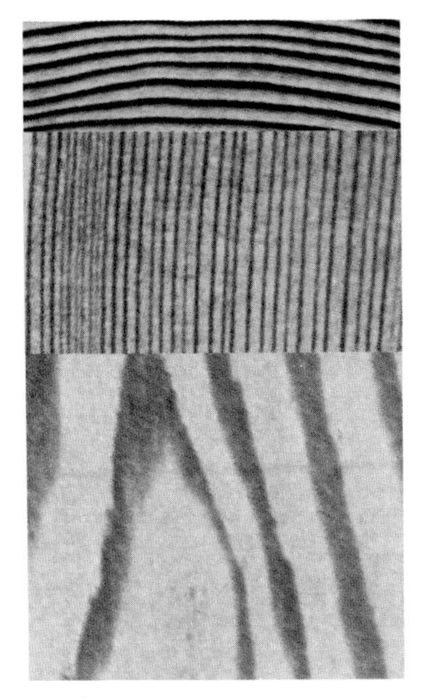
Douglas Fir*

Western White Pine*

Sugar Pine*

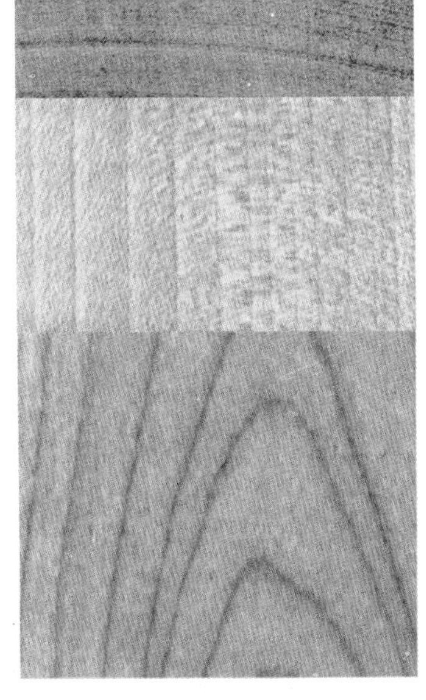
Yellow Birch

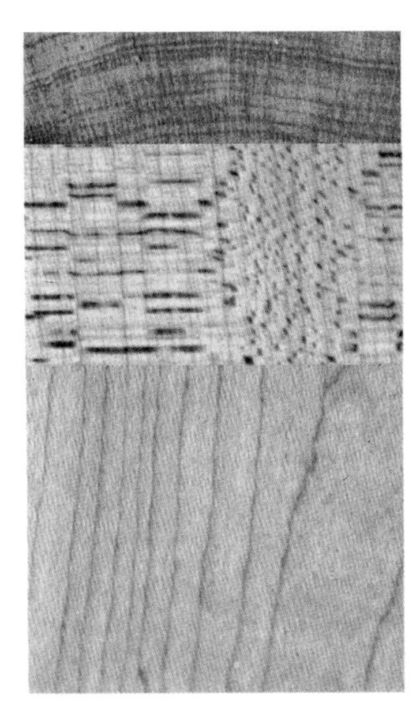
Sugar Maple

*Courtesy Southern Forest Products Association

Carefully selected and designed wood products combined with appropriate color creates a modern setting. (Armstrong Cork Company)

Wood can be combined with painted surfaces and touches of color to accent the natural wood. (Armstrong Cork Company)

Appropriately selected species of wood, properly finished, creates a setting for active living. (Armstrong Cork Company)

Wood is durable and is ideal for a recreation room. It requires a minimum of maintenance and provides a relaxing atmosphere. (Armstrong Cork Company)

The large laminated beams provide both beauty and strength. Wide open space without supporting walls is possible with laminated beams. (American Institute of Timber Construction)

Laminated beams may be of many different shapes to enhance the design and still give structural strength. (American Institute of Timber Construction)

Wood is used in many different ways in this room setting. Beautifully designed and harmonized wood furniture combined with room accessories adds to the warmth of the room. (Brandt Cabinet Works)

The warm wood furniture combined with color makes this an inviting room. The natural grain of the wood is visible through the transparent finish. (Brandt Cabinet Works)

The combinations of warm wood paneling, furniture, and colorful touches of accessories makes an attractive and comfortable home setting. (Masonite Corp.)

The furniture in this room is an example of good design in furniture construction. Notice the design of each piece complements the others. (Brandt Cabinet Works)

Figure 16-31 shows a piece of furniture with flush doors and drawers. The web frames are exposed and the front member matches the exterior of the cabinet.

The web frames are fitted into a dado in plywood case construction. The dadoes should be marked and cut on both ends with the same setup. Figure 16-32 shows dado cuts in an end for the web frame. The dado cuts are about one-half the thickness of the stock, and the width is one-half the thickness of the frame. The thickness of the frame is reduced to fit the dado, Fig. 16-33. The web frames are all trimmed to the exact length and width before they are cut to fit the dado. A flush drawer has a front the same width and length as the body of the drawer, whereas a lip drawer has a front which overlaps 3/8″ on all sides, Fig. 16-34. A flush drawer fits into the front cabinet frame. The lip drawer fits into the

Fig. 16-31. A cabinet with flush doors and drawers.

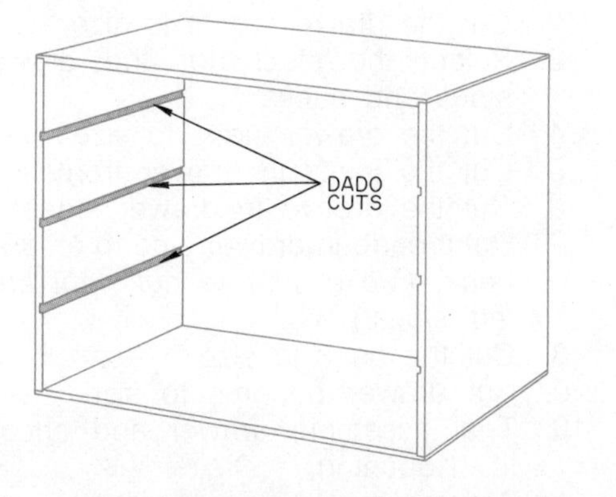

Fig. 16-32. Dado cuts in cabinet side to receive web panels.

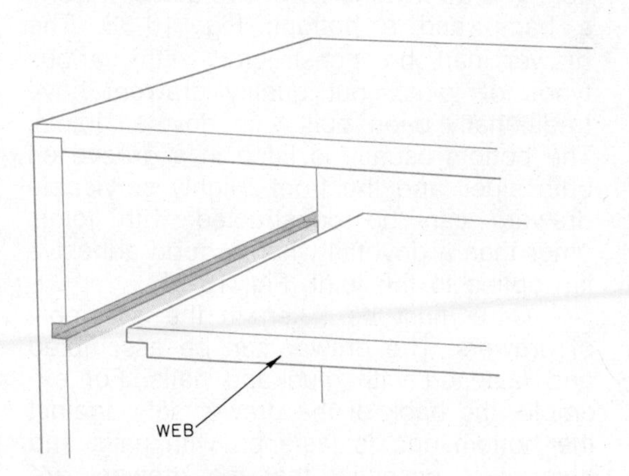

Fig. 16-33. Rabbeted web panel to fit cabinet side.

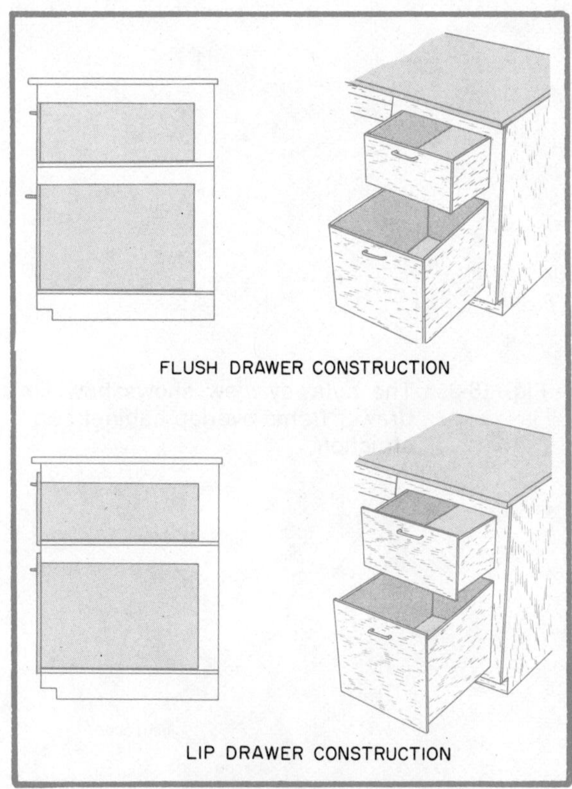

Fig. 16-34. Types of drawer construction.

front cabinet frame, but the lip partially covers the cabinet framing, Fig. 16-35.

Box construction previously discussed in this chapter is used for drawer construction. A drawer consists of two sides, a front, a back, and a bottom, Fig. 16-36. The drawer can be constructed with various types of joints, but quality drawers have traditionally been built with dovetail joints. The bottom usually is fitted in a groove on both sides and the front. Highly servicable drawers can be constructed with joints other than a dovetail when a good adhesive is applied to the joint, Fig. 16-37.

Care must be taken in the assembly of drawers. The drawer can be assembled and fastened with glue and nails. For example, the back of the drawer sets against the bottom and is fastened with nails and glue. It is essential that the drawers are square. A drawer which is out-of-square is very difficult to fit into a cabinet.

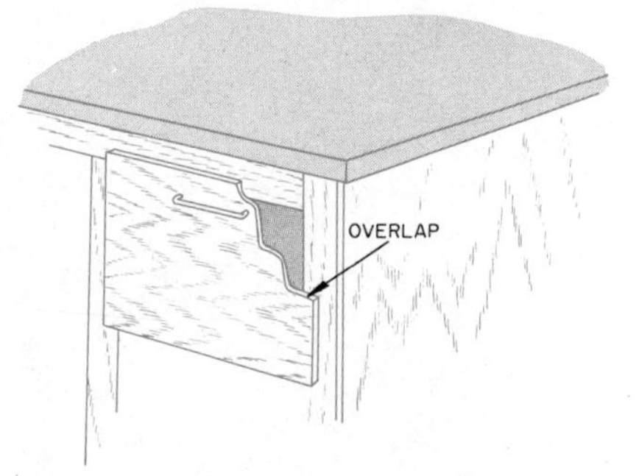

Fig. 16-35. The cutaway view shows how lip drawer fronts overlap cabinet construction.

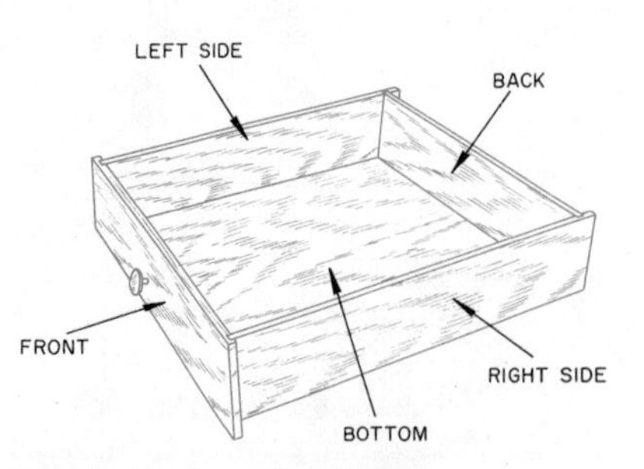

Fig. 16-36. Parts of a drawer.

The drawers are generally constructed after the cabinet in which they fit has been assembled. The drawers are fitted to the opening in the cabinet. The first step in drawer construction is to select the drawer fronts, as they should match the cabinet. When several drawers are constructed for the same cabinet, the grain pattern and color of the stock should be carefully selected and arranged.

Drawer fronts are usually constructed from stock ¾″ thick. The drawer sides can be cut from stock ½″ thick. Clear stock should be selected for the sides and back of the drawer. Pine, poplar, and oak are good woods for drawer construction. The bottoms of the drawer can be cut from ⅛″ or ¼″ plywood or hardboard.

FITTING THE DRAWER

A flush drawer front should be 1/16″ narrower and ⅛″ shorter than the opening in the cabinet. The lip drawer front should be ¾″ longer and wider than the cabinet opening. The groove in the drawer side is cut ⅜″ up from the bottom.

CONSTRUCTING A DRAWER

When several drawers of the same size are constructed, they can be semi-mass-produced. The procedure for constructing several drawers may be as follows:

1. Select the drawer fronts.
2. Cut the drawer fronts to size.
3. Select the stock for the drawer sides and backs.
4. Cut the drawer sides to size.
5. Cut the joints in drawer fronts.
6. Cut the groove in drawer sides.
7. Cut a dado in drawer side to receive back. (*Make sure to cut right and left sides.*)
8. Cut the back to size.
9. Cut drawer bottoms to size.
10. Trial assemble drawer and check fit in opening.
11. Assemble drawers. (Check for squareness.)

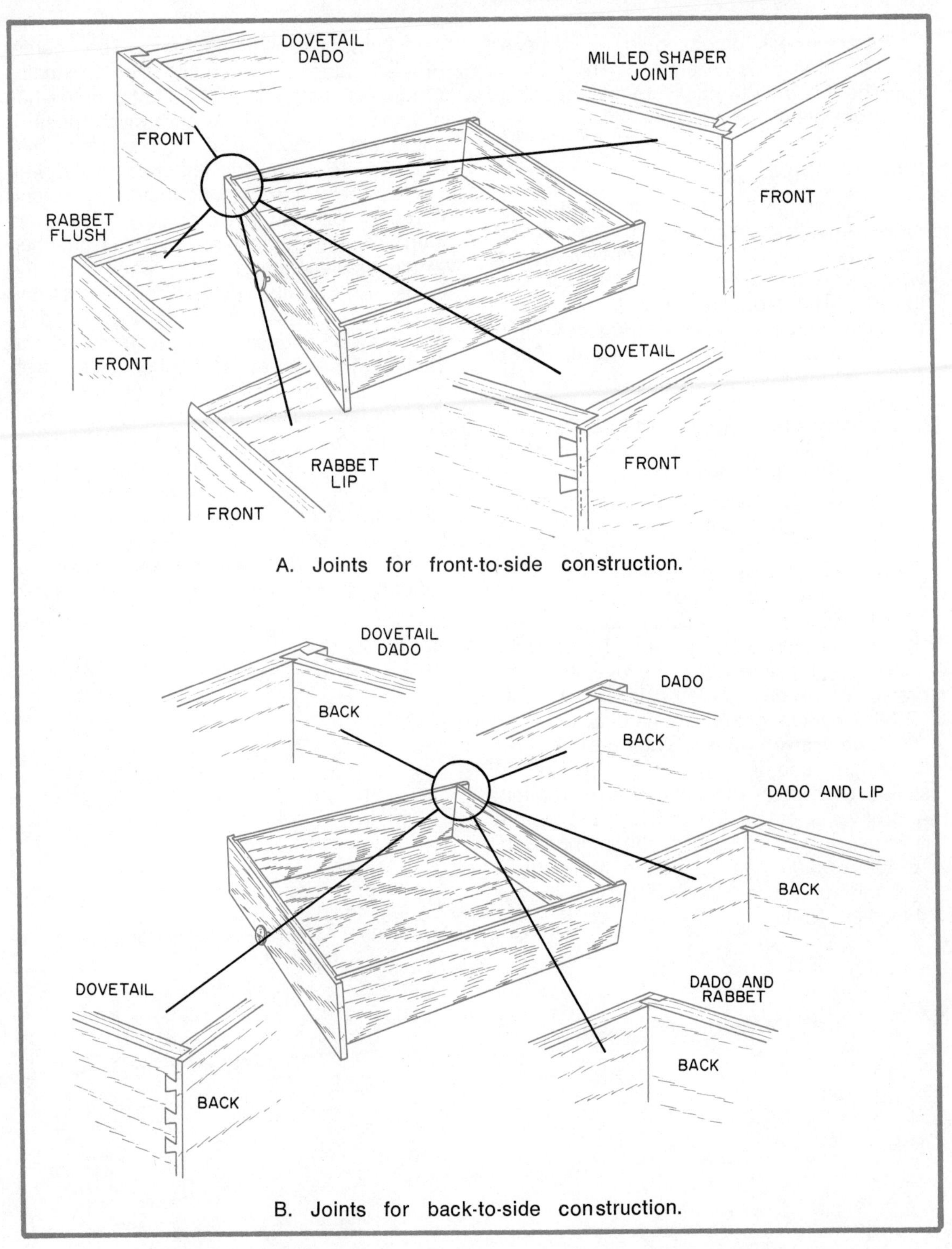

A. Joints for front-to-side construction.

B. Joints for back-to-side construction.

Fig. 16-37. Joints which can be used in drawer construction.

The excess glue should be removed after the drawer is assembled. The drawers may be fitted individually to the openings to insure a better fit.

DRAWER GUIDES

Drawer guides allow the drawer to move smoothly and freely in the cabinet. There are several types of drawer guides which can be either constructed or purchased. The type of guide which should be used depends on the method of cabinet construction. The simplest drawer guide system is the corner guide. It serves as both a runner and a guide, Fig. 16-38.

Corner drawer guide. The corner guide is constructed from two pieces of stock long enough to extend the depth of the cabinet. A rabbet is cut on the corner of each piece. The guide is attached with screws and/or glue. It is best to first attach the guide to the cabinet with screws so that it can be adjusted if it does not fit properly. If the drawer does close properly, the screws can be removed and glue may be applied. A top guide may be needed if the drawer drops down when it is opened, Fig. 16-39. The top guide is called a *kicker.*

Side drawer guides. Side drawer guides are often used if web frames are not installed as part of the cabinet construction. A groove is cut in the drawer side before it is assembled to receive the runner, Fig. 16-40. A dado can be cut in the cabinet and the runner is attached to the drawer side, Fig. 16-41.

Center drawer guides. The center drawer guide is frequently used in quality furniture construction. The center guide has a minimum of friction and gives positive alignment. It consists of a guide and a runner, Fig. 16-42 and is attached to the web frame. The guide is constructed by cutting a recess or plow in a piece of stock the length of the guide. This piece is attached to the drawer bottom. The runner is cut the same size as the plowed portion of the guide. It is positioned on the web frame so the drawer aligns with the front of the cabinet. The runner is attached to the web frame with screws. The front of the runner can be tapered on the sides slightly about 2″ back from the front edge, making it easier to replace the drawer when it is removed. The drawer sides should be in contact with the web frame when the guide is properly installed.

Suspended drawer guides. The drawer support and guides can be suspended

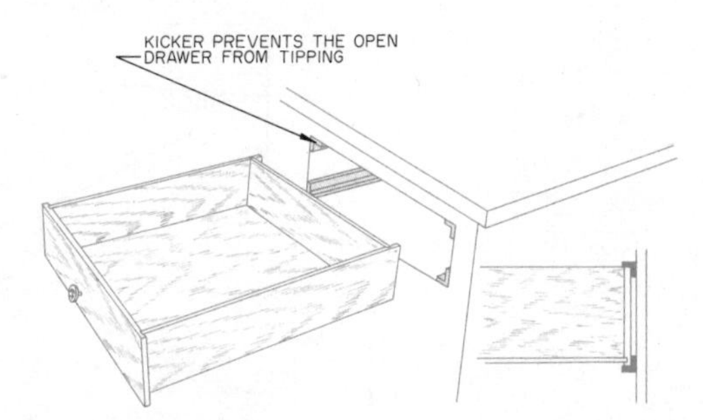

Fig. 16-39. A corner drawer guide with a top kicker.

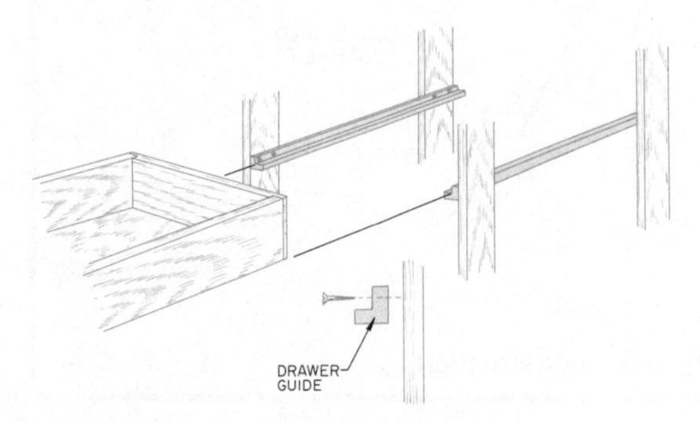

Fig. 16-38. A corner drawer guide.

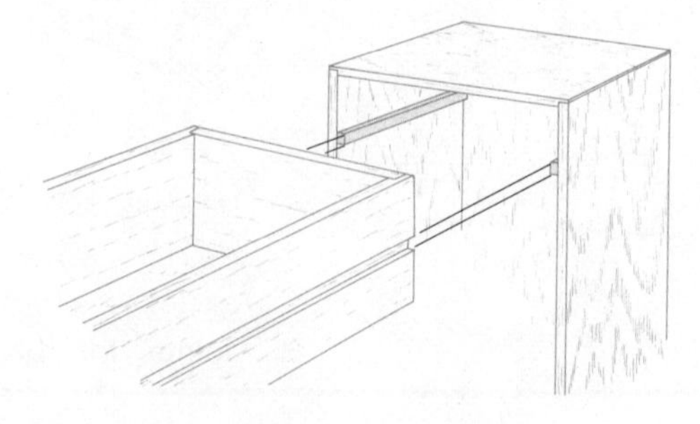

Fig. 16-40. Side drawer guide system.

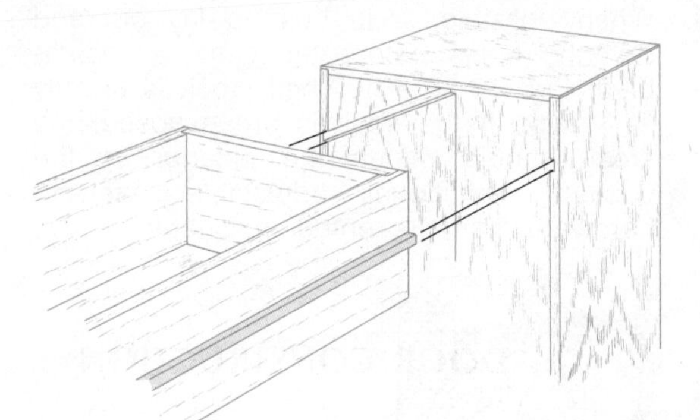

Fig. 16-41. Alternate method of constructing side drawer guides.

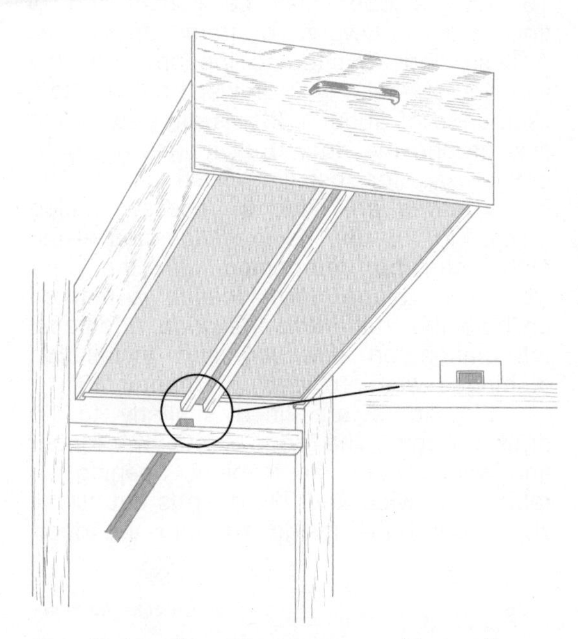

Fig. 16-42. A center drawer guide system.

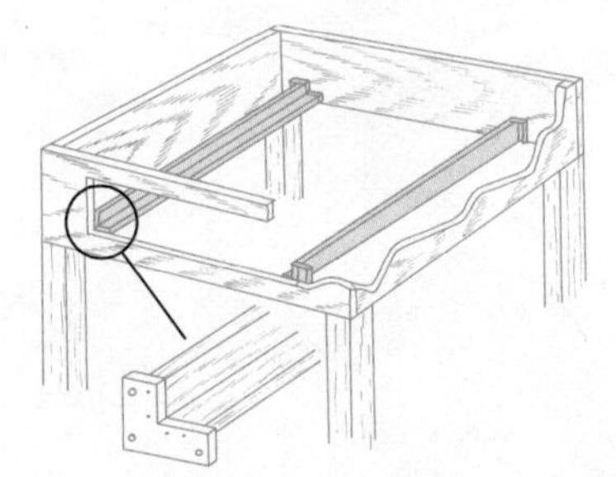

Fig. 16-43. Suspended drawer guides.

from a top such as in leg and rail construction, Fig. 16-43. Thus, the drawer is both supported and guided by the same suspension method. This method is satisfactory for shallow drawers of less than 4″ deep. Drawer glides are waxed to reduce the friction.

Commercial drawer guides. There are numerous commercial drawer guide systems which may be purchased, Fig. 16-44.

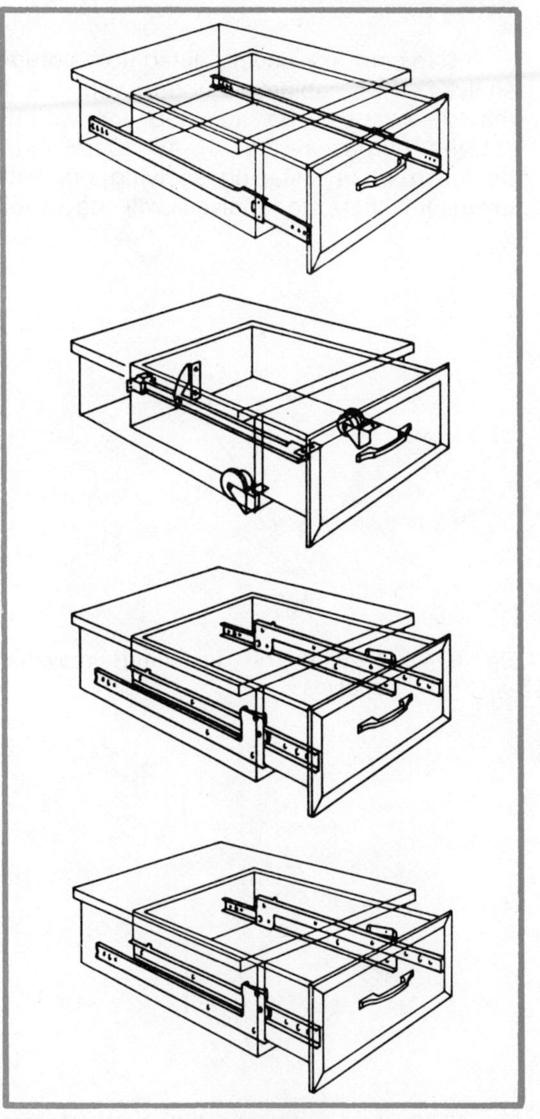

Fig. 16-44. A commercial drawer guide. (Amerock Corp.)

The roller supports combined with bearings are excellent for heavy drawers which may be used frequently, such as in dressers, kitchen cabinets, wooden file drawers, and others.

SHELF SUPPORTS

Cabinets are usually fitted for storage shelves. The cabinet is more versatile if the shelves can be adjusted for various storage purposes, Fig. 16-45. Care must be taken when installing shelving brackets to insure that the shelves will be level.

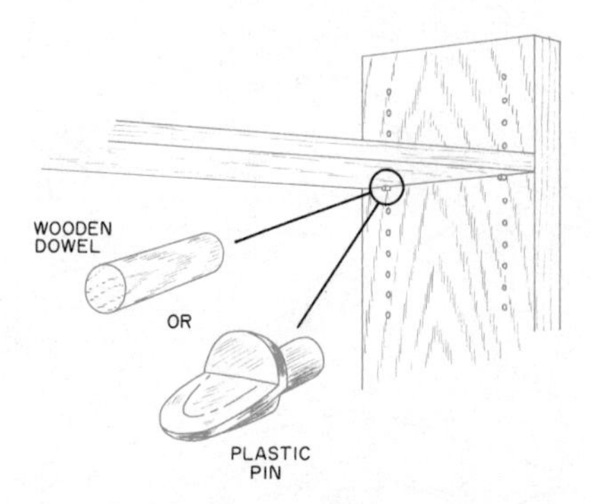

Fig. 16-45. Method of installing movable shelves.

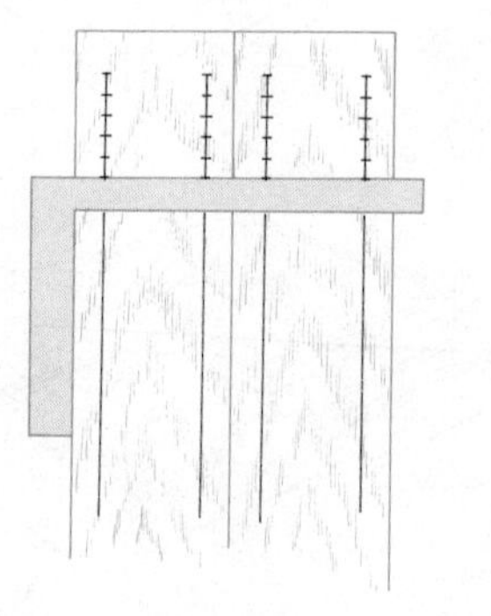

Fig. 16-46. Lay out and drill holes before cabinet assembly.

When possible, it is best to lay out and drill the holes before the case is assembled, Fig. 16-46. The shelf stock is usually ¾″ thick. If the shelves are exposed, the stock should match the exterior of the cabinet. Additional adjustable shelving hardware was presented in Chapter 13.

DOOR CONSTRUCTION

Doors installed on most furniture are either of flush or lip construction, Fig. 16-47. The door may be constructed of solid stock, plywood, or frame and panel. A flush door is fitted to the opening very similarly to the way a drawer front is fitted. Clearance of about 1/32″ is allowed all the way around the door. Butt hinges are generally used for installing flush doors. The hinge leaves are fitted in recesses called *gains*, Fig. 16-48. The exact location of the hinge can be determined with the door wedged in place. This location is marked on both the frame and the door. Additional information on selection and installation of hinges was provided in Chapter 13.

Lip doors are fitted similarly to lip drawer fronts. The door is cut ¾″ longer and wider than the cabinet opening. A rabbet ⅜″ wide and ⅜″ deep is cut all the way around the inside edge of the door.

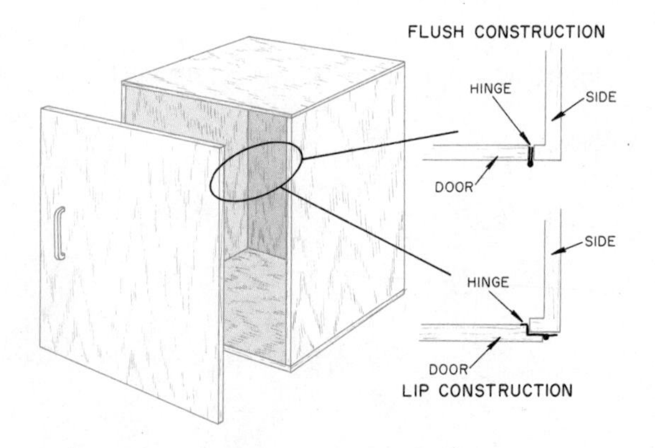

Fig. 16-47. Types of doors on cabinets.

The lip door is hung with a semi-concealed hinge, Fig. 16-49.

Cabinet doors are usually 3/4″ thick. The panel frame door is constructed in the same manner as shown in Figs. 16-23 and 16-24.

SLIDING DOORS

Sliding doors are easy to construct and install. They complement some types of furniture styles, and are often used when limited space does not allow a swinging door. Standard sheet stock such as plywood or hardboard make ideal sliding door panels, or plate glass can be used for the sliding door in display cabinets. Plastic sliding door tracks should be used for glass sliding doors to reduce the friction sometimes prevalent with heavy glass panels.

There are several techniques which may be used for installing sliding doors. A groove cut in the top and bottom of the cabinet will serve as a guide. The top groove is cut twice as deep as the lower groove, making it possible to remove the sliding panel by raising it in the upper guide, Fig. 16-50. When thick sliding panels are used, a rabbet cut on the inside edge of each panel reduces the distance between the doors. The rabbet on the bottom panel is cut deeper than the groove to reduce friction and allow free movement, Fig. 16-51.

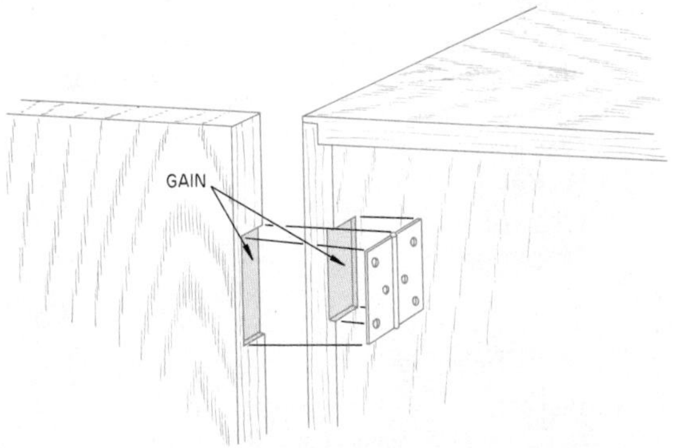

GAIN

Fig. 16-48. Hinges on flush doors are placed in gains.

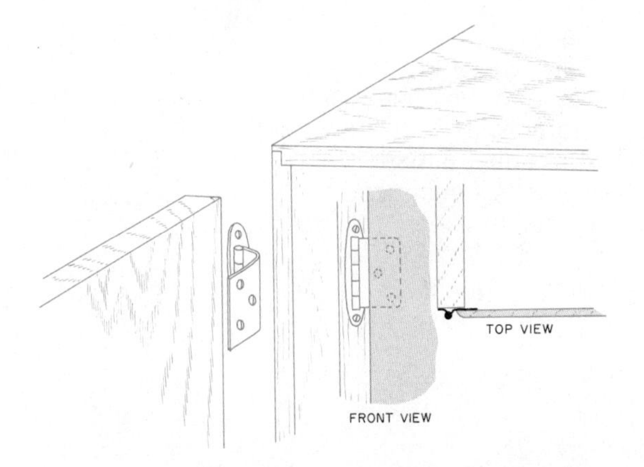

Fig. 16-49. Installation of semi-concealed hinge on a lip door.

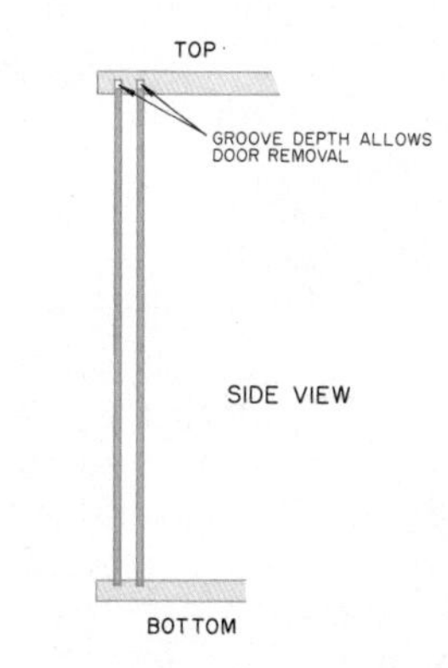

Fig. 16-50. Sliding doors installed in groove cuts in top and bottom of a cabinet.

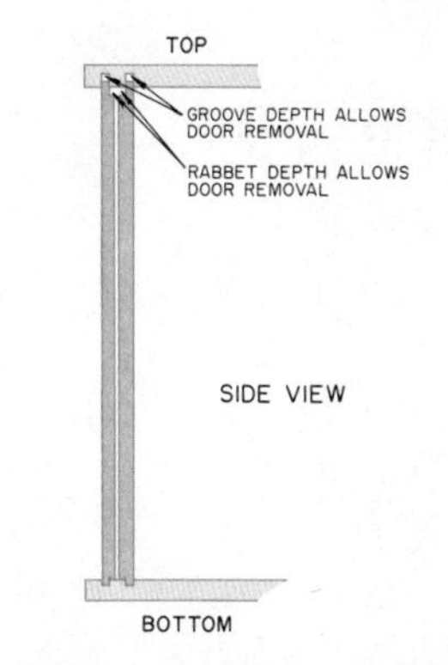

Fig. 16-51. The clearance between thick doors can be reduced by rabbeting the door.

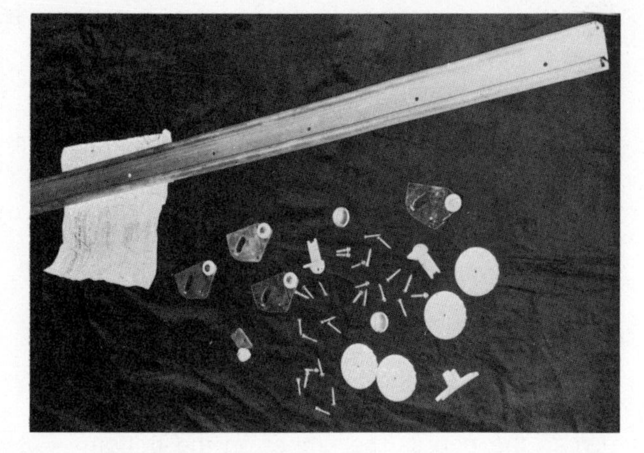

Fig. 16-52. Manufactured sliding door track parts.

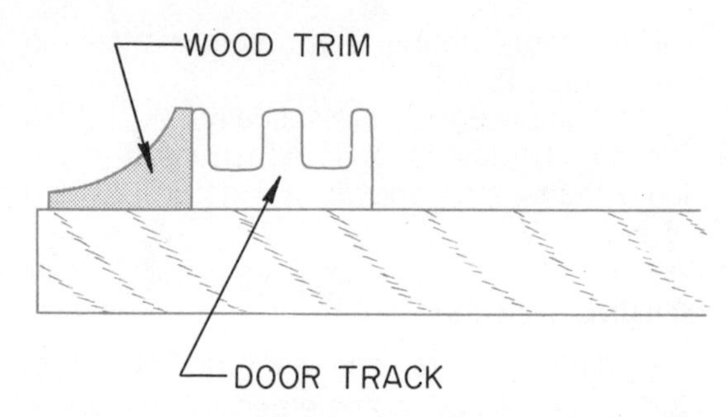

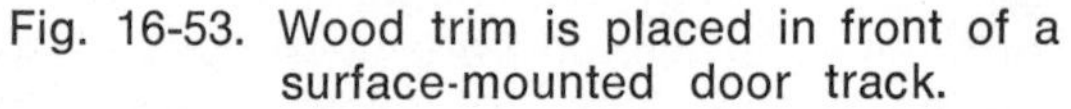

Fig. 16-53. Wood trim is placed in front of a surface-mounted door track.

Specially-designed plastic and aluminum sliding door tracks are also available, Fig. 16-52. Instructions for use are given by the manufacturer. The track can be fitted in a groove cut in the top and bottom of the cabinet. The upper track is deeper than the lower track so the doors can be removed. A piece of trim can be used to cover the exposed track when it is surface-mounted, Fig. 16-53. The track is available for different door thicknesses. The most common sizes of tracks are for doors 1/4″ and 3/4″ thick.

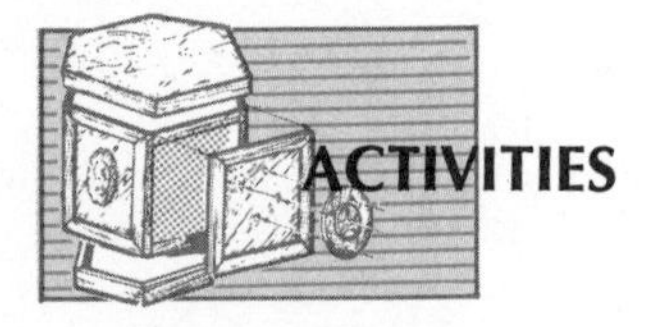

ACTIVITIES

PROBLEM

List the most appropriate construction method for building the following products. Remember, more than one technique might be appropriate for some products.

- End table
- Nightstand
- Kitchen cabinet
- Bathroom vanity
- Buffet
- Drawer
- Bookcase

VOCABULARY

Stiles
Rails
Case construction
Leg and rail construction
Frame with cover construction
Box construction
Carcass construction
Cabinetmaker
Dowel joint
Drawer guide
Gain

End Table

Nightstand

Kitchen Cabinet

Bathroom Vanity

Buffet

Drawer

Bookcase

INTRODUCTION

Many materials are formed by pouring them into a mold in a molten state (softened or liquefied). For example, concrete is placed in a form and allowed to cure; waffles are formed and cooked in a mold. Metal can also be poured into a mold and formed in the shape of the desired object. The process of shaping molten metal in a mold is called *casting*. The mold is generally formed in a special foundry sand.

A cavity in the sand is formed around a *pattern*. The pattern is a full-sized model of the finished object to be produced by casting. The pattern can be formed from wood, metal, plaster-of-paris or wax. Wood

Fig. 17-1. A skilled patternmaker constructs a foundry pattern.

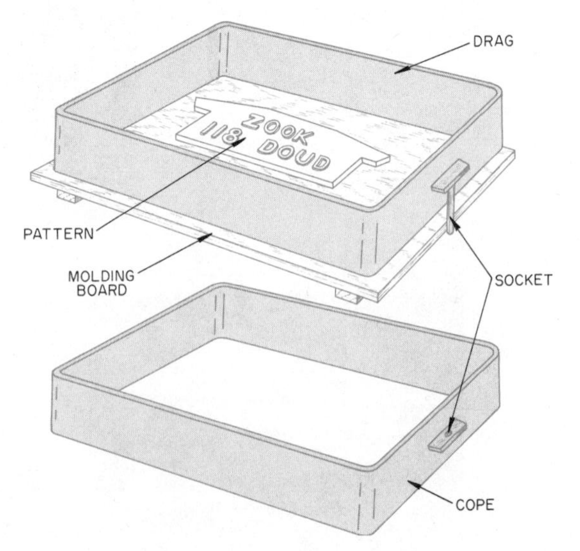

Fig. 17-2. A foundry flask for making a sand mold.

is one of the best materials as it can be shaped easily by sawing, sanding, and carving. Wood also provides a smooth surface on the finished casting. The person who makes patterns is a highly-skilled craftsman called a *patternmaker*, Fig. 17-1.

The pattern is used for making the sand mold. It is placed in a box-like case called a *flask* which consists of two parts for easy removal of the mold. The lower part is called the *drag* and upper part the *cope*, Fig. 17-2. Alignment pins are placed on the ends of the flask.

MAKING A MOLD

The pattern is placed on a board called a *molding board*. The drag portion of the flask is placed on the molding board with the pattern in the center, Fig. 17-3. Fine sand passed through a screen called a *riddle* is placed directly over the pattern. The drag is filled with more sand and packed solid using a *rammer*, Fig. 17-4.

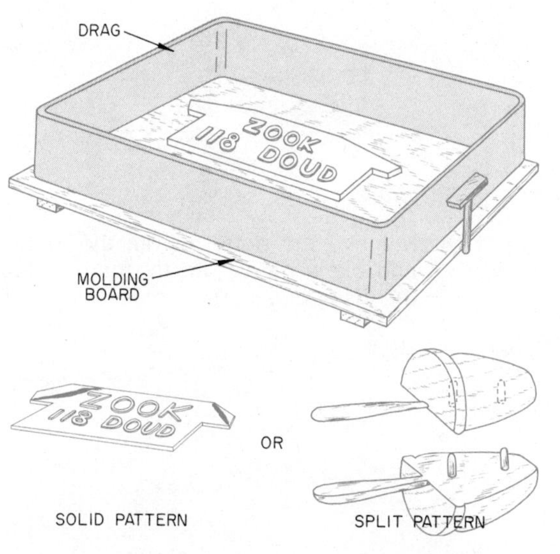

Fig. 17-3. Placement of the pattern in the drag.

After the sand has been packed to the top of the drag, the excess is removed by striking with a *strike bar*, Fig. 17-5.

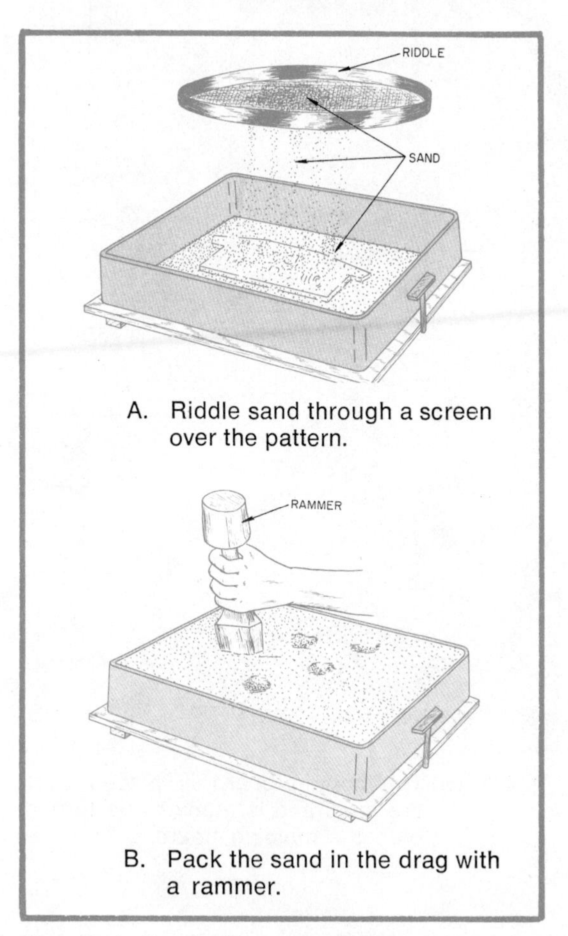

A. Riddle sand through a screen over the pattern.

B. Pack the sand in the drag with a rammer.

Fig. 17-4. Forming a sand mold.

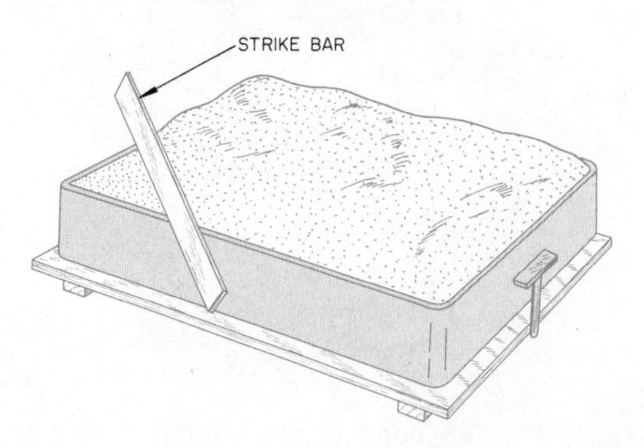

Fig. 17-5. Strike off excess sand with a strike bar.

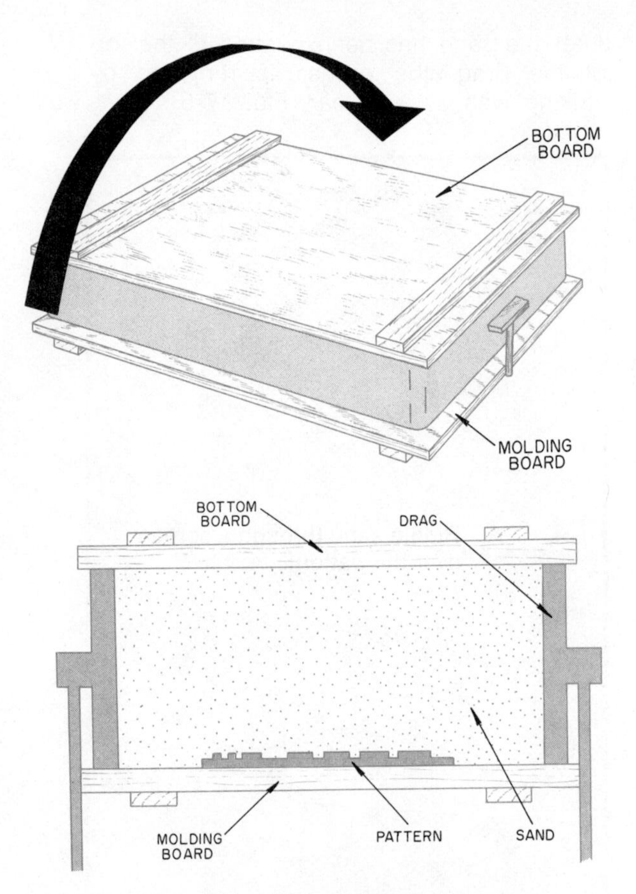

Fig. 17-6. The bottom board is placed over the mold and is ready to be turned onto the molding board.

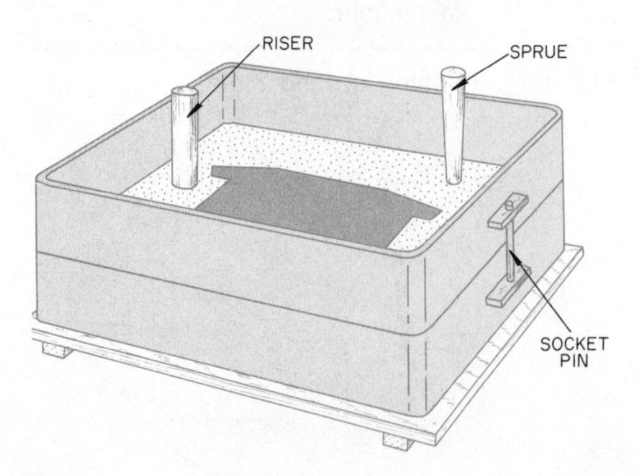

Fig. 17-7. The cope is set on the drag and the sprue and riser pins are placed in position.

A bottom board is placed on the top of the drag. The bottom board and molding board are held together firmly and the drag is turned over on the molding bench, Fig. 17-6. The pins on the drag are now pointed upward. The molding board is removed and the bottom of the pattern is exposed. The cope is placed on top of the drag, and aligned with *socket pins*, Fig. 17-7. Sand is riddled on the top of the pattern and rammed firmly. The same procedure is used for placing the sand in the cope as was used for the drag. The *sprue* and *riser pins* are removed, the cope is lifted off the drag and carefully laid on its side, and the pattern is removed from the drag. It must be removed carefully in a quick straight-up motion to prevent damage to the mold, Fig. 17-8. A gate is cut from the riser and sprue hole to the pattern. The molten metal

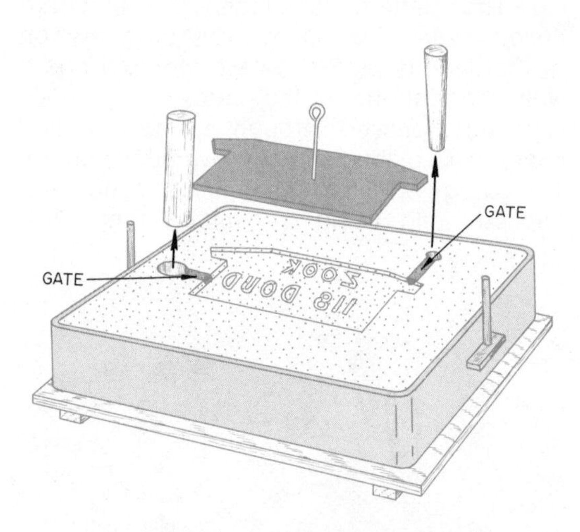

Fig. 17-8. Remove the pattern from the mold and cut a gate from the pattern to the riser and sprue holes.

Fig. 17-9. A solid pattern on which one side is flat.

flows to the cavity created by the pattern through the sprue hole, and some of the gases escape through the riser hole. The cope is replaced on the drag. Molten metal is then poured into the sprue hole.

TYPES OF FOUNDRY PATTERNS

There are two basic types of patterns which are commonly used for foundry castings: (1) *solid pattern* and (2) *split pattern*. The solid pattern is made as one piece and may be used to cast objects with one flat surface, Fig. 17-9. An irregularly shaped object requires a split pattern, made in two halves, Fig. 17-10. The two halves are aligned with dowel pins.

The sides of a pattern must be tapered so it can be removed from the mold. This taper is called *draft*, Fig. 17-11A. It is called *positive draft* when it is possible to remove the pattern from the mold. When it cannot be removed without breaking the sand mold, it is called *negative draft*, Fig. 17-11B. A pattern with a draft of one or two degrees can usually be removed.

ALLOWANCES IN PATTERN MAKING

The molten metal shrinks as it cools. The pattern is made slightly larger than the size of the finished casting to allow for shrinkage. A special rule called a *shrink rule* is used by the patternmaker which is 1/8″ to 1/4″ longer per foot than an ordinary rule. Various types of metal shrink more than others. Iron shrinks about 1/8″ per foot and aluminum about 1/4″ per foot. These allowances must be considered prior to construction of the pattern. Additional size must also be allowed if the casting will be machined.

CONSTRUCTING A ONE-PIECE PATTERN

The pattern is constructed from a high-quality piece of pine or mahogany the size specified on the working drawing. If letters are used as part of the pattern,

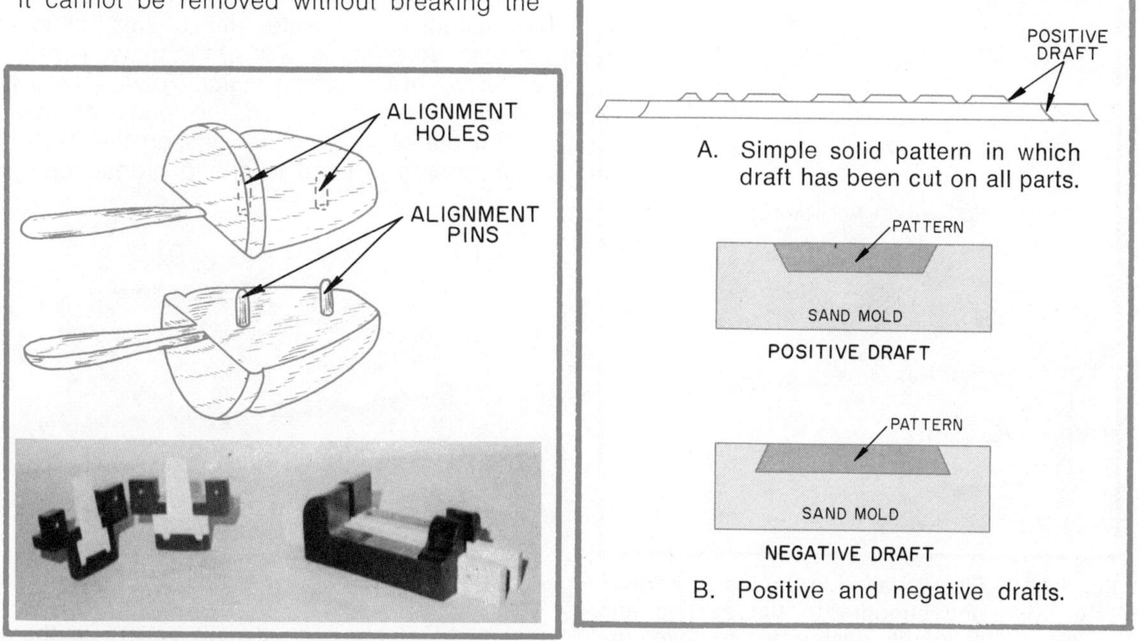

Fig. 17-10. Split patterns.

Fig. 17-11. Pattern drafts.

they can be cut from thin plywood or tempered hardboard. The draft is cut on the letters by tilting the jig saw table to a 2° angle. All of the cutting is done on the same side of the blade to give the proper draft on all sides. The letters are attached to the backing board with water-resistant glue. Small brads may also be used with glue.

The draft is cut on the backing board before the letters are attached. All parts of the pattern must be sanded smooth. A sealer coat of shellac or lacquer is applied to the entire pattern. Precast metal letters can also be used for patternmaking.

A deep pattern with square inside corners is difficult to remove from the mold. The inside corners are both stronger and easier to mold if they are rounded. The radius which makes the corner round is called a *fillet,* Fig. 17-12. A wax material is often used to form the fillet. The wax is smoothed in place with a special fillet iron which is heated and applied to the wax.

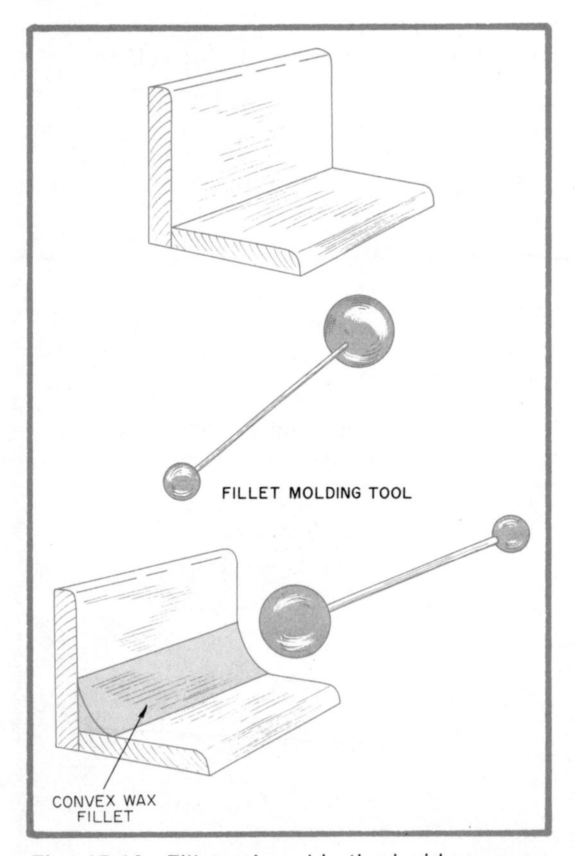

Fig. 17-12. Fillets placed in the inside corners add strength to the casting and make it easier to remove the pattern.

SPLIT-PATTERN MOLDING

A split pattern consists of two parts. The *parting line* is between the cope and the drag. Half of the pattern is in the cope and the other half is in the drag, Fig. 17-13. The parting line on the pattern provides a flat surface against the molding board when the mold is rammed. The second half of the pattern is put in place when the drag is turned over and the cope is placed on top.

Many split patterns are turned on the lathe after the parting line is determined and the alignment dowels are installed, Fig. 17-14. The stock is held together by glue while it is turned in the lathe.

MATCH PLATE PATTERNS

Sometimes split patterns are attached permanently to a plate. Half of the pattern is mounted on each side of the plate. These patterns are called *match plates,* Fig. 17-15. Holes drilled on the ends of the match plates fit over the pins on the flask. This method is used to speed production.

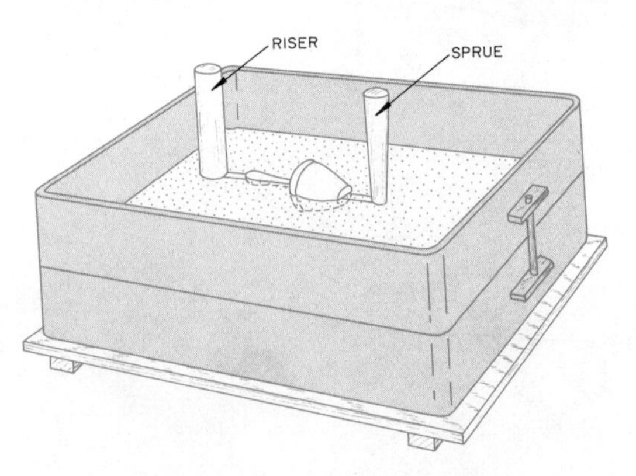

Fig. 17-13. Placement of split pattern in the flask.

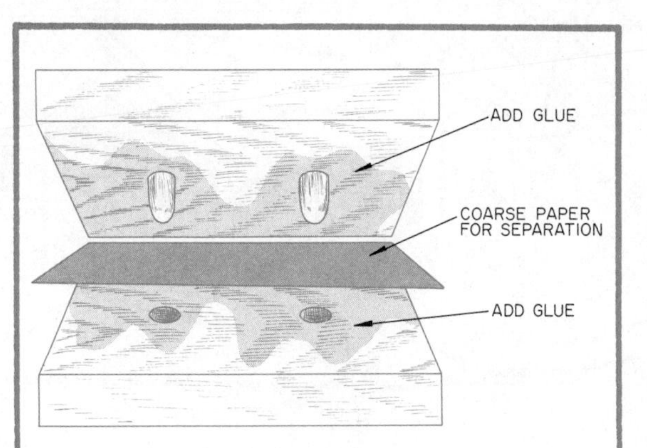

A. Glue both pieces of the pattern mold and place coarse paper between them.

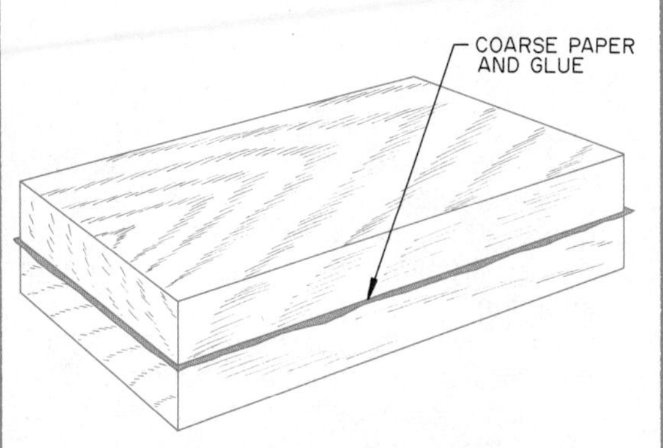

B. Set two pieces together and allow to dry before turning.

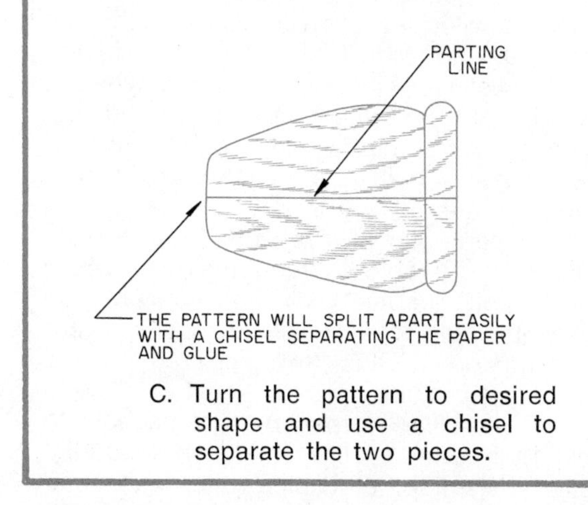

C. Turn the pattern to desired shape and use a chisel to separate the two pieces.

Fig. 17-14. Procedure for turning a split pattern.

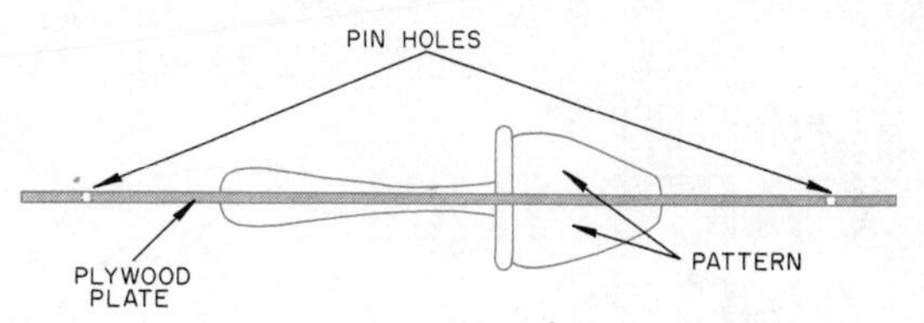

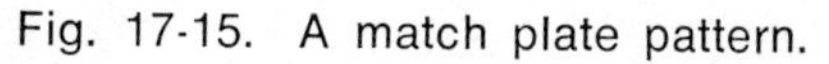
Fig. 17-15. A match plate pattern.

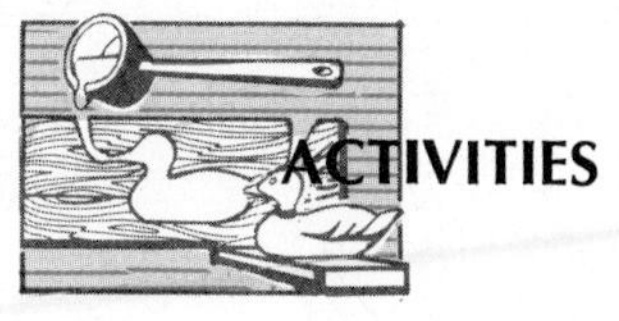

ACTIVITIES

PROBLEM

1. Design a one-piece foundry pattern for a lamp base, ash tray, book end, or some product idea of your selection.
2. Design a two-piece foundry pattern for a lamp base, ash tray, book end, or some product idea of your selection.

The foundry pattern you have designed should be a detailed drawing. Show the draft angle with a selection view. Use a shrink rule for accuracy. Specify the material and operations for constructing a finished pattern.

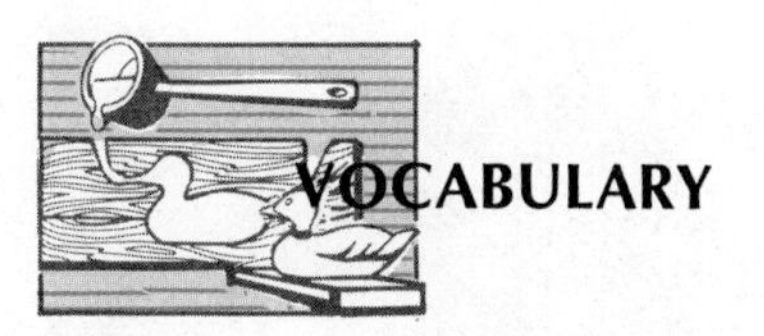

VOCABULARY

Casting
Patternmaker
Flask
Cope
Drag
Split-pattern
Sprue
Riser
Draft
Match plate pattern

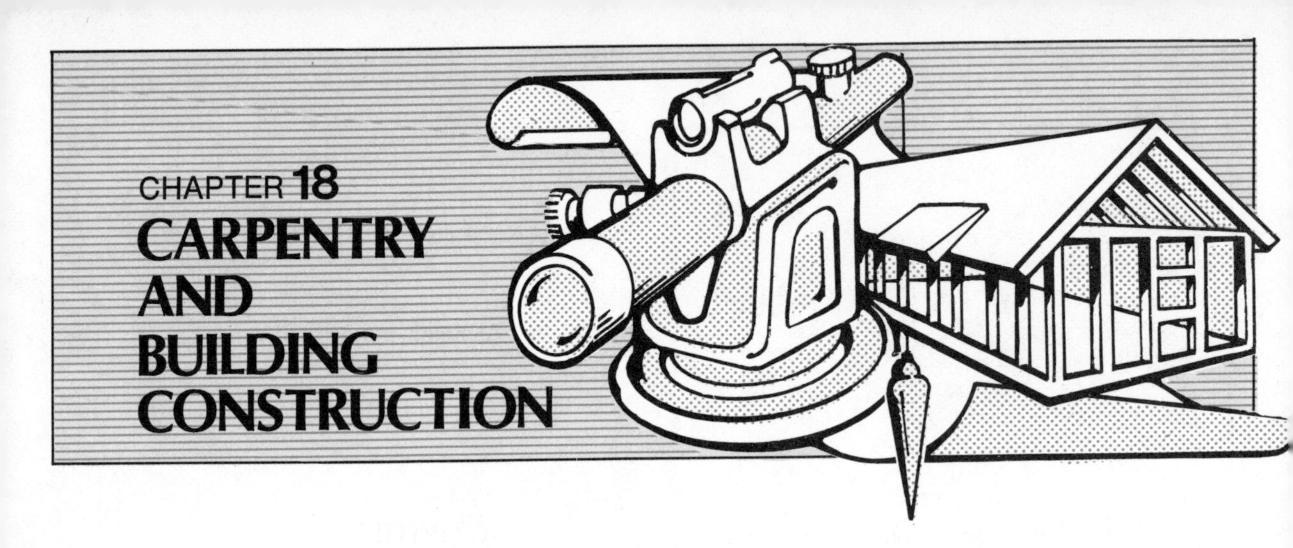

CHAPTER 18 CARPENTRY AND BUILDING CONSTRUCTION

INTRODUCTION

Carpentry is the largest single skilled trade in the woodworking industry. The demand for skilled carpenters is great. It is projected that the need for carpenters in the United States will continue to increase due to the population growth. Today, even in an age of miracle-man building materials, wood remains the most important. This chapter will deal with wood frame construction common to residential construction.

Fig. 18-1. A form constructed by a carpenter is filled with cement.

The carpenter performs many jobs related to the construction industry. Carpentry also includes the erection of forms in which concrete is placed for bridges, foundations, and buildings, Fig. 18-1. Carpentry requires many personal skills and the knowledge of various construction methods and materials. A skilled craftsman must know how to use many types of hand tools, portable power tools, and stationary power equipment. Today, eight of ten houses built are constructed with a wood frame.

A carpenter always works from a plan. He must be able to read the plan which has been prepared by an architect. He must be able to estimate the cost of labor and materials, and the time required to build a structure.

The skilled carpenter must be able to build the forms for the foundation, place the sills, joists, floor, erect studs and rafters, and apply roof sheathing, siding and lay shingles. This aspect of building is regarded as rough or framing carpentry. The finish carpentry consists of hanging doors, trimming windows, laying finish floors, and building cabinets. A master craftsman can perform all of the carpentry tasks involved in erecting a structure.

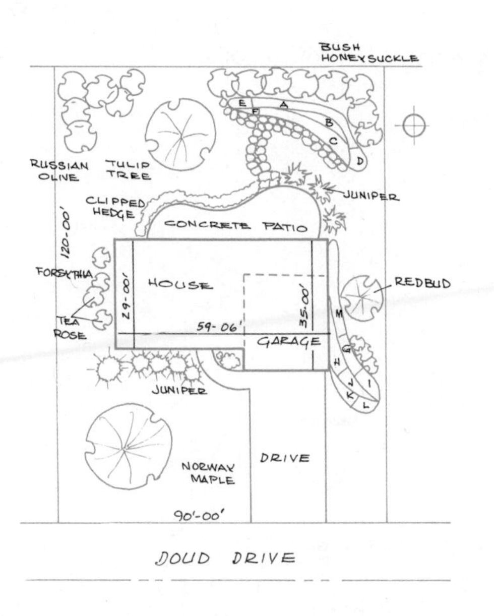

Fig. 18-2. A plot plan for a house.

The carpenter works in conjunction with other skilled tradesmen such as electricians, plumbers, plasterers, masons, and painters. It is helpful if the carpenter has a general understanding of the related trades.

HOUSE PLANS

A drawing of the proposed house is the first step towards the final construction. The drawings consist of a plot plan, floor plans, exterior elevations, structural sections, and details.

Plot plans. The plot plan shows how the structure will be placed on the building lot, the distance the house will be set back from the street, and the distance from the other property lines, Fig. 18-2. The plot plan also gives the elevation of the finished grade line and may also show the landscape plan.

Floor plan. The floor plan is the most important single drawing for house planning. It appears as a horizontal section of the structure and as a section just below the door and window level, Fig. 18-3. The

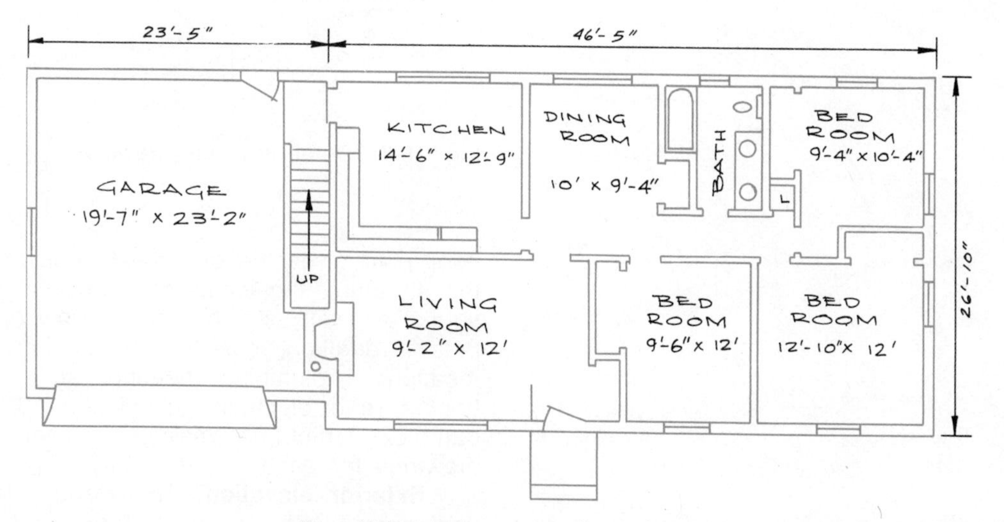

Fig. 18-3. A floor plan for a home.

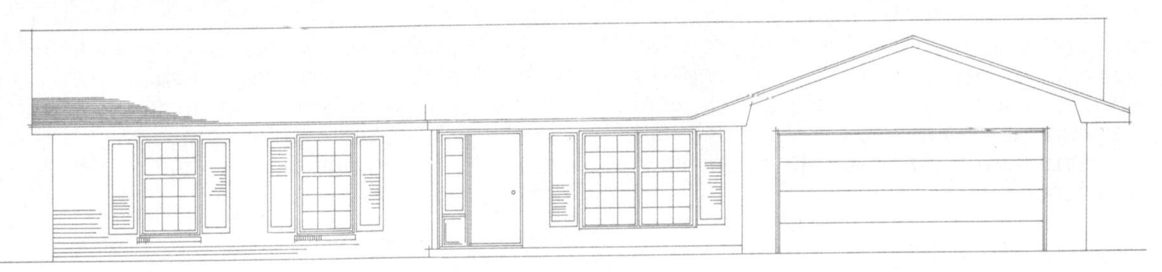

Fig. 18-4. A front exterior elevation view of a house.

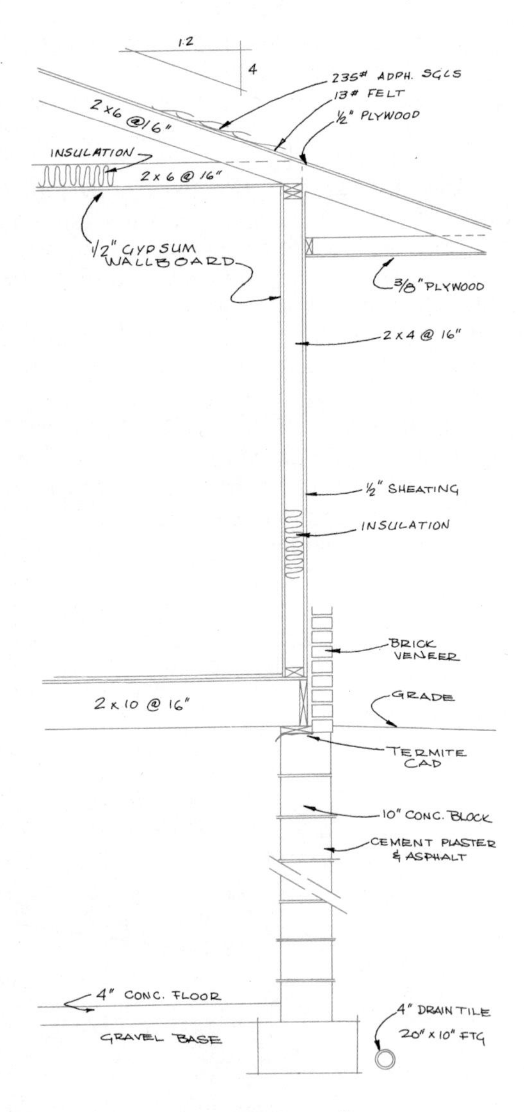

Fig. 18-5. A typical wall section.

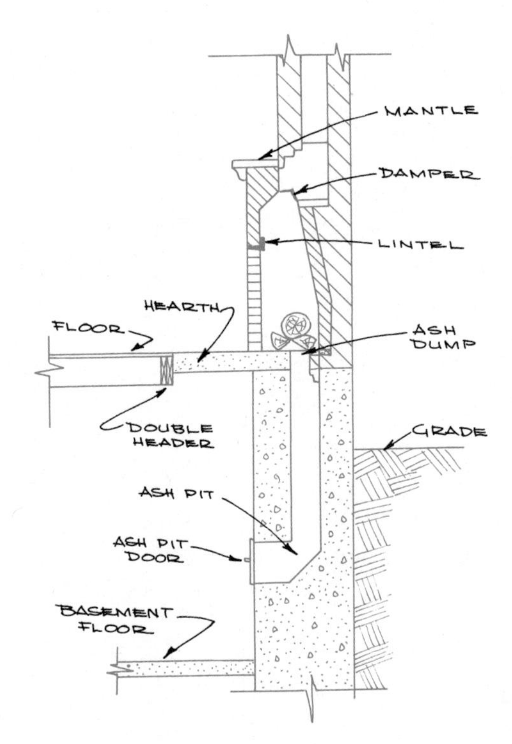

Fig. 18-6. Detail fireplace drawing.

floor plan gives the outside dimensions of the structure, the location of the wall partitions, as well as door and window openings. In addition, most floor plans show the location of plumbing, heating, air conditioning, and electrical installations. Large complex structures may have separate drawings for each type of utility.

Exterior elevation. An exterior elevation drawing shows the exterior appearance of the structure upon completion. The type of building material for the exterior is given on the elevation drawing, Fig. 18-4.

Structural section drawings. A structural section drawing is a view of a typical architectural or structural feature of the building. The section drawing is typical of the structural method and materials used throughout the building. A typical wall section from the foundation to the roof is shown in Fig. 18-5.

Detail drawings. These drawings show in greater detail special features of a structure. The detail drawing is generally more complete and drawn in a larger scale. Fireplaces and special built-in cabinets or storage areas are often shown as separate detail drawings, Fig. 18-6.

READING BLUEPRINTS

Architectural drawings such as house plans are composed of symbols. Thorough understanding of a plan requires that one know what the symbols represent. Various symbols are used to identify materials specified from the foundation to the roof, Fig. 18-7.

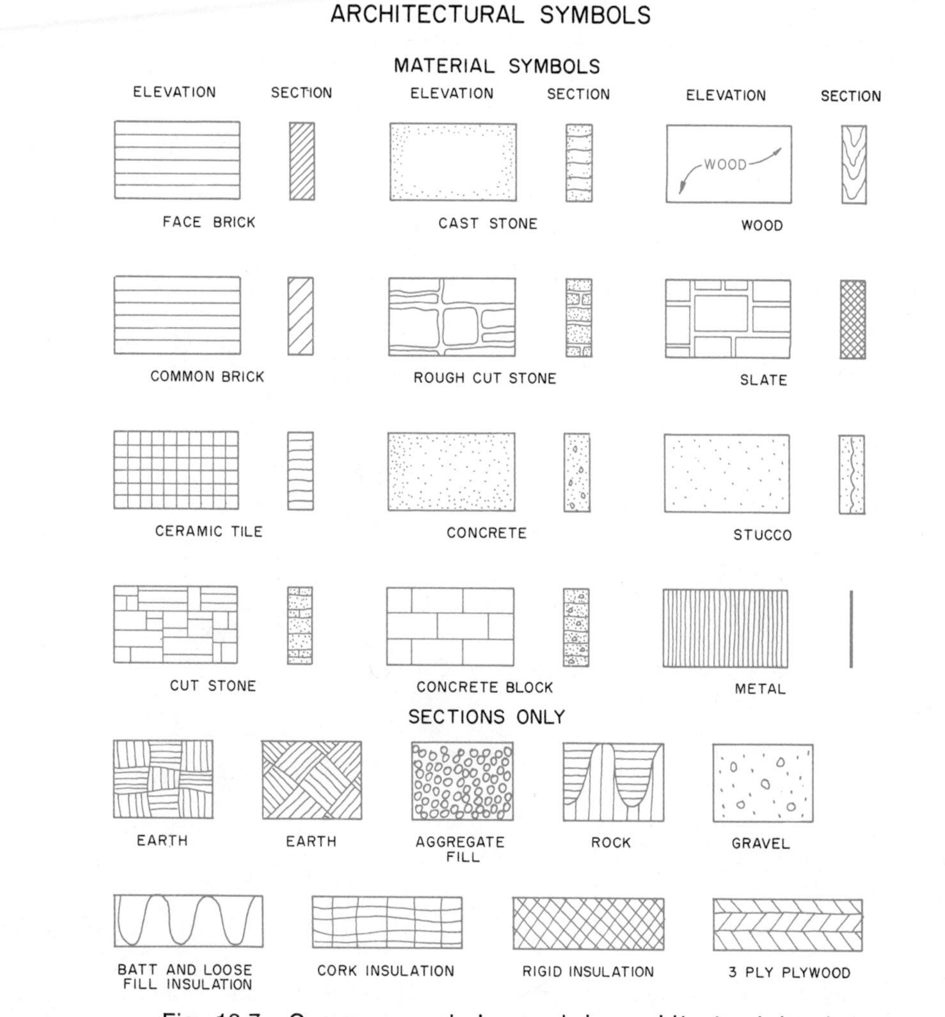

Fig. 18-7. Common symbols used in architectural drawing.

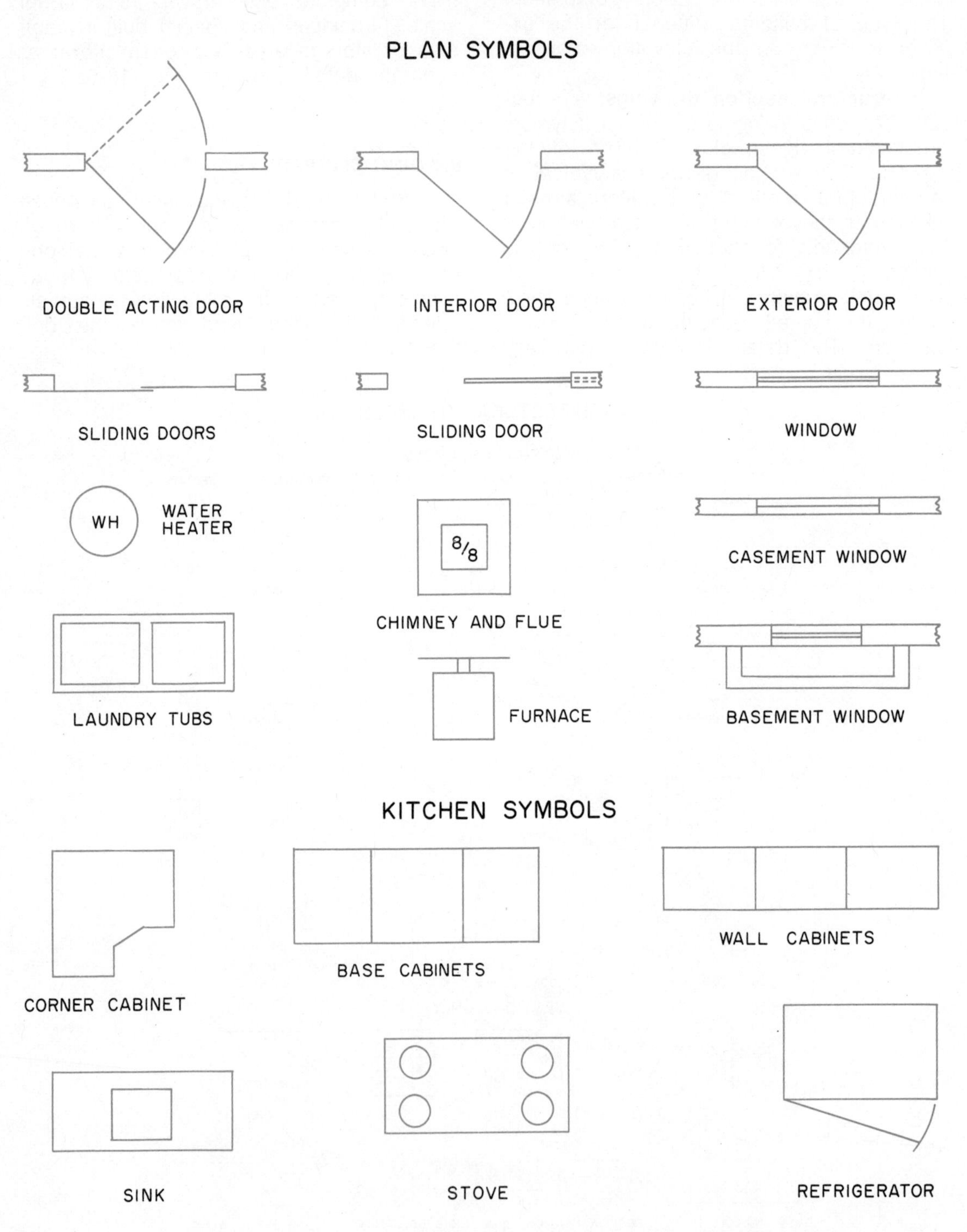

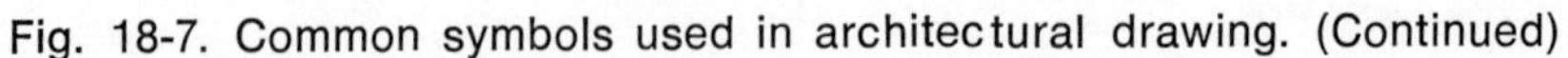
Fig. 18-7. Common symbols used in architectural drawing. (Continued)

ARCHITECTURAL SYMBOLS

BATHROOM SYMBOLS

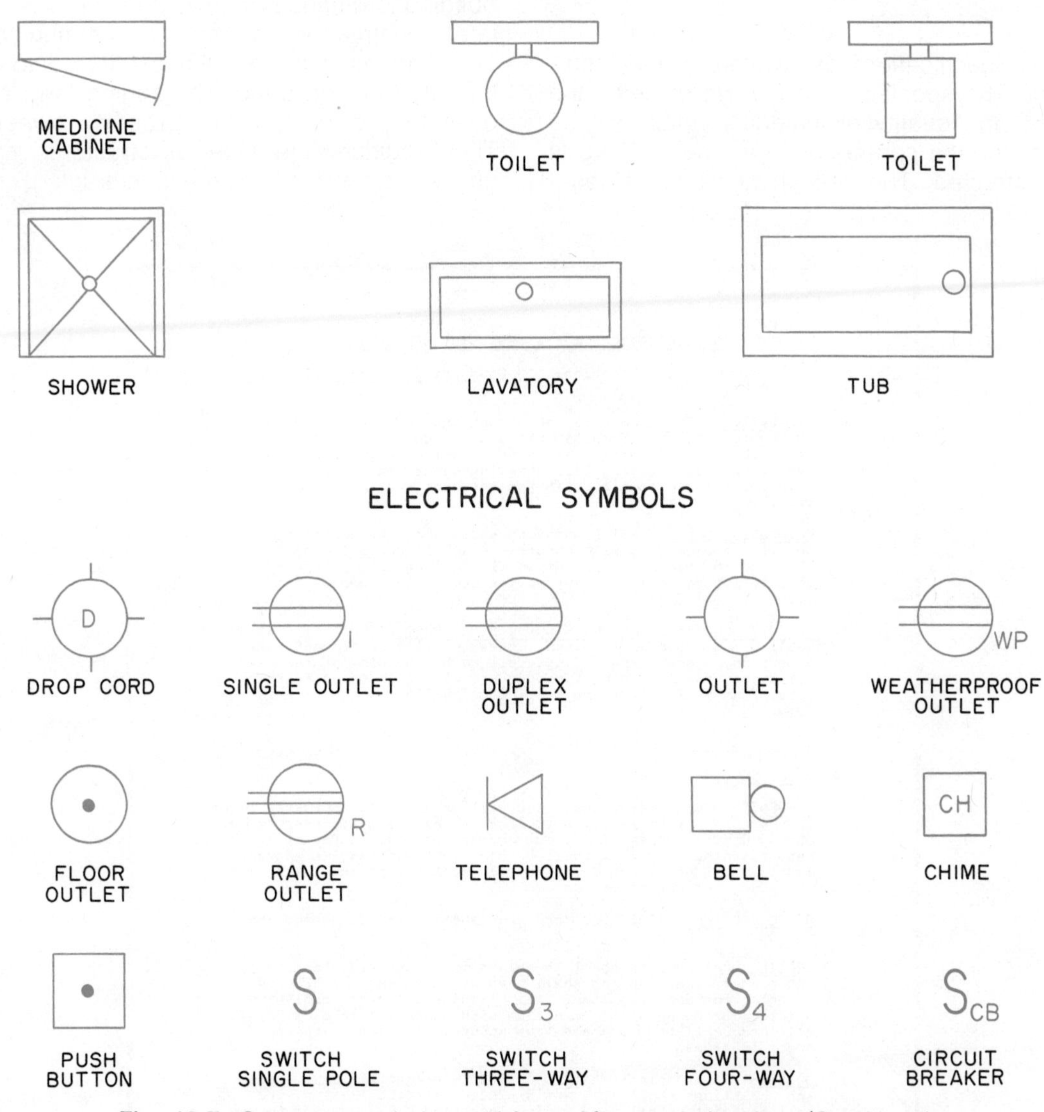

Fig. 18-7. Common symbols used in architectural drawing. (Continued)

SPECIFICATIONS

Specifications are written by the architect. The specifications explain in detail the type and quality of materials which will be used by the contractor when the building is constructed. The specifications are a legal contract between the buyer and the contractor. The acceptable quality of construction as well as materials is often written in the specifications, Fig. 18-8. Local building ordinances and zoning laws regulate construction in various communities. This ordinance is sometimes referred to as the city building code. The zoning laws are designed to protect the property owners. They regulate the type of structures and businesses which may be constructed and

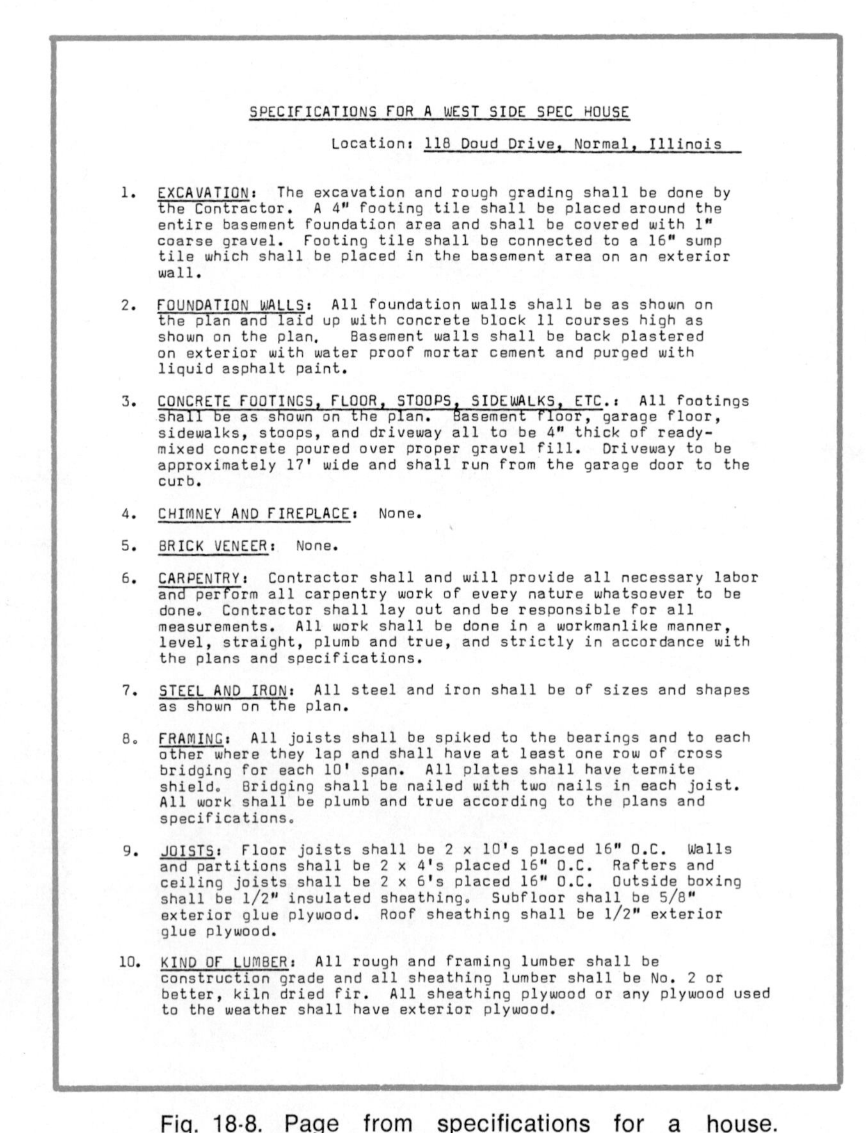

SPECIFICATIONS FOR A WEST SIDE SPEC HOUSE

Location: 118 Doud Drive, Normal, Illinois

1. EXCAVATION: The excavation and rough grading shall be done by the Contractor. A 4" footing tile shall be placed around the entire basement foundation area and shall be covered with 1" coarse gravel. Footing tile shall be connected to a 16" sump tile which shall be placed in the basement area on an exterior wall.

2. FOUNDATION WALLS: All foundation walls shall be as shown on the plan and laid up with concrete block 11 courses high as shown on the plan. Basement walls shall be back plastered on exterior with water proof mortar cement and purged with liquid asphalt paint.

3. CONCRETE FOOTINGS, FLOOR, STOOPS, SIDEWALKS, ETC.: All footings shall be as shown on the plan. Basement floor, garage floor, sidewalks, stoops, and driveway all to be 4" thick of ready-mixed concrete poured over proper gravel fill. Driveway to be approximately 17' wide and shall run from the garage door to the curb.

4. CHIMNEY AND FIREPLACE: None.

5. BRICK VENEER: None.

6. CARPENTRY: Contractor shall and will provide all necessary labor and perform all carpentry work of every nature whatsoever to be done. Contractor shall lay out and be responsible for all measurements. All work shall be done in a workmanlike manner, level, straight, plumb and true, and strictly in accordance with the plans and specifications.

7. STEEL AND IRON: All steel and iron shall be of sizes and shapes as shown on the plan.

8. FRAMING: All joists shall be spiked to the bearings and to each other where they lap and shall have at least one row of cross bridging for each 10' span. All plates shall have termite shield. Bridging shall be nailed with two nails in each joist. All work shall be plumb and true according to the plans and specifications.

9. JOISTS: Floor joists shall be 2 x 10's placed 16" O.C. Walls and partitions shall be 2 x 4's placed 16" O.C. Rafters and ceiling joists shall be 2 x 6's placed 16" O.C. Outside boxing shall be 1/2" insulated sheathing. Subfloor shall be 5/8" exterior glue plywood. Roof sheathing shall be 1/2" exterior glue plywood.

10. KIND OF LUMBER: All rough and framing lumber shall be construction grade and all sheathing lumber shall be No. 2 or better, kiln dried fir. All sheathing plywood or any plywood used to the weather shall have exterior plywood.

Fig. 18-8. Page from specifications for a house.

operated in a given vicinity. Zoning regulations in most cities appear as follows: (1) single family dwellings, (2) multi-family dwellings, (3) apartment complexes, (4) commercial buildings, (5) light industrial buildings and, (6) heavy industrial plants. The local zoning regulations can be changed as the city grows by action of the governing body of that city. Public hearings and pulic acceptance are required before zoning can be changed.

The city building code regulates the method of construction and the type of materials. This includes electrical, plumbing, and heating installations, Fig. 18-9. It also specifies how the house will be located on the property, that is, the distance from the property lines to the public street.

A building permit is required in most cities before construction is started. The permit to build is issued after the local building authorities have reviewed the house plans and specifications. When the specifications comply with the local ordinances, the permit is issued, and should always be displayed on the site, Fig. 18-10. The building authorities inspect the structure at various stages of construction. If the materials and construction methods do not comply with the ordinances and specifications, the work is stopped.

LOCATING THE STRUCTURE

After the plans are complete and a building permit is obtained, the structure is located on the building lot. The building lot should be suited to the house size and architectural design. The process of loca-

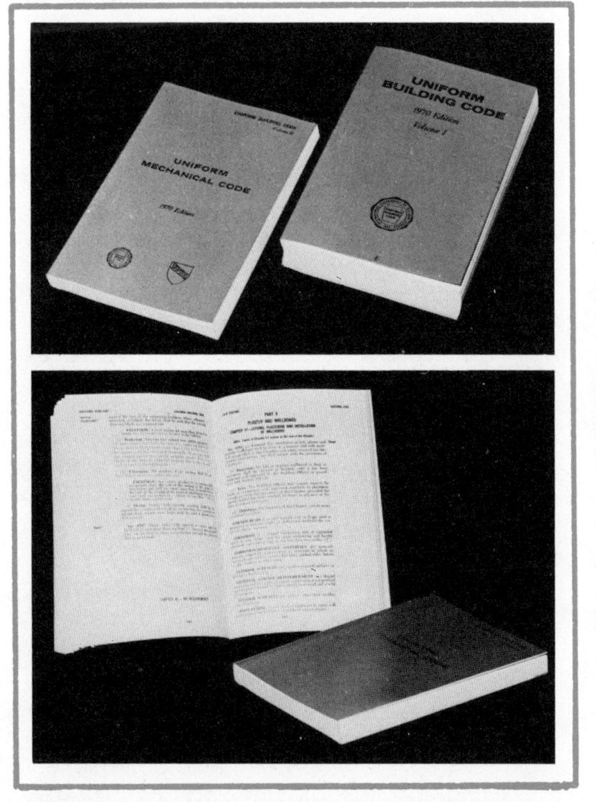

Fig. 18-9. Typical code books for locality in which structure is built.

Fig. 18-10. Building permit displayed on site of construction.

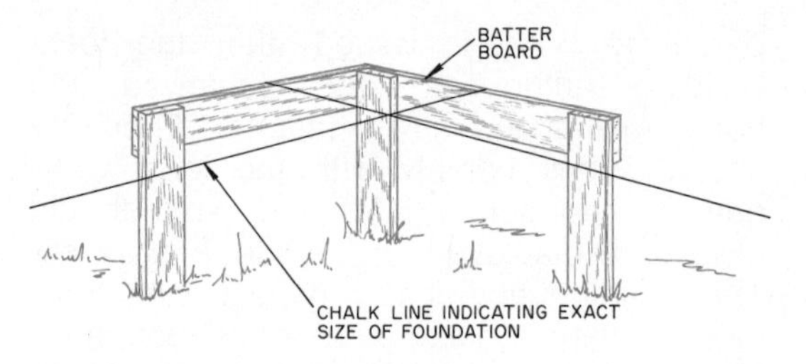

Fig. 18-11. Batter boards for locating and squaring a structure.

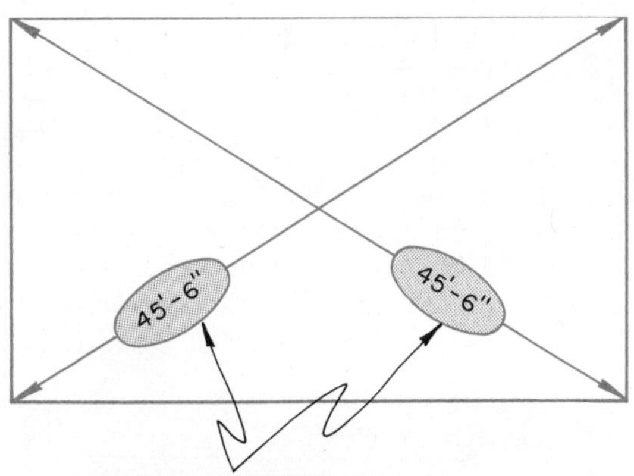

Fig. 18-12. Checking squareness by measuring diagonally.

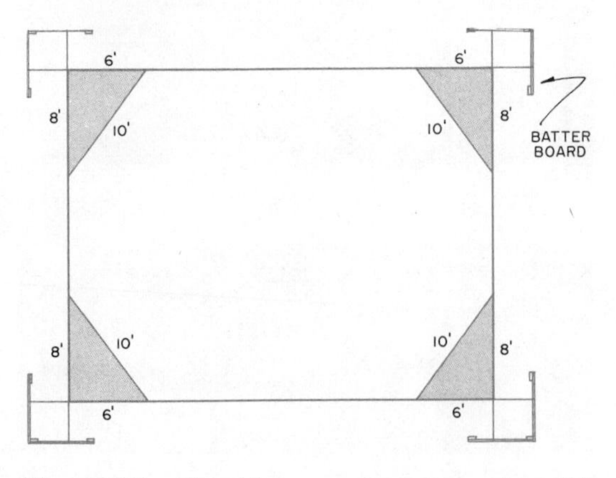

Fig. 18-13. Checking squareness using triangular measurements.

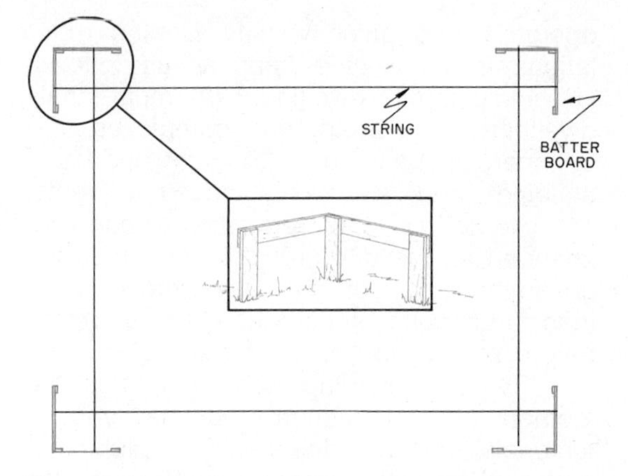

Fig. 18-14. Batter boards placed with string layout for a structure with squared corners.

ting the house is usually referred to as "staking out" the house. Generally the house is staked-out by first locating the edge. The location of the front from the street is usually determined by the local building code or the existing houses on either side. The distance from the street is measured and located with stakes. A line is stretched across the full distance of the lot. Stakes are set along the line to represent the front corners of the house. They are driven flush with the ground. A nail driven in the top of the corner stakes gives a more exact location.

Batter boards are set at each corner at least 4′ behind the approximate corner locations. The batter boards consist of 2×4 stakes driven in the ground, which are then connected with 1×6 boards to form a right angle to each corner, Fig. 18-11. The batter boards are set level as they are attached to the stakes. Carpenter's twine is stretched over the batter boards to locate the footings. The staking process can be checked for squareness by checking the diagonal measurements. The diagonal measurements will be the same for both diagonals if the structure is square, Fig. 18-12. The squareness can also be checked by using the triangle method, Fig. 18-13. A triangle with sides of 6′0″, 8′0″, and a

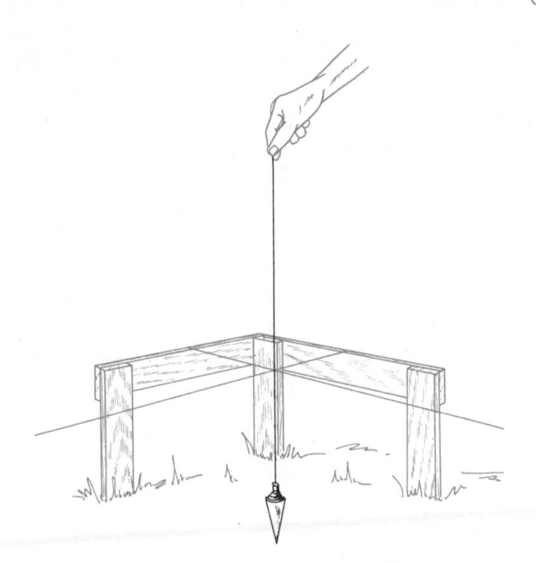

Fig. 18-15. Locating ground points with a plumb bob.

Fig. 18-16. Excavating a basement opening using heavy equipment.

hypotenuse of 10′0″ will result in a right angle. Figure 18-14 shows the structure completely staked with the string stretched to form a structure with square corners.

A plumb bob is suspended from the points at which the strings intersect to form the corners to the ground level. A stake is placed at this point, locating the outside corner of the footing, Fig. 18-15.

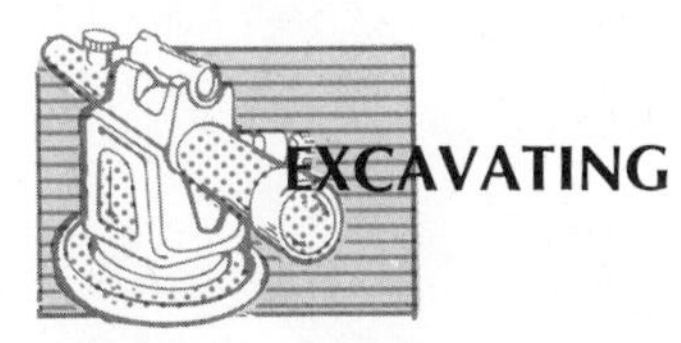

EXCAVATING

Earthmoving is the next step after the building structure has been located on the property. Heavy earthmoving equipment such as shown in Fig. 18-16 is used for excavating a basement. The top soil is first removed and stockpiled, and because it is scarce and expensive, it will be spread on the property when the lawn is prepared. The excess fill is either hauled from the site or piled far enough from the future structure so it will not obstruct the work. When the house does not have a basement, a narrow trench is dug around the perimeter of the structure. This trench is dug to below the frost line. A trenching machine such as shown in Fig. 18-17 is used for this type of excavation.

Fig. 18-17. Excavating for a slab foundation using a trenching machine. (Clark Equipment Co.)

FOOTINGS AND FOUNDATIONS

The footings and foundations are important parts of any permanent structure. The footings must be below the frost level and large enough to support the building. Inadequate footing will cause excessive settling of the building. The footings are the enlarged base below the foundation wall which distribute the weight of the structure to the soil. Thus, the size of the footing is determined by the weight of the structure and the soil conditions. The ability of the soil to support a structure is tested by engineers before constructing large buildings.

The foundation rests on the footings and extends to the sills on which the floor joists are supported. The depth of the footing is usually regulated by local building codes, and is often 4′ or more in cold northern sections of the United States. This depth protects the foundation from frost action. Footings are generally of poured concrete. The width of the footing should be at least twice the width of the concrete foundation wall which it will support, and the thickness of the footing equal to the width of the foundation wall. The width of the wall foundation would be equal to the proposed wall thickness of the structure it will support, Fig. 18-18. Local codes and practices should be followed for designing foundations and footings.

Drain tile is often placed on the outside of the footing. It is connected to a drain system and carries excess water away from the foundation wall. The foundation wall may be either solid poured concrete or concrete block. When concrete blocks are used, a thin coating of masonry mortar is applied to the exterior surface. The masonry plaster is then coated with an asphalt material to provide additional waterproofing.

Construction costs can be reduced by building houses on concrete slabs without basements. Special consideration must be given to insulation and vapor barriers for this method of construction. Figure 18-19 shows a typical footing and foundation design for concrete slab floor construction.

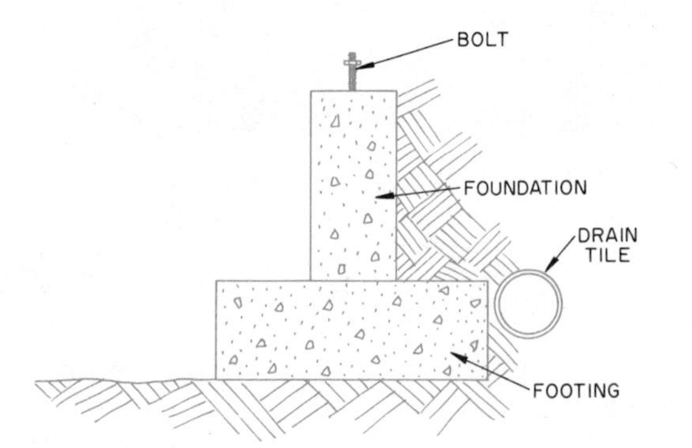

Fig. 18-18. Typical footing for houses with basements.

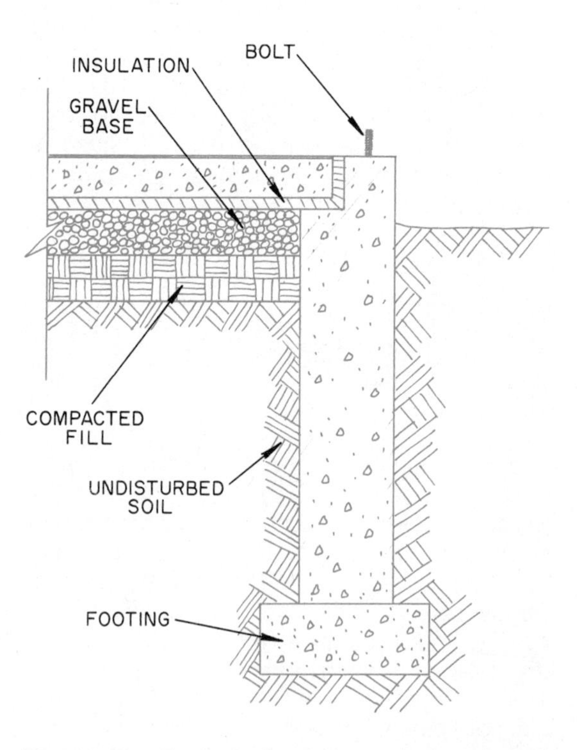

Fig. 18-19. Typical footing and foundation for slab foundation construction.

The concrete floor is then covered with carpet or floor tile.

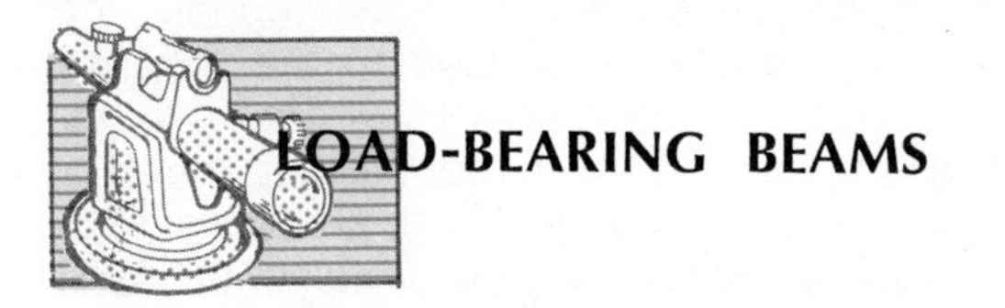

LOAD-BEARING BEAMS

Most houses require a center beam support to carry the load-bearing floor joists. Posts or steel columns are spaced about every 8′ along the support beam. A special concrete pad called a *pier* supports the post, Fig. 18-20.

The center beam or *girders* may be either of steel or wood, Fig. 18-21. Wood beams may be either solid or build-up. Build-up beams are generally more satisfactory as they are constructed of drier, more stable wood. The build-up beam is constructed of two or more pieces of 2″ dimension lumber laminated together with nails.

SILL AND FLOOR CONSTRUCTION

The sill is placed on the foundation wall. It is secured to the foundation by the

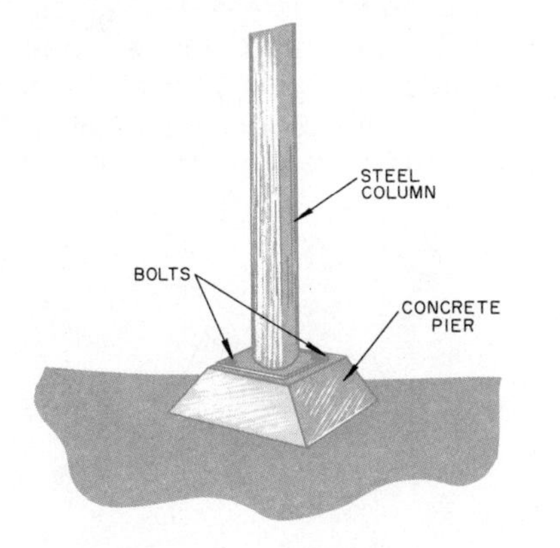

Fig. 18-20. Pier to support a steel column or post.

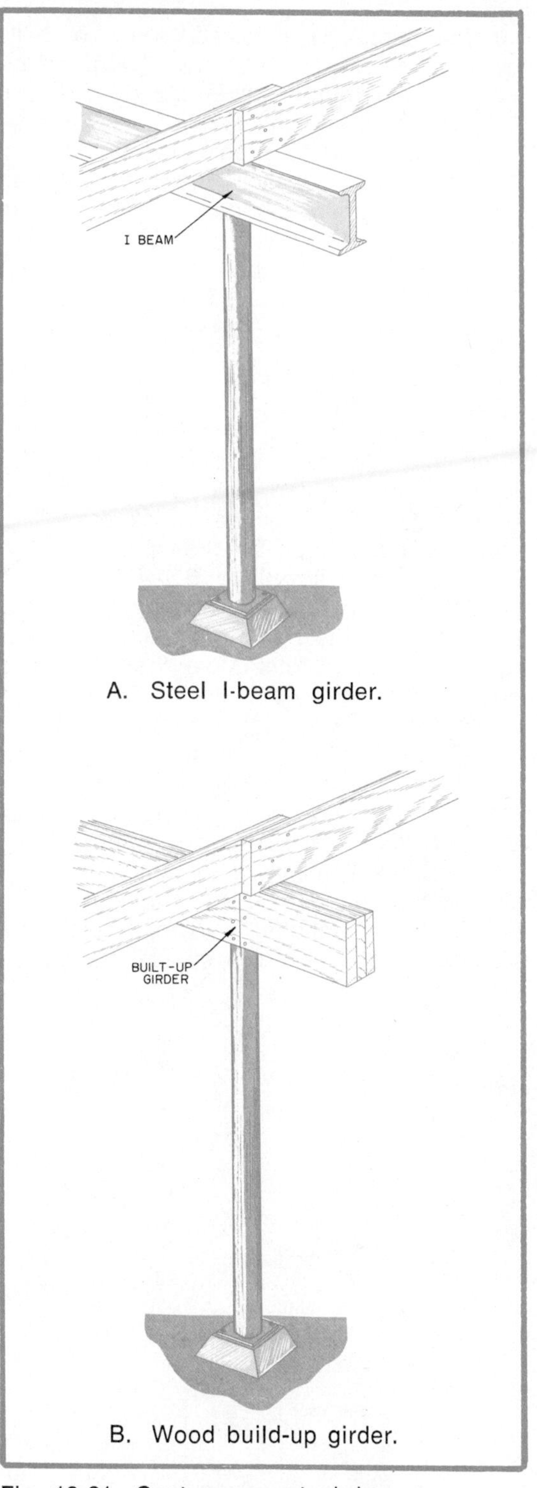

A. Steel I-beam girder.

B. Wood build-up girder.

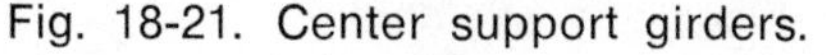

Fig. 18-21. Center support girders.

anchor bolts. Nuts placed on the bolts anchor the sill firmly in place, Fig. 18-22. A piece of thin metal is placed between the wood sill and the concrete to serve as a termite shield to protect the wood from termites. An insulation material wrapped in treated paper is placed on top of the foundation wall. This is called sill-sealer and provides a weather seal.

FLOOR JOISTS

Floor joists are placed on the sill. The spacing of the floor joists is 16″ on-center (o.c.) for most types of construction. The strength of the joist is determined by the loads which the floor will carry. Joists are generally of 2″ nominal thickness. The width is determined by the span and the load. Common sizes are 8″, 10″, or 12″ widths for most house construction. The distance the joists are spaced apart also determines the width. Most local building codes specify the minimum size joists which may be used for a particular span.

The layout for the joists starts at one end of the structure. Equal spacing, generally of 16″ o.c., is used the full length of the building as standard spacing of 16″ o.c. requires a minimum of cutting for installation of standard 4′ × 8′ building materials. Joists are often curved or crowned on one edge. The crown should always be placed on top when fastened to the sill. A header joist is placed on both ends of the joists. The joists are nailed to the header joist and toe-nailed to the sill. The joists are spliced on the center girder support, either butted together or overlapped, Fig. 18-21. When the joists are butted, a scab is placed on one side. If the joist overlaps on the girder, the two ends are nailed together. It is a good policy to double the floor joist under the load bearing partition walls.

Bridging. Bridging is placed between the floor joists. This helps distribute the load and prevents movement. The two most common methods of bridging include solid and crossed bridging, Fig. 18-24. Solid bridging consists of placing solid lumber of the same width and thickness of the joist between each joist. Cross bridging is constructed from 1″ × 3″ boards cut at an angle and nailed between the joist. Metal bridging is also available. Generally one row of bridging placed in the center of the joist span is adequate.

SUBFLOOR

The *subfloor* is placed over the joists. It forms a solid deck and serves as a base for the finish flooring. The subfloor material is either surfaced lumber or plywood sheet stock. Plywood is becoming the most popular because it takes less time to install. When lumber is used, it is generally less than 8″ wide and not less than ¾″ thick. Boards may be installed at right angles to the joists or diagonally. If the boards are laid diagonally, the finished

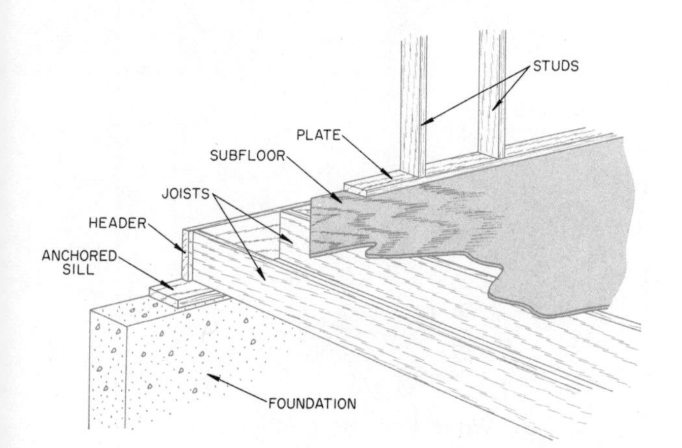

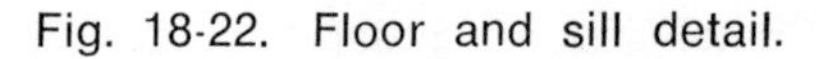

Fig. 18-22. Floor and sill detail.

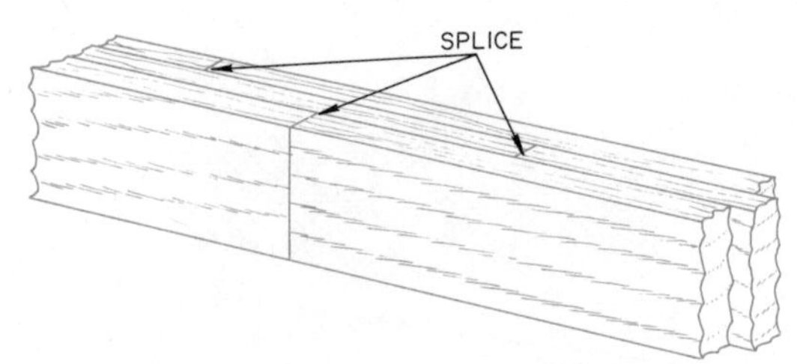

Fig. 18-23. Splicing a center girder.

floor may be installed in either direction. Plywood must be at least ½″ in thickness for subflooring, Fig. 18-25. Stock of either ⅝″ or ¾″ thickness will provide a firmer deck.

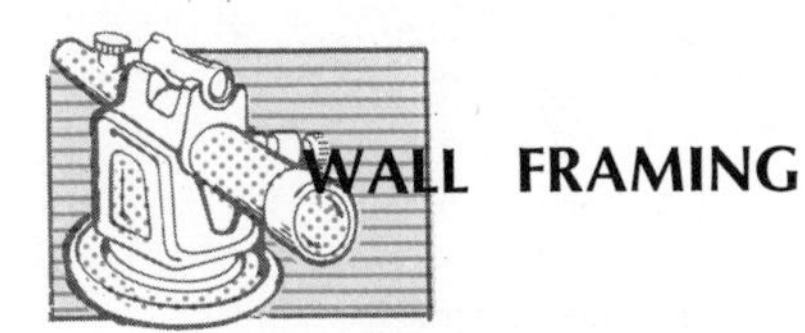

WALL FRAMING

There are several framing methods common to house construction. The *platform* construction technique is most often used for single story construction. The wall framing is erected above the subfloor. The walls are assembled on the subfloor as a unit and raised into place, Fig. 18-26. The layout is made for the studs, window and door headers, cripple studs, and window sills, then fastened together before the wall is raised into position.

The layout for the position of the studs, window openings, and door openings is made on the top plates and sole

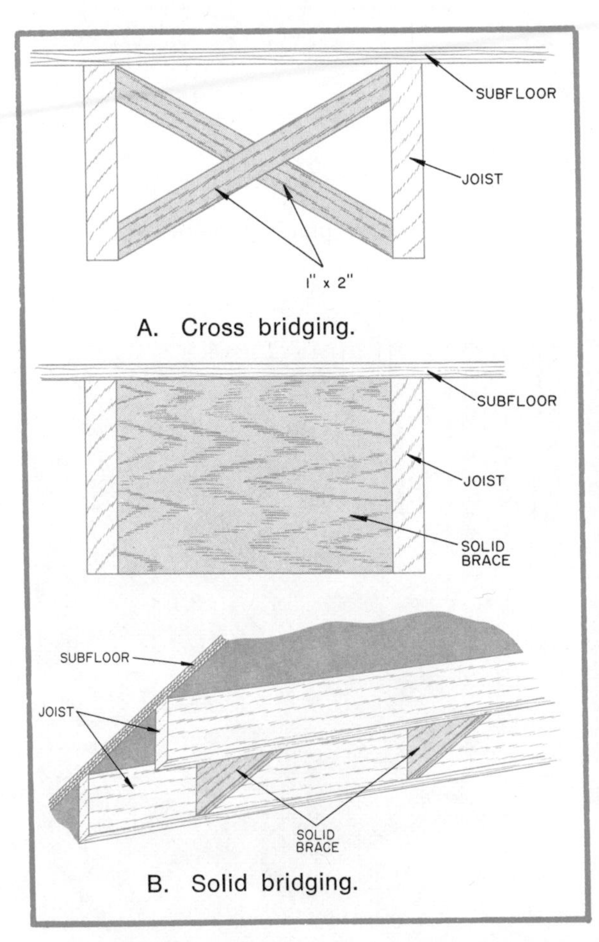

Fig. 18-24. Common floor bridging between joists.

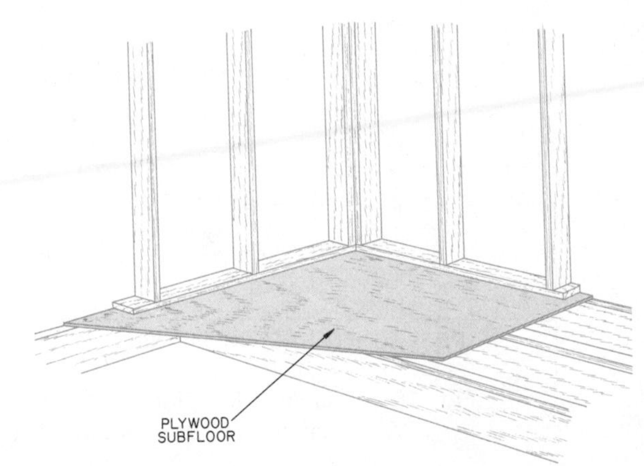

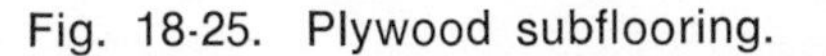

Fig. 18-25. Plywood subflooring.

Fig. 18-26. Placing a preconstructed wall section.

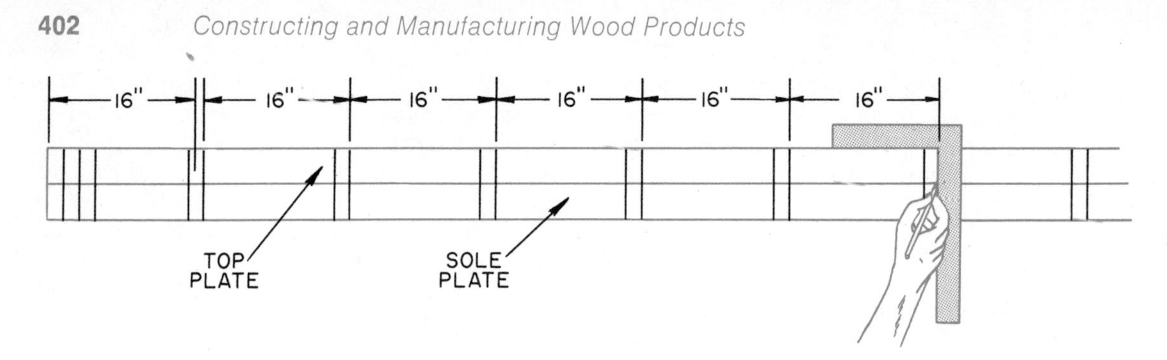

Fig. 18-27. Layout for wall framing on top and sole plates.

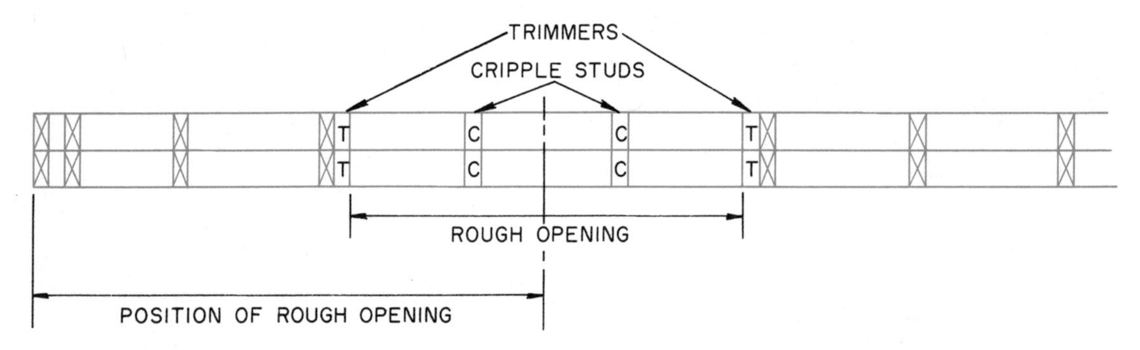

Fig. 18-28. Layout for placement of studs on top and sole plates requires locating rough openings first.

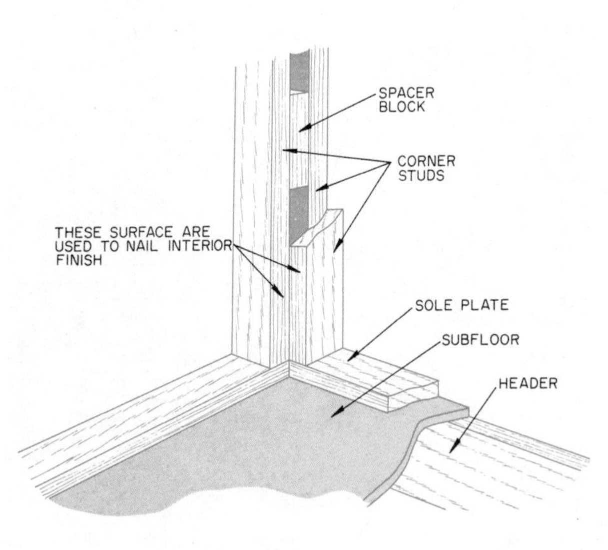

Fig. 18-29. Corner stud detail.

Fig. 18-30. Nailing the wall section together.

plates, Fig. 18-27. A line is placed across both plates with a square. An "X" is placed on one side of the line. This indicates on which side of the line studs will be placed. A "T" is used to show the position of trimmer studs as the layout is made; and "C" indicates a cripple stud, Fig. 18-28. The center of the rough opening is first located and the width of the opening is measured from the center of the opening. The window and door manufacturers supply the exact rough opening measurements needed for installation of a particular door or window.

Extra studs are placed at the corner of the wall section. These are needed to nail the interior wall covering material in place, and are given in a detail drawing, Fig. 18-29. The top and sole plates are then nailed to the vertical members or studs, Fig. 18-30. Sixteen penny nails are used for nailing the frames, with the studs placed on 16″ centers. The wall frame is constructed of 2×4's. The length of the 2×4's are such that the distance from the finished floor to the ceiling is slightly greater than 8′. The height over 8′ allows room for installing standard materials which are 8′ long, so usually 1″ of extra height is adequate.

INSTALLING HEADERS IN LIGHT FRAME CONSTRUCTION

The typical wall frame in Fig. 18-31 shows headers placed at the top of the wall section above window and door openings. The header helps support the weight of the roof over the openings. The headers consist of two members nailed together with 3/8″ spacer placed between them to make the header the same thickness as the width of the studs. The size of the header members is determined by the width of the opening. The following sizes are minimums established by the Federal Housing Administration (F.H.A.). Spans up to 4′-0″ require two 2×4's, spans up to 6′-0″ require two 2×6's, spans up to 8′-0″ require two 2×8's and spans up to 10′-0″ require two 2 × 10's. The members are always placed on edge. A header constructed of 2 × 12's is commonly used over all openings to eliminate the need for cripple studs, Fig. 18-32.

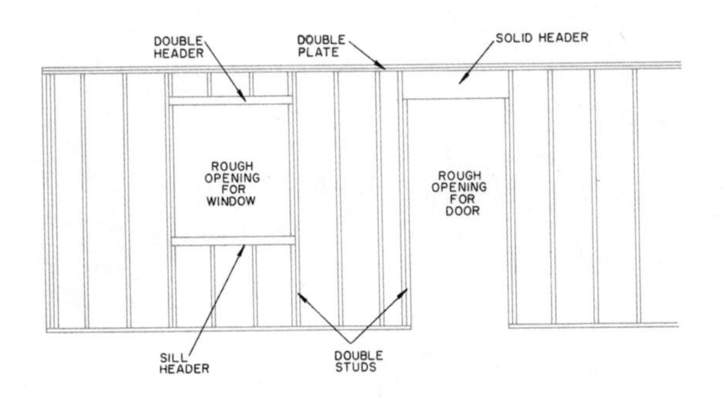

Fig. 18-31. A typical exterior wall frame.

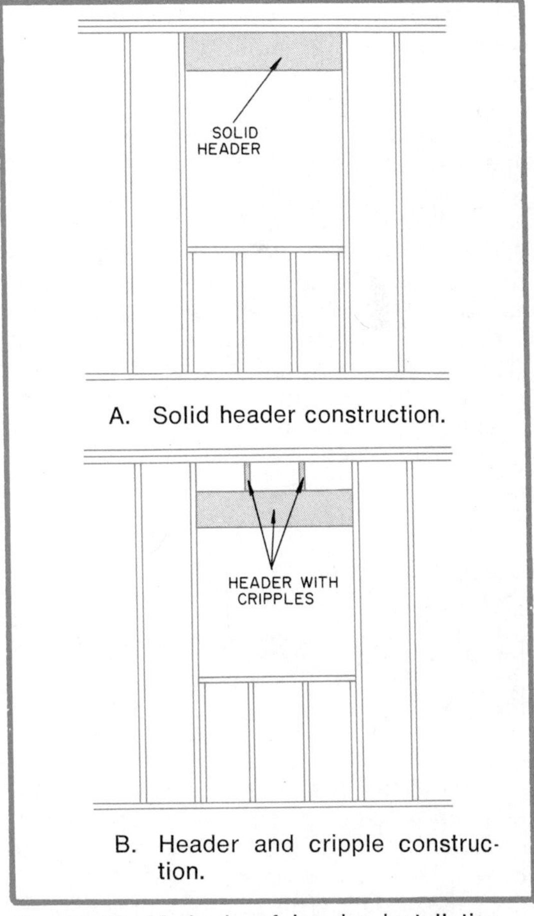

A. Solid header construction.

B. Header and cripple construction.

Fig. 18-32. Methods of header installation.

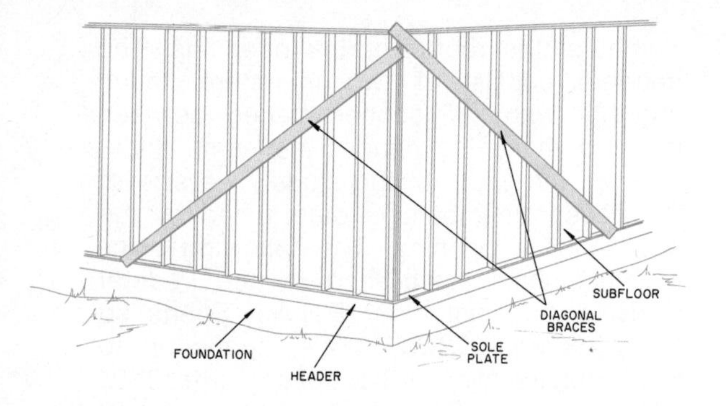

Fig. 18-33. Diagonal bracing stabilizes the frame before sheathing is applied.

The top of the rough opening is always the same height regardless of the window and door size in standard framing. The height of the opening is 6′-10″. The location and size of the rough openings are given on the architectural plan.

The wall is positioned and nailed to the subfloor. It is then plumbed (placed vertical) and braced. The bracing is left until all the walls are in place and the exterior is covered with sheathing, Fig. 18-33. The exterior sheathing is the material nailed to the outside of the studs to enclose the structure. Various materials such as boards, plywood, or composition board may be used for sheathing. Sometimes the sheathing is applied before the walls are raised into place.

INTERIOR PARTITIONS

Partitions are either bearing or nonbearing. A bearing partition supports part of the weight of the ceiling and roof; a nonbearing wall serves only to divide the structure into rooms. The position of the partitions is given on the floor plan. The partition walls are constructed in the same manner as exterior walls, and are framed from 2×4 lumber. The header in non-bearing walls can generally be constructed from 2×4's regardless of the size of the opening. An extra stud as shown in Fig. 18-34 is placed in the outside wall to provide nailing at the intersection of the partition.

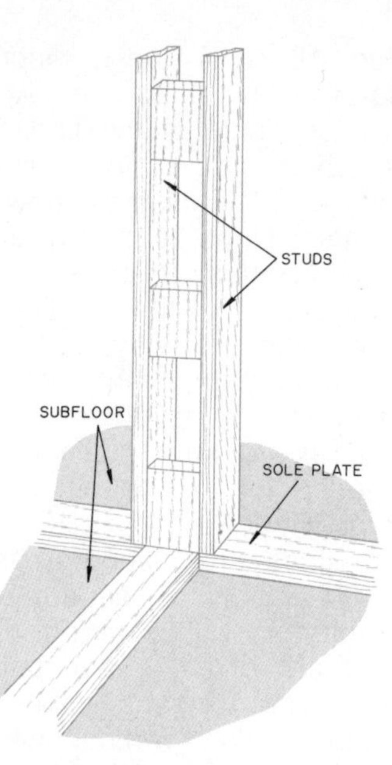

Fig. 18-34. Placement of the studs for joining interior partition walls.

The top plate is doubled after the partitions are erected. The double plate ties the walls together as it laps the adjacent wall and staggers the joints. It also gives the wall greater support for the roof.

WALL SHEATHING

The wall sheathing placed on the exterior side of the studs encloses the framed structure. The most common types of sheathing used include boards, plywood, structural insulation board, and gypsum sheathing. The manufactured sheet materials are used extensively as they can be installed in a minimum of time, Fig. 18-35. Plywood and insulated sheathing in 4′ × 8′ sheets are applied vertically. Boards are applied either horizontally or diagonally.

The sheathing adds strength to the structure and provides insulating qualities. The sheet materials used for sheathing may also serve as the exterior siding. This practice is generally limited to mild climates.

CONVENTIONAL CEILING AND ROOF CONSTRUCTION

Ceiling and roof construction varies greatly depending on the roof design. Until recent years the roof was framed of separate members. The members consisted of ceiling joists and rafters combined with the necessary braces and ties depending upon the roof design, Fig. 18-36. This method is sometimes referred to as conventional roof construction. The ceiling joists are placed on the bearing partitions and outside walls like floor joists. They are generally 2×6's placed 16″ o.c. for single story dwellings. The ceiling joist supports the finished ceiling. They are securely nailed in the center and on the outer plate. The ceiling joists are normally installed to span the narrow width of the structure, and are spliced on the load bearing walls.

TYPES OF ROOF CONSTRUCTION

The gable roof is the most common type of sloped roof. It is widely used because it is easy to construct and gives a pleasing appearance. The style of the house usually determines the type of roof,

Fig. 18-35. Applying exterior sheathing.

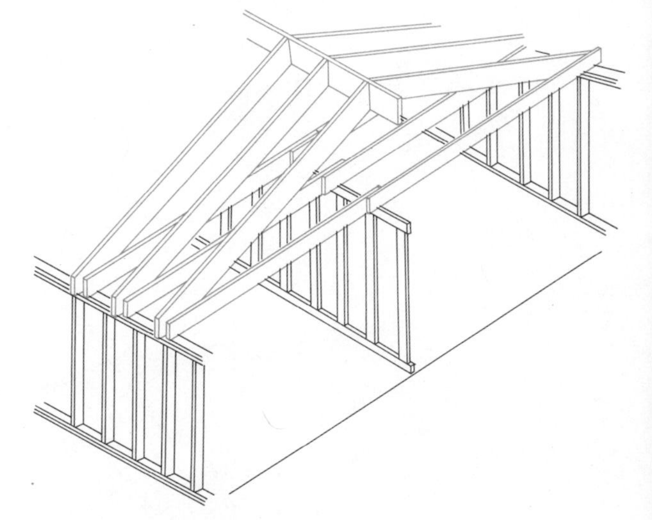

Fig. 18-36. Placement of ceiling joist for conventionally framed roof.

Fig. 18-37. The slope of the roof is measured by the number of inches of rise per running foot. A roof which raises 4″ for every 12″ of run would be referred to as a 4/12 or ⅓ pitch roof. Figure 18-38 shows the framing members of a gable roof and the way the roof pitch is shown on a drawing.

SHED

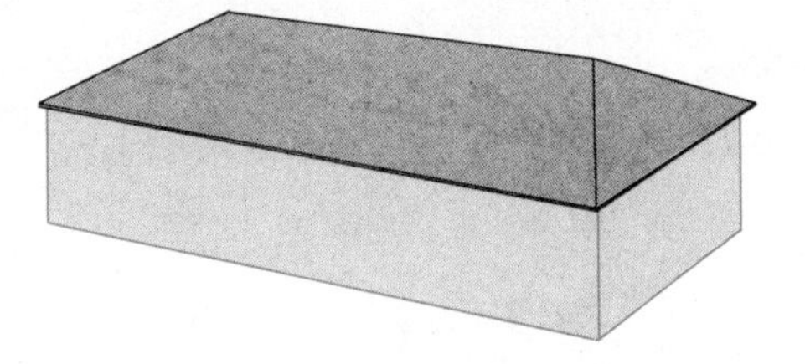

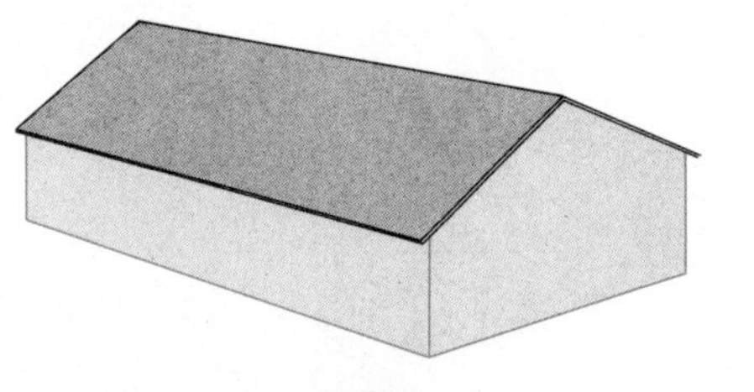

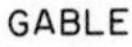

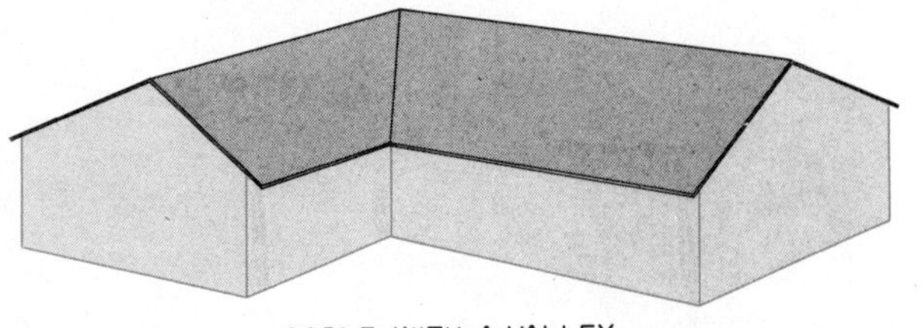

GABLE WITH A VALLEY

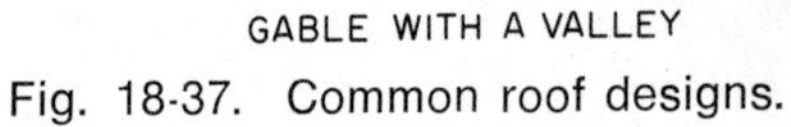

Fig. 18-37. Common roof designs.

LAYOUT FOR A COMMON RAFTER

A common rafter extends from the plate to the ridge board as shown in Fig. 18-39. The length of the rafter is measured along an imaginary line parallel to the edge of the rafter. This line intersects with the inside corner of the *birds mouth* (notch which fits on the plate).

The length of a rafter and angle of the cuts can be marked using a framing or rafter square, and is determined by the step method or from the rafter table printed on the square. When the step method is used, the rise of the roof such as 4 is located on the tongue. The run, which is always 12, is located on the blade of the square, Fig. 18-40.

The layout is started at the top or ridge. The square is positioned and the angle of the ridge cut is marked. This angle is called the plumb cut. A mark is also made at the 12″ mark. The square is moved and the 4″ mark on the tongue is aligned

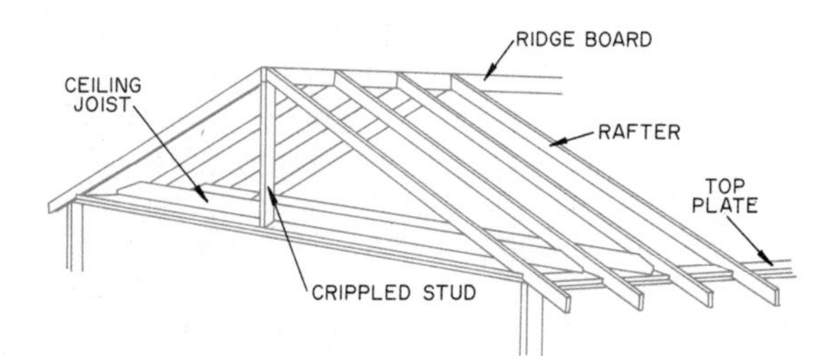

Fig. 18-38. Framing members of a gable roof.

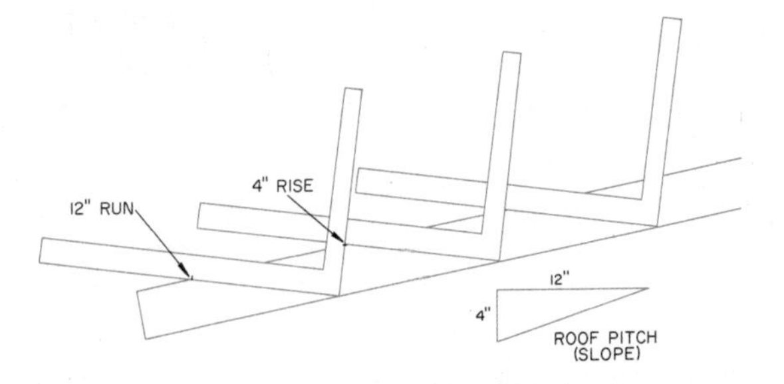

Fig. 18-39. Position of the rise and run on the rafter square for stepping off the rafter.

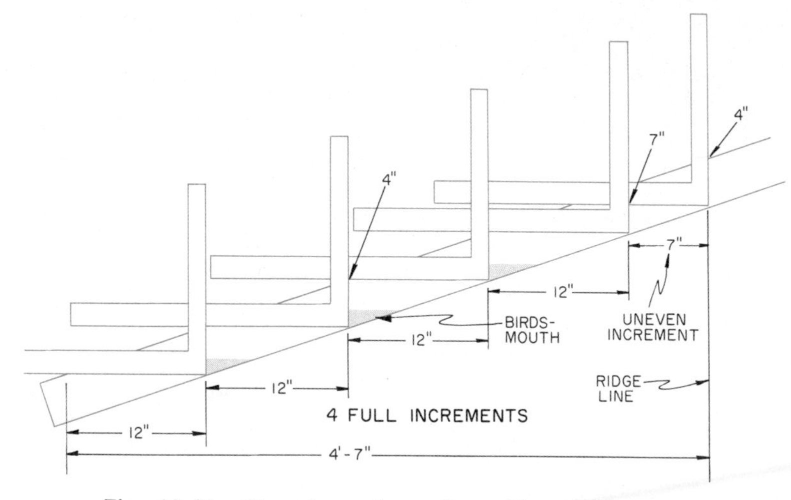

Fig. 18-40. Stepping off a rafter with a rafter square.

with the 12″ mark from the previous step, Fig. 18-41. Each step of the square is equal to one foot of the run. If the building is 10′ wide, five steps would be made with the square. The birds mouth to fit the plate is made by reversing the square. The 5″ and 12″ marks on the square are used to give the correct angle. The cut is to about one-third the width of the rafter. The length of each rafter is reduced one-half the thickness of the ridge board.

The overhang may be any desired length. If the end of the rafter tail is plumb, the rafter square is used to mark the angle. The same numbers on the square used for the rafter layout are used to make the plumb cut.

A rafter can also be laid out using the rafter table which is printed on the blade of the square. The length of the rafter for each foot of run is located on the first line in the table, Fig. 18-42. The number representing the rise is located on the blade of the square. For example, locate 5 on the blade. In the first line directly below the 5, locate the number 13. This indicates that for each foot of run, a rafter with a rise of 5″ would need to be 13″ long. The 13″ is multiplied by the run to give the total length of the rafter from the ridge to the outside of the

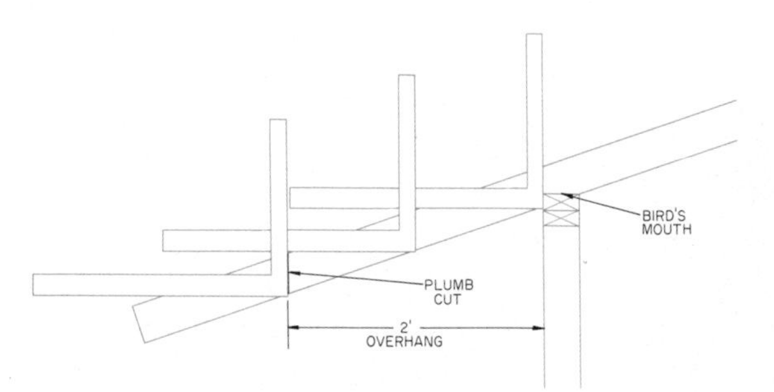

Fig. 18-41. Plumb cut and birds mouth layout on the rafter.

Fig. 18-42. Tables are printed on the rafter square.

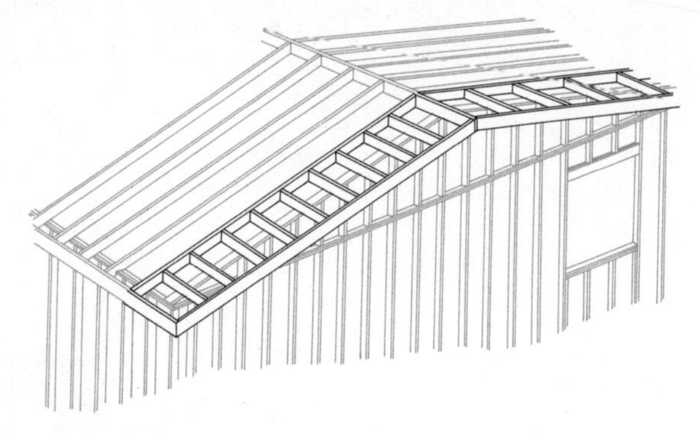

Fig. 18-43. Framing a gable overhang.

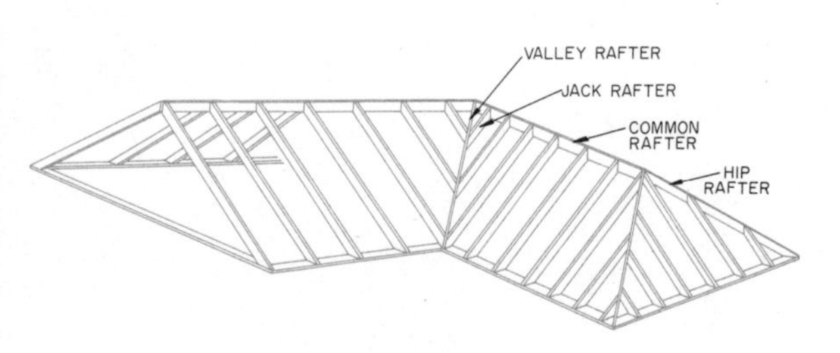

Fig. 18-44. This roof construction combines hip, valley, common, and jack rafters.

plate. For example, a building 24′ wide (span) would have a run of 12′. The 13″ from the table multiplied by 12 equals 286″ This is the length of the rafter from the plate to the ridge minus one-half the thickness of the ridge board.

Most gable roofs have an overhang on the end. Additional support is required for the overhanging portion. Figure 18-43 shows one method of framing the overhang. The ridge board is extended equal to the amount of the overhang. Lock-outs are attached to the rafters at intervals from the ridge to the rafter tail. The roof sheathing ties the overhanging frame together and gives additional support.

Studs are placed under the outside gable end rafters and extend to the plate. The top end of the studs are cut at an angle the same pitch as the roof. The exterior sheathing and siding is attached to the gable end studs.

HIP AND VALLEY RAFTERS

A hip roof such as shown in Fig. 18-44 consists of common rafters as well as hip and jack rafters. A hip rafter extends from the outside corner of the structure to the ridge board. Jack rafters extend from the plate to the hip rafter.

The rafter square is used for measuring and laying out hip, valley and jack rafters. The second line of the rafter table on the square gives the necessary information. The pitch of the roof such as 5/12 is located on the number scale along the blade of the square. Below the 5 listed on the second line is the number 17.69, meaning that the rafter is 17.69″ long for each 12″ of run. The total length of the rafter is determined by multiplying 17.69 by the number of feet in the runs. The length of the tail is added and one-half the thickness of the ridge board is subtracted.

The plumb cut and birds mouth cut are laid out using the rise of the roof and the 17″ mark on the blade of the square. The 17″ mark is used because the hip rafter forms the hypotenuse of a right angle triangle with equal sides. The hypotenuse, when figured geometrically, equals 16.97″ but is normally rounded to 17″.

A side cut must also be made on the top and bottom of hip and valley rafters. The necessary information is given on the bottom line of the table on the square. The rise per foot is located on the blade and the angle of the cut is given directly below on the bottom line. For example, if the rise is 5″, the number in the table is 11½. The layout for the top and bottom cut is made with the square. The number 11½ is used on the tongue and 12 is used on the blade. The side cut layout intersects with the layouts made for the top and bottom cut.

Jack rafters are similar to common rafters, but are shorter. They intersect with either the hip rafter or valley rafter rather than extending from the ridge to the plate. The third and fourth lines on the rafter table are for laying out jack rafters. The first jack rafter down the ridge is shorter than a common rafter the amount specified in the table. The length of each jack rafter is shortened by this specified amount, plus one-half the thickness of the ridge board measured at a 45° angle. The bottom cut is the same as for common rafters, and a cheek cut is placed on top of the jack rafter to fit the hip rafter. The jack rafters from opposite sides bearing on the hip rafter intersect at the same point.

TRUSS ROOFS

Roof trusses are being used extensively in house construction. The roof truss forms the entire framework for the roof and interior ceiling. They are designed to extend from one outer wall to the other. The truss is designed so it is not necessary to have load bearing walls on the interior of the structure, thus allowing for greater flexibility for interior planning.

The truss rafter saves material and the structure can be enclosed very quickly. The truss is fabricated and then positioned after the outside walls are erected.

Numerous designs have been used for trusses. The W-type truss is the one most commonly used in house construction, Fig. 18-45. Trusses are generally designed for placement on 24″ centers rather than the conventional 16″ centers.

ROOF SHEATHING

The roof sheathing is placed on the rafters or trusses. The sheathing material is generally either 1″ boards or plywood. When boards are used under asphalt shingles, they are laid tight together without spacing between. The width of the boards should be 8″ or less to minimize shrinkage problems, and should be placed on the rafters so the end joints are staggered, Fig. 18-46. The boards are usually narrower and spaced about 1½″ apart when wood shingles are used as roofing.

When plywood is used as roof sheathing, the face grain should be laid perpendicular to the rafters, Fig. 18-47. The end joints should be staggered at least one rafter. The minimum thickness which should be used is generally ⅜″. A smoother roof will result in using ½″ materials.

ROOF FINISH

A sloped roof is generally covered with either asphalt or wood shakes. A flat roof is finished with a built-up roof consisting of layers of asphalt-impregnated paper

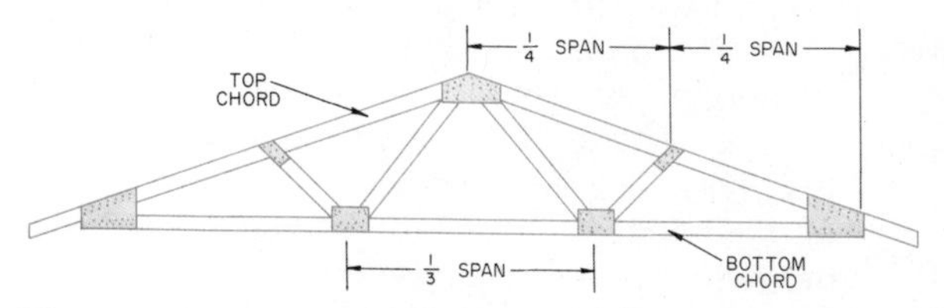

Fig. 18-45. W-type roof truss, commonly used in house construction.

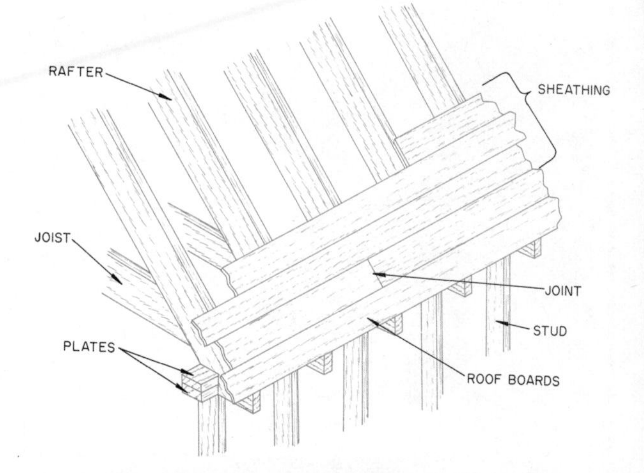

Fig. 18-46. Placement of board roof sheathing.

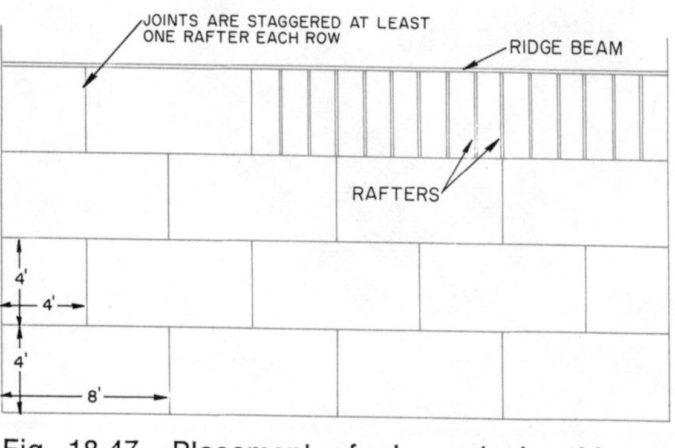

Fig. 18-47. Placement of plywood sheathing on roof rafters.

covered with hot pitch or roofing tar. The paper and pitch are applied in alternate layers. The final coat of pitch is covered with gravel or small stone.

Asphalt shingles are most generally applied in residential construction. They are coated with mineral granules and are available in various weights. A 240 lb. shingle is commonly used, meaning enough shingles to cover 100 square feet weigh 240 lbs. They are also available in many colors and shapes. Three-tab shingles are the most common and are widely used, as they can be applied rapidly, Fig. 18-48. They are applied over 15 lbs. (per 100 sq. ft.) asphalt building paper and either nailed or stapled in place.

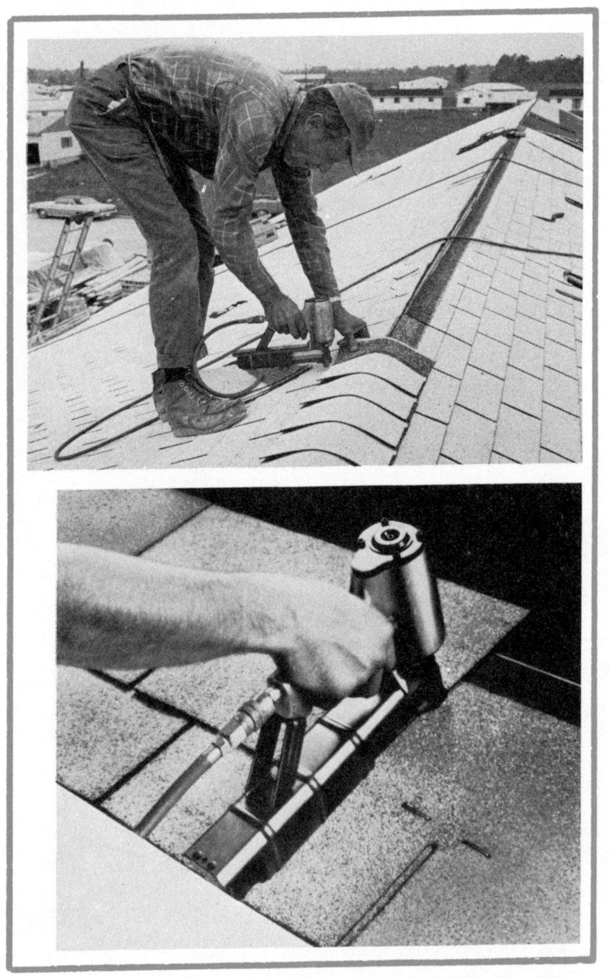

Fig. 18-48. Applying three-tab asphalt shingles. (Senco Products, Inc.)

Wood shingles are usually made from western red cedar or redwood. The shingle is cut from the heartwood of the tree. Both of these woods are highly resistant to decay. The first course extends about an inch below the fascia board and is doubled, Fig. 18-49. A space of about 1/8″ to 1/4″ is left between each shingle to allow expansion when the shingle becomes wet. The shingles are fastened with two rust-resistant nails placed about 3/4″ from the edge and 1 1/2″ above the bottom of the next course. Each course is laid with an exposure of between 3″ and 5″. The amount of exposure is determined by the roof pitch and the climate.

EXTERIOR DOORS

Doors are placed in the structure after the exterior walls and roof have been assem-

Fig. 18-49. Applying wood shingles. (Senco Products, Inc.)

bled. A door 1¾″ thick and 6′8″ high is generally installed. The main exterior entrance door is 3′ wide. Other side and passage doors are also 6′8″ high, but are usually 2′8″ wide.

The door is installed in a frame at least 1⅛″ thick. The sides of the frame called the *jamb* are rabbeted to serve as a stop for the door. The upper part of the frame called the header is also rabbeted. The bottom cross-member of the frame is called the sill. A threshold is attached to the sill after the finished floor is in place. The threshold covers the joint between the sill and the finished floor, as well as a weather stop to seal out drafts. The joists of the floor frame are generally trimmed away slightly so the sill will fit flush with the subfloor. The exterior portion of the door frame is trimmed with casing, Fig. 18-50. Combination doors of either wood or aluminum consisting of a screen and storm are usually mounted outside of the door frame.

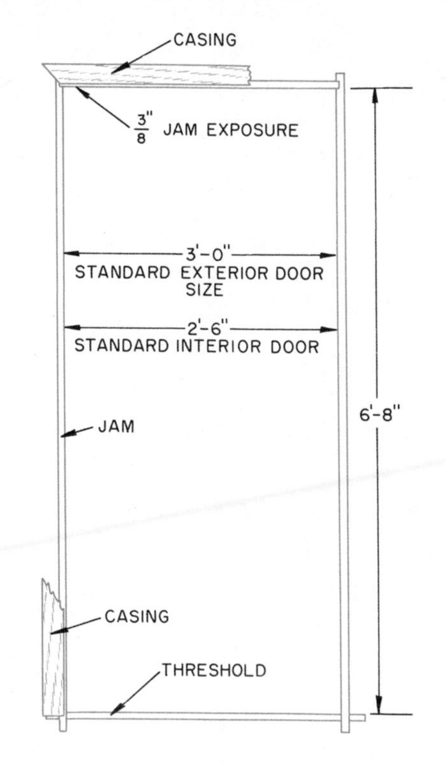

Fig. 18-50. Door assembly.

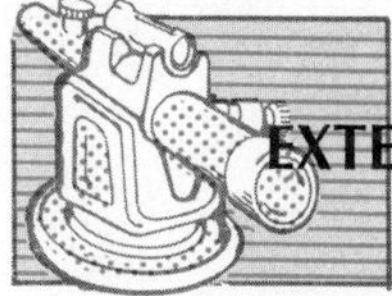

EXTERIOR WINDOWS

Various window styles are available with each having certain advantages and disadvantages. Among the most popular are double-hung, casement, horizontal sliding, and awning windows. The frames are made of either wood or metal. The heat loss is much less through wood than metal. Insulating glass is installed in some window sashes. It consists of two panels of glass spaced apart with the edges sealed, to be airtight. The air between the panes of glass serves as insulation. A storm window is not necessary when insulated glass is installed, Fig. 18-51.

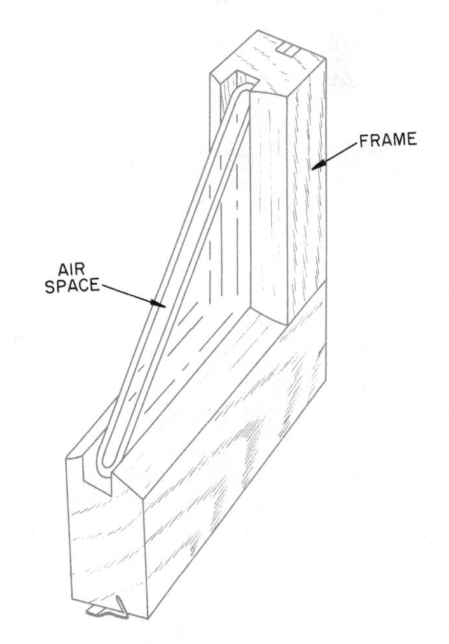

Fig. 18-51. Insulated glass installed in a window sash.

Double-hung and casement windows are widely used in residential construction. The window units are generally delivered from the factory assembled and ready to install. Windows are available in a wide range of standard sizes. The size and style

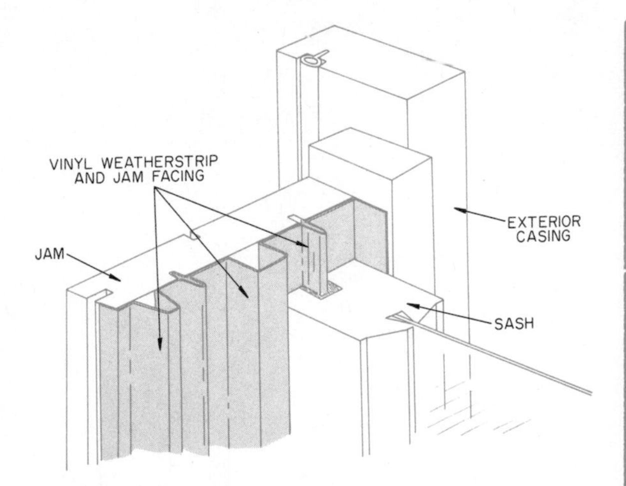

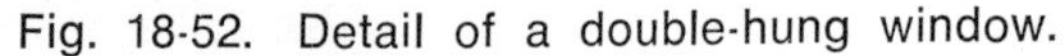
Fig. 18-52. Detail of a double-hung window.

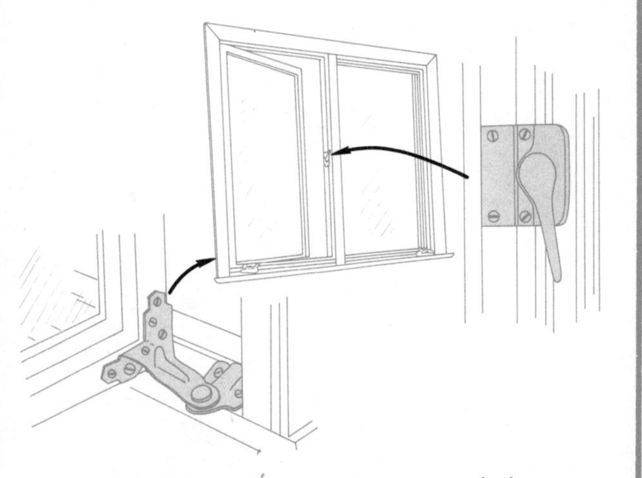
Fig. 18-53. Detail of a casement window.

A. Plumbing the side of the window.

B. Leveling the top of the window.

Fig. 18-54. Installing a wood frame window in a rough opening.

are generally determined by the architect and specified on the plan. Each sash is movable and slides up and down in the frame unit of a double-hung window, Fig. 18-52.

A casement window, Fig. 18-53, consists of a sash or several sashes hinged along the side of the sash to the jamb. It is designed to swing outward. The screen is mounted to the stationary frame. The major advantage of a casement window is the entire window area can be opened for ventilation. Insulating glass is generally installed in the sash of most casement windows.

The exterior trim is already attached to most window frames as they are manufactured. The window is installed in the rough opening which was provided during the framing process. The window is placed on the sill of the rough opening and leveled. The sides of the window are plumbed (checked with the level to see if they are perfectly straight up and down), Fig. 18-54. Wood frame windows are held securely in place with casing nails through exterior trim and driven into the stud framing. Special installation procedures supplied by the manufacturer should be observed.

EXTERIOR TRIM AND SIDING

After the structure is enclosed including the installation of the doors and windows, the siding and trim can be completed.

COMPLETING THE CORNICE

The cornice must first be enclosed. The cornice is the projection of the rafters over the plate that connects with the sidewalls. The cornice is formed on the sides on gable roofs, but is formed all the way around on a hip roof. A boxed cornice which conceals the ends of the rafters and roof sheathing is generally used in residential construction.

INSTALLING THE SOFFIT

The soffit may be installed horizontally or at the same angle as the roof, Fig. 18-55. Less material and labor is required when the soffit is placed at the same angle as the roof. The soffit is generally sheet material nailed on the bottom side of the rafters. A horizontal soffit is widely used and requires special framing. When horizontal soffit is installed, the ledger strip is placed on the exterior wall before the roof sheathing is placed on the rafters, Fig. 18-56. The fascia board is also attached to the rafters before the shingles are attached if a metal drip edge is placed under the shingles. A vent is installed in the soffit to provide ventilation in the attic.

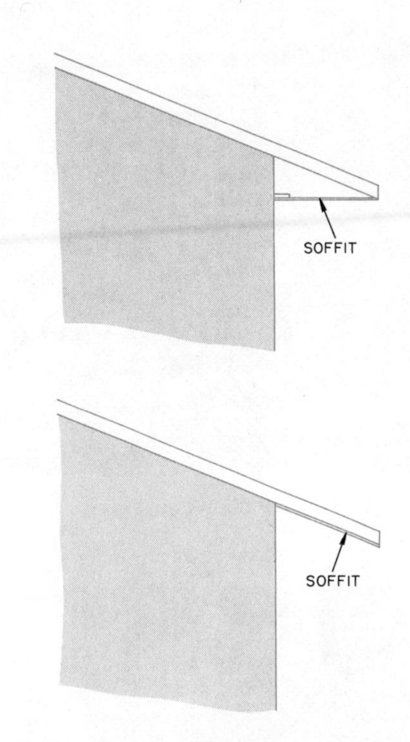

Fig. 18-55. Soffit construction methods on a sloped roof.

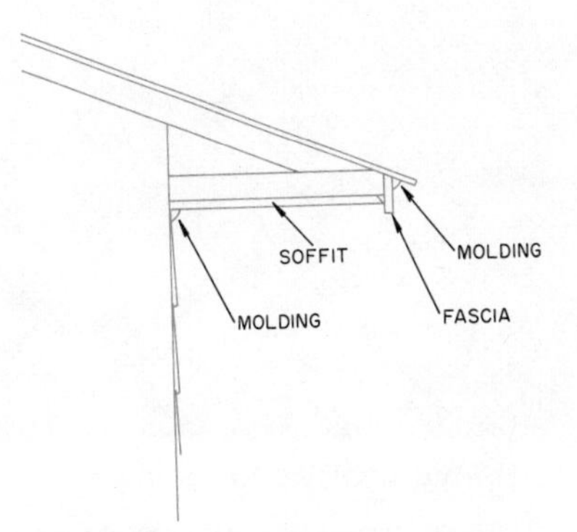

Fig. 18-56. Boxed horizontal soffit detail.

A. Horizontal wood siding.

B. Vertical wood siding.

Fig. 18-57. Wood siding.

APPLYING EXTERIOR SIDING

There are numerous types of exterior siding materials which can be installed. Wood siding has traditionally been used on American homes. It provides a pleasing appearance and has good insulating qualities. The siding may be placed either horizontally or vertically on the structure, Fig. 18-57.

The overlap on each piece of bevel siding is gauged so each piece is exposed the same amount. The distance from the bottom of the wall to soffit is measured and the total number of courses are figured. After the proper spacing has been determined, a small board with the location of each course is laid out. The gauge is called a "story pole." Figure 18-58 shows carpenters installing horizontal bevel siding. Metal corners are used on the inside and outside corners, Fig. 18-59. Rust-proof nails are used to apply all exterior siding.

Plywood and a variety of other sheet materials can also be used for exterior siding. Sheet stock 3/8" thick is the minimum thickness recommended for application on 16" stud spacing, and is always applied vertically. It is nailed around the perimeter of the sheet and on the studs. The sheet material must be the proper kind for exterior siding.

INSTALLATION OF UTILITY SERVICES

The service utilities including electrical, plumbing, air conditioning, and heating systems are installed after the exterior structure is enclosed. Workmen of the respective trades make these installations. The majority of the utilities are placed (roughed in) on the interior walls before the covering is applied to them, Fig. 18-60. The utilities are "hooked up" as the interior nears completion. Special trenches are dug from the street to the house for installing the utilities.

A. Applying starter strip.

B. Applying bevel siding to sheathing.

Fig. 18-58. Applying horizontal bevel siding.

THERMAL INSULATION

Insulation is placed in the outer walls and ceilings to minimize the loss of heat through the walls in cold weather. The insulation also reduces the flow of hot air

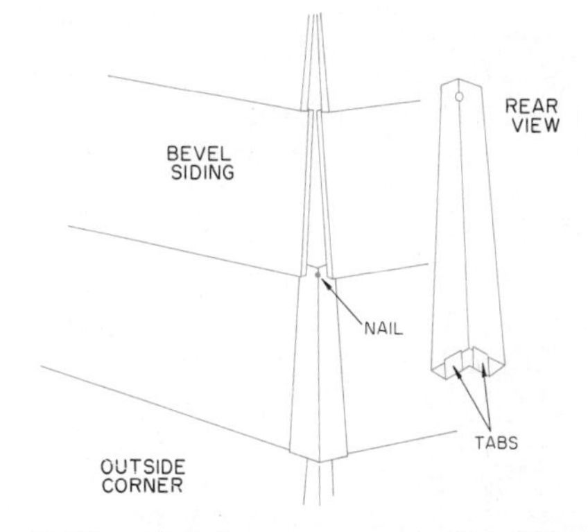

Fig. 18-59. Metal corner used with bevel siding.

A. Electrical utilities.

B. Plumbing utilities.

Fig. 18-60. Roughing in utilities during construction. (National Homes Corp.)

when the house is air-conditioned in hot weather. The structural materials such as wood and sheet materials have some insulating value. The air spaces between the studs also serve as an insulation.

Commercial insulation is generally placed in the walls and ceilings of structures which are artificially heated or cooled. There are numerous insulating materials including blankets and batts, expanded polystyrene, and poured loose fill, Fig. 18-61. The flexible batt and blanket insulation is available in rolls or packages in widths designed for 16″ and 24″ stud spacing. It is available in standard thicknesses of 1½″, 2″, and 4″. It is made from tiny glass filaments glued to a backing. The backing, which also serves as a vapor barrier, is placed towards the heated side of the wall. The vapor barrier prevents the flow of moist air from the inside out through the walls. If it is permitted to escape, the warm moist air will condense when it strikes cold air outside. Excessive condensation will damage exterior and interior finish and could also cause rotting in the wood frame. Aluminum foil backing material on fiberglass insulation serves as a combination reflective insulation and vapor barrier. Insulation materials can also be used inside interior walls to reduce noise transmission.

Fill insulation is poured or blown into place. It is best suited for placement in an unheated attic between the ceiling joists. Blanket and batten insulation is stapled in place between the ceiling joists or studs. Some blanket insulation has a side strip which holds it in place.

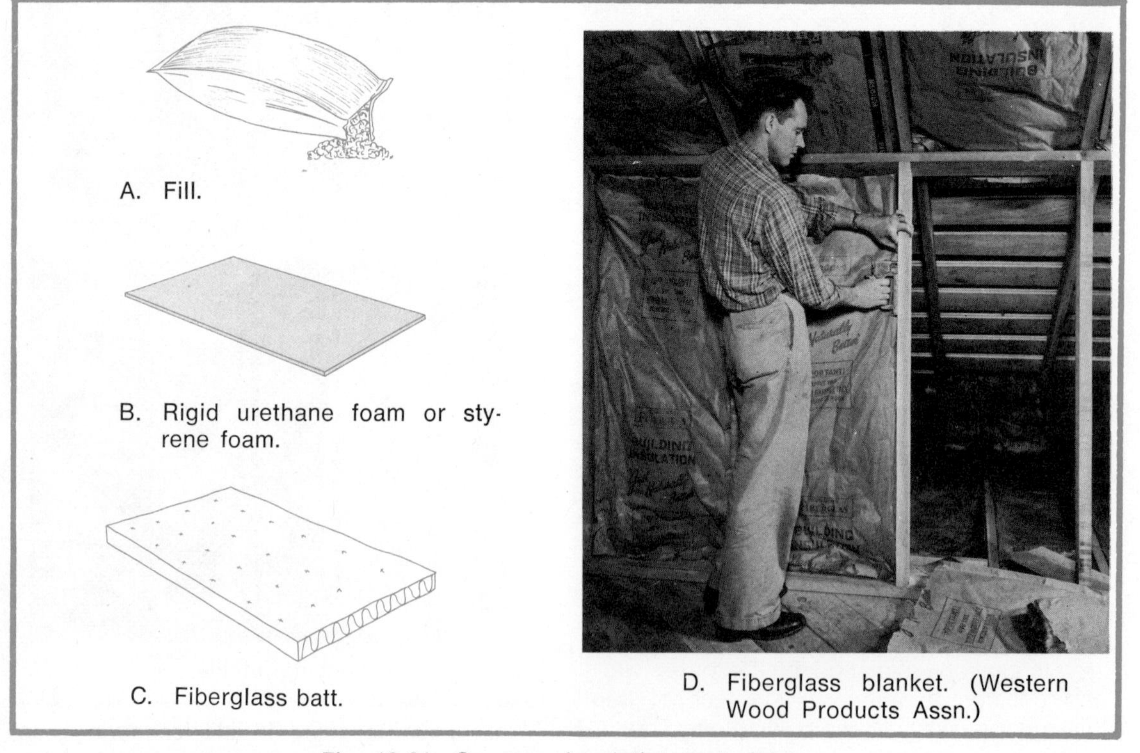

A. Fill.

B. Rigid urethane foam or styrene foam.

C. Fiberglass batt.

D. Fiberglass blanket. (Western Wood Products Assn.)

Fig. 18-61. Common insulation materials.

INTERIOR CEILING AND WALL FINISH

The interior walls of a house are generally covered with either gypsum wall board, plaster, or a wood paneling. The gypsum wall board is a sheet material composed of gypsum covered with paper on both sides. The process is commonly referred to as drywall construction. The wall board comes in a variety of standard sizes 4′ wide ranging in length from 8′ to 16′, and is either 3/8″ or 1/2″ thick. The 1/2″ material is recommended when only one layer is applied, Fig. 18-62. The sheets can be attached to the walls in either a vertical or horizontal position and held with nails. The joints are finished and concealed with special tape and joint compound.

Gypsum board is also used for plastered walls. Sheets of gypsum 3/8″ thick, 16″ wide and 4′ long are nailed on the studs and serve as the base on which the plaster is applied. Metal lath may be used instead of gypsum board for the base. It is often placed on inside corners which will undergo stress due to shrinkage of the framing lumber to reduce the amount of cracking. Metal corner beads of lath-like material are installed on outside corners and around door openings. Plaster grounds, which are strips of wood tacked around window and door openings, serve as guides or strike-off edges. The thickness of the plaster is regulated by the width of the standard door jamb to be installed in the openings, and is usually 5¼″.

Wood wall paneling is generally applied over wall board or plaster lath. This gives additional stability to the lightweight sheet material. The paneling is manufactured in 4′ × 8′ sheets either 3/16″ or ¼″ thick. It is applied to the wall with nails or adhesive, Fig. 18-63.

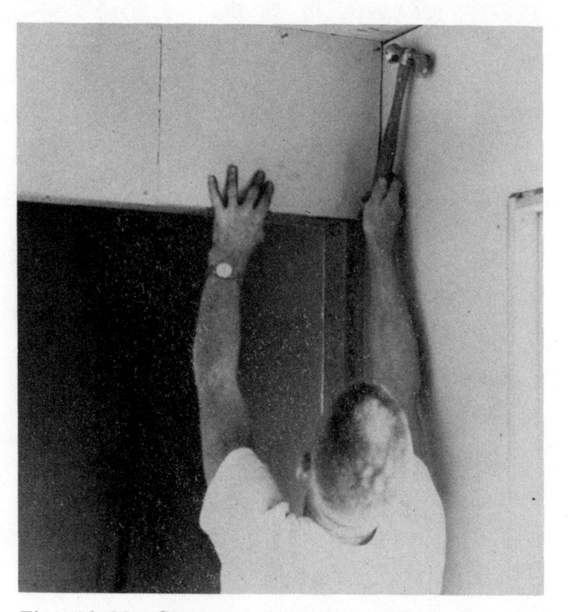

Fig. 18-62. Gypsum board applied to the studs in drywall construction.

Fig. 18-63. Installing wood paneling. (Hardwood Plywood Manufacturers Association and Georgia Pacific)

INTERIOR DOORS, FRAMES, AND TRIM

The interior doors in a house are 6′8″ and the width varies depending upon the room. Bedroom doors are generally a minimum of 2′6″ wide; bathroom doors may be as narrow as 2′4″. The width of the doors on wardrobe and linen closets varies considerably. The doors may be either swinging, sliding, or folding, Fig. 18-64.

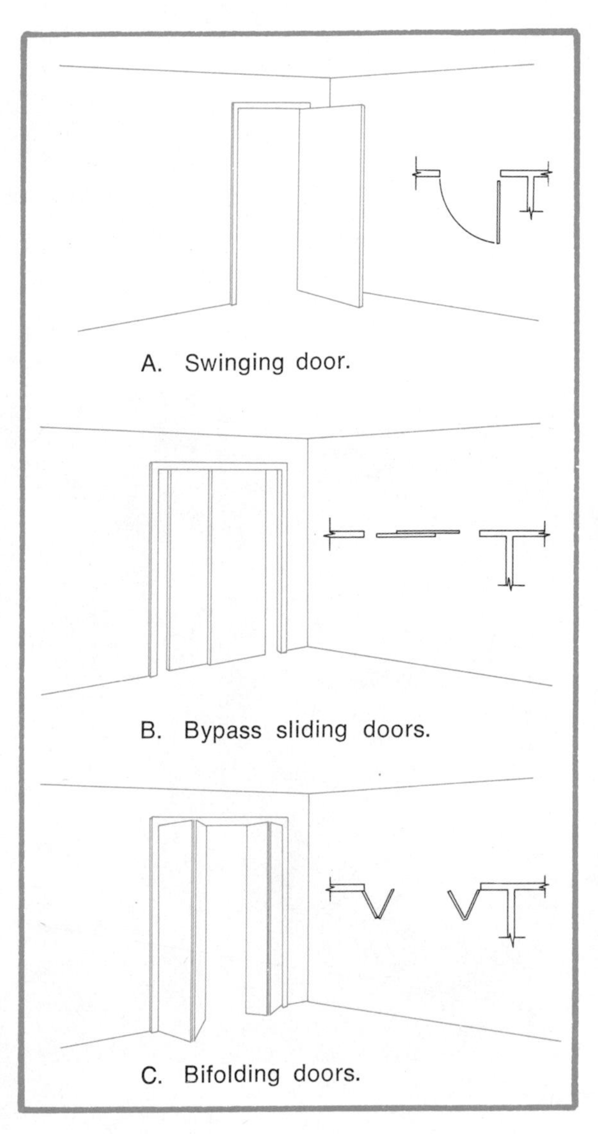

A. Swinging door.

B. Bypass sliding doors.

C. Bifolding doors.

Fig. 18-64. Interior doors commonly used.

INSTALLING INTERIOR DOORS

Prehung interior doors, which consist of the jamb fitted with a door ready for installation in the opening, are becoming widely accepted, Fig. 18-65. The door often consists of a two-piece jamb and has the casing attached to both sides at the factory.

When conventional doors are used, both sides of the jamb are trimmed with casing. The bottom of the casing is cut square and the top is mitered at a 45° angle.

The jamb is leveled and plumbed in the rough opening when the door is installed. The door is then bored for instal-

Fig. 18-65. A pre-hung door. (National Homes Corporation)

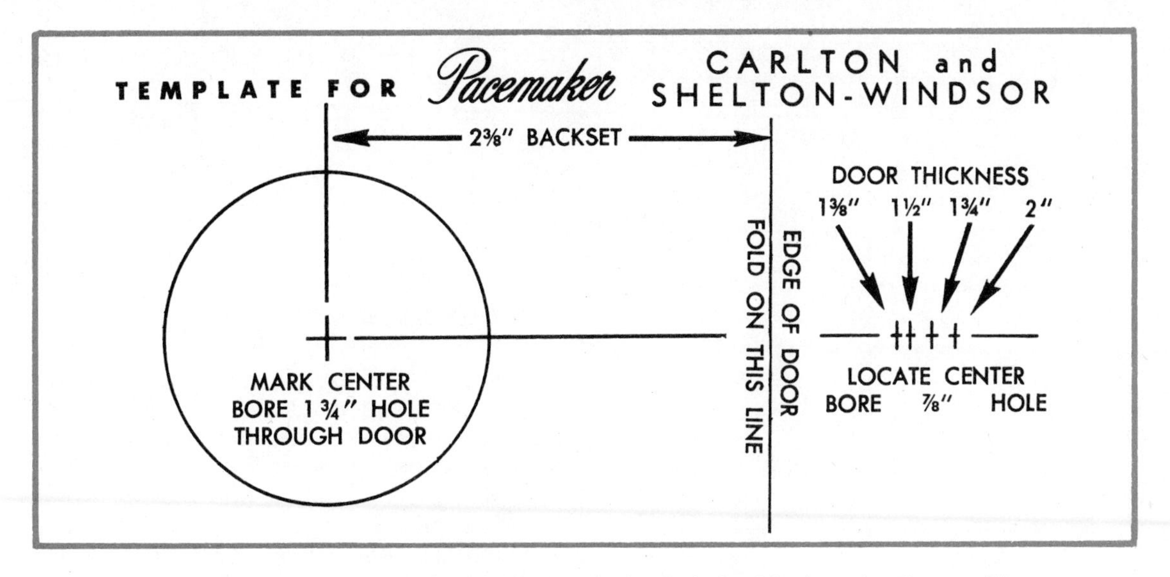

Fig. 18-66. A paper template is generally included with door hardware. (Harloc Products Corporation)

lation of the lockset, Fig. 18-66. A template is usually supplied with the lock for locating the holes. The knob is placed 36″ to 38″ from the floor.

INSTALLING INTERIOR WINDOW TRIM

The interior window trim should match the casing on the doors and in the rest of the room. This trim consists of the stool, apron, casing, and stop, Fig. 18-67. The *stool* is the first member cut and put in place. It laps the window sill and extends beyond the casing on both sides. Each end is notched against the finished wall so it will fit to the window sash and the wall. It is advisable to predrill the nail holes if hardwood is used to prevent splitting when the window trim is attached. The casing is mitered on the top corners and butted against the stool. The inner edge is placed about ¼″ from the inside of the window frame. The *stop* covers the joint between the jamb and the casing. The *apron* is nailed under the stool to completely finish the windows.

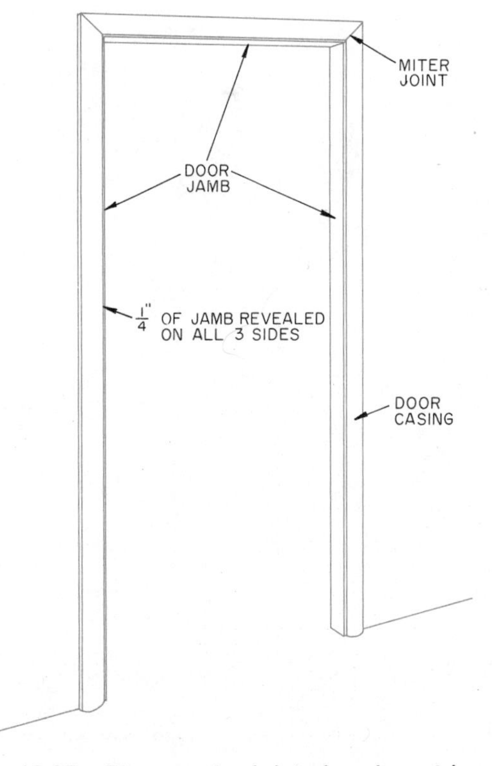

Fig. 18-67. Placement of interior door trim.

Figure 18-68 shows an alternate method of trimming windows when a stool and apron is not used. It is sometimes referred as *picture frame trimming* and is often used on casement windows. The casing is mitered on all four corners and extends all the way around the window frame. A stop strip is placed against the sash on all four sides.

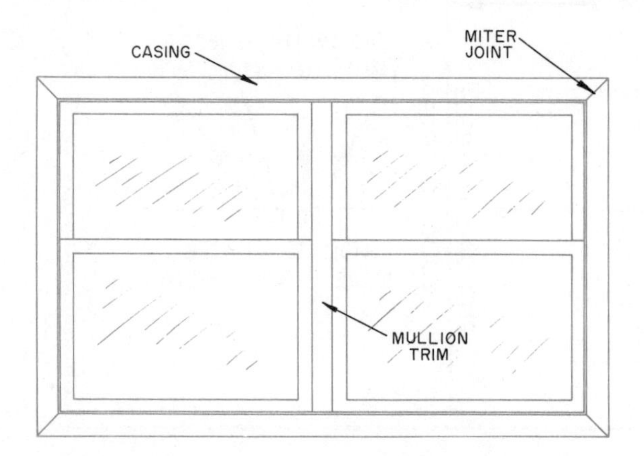

Fig. 18-68. Window trimmed with entire frame enclosed with casing.

INSTALLING BASE MOLDING

Base molding is placed between the finish floors and the finished walls to conceal the joint between the floor and the wall. A second piece called the base shoe is also sometimes used with the base, but when carpet is installed only the base molding is needed, Fig. 18-69. The outside corners are mitered and the inside ones are coped.

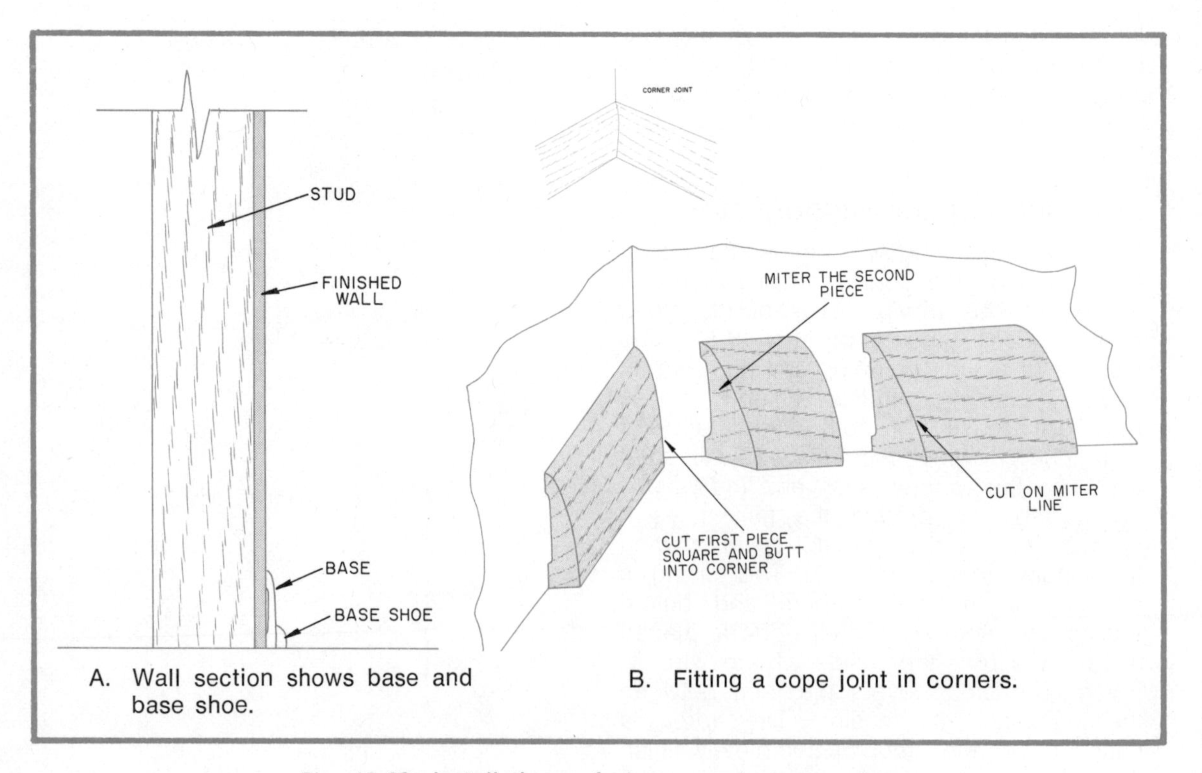

A. Wall section shows base and base shoe.

B. Fitting a cope joint in corners.

Fig. 18-69. Installation of base and base shoe.

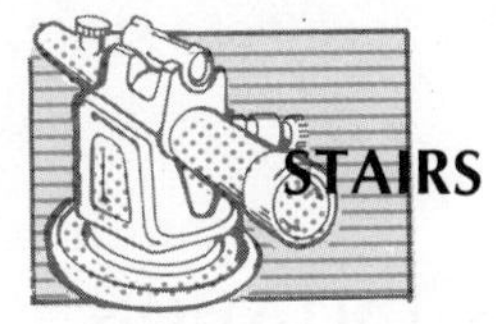

STAIRS

Stairs provide a means of moving from one floor level to another. The first consideration in the construction of stairs is safety. They must be designed so that there is ample head clearance of at least 6′8″. The stairwell is provided when the house is rough framed, Fig. 18-70. The rough opening should be approximately 10′ long. The width should provide a minimum of 2′8″ between the handrail and the finished wall on the opposite side.

A straight, continuous stairway as used in most houses consists of *stringers*, *risers*, and *treads*. The stringer is the main support member which carries the treads and risers, the tread is the horizontal member which serves as the step, and the riser is the vertical member which forms the back of each step. The size and number of treads and risers is based on the total run (horizontal length of the stairs) and total rise (vertical distance from one floor to the next).

A general rule of the thumb applied to stair construction is that the total of one tread and riser should equal between 17″ and 18″. The rise should equal 7″ to 8″ and the tread 9″ to 11″.

The number of risers required can be determined by dividing the total rise in inches by the desired height of the riser. For example, suppose the total rise is 7′10″ or 94″. This, divided by a desired rise of 7″ equals approximately 13.42, indicating that either 13 or 14 risers will be needed. The number of risers must be a whole number, so we will use 13 risers. Now, divide the 94″ by 13 to determine the definite height of each rise. This is 7.23″ which rounded to the nearest 1/8″, equals 7 1/4″. The slight difference between .23 and 1/4″ when rounded off will be insignificant if it is divided between several risers in the layout process.

A tread of either 9″ or 10″ may be used to stay within the total range of 17″ or 18″ for combined tread and riser distance. Let's assume a tread width of 10″. There will always be one less tread than riser. Since we have 13 risers, we will have 12 treads. The total run would equal 12 multiplied by the width of each tread of 10″, which equals 120″, or 10′ for the total run.

A framing square used in the same manner as for rafter layout can be used for laying out the stringer. From the previous example, the 10″ tread width would be located on the blade of the square. The 7 1/4″ mark is located on the tongue of the square. The spacing for each tread and riser is stepped off as shown in Fig. 18-71.

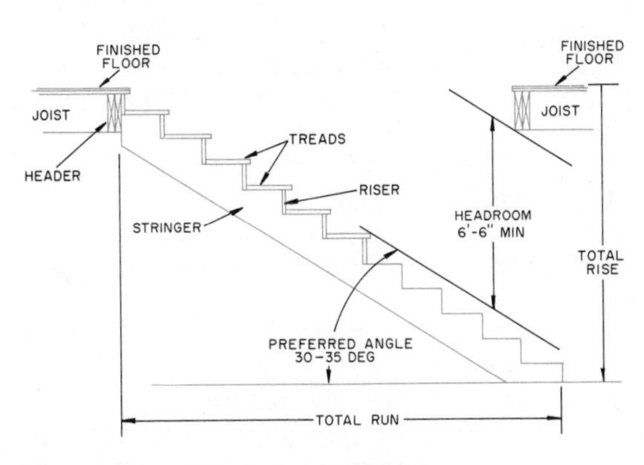

Fig. 18-70. Parts of a stairwell.

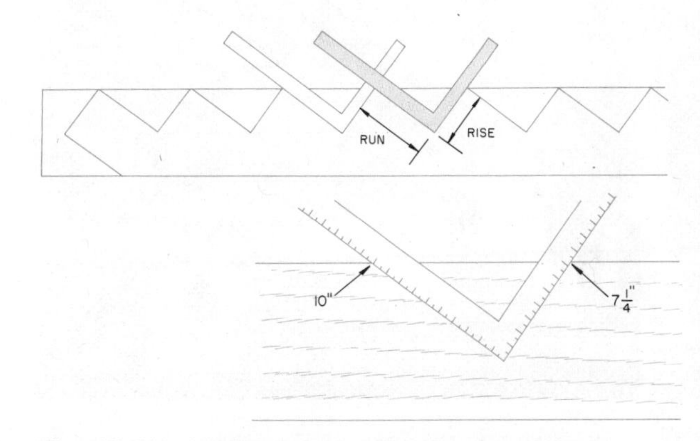

Fig. 18-71. Laying out the stair stringer using a framing square.

A 2″ × 12″ board is used for the stringer. It is best to install three stringers to carry the load of the stairs. The first stringer can be used as a pattern to mark the others for cutting.

FINISH FLOOR COVERINGS

A vast variety of floor coverings can be used as the finished floor. Linoleum, tile, hardwood, and carpet of many styles and types are most popular. An underlayment of plywood, particle board, or hardboard is installed over the subfloor to form a smooth, solid base for tile, linoleum, and carpet. The underlayment is placed in the opposite direction of the subfloor. When hardwood is installed, a layer of 15 lb. asphalt-saturated felt is placed between the subfloor and hardwood floor. Underlayment is not used with hardwood flooring. The flooring is tongue-and-grooved and the nails are driven on the tongue side so they are not exposed. Most hardwood flooring is prefinished and does not require sanding and finishing.

Fig. 18-72. Erection of a prefabricated house. (Senco Products, Inc.)

CABINET INSTALLATION

Most cabinets installed in new homes are manufactured in a factory rather than being constructed by the carpenter. It is generally less expensive for the carpenter to buy manufactured cabinets, since they are a standard component in house construction and because they are produced by mass-production methods. Cabinets of the desired design and size can be specified. However, some special cabinets are constructed by the carpenter. Various construction methods for cabinets and built-ins were discussed in Chapter 16.

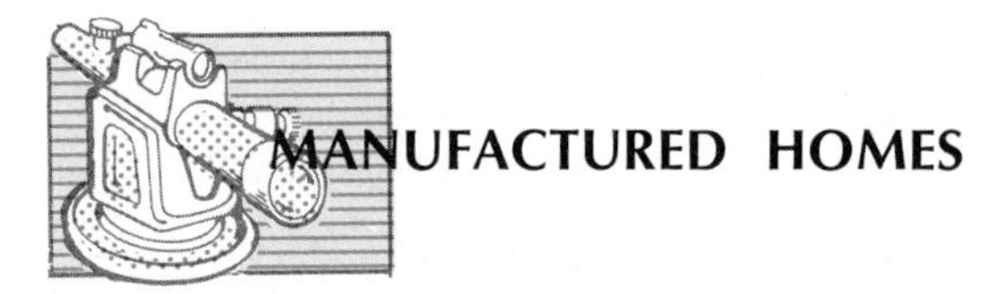

MANUFACTURED HOMES

The manufactured home industry has expanded rapidly in the past decade. A manufactured house is one which has been constructed in part in a factory. The component parts are then shipped to the building site where the house is erected. The production method used in the manufacture of other products is being applied to the construction industry. The manufactured house can be produced at a lower cost than the traditional custom-built homes, Fig. 18-72.

Generally, the entire house is not manufactured in a factory. Sometimes only the stock is cut to size and shape and the assembly is performed on the site. This would be regarded as a "precut" house. The studs, rafters, headers, joists, and other components are cut as prescribed by the plans and shipped to the site.

Roof trusses are often purchased as assembled components ready to put in place, Fig. 18-73. Truss rafters reduce the time required to erect an enclosed shell.

Prehung doors and window assemblies are manufactured components. They also reduce the time required to construct a house.

A prefabricated house is usually built in panels, Fig. 18-74. The panels include the assembled 2×4 studs and plates, complete with the exterior sheathing and windows in place. The panel is hauled by truck to the site and nailed to the floor deck. Some manufacturers construct the subfloor as panels to be placed on the foundation in units. The wall panels may be as long as 30′, complete with electrical wiring, insulation, and finished interior and exterior walls.

Transportation difficulties control the degree to which a house can be completed in the factory. A considerable amount of work remains to be completed on the site, even when a manufactured house is constructed. The excavation, foundation, and utility installation must be done on the site.

Special jigs and fixtures are used for fabricating the wall sections. Automatic nailing machines are used to fasten the studs in place. The sheathing is applied to the wall frame as it moves down the production line, Fig. 18-75. The windows are installed and the exterior siding may also be applied. The component pieces for truss rafters used on prefabricated houses are cut and placed in jigs for assembly. The rafters and wall sections are usually turned over by a hydraulic system.

Fig. 18-73. A truss rafter ready to be placed on a prefabricated house.

Fig. 18-74. A wall section fabricated in a factory. (Senco Products, Inc.)

Fig. 18-75. Sheathing can be applied to prefabricated wall sections. (Senco Products, Inc.)

MODULAR CONSTRUCTION

Homes are also being manufactured in factories to the degree of completion that they are being hauled to the erection site as a one-piece unit. Modular homes are built like a conventional home except it is done in a factory, Fig. 18-76. The size of the modular home is somewhat limited by the size which can be moved on the highways. To overcome this problem, two modules are often placed together after they arrive on the site, Fig. 18-77.

Many of the modular homes are being constructed from component parts as large as a total bathroom unit. The unit is constructed of plastic materials and is placed in the house before the sidewalls are enclosed.

The module can be placed on a foundation or basement wall. It is designed so that it can be raised from the transport to the foundation by means of a crane. It is also possible to produce two-story modules, by placing the completed second story unit on top of the first story, Fig. 18-78.

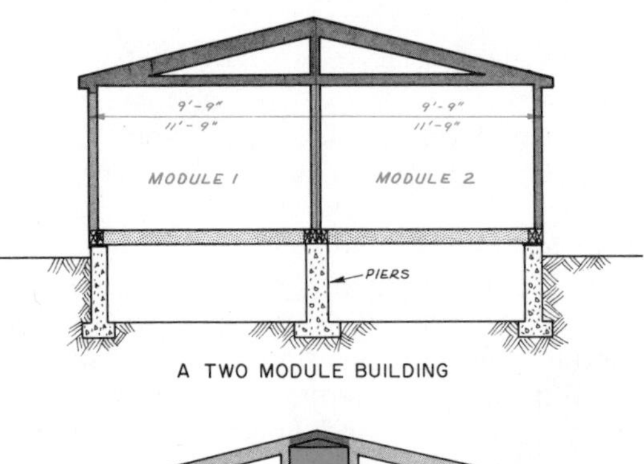

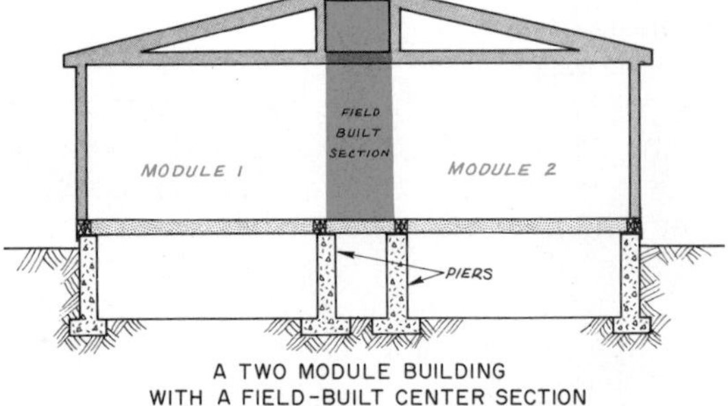

Fig. 18-77. Two modules are placed together on the construction site.

Fig. 18-76. A modular home.

Fig. 18-78. Second story is raised into position to complete modular structure.

ADVANTAGES OF MODULAR CONSTRUCTION

The greatest advantage of modular homes is the *cost*. The cost is less because the houses are constructed using modern factory mass-production techniques. In addition, the house can be erected on the site in as little as one day if the foundation has been prepared in advance.

ACTIVITIES

PROBLEM

Estimating is a very important part of the construction industry. Before a cost estimate can be made an estimate of the materials needed to construct the building must be made. A contractor estimates the amount of material needed for a structure from the plans or drawings.

Below is a floor plan for a double garage. On a sheet of paper, calculate the amount of the following materials which will be needed if framing materials are placed 16″ O.C.

A. 2 × 4 studs, cripples and trimmers
B. 2 × 4 plates
C. 4 × 8 sheets of wall sheathing
D. 2 × 12 headers

VOCABULARY

Plot plan
Exterior Elevation view
Drywall Construction
Specification
Building Ordinance
Excavation
Footing
Foundation
Load bearing wall
Subfloor
Soffit
Header
Fascia
Sheathing
Truss rafter
Modular Construction
Hip Rafter
Roof pitch
Tread
Riser

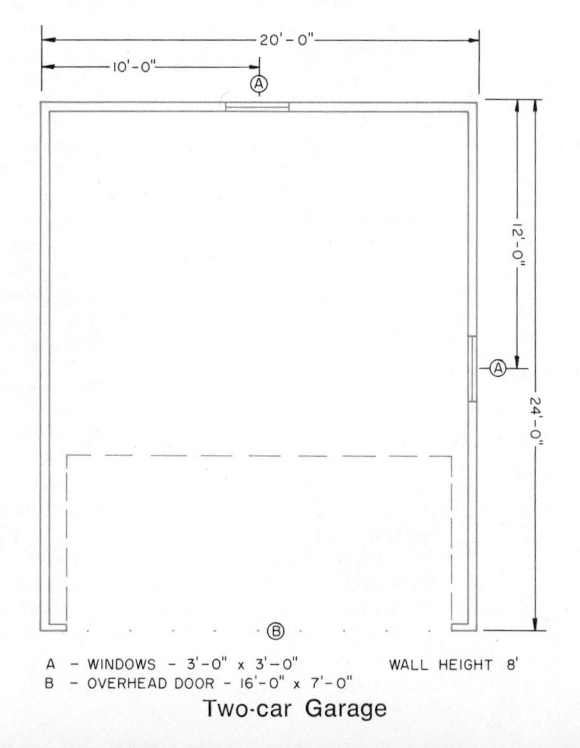

Two-car Garage

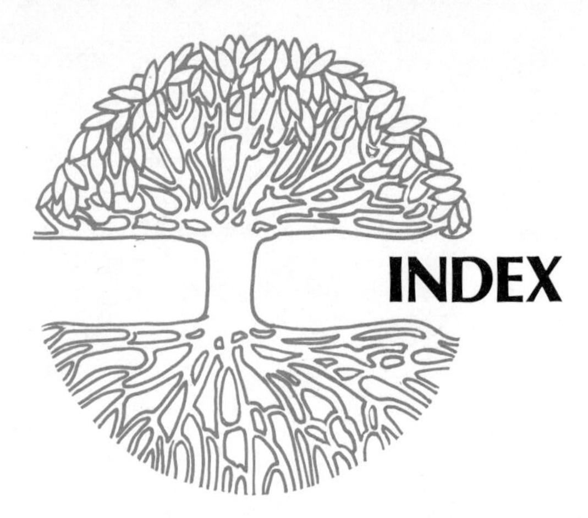

INDEX